Spon's Mechanical and Electrical Services Price Book

2002

Key titles from Spon Press:

Building Regulations Explained
J. Stephenson

Construction Scheduling, Cost Optimisation and Management
H. Adeli and A. Karim

Dictionary of Architectural and Building Technology
H. Cowan and P. Smith

Energy & Environment in Architecture
N. Baker and K. Steemers

Fire From First Principles
P. Stollard and J. Abrahams

Green Building Handbook Volumes 1 & 2
T. Woolley et al

Hazardous Building Materials 2nd Edition
S. Curwell and C. March

Profiled Sheet Roofing and Cladding
National Federation of Roofing Contractors

Refurbishment and Upgrading of Existing Buildings
D. Highfield

Spon's Building Costs Guide for Educational Premises
Barnsley and Partners

Spon's Construction Resource Handbook
B. Spain

Spon's House Improvement Price Book
B. Spain

Spon's Irish Construction Price Book
Franklin and Andrews

Spon's Railway Construction Price Book
Franklin and Andrews

Timber Decay in Buildings
B. Ridout

Understanding the Building Regulations
S. Polley

Understanding JCT Standard Building Contracts
D. Chappell

Understanding Quality Assurance in Construction
H. W. Chung

To order or obtain further information on any of the above or receive a full catalogue please contact:
The Marketing Department, Spon Press, 11 New Fetter Lane, London, EC4P 4EE. Tel: 020 7583 9855; Fax: 020 7842 2298

For a complete listing of all our titles please visit www.sponpress.com

Spon's Mechanical and Electrical Services Price Book

Edited by
MOTT GREEN & WALL
Building Services Cost Consultants
and Value Engineers

2002

Thirty third edition

London and New York

First edition 1968
Thirty third edition 2002
Published by Spon Press
11 New Fetter Lane, London EC4P 4EE

Simultaneously published in the USA and Canada
by Spon Press
29 West 35th Street, New York, NY 10001

Spon Press is an imprint of the Taylor & Francis Group

© 2002 Spon Press

Printed and bound in Great Britain by
TJ International Ltd, Padstow, Cornwall

All rights reserved. No part of this book may be reprinted or
reproduced or utilised in any form or by any electronic, mechanical,
or other means, now known or hereafter invented, including
photocopying and recording, or in any information storage or
retrieval system, without permission in writing from the publishers.

Publisher's note
This book has been produced from camera-ready copy supplied by
the authors

British Library Cataloguing in Publication Data
A catalogue record for this book is available from the British Library

Library of Congress Cataloging in Publication Data
A catalogue record has been requested

ISBN 0-415-24280-0
ISSN 0305-4543

Preface

The Thirty Third edition of *Spon's Mechanical and Electrical Services Price Book* has been reformatted to align more accurately with the engineering services format of SMM 7. This is Mott Green & Wall's second year of producing and editing *Spon's Mechanical and Electrical Services Price Book* and this year we have undertaken a thorough re-alignment of the book's layout. In addition, we have included new sections and descriptions to current industry trends.

This years book includes new items, additions to existing sections and revising whole sections, which includes the following. Rainwater installations, pressfit pipework and fittings, ductwork now reflects construction break points, Modular wiring, lighting control, emergency lighting, earthing.

Before referring to prices or other information in the book, readers are advised to study the `Directions' which precede each section of the Materials Costs/Measured Work Prices. As before, no allowance has been made in any of the sections for Value Added Tax.

Approximate Estimating sections give a broad scope of outline costs for various building types and all-in rates for certain elements and selected specialist activities; these are followed by elemental analysis and then by a quantified analysis of a specific building types.

The prime purpose of the Materials Costs/Measured Work Prices part is to provide industry average prices for mechanical and electrical services, giving a reasonably accurate indication of their likely cost. Supplementary information is included which will enable readers to make adjustments to suit their own requirements. It cannot be emphasised too strongly that it is not intended that prices should be used in the preparation of an actual tender without adjustment for the circumstances of the particular project such as productivity, locality, project size and current market conditions. Adjustments should be made to standard rates for time, location, local conditions, site constraints and any other factor likely to affect the costs of a specific scheme. Readers are referred to the build up of the gang rates, where allowances are included for supervision, labour related insurances, where the percentage allowances for overhead, profit and preliminaries are defined.

Readers are reminded of the service available in *Spon's Price Book Update*, where details of significant changes to the published information are given. The printed update is circulated free of charge every 3 months, until the publication of next year's *Price Book*, to those readers who have requested it. In order to receive this the coloured card bound in with this volume should be completed and returned.

As with previous editions the Editors invite the views of readers, critical or otherwise, which might usefully be considered when preparing future editions of this work.

While every effort is made to ensure the accuracy of the information given in this publication, neither the Editors nor Publishers in any way accept liability for loss of any kind resulting from the use made by any person of such information.

In conclusion, the Editors record their appreciation of the indispensable assistance received from many individuals and organisations in compiling this book.

MOTT GREEN & WALL
Building Services Cost Consultants
and Value Engineers
2nd Floor
Africa House
64-78 Kingsway
London WC2B 6NN

Telephone: 0207 836 0836
Facsimile: 0207 242 0394

e-mail: spons@mottgreenwall.co.uk

Visit www.pricebooks.co.uk

Contents

Preface	page v
Special Acknowledgements	page ix
Acknowledgements	page xi

PART ONE: APPROXIMATE ESTIMATING

Directions	3
Cost Indices	4
Outline Costs	6
Elemental and All-in-Rates	7
Elemental Costs	24

PART TWO: MATERIAL COSTS/MEASURED WORK PRICES

Mechanical Installations

Directions	38
R : Disposal Systems	
R10 : Rainwater Pipework/Gutters	44
R11 : Above Ground Drainage	64
S : Piped Supply System	
S10 : Cold Water	90
S11 : Hot Water	149
S32 : Natural Gas	153
S41 : Fuel Oil Storage/Distribution	155
S60 : Fire Hose Reels	156
S61 : Dry Risers	157
S63 : Sprinklers	158
S65 : Fire Hydrants	162
T : Mechanical/Cooling/Heating Systems	
T10 : Gas/Oil Fired Boilers	164
T13 : Packaged Steam Generators	177
T14 : Heat Pumps	178
T31 : Low Temperature Hot Water Heating	179
T33 : Steam Heating	267
T42 : Local Heating Units	271
T61 : Chilled Water	272
T70 : Local Cooling Units	274
U : Ventilation/Air Conditioning Systems	
U10 : Ductwork	275
U14 : Smoke Extract	402
U70 : Air Curtains	397

viii

Contents

Electrical Installations

Directions   page 426

V : Electrical Supply/Power/Lighting
V11 : HV Supply   429
V12 : LV Supply   440
V20 : LV Distribution   452
V21 : General Lighting   515
V22 : Small Power   523
V32 : Uninterrupted Power Supply   532
V40 : Emergency Lighting   534

W : Communications/Security/Control
W10 : Telecommunications   537
W20 : Radio/Television   538
W23 : Clocks   540
W30 : Data Transmission   542
W50 : Fire Detection and Alarm   549
W51 : Earthing and Bonding   552
W52 : Lightning Protection   553

PART THREE: RATES OF WAGES AND WORKING RULES

Mechanical Installations

Rates of Wages   557
Working Rules (HVCA JIB PMES)   563

Electrical Installations

Rates of Wages   595
Working Rules   602

PART FOUR: DAYWORK

Heating and Ventilating Industry   617
Electrical Industry   620
Building Industry Plant Hire Costs   624

Tables and Memoranda   637

Index   661

Special Acknowledgements

The editors wish to record their appreciation of the assistance given by the following organisations in the compilation of this edition.

ABBEY THERMAL INSULATION LTD.
23-24, Riverside House,
Lower Southend Road, Wickford, Essex SS11 8BB
Telephone: 01268 572116- Facsimile: 01268 572117
E-mail: general@abbeythermal.com

T. Clarke plc

Electrical Engineers & Contractors
Stanhope House
116-118 Walworth Road
London SE17 1JY
Tel: 020 7358 5000
Fax: 020 7701 6265
e-mail: info@tclarke.co.uk
Internet: www.tclarke.co.uk

Hampden Park Industrial Estate
Eastbourne
East Sussex
BN22 9AX
Tel : 01323 501234
Fax : 01323 508752

FUJITEC

Fujitec UK Limited
1 Newtons Court
Crossways Business Park
Dartford Kent DA2 6QL
Tel : 01322 424400
Fax : 01322 424401

Sulzer Infra (UK) Ltd
Westmead, Farnborough
Hants GU14 7LP
Tel: 01252 544400
Fax: 01252 378988

H&K

Hall & Kay Fire Engineering
Sefton Lodge
Clewer Hill Road
Windsor
Berkshire **SL4 4FT**
Tel: 01753-833444
Fax: 01753-833434
E-Mail: Windsor@hkfire.co.uk

cableship

Cableship Limited
The Hallmarks
146 Field End Road
Eastcote
Pinner
Middx HA5 1RJ
Tel : 020 8429 3333
Fax : 020 8429 4982

senior HARGREAVES
DUCTWORK SPECIALISTS

Lord Street, Bury, Lancashire. BL9 0RG
Tel: 0161 764 5082 • Fax: 0161 762 2336
E-Mail: sales@senior-hargreaves.co.uk

Special Acknowledgements

Power House
6 Power Road
Chiswick
London W4 5PY
Tel No: +44 (0) 20 8994 6462
Fax No: +44 (0) 20 8994 0678
E-Mail: howlondon@howfire.co.uk
www.howfire.co.uk

Yeoman House
63 Croydon Road
London SE20 7TS
Tel : 020 8778 9666   Fax : 020 8659 9386

Service House
145 Gosport Road
London E17 7LX
Tel : 020 8281 8000
Fax : 020 8281 8001

Drake & Scull House
51 Great North Road
Hatfield
Herts AL9 5EN
Tel : 01707 630300
Fax : 01707 630333

ASCO Power Technologies Ltd
Fourth Avenue
Globe Park
Marlow
Buckinghamshire. SL7 1YG
Tel: 01628 403872 Fax: 01628 403870
E-Mail: sales.uk@asco-pt.co.uk

Sauter Automation Limited
Inova House
Hampshire International Business Park
Crockford Lane, Chineham
Basingstoke, Hampshire
RG24 8WH
Tel: +44 (0) 1256 374400
Fax: +44 (0) 1256 374455
E-mail info@uk.sauter-bc.com

Acknowledgements

The editors wish to record their appreciation of the assistance given by many individuals and organisations in the compilation of this edition.

Manufacturers, Distributors and Sub-Contractors who have contributed this year include:-

APV-Vent Axia Industrial Division
Unit 2
Caledonia Way
Stretford Motorway Estate
Barton Dock Road
Manchester M32 0ZH
Fans
Tel: 0161 865 8421
Fax: 0161 865 0098

AVK/SEG (UK) Ltd
Power Systems House
Manor Courtyard
Hughenden Avenue
High Wycombe
Bucks HP13 5RE
Standby Generators
UPS
Tel: 01494 435600
Fax: 01494 435610
Power_Systems@AVK-SEG.co.uk
www.AVK-SEG.co.uk

Brights of London Ltd
Westgate Business Park
Westgate Carr Road
Pickering
N Yorks YO18 8LX
Clock Systems
Tel: (020) 8786 8466
Fax: (020) 8786 8477

Brush Transformers Ltd
P O Box 20
Falcon Works
Loughborough
Leics LE11 1HN
Transformers
Tel : 01509 617892
Fax : 01509 612819

BSS UK Ltd (Wirral Branch)
Rossmore Road East
Rossmore Road Industrial Estate
Ellesmere Port
South Wirral L65 3DD
Pipeline Fittings
Tel: 0151 355 8281
Fax: 0151 375 2001

Aqua-Gas (Valves and Fittings) Ltd
Donkin Valves Division
Derby Road
Chesterfield
S40 2EB
Gas Valves
Tel: 01246 558707
Fax: 01246 558712

AVO International
Archcliffe Road
Dover
Kent CT17 9EN
Air Distribution Equipment
Tel: 01304 502232
Fax: 01304 207460

S G Baldwin
Barholm Road
Tallington
Stamford
Lincolnshire PEG 4RL
Concrete Cable Protection
Tel: 01778 345455
Fax: 01778 345949

BATT Electrical Company
Unit 7
Flag Business Exchange
Vicarage Farm Road
Peterborough
Cambs PE1 5TX
Cables
Tel: 01733 558485
Fax: 01733 558485
www.batt.co.uk

Caradon Stelrad Ideal Boilers
PO Box 103
National Avenue
Kingston-upon-Hall
North Humberside
HU5 4JN
Boilers/Heating Products
Tel: 08708 400030
Fax: 08708 400059

Acknowledgements

Caradon MK Electric Ltd
The Arnold Centre
Paycocke Road
Basildon
Essex SS14 3EA
Cabling Systems, Switchgear &
Distribution, Wiring, Fittings
& Accessories
Tel: 01268 563000
Fax: 01268 563563
website: www.mkelectric.co.uk

Crane Fluid Systems
National Sales Centre
Nacton Road
Ipswich
Suffolk
1PE 9QH
Valves, Fittings, Actuators
Tel: 01473 270222
Fax: 01473 270393

Cumbria Heating Components
Block Two
Brymau Three Industrial Estate
River Lane
Saltney
Chester CH4 8RH
Domestic & Industrial Heating Equipment
Tel: 01244 671877
Fax: 01244 671305

Cutler Hammer Limited
Mill Street
Ottery St Mary
Devon
EX11 1AG
Cable Management Systems
Tel: 01404 812131
Fax: 01404 815471

Dorman-Smith Switchgear Ltd
Blackpool Road
Preston PR2 2DQ
Switchgear & Distribution
Tel: 01772 728271
Fax: 01772 726276

Friatec UK Ltd
Friatec House
Old Parkbury Lane
Colney Street
Saint Albans
Herts
AL2 2ED
PVCC Pipes and Fittings
Tel: 01923 857878
Fax : 01923 853434

CCT Pipefreezing Ltd
74 Southbridge Road
Croydon
Surrey CR0 1AE
Pipefreezing
Tel: (020) 8680 2230
Fax: (020) 8680 2234

Crabtree Electrical Industries Ltd
Lincoln Works
Walsall WS1 2DN
Switchgear & Distribution,
Wiring, Fittings & Accessories
Tel: 01922 721202
Fax: 01922 721321

Drainage Centre
Cray Avenue
St Mary Cray
Orpington
Kent BR5 3RH
C.I. Pipes
Tel: 020 8855 8121
Fax : 020 8556 2716

Dunham Bush
Downley Road
Havant
Hampshire
P09 2JD
Air Distribution, Heating
Tel: 02392 477700
Fax: 02392 450396

Durapipe S&LP
Walsall Road
Norton Canes
Cannock
Staffs WS11 3NS
Thermoplastic Piping
Tel: 01543 279909
Fax: 01543 279450
e-mail: marketing@durapipe-slp.co.uk

Geberit Ltd
New Hythe Business Park
Aylesford
Kent ME20 7PJ
UPVC/ABS/Polypropelene Pipework and Fittings
Tel: 01622 717811
Fax: 01622 716920
technical@geberit-terrain.demon.co.uk
www.geberit.co.uk

Acknowledgements

Furse and Company Limited
Wilford Road
Nottingham NG2 1EB
Lightning Protection
Tel: 0115 863471
Fax: 0115 9860071

GEC-Alsthom Installation Equipment
East Lancashire Road
Liverpool L10 5HB
Switchgear & Distribution
Tel: 0151 525 8371
Fax: 0151 523 7007

Hattersley, Newman, Hender Ltd
Burscough Road
Ormskirk
Lancashire L39 2XG
Valves
Tel: 01695 577199
Fax: 01695 578775
e-mail: uksales@hattersley-valves.co.uk

Hawker Siddeley Switchgear Ltd
P O Box 19
Falcon Works
Loughborough
Leicestershire LE11 1HL
Switchgear & Distribution
Tel: 01509 611311
Fax: 01509 610404

Horseley Bridge Tanks
27/29 Thornleigh Trading Estate
Blowers Green
Dudley
West Midlands DY2 8UB
Storage Tanks
Tel: 01384 459119
Fax: 01384 459117

JIB- For the Electrical Industry
Kingswood House
47/51 Sidcup Hill
Kent
DA14 6HP
Labour
Tel: (020) 8302 0031
Fax: (020) 8309 1103

JIB-Plumb & Mech Eng Industry
The Joint Industry Board For Plumbing
Brook House
Brook Street
St. Neots
Huntingdon PE19 2HW
Labour
Tel: 01480 476925
Fax: 01480 403081

Engineering Appliances Ltd
Unit 11
Sunbury Cross Ind Est
Brooklands Close
Sunbury On Thames
TW16 7DX
Expansion Joints
Tel: 01932 788888
Fax: 01932 761263
e-mail: info@engineering-appliances.co

HRP Ltd
5 Rivington Court
Hardwick Grange
Warrington WA1 4RT
Air Conditioning
Tel: 01925 837688
Fax: 01925 831093

HVCA Publications
Old Mansion House
Eamont Bridge
Penrith
Cumbria
CA10 2BX
Labour
Tel: 01768 864771
Fax :01768 867138

IAC
IEC House
Moorside Road
Winchester
Hampshire
SO23 7US
Attenuators
Tel : 01962 873000
Fax : 01962 873102

Igranic Control Systems
Murdoch Road
Bedford
MK41 7PT
Circuit Breakers
Tel : 01234 267242
Fax : 01234 219061

IMI Yorkshire Fittings Ltd
P O Box 166
Leeds LS1 1RD
**Copper Pipes, Fittings, Fixings
& Sundries**
Tel : 0113 270 6945
Fax : 0113 270 5644

Kopex International Ltd
189 Bath Road
Slough
Berks SL1 4AR
Wiring, Fittings & Accessories
Tel: 01753 534931
Fax: 01753 693521

W Lucy & Co. Ltd
Eagle Works
Walton Well Road
Oxford OX2 6EE
Switchgear
Tel: 01865 311411
Fax: 01865 310504
e-mail : switchgear.sales@wlucy.co.uk
www.lucyswitchgear.com

Lancashire Fittings Limited
The Science Village
Claro Road
Harrogate
North Yorkshire HG1 4AF
Stainless Steel Fittings
Tel: 01423 522355
Fax: 01423 506111
e-mail: kenidle@lancs-fittings.co.uk

Marley Extrusions Ltd
Dickley Lane
Lenham
Maidstone
Kent ME17 2DE
UPVC, Pipes, Fittings, Fixings & Sundries
Tel: 01622 858888
Fax: 01622 858725
www.marley.co.uk

Mita (UK) Ltd
Bodelwyddan Business Park
Bodelwyddan
Denbighshire
LL18 5SX
Cabling Management Systems
Tel: 01745 586000
Fax: 01745 586015

Modular Wiring Systems
24 Watford Metro Centre
Dwight Road
Watford WD18 9XB
Modular Wiring
Tel: 01923 220777
Fax: 01923 800787
info@modwire.com
www.modwire.com

Jasun Filtration
Unit 2B/2c
Yeo Road
Bridgewater
Somerset
TA6 5NA
Air Filtration
Tel: 01278 452277
Fax: 01278 450873

Marshall Tufflex Ltd
Ponswood Industrial Estate
Hastings
East Sussex TN34 1YJ
Cabling Management Systems
Tel: 01424 427691
Fax: 01424 720670

MEM Ltd
Whitegate
Broadway, Chadderton
Oldham OL9 9QG
Wiring, Fittings & Accessories
Tel: 0161 652 1111
Fax: 0161 626 1709

Netco Ltd
4 Watt House
Pensnett Estate
Kingswinford
West Midlands DY6 8XZ
Trace Heating
Tel: 01384 400750
Fax: 01384 400314

Ozonair Limited
Quarrywood Industrial Estate
Aylesford
Maidstone
Kent ME20 7NB
Air Distribution Equipment
Tel: 01622 717861
Fax: 01622 719291

Pegler Ltd
St Catherines Avenue
Doncaster
South Yorkshire DN4 8DF
Radiators, Heaters & Controls, Valves
Tel: 01302 329777
Fax: 01302 730515

Philmac PTY Ltd
Diplocks Way
Hailsham
East Sussex
BN27 3JF
Pipes
Tel: 01323 847323

Acknowledgements

Myson RCM
Old Wolverton Road
Milton Keynes
Buckinghamshire MK12 5PT
Skirting Radiator
Tel: 01908 321155
Fax: 01908 317387

Pipeline Centre
116 London Road
Hailsham
East Sussex BN27 3AL
Plastic Pipeline
Tel : 01323 44233
Fax : 01323 847488

Pirelli Cables Ltd
Special Cables Division
PO Box 30
Chickenhall Lane
Eastleigh
Hants SO5 5XA
Cables
Tel: 01703 644544
Fax: 01703 649649

Potterton Myson Limited
Eastern Avenue
Team Valley Trading Estate
Gateshead
Newcastle upon Tyne NE11 0PG
Heating Products
Tel: 0191 491 7500
Fax: 0191 491 756

Raychem Ltd
Chemelex Division
Faraday Road
Dorcan Swindon
Wiltshire SN3 5HH
Leak Sensing Cable
Tel: 01793 572663
Fax: 01793 572189

Salamandre Plc
Hunts Rise
South Marston Park
South Marston
Swindon
Wiltshire SN3 4RE
Cabling Trunking Systems
Tel: 01793 828000
Fax: 01793 828597

Sauter Automation
86 Talbot Road
Old Trafford
Manchester M16 0PG
Butterfly Control Valves
Tel: 0161 8724791
Fax: 0161 8480855

Fax: 01323 844775
Schneider Electric Ltd
Stafford Park 5
Telford
Shropshire TF3 3BL
Busbar
Tel : 01952 209589
Fax : 01952 292238

Selkirk Manufacturing Ltd
Bassett House
High Street
Banstead
Surrey SM7 2LZ
Chimney Systems, Exhausts
Tel: 01737 353388
Fax: 01737 362501
e-mail: info@selkirk.co.uk

South Wales Transformers
Newport Road
Blackwood
Gwent NP2 2XP
Switchgear & Distribution
Tel: 01495 232100
Fax: 01495 232132
e-mail: sales@swtran.co.uk

Spirax-Sarco Ltd
Charlton House
Cheltenham
Gloucestershire
GL53 8ER
Traps and Valves
Tel: 01242 521361
Fax: 01242 573342

Sunvic Controls Ltd
Bellshill Road
Glasgow
G71 6NP
Heat Controls
Tel: 01698 812944
Fax: 01698 813637

Swifts of Scarborough Limited
Cayton Low Road
Eastfield
Scarborough
North Yorkshire
YO11 3BY
Cable Trays
Tel: 01723 583131
Fax: 01723 584625

xvi *Acknowledgements*

Thorn Lighting Ltd
3 King George Close
Eastern Avenue West
Romford
Essex RM7 7PP
Lighting
Tel: 01708 766033
Fax: 01708 766381

Trox (UK) Limited
Caxton Way
Thetford
Norfolk IP24 3SQ
Grilles & Diffusers
Tel : 01842 754545
Fax : 01842 763051
e-mail : trox@troxuk.co.uk
www.troxuk.co.uk

Vokes Ltd
Henley Park
Guildford
Surrey GU3 2AF
Air Filters
Tel: 01483 569971
Fax: 01483 235384
e-mail: vokes@btvinc.com

Vent-Axia Limited
Unit 2
Caledonia Way
Stretford Motorway Estate
Barton Dock Road
Manchester M32 0ZH
Ventilation/Cooling, Fans
Tel: 0161 865 8421
Fax: 0161 865 0098

Walsall Conduits Ltd
Dial Lane
West Bromwich
West Midlands B70 0EB
Cabling Management Systems
Tel: 0121 557 1171
Fax: 0121 557 5631

Weidmuller (Klippon Products) Ltd
Power Station Road
Sheerness
Kent ME12 3AB
Wiring Accessories
Tel: 01795 580999
Fax: 01795 580115

Wibe Ltd
Unit 8D, Castle Vale Ind. Estate
Maybrook Road
Minworth
Sutton Coldfield
West Midlands
B76 1AL
Cabling Support Systems
Tel: 0121 313 1010
Fax: 0121 313 1020

Woods of Colchester
Tufnell Way
Colchester
Essex CO4 5AR
Air Distribution, Fans, Anti-vibration mountings
Tel: 01206 544122
Fax: 01206 574434
e-mail: enquiriesponor.com

PART ONE

Approximate Estimating

Directions, *page 3*
Cost Indices, *page 4*
Outline Costs, *page 6*
Elemental and All-in-Rates, *page 7*
Elemental Costs, *page 23*

SPON'S PRICE BOOKS 2002

Free CD-ROM when you order any Spon's 2002 Price Book.
Use the CD-ROM to:
- produce tender documents
- customise data
- keyword search
- export to other major packages
- perform simple calculations.

updates available to download from the web
www.pricebooks.co.uk

Spon's Architects' and Builders' Price Book 2002
Davis Langdon & Everest

"Spon's Price Books have always been a 'Bible' in my work - now they have got even better! The CDs are not only quick but easy to use. The CD ROMs will really help me to get the most from my Spon's in my role as a Freelance Surveyor."
Martin Taylor, Isle of Lewis

New Features for 2002 include:
- A new section on Captial Allowances
- Inclusion of new items within a seperate Measured Works section

September 2001: 1024 pages
Hb & CD-ROM: 0-415-26216-X: £110.00

Spon's Mechanical and Electrical Services Price Book 2002
Mott Green & Wall

"An essential reference for everybody concerned with the calculation of costs of mechanical and electrical works." *Cost Engineer*

New Features for 2002 include:
- New sections on modular wiring, emergency lighting, lighting control, sprinkler pre fabricated pipework, UPVC rainwater and gutters, carbon steel pipework and fittings

September 2001: 584 pages
Hb & CD-ROM: 0-415-26222-4: £110.00

Spon's Landscapes and External Works Price Book 2002
Davis Langdon & Everest, in association with Landscape Projects

New Features for 2002 include:
- Fees for professional services
- Revised and updated sections on Cost Information and how to use this book
- Revisions and expansions of the Approximate Estimating section, together with direct links into the Measured Works Section

September 2001: 484 pages
Hb & CD-ROM: 0-415-26220-8: £80.00

Spon's Civil Engineering and Highway Works Price Book 2002
Davis Langdon & Everest

New Features for 2002 include:
- A revised and extended section on Land Remediation
- The Rail Track section now includes data on Permanent Way work with fully reviewed pricing
- Fully reviewed pricing for the Geotextiles section

September 2001: 688 pages
Hb & CD-ROM: 0-415-26218-6: £120.00

Return your orders to: Spon Press Customer Service Department, ITPS, Cheriton House, North Way, Andover, Hampshire, SP10 5BE · Tel: +44 (0) 1264 343071 · Fax: + 44 (0) 1264 343005 · Email: book.orders@tandf.co.uk
Postage & Packing: 5% of order value (min. charge £1, max. charge £10) for 3–5 days delivery · Option of next day delivery at an additional £6.50.

DIRECTIONS

Prices shown are average prices on a Fixed Price basis for typical buildings for completion during the second quarter, of 2002. Unless otherwise shown, they exclude external services and professional fees.

The information in this section has been arranged to follow more closely the order in which estimates may be developed:

a)   Cost Indices and Regional Variations - gives indices and variations to be applied to estimates in general

b)   Outline Costs - gives a range of data (based on a rate per square metre) for all-in engineering costs associated with a wide variety of building functions.

For certain specified building types these initial all-in costs are further subdivided to give a broad division between the main work elements.

c)   Elemental and All-in-Rates - given for a number of items and complete component parts i.e. boiler plant, ductwork and mains and switchgear together with some other items not covered in parts of this or other sections of the book i.e. lifts, escalators, kitchen equipment, medical gases.

d)   Elemental Costs - are shown for five building types - comprising two office blocks, one hotel, one leisure related building and a supermarket. In each case, a full analysis of engineering services costs is given to show the division between all elements and their relative costs to the total building area. A regional variation factor has been applied to bring these analyses to a common London base.

Prices should be applied to the total floor area of all storeys of the building under consideration. The area should be measured between the external walls without deduction for internal walls and staircases/lift shafts.

Although prices are reviewed in the light of recent tenders it has only been possible to provide a range of prices for each building type. This should serve to emphasise that these can only be average prices for typical requirements and that such prices can vary widely depending upon a number of features. Rates per square metre should not therefore be used indiscriminately and each case must be assessed on its merits.

The prices do not include for incidental builder's work nor for profit and attendance by a main contractor where the work is executed as a sub-contract: they do however include for preliminaries, profit and overheads for the services contractor. Capital contributions to statutory authorities and public undertakings and the cost of work carried out by them have been excluded.

COST INDICES

The following tables reflect the major changes in cost to contractors but do not necessarily reflect changes in tender levels. In addition to changes in labour and materials costs, tenders are affected by other factors such as the degree of competition in the particular industry and area where the work is to be carried out, the availability of labour and the general economic situation. This has meant in recent years that, when there has been an abundance of work, tender levels have often increased at a greater rate than can be accounted for by increases in basic labour and material costs and, conversely, when there is a shortage of work this has often resulted in keener tenders. Allowances for these factors are impossible to assess on a general basis and can only be based on experience and knowledge of the particular circumstances. In compiling the tables the cost of labour has been calculated on the basis of a notional gang as set out elsewhere in the book. The proportion of labour to materials has been assumed as follows: Mechanical Services - 30:70 / Electrical Services - 50:50 (1976 = 100)

Mechanical Services

Year	First Quarter	Second Quarter	Third Quarter	Fourth Quarter
1990	288	298	299	301
1991	312	316	317	319
1992	323	325	326	329
1993	333	334	336	338
1994	341	342	345	350
1995	357	358	360	365
1996	364	361	357	359
1997	361	356	358	363
1998	365	363	368	373
1999	368	363	370	384
2000	384	386	388	400
2001	401	402	P 406	P 400
2002	P 416	F 417	F 418	F 451

Electrical Services

Year	First Quarter	Second Quarter	Third Quarter	Fourth Quarter
1990	323	324	326	330
1991	350	350	350	350
1992	367	368	368	370
1993	373	377	378	378
1994	380	381	383	389
1995	397	397	398	399
1996	406	404	401	402
1997	411	411	411	411
1998	410	423	422	422
1999	433	432	431	446
2000	458	464	465	468
2001	485	485	486	P 487
2002	P 507	F 508	F 509	F 510

(P = Provisional)
(F = Forecast)

Approximate Estimating

COST INDICES

Regional Variations

Prices throughout this Book generally apply to work in the London area (see Directions at the beginning of the Mechanical Installations and Electrical Installations sections). Prices for mechanical and electrical services installations to vary from region to region, largely as a result of differing labour costs but also dependent on accessibility, urbanisation and local market conditions.

The following table to regional factors are intended to provide readers with indicative adjustments that may be made to the prices in the Book for locations outside of London. The figures are of necessity averages for regions and further adjustments should be considered for city centre or very isolated or other known local factors.

Greater London	1.00	North West	0.90
South East	0.95	Northern	0.91
South West	0.91	Scotland	0.90
East Midlands	0.89	Wales	0.88
West Midlands	0.90	Northern Ireland	0.80
East Anglia	0.91	Channel Islands	1.20
Yorkshire & Humberside	0.88		

6 *Approximate Estimating*

OUTLINE COSTS

SQUARE METRE RATES FOR BASIC SERVICES BY BUILDING TYPE
The undernoted examples indicate the range of rates within which normal engineering services excluding lifts would be contained for each of the more common building types.

	Square metre £
Industrial Buildings (CI/SfB 2)	
Factories	
owner occupation	64 to 161
Warehouses	
high bay for owner occupation	86 to 171
Administrative, public, commercial and office buildings; general (CI/SfB 3)	
Civic offices	
fully air conditioned	350 to 500
Offices for letting	
non air conditioned; Cat A	170 to 190
fully air conditioned; Cat A	280 to 410
Offices for owner occupation	
non air conditioned	210 to 280
fully air conditioned	310 to 500
Health and welfare facilities (CI/SfB 4)	
District general hospitals	393 to 556
Private hospitals	415 to 654
Refreshment, entertainment, recreation buildings (CI/SfB 5)	
Arts and drama centres	294 to 403
Theatres, large	382 to 519
Educational, scientific and information buildings (CI/SfB 7)	
Secondary/middle schools	163 to 241
Universities	
arts buildings	212 to 268
science buildings	
physics	213 to 336
biology	299 to 353
chemistry	336 to 395
Museums	
national	721 to 901
Residential buildings (CI/SfB 8)	
Local authority schemes	
two storey houses	71 to 97
medium rise flats	96 to 126

Approximate Estimating

ELEMENTAL AND ALL-IN-RATES

ELEMENTAL RATES FOR ALTERNATIVE MECHANICAL SERVICES SOLUTIONS FOR OFFICES

The undernoted examples, when considered within the context of office accommodation, indicate the range of rates for alternative design solutions for each of the basic mechanical services elements based on gross area.

	Square metre £
Sanitaryware and Above ground disposal installation	
Normal services for building up to 3,000 m²	4 to 12
Normal services for low rise building over 3,000 m²	10 to 15
Water installation	
Normal services for building up to 3,000 m²	5 to 15
Normal services for building over 3,000 m²	9 to 17
Heating installation; including gas installations	
LPHW installations for building up to 3,000 m²	55 to 65
LPHW installations for building over 3,000 m²	45 to 60
Air conditioning: including ventilation	
Comfort cooling, 2 pipe fan coil, for building up to 3,000m²	110 to 130
Comfort cooling, 2 pipe fan coil, for building over 3,000m²	105 to 120
Comfort cooling, 2 pipe variable refrigerant volume for building up to 3,000m²	90 to 110
Full air conditioning, fan coil for building up to 3,000m²	150 to 170
Full air conditioning, fan coil for building over 3,000m²	135 to 155
Full air conditioning, variable air volume for building over 3,000m²	150 to 180
Full air conditioning, fan assisted variable air volume for building over 3,000m²	150 to 180
Full air conditioning, 3 pipe variable refrigerant volume, speculative offices	115 to 135
Full air conditioning, concealed chilled beams for building over 3,000m²	155 to 185
Full air conditioning, chilled ceiling for building over 3,000m²	175 to 210
Full air conditioning, displacement for building over 3,000m²	105 to 140
Fire Protection	
Dry risers	1 to 2
Sprinkler installation	15 to 20
BMS Controls: including MCC panels and control cabling	
Full air conditioning, fan coil	20 to 30
Full air conditioning, variable air volume	20 to 25

8   *Approximate Estimating*

ELEMENTAL AND ALL-IN-RATES

ELEMENTAL RATES FOR AIR CONDITIONING FOR OFFICES

Air conditioning installations vary considerably according to the type of building they serve in addition to the delivery system selected. No two buildings will have precisely the same requirements and therefore the figures below should be adjusted accordingly to the users particular project.

The following information has been compiled to indicate the average cost of two different design solutions, these being Variable air volume and Fan coil. Brief specification notes are provided to enable the user to make his own adjustments to the cost of individual elements to take into account his own design criteria.

It has been assumed that the building is situated in London.

Approximate costs of air conditioning installations for two specimen office blocks with gross floor areas of 3,000 m² and 10,000 m² respectively to Category A specification.

The costs allow for all plant and equipment, distribution ductwork and pipework, BMS controls and all associated electrical work.

	Floor area	
	3,000 m²	*10,000 m²*
	£	£
Variable air volume system, per m²	239.00	189.00
Fan Coil system, per m²	217.00	174.00

Approximate Estimating

ELEMENTAL AND ALL-IN-RATES

ELEMENTAL RATES FOR AIR CONDITIONING FOR OFFICES *continued*

VARIABLE AIR VOLUME SYSTEM

Spec'n Ref.	Elements	Office block of 3,000 m²		Office block of 10,000 m²	
		Cost of element £	Cost of element per m² of floor area £	Cost of element £	Cost of element per m² of floor area £
A	Boilers and flue	17,300	5.76	47,250	4.73
	LTHW heating				
B	Primary Pipework	30,300	10.10	99,800	9.98
C	Secondary pipework	26,800	8.93	87,200	8.72
D	Convectors and/or radiators	12,000	4.00	29,400	2.94
	Chilled Water				
E	Air cooled chillers	41,300	13.77	106,000	10.60
F	Primary Pipework	113,000	37.67	227,000	22.70
	Ductwork				
G	Air Handling Units	82,000	27.33	314,500	31.45
H	Extract Fans	11,700	3.90	42,000	4.20
J	VAV Boxes	53,000	17.67	183,800	18.38
K	Ductwork	189,000	63.00	374,900	37.49
L	BMS Controls	109,000	36.30	286,600	28.66
M	Electrical work in connection	32,500	10.83	91,400	9.14
	Totals	717,900	239.30	1,889,850	188.99
		say	239.00	say	189.00

10 *Approximate Estimating*

ELEMENTAL AND ALL-IN-RATES

ELEMENTAL RATES FOR AIR CONDITIONING FOR OFFICES *continued*

FAN COIL SYSTEM

Spec'n Ref.	Elements	*Office block of 3,000 m²*		*Office block of 10,000 m²*	
		Cost of element £	*Cost of element per m² of floor area* £	*Cost of element* £	*Cost of element per m² of floor area* £
A	Boilers and flue	17,300	5.76	47,250	4.73
	LTHW heating				
B	Primary Pipework	30,300	10.10	99,800	9.98
C	Secondary pipework	26,800	8.93	87,200	8.72
D	Convectors and/or radiators	12,000	4.00	29,400	2.94
	Chilled Water				
E	Air cooled chillers	41,300	13.77	106,000	10.60
F	Primary Pipework	113,000	37.67	227,000	22.70
G	Secondary Pipework	62,400	20.80	158,000	15.80
	Ductwork				
H	Air Handling Units	41,000	13.67	158,000	15.80
J	Extract Fans	11,700	3.90	42,000	4.20
K	Fan Coil Units	66,600	22.20	156,500	15.65
L	Ductwork	88,200	29.40	250,000	25.00
M	BMS Controls	109,000	36.30	286,600	28.66
N	Electrical work in connection	32,500	10.83	91,400	9.14
	Totals	652,100	217.37	1,739,150	173.92
			say £ 217.00		say £ 174.00

Approximate Estimating

ELEMENTAL AND ALL-IN-RATES

ELEMENTAL RATES FOR AIR CONDITIONING FOR OFFICES *continued*

Brief Specification Notes

Ref.

A   Boilers and Flues: includes gas fired boilers, gas installation, boiler control system, stainless steel twin wall flue.

B   Primary pipework: includes primary pump sets, inertia bases, valves, supports, water installation, insulation and identification.

C   Secondary Pipework: includes secondary pump sets, inertia bases, pressurisation sets ,valves, supports, insulation and identification.

D   Convector and/or radiators: includes steel panel radiators in circulation areas and staircases, Door heaters in reception entrance doors.

E   Air-cooled chillers: includes R134a refrigerant, controls system, AV mounts.

F   Primary pipework: includes primary pump sets and inertia bases, valves, supports, insulation and identification.

G   Secondary Pipework: includes secondary pump sets and inertia bases, pressurisation sets, valves, supports, insulation and identification.

H   Air Handling units: includes inverter drives, filters, heating and cooling batteries, attenuators and AV mounts

J   Extract Fans: includes attenuators and AV mounts.

K   VAV/FCU: includes LTHW/chilled 2 pipe or 4 pipe batteries, supports

L   Ductwork: DW 144 includes fittings and supports, volume control/ fire/smoke dampers, grilles and diffusers, insulation.

M   BMS controls: analogue addressable includes MCC panels, head end PC, intelligent outstations, valves, sensors and control cabling

N   Electrical work in connection: Electrical supplies to MCC panels and mechanical plant.

12   *Approximate Estimating*

ELEMENTAL AND ALL-IN-RATES

ALL IN RATES FOR PRICING MECHANICAL APPROXIMATE QUANTITIES *continued*

HEAT SOURCE

BOILERS

	Cost Per kW £
Gas fired boilers including gas train and controls	16.00 - 21.00
Gas fired boilers including gas train, controls, flue, plantroom pipework, valves and insulation, pumps and pressurisation unit	60.00 - 115.00

Approximate Estimating   13

ELEMENTAL AND ALL-IN-RATES

ALL IN RATES FOR PRICING MECHANICAL APPROXIMATE QUANTITIES *continued*

SPACE HEATING AND AIR TREATMENT

DOMESTIC CENTRAL HEATING

Gas fired fan assisted balanced flue boiler, pump, HWS cylinder and thermostatic controls, 3 way valve, copper distribution pipework, fittings, valves and insulation, radiators and thermostatic valves, feed and expansion tank, room thermostat :

£

Heating for 2 bedroom flat	2,400.00 - 3,000.00
Heating for 2 bedroom house	2,800.00 - 3,500.00
Heating for 3 bedroom house	3,600.00 - 4,500.00
Heating for 4 bedroom house	4,400.00 - 5,500.00
Heating for 5 bedroom house	5,200.00 - 6,500.00

Gas fired fan assisted balanced flue combination boiler, pump, mains water connection, copper distribution pipework, fittings, valves and insulation, radiators and thermostatic valves, room thermostat :

£

Heating for 1 bedroom flat	2,000.00 - 2,600.00
Heating for 2 bedroom flat	2,200.00 - 2,800.00
Heating for 2 bedroom house	2,600.00 - 3,300.00
Heating for 3 bedroom house	3,500.00 - 4,300.00

Balanced flue oil fired boiler, oil storage tank and pump, pump, HWS cylinder, 3 way valve, copper distribution pipework, fittings, valves, insulation, radiators and thermostatic valves, feed and expansion tank, room thermostat :

Heating for 2 bedroom house	3,800.00 - 4,500.00
Heating for 3 bedroom house	4,600.00 - 5,500.00
Heating for 4 bedroom house	5,400.00 - 6,500.00
Heating for 5 bedroom house	6,200.00 - 7,500.00

14   *Approximate Estimating*

ELEMENTAL AND ALL-IN-RATES

ALL IN RATES FOR PRICING MECHANICAL APPROXIMATE QUANTITIES *continued*

	Cost per kW £
CHILLED WATER	
Air cooled R134a refrigerant chiller including control panel, anti vibration mountings	90.00 – 115.00
Air cooled R134a chiller including control panel, antivibration mountings, plantroom pipework, valves, insulation, pumps and pressurisation units	120.00 - 215.00

DUCTWORK

The rates below allow for ductwork and for all other labour and material in fabrication fittings, supports and jointing to equipment, stop and capped ends, elbows, bends, diminishing and transition pieces, regular and reducing couplings, branch diffuser and 'snap on' grille connections, ties, 'Ys', crossover spigots, etc., turning vanes, regulating dampers, access doors and openings, handholes, test holes and covers, blanking plates, flanges, stiffeners, tie rods and all supports and brackets fixed to structure.

	Per M^2 £
Rectangular galvanised mild steel ductwork as HVCA DW 144 up to 1000mm longest side	35.00 - 40.00
Rectangular galvanised mild steel ductwork as HVCA DW 144 up to 2500mm longest side	40.00 - 50.00
Rectangular galvanised mild steel ductwork as HVCA DW 144 3000mm longest side and above	55.00 - 60.00
Circular galvanised mild steel ductwork as HVCA DW 144	40.00 - 50.00
Flat oval galvanised mild steel ductwork as HVCA DW 144 up to 545mm wide	40.00 - 45.00
Flat oval galvanised mild steel ductwork as HVCA DW 144 up to 880mm wide	45.00 - 50.00
Flat oval galvanised mild steel ductwork as HVCA DW 144 up to 1785mm wide	55.00 - 60.00

Approximate Estimating   15

ELEMENTAL AND ALL-IN-RATES

ALL IN RATES FOR PRICING MECHANICAL APPROXIMATE QUANTITIES *continued*

PACKAGE AIR HANDLING UNITS

Cost per M^3
£

Air handling unit including LPHW pre-heater coil, pre-filter panel, LPHW heater coils, chilled water coil, filter panels, inverter drive, motorised volume control dampers, sound attenuation, flexible connections to ductwork and all anti-vibration mountings.   3500.00 - 5000.00

EXTRACT FANS

Cost per M^3
£

Extract fan including inverter drive, sound attenuation, flexible connections to ductwork and all anti-vibration mountings   1,000.00 - 2000.00

PROTECTIVE INSTALLATIONS

SPRINKLER INSTALLATION

Sprinkler equipment installation, pipework, valve sets, booster pumps and water storage   50,000.00 - 75,000.00

£
Price per sprinkler head; including pipework, valves and supports 150.00

Recommended maximum area coverage per sprinkler head:
     Extra light hazard, 21 m² of floor area
     Ordinary hazard, 12 m² of floor area
     Extra high hazard, 9 m² of floor area

HOSE REELS AND DRY RISERS

Wall mounted concealed hose reel with 36 metre hose including approximately 15 metres of pipework and isolating valve:

     Price per hose reel 1,200.00

100mm dry riser main including 2 way breeching valve and box,, 65mm landing valve, complete with padlock and leather strap and automatic air vent and drain valve.

£
     Price per landing 1,000.00

THERMAL INSULATION

These have not been detailed in this section for sizes and types, see 'Prices for Measured Work' section.

16   *Approximate Estimating*

ELEMENTAL AND ALL-IN-RATES

ALL IN RATES FOR PRICING ELECTRICAL APPROXIMATE QUANTITIES

HV/LV INSTALLATIONS

The cost of HV/LV equipment will vary according to the electricity suppliers requirements, the duty required and the actual location of the site. For estimating purposes the items indicated below are typically the equipment required in a HV substation incorporated into a building.

RING MAIN UNIT

	Cost per Unit
	£
Ring Main Unit , 11kv including electrical terminations	7,500.00 to 15,000.00

TRANSFORMERS.

	Cost per KVA
	£
Oil filled transformers, 11kv to 415v including electrical terminations	20.00
Cast Resin transformers, 11kv to 415v including electrical terminations	15.00

HV SWITCHGEAR

	Cost per Section
Cubicle section HV switchpanel, Form 4 type 6 including air circuit breakers, meters and electrical terminations	£
	20,000.00

LV SWITCHGEAR

	Cost per Isolator
	£
LV switchpanel, Form 3 including all isolators, fuses, meters and electrical terminations	1,500.00 - 2,500.00
LV switchpanel, Form 4 type 5 including all isolators, fuses, meters and electrical terminations	2,500.00 - 3,500.00

PACKAGED SUB-STATION

	£
Extra over cost for prefabricated packaged sub station housing excludes base and protective security fencing	20,000.00 – 25,000.00

Spon Press Marketing

Freepost SN926

11 New Fetter Lane

LONDON EC4B 4FH

FREE UPDATES

REGISTER TODAY

Available to download from the web: www.pricebooks.co.uk

All four Spon Price Books - Architects' and Builders', Civil Engineering and Highway Works, Landscape and External Works and Mechanical and Electrical Services - are supported by an updating service. Three updates are issued during the year, in November, February and May. Each gives details of changes in prices of materials, wage rates and other significant items, with regional price level adjustments for Northern Ireland, Scotland and Wales and regions of England.

As a purchaser of a Spon Price Book you are entitled to this updating service for the 2002 edition - free of charge. Simply complete this registration card and return it to us - or register via the website **www.pricebooks.co.uk**. The updates will also be available to download from the website.

The Updating service terminates with the publication of the next annual edition.

REGISTRATION CARD for Spon's Price Book Update   2002
Please print your details clearly

Name..
(Please indicate membership of professional bodies e.g. RICS, RIBA, CioB etc.)
Position/Department...
Organisation..
Address..
....
.... Postcode........................
Tel:... Fax:...
E-mail address... (Please print clearly)

FIND OUT MORE ABOUT SPON BOOKS
Visit www.sponpress.com for more details.

Spon's Price Books Online
☐ Please tick the box if you are interested in subscribing to Spon's price data online. We will contact you if the service becomes available.
Please tick the other areas of interest
☐ International Price Books   ☐ Occupational Safety and Health
☐ Architecture   ☐ Landscape
☐ Planning   ☐ Journals - inc. Journal of Architecture;
☐ Civil Engineering   Building Research and Information;
☐ Building & Construction Management   Construction Management and Economics
☐ Environmental Engineering

☐ Other - note any other areas of interest..
....
....

Approximate Estimating

ELEMENTAL AND ALL-IN-RATES

ALL IN RATES FOR PRICING ELECTRICAL APPROXIMATE QUANTITIES *continued*

HV/LV INSTALLATIONS *continued*

Standby Generating Sets

Diesel powered including control panel, flue, oil day tank
and attenuation

Approximate installed cost .... £150.00 per kVA

Uninterruptible Power Supply

Rotary UPS including control panel, automatic bypass, DC
Isolator and batteries

Approximate installed cost ....£400.00 - £ 500.00 per kVA

Approximate Estimating

ELEMENTAL AND ALL-IN-RATES

ALL IN RATES FOR PRICING ELECTRICAL APPROXIMATE QUANTITIES *continued*

SMALL POWER

Approximate prices for wiring of power points of not exceeding 20m, including accessories, wireways but excluding distribution boards.

	Per Point £
13 amp Accessories	
Wired in PVC insulated twin and earth cable in ring main circuit	
Domestic properties	52.00
Commercial properties	70.00
Industrial properties	70.00
Wired in PVC insulated twin and earth cable in radial circuit	
Domestic properties	65.00
Commercial properties	85.00
Industrial property	85.00
Wired in LSF insulated single cable in ring main circuit	
Commercial properties	75.00
Industrial property	75.00
Wired in LSF insulated single cable in radial circuit	
Commercial properties	95.00
Industrial property	95.00
45 amp wired in PVC insulated twin and earth cable	
Domestic properties	95.00

Low voltage power circuits

Three phase four wire radial circuit feeding an individual load, wired in LSF insulated single cable including wireways, isolator, *not exceeding 10 metres; in commercial properties.*

Cable size Mm^2	£
1.5	160.00
2.5	175.00
4	190.00
6	205.00
10	240.00
16	265.00

Three phase four core radial circuit feeding an individual load item, wired in LSF/SWA/XLPE insulated cable including terminations, isolator; clipped to surface, *not exceeding 10 metres in commercial properties.*

Cable size Mm^2	£
1.5	130.00
2.5	145.00
4	160.00
6	175.00
10	270.00
16	350.00

Approximate Estimating

ELEMENTAL AND ALL-IN-RATES

ALL IN RATES FOR PRICING ELECTRICAL APPROXIMATE QUANTITIES *continued*

LIGHTING

Approximate prices for wiring of lighting points including rose, wireways but excluding distribution boards, luminaires and switches.

	Per Point £
Final Circuits	
Wired in PVC insulated twin and earth cable	
Domestic properties	35.00
Commercial properties	45.00
Industrial properties	45.00
Wired in LSF insulated single cable	
Commercial properties	60.00
Industrial property	65.00

ELECTRICAL WORKS IN CONNECTION WITH MECHANICAL SERVICES.

The cost of electrical connections to mechanical services equipment will vary depending on the type of building and complexity of the equipment. Therefore a rate of £ 5.00 per m² of gross floor area should be a useful guide to allow for power wiring, isolators and associated wireways.

FIRE ALARMS

Fire alarm circuits
Fire alarm system wired in two core MICC insulated cables, including all terminations, supports and wireways.

	Cost per point £
Sounder	165.00
Break glass contact	145.00
Heat detector	210.00
Smoke detector	210.00

Fire alarm system wired in two core firetuff insulated cables, including all terminations, supports and wireways.

	Average cost per point £
Sounder	160.00
Break glass contact	140.00
Heat detector	200.00
Smoke detector	200.00

For costs for zone control panel, battery chargers and batteries, see 'Prices for Measured Work' section.

20 *Approximate Estimating*

ELEMENTAL AND ALL-IN-RATES

ALL IN RATES FOR PRICING SPECIALIST APPROXIMATE QUANTITIES *continued*

EXTERNAL LIGHTING

Estate road lighting
Post type road lighting lantern 70 watt CDM-T 3000k complete with 5m high column with hinged lockable door, control gear and cut-out including 2.5 mm two core butyal cable internal wiring, interconnections and earthing fed by 16 mm² four core XLPE/SWA /LSF cable and terminations. Approximate installed price *per metre road length* (based on 300 metres run) including time switch but excluding builder's work in connection

Columns erected on same side of road at 30 m intervals £35.73
Columns erected on both sides of road at 30 m intervals
in staggered formation £35.73

Bollard lighting
Bollard lighting fitting 26 watt TC-D 3500k including control gear, all internal wiring, interconnections, earthing and 25 metres of 2.5 mm² three core XLPE/SWA/LSF cable

Approximate installed price excluding builder's work in connection............................ £784.00 *each*

Outdoor flood lighting
Wall mounted outdoor flood light fitting complete with tungsten halogen lamp, mounting bracket, wire guard and all internal wiring; fixed to brickwork or concrete and connected.

Installed price 500 watt .... £110.00 *each*
Installed price 1000 watt .... £136.00 *each*

Pedestal mounted outdoor floor light fitting complete with1000 watt MBF/U lamp, mounting bracket, control gear, contained in weatherproof steel box, all internal wiring, interconnections and earthing; fixed to brickwork or concrete and connected
Approximate installed price excluding builder's work in connection .... £910.25 *each*

Approximate Estimating

ELEMENTAL AND ALL-IN-RATES

ALL IN RATES FOR PRICING SPECIALIST APPROXIMATE QUANTITIES *continued*

LIFT INSTALLATIONS

The cost of lift installations will vary depending upon a variety of circumstances. The following prices assume a car height of 2.2 metres, manufacturers standard finish to cars, brushed stainless steel 2 panel centre opening doors to BSEN81 part 1 & 2 and Lift Regulations 1997.

Passenger Lifts	8 Person	10 Person	13 Person	16 Person	21 Person	26 Person
Electrically operated AC drive serving 8 levels with directional collective controls and a speed of 1.6 m/s	114,000	116,400	120,000	124,800	132,000	114,000
Add to above for:						
Bottom motor room	6,300	6,300	6,300	7,560	7,560	6,300
Extra levels served	3,780	3,780	4,410	4,410	4,410	3,780
Increase speed from 1.6 to 2.0 m/s	756	756	1,008	1,008	1,260	756
Increase speed from 2.0 to 2.5 m/s	1,764	1,764	2,142	2,142	2,520	1,764
Enhanced finish to car	12,000	12,000	12,000	12,000	14,400	12,000
Glass back	2,160	2,520	3,000	3,600	3,600	2,160
Glass doors	17,238	17,238	19,422	19,422	19,422	17,238
Fire fighting control	5,040	5,040	5,040	5,040	5,040	5,040
Increase height of car by 200mm	504	504	630	630	630	504
Top hat section	1,008	1,134	1,260	1,890	1,890	1,008
Shaft lighting/small power	3,480	3,480	3,480	3,480	3,480	3,480
Motor room lighting/small power	1,200	1,200	1,320	1,440	1,440	1,200
Lifting beams	1,680	1,680	1,920	2,040	2,040	1,680
Shaft secondary steel work	5,280	5,400	5,640	5,760	5,760	5,280
Dust sealing - shaft	2,880	2,880	3,120	3,360	3,600	24,001
Dust sealing - machine room	720	720	1,200	1,200	1,200	720
Electrically operated AC drive gearless serving 20 levels with directional collective controls and a speed of 2.5 m/s	--	--	282,000	292,800	300,000	--
Add to above for:						
Extra levels served	--	--	7,200	7,200	7,200	--
Increase speed from 2.5 to 4.0 m/s	--	--	12,000	12,000	14,400	--
Increase speed from 4.0 to 6.0 m/s	--	--	2,400	2,400	4,800	--
Enhanced finish to car	--	--	12,000	14,400	16,800	--
Fire fighting control	--	--	6,000	6,000	6,000	--
Increase height of car by 200mm	--	--	630	630	630	--
Top hat section	--	--	1,260	1,890	1,890	--
Shaft lighting/small power	--	--	3,480	3,480	3,480	--
Motor room lighting/small power	--	--	1,320	1,440	1,440	--
Lifting beams	--	--	1,920	2,040	2,040	--
Shaft secondary steel work	--	--	5,640	5,760	5,760	--
Dust sealing - shaft	--	--	3,120	3,360	3,600	--
Dust sealing - machine room	--	--	1,200	1,200	1,200	--

22   *Approximate Estimating*

ELEMENTAL AND ALL-IN-RATES

ALL IN RATES FOR PRICING SPECIALIST APPROXIMATE QUANTITIES *continued*

LIFT INSTALLATIONS *continued*

Passenger Lifts	8 Person	10 Person	13 Person	16 Person	21 Person	26 Person
Oil hydraulic operated drive serving 5 levels with directional collective controls and a speed of 0.63 m/s	56,400	52,800	54,000	--	--	--
Add to above for:						
Glass back	2,160	2,520	3,000	--	--	--
Glass doors	17,238	17,238	19,422	--	--	--
Fire fighting control	3,780	3,780	3,780	--	--	--
Shaft lighting/small power	3,480	3,480	3,480	--	--	--
Motor room lighting/small power	1,200	1,200	1,200	--	--	--
Shaft secondary steel work	5,280	5,400	5,640	--	--	--
Dust sealing - shaft	2,880	2,880	3,120	--	--	--
Dust sealing - machine room	720	720	1,200	--	--	--
Machineroomless electrically operated AC drive serving 8 levels with directional collective controls and a speed of 1.0 m/s	52,500	52,750	56,250	--	--	--
Add to above for:						
Extra levels served	6,250	6,250	6,250	--	--	--
Fire fighting control	6,250	6,250	6,250	--	--	--
Increase height of car by 200mm	1,250	1,250	1,250	--	--	--
Top hat section	1,875	1,875	1,875	--	--	--
Shaft lighting/small power	3,480	3,480	3,480	--	--	--
Lifting beams	1,680	1,680	1,680	--	--	--
Shaft secondary steel work	5,000	5,000	5,500	--	--	--
Dust sealing - shaft	2,880	2,880	2,880	--	--	--

Approximate Estimating

ELEMENTAL AND ALL-IN-RATES

ALL IN RATES FOR PRICING SPECIALIST APPROXIMATE QUANTITIES *continued*

LIFT INSTALLATIONS *continued*

Goods Lifts	8 Person	10 Person	13 Person	16 Person	21 Person	26 Person
Electrically operated two speed serving 8 levels to take 1000 kg load, prime coated internal finish and a speed of 0.63 m/s	13,200	105,600	108,000	114,000	99,000	103,200
Add to above for:						
Bottom motor room	6,300	6,300	6,300	7,560	6,300	6,300
Extra levels served	3,780	3,780	4,410	4,410	3,675	3,780
Increase capacity from 1000 kgs load to 2000 kgs load	6,250	6,250	6,250	6,250	6,250	6,250
Increase capacity from 2000 kgs load to 3000 kgs load	6,250	6,250	6,250	6,250	6,250	6,250
Increased speed of travel from 1.0 to 1.6 metres per second	630	1,320	2,400	1,200	1,000	630
Painted internal finish	660	660	660	900	750	660
Stainless steel internal finish	3,840	3,840	5,040	5,040	4,800	3,840
Oil hydraulic serving 5 levels to take 1000kg load, prime coated internal finish and a speed of 0.63m/s	56,400	52,800	54,000	--	--	--
Add to above for:						
Painted internal finish	660	660	660	--	--	--
Stainless steel internal finish	3,780	3,840	3,840	--	--	--

ESCALATOR INSTALLATIONS

30Ø Pitch escalator with a rise of 3 to 6 metres with standard balustrades ......

800mm step width	72,000
1000mm step width	84,000
Add to above for	
Balustrade Lighting	7,800
Skirting Lighting	8,400
Emergency stop button pedestals	2,640
Truss cladding - Stainless steel..	24,000
Truss cladding - Spray painted steel	19,200

Approximate Estimating

ELEMENTAL COSTS

SUPERMARKET

Supermarket located in the South East with a total gross floor area of 4,000m², which includes a sales area of 2,350m² upon which this cost analysis is based.

The building is on one level and incorporates a main sales, coffee shop, bakery, offices and amenities areas and warehouse.

Cost Summary

El. Ref.	Element	Total Cost £	Cost /m² £
5A	Sanitaryware	6,000.00	1.50
5C	Disposal Installations		
	Soil and Waste	9,700.00	2.43
5D	Water Installations		
	Cold water services	27,500.00	6.88
	Hot water services	21,000.00	5.25
5E	Heat Source	71,500.00	17.88
5F	Space Heating and Air Treatment		
	Heating with ventilation with supplemental cooling via DX units	18,000.00	44.50
5G	Ventilating Services		
	Supply and extract system	45,000.00	11.25
5H	Electrical Installation		
	Generator	63,000.00	15.75
	LV supply/distribution	141,000.00	35.25
	General lighting	113,000.00	28.25
	Emergency lighting	28,400.00	7.10
	Small power	84,000.00	21.00
	Earthing and bonding	1,000.00	0.25
5I	Gas Installation		
	Gas mains services to plantroom	13,000.00	3.25
	Carried forward	642,100.00	160.53

Approximate Estimating

ELEMENTAL COSTS

El. Ref.	Element	Total Cost £	Cost /m² £
	Brought forward	642,100.00	160.53
5K	Protection		
	Sprinklers	105,000.00	26.25
5L	Communication Installation		
	Fire alarms and detection	23,000.00	5.75
	Telecommunications (Containment only)	21,000.00	5.25
	Public address	6,000.00	1.50
	CCTV	26,250.00	6.56
	Intruder alarm and detection	10,000.00	2.50
5M	Special Installations		
	BMS Installation	42,000.00	10.50
	Electrical services in connection	26,000.00	6.50
	Summary total	901,350.00	225.34

26 *Approximate Estimating*

ELEMENTAL COSTS

OFFICE BUILDING

Speculative office in Central London with a gross floor area of 13,211m², 4 pipe fan coil system, with roof mounted air cooled chillers, gas fired boilers, basement area shell and core retail.

Category ☐A☐ fit out nett area of 7,735m².

Cost Summary

El. Ref.	Element	Total Cost £	Cost /m² £
	SHELL AND CORE		
5A	Sanitaryware	71,000.00	5.39
5C	Disposal Installations		
	Rainwater	46,760.00	3.54
	Soil and Waste	133,900.00	10.06
5D	Water Installations		
	Hot and cold water services	117,800.00	8.92
5E	Space Heating and Air Treatment		
	LTHW Heating in plantroom and risers	175,752.00	13.30
	Chilled water in basement and risers	283,812.00	21.48
	Supply and extract ductwork in plantroom and risers	178,943.00	13.55
5G	Ventilating Services		
	Toilet extract ventilation	37,000.00	2.80
	Car park extract	29,100.00	2.20
	Miscellaneous ventilation systems	146,800.00	11.11
5H	Electrical Installation		
	Generator	38,702.00	2.92
	LV supply/distribution	254,820.00	19.29
	General lighting	95,158.00	7.20
	General power	28,992.00	2.19
	External lighting	4,900.00	0.37
5I	Gas Installation	6,500.00	0.49
	Carried forward	1,649,939	124.89

Approximate Estimating

ELEMENTAL COSTS

El. Ref.	Element	Total Cost £	Cost /m² £
	Brought forward	1,649,939.00	124.89
5K	Protection		
	Dry risers	16,091.00	1.21
	Hosereels	22,611.00	1.71
	Sprinklers	307,948.00	23.31
	Earthing and bonding	9,710.00	0.74
	Lightning protection	1,664.00	0.13
5L	Communication Installation		
	Fire alarms	91,969.00	6.96
	Voice and data (wireways)	16,091.00	1.22
	Security	58,122.00	4.40
5M	Special Installation		
	Building management systems	203,218.00	15.38
	Electrical services in connection	114,500.00	8.67
	Summary total (based on gross floor area).......................	2,491,863.00	188.62

El. Ref.	Element	Total Cost £	Cost /m² £
	CATEGORY 'A' FIT OUT		
5F	Space Heating and Air Treatment		
	LTHW Heating	125,815.00	16.27
	Chilled water	272,575.00	35.24
	Ductwork, including fan coil units	456,374.00	59.00
5H	Electrical Installation		
	General lighting	601,500.00	77.76
5L	Protection		
	Earthing and bonding	1,664.00	0.22
5M	Special Installations		
	Building management system.........................	116,100.00	15.01
	Electrical services in connection	22,600.00	2.92
	Summary total (based on nett area)................	1,596,628.00	206.42

Approximate Estimating

ELEMENTAL COSTS

BUSINESS PARK

New build office in South East part of a speculative business park consisting of two 3 storey existing buildings and 1 new build, fitted out to Category A specification. Four pipe FCU system, external remote chiller, BMS controlled with all three buildings linked with an area of 7,500m² gross and nett area of 6,000m².

Cost Summary

El. Ref.	Element	Total Cost £	Cost /m² £
	SHELL AND CORE		
5A	Sanitaryware	26,413.00	3.52
5C	Disposal Installations		
	Rainwater	16,715.00	2.23
	Soil and waste	50,933.00	6.79
5D	Water Installations		
	Cold water services	33,272.00	4.44
	Hot water services	7,726.00	1.03
5E	Heat Source	Included in 5F	
5F	Space Heating and Air Treatment		
	LTHW Heating; plantroom and risers	150,829.00	20.11
	Chilled water; plantroom and risers	191,514.00	25.54
5G	Ventilating Services		
	Toilet and miscellaneous ventilation	24,000.00	3.20
5H	Electrical Installation		
	LV supply/distribution	56,200.00	7.49
	General lighting	77,977.00	10.40
	General power	11,700.00	1.56
5I	Gas Installation	8,700.00	1.16
	Carried forward	655,979.00	87.46

Approximate Estimating 29

ELEMENTAL COSTS

El. Ref.	Element	Total Cost £	Cost /m² £
	Brought forward	655,979.00	87.46
5K	Protective Installation		
	Earthing and bonding	2,200.00	0.29
	Lightning protection	4,700.00	0.63
5L	Communication Installation		
	Fire alarms	33,272.00	4.44
	Security (wireways)	7,800.00	1.04
	Data and voice (wireways)	7,700.00	1.03
5M	Special Installation		
	Building management systems	225,022.00	30.00
	Electrical services in connection	4,500.00	0.60
	Summary total (based on gross floor area)........................	941,173.00	125.49

Approximate Estimating

ELEMENTAL COSTS

El. Ref.	Element	Total Cost £	Cost /m² £
	CATEGORY 'A' FIT OUT		
5C	Disposal Installation		
	FCU Condensate	27,754.00	4.62
5F	Space Heating and Air Treatment		
	LTHW Heating	93,273.00	15.55
	Chilled water	114,000.00	19.00
	Supply and extract ductwork	366,000.00	61.00
5H	Electrical Installation		
	General lighting	177,800.00	29.63
5K	Protective Installation		
	Earthing and bonding	2,838.00	0.47
5L	Communication Installation		
	Fire alarms	18,600.00	3.10
5M	Special Installations		
	Electrical services in connection	33,400.00	5.57
	Summary total (Based on nett area)	833,665.00	138.94

Approximate Estimating

ELEMENTAL COSTS

PERFORMING ARTS CENTRE

Performing Arts centre with a gross floor area of 8,203m², upon which this cost analysis is based.

The development comprises of dance studios and theatre auditorium. The theatre has all the necessary stage lighting, machinery and equipment required in a modern professional theatre (excluded from rates).

Cost Summary

El. Ref.	Element	Total Cost £	Cost /m² £
5A	Sanitaryware..	68,267.00	8.32
5C	Disposal Installations		
	Soil and Waste	34,133.00	4.16
5D	Water Installations		
	Cold water services	57,230.00	6.98
	Hot water services	54,045.00	6.59
5E	Heat Source...	95,574.00	11.65
5F	Space Heating and Air Treatment		
	Heating with cooling	551,823.00	67.27
5G	Ventilating Services		
	Extract systems.......................................	748,233.00	91.21
5H	Electrical Installation		
	LV supply/distribution	285,583.00	34.81
	General lighting	636,019.00	77.53
	Small power	167,254.00	20.39
5I	Gas Installation...	21,845.00	2.66
5K	Protection		
	Lighting protection....................................	6,827.00	0.83
5L	Communication Installation		
	Fire alarms and detection	136,534.00	16.64
	Voice and Data...	167,254.00	20.39
	Security..	153,600.00	16.46
5M	Special Installation		
	Building management systems......................	250,312.00	30.51
	Theatre systems..	119,467.00	14.56
	Summary total	3,551,000.00	432.89

32   *Approximate Estimating*

ELEMENTAL COSTS

Hospital

Combined Neuroscience/Cardiothoracic Hospital with a gross floor area of 23,000 m², upon which this cost analysis is based.

The development is on three floors plus lower ground floor split between Neuroscience Department with an area of 8,000 m² and Cardiothoracic Department with an area of 9,000 m². In addition there is a shared accommodation area with an area of 2,000 m² and back of house support area of 4,000 m².

Cost Summary

El. Ref.	Element	Total Cost £	Cost /m² £
5A	Sanitaryware	299,000.00	13.00
5C	Disposal Installations		
	Rainwater / Soil and Waste...	324,000.00	14.09
5D	Water Installations		
	Cold water services	622,000.00	27.04
	Hot water services	637,000.00	27.70
5E	Heat Source...	332,000.00	14.43
5F	Space Heating and Air Treatment.................................	3,780,000.00	164.35
5G	Ventilating Services		
	Toilet extract system	150,000.00	6.50
5H	Electrical Installation		
	HV supply/distribution	382,000.00	16.61
	Standby Power..	38,000.00	1.65
	UPS	74,000.00	3.21
	LV supply/distribution	447,000.00	40.00
	Lighting	920,000.00	22.21
	Small power	157,000.00	6.83
	Emergency power..	791,000.00	34.39
	Bedhead Trunking...	182,000.00	7.91
5I	Gas Installation..	17,000.00	0.74
5K	Protection		
	Earthing and bonding	56,000.00	2.44
	Dry Riser...	19,000.00	0.83
	Carried forward	9,227,000.00	403.93

Approximate Estimating

ELEMENTAL COSTS

El. Ref.	Element	Total Cost £	Cost /m² £
	Brought forward	9,227,000.00	403.93
5L	Communication Installation		
	Fire alarms and detection	483,000.00	21.00
	Voice and Data	277,000.00	12.04
	Television......	17,000.00	0.74
	Induction Loops..	7,000.00	0.30
	Nurse Call..	382,000.00	16.61
	Security..	104,000.00	4.52
5M	Special Installation		
	Building management systems......................................	843,000.00	36.65
	Clean Steam..	359,000.00	15.61
	Summary total	11,699,000.00	511.40

Approximate Estimating

ELEMENTAL COSTS

Building Management Installations

New office with a gross area of 10,000 m^2 and a nett lettable area of 7,200 m^2. Building comprises of 4 floors, basement and roof plantroom. System comprises monitoring and control of the mechanical services via direct digital control (DDC) outstations via a central operators terminal.

Whilst the shell and core installation is constant, we have identified three options for the Category A fit out.

Cost Summary

El. Ref.	Element	Total Cost £	Cost /Point £
	Shell and Core – 440 Points		
1.0	**Central Equipment** Operator Station Software Printers Laptop PC Modems, multiplexers etc............................	10,481.00	23.82
2.0	**Field Equipment** Network devices Valves/Actuators Sensing/Interfacing devices........................	32,482.00	74.50
3.0	**Cabling – Power/Control** Power – from MCC to equipment Control – from MCC/Outstations to equipment Cable ways..	83,857.00	190.59
4.0	**Motor Control Centres** Motor control centres including control equipment Field mount inverter drives.........................	52,426.00	119.15
5.0	**Programming** Central facility software Network devices software Graphics...	23,569.00	53.57
6.0	**On site testing and Commissioning** Operating stations & Equipment Software and graphics Power cabling Control cabling – point to point MCC's & inverters.....................................	17,653.00	40.12
7.0	**On site testing and Commissioning** Operating stations & Equipment Software and graphics MCC's & inverters.....................................	3,677.00	8.36
	Total for Shell and Core............................	224,145.00	509.42

Approximate Estimating

ELEMENTAL COSTS

El. Ref.	Element	Total Cost £	Cost /Point £
	Category A Fit Out		
	Option 1 - 4 pipe fan coil – 756 points		
1.0	**Field Equipment** Network devices Valves/actuators Sensing devices...	44,154.00	58.40
2.0	**Cabling** Power – from local isolator to DDC controller Control – from DDC controller to field equipment................	24,948.00	33.00
3.0	**Programming** Software – central facility Software – network devices Graphics...	13,811.00	18.27
4.0	**On site testing and commissioning** Equipment Programming/graphics Power and control cabling...	14,383.00	19.03
	Total Option 1 – four pipe fan coil................................	97,296.00	128.70
	Category A Fit Out		
	Option 2 - 2 pipe fan coil with electric heating – 756 points		
1.0	**Field Equipment** Network devices Valves/actuators/thyristors Sensing devices...	56,628.00	74.91
2.0	**Cabling** Power – from local isolator to DDC controller Control – from DDC controller to field equipment................	26,123.00	34.57
3.0	**Programming** Software – central facility Software – network devices Graphics...	13,811.00	18.27
4.0	**On site testing and commissioning** Equipment Programming/graphics Power and control cabling...	14,383.00	19.03
	Total Option 2 – 2 pipe fan coil with electric heating..........	110,945.00	146.78

Approximate Estimating

ELEMENTAL COSTS

El. Ref.	Element	Total Cost £	Cost /Point £
	Category A Fit Out		
	Option 3 – Chilled Beams with perimeter heating – 567 points		
1.0	**Field Equipment** Network devices Valves/actuators Sensing devices...	39,996.00	70.54
2.0	**Cabling** Power – from local isolator to DDC controller Control – from DDC controller to field equipment................	26,532.00	46.79
3.0	**Programming** Software – central facility Software – network devices Graphics...	13,811.00	24.36
4.0	**On site testing and commissioning** Equipment Programming/graphics Power and control cabling..	14,762.00	26.04
	Total Option 3 – Chilled beams with perimeter heating.......	95,101.00	167.73

El. Ref.	Element	Total Cost £	Cost /Point £
	Combined Shell & Core and Fit Out		
1.0	**Option 1 – 4 pipe fan coil – 1196 points** Shell and core Category A Fit out	321,441.00	268.76
2.0	**Option 2 – 2 pipe fan coil with electrical heating – 1196 points** Shell and core Category A Fit out	335,090.00	280.16
3.0	**Option 3 – Chilled beams with perimeter heating – 1007 points** Shell and core Category A Fit out	319,246.00	317.03

PART TWO

Material Costs/
Measured Work Prices

Mechanical Installations

R : Disposal Systems
R10 : Rainwater Pipework/Gutters — 44
R11 : Above Ground Drainage — 64

S : Piped Supply System
S10 : Cold Water — 90
S11 : Hot Water — 149
S32 : Natural Gas — 153
S41 : Fuel Oil Storage/Distribution — 155
S60 : Fire Hose Reels — 156
S61 : Dry Risers — 157
S63 : Sprinklers — 158
S65 : Fire Hydrants — 162

T : Mechanical/Cooling/Heating Systems
T10 : Gas/Oil Fired Boilers — 164
T13 : Packaged Steam Generators — 177
T14 : Heat Pumps — 178
T31 : Low Temperature Hot Water Heating — 179
T33 : Steam Heating — 267
T42 : Local Heating Units — 271
T61 : Chilled Water — 272
T70 : Local Cooling Units — 274

U : Ventilation/Air Conditioning Systems
U10 : Ductwork — 275
U14 : Smoke Extract — 402
U70 : Air Curtains — 397

Electrical Installations

V : Electrical Supply/Power/Lighting
V11 : HV Supply — 429
V12 : LV Supply — 440
V20 : LV Distribution — 452
V21 : General Lighting — 515
V22 : Small Power — 523
V32 : Uninterrupted Power Supply — 532
V40 : Emergency Lighting — 534

W : Telecommunications/Security/Control
W10 : Communications — 537
W20 : Radio/Television — 538
W23 : Clocks — 540
W30 : Data Transmission — 542
W50 : Fire Detection and Alarm — 549
W51 : Earthing and Bonding — 552
W52 : Lightning Protection — 553

Visit www.pricebooks.co.uk

Mechanical Installations

MATERIAL COSTS/MEASURED WORK PRICES

DIRECTIONS

The following explanations are given for each of the column headings and letter codes.

Unit	Prices for each unit are given as singular (1 metre, 1 nr).
Net price	Industry tender prices, plus nominal allowance for fixings, waste and applicable trade discounts.
Material cost	Net price plus percentage allowance for overheads and profit and preliminaries.
Labour constant	Gang norm (in man-hours) for each operation.
Labour cost	Labour constant multiplied by the appropriate all-in man-hour cost. (See also relevant Rates of Wages Section)
Measured work price	Material cost plus Labour cost.

MATERIAL COSTS

The Material Costs given are based at Second Quarter 2001 but exclude any charges in respect of VAT.

MEASURED WORK PRICES

These prices are intended to apply to new work in the London area. The prices are for reasonable quantities of work and the user should make suitable adjustments if the quantities are especially small or especially large. Adjustments may also be required for locality (e.g. outside London) and for the market conditions (e.g. volume of work on hand or on offer) at the time of use.

Keep your figures up to date, free of charge
Download updates from the web: www.pricebooks.co.uk

This section, and most of the other information in this Price Book, is brought up to date every three months with the Price Book Updates, until the next annual edition. The updates are available free to all Price Book purchasers.

To ensure you receive your free copies, either complete the reply card from the centre of the book and return it to us or register via the website www.pricebooks.co.uk

Material Costs/Measured Work Prices - Mechanical Installations   39

DIRECTIONS

MECHANICAL INSTALLATIONS
The labour rate has been based on average tendered rates per man hour and for Third Quarter 2001 plus allowances for all other emoluments and expenses. To this rate has been added 20% to cover site and head office overheads and preliminary items together with 5% for profit, resulting in an inclusive rate of £ 17.60 per man hour. The rate has been calculated on a working year of 2,016.40 hours; a detailed build-up of the rate is given at the end of these Directions.

DUCTWORK INSTALLATIONS
The labour rate has been based on an average tendered rate and for Third Quarter 2001. To this rate has been added 45.5% to cover shop, site and head office overheads and preliminary items together with 10% for profit, resulting in an inclusive rate of £ 21.83 per man hour. The rate per man hour has been calculated on a working year of 2,016.4 hours.

In calculating the 'Measured Work Prices' the following assumptions have been made:
(a)   That the work is carried out as a sub-contract under the Standard Form of Building Contract.
(b)   That, unless otherwise stated, the work is being carried out in open areas at a height which would not require more than simple scaffolding.
(c)   That the building in which the work is being carried out is no more than six storeys high.

Where these assumptions are not valid, as for example where work is carried out in ducts and similar confined spaces or in multi-storey structures when additional time is needed to get to and from upper floors, then an appropriate adjustment must be made to the prices. Such adjustment will normally be to the labour element only.

Material Costs/Measured Work Prices - Mechanical Installations

DIRECTIONS

LABOUR RATE - MECHANICAL ENGINEERING

The annual cost of notional twelve man gang.

		FOREMAN	SENIOR CRAFTSMAN (+2 Welding skill)	SENIOR CRAFTSMAN	CRAFTSMAN	INSTALLER	MATE (Over 18)	SUB TOTALS
		1 NR	1 NR	2 NR	4 NR	2 NR	2 NR	
Hourly Rate from 3 September 2001		10.00	8.61	8.26	7.56	6.87	5.79	
Working hours per annum per man		1,702.40	1,702.40	1,702.40	1,702.40	1,702.40	1,702.40	
x Hourly rate x nr of men = £ per annum		17,024.00	14,657.66	28,123.65	51,480.58	23,390.98	19,713.79	154,390.66
Overtime Rate		14.75	13.73	12.19	11.16	10.14	8.54	
Overtime hours per annum per man		314	314	314	314	314	314	
x Hourly rate x nr of men = £ per annum		4631.5	4311.22	7655.32	14016.96	6367.92	5363.12	42,346.04
Total		21,655.50	18,968.88	35,778.97	65,497.54	29,758.90	25,076.91	196,736.70
Incentive schemes (insert percentage)	5.00%	1,082.78	948.44	1,788.95	3,274.88	1,487.94	1,253.85	9,836.83
Daily Travel Time Allowance (10-20 miles each way)		3.85	3.85	3.85	3.85	3.85	3.31	
Days per annum per man		224	224	224	224	224	224	
x nr of men = £ per annum		862.40	862.40	1,724.80	3,449.60	1,724.80	1,482.88	10,106.88
Daily Travel Fare (10-20 miles each way)		5.40	5.40	5.40	5.40	5.40	5.40	
Days per annum per man		224	224	224	224	224	224	
x nr of men = £ per annum		1,209.60	1,209.60	2,419.20	4,838.40	2,419.20	2,419.20	14,515.20
National Insurance Contributions								
Weekly gross pay (subject to NI) each		23600.68	20779.73	39292.72	72222.01	32971.64	27813.64	
% of NI Contributions		11.9	11.9	11.9	11.9	11.9	11.9	
£ Contributions/annum		2235.57	1899.87	3530.01	6302.77	2777.80	2164.00	18,910.01
Holiday Credit and Welfare contributions:								
Number of weeks		52	52	52	52	52	52	
Total weekly £ contribution each		53.02	49.68	49.68	46.29	44.60	42.90	
x nr of men = £ contributions/annum		2,757.04	2,583.36	5,166.72	9,628.32	4,638.40	4,461.60	29,235.44
Holiday Top-up Funding including overtime		9.21	4.54	2.53	1.89	0.00	0.00	
Cost		478.92	236.08	263.12	393.12	0.00	0.00	1,371.24

SUB-TOTAL			280,712.30
TRAINING (INCLUDING ANY TRADE REGISTRATIONS) - SAY		1.00%	2,807.12
SEVERANCE PAY AND SUNDRY COSTS - SAY		1.50%	4,252.79
EMPLOYER'S LIABILITY AND THIRD PARTY INSURANCE - SAY		2.00%	5,755.44
ANNUAL COST OF NOTIONAL GANG			293,527.65
MEN ACTUALLY WORKING = 10.5 THEREFORE ANNUAL COST PER PRODUCTIVE MAN			27,955.01
AVERAGE NR OF HOURS WORKED PER MAN 2016.4 THEREFORE ALL IN MAN HOUR			13.86
PRELIMINARY ITEMS - SAY		7.50%	1.04
SITE AND HEAD OFFICE OVERHEADS - SAY		12.50%	1.86
PROFIT - SAY		5.00%	0.84
THEREFORE INCLUSIVE MAN HOUR			17.60

Material Costs/Measured Work Prices - Mechanical Installations 41

DIRECTIONS

Notes:

(1)   The following assumptions have been made in the above calculations:-
      (a)   The working week of 38 hours i.e. the normal working week as defined by the National Agreement.
      (b)   The actual hours worked are five days of 9 hours each.
      (c)   A working year of 2016 hours.
      (d)   Five days in the year are lost through sickness or similar reason.
(2)   The incentive scheme addition of 5% is intended to reflect bonus schemes typically in use.
(3)   National insurance contributions are those effective from 6 April 2001.
(4)   Weekly Holiday Credit/Welfare Stamp values are those effective from 2 October 2000.
(5)   Rates base from 3 September 2001.
(6)   Fares (Waterloo to New Malden) effective from January 2001. Tel: 08757 487490.

Material Costs/Measured Work Prices - Mechanical Installations

DIRECTIONS

LABOUR RATE - DUCTWORK

The annual cost of notional eight man gang.

		FOREMAN	SENIOR CRAFTSMAN	CRAFTSMAN	INSTALLER	SUB TOTALS
		1 NR	1 NR	4 NR	2 NR	
Hourly Rate from 3 September 20001		10.00	8.26	7.56	6.87	
Working hours per annum per man		1,702.40	1,702.40	1,702.40	1,702.40	
x Hourly rate x nr of men = £ per annum		17,024.00	14,061.82	51,480.58	23,390.98	105,957.38
Overtime Rate		14.75	12.19	11.16	10.14	
Overtime hours per annum per man		314	314	314	314	
x hourly rate x nr of men = £ per annum		4631.5	3827.66	14016.96	6367.92	28,844.04
Total		21,655.50	17,889.48	65,497.54	29,758.90	134,801.42
Incentive schemes (insert percentage)	5.00%	1,082.78	894.47	3,274.88	1,487.94	6,740.07
Daily Travel Time Allowance (10-20 miles each way)		3.85	3.85	3.85	3.85	
Days per annum per man		224	224	224	224	
x nr of men = £ per annum		862.40	86240	3,449.60	1,724.80	6,899.20
Daily Travel Time Allowance (10-20 miles each way)		5.40	5.40	5.40	5.40	
Days per annum per man		224	224	224	224	
x nr of men = £ per annum		1,209.60	1,209.60	4,838.40	2,419.20	9,676.80
National Insurance Contributions:						
Weekly gross pay (subject to NI) each		23600.68	19646.36	72222.01	32971.64	
% of NI Contributions		11.9	11.9	11.9	11.9	
£ Contributions/annum		2235.57	1765.00	6302.77	2777.80	13,081.13
Holiday Credit and Welfare contributions						
Number of weeks		52	52	52	52	
Total weekly £ contribution each		53.02	49.68	46.29	44.60	
x nr of men = £ Contributions/annum		2,757.04	2,583.36	9,628.32	4,638.40	19,607.12
Holiday Top-up Funding including overtime		9.21	2.53	1.89	0.00	
Cost		478.92	131.56	393.12	0.00	1,003.60

SUB-TOTAL		191,809.34
TRAINING (INCLUDING ANY TRADE REGISTRATIONS) - SAY	1.00%	1,918.09
SEVERANCE PAY AND SUNDRY COSTS - SAY	1.50%	2,905.91
EMPLOYER'S LIABILITY AND THIRD PARTY INSURANCE - SAY	2.00%	3,932.67
ANNUAL COST OF NOTIONAL GANG		200,566.01

MEN ACTUALLY WORKING = 7.5   THEREFORE ANNUAL COST PER PRODUCTIVE MAN   26,742.13

AVERAGE NR OF HOURS WORKED PER MAN   2016.4   THEREFORE ALL IN MAN HOUR   13.26

12.50%	1.66
33.00%	4.92
10.00%	1.98
	21.83

Material Costs/Measured Work Prices - Mechanical Installations

DIRECTIONS

Notes:

(1)   The following assumptions have been made in the above calculations:-
       (a)   The working week of 38 hours i.e. the normal working week as defined by the National Agreement.
       (b)   The actual hours worked are five days of 9 hours each.
       (c)   A working year of 2016 hours.
       (d)   Five days in the year are lost through sickness or similar reason.
(2)   The incentive scheme addition of 5% is intended to reflect bonus schemes typically in use.
(3)   National insurance contributions are those effective from 6 April 2001.
(4)   Weekly Holiday Credit/Welfare Stamp values are those effective from 2 October 2000.
(5)   Rates are based from 3 September 2001.
(6)   Fares (Waterloo to New Malden) effective from January 2001. Tel: 08757 487490

44   *Material Costs/Measured Work Prices - Mechanical Installations*

R: DISPOSAL SYSTEMS

Item	Net Price £	Material £	Labour hours	Labour £	Unit	Total rate £
R10: RAINWATER PIPEWORK/GUTTERS						
PVC-U Gutters: push fit joints; fixed with brackets to backgrounds; BS 4576 BS EN 607						
Half Round Gutter						
75mm	1.78	2.16	0.69	12.14	m	**14.30**
110 mm	3.05	3.70	0.64	11.26	m	**14.96**
150mm	4.99	6.05	0.82	14.43	m	**20.49**
Extra Over fittings Half Round PVC-U Gutter						
Union						
75mm	0.83	1.00	0.19	3.34	nr	**4.35**
150mm	2.51	3.20	0.28	4.93	nr	**8.13**
Rainwater pipe outlets						
Running: 75 x 53mm dia	1.89	2.30	0.12	2.11	nr	**4.41**
Running: 100 x 68mm dia	2.39	2.91	0.12	2.11	nr	**5.02**
Running: 150 x 110mm dia	4.85	5.89	0.12	2.11	nr	**8.01**
Stop end: 100 x 68mm dia	2.39	2.91	0.12	2.11	nr	**5.02**
Internal Stop ends: short						
75mm	0.83	1.00	0.09	1.58	nr	**2.59**
100mm	0.83	1.00	0.09	1.58	nr	**2.59**
150mm	0.95	1.16	0.09	1.58	nr	**2.74**
External Stop ends: short						
75mm	0.83	1.00	0.09	1.58	nr	**2.59**
100mm	0.83	1.00	0.09	1.58	nr	**2.59**
150mm	0.95	1.16	0.09	1.58	nr	**2.74**
Angles						
75mm; 45 degree	2.24	2.72	0.20	3.52	nr	**6.24**
75mm; 90 degree	2.24	2.72	0.20	3.52	nr	**6.24**
100mm; 90 degree	2.36	3.01	0.20	3.52	nr	**6.53**
100mm; 120 degree	2.62	3.18	0.20	3.52	nr	**6.70**
100mm; 135 degree	2.62	3.18	0.20	3.52	nr	**6.70**
100mm; Prefabricated to special angle	11.17	13.56	0.23	4.05	nr	**17.61**
100mm; Prefabricated to raked angle	12.08	14.67	0.23	4.05	nr	**18.72**
150mm; 90 degree	4.43	5.39	0.20	3.52	nr	**8.91**
Gutter adaptors						
100mm; Stainless steel clip	1.26	1.53	0.16	2.82	nr	**4.35**
100mm; Cast iron spigot	3.48	4.23	0.23	4.05	nr	**8.28**
100mm; Cast iron socket	3.48	4.23	0.23	4.05	nr	**8.28**
100mm; Cast iron "ogee" spigot	3.53	4.28	0.23	4.05	nr	**8.33**
100mm; Cast iron "ogee" socket	3.53	4.28	0.23	4.05	nr	**8.33**
100mm; Half round to Square PVC-U	6.10	7.41	0.23	4.05	nr	**11.45**
100mm; Gutter overshoot guard	7.56	9.18	0.58	10.21	nr	**19.39**

Material Costs/Measured Work Prices - Mechanical Installations

R: DISPOSAL SYSTEMS

Item	Net Price £	Material £	Labour hours	Labour £	Unit	Total rate £
Brackets: including fixing to backgrounds						
75mm; Fascia	0.49	0.60	0.15	2.64	nr	3.24
100mm; Jointing	1.26	1.53	0.16	2.82	nr	4.34
100mm; Support	0.54	0.65	0.16	2.82	nr	3.47
150mm; Fascia	0.86	1.04	0.16	2.82	nr	3.86
Bracket supports: including fixing to backgrounds						
Side rafter	2.26	2.74	0.16	2.82	nr	5.56
Top rafter	2.26	2.74	0.16	2.82	nr	5.56
Rise and fall	2.18	2.65	0.16	2.82	nr	5.46
Square Gutter						
120mm	4.23	5.14	0.82	14.43	m	19.57
Extra Over fittings Square PVC-U Gutter						
Rainwater pipe outlets						
Running: 62mm square	2.61	3.17	0.12	2.11	nr	5.28
Stop end: 62mm square	2.24	2.73	0.12	2.11	nr	4.84
Stop ends: short						
External	1.27	1.55	0.09	1.58	nr	3.13
Angles						
90 degree	2.65	3.38	0.20	3.52	nr	6.90
120 degree	9.77	12.46	0.20	3.52	nr	15.98
135 degree	2.65	3.38	0.20	3.52	nr	6.90
Prefabricated to special angle	11.46	13.91	0.23	4.05	nr	17.96
Prefabricated to raked angle	11.75	14.27	0.23	4.05	nr	18.32
Gutter Adaptors						
Cast iron	6.09	7.40	0.23	4.05	nr	11.45
Half round asbestos	6.09	7.40	0.23	4.05	nr	11.45
Brackets: including fixing to backgrounds						
Jointing	1.58	1.92	0.16	2.82	nr	4.73
Support	0.66	0.80	0.16	2.82	nr	3.62
Bracket support: including fixing to backgrounds						
Side rafter	2.26	2.74	0.16	2.82	nr	5.56
Top rafter	2.26	2.74	0.16	2.82	nr	5.56
Rise and fall	2.18	2.65	0.16	2.82	nr	5.46
High Capacity Square Gutter						
137mm	7.15	8.68	0.82	14.43	m	23.11
Extra Over fittings High Capacity Square UPV-C						
Rainwater pipe outlets						
Running: 75mm square	7.59	9.22	0.12	2.11	nr	11.33
Running: 82mm dia	7.59	9.22	0.12	2.11	nr	11.33
Running: 110mm dia	6.64	8.06	0.12	2.11	nr	10.17

Material Costs/Measured Work Prices - Mechanical Installations

R: DISPOSAL SYSTEMS

Item	Net Price £	Material £	Labour hours	Labour £	Unit	Total rate £
R10: RAINWATER PIPEWORK/GUTTERS						
PVC-U Gutters: push fit joints; fixed with brackets to backgrounds; BS 4576 BS EN 607 (Continued)						
Screwed outlet adaptor						
75mm square pipe	3.99	4.85	0.23	4.05	nr	8.90
Stop ends: short						
External	0.83	1.00	0.09	1.58	nr	2.59
Angles						
90 degree	7.06	9.00	0.20	3.52	nr	12.52
135 degree	13.81	17.60	0.20	3.52	nr	21.12
Prefabricated to special angle	16.07	19.51	0.23	4.05	nr	23.56
Prefabricated to raked internal angle	27.97	33.97	0.23	4.05	nr	38.02
Prefabricated to raked external angle	27.97	33.97	0.23	4.05	nr	38.02
Brackets: including fixing to backgrounds						
Jointing	4.32	5.51	0.16	2.82	nr	8.33
Support	1.80	2.30	0.16	2.82	nr	5.11
Overslung	1.68	2.14	0.16	2.82	nr	4.96
Bracket supports: including fixing to backgrounds						
Side rafter	2.71	3.28	0.16	2.82	nr	6.10
Top rafter	2.71	3.28	0.16	2.82	nr	6.10
Rise and fall	3.82	4.64	0.16	2.82	nr	7.46
Deep Eliptical Gutter						
137mm	3.38	4.11	0.82	14.43	m	18.54
Extra Over fittings Deep Eliptical PVC-U Gutter						
Rainwater pipe outlets						
Running: 68mm dia	2.53	3.22	0.12	2.11	nr	5.33
Running: 82mm dia	2.53	3.07	0.12	2.11	nr	5.18
Stop end: 68mm dia	2.67	3.24	0.12	2.11	nr	5.35
Stop ends: short						
External	1.24	1.51	0.09	1.58	nr	3.10
Angles						
90 degree	2.78	3.54	0.20	3.52	nr	7.06
135 degree	2.78	3.54	0.20	3.52	nr	7.06
Prefabricated to special angle	7.90	10.07	0.23	4.05	nr	14.12
Gutter Adaptors						
Stainless steel clip	1.83	2.22	0.16	2.82	nr	5.04
Marley deepflow	3.99	4.84	0.23	4.05	nr	8.89
Brackets: including fixing to backgrounds						
Jointing	1.66	2.11	0.16	2.82	nr	4.93
Support	0.74	0.94	0.16	2.82	nr	3.75

Material Costs/Measured Work Prices - Mechanical Installations

R: DISPOSAL SYSTEMS

Item	Net Price £	Material £	Labour hours	Labour £	Unit	Total rate £
Bracket support: including fixing to backgrounds						
Side rafter	2.71	3.45	0.16	2.82	nr	**6.27**
Top rafter	2.71	3.45	0.16	2.82	nr	**6.27**
Rise and fall	3.82	4.88	0.16	2.82	nr	**7.69**
Ogee Profile Gutter						
122mm	3.75	4.56	0.82	14.43	m	**18.99**
Extra Over fittings Ogee Profile PVC-U Gutter						
Rainwater pipe outlets						
Running: 68mm dia	2.76	3.35	0.12	2.11	nr	**5.46**
Stop ends: short						
Internal/External: left or right hand	1.41	1.71	0.09	1.58	nr	**3.29**
Angles						
90 degree: internal or external	3.05	3.71	0.20	3.52	nr	**7.23**
135 degree: internal or external	3.05	3.71	0.20	3.52	nr	**7.23**
Brackets: including fixing to backgrounds						
Jointing	1.99	2.42	0.16	2.82	nr	**5.24**
Support	0.81	0.98	0.16	2.82	nr	**3.80**
Overslung	0.81	0.98	0.16	2.82	nr	**3.80**
PVC-U Rainwater pipe: dry push fit joints; fixed with brackets to backgrounds; BS 4576/ BS EN 607						
Pipe: circular						
53mm	3.86	4.69	0.61	10.74	m	**15.42**
68mm	5.67	6.89	0.61	10.74	m	**17.63**
Extra Over fittings Circular Pipework PVC-U						
Pipe coupler: PVC-U to PVC-U						
68mm	1.01	1.23	0.12	2.11	nr	**3.34**
Pipe coupler: PVC-U to Cast Iron						
68mm: to 3" cast iron	2.24	2.73	0.17	2.99	nr	**5.72**
68mm: to 3.3/4" cast iron	10.00	12.14	0.17	2.99	nr	**15.14**
Access pipe: single socket						
68mm	6.42	7.79	0.15	2.64	nr	**10.43**
Bend: short radius						
53mm: 67.5 degree	1.39	1.69	0.20	3.52	nr	**5.21**
68mm: 92.5 degree	1.75	2.13	0.20	3.52	nr	**5.65**
68mm: 112.5 degree	1.47	1.78	0.20	3.52	nr	**5.30**
Bend: long radius						
68mm: 112 degree	1.75	2.12	0.20	3.52	nr	**5.64**

Material Costs/Measured Work Prices - Mechanical Installations

R: DISPOSAL SYSTEMS

Item	Net Price £	Material £	Labour hours	Labour £	Unit	Total rate £
R10: RAINWATER PIPEWORK/GUTTERS						
Extra Over fittings Circular Pipework PVC-U (Continued)						
Branch						
68mm: 92 degree	8.11	10.34	0.23	4.05	nr	**14.39**
68mm: 112 degree	3.96	5.05	0.23	4.05	nr	**9.10**
Double branch						
68mm: 112 degree	16.88	20.49	0.24	4.22	nr	**24.72**
Shoe						
53mm	1.39	1.69	0.12	2.11	nr	**3.80**
68mm	1.19	1.44	0.12	2.11	nr	**3.55**
Rainwater head: including fixing to backgrounds						
68mm	6.49	7.88	0.29	5.10	nr	**12.98**
Pipe clip: including fixing to backgrounds						
68mm	0.81	0.98	0.16	2.82	nr	**3.80**
Pipe clip adjustable: including fixing to backgrounds						
53mm	0.78	0.95	0.16	2.82	nr	**3.77**
68mm	1.75	2.13	0.16	2.82	nr	**4.95**
Pipe clip drive in: including fixing to backgrounds						
68mm	1.98	2.41	0.16	2.82	nr	**5.22**
Pipe: square						
62mm	3.00	3.65	0.45	7.92	m	**11.57**
75mm	5.05	6.13	0.45	7.92	m	**14.05**
Extra Over fittings Square Pipework PVC-U						
Pipe coupler: PVC-U to PVC-U						
62mm	1.38	1.68	0.20	3.52	nr	**5.20**
75mm	1.80	2.19	0.20	3.52	nr	**5.71**
Square to circular adaptor: single socket						
62mm to 68mm	2.14	2.59	0.20	3.52	nr	**6.11**
Square to circular adaptor: single socket						
75mm to 62mm	2.71	3.29	0.20	3.52	nr	**6.81**
Access pipe						
62mm	8.83	10.72	0.16	2.82	nr	**13.54**
75mm	11.09	13.47	0.16	2.82	nr	**16.28**
Bends						
62mm: 92.5 degree	1.89	2.29	0.20	3.52	nr	**5.81**
62mm: 112.5 degree	1.56	1.89	0.20	3.52	nr	**5.41**
75mm: 112.5 degree	3.62	4.40	0.20	3.52	nr	**7.92**

Material Costs/Measured Work Prices - Mechanical Installations   49

R: DISPOSAL SYSTEMS

Item	Net Price £	Material £	Labour hours	Labour £	Unit	Total rate £
Bends: prefabricated special angle						
62mm	8.76	10.64	0.23	4.05	nr	**14.69**
75mm	12.25	14.87	0.23	4.05	nr	**18.92**
Offset						
62mm	2.83	3.44	0.20	3.52	nr	**6.96**
75mm	8.00	9.72	0.20	3.52	nr	**13.24**
Offset: prefabricated special angle						
62mm	8.80	10.68	0.23	4.05	nr	**14.73**
Shoe						
62mm	1.47	1.78	0.12	2.11	nr	**3.89**
75mm	1.86	2.26	0.12	2.11	nr	**4.37**
Branch						
62mm	4.52	5.49	0.23	4.05	nr	**9.54**
75mm	11.94	14.50	0.23	4.05	nr	**18.55**
Double branch						
62mm	16.27	19.75	0.24	4.22	nr	**23.98**
Rainwater head						
62mm	5.82	7.07	0.29	5.10	nr	**12.18**
75mm	21.34	25.91	3.45	60.69	nr	**86.60**
Pipe clip: including fixing to backgrounds						
62mm	0.75	0.92	0.16	2.82	nr	**3.73**
75mm	1.51	1.83	0.16	2.82	nr	**4.65**
Pipe clip adjustable: including fixing to backgrounds						
62mm	2.32	2.81	0.16	2.82	nr	**5.63**
PVC-U Rainwater pipe: solvent welded joints; fixed with brackets to backgrounds; BS 4576/ BS EN 607						
Pipe: circular						
82mm	3.96	4.80	0.35	6.16	m	**10.96**
Extra Over fittings Circular pipework PVC-U						
Pipe coupler: PVC-U to PVC-U						
82mm	2.51	3.05	0.21	3.70	nr	**6.75**
Access pipe						
82mm	16.30	19.79	0.23	4.05	nr	**23.84**
Bend						
82mm: 92, 112.5 and 135 degree	7.30	8.87	0.29	5.10	nr	**13.97**

Material Costs/Measured Work Prices - Mechanical Installations

R: DISPOSAL SYSTEMS

Item	Net Price £	Material £	Labour hours	Labour £	Unit	Total rate £
R10: RAINWATER PIPEWORK/GUTTERS						
PVC-U Rainwater pipe: solvent welded joints; fixed with brackets to backgrounds; BS 4576/ BS EN 607 (Continued)						
Shoe						
82mm	5.43	6.60	0.29	5.10	nr	**11.70**
110mm	6.84	8.31	0.32	5.63	nr	**13.94**
Branch						
82mm: 92, 112.5 and 135 degree	8.74	10.61	0.35	6.16	nr	**16.77**
Rainwater head						
82mm	11.52	13.98	0.58	10.21	nr	**24.19**
110mm	10.50	12.75	0.58	10.21	nr	**22.96**
Pipe clip: galvanised; including fixing to backgrounds						
82mm	1.72	2.09	0.58	10.21	nr	**12.29**
Pipe clip: galvanised plastic coated; including fixing to backgrounds						
82mm	2.38	2.89	0.58	10.21	nr	**13.09**
Pipe clip: PVC-U including fixing to backgrounds						
82mm	1.28	1.56	0.58	10.21	nr	**11.76**
Pipe clip: PVC-U adjustable: including fixing to backgrounds						
82mm	2.35	2.86	0.58	10.21	nr	**13.06**
Roof Outlets: 178 dia; Flat						
50mm	10.84	13.17	1.15	20.24	nr	**33.41**
82mm	10.84	13.17	1.15	20.24	nr	**33.41**
Roof Outlets: 178mm dia; Domed						
50mm	10.84	13.17	1.15	20.24	nr	**33.41**
82mm	10.84	13.17	1.15	20.24	nr	**33.41**
Roof Outlets: 406mm dia; Flat						
82mm	21.22	25.76	1.15	20.24	nr	**46.00**
110mm	21.22	25.76	1.15	20.24	nr	**46.00**
Roof Outlets: 406mm dia; Domed						
82mm	21.22	25.76	1.15	20.24	nr	**46.00**
110mm	21.22	25.76	1.15	20.24	nr	**46.00**
Roof Outlets: 406mm dia; Inverted						
82mm	46.70	56.71	1.15	20.24	nr	**76.95**
110mm	46.70	56.71	1.15	20.24	nr	**76.95**
Roof Outlets: 406mm dia; Vent Pipe						
82mm	30.67	37.25	1.15	20.24	nr	**57.49**
110mm	30.67	37.25	1.15	20.24	nr	**57.49**

Material Costs/Measured Work Prices - Mechanical Installations

R: DISPOSAL SYSTEMS

Item	Net Price £	Material £	Labour hours	Labour £	Unit	Total rate £
Balcony outlets: screed						
82mm	18.04	21.91	1.15	20.24	nr	**42.15**
Balcony outlets:asphalt						
82mm	18.06	21.93	1.15	20.24	nr	**42.17**
Adaptors						
82mm x 62mm square pipe	1.41	1.71	0.21	3.70	nr	**5.40**
82mm x 68mm circular pipe	1.30	1.58	0.21	3.70	nr	**5.27**

For 110mm diameter pipework and fittings see R11: Above Ground Drainage

Material Costs/Measured Work Prices - Mechanical Installations

R: DISPOSAL SYSTEMS

Item	Net Price £	Material £	Labour hours	Labour £	Unit	Total rate £
R10: RAINWATER PIPEWORK/GUTTERS						
Cast iron gutters: mastic and bolted joints; BS 460; fixed with brackets to backgrounds						
Half Round Gutter						
100 mm	8.05	9.78	0.85	14.96	m	**24.74**
115 mm	8.34	10.12	0.97	17.07	m	**27.19**
125 mm	9.52	11.55	0.97	17.07	m	**28.63**
150 mm	15.66	19.01	1.12	19.71	m	**38.73**
Extra over fittings Half Round Gutter Cast Iron BS 460						
Union						
100 mm	3.36	4.09	0.39	6.86	nr	**10.95**
115 mm	4.10	4.97	0.48	8.45	nr	**13.42**
125 mm	4.73	5.74	0.48	8.45	nr	**14.19**
150 mm	5.31	6.45	0.55	9.68	nr	**16.13**
Stop end; internal						
100 mm	1.71	2.08	0.12	2.11	nr	**4.19**
115 mm	2.22	2.69	0.15	2.64	nr	**5.33**
125 mm	2.22	2.69	0.15	2.64	nr	**5.33**
150 mm	2.95	3.59	0.20	3.52	nr	**7.11**
Stop end; external						
100 mm	1.71	2.08	0.12	2.11	nr	**4.19**
115 mm	2.22	2.69	0.15	2.64	nr	**5.33**
125 mm	2.22	2.69	0.15	2.64	nr	**5.33**
150 mm	2.95	3.59	0.20	3.52	nr	**7.11**
90 degree angle; single socket						
100 mm	5.09	6.17	0.39	6.86	nr	**13.04**
115 mm	5.24	6.36	0.43	7.57	nr	**13.93**
125 mm	6.18	7.51	0.43	7.57	nr	**15.07**
150 mm	11.28	13.70	0.50	8.80	nr	**22.50**
90 degree angle; double socket						
100 mm	6.17	7.50	0.39	6.86	nr	**14.36**
115 mm	6.55	7.95	0.43	7.57	nr	**15.52**
125 mm	8.47	10.29	0.43	7.57	nr	**17.86**
135 degree angle; single socket						
100 mm	4.73	5.74	0.39	6.86	nr	**12.60**
115 mm	5.24	6.36	0.43	7.57	nr	**13.93**
125 mm	7.73	9.39	0.43	7.57	nr	**16.96**
150 mm	10.34	12.55	0.50	8.80	nr	**21.35**
Running outlet						
65 mm outlet						
100 mm	4.96	6.02	0.39	6.86	nr	**12.88**
115 mm	5.40	6.56	0.43	7.57	nr	**14.13**
125 mm	6.18	7.51	0.43	7.57	nr	**15.07**

Material Costs/Measured Work Prices - Mechanical Installations

R: DISPOSAL SYSTEMS

Item	Net Price £	Material £	Labour hours	Labour £	Unit	Total rate £
Running outlet						
75 mm outlet						
100 mm	4.96	6.02	0.39	6.86	nr	**12.88**
115 mm	5.40	6.56	0.43	7.57	nr	**14.13**
125 mm	6.18	7.51	0.43	7.57	nr	**15.07**
150 mm	10.51	12.76	0.50	8.80	nr	**21.56**
100 mm outlet						
150 mm	10.51	12.76	0.50	8.80	nr	**21.56**
Stop end outlet; socket						
65 mm outlet						
100 mm	3.94	4.79	0.39	6.86	nr	**11.65**
115 mm	4.34	5.27	0.43	7.57	nr	**12.84**
75 mm outlet						
125 mm	5.50	6.68	0.43	7.57	nr	**14.25**
150 mm	10.51	12.76	0.50	8.80	nr	**21.56**
100mm outlet						
150 mm	10.51	12.76	0.50	8.80	nr	**21.56**
Stop end outlet; spigot						
65 mm outlet						
100 mm	3.94	4.79	0.39	6.86	nr	**11.65**
115 mm	4.34	5.27	0.43	7.57	nr	**12.84**
75 mm outlet						
125 mm	5.50	6.68	0.43	7.57	nr	**14.25**
150 mm	10.51	12.76	0.50	8.80	nr	**21.56**
100mm outlet						
150 mm	10.51	12.76	0.50	8.80	nr	**21.56**
Brackets; fixed to backgrounds						
Fascia						
100 mm	1.38	1.67	0.16	2.82	nr	**4.49**
115 mm	1.38	1.67	0.16	2.82	nr	**4.49**
125 mm	1.38	1.67	0.16	2.82	nr	**4.49**
150 mm	1.74	2.12	0.16	2.82	nr	**4.93**
Rise and Fall						
100 mm	2.06	2.50	0.39	6.86	nr	**9.36**
115 mm	2.06	2.50	0.39	6.86	nr	**9.36**
125 mm	2.46	2.99	0.39	6.86	nr	**9.85**
150 mm	2.89	3.50	0.39	6.86	nr	**10.37**
Top rafter						
100 mm	1.32	1.60	0.16	2.82	nr	**4.42**
115 mm	1.32	1.60	0.16	2.82	nr	**4.42**
125 mm	1.36	1.65	0.16	2.82	nr	**4.47**
150 mm	2.06	2.50	0.16	2.82	nr	**5.31**

DAVIS LANGDON & EVEREST
Authors of Spon's Price Books

Davis Langdon & Everest is an independent practice of Chartered Quantity Surveyors, with some 1000 staff in 20 UK offices and, through Davis Langdon & Seah International, some 2,500 staff in 85 offices worldwide.

DLE manages client requirements, controls risk, manages cost and maximises value for money, throughout the course of construction projects, always aiming to be - and to deliver - the best.

TYPICAL PROJECT STAGES, DLE INTEGRATED SERVICES AND THEIR EFFECT:

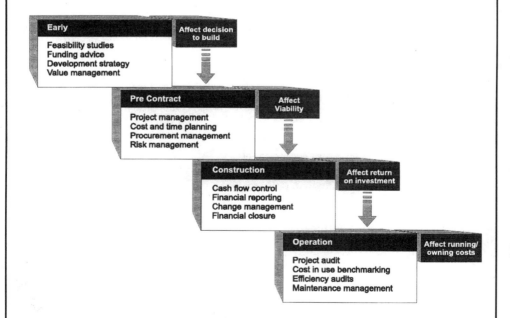

EUROPE ⇨ ASIA - AUSTRALIA - AFRICA - AMERICA
GLOBAL REACH - LOCAL DELIVERY

DAVIS LANGDON & EVEREST

LONDON
Princes House
39 Kingsway
London
WC2B 6TP
Tel : (020) 7497 9000
Fax : (020) 7497 8858
Email:rob.smith@davislangdon-uk.com

BIRMINGHAM
29 Woodbourne Road
Harborne
Birmingham
B17 8BY
Tel : (0121) 4299511
Fax : (0121) 4292544
Email:richard.d.taylor@davislangdon-uk.com

BRISTOL
St Lawrence House
29/31 Broad Street
Bristol
BS1 2HF
Tel : (0117) 9277832
Fax : (0117) 9251350
Email:alan.trolley@davislangdon-uk.com

CAMBRIDGE
36 Storey's Way
Cambridge
CB3 0DT
Tel : (01223) 351258
Fax : (01223) 321002
Email:stephen.bugg@davislangdon-uk.com

CARDIFF
4 Pierhead Street
Capital Waterside
Cardiff
CF10 4QP
Tel : (029) 20497497
Fax : (029) 20497111
Email:paul.edwards@davislangdon-uk.com

EDINBURGH
74 Great King Street
Edinburgh
EH3 6QU
Tel : (0131) 557 5306
Fax : (0131) 557 5704
Email:ian.mcandie@davislangdon-uk.com

GATESHEAD
11 Regent Terrace
Gateshead
Tyne and Wear
NE8 1LU
Tel : (0191) 477 3844
Fax : (0191) 490 1742
Email:gary.lockey@davislangdon-uk.com

GLASGOW
Cumbrae House
15 Carlton Court
Glasgow
G5 9JP
Tel : (0141) 429 6677
Fax : (0141) 429 2255
Email:hugh.fisher@davislangdon-uk.com

LEEDS
Duncan House
14 Duncan Street
Leeds
LS1 6DL
Tel : (0113) 2432481
Fax : (0113) 2424601
Email:tony.brennan@davislangdon-uk.com

LIVERPOOL
Cunard Building
Water Street
Liverpool L3 1JR
Tel : (0151) 2361992
Fax : (0151) 2275401
Email:john.davenport@davislangdon-uk.com

MANCHESTER
Cloister House
Riverside
New Bailey Street
Manchester M3 5AG
Tel : (0161) 819 7600
Fax : (0161) 819 1818
Email:paul.stanion@davislangdon-uk.com

NORWICH
63 Thorpe Road
Norwich
NR1 1UD
Tel : (01603) 628194
Fax : (01603) 615928
Email:michael.ladbrook@davislangdon-uk.com

MILTON KEYNES
Everest House
Rockingham Drive
Linford Wood
Milton Keynes MK14 6LY
Tel : (01908) 304700
Fax : (01908) 660059
Email:kevin.sims@davislangdon-uk.com

OXFORD
Avalon House
Marcham Road
Abingdon
Oxford OX14 1TZ
Tel : (01235) 555025
Fax : (01235) 554909
Email:paul.coomber@davislangdon-uk.com

PETERBOROUGH
Charterhouse
66 Broadway
Peterborough
PE1 1SU
Tel: (01733) 343625
Fax: (01733) 349177
Email:colin.harrison@davislangdon-uk.com

PLYMOUTH
3 Russell Court
St Andrew Street
Plymouth PL1 2AX
Tel : (01752) 668372
Fax : (01752) 221219
Email:gareth.steventon@davislangdon-uk.com

PORTSMOUTH
St Andrews Court
St Michaels Road
Portsmouth
PO1 2PR
Tel : (023) 92815218
Fax : (023) 92827156
Email:brian.bartholomew@davislangdon-uk.com

SOUTHAMPTON
Brunswick House
Brunswick Place
Southampton SO15 2AP
Tel : (023) 80333438
Fax : (023) 80226099
Email:richard.pitman@davislangdon-uk.com

DAVIS LANGDON CONSULTANCY
Princes House
39 Kingsway
London WC2B 6TP
Tel : (020) 7379 3322
Fax : (020) 7379 3030
Email:jim.meikle@davislangdon-uk.com

MOTT GREEN & WALL
Africa House
64-78 Kingsway
London
WC2B 6NN
Tel : (020) 7836 0836
Fax : (020) 7242 0394
Email:barry.nugent@mottgreenwall.co.uk

SCHUMANN SMITH
14th Flr, Southgate House
St Georges Way
Stevenage
Hertfordshire SG1 1HG
Tel : (01438) 742642
Fax : (01438) 742632
Email:nschumann@schumannsmith.com

NBW CROSHER & JAMES
1-5 Exchange Court
Strand
London WC2R 0PQ
Tel : (020) 7845 0600
Fax : (020) 7845 0601
Email:sandersr@nbwcrosherjames.com

NBW CROSHER & JAMES
102 New Street
Birmingham
B2 4HQ
Tel : (0121) 632 3600
Fax : (0121) 632 3601
Email:whittakerr@bir.nbwcrosherjames.com

NBW CROSHER & JAMES
5 Coates Crescent
Edinburgh
EH3 9AL
Tel : (0131) 220 4225
Fax : (0131) 220 4226
Email:mcfarlanel@nbwcrosherjames.com

with offices throughout Europe, the Middle East, Asia, Australia, Africa and the USA which together form

DAVIS LANGDON & SEAH INTERNATIONAL

Material Costs/Measured Work Prices - Mechanical Installations

R: DISPOSAL SYSTEMS

Item	Net Price £	Material £	Labour hours	Labour £	Unit	Total rate £
R10: RAINWATER PIPEWORK/GUTTERS						
Cast iron gutters: mastic and bolted joints; BS 460; fixed with brackets to backgrounds (Continued)						
Side rafter						
100 mm	1.32	1.60	0.16	2.82	nr	**4.42**
115 mm	1.32	1.60	0.16	2.82	nr	**4.42**
125 mm	1.36	1.65	0.16	2.82	nr	**4.47**
150 mm	2.06	2.50	0.16	2.82	nr	**5.31**
Half Round; 3 mm thick Double Beaded Gutter						
100 mm	7.82	9.49	0.85	14.96	m	**24.45**
115 mm	8.24	10.00	0.85	14.96	m	**24.96**
125 mm	9.23	11.21	0.97	17.07	m	**28.29**
Extra over fittings Half Round 3mm thick Gutter BS 460						
Union						
100 mm	3.36	4.08	0.38	6.69	nr	**10.76**
115 mm	4.07	4.95	0.38	6.69	nr	**11.63**
125 mm	4.67	5.67	0.43	7.57	nr	**13.23**
Stop end; internal						
100 mm	1.53	1.86	0.12	2.11	nr	**3.97**
115 mm	2.18	2.64	0.12	2.11	nr	**4.76**
125 mm	2.22	2.70	0.15	2.64	nr	**5.34**
Stop end; external						
100 mm	1.53	1.86	0.12	2.11	nr	**3.97**
115 mm	2.18	2.64	0.12	2.11	nr	**4.76**
125 mm	2.18	2.64	0.15	2.64	nr	**5.28**
90 degree angle; single socket						
100 mm	5.19	6.30	0.38	6.69	nr	**12.99**
115 mm	5.36	6.51	0.38	6.69	nr	**13.20**
125 mm	6.52	7.92	0.43	7.57	nr	**15.49**
135 degree angle; single socket						
100 mm	5.19	6.30	0.38	6.69	nr	**12.99**
115 mm	5.36	6.51	0.38	6.69	nr	**13.20**
125 mm	6.52	7.92	0.43	7.57	nr	**15.49**
Running outlet						
65 mm outlet						
100 mm	5.19	6.30	0.38	6.69	nr	**12.99**
115 mm	5.25	6.38	0.38	6.69	nr	**13.07**
125 mm	6.39	7.76	0.43	7.57	nr	**15.33**
75 mm outlet						
115 mm	5.36	6.51	0.38	6.69	nr	**13.20**
125 mm	6.39	7.76	0.43	7.57	nr	**15.33**

Material Costs/Measured Work Prices - Mechanical Installations

R: DISPOSAL SYSTEMS

Item	Net Price £	Material £	Labour hours	Labour £	Unit	Total rate £
Stop end outlet; socket						
65 mm outlet						
100 mm	4.02	4.88	0.38	6.69	nr	**11.57**
115 mm	4.89	5.93	0.38	6.69	nr	**12.62**
125 mm	6.39	7.76	0.43	7.57	nr	**15.33**
75 mm outlet						
125 mm	5.48	6.66	0.43	7.57	nr	**14.22**
Stop end outlet; spigot						
65 mm outlet						
100 mm	4.02	4.88	0.38	6.69	nr	**11.57**
115 mm	4.89	5.93	0.38	6.69	nr	**12.62**
125 mm	5.60	6.79	0.43	7.57	nr	**14.36**
Brackets; fixed to backgrounds						
Fascia						
100 mm	1.38	1.67	0.16	2.82	nr	**4.49**
115 mm	1.38	1.67	0.16	2.82	nr	**4.49**
125 mm	1.38	1.67	0.16	2.82	nr	**4.49**
Deep Half Round Gutter						
100 x 75 mm	15.80	19.18	0.85	14.96	m	**34.14**
125 x 75 mm	20.12	24.43	0.97	17.07	m	**41.50**
Extra over fittings Deep Half Round Gutter BS 460						
Union						
100 x 75 mm	5.51	6.69	0.38	6.69	nr	**13.38**
125 x 75 mm	5.95	7.22	0.43	7.57	nr	**14.79**
Stop end; internal						
100 x 75 mm	4.83	5.86	0.12	2.11	nr	**7.97**
125 x 75 mm	5.95	7.22	0.15	2.64	nr	**9.86**
Stop end; external						
100 x 75 mm	4.83	5.86	0.12	2.11	nr	**7.97**
125 x 75 mm	5.95	7.22	0.15	2.64	nr	**9.86**
90 degree angle; single socket						
100 x 75 mm	14.01	17.02	0.38	6.69	nr	**23.71**
125 x 75 mm	17.79	21.60	0.43	7.57	nr	**29.17**
135 degree angle; single socket						
100 x 75 mm	14.01	17.02	0.38	6.69	nr	**23.71**
125 x 75 mm	17.79	21.60	0.43	7.57	nr	**29.17**
Running outlet						
65 mm outlet						
100 x 75 mm	14.01	17.02	0.38	6.69	nr	**23.71**
125 x 75 mm	17.79	21.60	0.43	7.57	nr	**29.17**
75 mm outlet						
100 x 75 mm	14.01	17.02	0.38	6.69	nr	**23.71**
125 x 75 mm	17.79	21.60	0.43	7.57	nr	**29.17**

56   *Material Costs/Measured Work Prices - Mechanical Installations*

R: DISPOSAL SYSTEMS

Item	Net Price £	Material £	Labour hours	Labour £	Unit	Total rate £
R10: RAINWATER PIPEWORK/GUTTERS						
Cast iron gutters: mastic and bolted joints; BS 460; fixed with brackets to backgrounds (Continued)						
Stop end outlet; socket						
65 mm outlet						
100 x 75 mm	9.39	11.40	0.38	6.69	nr	**18.09**
75 mm outlet						
100 x 75 mm	9.39	11.40	0.38	6.69	nr	**18.09**
125 x 75 mm	12.00	14.58	0.43	7.57	nr	**22.15**
Stop end outlet; spigot						
65 mm outlet						
100 x 75 mm	9.39	11.40	0.38	6.69	nr	**18.09**
75 mm outlet						
100 x 75 mm	9.39	11.40	0.38	6.69	nr	**18.09**
125 x 75 mm	12.00	14.58	0.43	7.57	nr	**22.15**
Brackets; fixed to backgrounds						
Fascia						
100 x 75 mm	4.83	5.86	0.16	2.82	nr	**8.68**
125 x 75 mm	5.95	7.22	0.16	2.82	nr	**10.04**
Ogee Gutter						
100 mm	12.27	14.90	0.85	14.96	m	**29.86**
115 mm	14.14	17.17	0.97	17.07	m	**34.24**
125 mm	14.54	17.66	0.97	17.07	m	**34.73**
Extra over fittings Ogee Cast Iron Gutter BS 460						
Union						
100 mm	3.23	3.92	0.38	6.69	nr	**10.61**
115 mm	3.36	4.09	0.43	7.57	nr	**11.65**
125 mm	4.22	5.13	0.43	7.57	nr	**12.70**
Stop end; internal						
100 mm	1.51	1.83	0.12	2.11	nr	**3.94**
115 mm	1.99	2.42	0.15	2.64	nr	**5.06**
125 mm	1.99	2.42	0.15	2.64	nr	**5.06**
Stop end; external						
100 mm	1.51	1.83	0.12	2.11	nr	**3.94**
115 mm	1.99	2.42	0.15	2.64	nr	**5.06**
125 mm	1.99	2.42	0.15	2.64	nr	**5.06**
90 degree angle; internal						
100 mm	5.31	6.45	0.38	6.69	nr	**13.14**
115 mm	5.75	6.99	0.43	7.57	nr	**14.56**
125 mm	6.28	7.63	0.43	7.57	nr	**15.19**

Material Costs/Measured Work Prices - Mechanical Installations

R: DISPOSAL SYSTEMS

Item	Net Price £	Material £	Labour hours	Labour £	Unit	Total rate £
90 degree angle; external						
100 mm	5.41	6.57	0.38	6.69	nr	**13.26**
115 mm	5.88	7.14	0.43	7.57	nr	**14.70**
125 mm	6.41	7.78	0.43	7.57	nr	**15.35**
135 degree angle; internal						
100 mm	5.31	6.45	0.38	6.69	nr	**13.14**
115 mm	5.75	6.99	0.43	7.57	nr	**14.56**
125 mm	6.28	7.63	0.43	7.57	nr	**15.19**
135 degree angle; external						
100 mm	5.31	6.45	0.38	6.69	nr	**13.14**
115 mm	5.75	6.99	0.43	7.57	nr	**14.56**
125 mm	6.41	7.78	0.43	7.57	nr	**15.35**
Running outlet						
65 mm outlet						
100 mm	14.01	17.02	0.38	6.69	nr	**23.71**
115 mm	17.79	21.60	0.43	7.57	nr	**29.17**
125 mm	14.01	17.02	0.43	7.57	nr	**24.59**
75 mm outlet						
125 mm	14.01	17.02	0.43	7.57	nr	**24.59**
Stop end outlet; socket						
65 mm outlet						
100 mm	9.39	11.40	0.38	6.69	nr	**18.09**
115 mm	9.39	11.40	0.43	7.57	nr	**18.97**
125 mm	12.00	14.58	0.43	7.57	nr	**22.15**
75 mm outlet						
125 mm	12.00	14.58	0.43	7.57	nr	**22.15**
Stop end outlet; spigot						
65 mm outlet						
100 mm	9.39	11.40	0.38	6.69	nr	**18.09**
115 mm	9.39	11.40	0.43	7.57	nr	**18.97**
125 mm	12.00	14.58	0.43	7.57	nr	**22.15**
75 mm outlet						
125 mm	12.00	14.58	0.43	7.57	nr	**22.15**
Brackets; fixed to backgrounds						
Fascia						
100 mm	4.83	5.86	0.16	2.82	nr	**8.68**
115 mm	5.95	7.22	0.16	2.82	nr	**10.04**
125 mm	5.95	7.22	0.16	2.82	nr	**10.04**
Notts Ogee Gutter						
115 mm	17.69	21.48	0.85	14.96	m	**36.44**
Extra over fittings Notts Ogee Cast Iron Gutter BS 460						
Union						
115 mm	5.52	6.70	0.38	6.69	nr	**13.39**
Stop end; internal						
115 mm	4.71	5.72	0.16	2.82	nr	**8.54**

58   *Material Costs/Measured Work Prices - Mechanical Installations*

R: DISPOSAL SYSTEMS

Item	Net Price £	Material £	Labour hours	Labour £	Unit	Total rate £
R10: RAINWATER PIPEWORK/GUTTERS						
Cast iron gutters: mastic and bolted joints; BS 460; fixed with brackets to backgrounds (Continued)						
Stop end; external						
115 mm	4.71	5.72	0.16	2.82	nr	**8.54**
90 degree angle; internal						
115 mm	13.18	16.00	0.43	7.57	nr	**23.57**
90 degree angle; external						
115 mm	13.18	16.00	0.43	7.57	nr	**23.57**
135 degree angle; internal						
115 mm	13.18	16.00	0.43	7.57	nr	**23.57**
135 degree angle; external						
115 mm	13.18	16.00	0.43	7.57	nr	**23.57**
Running outlet						
65 mm outlet						
115 mm	15.81	19.20	0.43	7.57	nr	**26.77**
75 mm outlet						
115 mm	13.20	16.03	0.43	7.57	nr	**23.60**
Stop end outlet; socket						
65 mm outlet						
115 mm	10.44	12.67	0.43	7.57	nr	**20.24**
Stop end outlet; spigot						
65 mm outlet						
115 mm	12.53	15.22	0.43	7.57	nr	**22.78**
Brackets; fixed to backgrounds						
Fascia						
115 mm	4.71	5.72	0.16	2.82	nr	**8.54**
No 46 moulded Gutter						
100 x 75 mm	12.79	15.53	0.85	14.96	m	**30.49**
125 x 100 mm	18.41	22.35	0.97	17.07	m	**39.42**
Extra over fittings						
Union						
100 x 75 mm	8.02	9.73	0.38	6.69	nr	**16.42**
125 x 100 mm	9.00	10.93	0.43	7.57	nr	**18.49**
Stop end; internal						
100 x 75 mm	4.93	5.98	0.12	2.11	nr	**8.09**
125 x 100 mm	6.39	7.76	0.15	2.64	nr	**10.40**
Stop end; external						
100 x 75 mm	4.93	5.98	0.12	2.11	nr	**8.09**
125 x 100 mm	6.39	7.76	0.15	2.64	nr	**10.40**

Material Costs/Measured Work Prices - Mechanical Installations

R: DISPOSAL SYSTEMS

Item	Net Price £	Material £	Labour hours	Labour £	Unit	Total rate £
90 degree angle; internal						
100 x 75 mm	12.92	15.69	0.38	6.69	nr	**22.38**
125 x 100 mm	18.57	22.55	0.43	7.57	nr	**30.11**
90 degree angle; external						
100 x 75 mm	12.92	15.69	0.38	6.69	nr	**22.38**
125 x 100 mm	18.57	22.55	0.43	7.57	nr	**30.11**
135 degree angle; internal						
100 x 75 mm	12.92	15.69	0.38	6.69	nr	**22.38**
125 x 100 mm	18.57	22.55	0.43	7.57	nr	**30.11**
135 degree angle; external						
100 x 75 mm	12.92	15.69	0.38	6.69	nr	**22.38**
125 x 100 mm	18.57	22.55	0.43	7.57	nr	**30.11**
Running outlet						
65 mm outlet						
100 x 75 mm	12.92	15.69	0.38	6.69	nr	**22.38**
125 x 100 mm	18.57	22.55	0.43	7.57	nr	**30.11**
75 mm outlet						
100 x 75 mm	12.92	15.69	0.38	6.69	nr	**22.38**
125 x 100 mm	18.57	22.55	0.43	7.57	nr	**30.11**
100 mm outlet						
100 x 75 mm	18.57	22.55	0.38	6.69	nr	**29.23**
125 x 100 mm	18.57	22.55	0.43	7.57	nr	**30.11**
100 x 75 mm outlet						
125 x 100 mm	18.57	22.55	0.43	7.57	nr	**30.11**
Stop end outlet; socket						
65 mm outlet						
100 x 75 mm	10.25	12.44	0.38	6.69	nr	**19.13**
75 mm outlet						
125 x 100 mm	12.64	15.35	0.43	7.57	nr	**22.92**
Stop end outlet; spigot						
65 mm outlet						
100 x 75 mm	10.25	12.44	0.38	6.69	nr	**19.13**
75 mm outlet						
125 x 100 mm	12.64	15.35	0.43	7.57	nr	**22.92**
Brackets; fixed to backgrounds						
Fascia						
100 x 75 mm	2.61	3.17	0.16	2.82	nr	**5.99**
125 x 100 mm	2.61	3.17	0.16	2.82	nr	**5.99**
Box Gutter						
100 x 75 mm	22.65	27.51	0.85	14.96	m	**42.47**
Extra over fittings Box Cast Iron Gutter BS 460						
Union						
100 x 75 mm	3.66	4.45	0.38	6.69	nr	**11.13**

Material Costs/Measured Work Prices - Mechanical Installations

R: DISPOSAL SYSTEMS

Item	Net Price £	Material £	Labour hours	Labour £	Unit	Total rate £
R10: RAINWATER PIPEWORK/GUTTERS						
Cast iron gutters: mastic and bolted joints; BS 460; fixed with brackets to backgrounds (Continued)						
Stop end; external						
100 x 75 mm	2.83	3.44	0.12	2.11	nr	**5.55**
90 degree angle						
100 x 75 mm	9.75	11.84	0.38	6.69	nr	**18.53**
135 degree angle						
100 x 75 mm	13.18	16.00	0.38	6.69	nr	**22.69**
Running outlet						
65 mm outlet						
100 x 75 mm	9.75	11.84	0.38	6.69	nr	**18.53**
75 mm outlet						
100 x 75 mm	9.75	11.84	0.38	6.69	nr	**18.53**
100 x 75 mm outlet						
100 x 75 mm	9.75	11.84	0.38	6.69	nr	**18.53**
Brackets; fixed to backgrounds						
Fascia						
100 x 75 mm	2.87	3.48	0.16	2.82	nr	**6.30**
Cast iron rainwater pipe; dry joints; BS 460; fixed to backgrounds						
Circular						
Plain sockets						
65mm	16.35	19.86	0.69	12.14	m	**32.00**
75 mm	16.37	19.88	0.69	12.14	m	**32.03**
100 mm	20.91	25.39	0.69	12.14	m	**37.54**
Eared sockets						
65mm	13.12	15.94	0.62	10.91	m	**26.85**
75 mm	13.12	15.94	0.62	10.91	m	**26.85**
100 mm	17.62	21.39	0.62	10.91	m	**32.30**
Extra over fittings Circular Cast Iron Pipework BS 460						
Loose sockets						
Plain socket						
65mm	3.13	3.80	0.23	4.05	nr	**7.85**
75 mm	3.13	3.80	0.23	4.05	nr	**7.85**
100 mm	4.51	5.47	0.23	4.05	nr	**9.52**
Eared socket						
65mm	4.20	5.10	0.29	5.10	nr	**10.21**
75 mm	4.20	5.10	0.29	5.10	nr	**10.21**
100 mm	5.66	6.88	0.29	5.10	nr	**11.98**

Material Costs/Measured Work Prices - Mechanical Installations

R: DISPOSAL SYSTEMS

Item	Net Price £	Material £	Labour hours	Labour £	Unit	Total rate £
Shoe; front projection						
Plain socket						
65mm	10.05	12.20	0.23	4.05	nr	**16.25**
75 mm	10.05	12.20	0.23	4.05	nr	**16.25**
100 mm	13.54	16.44	0.23	4.05	nr	**20.49**
Eared socket						
65mm	11.59	14.07	0.29	5.10	nr	**19.17**
75 mm	11.59	14.07	0.29	5.10	nr	**19.17**
100 mm	15.10	18.34	0.29	5.10	nr	**23.44**
Access Pipe						
65mm	18.06	21.94	0.23	4.05	nr	**25.98**
75 mm	18.97	23.04	0.23	4.05	nr	**27.08**
100 mm	33.14	40.24	0.23	4.05	nr	**44.29**
100 mm; eared	37.38	45.39	0.29	5.10	nr	**50.49**
Bends; any degree						
65mm	7.09	8.62	0.23	4.05	nr	**12.66**
75 mm	8.62	10.46	0.23	4.05	nr	**14.51**
100 mm	12.17	14.78	0.23	4.05	nr	**18.83**
Branch						
92.5 degrees						
65mm	13.68	16.61	0.29	5.10	nr	**21.72**
75 mm	15.08	18.31	0.29	5.10	nr	**23.42**
100 mm	17.92	21.76	0.29	5.10	nr	**26.86**
112.5 degrees						
65mm	10.96	13.31	0.29	5.10	nr	**18.42**
75 mm	12.07	14.66	0.29	5.10	nr	**19.77**
135 degrees						
65mm	13.68	16.61	0.29	5.10	nr	**21.72**
75 mm	15.08	18.31	0.29	5.10	nr	**23.42**
Offsets						
75 to 150 mm projection						
65mm	10.86	13.19	0.25	4.40	nr	**17.59**
75 mm	10.86	13.19	0.25	4.40	nr	**17.59**
100 mm	20.49	24.88	0.25	4.40	nr	**29.28**
225 mm projection						
65mm	12.64	15.35	0.25	4.40	nr	**19.75**
75 mm	12.64	15.35	0.25	4.40	nr	**19.75**
100 mm	24.81	30.13	0.25	4.40	nr	**34.53**
305 mm projection						
65mm	14.80	17.97	0.25	4.40	nr	**22.37**
75 mm	15.54	18.87	0.25	4.40	nr	**23.27**
100 mm	24.81	30.13	0.25	4.40	nr	**34.53**
380 mm projection						
65mm	29.54	35.88	0.25	4.40	nr	**40.28**
75 mm	29.54	35.88	0.25	4.40	nr	**40.28**
100 mm	40.32	48.96	0.25	4.40	nr	**53.36**

Material Costs/Measured Work Prices - Mechanical Installations

R: DISPOSAL SYSTEMS

Item	Net Price £	Material £	Labour hours	Labour £	Unit	Total rate £
R10: RAINWATER PIPEWORK/GUTTERS						
Cast iron rainwater pipe; dry joints; BS 460; fixed to backgrounds (Continued)						
455 mm projection						
65mm	34.58	41.99	0.25	4.40	nr	**46.40**
75 mm	34.58	41.99	0.25	4.40	nr	**46.40**
100 mm	49.02	59.52	0.25	4.40	nr	**63.92**
Bracket; fixed to backgrounds						
65mm	4.07	4.94	0.29	5.10	nr	**10.04**
75 mm	4.09	4.96	0.29	5.10	nr	**10.07**
100 mm	4.14	5.03	0.29	5.10	nr	**10.13**
Wall spacer plate; eared pipework						
65mm	2.82	3.43	0.16	2.82	nr	**6.25**
75 mm	2.89	3.50	0.16	2.82	nr	**6.32**
100 mm	2.93	3.56	0.16	2.82	nr	**6.38**
Rectangular						
Plain socket						
100 x 75 mm	67.49	81.96	1.04	18.30	m	**100.26**
Eared Socket						
100 x 75 mm	68.39	83.05	1.16	20.42	m	**103.46**
Extra over fittings Rectangular Cast Iron Pipework BS 460						
Loose socket						
100 x 75 mm; plain	13.13	15.95	0.23	4.05	nr	**19.99**
100 x 75 mm; eared	21.76	26.43	0.29	5.10	nr	**31.53**
Shoe; front						
100 x 75 mm; plain	34.70	42.13	0.23	4.05	nr	**46.18**
100 x 75 mm; eared	68.39	83.05	0.29	5.10	nr	**88.15**
Shoe; side						
100 x 75 mm; plain	34.70	42.13	0.23	4.05	nr	**46.18**
100 x 75 mm; eared	68.39	83.05	0.29	5.10	nr	**88.15**
Bends; side; any degree						
100 x 75 mm; plain	34.70	42.13	0.25	4.40	nr	**46.53**
100 x 75 mm; 135 degree; plain	33.40	40.55	0.25	4.40	nr	**44.95**
Bends; side; any degree						
100 x 75 mm; eared	68.39	83.05	0.25	4.40	nr	**87.45**
Bends; front; any degree						
100 x 75 mm; plain	34.70	42.13	0.25	4.40	nr	**46.53**
100 x 75 mm; eared	68.39	83.05	0.25	4.40	nr	**87.45**

Material Costs/Measured Work Prices - Mechanical Installations   63

R: DISPOSAL SYSTEMS

Item	Net Price £	Material £	Labour hours	Labour £	Unit	Total rate £
Offset; side;						
Plain socket						
75mm projection	43.06	52.29	0.25	4.40	nr	**56.69**
115mm projection	44.79	54.39	0.25	4.40	nr	**58.79**
225mm projection	55.80	67.76	0.25	4.40	nr	**72.16**
305mm projection	64.31	78.09	0.25	4.40	nr	**82.49**
Eared socket						
150mm projection	55.30	67.15	0.25	4.40	nr	**71.55**
Offset; Front						
Plain socket						
75mm projection	32.14	39.03	0.25	4.40	nr	**43.43**
150mm projection	35.52	43.13	0.25	4.40	nr	**47.53**
225mm projection	44.82	54.43	0.25	4.40	nr	**58.83**
305mm projection	53.33	64.76	0.25	4.40	nr	**69.16**
Eared socket						
75mm projection	41.12	49.94	0.25	4.40	nr	**54.34**
150mm projection	35.52	43.13	0.25	4.40	nr	**47.53**
225mm projection	58.19	70.66	0.25	4.40	nr	**75.06**
305mm projection	62.20	75.53	0.25	4.40	nr	**79.93**
Offset; plinth						
115mm projection; Plain	42.72	51.88	0.25	4.40	nr	**56.28**
115mm projection; Eared	33.81	41.06	0.25	4.40	nr	**45.46**
Bracket; fixed to backgrounds						
100 x 75mm; build in holdabat	17.12	20.79	0.35	6.16	nr	**26.95**
100 x 75mm; trefoil earband	13.38	16.25	0.29	5.10	nr	**21.36**
100 x 75mm; plain earband	12.94	15.71	0.29	5.10	nr	**20.82**
Rainwater Heads						
Flat hopper						
210 x 160 x 185 mm; 65 mm outlet	9.05	10.99	0.40	7.04	nr	**18.03**
210 x 160 x 185 mm; 75 mm outlet	10.28	12.49	0.40	7.04	nr	**19.53**
250 x 215 x 215 mm; 100 mm outlet	22.78	27.67	0.40	7.04	nr	**34.71**
Flat rectangular						
225 x 125 x 125 mm; 65 mm outlet	15.82	19.21	0.40	7.04	nr	**26.25**
225 x 125 x 125 mm; 75 mm outlet	15.82	19.21	0.40	7.04	nr	**26.25**
280 x 150 x 130 mm; 100 mm outlet	21.84	26.52	0.40	7.04	nr	**33.56**
Rectangular						
250 x 180 x 175mm; 75 mm outlet	22.71	27.57	0.40	7.04	nr	**34.61**
250 x 180 x 175mm; 100 mm outlet	20.45	24.84	0.40	7.04	nr	**31.88**
300 x 250 x 200mm; 65 mm outlet	40.32	48.96	0.40	7.04	nr	**56.01**
300 x 250 x 200mm; 75 mm outlet	40.32	48.96	0.40	7.04	nr	**56.01**
300 x 250 x 200mm; 100 mm outlet	40.32	48.96	0.40	7.04	nr	**56.01**
300 x 250 x 200mm; 100 x 75 mm outlet	40.32	48.96	0.40	7.04	nr	**56.01**
Castellated rectangular						
250 x 180 x 175mm; 65 mm outlet	20.45	24.84	0.40	7.04	nr	**31.88**

Material Costs/Measured Work Prices - Mechanical Installations

R: DISPOSAL SYSTEMS

Item	Net Price £	Material £	Labour hours	Labour £	Unit	Total rate £
R11: ABOVE GROUND DRAINAGE						
Pricing note: degree angles are only indicated where material prices differ						
PVC-U overflow pipe; solvent welded joints; fixed with clips to backgrounds						
Pipe						
19mm	1.46	1.77	0.21	3.70	m	**5.47**
Extra Over fittings Overflow pipework PVC-U						
Straight coupler						
19mm	0.61	0.74	0.17	2.99	nr	**3.73**
Bend						
19mm: 91.25 degree	0.72	0.87	0.17	2.99	nr	**3.87**
19mm: 135 degree	0.73	0.89	0.17	2.99	nr	**3.88**
Tee						
19mm	0.78	0.95	0.18	3.17	nr	**4.12**
Reverse nut connector						
19mm	0.78	0.95	0.15	2.64	nr	**3.59**
BSP adaptor: solvent welded socket to threaded socket						
19mm x 3/4"	0.99	1.20	0.14	2.46	nr	**3.67**
Straight tank connector						
19mm	0.95	1.15	0.21	3.70	nr	**4.85**
32mm	1.77	2.15	0.28	4.93	nr	**7.08**
40mm	1.93	2.34	0.30	5.28	nr	**7.62**
Bent tank connector						
19mm	1.13	1.37	0.21	3.70	nr	**5.07**
Tundish						
19mm	25.36	30.79	0.38	6.69	nr	**37.48**
Pipe clip: including fixing to backgrounds						
19mm	0.78	0.95	0.18	3.17	nr	**4.12**
MuPVC waste pipe; solvent welded joints; fixed with clips to backgrounds; BS 5255						
Pipe						
32mm	0.96	1.17	0.23	4.05	m	**5.21**
40mm	1.15	1.40	0.23	4.05	m	**5.44**
50mm	1.52	1.85	0.26	4.58	m	**6.42**

Material Costs/Measured Work Prices - Mechanical Installations

R: DISPOSAL SYSTEMS

Item	Net Price £	Material £	Labour hours	Labour £	Unit	Total rate £
Extra Over fittings Waste Pipework MuPVC						
Screwed access plug						
32mm	0.47	0.57	0.18	3.17	nr	3.74
40mm	0.47	0.57	0.18	3.17	nr	3.74
50mm	1.10	1.34	0.25	4.40	nr	5.74
Straight coupling						
32mm	0.47	0.57	0.27	4.75	nr	5.32
40mm	0.47	0.57	0.27	4.75	nr	5.32
50mm	2.20	2.67	0.27	4.75	nr	7.42
Expansion coupling						
32mm	1.10	1.34	0.27	4.75	nr	6.09
40mm	1.10	1.34	0.27	4.75	nr	6.09
50mm	2.20	2.67	0.27	4.75	nr	7.42
MuPVC to Copper coupling						
32mm	1.10	1.34	0.27	4.75	nr	6.09
40mm	1.10	1.34	0.27	4.75	nr	6.09
50mm	1.10	1.34	0.27	4.75	nr	6.09
Spigot and socket coupling						
32mm	1.10	1.34	0.27	4.75	nr	6.09
40mm	1.10	1.34	0.27	4.75	nr	6.09
50mm	2.20	2.67	0.27	4.75	nr	7.42
Union						
32mm	1.10	1.34	0.28	4.93	nr	6.26
40mm	1.10	1.34	0.28	4.93	nr	6.26
50mm	1.10	1.34	-	-	nr	1.34
Reducer: socket						
32 x 19mm	0.47	0.57	-	-	nr	0.57
40 x 32mm	0.47	0.57	0.27	4.75	nr	5.32
50 x 32mm	0.47	0.57	0.27	4.75	nr	5.32
50 x 40mm	1.10	1.34	0.27	4.75	nr	6.09
Reducer: level invert						
40 x 32mm	0.47	0.57	0.27	4.75	nr	5.32
50 x 32mm	0.47	0.57	0.27	4.75	nr	5.32
50 x 40mm	1.10	1.34	0.27	4.75	nr	6.09
Swept bend						
32mm	0.47	0.57	0.27	4.75	nr	5.32
32mm: 165 degree	0.47	0.57	0.27	4.75	nr	5.32
40mm	0.47	0.57	0.27	4.75	nr	5.32
40mm: 165 degree	0.47	0.57	0.27	4.75	nr	5.32
50mm	1.10	1.34	0.30	5.28	nr	6.62
50mm: 165 degree	0.47	0.57	0.30	5.28	nr	5.85
Knuckle bend						
32mm	0.47	0.57	0.27	4.75	nr	5.32
40mm	0.47	0.57	0.27	4.75	nr	5.32

Material Costs/Measured Work Prices - Mechanical Installations

R: DISPOSAL SYSTEMS

Item	Net Price £	Material £	Labour hours	Labour £	Unit	Total rate £
R11: ABOVE GROUND DRAINAGE						
MuPVC waste pipe; solvent welded joints; fixed with clips to backgrounds; BS 5255 (Continued)						
Spigot and socket bend						
32mm	0.47	0.57	0.27	4.75	nr	**5.32**
32mm: 150 degree	0.47	0.57	0.27	4.75	nr	**5.32**
40mm	0.47	0.57	0.27	4.75	nr	**5.32**
50mm	0.47	0.57	0.30	5.28	nr	**5.85**
Swept tee						
32mm: 91.25 degree	0.47	0.57	0.31	5.46	nr	**6.03**
32mm: 135 degree	0.47	0.57	0.31	5.46	nr	**6.03**
40mm: 91.25 degree	0.47	0.57	0.31	5.46	nr	**6.03**
40mm: 135 degree	0.47	0.57	0.31	5.46	nr	**6.03**
50mm	1.10	1.34	0.31	5.46	nr	**6.79**
Swept cross						
40mm: 91.25 degree	3.31	4.02	0.31	5.46	nr	**9.48**
50mm: 91.25 degree	5.03	6.11	0.43	7.57	nr	**13.68**
50mm: 135 degree	5.03	6.11	0.31	5.46	nr	**11.56**
Male iron adaptor						
32mm	0.47	0.57	0.28	4.93	nr	**5.50**
40mm	0.47	0.57	0.28	4.93	nr	**5.50**
Female iron adaptor						
32mm	0.47	0.57	0.28	4.93	nr	**5.50**
40mm	0.47	0.57	0.28	4.93	nr	**5.50**
50mm	0.47	0.57	0.31	5.46	nr	**6.03**
Reverse nut adaptor						
32mm	0.47	0.57	-	-	nr	**0.57**
40mm	0.47	0.57	-	-	nr	**0.57**
Automatic air admittance valve						
32mm	8.24	10.01	0.27	4.75	nr	**14.76**
40mm	8.24	10.01	0.28	4.93	nr	**14.93**
50mm	8.24	10.01	0.31	5.46	nr	**15.46**
MuPVC to metal adpator: including heat shrunk joint to metal						
50mm	3.50	4.25	0.38	6.69	nr	**10.94**
Caulking bush: including joint to metal						
32mm	1.96	2.38	0.31	5.46	nr	**7.84**
40mm	1.96	2.38	0.31	5.46	nr	**7.84**
50mm	1.96	2.38	0.32	5.63	nr	**8.01**

Material Costs/Measured Work Prices - Mechanical Installations

R: DISPOSAL SYSTEMS

Item	Net Price £	Material £	Labour hours	Labour £	Unit	Total rate £
Weathering apron						
50mm	1.31	1.59	0.65	11.44	nr	**13.03**
Vent Cowl						
50mm	1.30	1.58	0.19	3.34	nr	**4.92**
Pipe clip: including fixing to backgrounds						
32mm	0.21	0.26	0.13	2.29	nr	**2.54**
40mm	0.27	0.33	0.13	2.29	nr	**2.62**
50mm	0.50	0.61	0.13	2.29	nr	**2.90**
Pipe clip: expansion: including fixing to backgrounds						
32mm	0.24	0.29	0.13	2.29	nr	**2.58**
40mm	0.31	0.38	0.13	2.29	nr	**2.66**
50mm	2.28	2.77	0.13	2.29	nr	**5.06**
Pipe clip: metal; including fixing to backgrounds						
32mm	0.93	1.13	0.13	2.29	nr	**3.42**
40mm	1.22	1.48	0.13	2.29	nr	**3.77**
50mm	2.28	2.77	0.13	2.29	nr	**5.06**
ABS waste pipe; solvent welded joints; fixed with clips to backgrounds; BS 5255						
Pipe						
32mm	0.96	1.17	0.23	4.05	m	**5.21**
40mm	1.15	1.40	0.23	4.05	m	**5.44**
50mm	1.52	1.85	0.26	4.58	m	**6.42**
Extra Over fittings Waste Pipework ABS						
Screwed access plug						
32mm	0.47	0.57	0.18	3.17	nr	**3.74**
40mm	0.47	0.57	0.18	3.17	nr	**3.74**
50mm	1.10	1.34	0.25	4.40	nr	**5.74**
Straight coupling						
32mm	0.47	0.57	0.27	4.75	nr	**5.32**
40mm	0.47	0.57	0.27	4.75	nr	**5.32**
50mm	1.10	1.34	0.27	4.75	nr	**6.09**
Expansion coupling						
32mm	1.10	1.34	0.27	4.75	nr	**6.09**
40mm	1.10	1.34	0.27	4.75	nr	**6.09**
50mm	2.20	2.67	0.27	4.75	nr	**7.42**

Material Costs/Measured Work Prices - Mechanical Installations

R: DISPOSAL SYSTEMS

Item	Net Price £	Material £	Labour hours	Labour £	Unit	Total rate £
R11: ABOVE GROUND DRAINAGE						
ABS waste pipe; solvent welded joints; fixed with clips to backgrounds; BS 5255 (Continued)						
ABS to Copper coupling						
32mm	1.10	1.34	0.27	4.75	nr	6.09
40mm	1.10	1.34	0.27	4.75	nr	6.09
50mm	2.20	2.67	0.27	4.75	nr	7.42
Reducer: socket						
40 x 32mm	0.47	0.57	0.27	4.75	nr	5.32
50 x 32mm	0.47	0.57	0.27	4.75	nr	5.32
50 x 40mm	1.10	1.34	0.27	4.75	nr	6.09
Swept bend						
32mm	0.47	0.57	0.27	4.75	nr	5.32
40mm	0.47	0.57	0.27	4.75	nr	5.32
50mm	1.10	1.34	0.30	5.28	nr	6.62
Knuckle bend						
32mm	0.47	0.57	0.27	4.75	nr	5.32
40mm	0.47	0.57	0.27	4.75	nr	5.32
Swept tee						
32mm	0.47	0.57	0.31	5.46	nr	6.03
40mm	0.47	0.57	0.31	5.46	nr	6.03
50mm	1.10	1.34	0.31	5.46	nr	6.79
Swept cross						
40mm	3.31	4.02	0.23	4.05	nr	8.07
50mm	5.03	6.11	0.43	7.57	nr	13.68
Male iron adaptor						
32mm	0.47	0.57	0.28	4.93	nr	5.50
40mm	0.47	0.57	0.28	4.93	nr	5.50
Female iron adapator						
32mm	0.47	0.57	0.28	4.93	nr	5.50
40mm	0.47	0.57	0.28	4.93	nr	5.50
50mm	1.10	1.34	0.31	5.46	nr	6.79
Tank connectors						
32mm	1.79	2.17	0.29	5.10	nr	7.28
40mm	1.95	2.37	0.29	5.10	nr	7.47
Caulking bush: including joint to pipework						
50mm	1.79	2.17	0.50	8.80	nr	10.97
Pipe clip: including fixing to backgrounds						
32mm	0.21	0.25	0.17	2.99	nr	3.24
40mm	0.27	0.33	0.17	2.99	nr	3.32
50mm	0.51	0.62	0.17	2.99	nr	3.62

Material Costs/Measured Work Prices - Mechanical Installations

R: DISPOSAL SYSTEMS

Item	Net Price £	Material £	Labour hours	Labour £	Unit	Total rate £
Pipe clip: expansion: including fixing to backgrounds						
32mm	0.24	0.29	0.17	2.99	nr	**3.28**
40mm	0.29	0.35	0.17	2.99	nr	**3.34**
50mm	0.73	0.88	0.17	2.99	nr	**3.88**
Pipe clip: metal; including fixing to backgrounds						
32mm	0.94	1.14	0.17	2.99	nr	**4.14**
40mm	1.12	1.36	0.17	2.99	nr	**4.35**
50mm	1.43	1.73	0.17	2.99	nr	**4.72**
Polypropylene waste pipe; push fit joints; fixed with clips to backgrounds; BS 5254						
Pipe						
32mm	0.78	0.95	0.21	3.70	m	**4.64**
40mm	0.90	1.09	0.21	3.70	m	**4.79**
50mm	1.42	1.72	0.38	6.69	m	**8.41**
Extra Over fittings Waste Pipework Polypropylene						
Screwed access plug						
32mm	0.49	0.59	0.16	2.82	nr	**3.41**
40mm	0.49	0.59	0.16	2.82	nr	**3.41**
50mm	0.85	1.03	0.20	3.52	nr	**4.55**
Straight coupling						
32mm	0.49	0.59	0.19	3.34	nr	**3.94**
40mm	0.49	0.59	0.19	3.34	nr	**3.94**
50mm	0.85	1.03	0.20	3.52	nr	**4.55**
Universal waste pipe coupler						
32mm dia.	0.55	0.67	0.20	3.52	nr	**4.19**
40mm dia.	0.60	0.73	0.20	3.52	nr	**4.25**
Reducer						
40 x 32mm	0.49	0.59	0.19	3.34	nr	**3.94**
50 x 32mm	0.49	0.59	0.19	3.34	nr	**3.94**
50 x 40mm	0.85	1.03	0.20	3.52	nr	**4.55**
Swept bend						
32mm	0.49	0.59	0.19	3.34	nr	**3.94**
40mm	0.49	0.59	0.19	3.34	nr	**3.94**
50mm	0.85	1.03	0.20	3.52	nr	**4.55**
Knuckle bend						
32mm	0.49	0.59	0.19	3.34	nr	**3.94**
40mm	0.49	0.59	0.19	3.34	nr	**3.94**
50mm	0.85	1.03	0.20	3.52	nr	**4.55**
Spigot and socket bend						
32mm	0.49	0.59	0.19	3.34	nr	**3.94**
40mm	0.49	0.59	0.19	3.34	nr	**3.94**

Material Costs/Measured Work Prices - Mechanical Installations

R: DISPOSAL SYSTEMS

Item	Net Price £	Material £	Labour hours	Labour £	Unit	Total rate £
R11: ABOVE GROUND DRAINAGE						
Polypropylene waste pipe; push fit joints; fixed with clips to backgrounds; BS 5254 (Continued)						
Swept tee						
32mm	0.49	0.59	0.22	3.87	nr	4.47
40mm	0.49	0.59	0.22	3.87	nr	4.47
50mm	0.88	1.07	0.23	4.05	nr	5.12
Male iron adaptor						
32mm	0.49	0.59	0.13	2.29	nr	2.88
40mm	0.49	0.59	0.19	3.34	nr	3.94
50mm	1.30	1.58	0.15	2.64	nr	4.22
Tank connector						
32mm	0.49	0.59	0.24	4.22	nr	4.82
40mm	0.49	0.59	0.24	4.22	nr	4.82
50mm	0.80	0.97	0.35	6.16	nr	7.13
Pipe clip: saddle; including fixing to backgrounds						
32mm	0.16	0.19	0.17	2.99	nr	3.19
40mm	0.16	0.19	0.17	2.99	nr	3.19
Pipe clip: including fixing to backgrounds						
50mm	0.40	0.49	0.17	2.99	nr	3.48
Polypropylene traps; including fixing to appliance and connection to pipework; BS 3943						
Tubular P trap; 75mm seal						
32mm dia.	1.57	1.91	0.20	3.52	nr	5.43
40mm dia.	1.73	2.10	0.20	3.52	nr	5.62
Tubular S trap; 75mm seal						
32mm dia.	1.63	1.98	0.20	3.52	nr	5.50
40mm dia.	1.79	2.17	0.20	3.52	nr	5.69
Running tubular P trap; 75mm seal						
32mm dia.	1.57	1.91	0.20	3.52	nr	5.43
40mm dia.	1.73	2.10	0.20	3.52	nr	5.62
Running tubular S trap; 75mm seal						
32mm dia.	1.63	1.98	0.20	3.52	nr	5.50
40mm dia.	1.79	2.17	0.20	3.52	nr	5.69
Spigot and socket bend; convertor from P to S Trap						
32mm	1.00	1.21	0.20	3.52	nr	4.73
40mm	1.12	1.36	0.21	3.70	nr	5.06
Bottle P trap; 75mm seal						
32mm dia.	1.36	1.65	0.20	3.52	nr	5.17
40mm dia.	1.57	1.91	0.20	3.52	nr	5.43

Material Costs/Measured Work Prices - Mechanical Installations

R: DISPOSAL SYSTEMS

Item	Net Price £	Material £	Labour hours	Labour £	Unit	Total rate £
Bottle S trap; 75mm seal						
32mm dia.	1.49	1.81	0.20	3.52	nr	**5.33**
40mm dia.	1.73	2.10	0.25	4.40	nr	**6.50**
Bottle P trap; resealing; 75mm seal						
32mm dia.	2.03	2.46	0.20	3.52	nr	**5.99**
40mm dia.	2.36	2.87	0.25	4.40	nr	**7.27**
Bottle S trap; resealing; 75mm seal						
32mm dia.	2.03	2.46	0.20	3.52	nr	**5.99**
40mm dia.	2.36	2.87	0.25	4.40	nr	**7.27**
Bath trap, low level; 38mm seal						
40mm dia.	1.93	2.34	0.25	4.40	nr	**6.74**
Bath trap, low level; 38mm seal complete with overflow hose						
40mm dia.	4.05	4.92	0.25	4.40	nr	**9.32**
Bath trap; 75mm seal complete with overlow hose						
40mm dia.	2.30	2.79	0.25	4.40	nr	**7.19**
Bath trap; 75mm seal complete with overflow hose and overflow outlet						
40mm dia.	4.45	5.40	0.20	3.52	nr	**8.92**
Bath trap; 75mm seal complete with overflow hose, overflow outlet and ABS chrome waste						
40mm dia.	4.45	5.40	0.20	3.52	nr	**8.92**
Washing machine trap; 75mm seal including stand pipe						
40mm dia.	2.89	3.51	0.25	4.40	nr	**7.91**
Washing machine standpipe						
40mm dia.	3.55	4.31	0.25	4.40	nr	**8.71**
Plastic unslotted chrome plated basin/sink waste including plug						
32mm	2.65	3.22	0.34	5.98	nr	**9.20**
40mm	3.27	3.97	0.34	5.98	nr	**9.96**
Plastic slotted chrome plated basin/sink waste including plug						
32mm	2.73	3.31	0.34	5.98	nr	**9.30**
40mm	3.27	3.97	0.34	5.98	nr	**9.96**
Bath overflow outlet; plastic; white						
42mm	5.50	6.68	0.37	6.51	nr	**13.19**
Bath overlow outlet; plastic; chrome plated						
42mm	6.15	7.47	0.37	6.51	nr	**13.98**

Material Costs/Measured Work Prices - Mechanical Installations

R: DISPOSAL SYSTEMS

Item	Net Price £	Material £	Labour hours	Labour £	Unit	Total rate £
R11: ABOVE GROUND DRAINAGE						
Polypropylene traps; including fixing to appliance and connection to pipework; BS 3943 (Continued)						
Combined cistren and bath overflow outlet; plastic; white						
42mm	10.74	13.04	0.39	6.86	nr	**19.91**
Combined cistern and bath overlow outlet; plastic; chrome plated						
42mm	11.69	14.20	0.39	6.86	nr	**21.06**
Cistern overflow outlet; plastic; white						
42mm	5.50	6.68	0.15	2.64	nr	**9.32**
Cistern overlow outlet; plastic; chrome plated						
42mm	6.15	7.47	0.15	2.64	nr	**10.11**

Material Costs/Measured Work Prices - Mechanical Installations

R: DISPOSAL SYSTEMS

Item	Net Price £	Material £	Labour hours	Labour £	Unit	Total rate £
PVC-U Soil and Waste pipe; solvent welded joints; fixed with clips to backgrounds; BS 4514/ BS EN 607						
Pipe						
82mm	5.57	6.76	0.35	6.16	m	**12.92**
110mm	5.86	7.12	0.41	7.22	m	**14.33**
160mm	15.10	18.34	0.51	8.98	m	**27.31**
Extra Over fittings Solvent Welded Pipework PVC-U						
Straight coupling						
82mm	2.16	2.62	0.21	3.70	nr	**6.32**
110mm	2.70	3.28	0.22	3.87	nr	**7.15**
160mm	7.79	9.46	0.24	4.22	nr	**13.68**
Expansion coupling						
82mm	2.16	2.62	0.21	3.70	nr	**6.32**
110mm	2.70	3.28	0.22	3.87	nr	**7.15**
160mm	7.79	9.46	0.24	4.22	nr	**13.68**
Slip coupling; double ring socket						
82mm	2.16	2.62	0.21	3.70	nr	**6.32**
110mm	2.70	3.28	0.22	3.87	nr	**7.15**
160mm	7.79	9.46	0.24	4.22	nr	**13.68**
Puddle flanges						
110mm	63.17	76.71	0.45	7.92	nr	**84.63**
160mm	84.42	102.51	0.55	9.68	nr	**112.19**
Socket reducer						
82 to 50mm	3.28	3.98	0.18	3.17	nr	**7.15**
110 to 50mm	4.09	4.97	0.18	3.17	nr	**8.13**
110 to 82mm	4.09	4.97	0.22	3.87	nr	**8.84**
160 to 110mm	8.54	10.37	0.26	4.58	nr	**14.95**
Socket plugs						
82mm	3.07	3.73	0.15	2.64	nr	**6.37**
110mm	3.72	4.52	0.20	3.52	nr	**8.04**
160mm	6.85	8.32	0.27	4.75	nr	**13.07**
Access door; including cutting into pipe						
82mm	5.10	6.19	0.28	4.93	nr	**11.12**
110mm	6.02	7.31	0.34	5.98	nr	**13.29**
160mm	11.33	13.76	0.46	8.10	nr	**21.85**
Screwed access cap						
82mm	8.90	10.81	0.15	2.64	nr	**13.45**
110mm	8.90	10.81	0.20	3.52	nr	**14.33**
160mm	11.33	13.76	0.27	4.75	nr	**18.51**

Material Costs/Measured Work Prices - Mechanical Installations

R: DISPOSAL SYSTEMS

Item	Net Price £	Material £	Labour hours	Labour £	Unit	Total rate £
R11: ABOVE GROUND DRAINAGE						
PVC-U Soil and Waste pipe; solvent welded joints; fixed with clips to backgrounds; BS 4514/ BS EN 607 (Continued)						
Access pipe: spigot and socket						
110mm	8.90	10.81	0.22	3.87	nr	**14.68**
Access pipe: double socket						
110mm	8.90	10.81	0.22	3.87	nr	**14.68**
Swept bend						
82mm	5.42	6.58	0.29	5.10	nr	**11.69**
110mm	6.34	7.70	0.32	5.63	nr	**13.33**
160mm	15.79	19.17	0.49	8.62	nr	**27.80**
Bend; special angle						
82mm	5.42	6.58	0.29	5.10	nr	**11.69**
110mm	6.33	7.69	0.32	5.63	nr	**13.32**
160mm	15.79	19.17	0.49	8.62	nr	**27.80**
Spigot and socket bend						
82mm	5.14	6.24	0.26	4.58	nr	**10.82**
110mm	6.03	7.32	0.32	5.63	nr	**12.95**
110mm: 135 degree	6.97	8.46	0.32	5.63	nr	**14.10**
160mm: 135 degree	15.01	18.23	0.44	7.74	nr	**25.97**
Variable bend: single socket						
110mm	12.36	15.01	0.33	5.81	nr	**20.82**
Variable bend: double socket						
110mm	12.36	15.01	0.33	5.81	nr	**20.82**
Access Bend						
110mm	17.58	21.35	0.33	5.81	nr	**27.16**
Single branch: two bosses						
82mm	7.57	9.19	0.35	6.16	nr	**15.35**
82mm: 104 degree	7.57	9.19	0.35	6.16	nr	**15.35**
110mm	8.38	10.18	0.42	7.39	nr	**17.57**
110mm: 135 degree	8.38	10.18	0.42	7.39	nr	**17.57**
160mm	37.20	45.17	0.50	8.80	nr	**53.97**
160mm: 135 degree	10.17	12.35	0.50	8.80	nr	**21.15**
Single branch; four bosses						
110mm	8.38	10.18	0.42	7.39	nr	**17.57**
Single access branch						
82mm	35.60	43.23	-	-	nr	**43.23**
110mm	20.09	24.40	-	-	nr	**24.40**
Unequal single branch						
160 x 160 x 110mm	20.10	24.41	0.50	8.80	nr	**33.21**
160 x 160 x 110mm: 135 degree	21.61	26.24	0.50	8.80	nr	**35.04**

Material Costs/Measured Work Prices - Mechanical Installations

R: DISPOSAL SYSTEMS

Item	Net Price £	Material £	Labour hours	Labour £	Unit	Total rate £
Double Branch						
110mm	8.38	10.18	0.42	7.39	nr	**17.57**
110mm: 135 degree	8.38	10.18	0.42	7.39	nr	**17.57**
Corner Branch						
110mm	8.38	10.18	0.42	7.39	nr	**17.57**
Unequal double Branch						
160 x 160 x 110mm	21.61	26.24	0.50	8.80	nr	**35.04**
Single boss pipe; single socket						
110 x 110 x 32mm	3.28	3.98	0.24	4.22	nr	**8.21**
110 x 110 x 40mm	3.28	3.98	0.24	4.22	nr	**8.21**
110 x 110 x 50mm	3.28	3.98	0.24	4.22	nr	**8.21**
Single boss pipe; triple socket						
110 x 110 x 40mm	3.28	3.98	0.24	4.22	nr	**8.21**
Waste boss; including cutting into pipe						
82 to 32mm	3.08	3.74	0.29	5.10	nr	**8.84**
82 to 40mm	3.08	3.74	0.29	5.10	nr	**8.84**
110 to 32mm	3.08	3.74	0.29	5.10	nr	**8.84**
110 to 40mm	3.08	3.74	0.29	5.10	nr	**8.84**
110 to 50mm	3.20	3.89	0.29	5.10	nr	**8.99**
160 to 32mm	4.37	5.31	0.30	5.28	nr	**10.59**
160 to 40mm	4.37	5.31	0.35	6.16	nr	**11.47**
160 to 50mm	4.37	5.31	0.40	7.04	nr	**12.35**
Self locking waste boss; including cutting into pipe						
110 to 32mm	4.12	5.00	0.30	5.28	nr	**10.28**
110 to 40mm	4.31	5.23	0.30	5.28	nr	**10.51**
110 to 50mm	4.88	5.93	0.30	5.28	nr	**11.21**
Adaptor saddle; including cutting to pipe						
82 to 32mm	1.86	2.26	0.29	5.10	nr	**7.36**
110 to 40mm	2.32	2.82	0.29	5.10	nr	**7.92**
160 to 50mm	4.19	5.09	0.29	5.10	nr	**10.19**
Branch boss adaptor						
32mm	1.17	1.42	0.26	4.58	nr	**6.00**
40 mm	1.17	1.42	0.26	4.58	nr	**6.00**
50 mm	1.68	2.04	0.26	4.58	nr	**6.62**
Branch boss adaptor bend						
32mm	1.17	1.42	0.26	4.58	nr	**6.00**
40 mm	1.17	1.42	0.26	4.58	nr	**6.00**
50 mm	1.68	2.04	0.26	4.58	nr	**6.62**
Automatic air admittance valve						
82 to 110mm	18.82	22.85	0.19	3.34	nr	**26.20**

Material Costs/Measured Work Prices - Mechanical Installations

R: DISPOSAL SYSTEMS

Item	Net Price £	Material £	Labour hours	Labour £	Unit	Total rate £
R11: ABOVE GROUND DRAINAGE						
PVC-U Soil and Waste pipe; solvent welded joints; fixed with clips to backgrounds; BS 4514/ BS EN 607 (Continued)						
PVC-U to metal adpator: including heat shrunk joint to metal						
110mm	5.14	6.24	0.57	10.03	nr	**16.27**
Caulking bush: including joint to pipework						
82mm	5.07	6.16	0.46	8.10	nr	**14.25**
110mm	5.07	6.16	0.46	8.10	nr	**14.25**
Vent cowl						
82mm	3.05	3.70	0.13	2.29	nr	**5.99**
110mm	3.30	4.01	0.13	2.29	nr	**6.30**
160mm	9.30	11.29	0.13	2.29	nr	**13.58**
Weathering apron; to lead slates						
82mm	1.59	1.93	1.15	20.24	nr	**22.17**
110mm	1.82	2.21	1.15	20.24	nr	**22.45**
160mm	5.48	6.65	1.15	20.24	nr	**26.90**
Weathering apron; to asphalt						
82mm	6.71	8.15	1.10	19.36	nr	**27.51**
110mm	6.71	8.15	1.10	19.36	nr	**27.51**
Weathering slate; flat; 406 x 406mm						
82mm	18.16	22.05	1.04	18.30	nr	**40.36**
110mm	18.16	22.05	1.04	18.30	nr	**40.36**
Weathering slate; flat; 457 x 457mm						
82mm	18.16	22.05	1.04	18.30	nr	**40.36**
110mm	18.16	22.05	1.04	18.30	nr	**40.36**
Weathering slate; angled; 610 x 610mm						
82mm	25.15	30.54	1.04	18.30	nr	**48.84**
110mm	25.15	30.54	1.04	18.30	nr	**48.84**
Galvanised steel pipe clip: including fixing to backgrounds						
82mm	1.69	2.05	0.18	3.17	nr	**5.22**
110mm	1.69	2.05	0.18	3.17	nr	**5.22**
160mm	4.21	5.11	0.18	3.17	nr	**8.28**
Plastic coated steel pipe clip: including fixing to backgrounds						
82mm	2.19	2.66	0.18	3.17	nr	**5.83**
110mm	2.28	2.77	0.18	3.17	nr	**5.94**
160mm	4.43	5.38	0.18	3.17	nr	**8.55**

Material Costs/Measured Work Prices - Mechanical Installations 77

R: DISPOSAL SYSTEMS

Item	Net Price £	Material £	Labour hours	Labour £	Unit	Total rate £
Plastic pipe clip: including fixing to backgrounds						
82mm	1.18	1.43	0.18	3.17	nr	**4.60**
110mm	1.23	1.49	0.18	3.17	nr	**4.66**
Plastic coated steel pipe clip: adjustable; including fixing to backgrounds						
82mm	2.31	2.81	0.20	3.52	nr	**6.33**
110mm	2.31	2.81	0.20	3.52	nr	**6.33**
Galvanised steel pipe clip: drive in; including fixing to backgrounds						
110mm	3.63	4.41	0.22	3.87	nr	**8.28**
PVC-U Soil and Waste pipe; ring seal joints; fixed with clips to backgrounds; BS 4514/BS EN 607						
Pipe						
82mm dia.	1.85	2.25	0.35	6.16	m	**8.41**
110mm dia.	1.95	2.37	0.41	7.22	m	**9.59**
160mm dia.	5.03	6.11	0.51	8.98	m	**15.08**
Extra Over fittings Ring Seal Pipework PVC-U						
Straight coupling						
82mm	1.37	1.66	0.21	3.70	nr	**5.36**
110mm	1.52	1.85	0.22	3.87	nr	**5.72**
160mm	4.96	6.02	0.24	4.22	nr	**10.25**
Straight coupling; double socket						
82mm	2.16	2.62	0.21	3.70	nr	**6.32**
110mm	2.70	3.28	0.22	3.87	nr	**7.15**
160mm	7.79	9.46	0.24	4.22	nr	**13.68**
Reducer; socket						
82 to 50mm	3.41	4.14	0.15	2.64	nr	**6.78**
110 to 50mm	5.49	6.67	0.15	2.64	nr	**9.31**
110 to 82mm	4.39	5.33	0.19	3.34	nr	**8.68**
160 to 110	8.90	10.81	0.31	5.46	nr	**16.26**
Access Cap						
82mm	5.57	6.76	0.15	2.64	nr	**9.40**
110mm	5.57	6.76	0.17	2.99	nr	**9.76**
Access Cap; pressure plug						
160mm	16.14	19.60	0.33	5.81	nr	**25.41**
Access pipe						
82mm	8.42	10.22	0.22	3.87	nr	**14.10**
110mm	12.09	14.68	0.22	3.87	nr	**18.55**
160mm	26.17	31.78	0.24	4.22	nr	**36.00**

R: DISPOSAL SYSTEMS

Item	Net Price £	Material £	Labour hours	Labour £	Unit	Total rate £
R11: ABOVE GROUND DRAINAGE						
PVC-U Soil and Waste pipe; ring seal joints; fixed with clips to backgrounds; BS 4514/BS EN 607 (Continued)						
Bend						
82mm	6.31	7.66	0.29	5.10	nr	**12.77**
82mm; adjustable radius	6.31	7.66	0.29	5.10	nr	**12.77**
110mm	7.40	8.99	0.32	5.63	nr	**14.62**
110mm; adjustable radius	7.40	8.99	0.32	5.63	nr	**14.62**
160mm	18.41	22.36	0.49	8.62	nr	**30.98**
160mm; adjustable radius	7.40	8.99	0.49	8.62	nr	**17.61**
Bend; spigot and socket						
110mm	7.40	8.99	0.32	5.63	nr	**14.62**
Bend; offset						
82mm	6.18	7.50	0.21	3.70	nr	**11.20**
110mm	7.37	8.95	0.32	5.63	nr	**14.58**
160mm	19.92	24.19	-	-	nr	**24.19**
Bend; access						
110mm	15.42	18.72	0.33	5.81	nr	**24.53**
Single branch						
82mm	8.99	10.92	0.35	6.16	nr	**17.08**
110mm	11.32	13.75	0.42	7.39	nr	**21.14**
110mm; 45 degree	12.00	14.57	0.31	5.46	nr	**20.03**
160mm	37.09	45.04	0.50	8.80	nr	**53.84**
Single branch; access						
82mm	8.99	10.92	0.35	6.16	nr	**17.08**
110mm	11.51	13.98	0.42	7.39	nr	**21.37**
Unequal single branch						
160 x 160 x 110mm	22.03	26.75	0.50	8.80	nr	**35.55**
160 x 160 x 110mm; 45 degree	22.03	26.75	0.50	8.80	nr	**35.55**
Double branch; 4 bosses						
110mm	17.66	21.44	0.49	8.62	nr	**30.07**
Corner branch; 2 bosses						
110mm	41.85	50.82	0.49	8.62	nr	**59.44**
Multibranch; 4 bosses						
110mm	33.54	40.73	0.52	9.15	nr	**49.88**
Boss Branch						
110 x 32mm	2.37	2.88	0.34	5.98	nr	**8.86**
110 x 40mm	2.37	2.88	0.34	5.98	nr	**8.86**

Material Costs/Measured Work Prices - Mechanical Installations

R: DISPOSAL SYSTEMS

Item	Net Price £	Material £	Labour hours	Labour £	Unit	Total rate £
Strap on boss						
110 x 32mm	3.08	3.74	0.30	5.28	nr	**9.02**
110 x 40mm	3.08	3.74	0.30	5.28	nr	**9.02**
110 x 50mm	3.08	3.74	0.30	5.28	nr	**9.02**
Patch boss						
82 x 32mm	2.52	3.06	0.31	5.46	nr	**8.52**
82 x 40mm	2.52	3.06	0.31	5.46	nr	**8.52**
82 x 50mm	2.52	3.06	0.31	5.46	nr	**8.52**
Boss Pipe; collar 4 boss						
110mm	19.79	24.03	0.35	6.16	nr	**30.19**
Boss adaptor; rubber; push fit						
32mm	1.17	1.42	0.26	4.58	nr	**6.00**
40 mm	1.17	1.42	0.26	4.58	nr	**6.00**
50 mm	1.68	2.04	0.26	4.58	nr	**6.62**
WC connector; cap and seal; solvent socket						
110mm	4.79	5.82	0.23	4.05	nr	**9.86**
110mm; 90 degree	7.62	9.25	0.27	4.75	nr	**14.01**
Vent terminal						
82mm	3.05	3.70	0.13	2.29	nr	**5.99**
110mm	3.30	4.01	0.13	2.29	nr	**6.30**
160mm	9.30	11.29	0.13	2.29	nr	**13.58**
Weathering slate; inclined; 610 x 610mm						
82mm	25.15	30.54	1.04	18.30	nr	**48.84**
110mm	25.15	30.54	1.04	18.30	nr	**48.84**
Weathering slate; inclined; 450 x 450mm						
82mm	25.15	30.54	1.04	18.30	nr	**48.84**
110mm	25.15	30.54	1.04	18.30	nr	**48.84**
Weathering slate; flat; 400 x 400mm						
82mm	18.16	22.05	1.04	18.30	nr	**40.36**
110mm	18.16	22.05	1.04	18.30	nr	**40.36**
Air admittance valve						
82mm	18.82	22.85	0.19	3.34	nr	**26.20**
110mm	18.82	22.85	0.19	3.34	nr	**26.20**

R: DISPOSAL SYSTEMS

Item	Net Price £	Material £	Labour hours	Labour £	Unit	Total rate £
R11: ABOVE GROUND DRAINAGE						
Cast iron pipe; nitrile rubber gasket joint with continuity clip BS 416/6087; fixed to backgrounds						
Pipe						
50 mm	8.14	9.89	0.25	4.40	m	**14.29**
75 mm	9.11	11.07	0.45	7.92	m	**18.99**
100 mm	11.02	13.38	0.60	10.56	m	**23.94**
150 mm	23.00	27.93	0.70	12.33	m	**40.25**
Extra over fittings Nitrile Gasket Cast Iron Pipework BS 416/6087						
Standard coupling						
50 mm	4.13	5.02	0.50	8.80	nr	**13.82**
75 mm	4.57	5.56	0.60	10.56	nr	**16.12**
100 mm	5.96	7.24	0.67	11.80	nr	**19.03**
150 mm	11.92	14.47	0.83	14.62	nr	**29.09**
Conversion coupling						
65 x 75 mm	3.68	4.47	0.60	10.56	nr	**15.04**
70 x 75 mm	3.68	4.47	0.60	10.56	nr	**15.04**
90 x 100 mm	4.75	5.77	0.67	11.80	nr	**17.56**
Access pipe; round door						
50 mm	19.64	23.85	0.41	7.22	nr	**31.07**
75 mm	20.31	24.66	0.46	8.10	nr	**32.76**
100 mm	23.11	28.07	0.67	11.81	nr	**39.88**
150 mm	41.50	50.39	0.83	14.62	nr	**65.01**
Access pipe; square door						
100 mm	36.73	44.60	0.67	11.81	nr	**56.41**
150 mm	60.48	73.44	0.83	14.62	nr	**88.06**
Taper reducer						
75 mm	9.57	11.62	0.60	10.56	nr	**22.18**
100 mm	12.72	15.45	0.67	11.80	nr	**27.25**
150 mm	24.79	30.10	0.83	14.62	nr	**44.72**
Blank cap						
50 mm	5.07	6.15	0.24	4.22	nr	**10.38**
75 mm	5.74	6.97	0.26	4.58	nr	**11.55**
100 mm	6.88	8.35	0.32	5.63	nr	**13.99**
150 mm	12.44	15.11	0.40	7.04	nr	**22.15**
Blank cap; 50 mm screwed tapping						
75 mm	8.22	9.98	0.26	4.58	nr	**14.56**
100 mm	9.62	11.68	0.32	5.63	nr	**17.31**
150 mm	15.18	18.43	0.40	7.04	nr	**25.48**
Universal connector						
50 x 56/48/40 mm	2.98	3.62	0.33	5.81	nr	**9.43**

Material Costs/Measured Work Prices - Mechanical Installations

R: DISPOSAL SYSTEMS

Item	Net Price £	Material £	Labour hours	Labour £	Unit	Total rate £
Change piece; BS416						
100 mm	6.31	7.66	0.47	8.27	nr	**15.93**
WC connector						
100 mm	8.79	10.67	0.49	8.62	nr	**19.29**
Boss pipe; 2 " BSPT socket						
50 mm	17.78	21.60	0.58	10.21	nr	**31.80**
75 mm	18.12	22.00	0.65	11.44	nr	**33.44**
100 mm	19.67	23.88	0.79	13.90	nr	**37.79**
150 mm	34.29	41.63	0.86	15.14	nr	**56.77**
Boss pipe; 2 " BSPT socket; 135 degree						
100 mm	26.82	32.57	0.79	13.90	nr	**46.48**
Boss pipe; 2 x 2 " BSPT socket; opposed						
75 mm	22.48	27.29	0.65	11.44	nr	**38.73**
100 mm	26.82	32.57	0.79	13.90	nr	**46.48**
Boss pipe; 2 x 2 " BSPT socket; in line						
100 mm	27.71	33.65	0.79	13.90	nr	**47.55**
Boss pipe; 2 x 2 " BSPT socket; 90 degree						
100 mm	26.82	32.57	0.79	13.90	nr	**46.48**
Bend; short radius						
50 mm	7.29	8.85	0.50	8.80	nr	**17.65**
75 mm	7.29	8.85	0.60	10.56	nr	**19.41**
100 mm	10.09	12.25	0.67	11.80	nr	**24.05**
100 mm; 11 degree	16.75	20.35	0.67	11.80	nr	**32.14**
100 mm; 67 degree	16.75	20.35	0.67	11.80	nr	**32.14**
150 mm	18.02	21.89	0.83	14.62	nr	**36.51**
Access bend; short radius						
50 mm	19.97	24.25	0.50	8.80	nr	**33.05**
75 mm	20.63	25.06	0.60	10.56	nr	**35.62**
100 mm	21.34	25.91	0.67	11.80	nr	**37.71**
100 mm; 45 degree	25.32	30.74	0.67	11.80	nr	**42.54**
150 mm	30.31	36.81	0.83	14.62	nr	**51.43**
150 mm; 45 degree	41.23	50.07	0.83	14.62	nr	**64.69**
Long radius bend						
75 mm	17.44	21.17	0.60	10.56	nr	**31.74**
100 mm	16.34	19.84	0.67	11.80	nr	**31.64**
100 mm; 5 degree	16.75	20.35	0.67	11.80	nr	**32.14**
150 mm	33.71	40.94	0.83	14.62	nr	**55.56**
150 mm; 22.5 degree	41.69	50.63	0.83	14.62	nr	**65.25**
Access bend; long radius						
75 mm	25.56	31.03	0.60	10.56	nr	**41.60**
100 mm	21.34	25.91	0.67	11.80	nr	**37.71**
150 mm	30.31	36.81	0.83	14.62	nr	**51.43**
Long tail bend						
100 x 250 mm long	13.04	15.83	0.70	12.33	nr	**28.15**
100 x 815 mm long	32.03	38.89	0.70	12.33	nr	**51.22**

Material Costs/Measured Work Prices - Mechanical Installations

R: DISPOSAL SYSTEMS

Item	Net Price £	Material £	Labour hours	Labour £	Unit	Total rate £
R11: ABOVE GROUND DRAINAGE						
Cast iron pipe; nitrile rubber gasket joint with continuity clip BS 416/6087; fixed to backgrounds (Continued)						
Offset						
75 mm projection						
75 mm	12.25	14.87	0.53	9.33	nr	**24.20**
100 mm	14.17	17.20	0.66	11.62	nr	**28.82**
115 mm projection						
75 mm	16.57	20.13	0.53	9.33	nr	**29.45**
100 mm	17.06	20.71	0.66	11.62	nr	**32.33**
150 mm projection						
75 mm	11.73	14.24	0.53	9.33	nr	**23.57**
100 mm	19.52	23.70	0.66	11.62	nr	**35.32**
225 mm projection						
100 mm	20.75	25.20	0.66	11.62	nr	**36.81**
300 mm projection						
100 mm	22.20	26.96	0.66	11.62	nr	**38.57**
Branch; equal and unequal						
50 mm	10.97	13.32	0.78	13.73	nr	**27.05**
75 mm	10.97	13.32	0.85	14.97	nr	**28.29**
100 mm	15.59	18.94	1.00	17.60	nr	**36.54**
150 mm	36.60	44.44	1.20	21.13	nr	**65.57**
150 x 100 mm; 87.5 degree	45.22	54.91	1.21	21.21	nr	**76.12**
150 x 100 mm; 45 degree	45.22	54.91	1.21	21.21	nr	**76.12**
Branch; 2" BSPT screwed socket						
100 mm	22.79	27.68	1.00	17.60	nr	**45.28**
Branch; long tail						
100 x 915 mm long	46.45	56.41	1.00	17.60	nr	**74.01**
Access branch; equal and unequal						
50 mm	25.92	31.47	0.78	13.73	nr	**45.20**
75 mm	26.92	32.69	0.85	14.97	nr	**47.65**
100 mm	26.85	32.61	1.02	17.96	nr	**50.57**
150 mm	50.60	61.45	1.20	21.13	nr	**82.58**
150 x 100 mm; 87.5 degree	54.56	66.26	1.20	21.13	nr	**87.39**
150 x 100 mm; 45 degree	54.56	66.26	1.20	21.13	nr	**87.39**
Parallel branch						
100 mm	29.51	35.84	1.00	17.60	nr	**53.44**
Double branch						
75 mm	27.96	33.95	0.95	16.72	nr	**50.67**
100 mm	19.29	23.43	1.30	22.89	nr	**46.31**
150 x 100 mm	77.63	94.27	1.56	27.46	nr	**121.73**

Material Costs/Measured Work Prices - Mechanical Installations   83

R: DISPOSAL SYSTEMS

Item	Net Price £	Material £	Labour hours	Labour £	Unit	Total rate £
Double access branch						
100 mm	30.54	37.08	1.43	25.18	nr	**62.26**
Corner branch						
100 mm	37.90	46.02	1.30	22.89	nr	**68.90**
Puddle flange; grey epoxy coated						
100 mm	14.54	17.65	1.00	17.60	nr	**35.25**
WC connector						
100 mm	14.96	18.17	0.66	11.62	nr	**29.79**
Roof vent connector; asphalt						
75 mm	17.33	21.04	0.90	15.84	nr	**36.88**
100 mm	13.39	16.26	0.97	17.07	nr	**33.33**
P trap						
100 mm	23.04	27.98	1.00	17.60	nr	**45.58**
P trap with access						
50 mm	21.98	26.69	0.77	13.55	nr	**40.25**
75 mm	22.32	27.10	0.90	15.84	nr	**42.94**
100 mm	23.04	27.98	1.16	20.42	nr	**48.40**
150 mm	72.46	87.99	1.77	31.15	nr	**119.15**
Bellmouth gully inlet						
100 mm	23.39	28.40	1.08	19.01	nr	**47.41**
Balcony gully inlet						
100 mm	26.93	32.71	1.08	19.01	nr	**51.71**
Roof outlet						
Flat grate						
75 mm	39.63	48.12	0.83	14.61	nr	**62.73**
100 mm	45.65	55.43	1.08	19.01	nr	**74.44**
Dome grate						
75 mm	40.58	49.27	0.83	14.61	nr	**63.88**
100 mm	46.57	56.55	1.08	19.01	nr	**75.56**
Top Hat						
100 mm	69.87	84.84	1.08	19.01	nr	**103.85**
Brackets; fixed to backgrounds						
50 mm	2.39	2.90	0.15	2.64	nr	**5.54**
75 mm	2.39	2.90	0.18	3.17	nr	**6.06**
100 mm	2.65	3.21	0.18	3.17	nr	**6.38**
150 mm	4.66	5.66	0.20	3.52	nr	**9.18**

R: DISPOSAL SYSTEMS

Item	Net Price £	Material £	Labour hours	Labour £	Unit	Total rate £
R11: ABOVE GROUND DRAINAGE						
Cast iron pipe; EPDM rubber gasket joint with continuity clip; BS EN877; fixed to backgrounds						
Pipe						
50 mm	6.62	8.28	0.25	4.40	m	**12.68**
70 mm	7.45	9.32	0.45	7.92	m	**17.24**
100 mm	8.82	11.04	0.60	10.56	m	**21.60**
125 mm	14.05	17.57	0.65	11.44	m	**29.01**
150 mm	17.30	21.64	0.70	12.32	m	**33.96**
200 mm	34.18	42.75	1.14	20.06	m	**62.82**
250 mm	40.82	51.06	1.25	22.00	m	**73.06**
300 mm	50.00	62.53	1.53	26.93	m	**89.46**
Extra over fittings EDPM Rubber Jointed Cast Iron Pipework BS EN 877						
Coupling						
50 mm	2.05	2.49	0.50	8.80	nr	**11.29**
70 mm	2.26	2.75	0.60	10.56	nr	**13.31**
100 mm	2.95	3.58	0.67	11.80	nr	**15.38**
125 mm	3.66	4.45	0.75	13.21	nr	**17.65**
150mm	5.89	7.15	0.83	14.62	nr	**21.77**
200 mm	13.18	16.01	1.21	21.30	nr	**37.30**
250 mm	17.07	20.72	1.33	23.41	nr	**44.13**
300 mm	20.67	25.10	1.63	28.69	nr	**53.79**
Push fit joint						
Plain socket						
50 mm	3.82	4.64	0.25	4.40	nr	**9.04**
70 mm	3.82	4.64	0.25	4.40	nr	**9.04**
100 mm	4.39	5.33	0.25	4.40	nr	**9.73**
Eared socket						
50 mm	3.94	4.78	0.25	4.40	nr	**9.18**
70 mm	3.94	4.78	0.25	4.40	nr	**9.18**
100 mm	6.58	7.98	0.25	4.40	nr	**12.39**
Slip socket						
50 mm	6.82	8.28	0.25	4.40	nr	**12.68**
70 mm	6.82	8.28	0.25	4.40	nr	**12.68**
100 mm	7.97	9.68	0.25	4.40	nr	**14.08**
Stack support pipe						
70 mm	12.72	15.45	0.74	13.02	nr	**28.48**
100 mm	15.46	18.78	0.88	15.49	nr	**34.27**
125 mm	17.56	21.32	1.00	17.60	nr	**38.92**
150 mm	24.17	29.35	1.19	20.95	nr	**50.30**
200 mm	35.60	43.23	1.31	23.06	nr	**66.28**

Material Costs/Measured Work Prices - Mechanical Installations

R: DISPOSAL SYSTEMS

Item	Net Price £	Material £	Labour hours	Labour £	Unit	Total rate £
Access pipe; round door						
50 mm	13.56	16.47	0.54	9.50	nr	**25.98**
70 mm	14.54	17.66	0.64	11.26	nr	**28.92**
100 mm	16.91	20.54	0.60	10.56	nr	**31.10**
150 mm	31.70	38.50	0.83	14.61	nr	**53.11**
Access pipe; square door						
100 mm	27.20	33.03	0.60	10.56	nr	**43.59**
125 mm	74.59	90.57	0.67	11.79	nr	**102.36**
150 mm	81.45	98.91	0.71	12.50	nr	**111.41**
200 mm	92.60	112.44	1.21	21.30	nr	**133.74**
250 mm	138.34	167.99	1.31	23.06	nr	**191.05**
300 mm	171.32	208.04	1.43	25.17	nr	**233.21**
Taper reducer						
70 mm	7.52	9.13	0.51	8.98	nr	**18.11**
100 mm	8.69	10.56	0.58	10.21	nr	**20.76**
125 mm	9.42	11.44	0.64	11.26	nr	**22.70**
150 mm	16.00	19.43	0.67	11.79	nr	**31.22**
200 mm	25.93	31.48	1.15	20.24	nr	**51.72**
250 mm	54.57	66.27	1.25	22.00	nr	**88.27**
300 mm	73.98	89.83	1.37	24.11	nr	**113.95**
Blank cap						
50 mm	3.48	4.23	0.24	4.22	nr	**8.45**
70 mm	3.77	4.58	0.26	4.58	nr	**9.15**
100 mm	4.79	5.81	0.32	5.63	nr	**11.44**
125 mm	6.15	7.47	0.35	6.16	nr	**13.63**
150 mm	8.42	10.22	0.40	7.04	nr	**17.26**
200 mm	24.40	29.62	0.60	10.56	nr	**40.18**
250 mm	39.98	48.55	0.65	11.44	nr	**59.99**
300 mm	49.87	60.55	0.72	12.67	nr	**73.22**
Blank cap; 50 mm screwed tapping						
70 mm	5.63	6.84	0.26	4.58	nr	**11.42**
100 mm	6.58	7.99	0.32	5.63	nr	**13.63**
150 mm	10.25	12.45	0.40	7.04	nr	**19.49**
Universal connector; EPDM rubber						
50 x 56/48/40 mm	2.98	3.62	0.30	5.28	nr	**8.90**
Blank end; push fit						
100 x 38/32 mm	6.58	7.99	0.39	6.86	nr	**14.86**
Boss pipe; 2 " BSPT socket						
50 mm	12.11	14.70	0.54	9.50	nr	**24.21**
75 mm	12.53	15.21	0.64	11.26	nr	**26.48**
100 mm	15.69	19.05	0.78	13.73	nr	**32.78**
150 mm	27.74	33.68	1.09	19.18	nr	**52.87**
Boss pipe; 2 x 2 " BSPT socket; opposed						
100 mm	18.55	22.52	0.78	13.73	nr	**36.25**
Boss pipe; 2 x 2 " BSPT socket; 90 degree						
100 mm	18.55	22.52	0.78	13.73	nr	**36.25**

Material Costs/Measured Work Prices - Mechanical Installations

R: DISPOSAL SYSTEMS

Item	Net Price £	Material £	Labour hours	Labour £	Unit	Total rate £
R11: ABOVE GROUND DRAINAGE						
Cast iron pipe; EPDM rubber gasket joint with continuity clip; BS EN877; fixed to backgrounds (Continued)						
Manifold connector						
100 mm	22.96	27.88	0.78	13.73	nr	41.61
Bend; short radius						
50mm	7.66	9.30	0.50	8.80	nr	18.10
70mm	8.53	10.35	0.60	10.56	nr	20.91
100mm	10.64	12.91	0.67	11.80	nr	24.71
125 mm	15.71	19.08	0.78	13.73	nr	32.81
150mm	20.28	24.62	0.83	14.62	nr	39.24
200 mm; 45 degree	51.68	62.75	1.21	21.30	nr	84.05
250 mm; 45 degree	83.55	101.45	1.31	23.06	nr	124.51
300 mm; 45 degree	110.83	134.58	1.43	25.17	nr	159.75
Access bend; short radius						
70 mm	12.28	14.91	0.64	11.26	nr	26.18
100mm	17.24	20.94	0.78	13.73	nr	34.66
150mm	29.38	35.67	0.83	14.62	nr	50.29
Bend; long radius bend						
100mm	17.12	20.79	0.67	11.80	nr	32.58
100 mm; 22 degree	15.44	18.74	0.78	13.73	nr	32.47
150mm	42.94	52.14	0.83	14.62	nr	66.76
Access bend; long radius						
100mm	20.55	24.96	0.67	11.80	nr	36.75
Bend ; long tail						
100 mm	14.51	17.62	0.78	13.73	nr	31.34
Bend ; long tail double						
70 mm	19.17	23.28	0.64	11.26	nr	34.54
100 mm	20.52	24.92	0.78	13.73	nr	38.65
Bend; air pipe						
100 mm	18.55	22.53	0.78	13.73	nr	36.26
Offset						
75 mm projection						
100 mm	10.24	12.43	0.78	13.73	nr	26.16
130 mm projection						
50 mm	8.08	9.81	0.54	9.50	nr	19.31
70 mm	11.40	13.85	0.64	11.26	nr	25.11
100 mm	14.96	18.16	0.78	13.73	nr	31.89
125 mm	17.96	21.81	0.78	13.73	nr	35.54

Material Costs/Measured Work Prices - Mechanical Installations

R: DISPOSAL SYSTEMS

Item	Net Price £	Material £	Labour hours	Labour £	Unit	Total rate £
Branch; equal and unequal						
50 mm	11.86	14.40	0.78	13.73	nr	28.13
70 mm	12.80	15.55	0.85	14.97	nr	30.51
100 mm	16.52	20.06	1.00	17.60	nr	37.66
125 mm	27.54	33.44	1.16	20.42	nr	53.85
150 mm	36.01	43.73	1.37	24.11	nr	67.84
200 mm	89.10	108.19	1.51	26.58	nr	134.77
250 mm	162.12	196.86	1.63	28.69	nr	225.55
300mm	204.73	248.60	1.77	31.15	nr	279.75
Branch; radius; equal and unequal						
70 mm	14.01	17.01	0.79	13.90	nr	30.91
100 mm	18.05	21.92	0.96	16.90	nr	38.82
125 mm	30.85	37.46	1.16	20.42	nr	57.88
150 mm	39.68	48.18	1.37	24.11	nr	72.30
200 mm	99.01	120.22	1.51	26.58	nr	146.80
Branch; long tail						
100 mm	36.56	44.40	0.96	16.90	nr	61.29
Access branch; radius; equal and unequal						
70 mm	17.91	21.75	0.79	13.90	nr	35.65
100 mm	23.93	29.06	0.96	16.90	nr	45.96
150 mm	49.59	60.22	1.20	21.13	nr	81.35
Double branch; equal and unequal						
100mm	22.29	27.06	1.30	22.89	nr	49.95
100 mm; 69 degree	24.39	29.62	1.30	22.89	nr	52.50
150 mm	54.38	66.03	1.37	24.11	nr	90.14
200 mm	99.01	120.22	1.51	26.58	nr	146.80
Double branch; radius; equal and unequal						
100 mm	22.29	27.06	1.30	22.89	nr	49.95
150 mm	54.38	66.03	1.37	24.11	nr	90.14
200 mm	99.01	120.22	1.51	26.58	nr	146.80
Corner branch						
100 mm	31.89	38.73	1.30	22.89	nr	61.62
Corner branch; long tail						
100 mm	31.89	38.73	1.30	22.89	nr	61.62
Roof vent connector; asphalt						
100 mm	13.35	16.21	0.78	13.73	nr	29.94
Roof vent connector; felt						
100 mm	29.36	35.65	0.78	13.73	nr	49.38
Movement connector						
100 mm	20.34	24.69	0.78	13.73	nr	38.42
150 mm	38.09	46.25	0.78	13.73	nr	59.98
Expansion plugs						
70 mm	5.63	6.84	0.32	5.63	nr	12.47
100 mm	6.58	7.99	0.39	6.86	nr	14.86
150 mm	10.25	12.45	0.55	9.68	nr	22.13

Material Costs/Measured Work Prices - Mechanical Installations

R: DISPOSAL SYSTEMS

Item	Net Price £	Material £	Labour hours	Labour £	Unit	Total rate £
R11: ABOVE GROUND DRAINAGE						
Cast iron pipe; EPDM rubber gasket joint with continuity clip; BS EN877; fixed to backgrounds (Continued)						
P trap						
100mm dia.	17.49	21.23	1.00	17.60	nr	**38.84**
P trap with access						
50 mm	15.48	18.80	0.54	9.50	nr	**28.30**
70 mm	15.69	19.05	0.64	11.26	nr	**30.32**
100mm	17.49	21.23	1.00	17.60	nr	**38.84**
150 mm	31.89	38.72	1.09	19.18	nr	**57.90**
Branch trap						
100 mm	43.40	52.70	1.17	20.59	nr	**73.29**
Balcony gully inlet						
100 mm	26.93	32.71	1.00	17.60	nr	**50.31**
Roof outlet						
Flat grate						
70 mm	39.63	48.12	1.00	17.60	nr	**65.72**
100 mm	45.65	55.43	1.00	17.60	nr	**73.03**
Dome grate						
70 mm	40.58	49.27	1.00	17.60	nr	**66.88**
100 mm	-	-	1.00	17.60	nr	**17.60**
Top Hat						
100 mm	69.87	84.84	1.00	17.60	nr	**102.44**
Floor Drains; for Cast Iron Pipework BS 416 and BS EN877						
Adjustable clamp plate body						
100 mm; 165mm nickel bronze grate and frame	29.20	35.46	0.50	8.80	nr	**44.26**
100 mm; 165mm nickel bronze rodding eye	34.32	41.68	0.50	8.80	nr	**50.48**
100 mm; 150 x 150mm nickel bronze grate and frame	31.06	37.71	0.50	8.80	nr	**46.52**
100 mm; 150 x 150 nickel bronze rodding eye	35.91	43.60	0.50	8.80	nr	**52.41**
Deck plate body						
100 mm; 165mm nickel bronze grate and frame	64.02	77.75	0.50	8.80	nr	**86.55**
100 mm; 165mm nickel bronze rodding eye	64.02	77.75	0.50	8.80	nr	**86.55**
100 mm; 150 x 150mm nickel bronze grate and frame	64.02	77.75	0.50	8.80	nr	**86.55**
100 mm; 150 x 150mm nickel bronze rodding eye	64.02	77.75	0.50	8.80	nr	**86.55**
Extra for						
100 mm; Srewed extension piece	12.72	15.45	0.30	5.28	nr	**20.73**
100 mm; Grating extension piece; screwed or spigot	9.21	11.19	0.30	5.28	nr	**16.47**
100 mm; Brewary trap	351.31	426.59	2.00	35.20	nr	**461.79**

Material Costs/Measured Work Prices - Mechanical Installations

R: DISPOSAL SYSTEMS

Item	Net Price £	Material £	Labour hours	Labour £	Unit	Total rate £
Brackets; fixed to backgrounds						
Ductile iron						
50mm	2.20	2.67	0.10	1.76	nr	4.43
70 mm	2.20	2.67	0.10	1.76	nr	4.43
100 mm	2.39	2.90	0.15	2.64	nr	5.54
150 mm	4.28	5.19	0.20	3.52	nr	8.71
200 mm	15.52	18.85	0.25	4.40	nr	23.25
Mild steel; vertical						
125 mm	2.73	3.32	0.15	2.64	nr	5.96
Mild steel; stand off						
250 mm	9.39	11.40	0.25	4.40	nr	15.80
300 mm	10.22	12.40	0.25	4.40	nr	16.80
Stack support; rubber seal						
70 mm	12.72	15.45	0.74	13.02	nr	28.48
100 mm	15.46	18.78	0.88	15.49	nr	34.27
125 mm	17.56	21.32	1.00	17.60	nr	38.92
150 mm	24.17	29.35	1.19	20.95	nr	50.30
200 mm	35.60	43.23	1.31	23.06	nr	66.28
Wall spacer plate; cast iron (eared sockets)						
50 mm	1.66	2.02	0.10	1.76	nr	3.78
70 mm	1.66	2.02	0.10	1.76	nr	3.78
100 mm	1.66	2.02	0.10	1.76	nr	3.78

Material Costs/Measured Work Prices - Mechanical Installations

S:PIPED SUPPLY SYSTEMS

Item	Net Price £	Material £	Labour hours	Labour £	Unit	Total rate £
S10 : COLD WATER						
Y10 - PIPELINES						
MEDIUM DENSITY POLYETHELENE - BLUE						
Pipes for water distribution; laid underground; electrofusion joints in the running length; BS 6572						
Coiled service pipe						
20mm dia.	0.19	0.24	0.37	6.51	m	**6.75**
25mm dia.	0.24	0.30	0.41	7.22	m	**7.52**
32mm dia.	0.40	0.50	0.47	8.27	m	**8.77**
50mm dia.	0.96	1.20	0.53	9.33	m	**10.53**
63mm dia.	1.53	1.91	0.60	10.56	m	**12.48**
Mains service pipe						
90mm dia.	4.14	5.18	0.90	15.84	m	**21.02**
110mm dia	6.98	8.48	1.10	19.36	m	**27.84**
125mm dia.	7.85	9.82	1.20	21.13	m	**30.95**
160mm dia.	14.08	17.10	1.48	26.05	m	**43.15**
180mm dia.	15.86	19.84	1.50	26.43	m	**46.26**
225mm dia	28.82	35.00	1.77	31.15	m	**66.15**
250mm dia.	31.97	39.99	1.75	30.82	m	**70.81**
315mm dia	47.04	57.12	1.90	33.44	m	**90.56**
Extra over fittings; MDPE - Blue; electrofusion joints						
Coupler						
20mm dia	3.08	3.74	0.36	6.34	nr	**10.08**
25mm dia	3.39	4.12	0.40	7.04	nr	**11.16**
32mm dia	3.65	4.43	0.44	7.74	nr	**12.18**
40mm dia	3.79	4.60	0.48	8.45	nr	**13.05**
50mm dia	3.91	4.75	0.52	9.15	nr	**13.90**
63mm dia.	4.16	5.05	0.58	10.21	nr	**15.26**
90mm dia.	6.14	7.46	0.67	11.79	nr	**19.25**
110mm dia	8.80	10.69	0.74	13.02	nr	**23.71**
125mm dia.	11.15	13.54	0.83	14.61	nr	**28.15**
160mm dia.	16.52	20.06	1.00	17.60	nr	**37.66**
180mm dia.	20.70	25.14	1.25	22.00	nr	**47.14**
225mm dia	35.82	43.50	1.35	23.76	nr	**67.26**
250mm dia.	48.51	58.91	1.50	26.40	nr	**85.31**
315mm dia	62.38	75.75	1.80	31.68	nr	**107.43**

Material Costs/Measured Work Prices - Mechanical Installations

S:PIPED SUPPLY SYSTEMS

Item	Net Price £	Material £	Labour hours	Labour £	Unit	Total rate £
Extra over fittings; MDPE - Blue; butt fused joints						
Cap						
25mm dia	6.49	7.88	0.20	3.52	nr	**11.40**
32mm dia	6.76	8.21	0.22	3.87	nr	**12.08**
40mm dia	7.08	8.60	0.24	4.22	nr	**12.82**
50mm dia	7.14	8.67	0.26	4.58	nr	**13.25**
63mm dia.	7.30	8.87	0.32	5.63	nr	**14.50**
90mm dia.	11.82	14.35	0.37	6.51	nr	**20.86**
110mm dia	16.57	20.12	0.40	7.04	nr	**27.16**
125mm dia.	19.24	23.36	0.46	8.10	nr	**31.46**
160mm dia	25.73	31.24	0.50	8.80	nr	**40.04**
180mm dia.	36.96	44.88	0.60	10.56	nr	**55.44**
225mm dia	48.18	58.51	0.68	11.97	nr	**70.47**
250mm dia	64.30	78.08	0.75	13.20	nr	**91.28**
315mm dia	79.20	96.17	0.90	15.84	nr	**112.01**
Reducer						
63 x 32mm dia	7.11	8.63	0.54	9.50	nr	**18.14**
63 x 50mm dia	7.79	9.46	0.60	10.56	nr	**20.02**
90 x 63mm dia.	8.57	10.41	0.67	11.79	nr	**22.20**
110 x 90mm dia	12.81	15.56	0.74	13.02	nr	**28.58**
125 x 90mm dia.	17.19	20.87	0.83	14.61	nr	**35.48**
125 x 110mm dia	21.50	26.11	1.00	17.60	nr	**43.71**
160 x 110mm dia	25.63	31.12	1.10	19.36	nr	**50.48**
180 x 125mm dia.	31.53	38.29	1.25	22.00	nr	**60.29**
225 x 160mm dia	43.61	52.96	1.40	24.64	nr	**77.60**
250 x 180mm dia	62.74	76.19	1.80	31.68	nr	**107.87**
315 x 250mm dia	77.58	94.21	2.40	42.24	nr	**136.45**
Bend; 45 degree						
50mm dia	8.91	10.82	0.50	8.80	nr	**19.62**
63mm dia.	10.73	13.02	0.58	10.21	nr	**23.23**
90mm dia.	15.07	18.30	0.67	11.79	nr	**30.09**
110mm dia	22.43	27.24	0.74	13.02	nr	**40.26**
125mm dia.	26.22	31.84	0.83	14.61	nr	**46.45**
160mm dia	39.85	48.39	1.00	17.60	nr	**65.99**
180mm dia.	57.65	70.01	1.25	22.00	nr	**92.01**
225mm dia	85.76	104.14	1.40	24.64	nr	**128.78**
250mm dia	112.04	136.05	1.80	31.68	nr	**167.73**
315mm dia	157.83	191.65	2.40	42.24	nr	**233.89**

Material Costs/Measured Work Prices - Mechanical Installations

S:PIPED SUPPLY SYSTEMS

Item	Net Price £	Material £	Labour hours	Labour £	Unit	Total rate £
S10 : COLD WATER						
Y10 - PIPELINES						
MEDIUM DENSITY POLYETHELENE - BLUE (Continued)						
Bend; 90 degree						
50mm dia	8.91	10.82	0.50	8.80	nr	**19.62**
63mm dia.	10.73	13.02	0.58	10.21	nr	**23.23**
90mm dia.	16.56	20.11	0.67	11.79	nr	**31.91**
110mm dia	22.80	27.69	0.74	13.02	nr	**40.71**
125mm dia.	27.10	32.91	0.83	14.61	nr	**47.51**
160mm dia	41.99	50.99	1.00	17.60	nr	**68.59**
180mm dia.	61.46	74.63	1.25	22.00	nr	**96.63**
225mm dia	86.89	105.51	1.40	24.64	nr	**130.15**
250mm dia	113.54	137.87	1.80	31.68	nr	**169.55**
315mm dia	161.14	195.67	2.40	42.24	nr	**237.91**
Equal tee						
50mm dia	8.93	10.84	0.70	12.32	nr	**23.16**
63mm dia.	10.52	12.78	0.75	13.20	nr	**25.98**
90mm dia.	19.20	23.31	0.87	15.31	nr	**38.63**
110mm dia	29.53	35.86	1.00	17.60	nr	**53.46**
125mm dia.	36.74	44.62	1.08	19.01	nr	**63.63**
160mm dia	44.40	53.91	1.35	23.76	nr	**77.68**
180mm dia.	62.30	75.65	1.63	28.69	nr	**104.34**
225mm dia	72.37	87.88	1.90	33.44	nr	**121.32**
250mm dia	98.88	120.07	2.70	47.52	nr	**167.59**
315mm dia	129.55	157.31	3.60	63.36	nr	**220.68**
Extra over plastic fittings, compression joints						
Straight connector						
20mm dia.	1.80	2.19	0.38	6.69	nr	**8.88**
25mm dia.	2.07	2.51	0.45	7.92	nr	**10.43**
32mm dia.	4.10	4.98	0.50	8.80	nr	**13.78**
50mm dia.	9.71	11.79	0.68	11.97	nr	**23.77**
63mm dia.	14.83	18.01	0.85	14.97	nr	**32.98**
Reducing connector						
25mm dia.	3.02	3.66	0.38	6.69	nr	**10.35**
32mm dia.	5.08	6.16	0.45	7.92	nr	**14.08**
50mm dia.	11.65	14.14	0.50	8.80	nr	**22.94**
63mm dia.	17.02	20.67	0.62	10.92	nr	**31.58**
Straight connector; polyethylene to MI						
20mm dia.	1.49	1.81	0.31	5.46	nr	**7.27**
25mm dia.	1.90	2.31	0.35	6.16	nr	**8.47**
32mm dia.	2.79	3.39	0.40	7.04	nr	**10.43**
50mm dia.	6.90	8.38	0.55	9.68	nr	**18.06**
63mm dia.	9.58	11.63	0.65	11.44	nr	**23.07**

Material Costs/Measured Work Prices - Mechanical Installations

S:PIPED SUPPLY SYSTEMS

Item	Net Price £	Material £	Labour hours	Labour £	Unit	Total rate £
Straight connector; polyethylene to FI						
20mm dia.	1.96	2.38	0.31	5.46	nr	7.84
25mm dia.	2.15	2.61	0.35	6.16	nr	8.77
32mm dia.	2.65	3.21	0.40	7.04	nr	10.25
50mm dia.	8.05	9.77	0.55	9.68	nr	19.45
63mm dia.	11.13	13.52	0.75	13.20	nr	26.72
Elbow						
20mm dia.	2.15	2.61	0.38	6.69	nr	9.30
25mm dia.	3.17	3.85	0.45	7.92	nr	11.77
32mm dia.	4.60	5.58	0.50	8.80	nr	14.38
50mm dia.	10.73	13.03	0.68	11.97	nr	25.00
63mm dia.	14.57	17.69	0.80	14.08	nr	31.77
Elbow; polyethylene to MI						
25mm dia.	2.65	3.21	0.35	6.16	nr	9.37
Elbow; polyethylene to FI						
20mm dia.	1.94	2.36	0.31	5.46	nr	7.82
25mm dia.	2.65	3.21	0.35	6.16	nr	9.37
32mm dia.	3.92	4.76	0.42	7.39	nr	12.16
50mm dia.	9.31	11.31	0.50	8.80	nr	20.11
63mm dia.	12.17	14.78	0.55	9.68	nr	24.46
Tank coupling						
25mm dia.	6.90	8.38	0.42	7.39	nr	15.77
Equal tee						
20mm dia.	3.14	3.81	0.53	9.33	nr	13.15
25mm dia.	4.77	5.79	0.55	9.68	nr	15.47
32mm dia.	6.08	7.39	0.64	11.27	nr	18.66
50mm dia.	14.19	17.23	0.75	13.20	nr	30.44
63mm dia.	21.20	25.75	0.87	15.32	nr	41.07
Equal tee; FI branch						
20mm dia.	2.82	3.42	0.45	7.92	nr	11.34
25mm dia.	4.39	5.33	0.50	8.80	nr	14.13
32mm dia.	5.60	6.80	0.60	10.56	nr	17.36
50mm dia.	19.40	23.56	0.68	11.97	nr	35.54
63mm dia.	21.63	26.26	0.81	14.26	nr	40.53
Equal tee; MI branch						
25mm dia.	5.03	6.11	0.50	8.80	nr	14.91

Material Costs/Measured Work Prices - Mechanical Installations

S:PIPED SUPPLY SYSTEMS

Item	Net Price £	Material £	Labour hours	Labour £	Unit	Total rate £
S10 : COLD WATER						
Y10 - PIPELINES						
ABS						
Pipes; solvent welded joints in the running length						
Class C (9 bar pressure)						
1" dia.	1.74	2.11	0.30	5.28	m	**7.39**
1 1/4" dia.	2.93	3.56	0.33	5.81	m	**9.37**
1 1/2" dia.	3.70	4.49	0.36	6.34	m	**10.83**
2" dia.	5.01	6.08	0.39	6.86	m	**12.95**
2 1/2" dia.	8.82	10.71	0.40	7.04	m	**17.75**
3" dia.	11.02	13.38	0.46	8.10	m	**21.48**
4" dia.	17.83	21.65	0.53	9.33	m	**30.98**
6" dia.	36.31	44.09	0.76	13.38	m	**57.47**
8" dia.	66.81	81.13	0.97	17.07	m	**98.20**
Class E (15 bar pressure)						
1/2" dia.	2.64	3.21	0.24	4.22	m	**7.43**
3/4" dia.	2.01	2.44	0.27	4.75	m	**7.19**
1" dia.	2.63	3.19	0.30	5.28	m	**8.47**
1 1/4" dia.	3.94	4.78	0.33	5.81	m	**10.59**
1 1/2" dia.	5.18	6.29	0.36	6.34	m	**12.63**
2" dia.	6.52	7.92	0.39	6.86	m	**14.78**
3" dia.	13.84	16.81	0.49	8.62	m	**25.43**
4" dia.	21.99	26.70	0.57	10.03	m	**36.73**
Extra Over fittings; solvent welded joints						
Cap						
1/2" dia.	0.53	0.64	0.16	2.82	nr	**3.46**
3/4" dia.	0.61	0.74	0.19	3.34	nr	**4.08**
1" dia.	0.70	0.85	0.22	3.87	nr	**4.72**
1 1/4" dia.	1.16	1.41	0.25	4.40	nr	**5.81**
1 1/2" dia.	1.80	2.19	0.28	4.93	nr	**7.11**
2" dia.	2.30	2.79	0.31	5.46	nr	**8.25**
3" dia.	6.88	8.35	0.36	6.34	nr	**14.69**
4" dia.	10.51	12.76	0.44	7.74	nr	**20.51**
Elbow 90 degree						
1/2" dia.	0.74	0.90	0.29	5.10	nr	**6.00**
3/4" dia.	0.89	1.08	0.34	5.98	nr	**7.07**
1" dia.	1.24	1.51	0.40	7.04	nr	**8.55**
1 1/4" dia.	2.08	2.53	0.45	7.92	nr	**10.45**
1 1/2" dia.	2.72	3.30	0.51	8.98	nr	**12.28**
2" dia.	4.13	5.02	0.56	9.86	nr	**14.87**
3" dia.	11.86	14.40	0.65	11.44	nr	**25.84**
4" dia.	17.71	21.51	0.80	14.08	nr	**35.59**
6" dia.	71.23	86.49	1.21	21.30	nr	**107.79**
8" dia.	113.92	138.33	1.45	25.52	nr	**163.85**

Material Costs/Measured Work Prices - Mechanical Installations

S:PIPED SUPPLY SYSTEMS

Item	Net Price £	Material £	Labour hours	Labour £	Unit	Total rate £
Elbow 45 degree						
1/2" dia.	1.41	1.71	0.29	5.10	nr	**6.82**
3/4" dia.	1.44	1.75	0.34	5.98	nr	**7.73**
1" dia.	1.80	2.19	0.40	7.04	nr	**9.23**
1 1/4" dia.	2.65	3.22	0.45	7.92	nr	**11.14**
1 1/2" dia.	3.29	4.00	0.51	8.98	nr	**12.97**
2" dia.	4.55	5.53	0.56	9.86	nr	**15.38**
3" dia.	10.72	13.02	0.65	11.44	nr	**24.46**
4" dia.	22.22	26.98	0.80	14.08	nr	**41.06**
6" dia.	46.06	55.93	1.21	21.30	nr	**77.23**
8" dia.	103.82	126.07	1.45	25.52	nr	**151.59**
Reducing bush						
3/4" x 1/2" dia.	0.65	0.79	0.42	7.39	nr	**8.18**
1" x 1/2" dia.	0.70	0.85	0.45	7.92	nr	**8.77**
1" x 3/4" dia.	0.81	0.98	0.45	7.92	nr	**8.90**
1 1/4" x 1" dia.	1.54	1.87	0.48	8.45	nr	**10.32**
1 1/2" x 3/4" dia.	1.24	1.51	0.51	8.98	nr	**10.48**
1 1/2" x 1" dia.	1.24	1.51	0.51	8.98	nr	**10.48**
1 1/2" x 1 1/4" dia.	1.73	2.10	0.51	8.98	nr	**11.08**
2" x 1" dia.	1.62	1.97	0.56	9.86	nr	**11.82**
2" x 1 1/4" dia.	1.62	1.97	0.56	9.86	nr	**11.82**
2" x 1 1/2" dia.	2.68	3.25	0.56	9.86	nr	**13.11**
3" x 1 1/2" dia.	4.55	5.53	0.65	11.44	nr	**16.97**
3" x 2" dia.	7.97	9.68	0.65	11.44	nr	**21.12**
4" x 3" dia.	11.92	14.47	0.80	14.08	nr	**28.56**
6" x 4" dia.	43.17	52.42	1.21	21.30	nr	**73.72**
Union						
1/2" dia.	2.93	3.56	0.34	5.98	nr	**9.54**
3/4" dia.	3.17	3.85	0.39	6.86	nr	**10.71**
1" dia.	4.26	5.17	0.43	7.57	nr	**12.74**
1 1/4" dia.	5.22	6.34	0.50	8.80	nr	**15.14**
1 1/2" dia.	7.19	8.73	0.57	10.03	nr	**18.76**
2" dia.	9.38	11.39	0.62	10.91	nr	**22.30**
Sockets						
1/2" dia.	0.55	0.67	0.34	5.98	nr	**6.65**
3/4" dia.	0.61	0.74	0.39	6.86	nr	**7.61**
1" dia.	0.70	0.85	0.43	7.57	nr	**8.42**
1 1/4" dia.	1.24	1.51	0.50	8.80	nr	**10.31**
1 1/2" dia.	1.48	1.80	0.57	10.03	nr	**11.83**
2" dia.	2.08	2.53	0.62	10.91	nr	**13.44**
3" dia.	8.39	10.19	0.70	12.32	nr	**22.51**
4" dia.	11.92	14.47	0.70	12.32	nr	**26.80**
6" dia.	29.77	36.15	1.26	22.18	nr	**58.33**
8" dia.	62.28	75.63	1.55	27.28	nr	**102.91**

Material Costs/Measured Work Prices - Mechanical Installations

S:PIPED SUPPLY SYSTEMS

Item	Net Price £	Material £	Labour hours	Labour £	Unit	Total rate £
S10 : COLD WATER						
Y10 - PIPELINES						
ABS (Continued)						
Barrel nipple						
1/2" dia.	1.02	1.24	0.34	5.98	nr	**7.22**
3/4" dia.	1.34	1.63	0.39	6.86	nr	**8.49**
1" dia.	1.73	2.10	0.43	7.57	nr	**9.67**
1 1/4" dia.	2.40	2.91	0.50	8.80	nr	**11.71**
1 1/2" dia.	2.82	3.42	0.57	10.03	nr	**13.46**
2" dia.	3.42	4.15	0.62	10.91	nr	**15.07**
3" dia.	9.01	10.94	0.70	12.32	nr	**23.26**
Tee 90 degree						
1/2" dia.	0.84	1.02	0.41	7.22	nr	**8.24**
3/4" dia.	1.16	1.41	0.47	8.27	nr	**9.68**
1" dia.	1.62	1.97	0.55	9.68	nr	**11.65**
1 1/4" dia.	2.33	2.83	0.64	11.26	nr	**14.09**
1 1/2" dia.	3.42	4.15	0.71	12.50	nr	**16.65**
2" dia.	5.22	6.34	0.78	13.73	nr	**20.07**
3" dia.	15.23	18.49	0.91	16.02	nr	**34.51**
4" dia.	22.36	27.15	1.12	19.71	nr	**46.86**
6" dia.	78.16	94.91	1.69	29.75	nr	**124.66**
8" dia.	127.68	155.04	2.03	35.73	nr	**190.77**
Full face flange						
1/2" dia.	9.06	11.00	0.10	1.76	nr	**12.76**
3/4" dia.	9.25	11.23	0.13	2.29	nr	**13.52**
1" dia.	9.91	12.03	0.15	2.64	nr	**14.67**
1 1/4" dia.	13.61	16.53	0.18	3.17	nr	**19.69**
1 1/2" dia.	15.54	18.87	0.21	3.70	nr	**22.57**
2" dia.	19.53	23.72	0.29	5.10	nr	**28.82**
3" dia.	36.03	43.75	0.37	6.51	nr	**50.26**
4" dia.	46.23	56.14	0.41	7.22	nr	**63.35**
PVC -U						
Pipes; solvent welded joints in the running length						
Class C (9 bar pressure)						
2" dia.	5.50	6.68	0.41	7.22	m	**13.89**
3" dia.	11.21	13.61	0.47	8.27	m	**21.88**
4" dia.	19.45	23.62	0.50	8.80	m	**32.42**
6" dia.	42.76	51.92	1.76	30.98	m	**82.90**
Class D (12 bar pressure)						
1 1/4" dia.	3.20	3.89	0.41	7.22	m	**11.10**
1 1/2" dia.	4.38	5.32	0.42	7.39	m	**12.71**
2" dia.	6.75	8.20	0.45	7.92	m	**16.12**
3" dia.	15.04	18.26	0.48	8.45	m	**26.71**
4" dia.	24.88	30.21	0.53	9.33	m	**39.54**
6" dia.	47.43	57.59	0.58	10.21	m	**67.80**

Material Costs/Measured Work Prices - Mechanical Installations

S:PIPED SUPPLY SYSTEMS

Item	Net Price £	Material £	Labour hours	Labour £	Unit	Total rate £
Class E (15 bar pressure)						
1/2" dia.	1.57	1.91	0.38	6.69	m	8.59
3/4" dia.	2.20	2.67	0.40	7.04	m	9.71
1" dia.	2.55	3.10	0.41	7.22	m	10.31
1 1/4" dia.	3.80	4.61	0.41	7.22	m	11.83
1 1/2" dia.	4.91	5.96	0.42	7.39	m	13.35
2" dia.	7.64	9.28	0.45	7.92	m	17.20
3" dia.	17.11	20.78	0.47	8.27	m	29.05
4" dia.	27.84	33.81	0.50	8.80	m	42.61
6" dia.	60.98	74.05	0.53	9.33	m	83.38
Class 7						
1/2" dia.	2.69	3.27	0.32	5.63	m	8.90
3/4" dia.	3.74	4.54	0.33	5.81	m	10.35
1" dia.	5.68	6.90	0.40	7.04	m	13.94
1 1/4" dia.	7.84	9.52	0.40	7.04	m	16.56
1 1/2" dia.	9.69	11.77	0.41	7.22	m	18.98
2" dia.	16.05	19.49	0.43	7.57	m	27.06
Extra Over fittings; solvent welded joints						
End cap						
1/2" dia.	0.49	0.59	0.17	2.99	nr	3.59
3/4" dia.	0.59	0.72	0.19	3.34	nr	4.06
1" dia.	0.66	0.80	0.22	3.87	nr	4.67
1 1/4" dia.	1.02	1.24	0.25	4.40	nr	5.64
1 1/2" dia.	1.73	2.10	0.28	4.93	nr	7.03
2" dia.	2.11	2.56	0.31	5.46	nr	8.02
3" dia.	6.49	7.88	0.36	6.34	nr	14.22
4" dia.	10.01	12.16	0.44	7.74	nr	19.90
6" dia.	24.19	29.37	0.67	11.79	nr	41.17
Socket						
1/2" dia.	0.53	0.64	0.31	5.46	nr	6.10
3/4" dia.	0.59	0.72	0.35	6.16	nr	6.88
1" dia.	0.69	0.84	0.42	7.39	nr	8.23
1 1/4" dia.	1.24	1.51	0.45	7.92	nr	9.43
1 1/2" dia.	1.44	1.75	0.51	8.98	nr	10.72
2" dia.	2.04	2.48	0.56	9.86	nr	12.33
3" dia.	7.83	9.51	0.65	11.44	nr	20.95
4" dia.	11.36	13.79	0.80	14.08	nr	27.88
6" dia.	28.50	34.61	1.21	21.30	nr	55.90
Reducing socket						
3/4 x 1/2" dia.	0.63	0.77	0.31	5.46	nr	6.22
1 x 3/4" dia.	0.78	0.95	0.35	6.16	nr	7.11
1 1/4 x 1" dia.	1.48	1.80	0.42	7.39	nr	9.19
1 1/2 x 1 1/4" dia.	1.66	2.02	0.45	7.92	nr	9.94
2 x 1 1/2" dia.	2.50	3.04	0.51	8.98	nr	12.01
3 x 2" dia.	7.62	9.25	0.56	9.86	nr	19.11
4 x 3" dia.	11.29	13.71	0.65	11.44	nr	25.15
6 x 4" dia.	41.12	49.93	0.80	14.08	nr	64.01
8 x 6" dia.	66.79	81.10	1.21	21.30	nr	102.40

Material Costs/Measured Work Prices - Mechanical Installations

S:PIPED SUPPLY SYSTEMS

Item	Net Price £	Material £	Labour hours	Labour £	Unit	Total rate £
S10 : COLD WATER						
Y10 - PIPELINES						
PVC-U (Continued)						
Elbow 90 degree						
1/2" dia.	0.70	0.85	0.31	5.46	nr	**6.31**
3/4" dia.	0.84	1.02	0.35	6.16	nr	**7.18**
1" dia.	1.16	1.41	0.42	7.39	nr	**8.80**
1 1/4" dia.	2.04	2.48	0.45	7.92	nr	**10.40**
1 1/2" dia.	2.65	3.22	0.45	7.92	nr	**11.14**
2" dia.	3.91	4.75	0.56	9.86	nr	**14.60**
3" dia.	11.29	13.71	0.65	11.44	nr	**25.15**
4" dia.	17.01	20.66	0.80	14.08	nr	**34.74**
6" dia.	67.29	81.71	1.21	21.30	nr	**103.01**
Elbow 45 degree						
1/2" dia.	1.34	1.63	0.31	5.46	nr	**7.08**
3/4" dia.	1.41	1.71	0.35	6.16	nr	**7.87**
1" dia.	1.73	2.10	0.45	7.92	nr	**10.02**
1 1/4" dia.	2.47	3.00	0.45	7.92	nr	**10.92**
1 1/2" dia.	3.10	3.76	0.51	8.98	nr	**12.74**
2" dia.	4.37	5.31	0.56	9.86	nr	**15.16**
3" dia.	10.30	12.51	0.65	11.44	nr	**23.95**
4" dia.	21.16	25.69	0.80	14.08	nr	**39.78**
6" dia.	43.66	53.02	1.21	21.30	nr	**74.31**
Bend 90 degree (long radius)						
3" dia.	31.46	38.20	0.65	11.44	nr	**49.64**
4" dia.	63.55	77.17	0.80	14.08	nr	**91.25**
6" dia.	139.66	169.59	1.21	21.30	nr	**190.89**
Bend 45 degree (long radius)						
1" dia.	4.16	5.05	0.45	7.92	nr	**12.97**
1 1/2" dia.	7.48	9.08	0.51	8.98	nr	**18.06**
2" dia.	12.21	14.83	0.56	9.86	nr	**24.68**
3" dia.	26.10	31.69	0.65	11.44	nr	**43.13**
4" dia.	50.79	61.67	0.80	14.08	nr	**75.75**
6" dia.	111.82	135.78	1.21	21.30	nr	**157.08**
Socket union						
1/2" dia.	2.72	3.30	0.34	5.98	nr	**9.29**
3/4" dia.	3.10	3.76	0.39	6.86	nr	**10.63**
1" dia.	4.02	4.88	0.45	7.92	nr	**12.80**
1 1/4" dia.	5.01	6.08	0.50	8.80	nr	**14.88**
1 1/2" dia.	6.88	8.35	0.57	10.03	nr	**18.39**
2" dia.	8.89	10.80	0.62	10.91	nr	**21.71**
3" dia.	33.08	40.17	0.70	12.32	nr	**52.49**
4" dia.	44.79	54.39	0.89	15.66	nr	**70.05**

Material Costs/Measured Work Prices - Mechanical Installations

S:PIPED SUPPLY SYSTEMS

Item	Net Price £	Material £	Labour hours	Labour £	Unit	Total rate £
Saddle plain						
2" x 1 1/4" dia.	6.98	8.48	0.42	7.39	nr	**15.87**
3" x 1 1/2" dia.	9.81	11.91	0.48	8.45	nr	**20.36**
4" x 2" dia.	11.07	13.44	0.68	11.97	nr	**25.41**
6" x 2" dia.	12.98	15.76	0.91	16.02	nr	**31.78**
Straight tank connector						
1/2" dia.	1.80	2.19	0.13	2.29	nr	**4.47**
3/4" dia.	2.04	2.48	0.14	2.46	nr	**4.94**
1" dia.	4.34	5.27	0.14	2.46	nr	**7.73**
1 1/4" dia.	11.00	13.36	0.16	2.82	nr	**16.17**
1 1/2" dia.	12.06	14.64	0.18	3.17	nr	**17.81**
2" dia.	14.46	17.56	0.24	4.22	nr	**21.78**
3" dia.	14.81	17.98	0.29	5.10	nr	**23.09**
Equal tee						
1/2" dia.	0.81	0.98	0.44	7.74	nr	**8.73**
3/4" dia.	1.02	1.24	0.48	8.45	nr	**9.69**
1" dia.	1.54	1.87	0.54	9.50	nr	**11.37**
1 1/4" dia.	2.18	2.65	0.70	12.32	nr	**14.97**
1 1/2" dia.	3.17	3.85	0.74	13.02	nr	**16.87**
2" dia.	5.01	6.08	0.80	14.08	nr	**20.16**
3" dia.	14.53	17.64	1.04	18.30	nr	**35.95**
4" dia.	21.30	25.86	1.28	22.53	nr	**48.39**
6" dia.	74.20	90.10	1.93	33.97	nr	**124.07**

PVC - C

Pipes; solvent welded in the running length

Item	Net Price £	Material £	Labour hours	Labour £	Unit	Total rate £
Pipe; 3m long; PN25						
16 x 2.0mm	1.92	2.33	0.20	3.52	m	**5.85**
20 x 2.3mm	3.15	3.82	0.20	3.52	m	**7.34**
25 x 2.8mm	4.24	5.15	0.20	3.52	m	**8.67**
32 x 3.6mm	6.14	7.46	0.20	3.52	m	**10.98**
Pipe; 5m long; PN25						
40 x 4.5mm	9.86	11.97	0.20	3.52	m	**15.49**
50 x 5.6mm	14.95	18.15	0.20	3.52	m	**21.67**
64 x 7.0mm	22.89	27.80	0.20	3.52	m	**31.32**

Material Costs/Measured Work Prices - Mechanical Installations

S:PIPED SUPPLY SYSTEMS

Item	Net Price £	Material £	Labour hours	Labour £	Unit	Total rate £
S10 : COLD WATER						
Y10 - PIPELINES						
PVC - C (Continued)						
Extra Over fittings; solvent welded joints						
Straight coupling; PN25						
16mm	0.45	0.54	0.20	3.52	nr	4.06
20mm	0.62	0.76	0.20	3.52	nr	4.28
25mm	0.79	0.96	0.20	3.52	nr	4.48
32mm	2.46	2.98	0.20	3.52	nr	6.50
40mm	3.15	3.82	0.20	3.52	nr	7.34
50mm	4.23	5.14	0.20	3.52	nr	8.66
63mm	7.45	9.05	0.20	3.52	nr	12.57
Elbow; 90 degree; PN25						
16mm	0.73	0.89	0.20	3.52	nr	4.41
20mm	1.12	1.36	0.20	3.52	nr	4.88
25mm	1.40	1.70	0.20	3.52	nr	5.23
32mm	2.93	3.56	0.20	3.52	nr	7.08
40mm	4.51	5.48	0.20	3.52	nr	9.00
50mm	6.27	7.61	0.20	3.52	nr	11.13
63mm	10.72	13.01	0.20	3.52	nr	16.53
Elbow; 45 degree; PN25						
16mm	0.73	0.89	0.20	3.52	nr	4.41
20mm	1.12	1.36	0.20	3.52	nr	4.88
25mm	1.40	1.70	0.20	3.52	nr	5.23
32mm	2.93	3.56	0.20	3.52	nr	7.08
40mm	4.51	5.48	0.20	3.52	nr	9.00
50mm	6.27	7.61	0.20	3.52	nr	11.13
63mm	10.72	13.01	0.20	3.52	nr	16.53
Reducer fitting; single stage reduction						
20/16mm	0.79	0.96	0.20	3.52	nr	4.48
25/20mm	0.96	1.16	0.20	3.52	nr	4.68
32/25mm	1.92	2.33	0.20	3.52	nr	5.85
40/32mm	2.54	3.09	0.20	3.52	nr	6.61
50/40mm	2.93	3.56	0.20	3.52	nr	7.08
63/50mm	4.45	5.40	0.20	3.52	nr	8.92
Equal tee; 90 degree; PN25						
16mm	1.24	1.50	0.20	3.52	nr	5.02
20mm	1.69	2.05	0.20	3.52	nr	5.57
25mm	2.14	2.59	0.20	3.52	nr	6.11
32mm	3.49	4.24	0.20	3.52	nr	7.76
40mm	6.04	7.33	0.20	3.52	nr	10.85
50mm	9.02	10.95	0.20	3.52	nr	14.47
63mm	15.22	18.48	0.20	3.52	nr	22.00

Material Costs/Measured Work Prices - Mechanical Installations

S:PIPED SUPPLY SYSTEMS

Item	Net Price £	Material £	Labour hours	Labour £	Unit	Total rate £
Cap; PN25						
16mm	0.57	0.69	0.20	3.52	nr	**4.21**
20mm	0.85	1.03	0.20	3.52	nr	**4.55**
25mm	1.12	1.36	0.20	3.52	nr	**4.88**
32mm	1.64	1.99	0.20	3.52	nr	**5.51**
40mm	2.26	2.75	0.20	3.52	nr	**6.27**
50mm	3.15	3.82	0.20	3.52	nr	**7.34**
63mm	5.01	6.09	0.20	3.52	nr	**9.61**

Material Costs/Measured Work Prices - Mechanical Installations

S:PIPED SUPPLY SYSTEMS

Item	Net Price £	Material £	Labour hours	Labour £	Unit	Total rate £
S10 : COLD WATER						
Y10 - PIPELINES						
SCREWED STEEL						
Galvanised steel pipes; screwed and socketed joints; BS 1387: 1985						
Galvanised; light						
15mm dia.	1.72	2.09	0.52	9.15	m	**11.24**
20mm dia.	2.13	2.59	0.55	9.68	m	**12.27**
25mm dia.	2.84	3.45	0.60	10.56	m	**14.01**
32mm dia.	3.53	4.29	0.67	11.79	m	**16.08**
40mm dia.	4.34	5.27	0.75	13.20	m	**18.47**
50mm dia.	5.68	6.90	0.85	14.96	m	**21.86**
65mm dia.	8.38	10.18	0.93	16.37	m	**26.54**
80mm dia.	10.08	12.24	1.07	18.83	m	**31.07**
100mm dia.	14.71	17.86	1.46	25.70	m	**43.56**
Galvanised; medium						
8mm dia.	1.57	1.91	0.51	8.98	m	**10.88**
10mm dia.	1.57	1.91	0.51	8.98	m	**10.88**
15mm dia.	1.54	1.87	0.52	9.15	m	**11.02**
20mm dia.	1.75	2.13	0.55	9.68	m	**11.81**
25mm dia.	2.44	2.96	0.60	10.56	m	**13.52**
32mm dia.	3.04	3.69	0.67	11.79	m	**15.48**
40mm dia.	3.54	4.30	0.75	13.20	m	**17.50**
50mm dia.	4.97	6.04	0.85	14.96	m	**21.00**
65mm dia.	6.84	8.31	0.93	16.37	m	**24.67**
80mm dia.	8.88	10.78	1.07	18.83	m	**29.62**
100mm dia.	12.71	15.43	1.46	25.70	m	**41.13**
125mm dia.	19.41	23.57	1.72	30.27	m	**53.84**
150mm dia.	22.86	27.76	1.96	34.50	m	**62.26**
Galvanised; heavy						
15mm dia.	1.82	2.21	0.52	9.15	m	**11.36**
20mm dia.	2.08	2.53	0.55	9.68	m	**12.21**
25mm dia.	2.96	3.59	0.60	10.56	m	**14.15**
32mm dia.	3.69	4.48	0.67	11.79	m	**16.27**
40mm dia.	4.31	5.23	0.75	13.20	m	**18.43**
50mm dia.	5.97	7.25	0.85	14.96	m	**22.21**
65mm dia.	8.20	9.96	0.93	16.37	m	**26.33**
80mm dia.	10.44	12.68	1.07	18.83	m	**31.51**
100mm dia.	14.73	17.89	1.46	25.70	m	**43.58**
125mm dia.	20.54	24.94	1.72	30.27	m	**55.22**
150mm dia.	24.33	29.54	1.96	34.50	m	**64.04**

Material Costs/Measured Work Prices - Mechanical Installations

S:PIPED SUPPLY SYSTEMS

Item	Net Price £	Material £	Labour hours	Labour £	Unit	Total rate £
Extra Over steel flanges, screwed and drilled; metric; BS 4504						
Screwed flanges; PN6						
15mm dia.	3.32	4.03	0.35	6.16	nr	**10.19**
20mm dia.	3.34	4.06	0.47	8.27	nr	**12.33**
25mm dia.	3.38	4.10	0.53	9.33	nr	**13.43**
32mm dia.	4.37	5.31	0.62	10.91	nr	**16.22**
40mm dia.	4.37	5.31	0.70	12.32	nr	**17.63**
50mm dia.	5.23	6.35	0.84	14.78	nr	**21.14**
65mm dia.	6.66	8.09	1.03	18.13	nr	**26.22**
80mm dia.	8.12	9.86	1.23	21.65	nr	**31.51**
100mm dia.	9.59	11.65	1.41	24.82	nr	**36.46**
125mm dia.	23.00	27.93	1.77	31.15	nr	**59.08**
150mm dia.	23.05	27.99	2.21	38.90	nr	**66.89**
Screwed flanges; PN16						
15mm dia.	3.21	3.90	0.35	6.16	nr	**10.06**
20mm dia.	3.22	3.91	0.47	8.27	nr	**12.18**
25mm dia.	3.29	4.00	0.53	9.33	nr	**13.32**
32mm dia.	4.48	5.44	0.62	10.91	nr	**16.35**
40mm dia.	4.49	5.45	0.70	12.32	nr	**17.77**
50mm dia.	5.30	6.44	0.84	14.78	nr	**21.22**
65mm dia.	6.48	7.87	1.03	18.13	nr	**26.00**
80mm dia.	8.40	10.20	1.23	21.65	nr	**31.85**
100mm dia.	9.71	11.79	1.41	24.82	nr	**36.61**
125mm dia.	22.89	27.80	1.77	31.15	nr	**58.95**
150mm dia.	23.62	28.68	2.21	38.90	nr	**67.58**
Extra Over steel flanges, screwed and drilled; imperial; BS 10						
Screwed flanges; Table E						
1/2" dia.	3.86	4.69	0.35	6.16	nr	**10.85**
3/4" dia.	3.88	4.71	0.47	8.27	nr	**12.98**
1" dia.	3.94	4.78	0.53	9.33	nr	**14.11**
1 1/4" dia.	4.10	4.98	0.62	10.91	nr	**15.89**
1 1/2" dia.	4.10	4.98	0.70	12.32	nr	**17.30**
2" dia.	4.14	5.03	0.84	14.78	nr	**19.81**
2 1/2" dia.	5.13	6.23	1.03	18.13	nr	**24.36**
3" dia.	5.57	6.76	1.23	21.65	nr	**28.41**
4" dia.	8.01	9.73	1.41	24.82	nr	**34.54**
5" dia.	14.81	17.98	1.77	31.15	nr	**49.14**
6" dia.	15.52	18.85	2.21	38.90	nr	**57.74**
Extra Over steel flange connections						
Bolted connection between pair of flanges; including gasket, bolts, nuts and washers						
50mm dia.	7.45	9.05	0.53	9.33	nr	**18.37**
65mm dia.	8.80	10.69	0.53	9.33	nr	**20.01**
80mm dia.	12.26	14.89	0.53	9.33	nr	**24.22**
100mm dia.	13.81	16.77	0.53	9.33	nr	**26.10**
125mm dia.	27.43	33.31	0.61	10.74	nr	**44.04**
150mm dia.	28.54	34.66	0.90	15.84	nr	**50.50**

Material Costs/Measured Work Prices - Mechanical Installations

S:PIPED SUPPLY SYSTEMS

Item	Net Price £	Material £	Labour hours	Labour £	Unit	Total rate £
S10 : COLD WATER						
Y10 - PIPELINES						
SCREWED STEEL (Continued)						
Extra Over heavy steel tubular fittings; BS 1387						
Long screw connection with socket and backnut						
15mm dia.	2.46	2.99	0.63	11.09	nr	**14.08**
20mm dia.	2.89	3.51	0.84	14.78	nr	**18.29**
25mm dia.	4.08	4.95	0.95	16.72	nr	**21.68**
32mm dia.	5.34	6.48	1.11	19.54	nr	**26.02**
40mm dia.	6.48	7.87	1.28	22.53	nr	**30.40**
50mm dia.	9.60	11.66	1.53	26.93	nr	**38.59**
65mm dia.	21.11	25.63	1.87	32.91	nr	**58.55**
80mm dia.	28.82	35.00	2.21	38.90	nr	**73.89**
100mm dia.	47.26	57.39	3.05	53.68	nr	**111.07**
Running nipple						
15mm dia.	0.60	0.73	0.50	8.80	nr	**9.53**
20mm dia.	0.75	0.91	0.68	11.97	nr	**12.88**
25mm dia.	0.92	1.12	0.77	13.55	nr	**14.67**
32mm dia.	1.30	1.58	0.90	15.84	nr	**17.42**
40mm dia.	1.74	2.11	1.03	18.13	nr	**20.24**
50mm dia.	2.67	3.24	1.23	21.65	nr	**24.89**
65mm dia.	7.06	8.57	1.50	26.40	nr	**34.97**
80mm dia.	10.40	12.63	1.78	31.33	nr	**43.96**
100mm dia.	11.04	13.41	2.38	41.89	nr	**55.30**
Barrel nipple						
15mm dia.	0.53	0.64	0.50	8.80	nr	**9.44**
20mm dia.	0.60	0.73	0.68	11.97	nr	**12.70**
25mm dia.	0.84	1.02	0.77	13.55	nr	**14.57**
32mm dia.	1.07	1.30	0.90	15.84	nr	**17.14**
40mm dia.	1.32	1.60	1.03	18.13	nr	**19.73**
50mm dia.	1.87	2.27	1.23	21.65	nr	**23.92**
65mm dia.	3.40	4.13	1.50	26.40	nr	**30.53**
80mm dia.	4.74	5.76	1.78	31.33	nr	**37.09**
100mm dia.	8.59	10.43	2.38	41.89	nr	**52.32**
125mm dia.	15.93	19.34	2.87	50.51	nr	**69.86**
150mm dia.	25.10	30.48	3.39	59.67	nr	**90.15**
Close taper nipple						
15mm dia.	0.87	1.06	0.50	8.80	nr	**9.86**
20mm dia.	0.96	1.17	0.68	11.97	nr	**13.13**
25mm dia.	1.32	1.60	0.77	13.55	nr	**15.16**
32mm dia.	1.80	2.19	0.90	15.84	nr	**18.03**
40mm dia.	2.24	2.72	1.03	18.13	nr	**20.85**
50mm dia.	3.25	3.95	1.23	21.65	nr	**25.60**
65mm dia.	6.23	7.57	1.50	26.40	nr	**33.97**
80mm dia.	8.30	10.08	1.78	31.33	nr	**41.41**
100mm dia.	15.75	19.13	2.38	41.89	nr	**61.02**

Material Costs/Measured Work Prices - Mechanical Installations

S:PIPED SUPPLY SYSTEMS

Item	Net Price £	Material £	Labour hours	Labour £	Unit	Total rate £
90 degree bend with socket						
15mm dia.	2.00	2.43	0.64	11.26	nr	**13.69**
20mm dia.	2.52	3.06	0.85	14.96	nr	**18.02**
25mm dia.	3.23	3.92	0.97	17.07	nr	**21.00**
32mm dia.	4.94	6.00	1.12	19.71	nr	**25.71**
40mm dia.	5.61	6.81	1.29	22.70	nr	**29.52**
50mm dia.	9.46	11.49	1.55	27.28	nr	**38.77**
65mm dia.	17.20	20.89	1.89	33.27	nr	**54.15**
80mm dia.	26.54	32.23	2.24	39.43	nr	**71.65**
100mm dia.	46.67	56.67	3.09	54.39	nr	**111.06**
125mm dia.	150.90	183.24	3.92	69.00	nr	**252.23**
150mm dia.	213.47	259.22	4.74	83.43	nr	**342.64**
Extra Over heavy steel fittings; BS 1740						
Plug						
15mm dia.	0.82	1.00	0.28	4.93	nr	**5.92**
20mm dia.	1.02	1.24	0.38	6.69	nr	**7.93**
25mm dia.	1.28	1.55	0.44	7.74	nr	**9.30**
32mm dia.	1.69	2.05	0.51	8.98	nr	**11.03**
40mm dia.	1.99	2.42	0.59	10.38	nr	**12.80**
50mm dia.	2.89	3.51	0.70	12.32	nr	**15.83**
65mm dia.	7.03	8.54	0.85	14.96	nr	**23.50**
80mm dia.	12.42	15.08	1.00	17.60	nr	**32.68**
100mm dia.	23.85	28.96	1.44	25.35	nr	**54.31**
Socket						
15mm dia.	0.46	0.56	0.64	11.26	nr	**11.82**
20mm dia.	0.54	0.66	0.85	14.96	nr	**15.62**
25mm dia.	0.70	0.85	0.97	17.07	nr	**17.92**
32mm dia.	1.02	1.24	1.12	19.71	nr	**20.95**
40mm dia.	1.31	1.59	1.29	22.70	nr	**24.30**
50mm dia.	1.96	2.38	1.55	27.28	nr	**29.66**
65mm dia.	3.35	4.07	1.89	33.27	nr	**37.33**
80mm dia.	4.69	5.70	2.24	39.43	nr	**45.12**
100mm dia.	8.94	10.86	3.09	54.39	nr	**65.24**
150mm dia.	31.56	38.32	4.74	83.43	nr	**121.75**
Elbow, female/female						
15mm dia.	2.97	3.61	0.64	11.26	nr	**14.87**
20mm dia.	3.62	4.40	0.85	14.96	nr	**19.36**
25mm dia.	4.91	5.96	0.97	17.07	nr	**23.04**
32mm dia.	9.82	11.92	1.12	19.71	nr	**31.64**
40mm dia.	9.82	11.92	1.29	22.70	nr	**34.63**
50mm dia.	12.67	15.39	1.55	27.28	nr	**42.67**
65mm dia.	28.55	34.67	1.89	33.27	nr	**67.93**
80mm dia.	41.96	50.95	2.24	39.43	nr	**90.38**
100mm dia.	79.18	96.15	3.09	54.39	nr	**150.53**

Material Costs/Measured Work Prices - Mechanical Installations

S:PIPED SUPPLY SYSTEMS

Item	Net Price £	Material £	Labour hours	Labour £	Unit	Total rate £
S10 : COLD WATER						
Y10 - PIPELINES						
SCREWED STEEL (Continued)						
Equal tee						
15mm dia.	2.83	3.44	0.91	16.02	nr	**19.45**
20mm dia.	3.48	4.23	1.22	21.47	nr	**25.70**
25mm dia.	5.61	6.81	1.40	24.64	nr	**31.45**
32mm dia.	10.41	12.64	1.62	28.51	nr	**41.15**
40mm dia.	10.41	12.64	1.86	32.74	nr	**45.38**
50mm dia.	14.95	18.15	2.21	38.90	nr	**57.05**
65mm dia.	40.75	49.48	2.72	47.87	nr	**97.36**
80mm dia.	44.96	54.59	3.21	56.50	nr	**111.09**
100mm dia.	91.97	111.68	4.44	78.15	nr	**189.83**
Extra Over malleable iron fittings; BS 143						
Cap						
15mm dia.	0.42	0.51	0.32	5.63	nr	**6.14**
20mm dia.	0.48	0.58	0.43	7.57	nr	**8.15**
25mm dia.	0.60	0.73	0.49	8.62	nr	**9.35**
32mm dia.	0.87	1.06	0.58	10.21	nr	**11.26**
40mm dia.	1.11	1.35	0.66	11.62	nr	**12.96**
50mm dia.	2.16	2.62	0.78	13.73	nr	**16.35**
65mm dia.	3.54	4.30	0.96	16.90	nr	**21.20**
80mm dia.	4.02	4.88	1.13	19.89	nr	**24.77**
100mm dia.	8.79	10.67	1.70	29.92	nr	**40.60**
Plain plug, hollow						
15mm dia.	0.33	0.40	0.28	4.93	nr	**5.33**
20mm dia.	0.42	0.51	0.38	6.69	nr	**7.20**
25mm dia.	0.52	0.63	0.44	7.74	nr	**8.38**
32mm dia.	0.72	0.87	0.51	8.98	nr	**9.85**
40mm dia.	1.19	1.45	0.59	10.38	nr	**11.83**
50mm dia.	1.68	2.04	0.70	12.32	nr	**14.36**
65mm dia.	2.62	3.18	0.85	14.96	nr	**18.14**
80mm dia.	3.90	4.74	1.00	17.60	nr	**22.34**
100mm dia.	7.18	8.72	1.44	25.35	nr	**34.06**
125mm dia.	24.51	29.76	1.98	34.85	nr	**64.61**
150mm dia.	29.66	36.02	2.53	44.53	nr	**80.55**
Plain plug, solid						
15mm dia.	0.87	1.06	0.29	5.10	nr	**6.16**
20mm dia.	0.87	1.06	0.38	6.69	nr	**7.74**
25mm dia.	1.29	1.57	0.44	7.74	nr	**9.31**
32mm dia.	1.56	1.89	0.51	8.98	nr	**10.87**
40mm dia.	2.10	2.55	0.59	10.38	nr	**12.93**
50mm dia.	2.76	3.35	0.70	12.32	nr	**15.67**
65mm dia.	3.38	4.10	0.85	14.96	nr	**19.06**
80mm dia.	5.04	6.12	1.00	17.60	nr	**23.72**

Material Costs/Measured Work Prices - Mechanical Installations

S:PIPED SUPPLY SYSTEMS

Item	Net Price £	Material £	Labour hours	Labour £	Unit	Total rate £
Elbow, male/female						
15mm dia.	0.54	0.66	0.64	11.26	nr	11.92
20mm dia.	0.72	0.87	0.85	14.96	nr	15.83
25mm dia.	1.19	1.45	0.97	17.07	nr	18.52
32mm dia.	1.98	2.40	1.12	19.71	nr	22.12
40mm dia.	2.94	3.57	1.29	22.70	nr	26.27
50mm dia.	3.78	4.59	1.55	27.28	nr	31.87
65mm dia.	7.51	9.12	1.89	33.27	nr	42.38
80mm dia.	10.26	12.46	2.24	39.43	nr	51.88
100mm dia.	17.93	21.77	3.09	54.39	nr	76.16
Elbow						
15mm dia.	0.48	0.58	0.64	11.26	nr	11.85
20mm dia.	0.66	0.80	0.85	14.96	nr	15.76
25mm dia.	1.02	1.24	0.97	17.07	nr	18.31
32mm dia.	1.68	2.04	1.12	19.71	nr	21.75
40mm dia.	2.81	3.41	1.29	22.70	nr	26.12
50mm dia.	3.30	4.01	1.55	27.28	nr	31.29
65mm dia.	6.51	7.91	1.89	33.27	nr	41.17
80mm dia.	9.55	11.60	2.24	39.43	nr	51.02
100mm dia.	16.41	19.93	3.09	54.39	nr	74.31
125mm dia.	35.16	42.69	4.44	78.15	nr	120.84
150mm dia.	65.46	79.49	5.79	101.91	nr	181.40
45 degree elbow						
15mm dia.	1.02	1.24	0.64	11.26	nr	12.50
20mm dia.	1.26	1.53	0.85	14.96	nr	16.49
25mm dia.	1.85	2.25	0.97	17.07	nr	19.32
32mm dia.	3.42	4.15	1.12	19.71	nr	23.87
40mm dia.	4.19	5.09	1.29	22.70	nr	27.79
50mm dia.	5.75	6.98	1.55	27.28	nr	34.26
65mm dia.	7.74	9.40	1.89	33.27	nr	42.66
80mm dia.	11.63	14.12	2.24	39.43	nr	53.55
100mm dia.	22.47	27.29	3.09	54.39	nr	81.67
150mm dia.	60.15	73.04	5.79	101.91	nr	174.95
Bend, male/female						
15mm dia.	0.84	1.02	0.64	11.26	nr	12.28
20mm dia.	1.23	1.49	0.85	14.96	nr	16.45
25mm dia.	1.80	2.19	0.97	17.07	nr	19.26
32mm dia.	2.58	3.13	1.12	19.71	nr	22.85
40mm dia.	3.78	4.59	1.29	22.70	nr	27.30
50mm dia.	7.10	8.62	1.55	27.28	nr	35.90
65mm dia.	10.88	13.21	1.89	33.27	nr	46.48
80mm dia.	14.74	17.90	2.24	39.43	nr	57.32
100mm dia.	36.50	44.32	3.09	54.39	nr	98.71
Bend, male						
15mm dia.	1.71	2.08	0.64	11.26	nr	13.34
20mm dia.	1.92	2.33	0.85	14.96	nr	17.29
25mm dia.	2.81	3.41	0.97	17.07	nr	20.48
32mm dia.	5.52	6.70	1.12	19.71	nr	26.42
40mm dia.	7.74	9.40	1.29	22.70	nr	32.10
50mm dia.	10.34	12.56	1.55	27.28	nr	39.84

Material Costs/Measured Work Prices - Mechanical Installations

S:PIPED SUPPLY SYSTEMS

Item	Net Price £	Material £	Labour hours	Labour £	Unit	Total rate £
S10 : COLD WATER						
Y10 - PIPELINES						
SCREWED STEEL (Continued)						
Bend, female						
15mm dia.	0.87	1.06	0.64	11.26	nr	**12.32**
20mm dia.	1.23	1.49	0.85	14.96	nr	**16.45**
25mm dia.	1.74	2.11	0.97	17.07	nr	**19.19**
32mm dia.	2.97	3.61	1.12	19.71	nr	**23.32**
40mm dia.	3.53	4.29	1.29	22.70	nr	**26.99**
50mm dia.	5.57	6.76	1.55	27.28	nr	**34.04**
65mm dia.	10.88	13.21	1.89	33.27	nr	**46.48**
80mm dia.	16.12	19.57	2.24	39.43	nr	**59.00**
100mm dia.	33.81	41.06	3.09	54.39	nr	**95.44**
125mm dia.	86.43	104.95	4.44	78.15	nr	**183.10**
150mm dia.	132.02	160.31	5.79	101.91	nr	**262.22**
Return bend						
15mm dia.	3.45	4.19	0.64	11.26	nr	**15.45**
20mm dia.	5.57	6.76	0.85	14.96	nr	**21.72**
25mm dia.	6.95	8.44	0.97	17.07	nr	**25.51**
32mm dia.	9.70	11.78	1.12	19.71	nr	**31.49**
40mm dia.	11.57	14.05	1.29	22.70	nr	**36.75**
50mm dia.	17.64	21.42	1.55	27.28	nr	**48.70**
Equal socket, parallel thread						
15mm dia.	0.45	0.55	0.64	11.26	nr	**11.81**
20mm dia.	0.54	0.66	0.85	14.96	nr	**15.62**
25mm dia.	0.72	0.87	0.97	17.07	nr	**17.95**
32mm dia.	1.23	1.49	1.12	19.71	nr	**21.21**
40mm dia.	1.68	2.04	1.29	22.70	nr	**24.74**
50mm dia.	2.52	3.06	1.55	27.28	nr	**30.34**
65mm dia.	4.22	5.12	1.89	33.27	nr	**38.39**
80mm dia.	5.80	7.04	2.24	39.43	nr	**46.47**
100mm dia.	9.85	11.96	3.09	54.39	nr	**66.35**
Concentric reducing socket						
20 x 15mm dia.	0.66	0.80	0.76	13.38	nr	**14.18**
25 x 15mm dia.	0.87	1.06	0.86	15.14	nr	**16.19**
25 x 20mm dia.	0.81	0.98	0.86	15.14	nr	**16.12**
32 x 25mm dia.	1.38	1.68	1.01	17.78	nr	**19.45**
40 x 25mm dia.	1.62	1.97	1.16	20.42	nr	**22.38**
40 x 32mm dia.	1.80	2.19	1.16	20.42	nr	**22.60**
50 x 25mm dia.	3.12	3.79	1.38	24.29	nr	**28.08**
50 x 40mm dia.	2.52	3.06	1.38	24.29	nr	**27.35**
65 x 50mm dia.	4.40	5.34	1.69	29.75	nr	**35.09**
80 x 50mm dia.	5.48	6.65	2.00	35.20	nr	**41.86**
100 x 50mm dia.	10.93	13.27	2.75	48.40	nr	**61.67**
100 x 80mm dia.	10.15	12.33	2.75	48.40	nr	**60.73**
150 x 100mm dia.	26.75	32.48	4.10	72.16	nr	**104.65**

Material Costs/Measured Work Prices - Mechanical Installations

S:PIPED SUPPLY SYSTEMS

Item	Net Price £	Material £	Labour hours	Labour £	Unit	Total rate £
Eccentric reducing socket						
20 x 15mm dia.	1.17	1.42	0.76	13.38	nr	14.80
25 x 15mm dia.	3.33	4.04	0.86	15.14	nr	19.18
25 x 20mm dia.	3.77	4.58	0.86	15.14	nr	19.71
32 x 25mm dia.	4.32	5.25	1.01	17.78	nr	23.02
40 x 25mm dia.	4.93	5.99	1.16	20.42	nr	26.40
40 x 32mm dia.	2.67	3.24	1.16	20.42	nr	23.66
50 x 25mm dia.	2.69	3.27	1.38	24.29	nr	27.56
50 x 40mm dia.	3.21	3.90	1.38	24.29	nr	28.19
65 x 50mm dia.	5.48	6.65	1.69	29.75	nr	36.40
80 x 50mm dia.	8.91	10.82	2.00	35.20	nr	46.02
Hexagon bush						
20 x 15mm dia.	0.42	0.51	0.37	6.51	nr	7.02
25 x 15mm dia.	0.52	0.63	0.43	7.57	nr	8.20
25 x 20mm dia.	0.54	0.66	0.43	7.57	nr	8.22
32 x 25mm dia.	0.63	0.77	0.51	8.98	nr	9.74
40 x 25mm dia.	0.87	1.06	0.58	10.21	nr	11.26
40 x 32mm dia.	0.90	1.09	0.58	10.21	nr	11.30
50 x 25mm dia.	1.80	2.19	0.71	12.50	nr	14.68
50 x 40mm dia.	1.68	2.04	0.71	12.50	nr	14.54
65 x 50mm dia.	2.87	3.48	0.84	14.78	nr	18.27
80 x 50mm dia.	4.34	5.27	1.00	17.60	nr	22.87
100 x 50mm dia.	9.61	11.67	1.52	26.75	nr	38.42
100 x 80mm dia.	8.00	9.71	1.52	26.75	nr	36.47
150 x 100mm dia.	25.32	30.75	2.48	43.65	nr	74.40
Hexagon nipple						
15mm dia.	0.45	0.55	0.28	4.93	nr	5.47
20mm dia.	0.52	0.63	0.38	6.69	nr	7.32
25mm dia.	0.72	0.87	0.44	7.74	nr	8.62
32mm dia.	1.19	1.45	0.51	8.98	nr	10.42
40mm dia.	1.38	1.68	0.59	10.38	nr	12.06
50mm dia.	2.52	3.06	0.70	12.32	nr	15.38
65mm dia.	3.75	4.55	0.85	14.96	nr	19.51
80mm dia.	5.43	6.59	1.00	17.60	nr	24.19
100mm dia.	9.20	11.17	1.44	25.35	nr	36.52
150mm dia.	26.02	31.60	2.32	40.83	nr	72.43
Union, male/female						
15mm dia.	4.02	4.88	0.64	11.26	nr	16.15
20mm dia.	5.43	6.59	0.85	14.96	nr	21.55
25mm dia.	6.87	8.34	0.97	17.07	nr	25.41
32mm dia.	9.50	11.54	1.12	19.71	nr	31.25
40mm dia.	11.30	13.72	1.29	22.70	nr	36.43
50mm dia.	15.10	18.34	1.55	27.28	nr	45.62
65mm dia.	29.58	35.92	1.89	33.27	nr	69.18
80mm dia.	40.67	49.39	2.24	39.43	nr	88.81

Material Costs/Measured Work Prices - Mechanical Installations

S:PIPED SUPPLY SYSTEMS

Item	Net Price £	Material £	Labour hours	Labour £	Unit	Total rate £
S10 : COLD WATER						
Y10 - PIPELINES						
SCREWED STEEL (Continued)						
Union, female						
15mm dia.	3.85	4.68	0.64	11.26	nr	**15.94**
20mm dia.	4.30	5.22	0.85	14.96	nr	**20.18**
25mm dia.	5.64	6.85	0.97	17.07	nr	**23.92**
32mm dia.	7.80	9.47	1.12	19.71	nr	**29.18**
40mm dia.	9.45	11.48	1.29	22.70	nr	**34.18**
50mm dia.	11.98	14.55	1.55	27.28	nr	**41.83**
65mm dia.	23.99	29.13	1.89	33.27	nr	**62.40**
80mm dia.	39.17	47.56	2.24	39.43	nr	**86.99**
100mm dia.	73.67	89.46	3.09	54.39	nr	**143.84**
Union elbow, male/female						
15mm dia.	3.29	4.00	0.64	11.26	nr	**15.26**
20mm dia.	4.34	5.27	0.85	14.96	nr	**20.23**
25mm dia.	5.67	6.89	0.97	17.07	nr	**23.96**
Twin elbow						
15mm dia.	2.28	2.77	0.91	16.02	nr	**18.79**
20mm dia.	2.52	3.06	1.22	21.47	nr	**24.53**
25mm dia.	4.08	4.95	1.39	24.47	nr	**29.42**
32mm dia.	7.22	8.77	1.62	28.51	nr	**37.28**
40mm dia.	9.15	11.11	1.86	32.74	nr	**43.85**
50mm dia.	11.74	14.26	2.21	38.90	nr	**53.15**
65mm dia.	18.98	23.05	2.72	47.87	nr	**70.92**
80mm dia.	32.34	39.27	3.21	56.50	nr	**95.77**
Equal tee						
15mm dia.	0.66	0.80	0.91	16.02	nr	**16.82**
20mm dia.	0.96	1.17	1.22	21.47	nr	**22.64**
25mm dia.	1.38	1.68	1.39	24.47	nr	**26.14**
32mm dia.	2.28	2.77	1.62	28.51	nr	**31.28**
40mm dia.	3.12	3.79	1.86	32.74	nr	**36.53**
50mm dia.	4.49	5.45	2.21	38.90	nr	**44.35**
65mm dia.	9.37	11.38	2.72	47.87	nr	**59.25**
80mm dia.	10.93	13.27	3.21	56.50	nr	**69.77**
100mm dia.	19.81	24.06	4.44	78.15	nr	**102.20**
125mm dia.	48.57	58.98	5.38	94.69	nr	**153.67**
150mm dia.	77.41	94.00	6.31	111.06	nr	**205.06**

Material Costs/Measured Work Prices - Mechanical Installations

S:PIPED SUPPLY SYSTEMS

Item	Net Price £	Material £	Labour hours	Labour £	Unit	Total rate £
Tee reducing on branch						
20 x 15mm dia.	0.87	1.06	1.22	21.47	nr	**22.53**
25 x 15mm dia.	1.19	1.45	1.39	24.47	nr	**25.91**
25 x 20mm dia.	1.26	1.53	1.39	24.47	nr	**26.00**
32 x 25mm dia.	2.22	2.70	1.62	28.51	nr	**31.21**
40 x 25mm dia.	2.94	3.57	1.86	32.74	nr	**36.31**
40 x 32mm dia.	3.83	4.65	1.86	32.74	nr	**37.39**
50 x 25mm dia.	4.00	4.86	2.21	38.90	nr	**43.75**
50 x 40mm dia.	5.39	6.55	2.21	38.90	nr	**45.44**
65 x 50mm dia.	8.32	10.10	2.72	47.87	nr	**57.98**
80 x 50mm dia.	11.26	13.67	3.21	56.50	nr	**70.17**
100 x 50mm dia.	16.41	19.93	4.44	78.15	nr	**98.07**
100 x 80mm dia.	25.32	30.75	4.44	78.15	nr	**108.89**
150 x 100mm dia.	57.01	69.23	6.31	111.06	nr	**180.29**
Equal pitcher tee						
15mm dia.	1.80	2.19	0.91	16.02	nr	**18.20**
20mm dia.	2.22	2.70	1.22	21.47	nr	**24.17**
25mm dia.	3.33	4.04	1.39	24.47	nr	**28.51**
32mm dia.	4.55	5.53	1.62	28.51	nr	**34.04**
40mm dia.	7.05	8.56	1.86	32.74	nr	**41.30**
50mm dia.	9.89	12.01	2.21	38.90	nr	**50.91**
65mm dia.	14.06	17.07	2.72	47.87	nr	**64.95**
80mm dia.	19.33	23.47	3.21	56.50	nr	**79.97**
100mm dia.	43.51	52.83	4.44	78.15	nr	**130.98**
Cross						
15mm dia.	1.56	1.89	1.00	17.60	nr	**19.50**
20mm dia.	2.34	2.84	1.33	23.41	nr	**26.25**
25mm dia.	2.97	3.61	1.51	26.58	nr	**30.18**
32mm dia.	3.89	4.72	1.77	31.15	nr	**35.88**
40mm dia.	5.24	6.36	2.02	35.55	nr	**41.92**
50mm dia.	8.15	9.90	2.42	42.59	nr	**52.49**
65mm dia.	11.63	14.12	2.97	52.27	nr	**66.40**
80mm dia.	15.47	18.79	3.50	61.60	nr	**80.39**
100mm dia.	28.13	34.16	4.84	85.19	nr	**119.35**

Material Costs/Measured Work Prices - Mechanical Installations

S:PIPED SUPPLY SYSTEMS

Item	Net Price £	Material £	Labour hours	Labour £	Unit	Total rate £
S10 : COLD WATER						
Y10 - PIPELINES						
COPPER						
Microbore copper pipe; capillary or compression joints in the running length; BS 2871						
Table W						
6mm dia.	0.48	0.58	0.40	7.04	m	**7.62**
8mm dia.	0.54	0.66	0.40	7.04	m	**7.70**
10mm dia.	0.65	0.79	0.41	7.22	m	**8.01**
Table W; plastic coated gas and cold water service pipe for corrosive and aggresive environments						
6mm dia.	0.68	0.87	0.44	7.74	m	**8.61**
8mm dia.	0.77	0.94	0.44	7.74	m	**8.68**
10mm dia.	1.06	1.29	0.48	8.45	m	**9.74**
Table W; profiled plastic coated central heating and hot water service pipe for heat loss reduction						
8mm dia.	0.81	0.98	0.44	7.74	m	**8.73**
10mm dia.	1.13	1.37	0.48	8.45	m	**9.82**
Microbore accessories						
Manifold connectors; side entry one way flow 22mm body						
4 x 8mm connections	7.18	8.72	0.59	10.38	nr	**19.10**
6 x 8mm connections	8.54	10.37	0.87	15.32	nr	**25.69**
2 x 10mm connections	4.69	5.70	0.33	5.81	nr	**11.50**
4 x 10mm connections	7.50	9.11	0.65	11.44	nr	**20.55**
Manifold connectors; linear flow 22mm body						
4 x 8mm connections	6.17	7.49	0.59	10.38	nr	**17.88**
4 x 10mm connections	7.95	9.65	0.65	11.44	nr	**21.09**
Manifold connectors; linear flow 28mm body						
6 x 8mm connections	8.86	10.76	0.87	15.32	nr	**26.08**

Material Costs/Measured Work Prices - Mechanical Installations

S:PIPED SUPPLY SYSTEMS

Item	Net Price £	Material £	Labour hours	Labour £	Unit	Total rate £
Copper pipe; capillary or compression joints in the running length; BS 2871						
Table X						
8mm dia.	0.57	0.69	0.36	6.34	m	**7.03**
10mm dia.	0.69	0.84	0.37	6.51	m	**7.35**
12mm dia.	0.80	0.97	0.39	6.86	m	**7.84**
15mm dia.	0.79	0.96	0.40	7.04	m	**8.00**
22mm dia.	1.54	1.87	0.47	8.27	m	**10.14**
28mm dia.	2.14	2.60	0.51	8.98	m	**11.57**
35mm dia.	4.84	5.88	0.58	10.21	m	**16.09**
42mm dia.	6.00	7.29	0.66	11.62	m	**18.90**
54mm dia.	7.85	9.53	0.72	12.67	m	**22.20**
67mm dia.	12.35	15.00	0.75	13.20	m	**28.20**
76mm dia.	17.60	21.37	0.76	13.38	m	**34.75**
108mm dia.	26.11	31.71	0.78	13.73	m	**45.43**
133mm dia.	33.40	40.56	1.05	18.48	m	**59.04**
159mm dia.	50.62	61.47	1.15	20.24	m	**81.71**

S:PIPED SUPPLY SYSTEMS

Item	Net Price £	Material £	Labour hours	Labour £	Unit	Total rate £
S10 : COLD WATER						
Y10 - PIPELINES						
COPPER (Continued)						
Table Y						
8mm dia.	0.74	0.90	0.38	6.69	m	**7.59**
10mm dia.	0.85	1.03	0.39	6.86	m	**7.90**
12mm dia.	1.12	1.36	0.41	7.22	m	**8.58**
15mm dia.	1.63	1.98	0.43	7.57	m	**9.55**
22mm dia.	2.99	3.63	0.50	8.80	m	**12.43**
28mm dia.	4.17	5.06	0.54	9.50	m	**14.57**
35mm dia.	5.59	6.79	0.62	10.91	m	**17.70**
42mm dia.	6.88	8.35	0.71	12.50	m	**20.85**
54mm dia.	11.90	14.45	0.78	13.73	m	**28.18**
67mm dia.	18.53	22.50	0.82	14.43	m	**36.93**
76mm dia.	22.78	27.66	0.60	10.56	m	**38.22**
108mm dia.	40.59	49.29	0.88	15.49	m	**64.78**
Table X; plastic coated gas and cold water service pipe for corrosive and aggresive environments						
15mm dia.	1.32	1.60	0.59	10.38	m	**11.99**
22mm dia.	2.54	3.08	0.68	11.97	m	**15.06**
28mm dia.	3.16	3.84	0.74	13.03	m	**16.87**
35mm dia.	6.39	7.76	0.85	14.97	m	**22.73**
42mm dia.	7.54	9.16	0.96	16.91	m	**26.06**
54mm dia.	9.26	11.24	1.06	18.66	m	**29.90**
Table Y; plastic coated gas and cold water service pipe for corrosive and aggresive environments						
15mm dia.	2.24	2.72	0.61	10.74	m	**13.46**
22mm dia.	3.91	4.75	0.69	12.14	m	**16.89**
28mm dia.	4.99	6.06	0.76	13.38	m	**19.44**
35mm dia.	7.67	9.31	0.87	15.31	m	**24.63**
42mm dia.	9.01	10.94	0.99	17.42	m	**28.37**
54mm dia.	14.30	17.36	1.09	19.18	m	**36.55**
Table X; profiled plastic coated central heating and hot water service pipe for heat loss reduction						
15mm dia.	1.40	1.70	0.59	10.38	m	**12.08**
22mm dia.	2.67	3.24	0.68	11.97	m	**15.22**
28mm dia.	3.32	4.03	0.74	13.03	m	**17.06**
35mm dia.	6.59	8.00	0.85	14.97	m	**22.97**
42mm dia.	7.74	9.40	0.96	16.91	m	**26.31**
54mm dia.	9.47	11.50	1.06	18.66	m	**30.16**
Table Y; profiled plastic coated central heating and hot water service pipe for heat loss reduction						
12mm dia.	1.43	1.74	0.61	10.74	m	**12.47**
15mm dia.	2.34	2.84	0.61	10.74	m	**13.58**
22mm dia.	4.08	4.95	0.69	12.14	m	**17.10**

Material Costs/Measured Work Prices - Mechanical Installations

S:PIPED SUPPLY SYSTEMS

Item	Net Price £	Material £	Labour hours	Labour £	Unit	Total rate £
Extra Over copper pipes; capillary fittings; BS 864						
Stop end						
15mm dia.	0.61	0.74	0.13	2.29	nr	**3.03**
22mm dia.	1.14	1.38	0.14	2.46	nr	**3.85**
28mm dia.	1.90	2.31	0.17	2.99	nr	**5.30**
35mm dia.	4.13	5.02	0.19	3.34	nr	**8.36**
42mm dia.	6.72	8.16	0.22	3.87	nr	**12.03**
54mm dia.	9.37	11.38	0.23	4.05	nr	**15.43**
Straight coupling; copper to copper						
6mm dia.	0.50	0.61	0.23	4.05	nr	**4.66**
8mm dia.	0.52	0.63	0.23	4.05	nr	**4.68**
10mm dia.	0.26	0.32	0.23	4.05	nr	**4.36**
15mm dia.	0.09	0.11	0.23	4.05	nr	**4.16**
22mm dia.	0.24	0.29	0.26	4.58	nr	**4.87**
28mm dia.	0.57	0.69	0.30	5.28	nr	**5.97**
35mm dia.	1.81	2.20	0.34	5.98	nr	**8.18**
42mm dia.	2.97	3.61	0.38	6.69	nr	**10.29**
54mm dia.	5.48	6.65	0.42	7.39	nr	**14.05**
67mm dia.	16.32	19.82	0.53	9.33	nr	**29.15**
Adaptor coupling; imperial to metric						
1/2" x 15mm dia.	1.31	1.59	0.27	4.75	nr	**6.34**
3/4" x 22mm dia.	1.15	1.40	0.31	5.46	nr	**6.85**
1" x 28mm dia.	2.27	2.76	0.36	6.34	nr	**9.09**
1 1/4" x 35mm dia.	3.72	4.52	0.41	7.22	nr	**11.73**
1 1/2" x 42mm dia.	4.73	5.74	0.46	8.10	nr	**13.84**
Reducing coupling						
15 x 10mm dia.	1.09	1.32	0.23	4.05	nr	**5.37**
22 x 10mm dia.	1.59	1.93	0.26	4.58	nr	**6.51**
22 x 15mm dia.	1.09	1.32	0.27	4.75	nr	**6.08**
28 x 15mm dia.	2.46	2.99	0.28	4.93	nr	**7.92**
28 x 22mm dia.	1.50	1.82	0.30	5.28	nr	**7.10**
35 x 28mm dia.	3.53	4.29	0.34	5.98	nr	**10.27**
42 x 35mm dia.	5.20	6.31	0.38	6.69	nr	**13.00**
54 x 35mm dia.	9.11	11.06	0.42	7.39	nr	**18.45**
54 x 42mm dia.	9.93	12.06	0.42	7.39	nr	**19.45**
Straight female connector						
15mm x 1/2" dia.	1.38	1.68	0.27	4.75	nr	**6.43**
22mm x 3/4" dia.	1.96	2.38	0.31	5.46	nr	**7.84**
28mm x 1" dia.	3.68	4.47	0.36	6.34	nr	**10.80**
35mm x 1 1/4" dia.	6.39	7.76	0.41	7.22	nr	**14.98**
42mm x 1 1/2" dia.	8.29	10.07	0.46	8.10	nr	**18.16**
54mm x 2" dia.	13.13	15.94	0.52	9.15	nr	**25.10**

Material Costs/Measured Work Prices - Mechanical Installations

S:PIPED SUPPLY SYSTEMS

Item	Net Price £	Material £	Labour hours	Labour £	Unit	Total rate £
S10 : COLD WATER						
Y10 - PIPELINES						
COPPER (Continued)						
Straight male connector						
15mm x 1/2" dia.	1.17	1.42	0.27	4.75	nr	**6.17**
22mm x 3/4" dia.	2.09	2.54	0.31	5.46	nr	**7.99**
28mm x 1" dia.	3.31	4.02	0.36	6.34	nr	**10.36**
35mm x 1 1/4" dia.	5.82	7.07	0.41	7.22	nr	**14.28**
42mm x 1 1/2" dia.	7.49	9.10	0.46	8.10	nr	**17.19**
54mm x 2" dia.	11.38	13.82	0.52	9.15	nr	**22.97**
67mm x 2 1/2" dia.	18.17	22.06	0.63	11.09	nr	**33.15**
Female reducing connector						
15mm x 3/4" dia.	3.35	4.07	0.27	4.75	nr	**8.82**
22mm x 1" dia.	5.77	7.01	0.31	5.46	nr	**12.46**
Male reducing connector						
15mm x 3/4" dia.	3.00	3.64	0.27	4.75	nr	**8.40**
22mm x 1" dia.	4.55	5.53	0.31	5.46	nr	**10.98**
Lead connector						
15mm dia.	1.02	1.24	0.28	4.93	nr	**6.17**
22mm dia.	1.53	1.86	0.32	5.63	nr	**7.49**
28mm dia.	2.08	2.53	0.37	6.51	nr	**9.04**
Flanged connector						
28mm dia.	21.42	26.01	0.36	6.34	nr	**32.35**
35mm dia.	27.12	32.93	0.41	7.22	nr	**40.15**
42mm dia.	32.41	39.36	0.46	8.10	nr	**47.45**
54mm dia.	48.99	59.49	0.52	9.15	nr	**68.64**
67mm dia.	60.49	73.45	0.61	10.74	nr	**84.19**
Tank connector						
15mm x 1/2" dia.	2.86	3.47	0.25	4.40	nr	**7.87**
22mm x 3/4" dia.	4.38	5.32	0.28	4.93	nr	**10.25**
28mm x 1" dia.	5.76	6.99	0.32	5.63	nr	**12.63**
35mm x 1 1/4" dia.	7.37	8.95	0.37	6.51	nr	**15.46**
42mm x 1 1/2" dia.	9.66	11.73	0.43	7.57	nr	**19.30**
54mm x 2" dia.	14.76	17.92	0.46	8.10	nr	**26.02**
Tank connector with long thread						
15mm x 1/2" dia.	3.72	4.52	0.30	5.28	nr	**9.80**
22mm x 3/4" dia.	5.30	6.44	0.33	5.81	nr	**12.24**
28mm x 1" dia.	6.53	7.93	0.39	6.86	nr	**14.79**

Material Costs/Measured Work Prices - Mechanical Installations

S:PIPED SUPPLY SYSTEMS

Item	Net Price £	Material £	Labour hours	Labour £	Unit	Total rate £
Reducer						
15 x 10mm dia.	0.37	0.45	0.23	4.05	nr	4.50
22 x 15mm dia.	0.43	0.52	0.26	4.58	nr	5.10
28 x 15mm dia.	1.24	1.51	0.28	4.93	nr	6.43
28 x 22mm dia.	0.95	1.15	0.30	5.28	nr	6.43
35 x 22mm dia.	3.44	4.18	0.34	5.98	nr	10.16
42 x 22mm dia.	6.20	7.53	0.36	6.34	nr	13.86
42 x 35mm dia.	4.81	5.84	0.38	6.69	nr	12.53
54 x 35mm dia.	10.06	12.22	0.40	7.04	nr	19.26
54 x 42mm dia.	8.68	10.54	0.42	7.39	nr	17.93
67 x 54mm dia.	11.80	14.33	0.53	9.33	nr	23.66
Adaptor; copper to female iron						
15mm x 1/2" dia.	2.33	2.83	0.27	4.75	nr	7.58
22mm x 3/4" dia.	3.54	4.30	0.31	5.46	nr	9.75
28mm x 1" dia.	5.00	6.07	0.36	6.34	nr	12.41
35mm x 1 1/4" dia.	9.02	10.95	0.41	7.22	nr	18.17
42mm x 1 1/2" dia.	11.37	13.81	0.46	8.10	nr	21.90
54mm x 2" dia.	13.69	16.62	0.52	9.15	nr	25.78
Adaptor; copper to male iron						
15mm x 1/2" dia.	2.37	2.88	0.27	4.75	nr	7.63
22mm x 3/4" dia.	3.03	3.68	0.31	5.46	nr	9.14
28mm x 1" dia.	5.06	6.14	0.36	6.34	nr	12.48
35mm x 1 1/4" dia.	7.37	8.95	0.41	7.22	nr	16.17
42mm x 1 1/2" dia.	10.19	12.37	0.46	8.10	nr	20.47
54mm x 2" dia.	13.69	16.62	0.52	9.15	nr	25.78
Union coupling						
15mm dia.	3.21	3.90	0.41	7.22	nr	11.11
22mm dia.	5.14	6.24	0.45	7.92	nr	14.16
28mm dia.	7.50	9.11	0.51	8.98	nr	18.08
35mm dia.	9.83	11.94	0.64	11.26	nr	23.20
42mm dia.	14.37	17.45	0.68	11.97	nr	29.42
54mm dia.	27.34	33.20	0.78	13.73	nr	46.93
67mm dia.	46.29	56.21	0.96	16.90	nr	73.11
Elbow						
15mm dia.	0.18	0.22	0.23	4.05	nr	4.27
22mm dia.	0.42	0.51	0.26	4.58	nr	5.09
28mm dia.	0.90	1.09	0.31	5.46	nr	6.55
35mm dia.	3.90	4.74	0.35	6.16	nr	10.90
42mm dia.	6.44	7.82	0.41	7.22	nr	15.04
54mm dia.	13.29	16.14	0.44	7.74	nr	23.88
67mm dia.	33.85	41.10	0.54	9.50	nr	50.61
Backplate elbow						
15mm dia.	2.46	2.99	0.51	8.98	nr	11.96
22mm dia.	5.17	6.28	0.54	9.50	nr	15.78
Overflow bend						
15mm dia.	6.09	7.40	0.23	4.05	nr	11.44
22mm dia.	7.29	8.85	0.26	4.58	nr	13.43

Material Costs/Measured Work Prices - Mechanical Installations

S:PIPED SUPPLY SYSTEMS

Item	Net Price £	Material £	Labour hours	Labour £	Unit	Total rate £
S10 : COLD WATER						
Y10 - PIPELINES						
COPPER (Continued)						
Return bend						
15mm dia.	3.62	4.40	0.23	4.05	nr	**8.44**
22mm dia.	7.10	8.62	0.26	4.58	nr	**13.20**
28mm dia.	9.06	11.00	0.31	5.46	nr	**16.46**
Obtuse elbow						
15mm dia.	0.47	0.57	0.23	4.05	nr	**4.62**
22mm dia.	0.97	1.18	0.26	4.58	nr	**5.75**
28mm dia.	1.88	2.28	0.31	5.46	nr	**7.74**
35mm dia.	5.77	7.01	0.36	6.34	nr	**13.34**
42mm dia.	10.26	12.46	0.41	7.22	nr	**19.68**
54mm dia.	18.57	22.55	0.44	7.74	nr	**30.29**
67mm dia.	33.67	40.89	0.54	9.50	nr	**50.39**
Straight tap connector						
15mm x 1/2" dia.	1.40	1.70	0.13	2.29	nr	**3.99**
22mm x 3/4" dia.	2.77	3.36	0.14	2.46	nr	**5.83**
Bent tap connector						
15mm x 1/2" dia.	1.03	1.25	0.13	2.29	nr	**3.54**
22mm x 3/4" dia.	2.62	3.18	0.14	2.46	nr	**5.65**
Bent male union connector						
15mm x 1/2" dia.	4.56	5.54	0.41	7.22	nr	**12.75**
22mm x 3/4" dia.	5.93	7.20	0.45	7.92	nr	**15.12**
28mm x 1" dia.	8.49	10.31	0.51	8.98	nr	**19.29**
35mm x 1 1/4" dia.	13.83	16.79	0.64	11.26	nr	**28.06**
42mm x 1 1/2" dia.	22.51	27.33	0.68	11.97	nr	**39.30**
54mm x 2" dia.	35.55	43.17	0.78	13.73	nr	**56.90**
Bent female union connector						
15mm dia.	4.56	5.54	0.41	7.22	nr	**12.75**
22mm x 3/4" dia.	5.93	7.20	0.45	7.92	nr	**15.12**
28mm x 1" dia.	8.49	10.31	0.51	8.98	nr	**19.29**
35mm x 1 1/4" dia.	13.83	16.79	0.64	11.26	nr	**28.06**
42mm x 1 1/2" dia.	22.51	27.33	0.68	11.97	nr	**39.30**
54mm x 2" dia.	35.55	43.17	0.78	13.73	nr	**56.90**
Straight union adaptor						
15mm x 3/4" dia.	1.96	2.38	0.41	7.22	nr	**9.60**
22mm x 1" dia.	2.82	3.42	0.45	7.92	nr	**11.34**
28mm x 1 1/4" dia.	4.47	5.43	0.51	8.98	nr	**14.40**
35mm x 1 1/2" dia.	6.89	8.37	0.64	11.26	nr	**19.63**
42mm x 2" dia.	8.70	10.56	0.68	11.97	nr	**22.53**
54mm x 2 1/2" dia.	13.44	16.32	0.78	13.73	nr	**30.05**
67mm x 3" dia.	24.75	30.05	0.96	16.90	nr	**46.95**

Material Costs/Measured Work Prices - Mechanical Installations

S:PIPED SUPPLY SYSTEMS

Item	Net Price £	Material £	Labour hours	Labour £	Unit	Total rate £
Straight male union connector						
15mm x 1/2" dia.	3.88	4.71	0.41	7.22	nr	**11.93**
22mm x 3/4" dia.	5.05	6.13	0.45	7.92	nr	**14.05**
28mm x 1" dia.	7.51	9.12	0.51	8.98	nr	**18.10**
35mm x 1 1/4" dia.	10.81	13.13	0.64	11.26	nr	**24.39**
42mm x 1 1/2" dia.	17.01	20.66	0.68	11.97	nr	**32.62**
54mm x 2" dia.	24.43	29.67	0.78	13.73	nr	**43.39**
Straight female union connector						
15mm x 1/2" dia.	3.88	4.71	0.41	7.22	nr	**11.93**
22mm x 3/4" dia.	5.05	6.13	0.45	7.92	nr	**14.05**
28mm x 1" dia.	7.51	9.12	0.51	8.98	nr	**18.10**
35mm x 1 1/4" dia.	10.81	13.13	0.64	11.26	nr	**24.39**
42mm x 1 1/2" dia.	17.01	20.66	0.68	11.97	nr	**32.62**
54mm x 2" dia.	24.43	29.67	0.78	13.73	nr	**43.39**
Male nipple						
3/4 x 1/2" dia.	1.93	2.34	0.28	4.93	nr	**7.27**
1 x 3/4" dia.	2.22	2.70	0.32	5.63	nr	**8.33**
1 1/4 x 1" dia.	3.04	3.69	0.37	6.51	nr	**10.20**
1 1/2 x 1 1/4" dia.	4.49	5.45	0.42	7.39	nr	**12.84**
2 x 1 1/2" dia.	9.20	11.17	0.46	8.10	nr	**19.27**
2 1/2 x 2" dia.	12.28	14.91	0.56	9.86	nr	**24.77**
Female nipple						
3/4 x 1/2" dia.	1.93	2.34	0.28	4.93	nr	**7.27**
1 x 3/4" dia.	2.22	2.70	0.32	5.63	nr	**8.33**
1 1/4 x 1" dia.	3.04	3.69	0.37	6.51	nr	**10.21**
1 1/2 x 1 1/4" dia.	4.49	5.45	0.42	7.39	nr	**12.84**
2 x 1 1/2" dia.	9.20	11.17	0.46	8.10	nr	**19.27**
2 1/2 x 2" dia.	12.28	14.91	0.56	9.86	nr	**24.77**
Equal tee						
10mm dia.	1.01	1.23	0.25	4.40	nr	**5.63**
15mm dia.	0.32	0.39	0.36	6.34	nr	**6.72**
22mm dia.	1.00	1.21	0.39	6.86	nr	**8.08**
28mm dia.	2.48	3.01	0.43	7.57	nr	**10.58**
35mm dia.	6.21	7.54	0.57	10.03	nr	**17.57**
42mm dia.	9.98	12.12	0.60	10.56	nr	**22.68**
54mm dia.	20.12	24.43	0.65	11.44	nr	**35.87**
67mm dia.	43.81	53.20	0.78	13.73	nr	**66.93**
Female tee, reducing branch Fl						
15 x 15mm x 1/4" dia.	2.93	3.56	0.36	6.34	nr	**9.89**
22 x 22mm x 1/2" dia.	2.04	2.48	0.39	6.86	nr	**9.34**
28 x 28mm x 3/4" dia.	7.01	8.51	0.43	7.57	nr	**16.08**
35 x 35mm x 3/4" dia.	9.60	11.66	0.47	8.27	nr	**19.93**
42 x 42mm x 1/2" dia.	12.15	14.75	0.60	10.56	nr	**25.31**
Backplate tee						
15 x 15mm x 1/2" dia.	5.53	6.72	0.62	10.91	nr	**17.63**
Heater tee						
1/2 x 1/2" x 15mm dia.	4.97	6.04	0.36	6.34	nr	**12.37**

Material Costs/Measured Work Prices - Mechanical Installations

S:PIPED SUPPLY SYSTEMS

Item	Net Price £	Material £	Labour hours	Labour £	Unit	Total rate £
S10 : COLD WATER						
Y10 - PIPELINES						
COPPER (Continued)						
Union heater tee						
1/2 x 1/2" x 15mm dia.	6.68	8.11	0.36	6.34	nr	**14.45**
Sweep tee - equal						
15mm dia.	3.96	4.81	0.36	6.34	nr	**11.14**
22mm dia.	5.10	6.19	0.39	6.86	nr	**13.06**
28mm dia.	8.56	10.39	0.43	7.57	nr	**17.96**
35mm dia.	12.14	14.74	0.57	10.03	nr	**24.77**
42mm dia.	18.01	21.87	0.60	10.56	nr	**32.43**
54mm dia.	19.96	24.24	0.65	11.44	nr	**35.68**
67mm dia.	34.73	42.17	0.78	13.73	nr	**55.90**
Sweep tee - reducing						
22 x 22 x 15mm dia.	4.27	5.19	0.39	6.86	nr	**12.05**
28 x 28 x 22mm dia.	7.24	8.79	0.43	7.57	nr	**16.36**
35 x 35 x 22mm dia.	12.14	14.74	0.57	10.03	nr	**24.77**
Sweep tee - double						
15mm dia.	4.47	5.43	0.36	6.34	nr	**11.76**
22mm dia.	6.09	7.40	0.39	6.86	nr	**14.26**
28mm dia.	9.25	11.23	0.43	7.57	nr	**18.80**
Cross						
15mm dia.	5.92	7.19	0.48	8.45	nr	**15.64**
22mm dia.	6.62	8.04	0.53	9.33	nr	**17.37**
28mm dia.	9.49	11.52	0.61	10.74	nr	**22.26**
Extra Over copper pipes; high duty capillary fittings; BS 864						
Stop end						
15mm dia.	2.99	3.63	0.16	2.82	nr	**6.45**
Straight coupling; copper to copper						
15mm dia.	1.40	1.70	0.27	4.75	nr	**6.45**
22mm dia.	2.21	2.68	0.32	5.63	nr	**8.32**
28mm dia.	3.15	3.83	0.37	6.51	nr	**10.34**
35mm dia.	5.54	6.73	0.43	7.57	nr	**14.30**
42mm dia.	6.07	7.37	0.50	8.80	nr	**16.17**
54mm dia.	8.93	10.84	0.54	9.50	nr	**20.35**
Reducing coupling						
15 x 12mm dia.	2.62	3.18	0.27	4.75	nr	**7.93**
22 x 15mm dia.	3.03	3.68	0.32	5.63	nr	**9.31**
28 x 22mm dia.	4.17	5.06	0.37	6.51	nr	**11.58**
Straight female connector						
15mm x 1/2" dia.	3.41	4.14	0.32	5.63	nr	**9.77**
22mm x 3/4" dia.	3.85	4.68	0.36	6.34	nr	**11.01**
28mm x 1" dia.	5.68	6.90	0.42	7.39	nr	**14.29**

Material Costs/Measured Work Prices - Mechanical Installations

S:PIPED SUPPLY SYSTEMS

Item	Net Price £	Material £	Labour hours	Labour £	Unit	Total rate £
Straight male connector						
15mm x 1/2" dia.	3.32	4.03	0.32	5.63	nr	9.66
22mm x 3/4" dia.	3.85	4.68	0.36	6.34	nr	11.01
28mm x 1" dia.	5.68	6.90	0.42	7.39	nr	14.29
42mm x 1 1/2" dia.	11.07	13.44	0.53	9.33	nr	22.77
54mm x 2" dia.	17.99	21.85	0.62	10.91	nr	32.76
Reducer						
15 x 12mm dia.	1.72	2.09	0.27	4.75	nr	6.84
22 x 15mm dia.	1.69	2.05	0.32	5.63	nr	7.68
28 x 22mm dia.	3.03	3.68	0.37	6.51	nr	10.19
35 x 28mm dia.	3.85	4.68	0.43	7.57	nr	12.24
42 x 35mm dia.	4.95	6.01	0.50	8.80	nr	14.81
54 x 42mm dia.	8.00	9.71	0.39	6.86	nr	16.58
Straight union adaptor						
15mm x 3/4" dia.	2.73	3.31	0.27	4.75	nr	8.07
22mm x 1" dia.	3.71	4.51	0.32	5.63	nr	10.14
28mm x 1 1/4" dia.	4.89	5.94	0.37	6.51	nr	12.45
35mm x 1 1/2" dia.	8.87	10.77	0.43	7.57	nr	18.34
42mm x 2" dia.	11.23	13.64	0.50	8.80	nr	22.44
Bent union adaptor						
15mm x 3/4" dia.	7.12	8.65	0.27	4.75	nr	13.40
22mm x 1" dia.	9.59	11.65	0.32	5.63	nr	17.28
28mm x 1 1/4" dia.	12.92	15.69	0.37	6.51	nr	22.20
Adaptor; male copper to FI						
15mm x 1/2" dia.	5.13	6.23	0.27	4.75	nr	10.98
22mm x 3/4" dia.	5.24	6.36	0.32	5.63	nr	12.00
Union coupling						
15mm dia.	6.25	7.59	0.54	9.50	nr	17.09
22mm dia.	8.00	9.71	0.60	10.56	nr	20.27
28mm dia.	11.10	13.48	0.68	11.97	nr	25.45
35mm dia.	19.37	23.52	0.83	14.61	nr	38.13
42mm dia.	22.82	27.71	0.89	15.66	nr	43.38
Elbow						
15mm dia.	4.02	4.88	0.27	4.75	nr	9.63
22mm dia.	4.30	5.22	0.32	5.63	nr	10.85
28mm dia.	6.39	7.76	0.37	6.51	nr	14.27
35mm dia.	9.98	12.12	0.43	7.57	nr	19.69
42mm dia.	12.43	15.09	0.50	8.80	nr	23.89
54mm dia.	21.63	26.27	0.52	9.15	nr	35.42
Return bend						
28mm dia.	13.09	15.90	0.37	6.51	nr	22.41
35mm dia.	15.21	18.47	0.43	7.57	nr	26.04
Bent male union connector						
15mm x 1/2" dia.	9.20	11.17	0.54	9.50	nr	20.68
22mm x 3/4" dia.	12.39	15.05	0.60	10.56	nr	25.61
28mm x 1" dia.	22.51	27.33	0.68	11.97	nr	39.30

Material Costs/Measured Work Prices - Mechanical Installations

S:PIPED SUPPLY SYSTEMS

Item	Net Price £	Material £	Labour hours	Labour £	Unit	Total rate £
S10 : COLD WATER						
Y10 - PIPELINES						
COPPER (Continued)						
Composite flange						
22mm dia.	15.41	18.71	0.35	6.16	nr	**24.87**
28mm dia.	16.48	20.01	0.37	6.51	nr	**26.53**
35mm dia.	22.29	27.07	0.38	6.69	nr	**33.76**
42mm dia.	25.67	31.17	0.41	7.22	nr	**38.39**
54mm dia.	35.99	43.70	0.43	7.57	nr	**51.27**
Equal tee						
15mm dia.	4.62	5.61	0.44	7.74	nr	**13.35**
22mm dia.	5.82	7.07	0.47	8.27	nr	**15.34**
28mm dia.	7.67	9.31	0.53	9.33	nr	**18.64**
35mm dia.	13.09	15.90	0.70	12.32	nr	**28.22**
42mm dia.	16.68	20.25	0.84	14.78	nr	**35.04**
54mm dia.	26.27	31.90	0.79	13.90	nr	**45.80**
Reducing tee						
15 x 12mm dia.	6.16	7.48	0.44	7.74	nr	**15.22**
22 x 15mm dia.	7.41	9.00	0.47	8.27	nr	**17.27**
28 x 22mm dia.	10.56	12.82	0.53	9.33	nr	**22.15**
35 x 28mm dia.	13.89	16.87	0.73	12.85	nr	**29.72**
42 x 28mm dia.	21.39	25.97	0.84	14.78	nr	**40.76**
54 x 28mm dia.	33.79	41.03	1.01	17.78	nr	**58.81**
Extra Over copper pipes; compression fittings; BS 864						
Stop end						
15mm dia.	0.82	1.00	0.10	1.76	nr	**2.76**
22mm dia.	0.98	1.19	0.29	5.10	nr	**6.29**
28mm dia.	2.86	3.47	0.15	2.64	nr	**6.11**
Straight connector; copper to copper						
15mm dia.	0.52	0.63	0.18	3.17	nr	**3.80**
22mm dia.	0.91	1.10	0.21	3.70	nr	**4.80**
28mm dia.	2.67	3.24	0.24	4.22	nr	**7.47**
Straight connector; copper to imperial copper						
22mm dia.	2.17	2.63	0.21	3.70	nr	**6.33**
Male coupling; copper to MI (BSP)						
15mm dia.	0.48	0.58	0.19	3.34	nr	**3.93**
22mm dia.	0.78	0.95	0.23	4.05	nr	**5.00**
28mm dia.	1.52	1.85	0.26	4.58	nr	**6.42**
Male coupling with long thread and backnut						
15mm dia.	2.81	3.41	0.19	3.34	nr	**6.76**
22mm dia.	3.67	4.46	0.23	4.05	nr	**8.50**

Material Costs/Measured Work Prices - Mechanical Installations

S:PIPED SUPPLY SYSTEMS

Item	Net Price £	Material £	Labour hours	Labour £	Unit	Total rate £
Female coupling; copper to FI (BSP)						
15mm dia.	0.60	0.73	0.19	3.34	nr	**4.07**
22mm dia.	0.90	1.09	0.23	4.05	nr	**5.14**
28mm dia.	1.90	2.31	0.27	4.75	nr	**7.06**
Elbow						
15mm dia.	0.64	0.78	0.18	3.17	nr	**3.95**
22mm dia.	1.07	1.30	0.21	3.70	nr	**5.00**
28mm dia.	3.31	4.02	0.24	4.22	nr	**8.24**
Male elbow; copper to FI (BSP)						
15mm x 1/2" dia.	2.01	2.44	0.19	3.34	nr	**5.78**
22mm x 3/4" dia.	1.28	1.55	0.23	4.05	nr	**5.60**
28mm x 1" dia.	4.25	5.16	0.27	4.75	nr	**9.91**
Female elbow; copper to FI (BSP)						
15mm x 1/2" dia.	1.61	1.96	0.19	3.34	nr	**5.30**
22mm x 3/4" dia.	2.36	2.87	0.23	4.05	nr	**6.91**
28mm x 1" dia.	4.06	4.93	0.27	4.75	nr	**9.68**
Backplate elbow						
15mm x 1/2" dia.	2.79	3.39	0.50	8.80	nr	**12.19**
Reducing set; internal						
15mm dia.	0.83	1.01	0.19	3.34	nr	**4.35**
Tank coupling; long thread						
22mm dia.	2.33	2.83	0.46	8.10	nr	**10.93**
Tee equal						
15mm dia.	0.81	0.98	0.28	4.93	nr	**5.91**
22mm dia.	1.37	1.66	0.30	5.28	nr	**6.94**
28mm dia.	4.60	5.59	0.34	5.98	nr	**11.57**
Tee reducing						
22mm dia.	2.75	3.34	0.30	5.28	nr	**8.62**
Backplate tee						
15mm dia.	3.98	4.83	0.62	10.91	nr	**15.75**

Material Costs/Measured Work Prices - Mechanical Installations

S:PIPED SUPPLY SYSTEMS

Item	Net Price £	Material £	Labour hours	Labour £	Unit	Total rate £
S10 : COLD WATER						
Y10 - PIPELINES						
COPPER (Continued)						
Extra over fittings; silver brazed welded joints						
Reducer						
76 x 67mm dia	19.61	23.81	1.40	24.64	nr	48.45
108 x 76.1mm dia	30.27	36.76	1.80	31.68	nr	68.44
133 x 108mm dia	45.61	55.38	2.20	38.72	nr	94.11
159 x 133mm dia	60.57	73.55	2.60	45.76	nr	119.31
90 degree elbow						
76mm dia	64.67	78.53	1.60	28.16	nr	106.69
108mm dia	123.48	149.94	2.00	35.20	nr	185.14
133mm dia	244.99	297.49	2.40	42.24	nr	339.73
159mm dia	306.76	372.50	2.80	49.28	nr	421.78
45 degree elbow						
76mm dia	58.80	71.40	1.60	28.16	nr	99.56
108mm dia	103.89	126.15	2.00	35.20	nr	161.36
133mm dia	209.73	254.68	2.40	42.24	nr	296.92
159mm dia	274.40	333.20	2.80	49.28	nr	382.49
Equal tee						
76mm dia	57.01	69.23	2.40	42.24	nr	111.47
108mm dia	103.24	125.36	3.00	52.80	nr	178.17
133mm dia	256.57	311.55	3.60	63.36	nr	374.92
159mm dia	285.09	346.18	4.20	73.92	nr	420.11
Extra Over copper pipes; dezincification resistant compression fittings; BS 864						
Stop end						
15mm dia.	1.12	1.36	0.10	1.76	nr	3.12
22mm dia.	1.58	1.92	0.13	2.29	nr	4.21
28mm dia.	3.08	3.74	0.15	2.64	nr	6.38
35mm dia.	4.88	5.93	0.18	3.17	nr	9.09
42mm dia.	8.04	9.76	0.20	3.52	nr	13.28
Straight coupling; copper to copper						
15mm dia.	0.90	1.09	0.18	3.17	nr	4.26
22mm dia.	1.43	1.74	0.21	3.70	nr	5.43
28mm dia.	3.04	3.69	0.24	4.22	nr	7.92
35mm dia.	6.58	7.99	0.29	5.10	nr	13.09
42mm dia.	8.47	10.29	0.33	5.81	nr	16.09
54mm dia.	12.36	15.01	0.38	6.69	nr	21.70
Straight swivel connector; copper to imperial copper						
22mm dia.	2.93	3.56	0.20	3.52	nr	7.08

Material Costs/Measured Work Prices - Mechanical Installations

S:PIPED SUPPLY SYSTEMS

Item	Net Price £	Material £	Labour hours	Labour £	Unit	Total rate £
Male coupling; copper to MI (BSP)						
15mm x 1/2" dia.	0.78	0.95	0.19	3.34	nr	4.29
22mm x 3/4" dia.	1.24	1.51	0.23	4.05	nr	5.55
28mm x 1" dia.	2.70	3.28	0.26	4.58	nr	7.85
35mm x 1 1/4" dia.	4.70	5.71	0.32	5.63	nr	11.34
42mm x 1 1/2" dia.	7.50	9.11	0.37	6.51	nr	15.62
54mm x 2" dia.	9.43	11.45	0.57	10.03	nr	21.48
Male coupling with long thread and backnuts						
22mm dia.	3.69	4.48	0.23	4.05	nr	8.53
28mm dia.	4.10	4.98	0.24	4.22	nr	9.20
Female coupling; copper to FI (BSP)						
15mm x 1/2" dia.	0.95	1.15	0.19	3.34	nr	4.50
22mm x 3/4" dia.	1.47	1.78	0.23	4.05	nr	5.83
28mm x 1" dia.	2.82	3.42	0.27	4.75	nr	8.18
35mm x 1 1/4" dia.	5.55	6.74	0.32	5.63	nr	12.37
42mm x 1 1/2" dia.	7.65	9.29	0.37	6.51	nr	15.80
54mm x 2" dia.	11.21	13.61	0.42	7.39	nr	21.00
Elbow						
15mm dia.	1.07	1.30	0.18	3.17	nr	4.47
22mm dia.	1.67	2.03	0.21	3.70	nr	5.72
28mm dia.	3.90	4.74	0.24	4.22	nr	8.96
35mm dia.	8.77	10.65	0.29	5.10	nr	15.75
42mm dia.	12.21	14.83	0.33	5.81	nr	20.63
54mm dia.	20.66	25.09	0.38	6.69	nr	31.78
Male elbow; copper to MI (BSP)						
15mm x 1/2" dia.	1.73	2.10	0.19	3.34	nr	5.44
22mm x 3/4" dia.	1.92	2.33	0.23	4.05	nr	6.38
28mm x 1" dia.	3.51	4.26	0.27	4.75	nr	9.01
Female elbow; copper to FI (BSP)						
15mm x 1/2" dia.	1.82	2.21	0.19	3.34	nr	5.55
22mm x 3/4" dia.	2.62	3.18	0.23	4.05	nr	7.23
28mm x 1" dia.	4.33	5.26	0.27	4.75	nr	10.01
Backplate elbow						
15mm x 1/2" dia.	2.37	2.88	0.50	8.80	nr	11.68
Straight tap connector						
15mm dia.	1.63	1.98	0.13	2.29	nr	4.27
22mm dia.	3.11	3.78	0.15	2.64	nr	6.42
Tank coupling						
15mm dia.	2.22	2.70	0.19	3.34	nr	6.04
22mm dia.	2.46	2.99	0.23	4.05	nr	7.04
28mm dia.	4.84	5.88	0.27	4.75	nr	10.63
35mm dia.	7.51	9.12	0.32	5.63	nr	14.75
42mm dia.	12.21	14.83	0.37	6.51	nr	21.34
54mm dia.	17.39	21.12	0.31	5.46	nr	26.57

126 *Material Costs/Measured Work Prices - Mechanical Installations*

S:PIPED SUPPLY SYSTEMS

Item	Net Price £	Material £	Labour hours	Labour £	Unit	Total rate £
S10 : COLD WATER						
Y10 - PIPELINES						
COPPER (Continued)						
Reducing set; internal						
15mm dia.	1.03	1.25	0.19	3.34	nr	**4.59**
22mm dia.	1.79	2.17	0.23	4.05	nr	**6.22**
Tee equal						
15mm dia.	1.51	1.83	0.28	4.93	nr	**6.76**
22mm dia.	2.43	2.95	0.30	5.28	nr	**8.23**
28mm dia.	6.32	7.67	0.34	5.98	nr	**13.66**
35mm dia.	10.92	13.26	0.43	7.57	nr	**20.83**
42mm dia.	16.96	20.59	0.46	8.10	nr	**28.69**
54mm dia.	27.27	33.11	0.54	9.50	nr	**42.62**
Tee reducing						
22mm dia.	3.92	4.76	0.30	5.28	nr	**10.04**
28mm dia.	6.21	7.54	0.34	5.98	nr	**13.53**
35mm dia.	10.67	12.96	0.43	7.57	nr	**20.52**
42mm dia.	16.66	20.23	0.46	8.10	nr	**28.33**
54mm dia.	28.07	34.09	0.54	9.50	nr	**43.59**
Extra Over copper pipes; bronze one piece brazing flanges; metric						
Bronze flange; PN6						
15mm dia.	13.49	16.38	0.27	4.75	nr	**21.13**
22mm dia.	15.99	19.42	0.32	5.63	nr	**25.05**
28mm dia.	18.38	22.32	0.36	6.34	nr	**28.66**
35mm dia.	25.96	31.52	0.47	8.27	nr	**39.80**
42mm dia.	31.58	38.35	0.54	9.50	nr	**47.85**
54mm dia.	44.65	54.22	0.63	11.09	nr	**65.31**
67mm dia.	51.42	62.44	0.77	13.55	nr	**75.99**
76mm dia.	59.26	71.96	0.93	16.37	nr	**88.33**
108mm dia.	79.04	95.98	1.14	20.06	nr	**116.04**
133mm dia.	95.75	116.27	1.41	24.82	nr	**141.09**
159mm dia.	136.20	165.39	1.74	30.63	nr	**196.01**
Bronze flange; PN10						
15mm dia.	17.51	21.26	0.27	4.75	nr	**26.01**
22mm dia.	20.37	24.74	0.32	5.63	nr	**30.37**
28mm dia.	20.60	25.01	0.38	6.69	nr	**31.70**
35mm dia.	28.23	34.28	0.47	8.27	nr	**42.55**
42mm dia.	33.54	40.73	0.54	9.50	nr	**50.23**
54mm dia.	47.40	57.56	0.63	11.09	nr	**68.65**
67mm dia.	51.42	62.44	0.77	13.55	nr	**75.99**
76mm dia.	65.68	79.76	0.93	16.37	nr	**96.12**
108mm dia.	96.42	117.08	1.14	20.06	nr	**137.15**
133mm dia.	110.74	134.47	1.41	24.82	nr	**159.29**
159mm dia.	169.63	205.98	1.74	30.63	nr	**236.61**

Material Costs/Measured Work Prices - Mechanical Installations

S:PIPED SUPPLY SYSTEMS

Item	Net Price £	Material £	Labour hours	Labour £	Unit	Total rate £
Bronze flange; PN16						
15mm dia.	12.82	15.57	0.27	4.75	nr	20.32
22mm dia.	15.67	19.03	0.32	5.63	nr	24.66
28mm dia.	17.33	21.04	0.38	6.69	nr	27.73
35mm dia.	24.96	30.31	0.47	8.27	nr	38.58
42mm dia.	29.53	35.86	0.54	9.50	nr	45.36
54mm dia.	39.89	48.44	0.63	11.09	nr	59.53
67mm dia.	58.15	70.61	0.77	13.55	nr	84.16
76mm dia.	67.31	81.73	0.93	16.37	nr	98.10
108mm dia.	71.16	86.41	1.14	20.06	nr	106.47
133mm dia.	130.63	158.62	1.41	24.82	nr	183.44
159mm dia.	157.97	191.82	1.74	30.63	nr	222.45
Extra Over copper pipes; bronze blank flanges; metric						
Bronze blank flange; PN6						
15mm dia.	11.62	14.11	0.27	4.75	nr	18.86
22mm dia.	14.82	18.00	0.27	4.75	nr	22.75
28mm dia.	15.23	18.49	0.27	4.75	nr	23.25
35mm dia.	24.93	30.27	0.32	5.63	nr	35.90
42mm dia.	33.63	40.84	0.32	5.63	nr	46.47
54mm dia.	36.50	44.32	0.34	5.98	nr	50.31
67mm dia.	45.62	55.40	0.36	6.34	nr	61.73
76mm dia.	58.77	71.36	0.37	6.51	nr	77.88
108mm dia.	93.38	113.39	0.41	7.22	nr	120.61
133mm dia.	110.16	133.77	0.58	10.21	nr	143.98
159mm dia.	138.79	168.53	0.61	10.74	nr	179.27
Bronze blank flange; PN10						
15mm dia.	14.05	17.06	0.27	4.75	nr	21.81
22mm dia.	18.18	22.08	0.27	4.75	nr	26.83
28mm dia.	20.13	24.44	0.27	4.75	nr	29.20
35mm dia.	24.93	30.27	0.32	5.63	nr	35.90
42mm dia.	46.72	56.73	0.32	5.63	nr	62.36
54mm dia.	53.26	64.67	0.34	5.98	nr	70.66
67mm dia.	57.14	69.39	0.46	8.10	nr	77.48
76mm dia.	73.42	89.15	0.47	8.27	nr	97.43
108mm dia.	112.13	136.16	0.51	8.98	nr	145.14
133mm dia.	118.38	143.75	0.58	10.21	nr	153.96
159mm dia.	216.50	262.90	0.71	12.50	nr	275.39
Bronze blank flange; PN16						
15mm dia.	11.40	13.84	0.27	4.75	nr	18.60
22mm dia.	15.16	18.41	0.27	4.75	nr	23.16
28mm dia.	17.57	21.34	0.27	4.75	nr	26.09
35mm dia.	24.34	29.56	0.32	5.63	nr	35.19
42mm dia.	37.31	45.31	0.32	5.63	nr	50.94
54mm dia.	43.50	52.82	0.34	5.98	nr	58.81
67mm dia.	71.84	87.24	0.46	8.10	nr	95.33
76mm dia.	81.62	99.11	0.47	8.27	nr	107.38
108mm dia.	100.23	121.71	0.51	8.98	nr	130.69
133mm dia.	158.97	193.04	0.58	10.21	nr	203.25
159mm dia.	203.68	247.33	0.71	12.50	nr	259.83

Material Costs/Measured Work Prices - Mechanical Installations

S:PIPED SUPPLY SYSTEMS

Item	Net Price £	Material £	Labour hours	Labour £	Unit	Total rate £
S10 : COLD WATER						
Y10 - PIPELINES						
COPPER (Continued)						
Extra Over copper pipes; bronze screwed flanges; metric						
Bronze screwed flange; 6 BSP						
15mm dia.	12.04	14.62	0.35	6.16	nr	**20.78**
22mm dia.	13.89	16.87	0.47	8.27	nr	**25.14**
28mm dia.	14.57	17.69	0.52	9.15	nr	**26.84**
35mm dia.	19.76	23.99	0.62	10.91	nr	**34.91**
42mm dia.	23.56	28.61	0.70	12.32	nr	**40.93**
54mm dia.	31.75	38.55	0.84	14.78	nr	**53.34**
67mm dia.	39.68	48.18	1.03	18.13	nr	**66.31**
76mm dia.	48.53	58.93	1.22	21.47	nr	**80.40**
108mm dia.	75.83	92.08	1.41	24.82	nr	**116.90**
133mm dia.	89.73	108.96	1.75	30.80	nr	**139.76**
159mm dia.	115.05	139.71	2.21	38.90	nr	**178.60**
Bronze screwed flange; 10 BSP						
15mm dia.	14.49	17.60	0.35	6.16	nr	**23.76**
22mm dia.	16.85	20.46	0.47	8.27	nr	**28.73**
28mm dia.	18.66	22.66	0.52	9.15	nr	**31.81**
35mm dia.	26.68	32.40	0.62	10.91	nr	**43.31**
42mm dia.	32.38	39.32	0.70	12.32	nr	**51.64**
54mm dia.	45.75	55.55	0.84	14.78	nr	**70.34**
67mm dia.	53.44	64.89	1.03	18.13	nr	**83.02**
76mm dia.	61.07	74.16	1.22	21.47	nr	**95.63**
108mm dia.	80.53	97.79	1.41	24.82	nr	**122.60**
133mm dia.	97.38	118.25	1.75	30.80	nr	**149.05**
159mm dia.	172.51	209.48	2.21	38.90	nr	**248.38**
Bronze screwed flange; 16 BSP						
15mm dia.	14.49	17.60	0.35	6.16	nr	**23.76**
22mm dia.	16.85	20.46	0.47	8.27	nr	**28.73**
28mm dia.	18.66	22.66	0.52	9.15	nr	**31.81**
35mm dia.	26.68	32.40	0.62	10.91	nr	**43.31**
42mm dia.	32.38	39.32	0.70	12.32	nr	**51.64**
54mm dia.	45.75	55.55	0.84	14.78	nr	**70.34**
67mm dia.	62.73	76.17	1.03	18.13	nr	**94.30**
76mm dia.	74.64	90.64	1.22	21.47	nr	**112.11**
108mm dia.	101.47	123.22	1.41	24.82	nr	**148.03**
133mm dia.	166.36	202.01	1.75	30.80	nr	**232.81**
159mm dia.	216.40	262.77	2.21	38.90	nr	**301.67**

Material Costs/Measured Work Prices - Mechanical Installations

S:PIPED SUPPLY SYSTEMS

Item	Net Price £	Material £	Labour hours	Labour £	Unit	Total rate £
Extra Over copper pipes; labours						
Made bend						
15mm dia.	-	-	0.26	4.58	nr	**4.58**
22mm dia.	-	-	0.28	4.93	nr	**4.93**
28mm dia.	-	-	0.31	5.46	nr	**5.46**
35mm dia.	-	-	0.42	7.39	nr	**7.39**
42mm dia.	-	-	0.51	8.98	nr	**8.98**
54mm dia.	-	-	0.58	10.21	nr	**10.21**
67mm dia.	-	-	0.69	12.14	nr	**12.14**
76mm dia.	-	-	0.80	14.08	nr	**14.08**
Bronze butt weld						
15mm dia.	-	-	0.25	4.40	nr	**4.40**
22mm dia.	-	-	0.31	5.46	nr	**5.46**
28mm dia.	-	-	0.37	6.51	nr	**6.51**
35mm dia.	-	-	0.49	8.62	nr	**8.62**
42mm dia.	-	-	0.58	10.21	nr	**10.21**
54mm dia.	-	-	0.72	12.67	nr	**12.67**
67mm dia.	-	-	0.88	15.49	nr	**15.49**
76mm dia.	-	-	1.08	19.01	nr	**19.01**
108mm dia.	-	-	1.37	24.11	nr	**24.11**
133mm dia.	-	-	1.73	30.45	nr	**30.45**
159mm dia.	-	-	2.03	35.73	nr	**35.73**
PRESSFIT						
Mechanical pressfit joints; butyl rubber O ring						
Coupler						
15mm dia	0.40	0.49	0.36	6.34	nr	**6.82**
22mm dia	0.64	0.78	0.36	6.34	nr	**7.11**
28mm dia	1.30	1.58	0.44	7.74	nr	**9.32**
35mm dia	2.10	2.55	0.44	7.74	nr	**10.29**
42mm dia	3.28	3.98	0.52	9.15	nr	**13.14**
54mm dia	5.48	6.65	0.60	10.56	nr	**17.21**
Stop end						
22mm dia	1.42	1.72	0.18	3.17	nr	**4.89**
28mm dia	2.37	2.88	0.22	3.87	nr	**6.75**
35mm dia	4.59	5.57	0.22	3.87	nr	**9.45**
42mm dia	7.51	9.12	0.26	4.58	nr	**13.70**
54mm dia	9.93	12.06	0.30	5.28	nr	**17.34**
Reducer						
22 x 15mm dia	0.89	1.08	0.36	6.34	nr	**7.42**
28 x 15mm dia	2.94	3.57	0.40	7.04	nr	**10.61**
28 x 22mm dia	2.24	2.72	0.40	7.04	nr	**9.76**
35 x 22mm dia	3.80	4.61	0.40	7.04	nr	**11.65**
35 x 28mm dia	2.95	3.58	0.44	7.74	nr	**11.33**
42 x 22mm dia	6.85	8.32	0.44	7.74	nr	**16.06**
42 x 28mm dia	6.45	7.83	0.48	8.45	nr	**16.28**
42 x 35mm dia	5.30	6.44	0.48	8.45	nr	**14.88**
54 x 35mm dia	10.59	12.86	0.52	9.15	nr	**22.01**
54 x 42mm dia	9.14	11.10	0.56	9.86	nr	**20.96**

130   *Material Costs/Measured Work Prices - Mechanical Installations*

S:PIPED SUPPLY SYSTEMS

Item	Net Price £	Material £	Labour hours	Labour £	Unit	Total rate £
S10 : COLD WATER						
Y10 - PIPELINES						
COPPER (Continued)						
90 degree elbow						
15mm dia	0.49	0.59	0.36	6.34	nr	**6.93**
22mm dia	0.94	1.14	0.36	6.34	nr	**7.48**
28mm dia	2.01	2.44	0.44	7.74	nr	**10.19**
35mm dia	4.10	4.98	0.44	7.74	nr	**12.72**
42mm dia	6.78	8.23	0.52	9.15	nr	**17.39**
54mm dia	12.59	15.29	0.60	10.56	nr	**25.85**
45 degree elbow						
15mm dia	0.74	0.90	0.36	6.34	nr	**7.23**
22mm dia	1.28	1.55	0.36	6.34	nr	**7.89**
28mm dia	2.33	2.83	0.44	7.74	nr	**10.57**
35mm dia	5.76	6.99	0.44	7.74	nr	**14.74**
42mm dia	10.80	13.11	0.52	9.15	nr	**22.27**
54mm dia	19.55	23.74	0.60	10.56	nr	**34.30**
Equal tee						
15mm dia	0.71	0.86	0.54	9.50	nr	**10.37**
22mm dia	1.55	1.88	0.54	9.50	nr	**11.39**
28mm dia	3.21	3.90	0.66	11.62	nr	**15.51**
35mm dia	6.54	7.94	0.66	11.62	nr	**19.56**
42mm dia	11.03	13.39	0.78	13.73	nr	**27.12**
54mm dia	19.06	23.14	0.90	15.84	nr	**38.99**
Reducing tee						
22 x 15mm dia	1.76	2.14	0.54	9.50	nr	**11.64**
28 x 15mm dia	5.07	6.16	0.62	10.91	nr	**17.07**
28 x 22mm dia	4.88	5.93	0.62	10.91	nr	**16.84**
35 x 22mm dia	7.15	8.68	0.62	10.91	nr	**19.59**
35 x 28mm dia	8.31	10.09	0.62	10.91	nr	**21.00**
42 x 28mm dia	15.66	19.02	0.70	12.32	nr	**31.34**
42 x 35mm dia	17.84	21.66	0.70	12.32	nr	**33.98**
54 x 35mm dia	28.65	34.79	0.82	14.43	nr	**49.22**
54 x 42mm dia	28.32	34.39	0.82	14.43	nr	**48.82**
Male iron connector; BSP thread						
15mm dia	1.74	2.11	0.18	3.17	nr	**5.28**
22mm dia	2.92	3.55	0.18	3.17	nr	**6.71**
28mm dia	5.74	6.97	0.22	3.87	nr	**10.84**
35mm dia	7.35	8.93	0.22	3.87	nr	**12.80**
42mm dia	8.67	10.53	0.26	4.58	nr	**15.10**
54mm dia	13.77	16.72	0.30	5.28	nr	**22.00**

Material Costs/Measured Work Prices - Mechanical Installations

S:PIPED SUPPLY SYSTEMS

Item	Net Price £	Material £	Labour hours	Labour £	Unit	Total rate £
90 degree elbow; male iron BSP thread						
15mm dia	2.67	3.24	0.36	6.34	nr	**9.58**
22mm dia	4.17	5.06	0.36	6.34	nr	**11.40**
28mm dia	6.26	7.60	0.44	7.74	nr	**15.35**
35mm dia	8.04	9.76	0.44	7.74	nr	**17.51**
42mm dia	10.96	13.31	0.52	9.15	nr	**22.46**
54mm dia	16.12	19.57	0.60	10.56	nr	**30.13**
Female iron connector; BSP thread						
15mm dia	1.74	2.11	0.18	3.17	nr	**5.28**
22mm dia	2.26	2.74	0.18	3.17	nr	**5.91**
28mm dia	4.33	5.26	0.22	3.87	nr	**9.13**
35mm dia	8.06	9.79	0.22	3.87	nr	**13.66**
42mm dia	8.72	10.59	0.26	4.58	nr	**15.16**
54mm dia	13.83	16.79	0.30	5.28	nr	**22.07**
90 degree elbow; female iron BSP thread						
15mm dia	2.57	3.12	0.36	6.34	nr	**9.46**
22mm dia	3.77	4.58	0.36	6.34	nr	**10.91**
28mm dia	6.26	7.60	0.44	7.74	nr	**15.35**
35mm dia	8.04	9.76	0.44	7.74	nr	**17.51**
42mm dia	10.96	13.31	0.52	9.15	nr	**22.46**
54mm dia	16.12	19.57	0.60	10.56	nr	**30.13**

Material Costs/Measured Work Prices - Mechanical Installations

S:PIPED SUPPLY SYSTEMS

Item	Net Price £	Material £	Labour hours	Labour £	Unit	Total rate £
S10 : COLD WATER						
Y10 - PIPELINES						
STAINLESS STEEL						
Stainless steel pipes; capillary or compression joints; BS 4127						
Grade 304; satin finish						
15mm dia.	1.67	2.03	0.41	7.22	m	9.24
22mm dia.	2.34	2.84	0.51	8.98	m	11.82
28mm dia.	3.18	3.86	0.58	10.21	m	14.07
35mm dia.	4.82	5.85	0.65	11.44	m	17.30
42mm dia.	6.12	7.43	0.71	12.50	m	19.93
54mm dia.	8.51	10.33	0.80	14.08	m	24.41
Grade 316 satin finish						
15mm dia.	2.15	2.61	0.61	10.74	m	13.35
22mm dia.	3.11	3.78	0.76	13.39	m	17.16
28mm dia.	4.75	5.77	0.87	15.32	m	21.09
35mm dia.	5.60	6.80	0.98	17.26	m	24.06
42mm dia.	7.19	8.73	1.06	18.66	m	27.39
54mm dia.	9.77	11.86	1.16	20.42	m	32.28
Extra Over stainless steel pipes; capillary fittings						
Straight coupling						
15mm dia.	0.70	0.85	0.25	4.40	nr	5.25
22mm dia.	1.14	1.38	0.28	4.93	nr	6.31
28mm dia.	1.49	1.81	0.33	5.81	nr	7.62
35mm dia.	3.45	4.19	0.37	6.51	nr	10.70
42mm dia.	3.98	4.83	0.42	7.39	nr	12.23
54mm dia.	5.98	7.26	0.45	7.92	nr	15.18
45 degree bend						
15mm dia.	2.74	3.33	0.25	4.40	nr	7.73
22mm dia.	4.05	4.92	0.30	5.22	nr	10.14
28mm dia.	5.86	7.12	0.33	5.81	nr	12.92
35mm dia.	8.10	9.84	0.37	6.51	nr	16.35
42mm dia.	10.75	13.05	0.42	7.39	nr	20.45
54mm dia.	15.11	18.35	0.45	7.92	nr	26.27
90 degree bend						
15mm dia.	1.94	2.36	0.28	4.93	nr	7.28
22mm dia.	2.63	3.19	0.28	4.93	nr	8.12
28mm dia.	3.71	4.51	0.33	5.81	nr	10.31
35mm dia.	12.29	14.92	0.37	6.51	nr	21.44
42mm dia.	16.91	20.53	0.42	7.39	nr	27.93
54mm dia.	22.93	27.84	0.45	7.92	nr	35.76

Material Costs/Measured Work Prices - Mechanical Installations

S:PIPED SUPPLY SYSTEMS

Item	Net Price £	Material £	Labour hours	Labour £	Unit	Total rate £
Reducer						
22 x 15mm dia.	4.48	5.44	0.28	4.93	nr	**10.37**
28 x 22mm dia.	4.82	5.85	0.33	5.81	nr	**11.66**
35 x 28mm dia.	6.12	7.43	0.37	6.51	nr	**13.95**
42 x 35mm dia.	6.60	8.01	0.42	7.39	nr	**15.41**
54 x 42mm dia.	19.57	23.76	0.48	8.46	nr	**32.23**
Tap connector						
15mm dia.	9.48	11.51	0.13	2.29	nr	**13.80**
22mm dia.	12.51	15.19	0.14	2.46	nr	**17.66**
28mm dia.	17.31	21.02	0.17	2.99	nr	**24.01**
Tank connector						
15mm dia.	12.24	14.86	0.13	2.29	nr	**17.15**
22mm dia.	18.21	22.11	0.13	2.29	nr	**24.40**
28mm dia	25.30	30.72	-	-	nr	**30.72**
35mm dia.	32.40	39.34	0.18	3.17	nr	**42.51**
42mm dia.	42.79	51.96	0.21	3.70	nr	**55.66**
54mm dia.	54.49	66.17	0.24	4.22	nr	**70.39**

Material Costs/Measured Work Prices - Mechanical Installations

S:PIPED SUPPLY SYSTEMS

Item	Net Price £	Material £	Labour hours	Labour £	Unit	Total rate £
S10 : COLD WATER						
Y10 - PIPELINES						
STAINLESS STEEL (Continued)						
Tee equal						
15mm dia.	3.49	4.24	0.37	6.51	nr	**10.75**
22mm dia.	4.34	5.27	0.40	7.04	nr	**12.31**
28mm dia.	5.24	6.36	0.45	7.92	nr	**14.28**
35mm dia.	12.60	15.30	0.59	10.39	nr	**25.69**
42mm dia.	15.53	18.86	0.62	10.92	nr	**29.78**
54mm dia.	31.39	38.12	0.67	11.80	nr	**49.91**
Unequal tee						
22 x 15mm dia.	7.07	8.59	0.37	6.51	nr	**15.10**
28 x 15mm dia.	7.96	9.67	0.45	7.92	nr	**17.59**
28 x 22mm dia.	7.96	9.67	0.45	7.92	nr	**17.59**
35 x 22mm dia.	13.90	16.88	0.59	10.39	nr	**27.27**
35 x 28mm dia.	13.90	16.88	0.59	10.39	nr	**27.27**
42 x 28mm dia.	17.11	20.78	0.62	10.92	nr	**31.69**
42 x 35mm dia.	17.11	20.78	0.62	10.92	nr	**31.69**
54 x 35mm dia.	35.41	43.00	0.67	11.80	nr	**54.79**
54 x 42mm dia.	35.41	43.00	0.67	11.80	nr	**54.79**
Union, conical seat						
15mm dia.	15.69	19.05	0.25	4.40	nr	**23.45**
22mm dia.	24.70	29.99	0.28	4.93	nr	**34.92**
28mm dia.	31.93	38.77	0.33	5.81	nr	**44.58**
35mm dia.	41.91	50.89	0.37	6.51	nr	**57.41**
42mm dia.	52.86	64.19	0.42	7.39	nr	**71.58**
54mm dia.	72.59	88.15	0.45	7.92	nr	**96.07**
Union, flat seat						
15mm dia.	16.38	19.89	0.25	4.40	nr	**24.29**
22mm dia.	25.57	31.05	0.28	4.93	nr	**35.98**
28mm dia.	32.96	40.02	0.33	5.81	nr	**45.83**
35mm dia.	43.06	52.29	0.37	6.51	nr	**58.80**
42mm dia.	54.25	65.88	0.42	7.39	nr	**73.27**
54mm dia.	72.76	88.35	0.45	7.92	nr	**96.27**
Extra Over stainless steel pipes; compression fittings						
Straight coupling						
15mm dia.	10.64	12.92	0.18	3.17	nr	**16.09**
22mm dia.	20.26	24.60	0.22	3.87	nr	**28.47**
28mm dia.	27.27	33.11	0.25	4.40	nr	**37.51**
35mm dia.	42.11	51.13	0.30	5.28	nr	**56.41**
42mm dia.	49.15	59.68	0.40	7.04	nr	**66.72**

Material Costs/Measured Work Prices - Mechanical Installations   135

S:PIPED SUPPLY SYSTEMS

Item	Net Price £	Material £	Labour hours	Labour £	Unit	Total rate £
90 degree bend						
15mm dia.	13.41	16.28	0.18	3.17	nr	**19.45**
22mm dia.	26.63	32.34	0.22	3.87	nr	**36.21**
28mm dia.	36.33	44.12	0.25	4.40	nr	**48.52**
35mm dia.	73.54	89.30	0.33	5.81	nr	**95.11**
42mm dia.	107.49	130.53	0.35	6.16	nr	**136.69**
Reducer						
22 x 15mm dia.	14.48	17.58	0.28	4.93	nr	**22.51**
28 x 22mm dia.	20.89	25.37	0.28	4.93	nr	**30.29**
35 x 28mm dia.	36.13	43.87	0.30	5.28	nr	**49.15**
42 x 35mm dia.	41.06	49.86	0.37	6.51	nr	**56.37**
Stud coupling						
15mm dia.	9.25	11.23	0.42	7.39	nr	**18.62**
22mm dia.	14.98	18.19	0.25	4.40	nr	**22.59**
28mm dia.	24.18	29.36	0.25	4.40	nr	**33.76**
35mm dia.	33.88	41.14	0.37	6.51	nr	**47.65**
42mm dia.	43.85	53.25	0.42	7.39	nr	**60.64**
Equal tee						
15mm dia.	18.88	22.93	0.37	6.51	nr	**29.44**
22mm dia.	39.01	47.37	0.40	7.04	nr	**54.41**
28mm dia.	53.33	64.76	0.45	7.92	nr	**72.68**
35mm dia.	106.07	128.80	0.59	10.39	nr	**139.19**
42mm dia.	147.07	178.59	0.62	10.92	nr	**189.50**
Running tee						
15mm dia.	23.24	28.22	0.37	6.51	nr	**34.73**
22mm dia.	41.86	50.83	0.40	7.04	nr	**57.87**
28mm dia.	70.87	86.06	0.59	10.39	nr	**96.45**
PRESS FIT						
Press fit jointing system; butyl rubber O ring mechanical joint						
Pipework						
15mm dia	1.37	1.66	0.46	8.10	m	**9.76**
22mm dia	2.64	3.21	0.48	8.45	m	**11.65**
28mm dia	3.38	4.10	0.52	9.15	m	**13.26**
35mm dia	6.43	7.81	0.56	9.86	m	**17.66**
42mm dia	7.74	9.40	0.58	10.21	m	**19.61**
54mm dia	9.80	11.90	0.66	11.62	m	**23.52**
Coupling						
15mm dia	0.24	0.29	0.36	6.34	nr	**6.63**
22mm dia	0.62	0.75	0.36	6.34	nr	**7.09**
28mm dia	1.25	1.52	0.44	7.74	nr	**9.26**
35mm dia	2.02	2.45	0.44	7.74	nr	**10.20**
42mm dia	3.15	3.83	0.52	9.15	nr	**12.98**
54mm dia	5.27	6.40	0.60	10.56	nr	**16.96**

Material Costs/Measured Work Prices - Mechanical Installations

S:PIPED SUPPLY SYSTEMS

Item	Net Price £	Material £	Labour hours	Labour £	Unit	Total rate £
S10 : COLD WATER						
Y10 - PIPELINES						
STAINLESS STEEL (Continued)						
Stop end						
22mm dia	1.37	1.66	0.18	3.17	nr	**4.83**
28mm dia	2.28	2.77	0.22	3.87	nr	**6.64**
35mm dia	4.41	5.36	0.22	3.87	nr	**9.23**
42mm dia	7.22	8.77	0.26	4.58	nr	**13.34**
54mm dia	9.55	11.60	0.30	5.28	nr	**16.88**
Reducer						
22 x 15mm dia	0.86	1.04	0.36	6.34	nr	**7.38**
28 x 15mm dia	2.83	3.44	0.40	7.04	nr	**10.48**
28 x 22mm dia	2.16	2.62	0.40	7.04	nr	**9.66**
35 x 22mm dia	3.65	4.43	0.40	7.04	nr	**11.47**
35 x 28mm dia	2.84	3.45	0.44	7.74	nr	**11.19**
42 x 35mm dia	5.10	6.19	0.48	8.45	nr	**14.64**
54 x 42mm dia	8.79	10.67	0.56	9.86	nr	**20.53**
90 degree bend						
15mm dia	0.47	0.57	0.36	6.34	nr	**6.91**
22mm dia	0.90	1.09	0.36	6.34	nr	**7.43**
28mm dia	1.93	2.34	0.44	7.74	nr	**10.09**
35mm dia	3.94	4.78	0.44	7.74	nr	**12.53**
42mm dia	6.52	7.92	0.52	9.15	nr	**17.07**
54mm dia	12.10	14.69	0.60	10.56	nr	**25.25**
45 degree bend						
15mm dia	0.71	0.86	0.36	6.34	nr	**7.20**
22mm dia	1.23	1.49	0.36	6.34	nr	**7.83**
28mm dia	2.24	2.72	0.44	7.74	nr	**10.46**
35mm dia	5.54	6.73	0.44	7.74	nr	**14.47**
42mm dia	10.38	12.60	0.52	9.15	nr	**21.76**
54mm dia	18.79	22.82	0.60	10.56	nr	**33.38**
Equal tee						
15mm dia	0.68	0.83	0.54	9.50	nr	**10.33**
22mm dia	1.49	1.81	0.54	9.50	nr	**11.31**
28mm dia	3.08	3.74	0.66	11.62	nr	**15.36**
35mm dia	6.28	7.63	0.66	11.62	nr	**19.24**
42mm dia	10.60	12.87	0.78	13.73	nr	**26.60**
54mm dia	18.33	22.26	0.90	15.84	nr	**38.10**
Reducing tee						
22 x 15mm dia	1.69	2.05	0.54	9.50	nr	**11.56**
28 x 15mm dia	4.88	5.93	0.62	10.91	nr	**16.84**
28 x 22mm dia	4.69	5.70	0.62	10.91	nr	**16.61**
35 x 22mm dia	6.87	8.34	0.62	10.91	nr	**19.25**
35 x 28mm dia	7.99	9.70	0.62	10.91	nr	**20.61**
42 x 28mm dia	15.06	18.29	0.70	12.32	nr	**30.61**
42 x 35mm dia	17.15	20.83	0.70	12.32	nr	**33.15**
54 x 35mm dia	27.54	33.44	0.82	14.43	nr	**47.87**
54 x 42mm dia	27.23	33.07	0.82	14.43	nr	**47.50**

Material Costs/Measured Work Prices - Mechanical Installations   137

S:PIPED SUPPLY SYSTEMS

Item	Net Price £	Material £	Labour hours	Labour £	Unit	Total rate £
FIXINGS						
For copper pipes						
Saddle band						
6mm dia.	0.05	0.06	0.11	1.94	nr	**2.00**
8mm dia.	0.05	0.06	0.12	2.11	nr	**2.17**
10mm dia.	0.05	0.06	0.12	2.11	nr	**2.17**
12mm dia.	0.05	0.06	0.12	2.11	nr	**2.17**
15mm dia.	0.06	0.07	0.13	2.29	nr	**2.36**
22mm dia.	0.06	0.07	0.13	2.29	nr	**2.36**
28mm dia.	0.07	0.09	0.16	2.82	nr	**2.90**
35mm dia.	0.10	0.12	0.18	3.17	nr	**3.29**
42mm dia.	0.22	0.27	0.21	3.70	nr	**3.96**
54mm dia.	0.30	0.36	0.21	3.70	nr	**4.06**
Single spacing clip						
15mm dia.	0.07	0.09	0.14	2.46	nr	**2.55**
22mm dia.	0.07	0.09	0.15	2.64	nr	**2.73**
28mm dia.	0.16	0.19	0.17	2.99	nr	**3.19**
Two piece spacing clip						
8mm dia. Bottom	0.05	0.06	0.11	1.94	nr	**2.00**
8mm dia. Top	0.05	0.06	0.11	1.94	nr	**2.00**
12mm dia. Bottom	0.05	0.06	0.13	2.29	nr	**2.35**
12mm dia. Top	0.05	0.06	0.13	2.29	nr	**2.35**
15mm dia. Bottom	0.06	0.07	0.13	2.29	nr	**2.36**
15mm dia. Top	0.06	0.07	0.13	2.29	nr	**2.36**
22mm dia. Bottom	0.06	0.07	0.14	2.46	nr	**2.54**
22mm dia. Top	0.06	0.07	0.14	2.46	nr	**2.54**
28mm dia. Bottom	0.06	0.07	0.16	2.82	nr	**2.89**
28mm dia. Top	0.10	0.12	0.16	2.82	nr	**2.94**
35mm dia. Bottom	0.07	0.09	0.21	3.70	nr	**3.78**
35mm dia. Top	0.15	0.18	0.21	3.70	nr	**3.88**
42mm dia. Bottom	0.16	0.19	0.21	3.70	nr	**3.89**
42mm dia. Top	0.27	0.33	0.21	3.70	nr	**4.02**
54mm dia. Bottom	0.19	0.23	0.21	3.70	nr	**3.93**
54mm dia. Top	0.36	0.44	0.21	3.70	nr	**4.13**
Single pipe bracket						
15mm dia.	0.59	0.72	0.14	2.46	nr	**3.18**
22mm dia.	0.67	0.81	0.14	2.46	nr	**3.28**
28mm dia.	0.80	0.97	0.17	2.99	nr	**3.96**
Single pipe bracket for building in						
22mm dia.	2.28	2.77	0.06	1.06	nr	**3.82**
28mm dia.	2.59	3.15	0.06	1.06	nr	**4.20**

Material Costs/Measured Work Prices - Mechanical Installations

S:PIPED SUPPLY SYSTEMS

Item	Net Price £	Material £	Labour hours	Labour £	Unit	Total rate £
S10 : COLD WATER						
Y10 - PIPELINES						
FIXINGS (Continued)						
Single pipe ring						
15mm dia.	0.98	1.19	0.26	4.58	nr	5.77
22mm dia.	1.04	1.26	0.26	4.58	nr	5.84
28mm dia.	1.25	1.52	0.31	5.46	nr	6.97
35mm dia.	1.33	1.61	0.32	5.63	nr	7.25
42mm dia.	1.46	1.77	0.32	5.63	nr	7.41
54mm dia.	1.75	2.13	0.34	5.98	nr	8.11
67mm dia.	4.09	4.97	0.35	6.16	nr	11.13
76mm dia.	5.15	6.25	0.42	7.39	nr	13.65
108mm dia.	7.98	9.69	0.42	7.39	nr	17.08
Double pipe ring						
15mm dia.	1.15	1.40	0.26	4.58	nr	5.97
22mm dia.	1.23	1.49	0.26	4.58	nr	6.07
28mm dia.	1.66	2.02	0.31	5.46	nr	7.47
35mm dia.	1.78	2.16	0.32	5.63	nr	7.79
42mm dia.	1.95	2.37	0.32	5.63	nr	8.00
54mm dia.	2.48	3.01	0.34	5.98	nr	9.00
67mm dia.	4.52	5.49	0.35	6.16	nr	11.65
76mm dia.	6.55	7.95	0.42	7.39	nr	15.35
108mm dia.	12.39	15.05	0.42	7.39	nr	22.44
Wall bracket						
15mm dia.	1.16	1.41	0.05	0.88	nr	2.29
22mm dia.	1.36	1.65	0.05	0.88	nr	2.53
28mm dia.	1.66	2.02	0.05	0.88	nr	2.90
35mm dia.	2.06	2.50	0.05	0.88	nr	3.38
42mm dia.	2.78	3.38	0.05	0.88	nr	4.26
54mm dia.	3.54	4.30	0.05	0.88	nr	5.18
Hospital bracket						
15mm dia.	1.54	1.87	0.26	4.58	nr	6.45
22mm dia.	1.63	1.98	0.26	4.58	nr	6.56
28mm dia.	2.00	2.43	0.31	5.46	nr	7.88
35mm dia.	2.18	2.65	0.32	5.63	nr	8.28
42mm dia.	3.34	4.06	0.32	5.63	nr	9.69
54mm dia.	4.40	5.34	0.34	5.98	nr	11.33
Screw on backplate, female						
15mm dia.	0.70	0.85	0.26	4.58	nr	5.43
22mm dia.	0.70	0.85	0.26	4.58	nr	5.43
28mm dia.	0.80	0.97	0.31	5.46	nr	6.43
35mm dia.	0.83	1.01	0.32	5.63	nr	6.64
42mm dia.	1.12	1.36	0.32	5.63	nr	6.99
54mm dia.	1.26	1.53	0.34	5.98	nr	7.51
67mm dia.	1.26	1.53	0.35	6.16	nr	7.69
76mm dia.	1.75	2.13	0.42	7.39	nr	9.52
108mm dia.	1.75	2.13	0.42	7.39	nr	9.52

Material Costs/Measured Work Prices - Mechanical Installations   139

S:PIPED SUPPLY SYSTEMS

Item	Net Price £	Material £	Labour hours	Labour £	Unit	Total rate £
Screw on backplate, male						
15mm dia.	0.64	0.78	0.26	4.58	nr	**5.35**
22mm dia.	0.64	0.78	0.26	4.58	nr	**5.35**
28mm dia.	0.72	0.87	0.31	5.46	nr	**6.33**
35mm dia.	0.76	0.92	0.32	5.63	nr	**6.56**
42mm dia.	0.99	1.20	0.32	5.63	nr	**6.83**
54mm dia.	1.14	1.38	0.34	5.98	nr	**7.37**
67mm dia.	1.14	1.38	0.35	6.16	nr	**7.54**
76mm dia.	1.72	2.09	0.42	7.39	nr	**9.48**
108mm dia.	1.72	2.09	0.42	7.39	nr	**9.48**
Pipe joist clips, single						
15mm dia.	0.47	0.57	0.08	1.41	nr	**1.98**
22mm dia.	0.47	0.57	0.08	1.41	nr	**1.98**
Pipe joist clips, double						
15mm dia.	0.66	0.80	0.08	1.41	nr	**2.21**
22mm dia.	0.66	0.80	0.08	1.41	nr	**2.21**

Material Costs/Measured Work Prices - Mechanical Installations

S:PIPED SUPPLY SYSTEMS

Item	Net Price £	Material £	Labour hours	Labour £	Unit	Total rate £
S10 : COLD WATER						
Y11 - PIPELINE ANCILLAIRES						
VALVES						
Regulators						
Gunmetal; self-acting two port thermostat; single seat; screwed; normally closed; with adjustable or fixed bleed device						
25mm dia.	243.90	296.17	1.46	25.70	nr	321.87
32mm dia.	251.32	305.18	1.45	25.52	nr	330.70
40mm dia.	268.62	326.19	1.55	27.29	nr	353.48
50mm dia.	323.01	392.23	1.68	29.57	nr	421.80
Self acting temperature regulator for storage calorifier; integral sensing element and pocket; screwed ends						
15mm dia.	269.20	326.89	1.32	23.23	nr	350.12
25mm dia.	295.49	358.81	1.52	26.75	nr	385.56
32mm dia.	381.59	463.37	1.79	31.51	nr	494.88
40mm dia.	466.80	566.83	1.99	35.03	nr	601.86
50mm dia.	545.65	662.59	2.26	39.78	nr	702.36
Self acting temperature regulator for storage calorifier; integral sensing element and pocket; flanged ends; bolted connection						
15mm dia.	395.19	479.88	0.61	10.74	nr	490.62
25mm dia.	452.29	549.22	0.72	12.67	nr	561.89
32mm dia.	570.13	692.30	0.94	16.54	nr	708.85
40mm dia.	675.27	819.98	1.03	18.13	nr	838.11
50mm dia.	784.04	952.05	1.18	20.77	nr	972.82
Chrome plated thermostatic mixing valves including non-return valves and inlet swivel connections with strainers; copper compression fittings						
15mm dia.	99.70	121.07	0.69	12.14	nr	133.22
Chrome plated thermostatic mixing valves including non-return valves and inlet swivel connections with angle pattern combined isolating valves and strainers; copper compression fittings						
15mm dia.	104.24	126.57	0.69	12.14	nr	138.72
Gunmetal thermostatic mixing valves including non-return valves and inlet swivel connections with strainers; copper compression fittings						
15mm dia.	90.52	109.91	0.69	12.14	nr	122.06

Material Costs/Measured Work Prices - Mechanical Installations

S:PIPED SUPPLY SYSTEMS

Item	Net Price £	Material £	Labour hours	Labour £	Unit	Total rate £
Gunmetal thermostatic mixing valves including non-return valves and inlet swivel connections with angle pattern combined isolating valves and strainers; copper compression fittings						
15mm dia.	95.17	115.57	0.69	12.14	nr	**127.71**
Ball float valves						
Bronze, equilibrium; copper float; working pressure cold services up to 16 bar; flanged ends; BS 4504 Table 16/21; bolted connections						
25mm dia.	261.72	317.81	1.04	18.30	nr	**336.11**
32mm dia.	363.09	440.90	1.22	21.47	nr	**462.37**
40mm dia.	413.42	502.02	1.38	24.29	nr	**526.30**
50mm dia.	679.51	825.13	1.66	29.22	nr	**854.35**
65mm dia.	804.84	977.32	1.93	33.97	nr	**1011.29**
80mm dia.	985.39	1196.56	2.16	38.02	nr	**1234.58**
Heavy, equilibrium; with long tail and backnut; copper float; screwed for iron						
25mm dia.	79.31	96.31	1.58	27.81	nr	**124.12**
32mm dia.	152.16	184.77	1.78	31.33	nr	**216.10**
40mm dia.	158.39	192.33	1.90	33.44	nr	**225.77**
50mm dia.	262.52	318.78	2.65	46.64	nr	**365.42**
Brass, ball valve; BS 1212; copper float; screwed						
15mm dia.	7.16	8.69	0.25	4.40	nr	**13.09**
22mm dia	12.07	14.66	-	-	nr	**14.66**
28mm dia	45.45	55.19	-	-	nr	**55.19**
Gate valves						
DZR copper alloy wedge non-rising stem; capillary joint to copper						
15mm dia.	5.27	6.40	0.84	14.78	nr	**21.18**
22mm dia.	6.22	7.55	1.01	17.78	nr	**25.33**
28mm dia.	8.61	10.46	1.19	20.95	nr	**31.40**
35mm dia.	12.91	15.68	1.38	24.29	nr	**39.97**
42mm dia.	18.17	22.06	1.62	28.51	nr	**50.58**
54mm dia.	25.17	30.56	1.94	34.15	nr	**64.71**
Cocks; capillary joints to copper						
Stopcock; brass head with gun metal body						
15mm dia.	2.49	3.02	0.45	7.92	nr	**10.94**
22mm dia.	4.66	5.66	0.46	8.10	nr	**13.75**
28mm dia.	13.26	16.10	0.54	9.50	nr	**25.61**
Lockshield stop cocks; brass head with gun metal body						
15mm dia.	6.12	7.43	0.45	7.92	nr	**15.35**
22mm dia.	8.80	10.69	0.46	8.10	nr	**18.78**
28mm dia.	15.58	18.92	0.54	9.50	nr	**28.42**

Material Costs/Measured Work Prices - Mechanical Installations

S:PIPED SUPPLY SYSTEMS

Item	Net Price £	Material £	Labour hours	Labour £	Unit	Total rate £
S10 : COLD WATER						
Y11 - PIPELINE ANCILLAIRES						
VALVES (Continued)						
DZR stopcock; brass head with gun metal body						
15mm dia.	6.26	7.60	0.45	7.92	nr	**15.52**
22mm dia.	10.87	13.20	0.46	8.10	nr	**21.30**
28mm dia.	18.10	21.98	0.54	9.50	nr	**31.48**
Gunmetal stopcock						
35mm dia.	28.39	34.47	0.69	12.14	nr	**46.62**
42mm dia.	37.71	45.79	0.71	12.50	nr	**58.29**
54mm dia.	56.32	68.39	0.81	14.26	nr	**82.65**
Lockshield DZR stopcock; brass head gun metal body						
15mm dia.	8.27	10.04	0.45	7.92	nr	**17.96**
22mm dia.	12.29	14.92	0.46	8.10	nr	**23.02**
28mm dia.	20.44	24.82	0.54	9.50	nr	**34.32**
Double union stopcock						
15mm dia.	8.79	10.67	0.60	10.56	nr	**21.23**
22mm dia.	12.26	14.89	0.60	10.56	nr	**25.45**
28mm dia.	21.65	26.29	0.69	12.14	nr	**38.43**
Double union lockshield stopcock						
15mm dia.	8.53	10.36	0.60	10.56	nr	**20.92**
22mm dia.	11.97	14.54	0.61	10.74	nr	**25.27**
28mm dia.	21.33	25.90	0.69	12.14	nr	**38.05**
Double union DZR stopcock						
15mm dia.	13.19	16.02	0.60	10.56	nr	**26.58**
22mm dia.	16.22	19.70	0.61	10.74	nr	**30.43**
28mm dia.	30.00	36.43	0.69	12.14	nr	**48.57**
Double union lockshield DZR stopcock						
15mm dia.	13.97	16.96	0.60	10.56	nr	**27.52**
22mm dia.	17.52	21.27	0.61	10.74	nr	**32.01**
28mm dia.	32.33	39.26	0.69	12.14	nr	**51.40**
Double union gun metal stopcock						
35mm dia.	49.99	60.70	0.63	11.09	nr	**71.79**
42mm dia.	68.61	83.31	0.67	11.79	nr	**95.11**
54mm dia.	107.93	131.06	0.85	14.96	nr	**146.02**
Stopcock with easy clean cover						
15mm dia.	6.43	7.81	0.60	10.56	nr	**18.37**
22mm dia.	9.12	11.07	0.60	10.56	nr	**21.63**
28mm dia.	16.04	19.48	0.66	11.62	nr	**31.09**
Union stopcock with easy clean cover						
15mm dia.	7.82	9.50	0.60	10.56	nr	**20.06**
22mm dia.	11.96	14.52	0.61	10.74	nr	**25.26**
28mm dia.	18.94	23.00	0.69	12.14	nr	**35.14**

Material Costs/Measured Work Prices - Mechanical Installations

S:PIPED SUPPLY SYSTEMS

Item	Net Price £	Material £	Labour hours	Labour £	Unit	Total rate £
Double union stopcock with easy clean cover						
15mm dia.	9.37	11.38	0.60	10.56	nr	**21.94**
22mm dia.	13.00	15.79	0.61	10.74	nr	**26.52**
28mm dia.	22.82	27.71	0.69	12.14	nr	**39.85**
Combined stopcock and drain						
15mm dia.	13.90	16.88	0.67	11.79	nr	**28.67**
22mm dia.	17.08	20.74	0.68	11.97	nr	**32.71**
Combined DZR stopcock and drain						
15mm dia.	18.36	22.29	0.67	11.79	nr	**34.09**
Gate valve						
DZR copper alloy wedge non-rising stem; compression joint to copper						
15mm dia.	5.27	6.40	0.84	14.78	nr	**21.18**
22mm dia.	6.22	7.55	1.01	17.78	nr	**25.33**
28mm dia.	8.61	10.46	1.19	20.95	nr	**31.40**
35mm dia.	12.91	15.68	1.38	24.29	nr	**39.97**
42mm dia.	18.17	22.06	1.62	28.51	nr	**50.58**
54mm dia.	25.17	30.56	1.94	34.15	nr	**64.71**
Cocks; compression joints to copper						
Stopcock; brass head gun metal body						
15mm dia.	3.24	3.93	0.42	7.39	nr	**11.33**
22mm dia.	5.72	6.95	0.42	7.39	nr	**14.34**
28mm dia.	14.89	18.08	0.45	7.92	nr	**26.00**
Lockshield stopcock; brass head gun metal body						
15mm dia.	7.07	8.59	0.42	7.39	nr	**15.98**
22mm dia.	9.97	12.11	0.42	7.39	nr	**19.50**
28mm dia.	19.18	23.29	0.45	7.92	nr	**31.21**
DZR Stopcock						
15mm dia.	7.57	9.19	0.38	6.69	nr	**15.88**
22mm dia.	12.43	15.09	0.39	6.86	nr	**21.96**
28mm dia.	20.57	24.98	0.40	7.04	nr	**32.02**
35mm dia.	37.86	45.97	0.52	9.15	nr	**55.13**
42mm dia.	54.21	65.83	0.54	9.50	nr	**75.33**
54mm dia.	73.83	89.65	0.63	11.09	nr	**100.74**
DZR Lockshield stopcock						
15mm dia.	8.70	10.56	0.38	6.69	nr	**17.25**
22mm dia.	13.64	16.56	0.39	6.86	nr	**23.43**
Stopcock with easy clean cover						
15mm dia.	7.82	9.50	0.42	7.39	nr	**16.89**
22mm dia.	10.97	13.32	0.42	7.39	nr	**20.71**
28mm dia.	19.68	23.90	0.45	7.92	nr	**31.82**

Material Costs/Measured Work Prices - Mechanical Installations

S:PIPED SUPPLY SYSTEMS

Item	Net Price £	Material £	Labour hours	Labour £	Unit	Total rate £
S10 : COLD WATER						
Y11 - PIPELINE ANCILLAIRES						
VALVES (Continued)						
Combined stop/draincock						
15mm dia.	10.92	13.26	0.22	3.87	nr	**17.13**
22mm dia.	14.07	17.09	0.45	7.92	nr	**25.01**
DZR Combined stop/draincock						
15mm dia.	20.63	25.05	0.41	7.22	nr	**32.27**
22mm dia.	19.65	23.86	0.42	7.39	nr	**31.25**
Stopcock to polyethylene						
15mm dia.	6.94	8.43	0.38	6.69	nr	**15.12**
20mm dia.	11.13	13.52	0.39	6.86	nr	**20.38**
25mm dia.	14.50	17.61	0.40	7.04	nr	**24.65**
Draw off coupling						
15mm dia.	4.72	5.73	0.38	6.69	nr	**12.42**
DZR Draw off coupling						
15mm dia.	6.50	7.89	0.38	6.69	nr	**14.58**
22mm dia.	7.53	9.14	0.39	6.86	nr	**16.01**
Draw off elbow						
15mm dia.	4.82	5.85	0.38	6.69	nr	**12.54**
22mm dia.	5.53	6.72	0.39	6.86	nr	**13.58**
Lockshield drain cock						
15mm dia.	3.31	4.02	0.41	7.22	nr	**11.24**
Check valves						
DZR copper alloy and bronze, WRc approved cartridge double check valve; BS 6282; working pressure cold services up to 10 bar at 65 degrees Celsius; screwed ends						
32mm dia.	47.60	57.80	1.38	24.29	nr	**82.09**
40mm dia.	64.44	78.25	1.62	28.51	nr	**106.76**
50mm dia.	87.07	105.73	1.94	34.15	nr	**139.87**

Material Costs/Measured Work Prices - Mechanical Installations 145

S:PIPED SUPPLY SYSTEMS

Item	Net Price £	Material £	Labour hours	Labour £	Unit	Total rate £
S10 : COLD WATER						
Y20 - PUMPS						
Pressurised cold water supply set; packaged; fully automatic; pumps with 3 phase motors; membrane tank; valves; control panel; inter connecting pipework; fixed on steel frame; includes fixing; electrical work elsewhere						
Cold water supply set						
3 litre/sec maximum delivery; 600 kN/m2 maximum head	4835.85	5872.17	10.38	182.70	nr	**6054.87**
7 litre/sec maximum delivery; 600 kN/m2 maximum head	9239.87	11219.97	10.38	182.70	nr	**11402.67**
16 litre/sec maximum delivery; 600 kN/m2maximum head	15360.08	18651.75	10.38	182.70	nr	**18834.44**
Automatic sump pump; self contained, totally enclosed electric motor, float switch and gear; includes fixing; electrical work elsewhere						
Small portable						
Single phase	104.91	127.39	1.83	32.21	nr	**159.60**
Fixed installation						
Single phase	1118.27	1357.92	3.23	56.85	nr	**1414.77**
Three phase	342.70	416.14	2.00	35.20	nr	**451.34**

Material Costs/Measured Work Prices - Mechanical Installations

S:PIPED SUPPLY SYSTEMS

Item	Net Price £	Material £	Labour hours	Labour £	Unit	Total rate £
S10 : COLD WATER						
Y21 - TANKS						
Cisterns; fibreglass; complete with ballvalve, fixing plate and fitted covers						
Rectangular						
70 litres capacity	131.17	159.28	1.33	23.41	nr	**182.69**
110 litres capacity	148.57	180.41	1.40	24.64	nr	**205.05**
170 litres capacity	163.31	198.31	1.61	28.34	nr	**226.64**
280 litres capacity	264.11	320.71	1.61	28.34	nr	**349.05**
420 litres capacity	344.28	418.06	1.99	35.03	nr	**453.08**
710 litres capacity	469.76	570.43	3.31	58.26	nr	**628.69**
840 litres capacity	542.04	658.20	3.60	63.36	nr	**721.56**
1590 litres capacity	817.46	992.64	13.32	234.44	nr	**1227.08**
2275 litres capacity	982.36	1192.88	20.18	355.18	nr	**1548.06**
3365 litres capacity	1648.26	2001.48	24.50	431.22	nr	**2432.70**
4545 litres capacity	1758.47	2135.31	29.91	526.44	nr	**2661.75**
Cisterns; polypropylene; complete with ball valve, fixing plate and cover; includes hoisting and placing in position						
Rectangular						
18 litres capacity	3.65	4.43	1.00	17.60	nr	**22.03**
68 litres capacity	16.34	19.85	1.00	17.60	nr	**37.45**
91 litres capacity	16.67	20.24	1.00	17.60	nr	**37.84**
114 litres capacity	22.15	26.89	1.00	17.60	nr	**44.49**
182 litres capacity	39.60	48.08	1.00	17.60	nr	**65.69**
227 litres capacity	39.91	48.47	1.00	17.60	nr	**66.07**
Circular						
114 litres capacity	18.61	22.60	1.00	17.60	nr	**40.20**
227 litres capacity	28.24	34.29	1.00	17.60	nr	**51.89**
318 litres capacity	75.79	92.03	1.00	17.60	nr	**109.63**
455 litres capacity	86.13	104.59	1.00	17.60	nr	**122.19**

Material Costs/Measured Work Prices - Mechanical Installations

S:PIPED SUPPLY SYSTEMS

Item	Net Price £	Material £	Labour hours	Labour £	Unit	Total rate £
S10 : COLD WATER						
Y50 -THERMAL INSULATION						
Flexible closed cell walled insulation; Class 1/Class O; adhesive joints; including around fittings						
6mm wall thickness						
15mm diameter	0.60	0.77	0.15	2.64	m	**3.41**
22mm diameter	0.70	0.91	0.15	2.64	m	**3.55**
28mm diameter	0.89	1.15	0.15	2.64	m	**3.79**
9mm wall thickness						
15mm diameter	0.63	0.81	0.15	2.64	m	**3.45**
22mm diameter	0.76	0.98	0.15	2.64	m	**3.62**
28mm diameter	0.84	1.08	0.15	2.64	m	**3.72**
35mm diameter	0.97	1.26	0.15	2.64	m	**3.90**
42mm diameter	1.13	1.46	0.15	2.64	m	**4.10**
54mm diameter	1.62	2.10	0.15	2.64	m	**4.74**
13mm wall thickness						
15mm diameter	0.82	1.06	0.15	2.64	m	**3.70**
22mm diameter	0.99	1.28	0.15	2.64	m	**3.92**
28mm diameter	1.21	1.57	0.15	2.64	m	**4.21**
35mm diameter	1.31	1.70	0.15	2.64	m	**4.34**
42mm diameter	1.56	2.02	0.15	2.64	m	**4.66**
54mm diameter	1.94	2.51	0.15	2.64	m	**5.15**
67mm diameter	3.06	3.96	0.15	2.64	m	**6.60**
76mm diameter	3.54	4.59	0.15	2.64	m	**7.23**
108mm diameter	5.19	6.72	0.15	2.64	m	**9.36**
19mm wall thickness						
15mm diameter	1.36	1.77	0.15	2.64	m	**4.41**
22mm diameter	1.66	2.15	0.15	2.64	m	**4.79**
28mm diameter	2.28	2.95	0.15	2.64	m	**5.59**
35mm diameter	2.66	3.44	0.15	2.64	m	**6.08**
42mm diameter	3.14	4.07	0.15	2.64	m	**6.71**
54mm diameter	4.00	5.18	0.15	2.64	m	**7.82**
67mm diameter	4.78	6.20	0.15	2.64	m	**8.84**
76mm diameter	5.50	7.13	0.22	3.87	m	**11.00**
108mm diameter	8.11	10.51	0.22	3.87	m	**14.38**
25mm wall thickness						
15mm diameter	2.71	3.51	0.15	2.64	m	**6.15**
22mm diameter	2.99	3.87	0.15	2.64	m	**6.51**
28mm diameter	3.39	4.39	0.15	2.64	m	**7.03**
35mm diameter	3.76	4.87	0.15	2.64	m	**7.51**
42mm diameter	4.01	5.20	0.15	2.64	m	**7.84**
54mm diameter	4.74	6.13	0.15	2.64	m	**8.77**
67mm diameter	5.78	7.48	0.15	2.64	m	**10.12**
76mm diameter	6.71	8.69	0.22	3.87	m	**12.56**

S:PIPED SUPPLY SYSTEMS

Item	Net Price £	Material £	Labour hours	Labour £	Unit	Total rate £
32mm wall thickness						
15mm diameter	3.43	4.44	0.15	2.64	m	**7.08**
22mm diameter	3.76	4.87	0.15	2.64	m	**7.51**
28mm diameter	4.38	5.67	0.15	2.64	m	**8.31**
35mm diameter	4.44	5.74	0.15	2.64	m	**8.38**
42mm diameter	5.16	6.68	0.15	2.64	m	**9.32**
54mm diameter	6.50	8.42	0.15	2.64	m	**11.06**
76mm diameter	9.72	12.59	0.22	3.87	m	**16.46**

For mineral fibre insulation rates see section T31 - Low Temperature Hot Water Heating

Material Costs/Measured Work Prices - Mechanical Installations

S:PIPED SUPPLY SYSTEMS

Item	Net Price £	Material £	Labour hours	Labour £	Unit	Total rate £
S11 - HOT WATER						
Y10 - PIPELINES						
For prices for pipework refer to Section						
S10 - Cold Water						
Y11 - PIPELINE ANCILLARIES						
For prices for ancillaries refer to Section						
S10 - Cold Water						
Y23 - STORAGE						
CYLINDERS/CALORIFIERS						
CYLINDERS						
Insulated copper storage cylinders; BS						
699; includes placing in position						
Grade 3 (maximum 10m working head)						
BS size 6; 115 litres capacity; 400mm dia.;						
1050mm height	61.65	74.86	1.50	26.43	nr	**101.29**
BS size 7; 120 litres capacity; 450mm dia.;						
900mm height	72.89	88.51	2.00	35.20	nr	**123.71**
BS size 8; 144 litres capacity; 450mm dia.;						
1050mm height	68.26	82.89	2.80	49.30	nr	**132.19**
Grade 4 (maximum 6m working head)						
BS size 2; 96 litres capacity; 400mm dia.; 900mm						
height	57.68	70.04	1.50	26.43	nr	**96.47**
BS size 7; 120 litres capacity; 450mm dia.;						
900mm height	47.73	57.96	1.50	26.43	nr	**84.39**
BS size 8; 144 litres capacity; 450mm dia.;						
1050mm height	59.88	72.71	1.50	26.43	nr	**99.14**
BS size 9; 166 litres capacity; 450mm dia.;						
1200mm height	87.56	106.32	1.50	26.43	nr	**132.75**
Storage cylinders; brazed copper						
construction; to BS 699; screwed						
bosses; includes placing in position						
Tested to 2.2 bar, 15m maximum head						
144 litres	273.11	331.64	3.00	52.86	nr	**384.49**
160 litres	308.74	374.90	3.00	52.86	nr	**427.76**
200 litres	318.24	386.44	3.76	66.17	nr	**452.61**
255 litres	362.18	439.80	3.76	66.17	nr	**505.96**
290 litres	489.23	594.07	3.76	66.17	nr	**660.24**
370 litres	560.48	680.59	4.50	79.28	nr	**759.87**
450 litres	763.49	927.11	5.00	88.00	nr	**1015.11**

150 **Material Costs/Measured Work Prices - Mechanical Installations**

S:PIPED SUPPLY SYSTEMS

Item	Net Price £	Material £	Labour hours	Labour £	Unit	Total rate £
S11 - HOT WATER						
Y23 - STORAGE CYLINDERS/CALORIFIERS CYLINDERS (Continued)						
Tested to 2.55 bar, 17m maximum head						
550 litres	826.71	1003.87	5.00	88.00	nr	**1091.88**
700 litres	966.51	1173.63	6.02	106.03	nr	**1279.66**
800 litres	1154.95	1402.46	6.54	115.04	nr	**1517.49**
900 litres	1209.66	1468.89	8.00	140.81	nr	**1609.70**
1000 litres	1276.53	1550.09	8.00	140.81	nr	**1690.90**
1250 litres	1398.10	1697.71	13.16	231.59	nr	**1929.30**
1500 litres	2127.54	2583.47	15.15	266.68	nr	**2850.15**
2000 litres	2553.23	3100.39	17.24	303.46	nr	**3403.85**
3000 litres	3586.43	4355.00	24.39	429.29	nr	**4784.29**
Indirect cylinders; copper; bolted top; up to 5 tappings for connections; BS 1586; includes placing in position						
Grade 3, tested to 1.45 bar, 10m maximum head						
74 litres capacity	124.40	151.06	1.50	26.43	nr	**177.49**
96 litres capacity	126.66	153.80	1.50	26.43	nr	**180.23**
114 litres capacity	130.06	157.93	1.50	26.43	nr	**184.36**
117 litres capacity	134.98	163.91	2.00	35.20	nr	**199.11**
140 litres capacity	139.10	168.91	2.50	44.00	nr	**212.91**
162 litres capacity	194.52	236.21	3.00	52.86	nr	**289.06**
190 litres capacity	212.61	258.17	3.51	61.76	nr	**319.93**
245 litres capacity	248.81	302.13	3.80	66.92	nr	**369.05**
280 litres capacity	441.06	535.58	4.00	70.40	nr	**605.98**
360 litres capacity	477.25	579.52	4.50	79.28	nr	**658.81**
440 litres capacity	554.15	672.90	4.50	79.28	nr	**752.19**
Grade 2, tested to 2.2 bar, 15m maximum head						
117 litres capacity	179.81	218.34	2.00	35.20	nr	**253.54**
140 litres capacity	195.65	237.58	2.50	44.00	nr	**281.58**
162 litres capacity	223.92	271.91	2.80	49.30	nr	**321.21**
190 litres capacity	260.12	315.86	3.00	52.86	nr	**368.72**
245 litres capacity	314.40	381.78	4.00	70.40	nr	**452.18**
280 litres capacity	502.13	609.74	4.00	70.40	nr	**680.14**
360 litres capacity	554.15	672.90	4.50	79.28	nr	**752.19**
440 litres capacity	655.93	796.50	4.50	79.28	nr	**875.78**
Grade 1, tested 3.65 bar, 25m maximum head						
190 litres capacity	386.78	469.67	3.00	52.86	nr	**522.52**
245 litres capacity	439.92	534.19	3.00	52.86	nr	**587.05**
280 litres capacity	622.01	755.31	4.00	70.40	nr	**825.71**
360 litres capacity	787.13	955.81	4.50	79.28	nr	**1035.09**
440 litres capacity	955.62	1160.41	4.50	79.28	nr	**1239.69**

Material Costs/Measured Work Prices - Mechanical Installations

S:PIPED SUPPLY SYSTEMS

Item	Net Price £	Material £	Labour hours	Labour £	Unit	Total rate £
Indirect cylinders, including manhole; BS 853						
Grade 3, tested to 1.5 bar, 10m maximum head						
550 litres capacity	831.22	1009.35	5.21	91.67	nr	**1101.02**
700 litres capacity	920.28	1117.50	6.02	106.03	nr	**1223.53**
800 litres capacity	1068.72	1297.75	6.54	115.04	nr	**1412.78**
1000 litres capacity	1335.90	1622.18	7.04	123.95	nr	**1746.13**
1500 litres capacity	1543.70	1874.51	10.00	176.01	nr	**2050.52**
2000 litres capacity	2137.44	2595.49	16.13	283.88	nr	**2879.38**
Grade 2, tested to 2.55 bar, 15m maximum head						
550 litres capacity	921.56	1119.05	5.21	91.67	nr	**1210.72**
700 litres capacity	1153.40	1400.57	6.02	106.03	nr	**1506.60**
800 litres capacity	1217.15	1477.99	6.54	115.04	nr	**1593.02**
1000 litres capacity	1506.95	1829.89	7.04	123.95	nr	**1953.84**
1500 litres capacity	1854.71	2252.17	10.00	176.01	nr	**2428.18**
2000 litres capacity	2318.39	2815.22	16.13	283.88	nr	**3099.10**
Grade 1, tested to 4 bar, 25m maximum head						
550 litres capacity	1072.25	1302.03	5.21	91.67	nr	**1393.70**
700 litres capacity	1217.15	1477.99	6.02	106.03	nr	**1584.01**
800 litres capacity	1304.09	1583.56	6.54	115.04	nr	**1698.59**
1000 litres capacity	1738.79	2111.41	7.04	123.95	nr	**2235.36**
1500 litres capacity	2086.54	2533.69	10.00	176.01	nr	**2709.69**
2000 litres capacity	2550.22	3096.73	16.13	283.88	nr	**3380.62**
Storage calorifiers; copper; heater battery capable of raising temperature of contents from 10 degree C to 65 degree C in one hour; static head not exceeding 1.35 bar; BS 853; includes fixing in position on cradles or legs						
Horizontal; primary LPHW at 82 degree C (on) 71 degree C (off)						
400 litres capacity	1244.01	1510.60	7.04	123.95	nr	**1634.55**
1000 litres capacity	1990.41	2416.95	8.00	140.81	nr	**2557.76**
2000 litres capacity	3980.84	4833.93	14.08	247.90	nr	**5081.83**
3000 litres capacity	4913.84	5966.88	25.00	440.02	nr	**6406.90**
4000 litres capacity	5971.25	7250.89	40.00	704.03	nr	**7954.92**
4500 litres capacity	6729.33	8171.43	50.00	880.04	nr	**9051.47**
Vertical; primary LPHW at 82 degree C (on) 71 degree C (off)						
400 litres capacity	1219.13	1480.39	7.04	123.95	nr	**1604.34**
1000 litres capacity	1959.33	2379.21	8.00	140.81	nr	**2520.02**
2000 litres capacity	3732.03	4531.80	14.08	247.90	nr	**4779.70**
3000 litres capacity	4665.05	5664.77	25.00	440.02	nr	**6104.79**
4000 litres capacity	5722.45	6948.77	40.00	704.03	nr	**7652.80**
4500 litres capacity	6468.86	7855.14	50.00	880.04	nr	**8735.18**

Material Costs/Measured Work Prices - Mechanical Installations

S:PIPED SUPPLY SYSTEMS

Item	Net Price £	Material £	Labour hours	Labour £	Unit	Total rate £
S11 - HOT WATER						
Y23 - STORAGE CYLINDERS/CALORIFIERS CYLINDERS (Continued)						
Storage calorifiers; galvanised mild steel; heater battery capable of raising temperature of contents from 10 degree C to 65 degree C in one hour; static head not exceeding 1.35 bar; BS 853; includes fixing in position on cradles or legs						
Horizontal; primary LPHW at 82 degree C (on) 71 degree C (off)						
400 litres capacity	1244.01	1510.60	7.04	123.95	nr	1634.55
1000 litres capacity	1990.41	2416.95	8.00	140.81	nr	2557.76
2000 litres capacity	3980.84	4833.93	14.08	247.90	nr	5081.83
3000 litres capacity	4913.84	5966.88	25.00	440.02	nr	6406.90
4000 litres capacity	5971.25	7250.89	40.00	704.03	nr	7954.92
4500 litres capacity	6729.33	8171.43	50.00	880.04	nr	9051.47
Vertical; primary LPHW at 82 degree C (on) 71 degree C (off)						
400 litres capacity	1219.13	1480.39	7.04	123.95	nr	1604.34
1000 litres capacity	1959.33	2379.21	8.00	140.81	nr	2520.02
2000 litres capacity	3732.03	4531.80	14.08	247.90	nr	4779.70
3000 litres capacity	4665.05	5664.77	25.00	440.02	nr	6104.79
4000 litres capacity	5722.45	6948.77	40.00	704.03	nr	7652.80
4500 litres capacity	6468.86	7855.14	50.00	880.04	nr	8735.18
Indirect cylinders; mild steel, welded throughout, galvanised; with bolted connections; includes placing in position						
3.2mm plate						
136 litres capacity	459.43	557.89	2.50	44.00	nr	601.89
159 litres capacity	494.78	600.81	2.80	49.30	nr	650.11
182 litres capacity	607.88	738.15	3.00	52.86	nr	791.00
227 litres capacity	749.24	909.80	3.00	52.86	nr	962.66
273 litres capacity	873.98	1061.27	4.00	70.40	nr	1131.68
364 litres capacity	1017.83	1235.95	4.50	79.28	nr	1315.23
455 litres capacity	1051.30	1276.59	5.00	88.00	nr	1364.60
683 litres capacity	1701.32	2065.91	6.02	106.03	nr	2171.94
910 litres capacity	1975.95	2399.40	7.04	123.95	nr	2523.35
Y50 - INSULATION **See sections S10 - Cold Water and T31 - Low Temperature Hot Water Heating**						

Material Costs/Measured Work Prices - Mechanical Installations

S:PIPED SUPPLY SYSTEMS

Item	Net Price £	Material £	Labour hours	Labour £	Unit	Total rate £
S32 : NATURAL GAS						
Y10 - PIPELINES						
MEDIUM DENSITY POLYETHELENE - YELLOW						
Pipe; laid underground; electrofusion joints in the running length; BS 6572						
Coiled service pipe						
20mm dia.	0.37	0.46	0.37	6.51	m	**6.98**
25mm dia.	0.48	0.60	0.41	7.22	m	**7.82**
32mm dia.	0.76	0.95	0.47	8.27	m	**9.23**
63mm dia.	2.91	3.65	0.60	10.56	m	**14.21**
90mm dia.	6.87	8.60	0.90	15.84	m	**24.44**
Mains service pipe						
63mm dia.	3.40	4.25	0.60	10.56	m	**14.81**
90mm dia.	4.55	5.69	0.90	15.84	m	**21.53**
125mm dia.	8.71	10.89	1.20	21.12	m	**32.02**
180mm dia.	17.72	22.17	1.50	26.40	m	**48.57**
250mm dia.	17.11	21.40	1.75	30.80	m	**52.20**
Extra over fittings, electrofusion joints						
Straight connector						
32mm dia.	2.39	2.90	0.47	8.27	nr	**11.18**
63mm dia.	4.48	5.44	0.58	10.21	nr	**15.65**
90mm dia.	7.09	8.61	0.67	11.79	nr	**20.40**
125mm dia.	12.05	14.63	0.83	14.61	nr	**29.24**
180mm dia.	23.05	27.99	1.25	22.00	nr	**49.99**
Reducing connector						
90 x 63mm dia.	9.72	11.80	0.67	11.79	nr	**23.60**
125 x 90mm dia.	19.53	23.71	0.83	14.61	nr	**38.32**
180 x 125mm dia.	35.79	43.46	1.25	22.00	nr	**65.46**
Bend; 45 degree						
90mm dia.	16.09	19.54	0.67	11.79	nr	**31.33**
125mm dia.	28.02	34.02	0.83	14.61	nr	**48.63**
180mm dia.	61.60	74.80	1.25	22.00	nr	**96.80**
Bend; 90 degree						
63mm dia.	11.64	14.13	0.58	10.21	nr	**24.34**
90mm dia.	17.68	21.47	0.67	11.79	nr	**33.26**
125mm dia.	28.94	35.14	0.83	14.61	nr	**49.75**
180mm dia.	65.70	79.78	1.25	22.00	nr	**101.79**

Material Costs/Measured Work Prices - Mechanical Installations

S:PIPED SUPPLY SYSTEMS

Item	Net Price £	Material £	Labour hours	Labour £	Unit	Total rate £
S32 : NATURAL GAS						
Y10 - PIPELINES						
MEDIUM DENSITY POLYETHELENE - YELLOW (Continued)						
Extra over malleable iron fittings, compression joints						
Straight connector						
20mm dia.	6.39	7.76	0.38	6.69	nr	14.45
25mm dia.	6.98	8.48	0.45	7.92	nr	16.40
32mm dia.	7.83	9.51	0.50	8.80	nr	18.31
63mm dia.	15.73	19.10	0.85	14.97	nr	34.06
Straight connector; polyethylene to MI						
20mm dia.	5.44	6.61	0.31	5.46	nr	12.06
25mm dia.	5.93	7.20	0.35	6.16	nr	13.36
32mm dia.	6.62	8.04	0.40	7.04	nr	15.08
63mm dia.	11.10	13.47	0.65	11.44	nr	24.92
Straight connector; polyethylene to FI						
20mm dia.	5.24	6.36	0.31	5.46	nr	11.82
25mm dia.	5.72	6.95	0.35	6.16	nr	13.11
32mm dia.	6.39	7.76	0.40	7.04	nr	14.80
63mm dia.	10.72	13.02	0.75	13.20	nr	26.22
Elbow						
20mm dia.	8.32	10.10	0.38	6.69	nr	16.79
25mm dia.	9.08	11.03	0.45	7.92	nr	18.95
32mm dia.	10.18	12.36	0.50	8.80	nr	21.16
63mm dia.	20.44	24.82	0.80	14.08	nr	38.90
Equal tee						
20mm dia.	9.68	11.75	0.53	9.33	nr	21.09
25mm dia.	11.27	13.69	0.55	9.68	nr	23.37
32mm dia.	14.22	17.27	0.64	11.27	nr	28.54
SCREWED STEEL						
For prices for steel pipework refer to Section T31 - Low Temperature Hot Water Heating						

NEW FROM SPON PRESS

The Architectural Expression of Environmental Control Systems

George Baird, Victoria University of Wellington, New Zealand

*The Architectural Expression of Environmental Control System*s examines the way project teams can approach the design and expression of both active and passive thermal environmental control systems in a more creative way. Using seminal case studies from around the world and interviews with the architects and environmental engineers involved, the book illustrates innovative responses to client, site and user requirements, focusing upon elegant design solutions to a perennial problem.

This book will inspire architects, building scientists and building services engineers to take a more creative approach to the design and expression of environmental control systems - whether active or passive, whether they influence overall building form or design detail.

March 2001: 276x219: 304pp
135 b+w photos, 40 colour, 90 line illustrations
Hb: 0-419-24430-1: £49.95

To Order: Tel: +44 (0) 8700 768853, or +44 (0) 1264 343071 Fax: +44 (0) 1264 343005, or
Post: Spon Press Customer Services, ITPS Andover, Hants, SP10 5BE, UK Email: book.orders@tandf.co.uk.

Postage & Packing: UK: 5% of order value (min. charge £1, max. charge £10) for 3-5 days delivery. Option of next day delivery at an additional £6.50. Europe: 10% of order value (min. charge £2.95, max. charge £20) for delivery surface post. Option of airmail at an additional £6.50. ROW: 15% of order value (min.charge£6.50, max. charge £30) for airmail delivery.

For a complete listing of all our titles visit: www.sponpress.com

NEW FROM SPON PRESS

Photovoltaics and Architecture

Edited by Randall Thomas, Max Fordham and Partners, UK
With a foreword by Amory Lovins

It has been said that the nineteenth century was the age of coal, the twentieth of oil, and the twenty-first will be the age of solar energy. *Photovoltaics and Architecture* describes vividly how buildings can contribute to this transition.

PVs are changing the form of our communities and buildings and encouraging designers to make the most use of solar energy. The challenges and potential are significant.

Photovoltaics and Architecture is the first book to set out the basic principles of PV design and examines their implications in a largely UK context. It will be of value to designers, clients and students.

March 2001: 297x210: 176pp
105 line illustrations, 63 b+w photos
Pb: 0-415-23182-5: £29.95

To Order: Tel: +44 (0) 8700 768853, or +44 (0) 1264 343071 Fax: +44 (0) 1264 343005, or
Post: Spon Press Customer Services, ITPS Andover, Hants, SP10 5BE, UK Email: book.orders@tandf.co.uk.

Postage & Packing: UK: 5% of order value (min. charge £1, max. charge £10) for 3-5 days delivery. Option of next day delivery at an additional £6.50. Europe: 10% of order value (min. charge £2.95, max. charge £20) for delivery surface post. Option of airmail at an additional £6.50. ROW: 15% of order value (min.charge£6.50, max. charge £30) for airmail delivery.

Material Costs/Measured Work Prices - Mechanical Installations

S:PIPED SUPPLY SYSTEMS

Item	Net Price £	Material £	Labour hours	Labour £	Unit	Total rate £
S41 : FUEL OIL STORAGE/DISTRIBUTION						
Y10 - PIPELINES						
For prices for pipework refer to Section						
T31 - Low Temperature Hot Water						
Heating						
Y21 - TANKS						
Fuel storage tanks; mild steel; with all						
necessary screwed bosses; oil resistant						
joint rings; includes placing in position						
Rectangular						
1360 litres (300 gallon) capacity; 2mm plate	242.05	293.92	12.03	211.74	nr	**505.66**
2730 litres (600 gallon) capacity; 2.5mm plate	323.42	392.73	18.60	327.37	nr	**720.10**
4550 litres (1000 gallon) capacity; 3mm plate	682.89	829.23	30.68	539.99	nr	**1369.23**
Fuel storage tanks; plastic; with all						
necessary screwed bosses; oil resistant						
joint rings; includes placing in position						
Cylindrical; horizontal						
1250 litres (285 gallon) capacity	189.52	230.13	3.73	65.65	nr	**295.79**
1350 litres (300 gallon) capacity	174.12	211.43	4.30	75.68	nr	**287.12**
2500 litres (550 gallon) capacity	297.31	361.02	4.88	85.89	nr	**446.92**
Cylindrical; vertical						
1365 litres (300 gallon) capacity	130.29	158.21	3.73	65.65	nr	**223.86**
2600 litres (570 gallon) capacity	200.18	243.08	4.88	85.89	nr	**328.97**
3635 litres (800 gallon) capacity	312.71	379.72	4.88	85.89	nr	**465.62**
5455 litres (1200 gallon) capacity	456.03	553.76	5.95	104.72	nr	**658.48**
10000 litres (2200 gallon) capacity	913.25	1108.96	7.68	135.17	nr	**1244.13**
Bunded tanks						
1135 litres (250 gallon) capacity	422.87	513.49	4.30	75.68	nr	**589.17**
1590 litres (350 gallon) capacity	501.04	608.41	4.88	85.89	nr	**694.30**
2500 litres (550 gallon) capacity	596.99	724.92	5.95	104.72	nr	**829.65**
5000 litres (1100 gallon) capacity	1216.48	1477.17	6.53	114.93	nr	**1592.10**

Material Costs/Measured Work Prices - Mechanical Installations

S:PIPED SUPPLY SYSTEMS

Item	Net Price £	Material £	Labour hours	Labour £	Unit	Total rate £
S60 : FIRE HOSE REELS						
Y10 - PIPELINES						
For prices for pipework refer to Section						
S10 - Cold Water						
Y11 - PIPELINE ANCILLARIES						
For prices for ancillaries refer to Section						
S10 - Cold Water						
Hose reels; automatic; connection to						
25mm screwed joint; reel with 30.5						
metres, 19mm rubber hose; suitable for						
working pressure up to 7 bar						
Reels						
Non-swing pattern	181.75	225.39	3.75	73.15	nr	**298.54**
Recessed non-swing pattern	191.25	237.17	3.75	73.15	nr	**310.32**
Swinging pattern	216.80	268.85	3.75	73.15	nr	**342.00**
Recessed swinging pattern	229.40	284.48	3.75	73.15	nr	**357.63**
Hose reels; manual; connection to 25mm						
screwed joint; reel with 30.5 metres,						
19mm rubber hose; suitable for working						
pressure up to 7 bar						
Reels						
Non-swing pattern	180.75	224.15	3.25	63.40	nr	**287.55**
Recessed non-swing pattern	188.25	233.45	3.25	63.40	nr	**296.85**
Swinging pattern	214.50	266.00	3.25	63.40	nr	**329.40**
Recessed swinging pattern	228.05	282.80	3.25	63.40	nr	**346.20**

Material Costs/Measured Work Prices - Mechanical Installations 157

S:PIPED SUPPLY SYSTEMS

Item	Net Price £	Material £	Labour hours	Labour £	Unit	Total rate £
S61 : DRY RISERS						
Y10 - PIPELINES **For prices for pipework refer to Section** **S10 - Cold Water**						
Y11 - PIPELINE ANCILLARIES						
VALVES						
Bronze/gunmetal inlet breeching for pumping in with 64mm dia. instantaneuos male coupling; with cap, chain and 25mm drain valve						
Double inlet with back pressure valve, screwed or flanged to steel	184.25	228.49	1.75	34.14	nr	**262.63**
Quadruple inlet with back pressure valve, flanged to steel	309.75	384.12	1.75	34.14	nr	**418.26**
Bronze/gunmetal gate type outlet valve with 64mm dia. instantaneous female coupling cap and chain; wheel head secured by padlock and leather strap						
Flanged BS table D inlet (bolted connection to counter flanges measured separately)	118.40	146.83	1.75	34.14	nr	**180.96**
Bronze/gunmetal landing type outlet - valve, with 64mm dia. instantaneous female coupling; cap and chain; wheelhead secured by padlock and leatherstrap, bolted connections to counter flanges measured separately						
Horizontal, flanged BS table D inlet	115.00	142.61	1.50	29.26	nr	**171.87**
Oblique, flanged BS table D inlet	115.00	142.61	1.50	29.26	nr	**171.87**
Air valve, screwed joint to steel						
25mm dia.	21.85	27.10	0.55	10.73	nr	**37.82**
INLET BOXES						
Steel dry riser inlet box with hinged wire glazed door suitably lettered (fixing by others)						
610 x 460 x 325mm; double inlet	209.00	259.18	3.00	58.52	nr	**317.70**
610 x 610 x 356mm; quadruple inlet	302.75	375.44	3.00	58.52	nr	**433.96**
Landing valve	137.50	170.51	3.00	58.52	nr	**229.03**

Material Costs/Measured Work Prices - Mechanical Installations

S:PIPED SUPPLY SYSTEMS

Item	Net Price £	Material £	Labour hours	Labour £	Unit	Total rate £
S63 : SPRINKLERS						
Y10 - PIPELINES						
Prefabricated black steel pipework; screwed joints, including all coupliings, unions and the like to BS 1387:1985; fixing to backgrounds						
Medium weight						
25mm dia	2.08	2.58	0.47	9.17	m	**11.75**
32mm dia	2.58	3.20	0.53	10.34	m	**13.54**
40mm dia	3.01	3.73	0.58	11.31	m	**15.05**
50mm dia	4.31	5.34	0.63	12.29	m	**17.63**
Extra over fittings						
Plug						
25mm dia	0.57	0.71	0.40	7.80	nr	**8.51**
32mm dia	0.81	1.00	0.44	8.58	nr	**9.59**
40mm dia	1.35	1.67	0.48	9.36	nr	**11.04**
50mm dia	1.89	2.34	0.56	10.92	nr	**13.27**
Reducer						
32mm dia	4.16	5.16	0.48	9.36	nr	**14.52**
40mm dia	4.71	5.84	0.55	10.73	nr	**16.57**
50mm dia	4.71	5.84	0.60	11.70	nr	**17.55**
Elbow; any degree						
25mm dia	2.13	2.64	0.44	8.58	nr	**11.22**
32mm dia	2.84	3.52	0.53	10.34	nr	**13.86**
40mm dia	4.00	4.96	0.60	11.70	nr	**16.66**
50mm dia	4.49	5.57	0.65	12.68	nr	**18.25**
Tee						
25mm dia	2.44	3.03	0.51	9.95	nr	**12.97**
32mm dia	3.40	4.22	0.54	10.53	nr	**14.75**
40mm dia	4.71	5.84	0.65	12.68	nr	**18.52**
50mm dia	6.24	7.74	0.78	15.22	nr	**22.95**
Cross tee						
25mm dia	3.34	4.14	1.16	22.63	nr	**26.77**
32mm dia	4.40	5.46	1.40	27.31	nr	**32.77**
40mm dia	5.92	7.34	1.60	31.21	nr	**38.55**
50mm dia	9.20	11.41	1.68	32.77	nr	**44.18**

Material Costs/Measured Work Prices - Mechanical Installations

S:PIPED SUPPLY SYSTEMS

Item	Net Price £	Material £	Labour hours	Labour £	Unit	Total rate £
S63 : SPRINKLERS						
Y10 - PIPELINES (Continued)						
Prefabricated black steel pipework; welded joints, including all couplings, unions and the like to BS 1387:1985; fixing to backgrounds						
Medium weight						
65mm dia	5.69	7.06	0.65	12.68	m	19.74
80mm dia	7.29	9.04	0.70	13.65	m	22.70
100mm dia	10.24	12.70	0.85	16.58	m	29.28
150mm dia	16.63	20.62	1.15	22.43	m	43.06
Extra over fittings						
Reducer						
65mm dia	10.89	13.50	2.70	52.67	nr	66.17
80mm dia	12.77	15.84	2.86	55.79	nr	71.63
100mm dia	18.51	22.95	3.22	62.81	nr	85.77
150mm dia	40.73	50.51	4.20	81.93	nr	132.44
Elbow; any degree						
65mm dia	13.70	16.99	3.06	59.69	nr	76.68
80mm dia	16.02	19.87	3.40	66.32	nr	86.19
100mm dia	26.39	32.73	3.70	72.18	nr	104.90
150mm dia	81.50	101.07	5.20	101.44	nr	202.50
Branch bend						
65mm dia	14.52	18.01	3.60	70.22	nr	88.23
80mm dia	18.11	22.46	3.80	74.13	nr	96.58
100mm dia	29.47	36.55	5.10	99.48	nr	136.03
150mm dia	90.84	112.65	7.50	146.30	nr	258.95
Prefabricated black steel pipe; victualic joints; including all couplings and the like to BS 1387: 1985; fixing to backgrounds						
Medium weight						
65mm dia	5.69	7.06	0.70	13.65	m	20.71
80mm dia	7.29	9.04	0.78	15.22	m	24.26
100mm dia	10.24	12.70	0.93	18.14	m	30.84
150mm dia	16.63	20.62	1.25	24.38	m	45.01
Extra over fittings						
Coupling						
65mm dia	7.05	8.74	0.26	5.07	nr	13.81
80mm dia	7.34	9.10	0.26	5.07	nr	14.17
100mm dia	8.82	10.94	0.32	6.24	nr	17.18
150mm dia	12.85	15.94	0.35	6.83	nr	22.76

Material Costs/Measured Work Prices - Mechanical Installations

S:PIPED SUPPLY SYSTEMS

Item	Net Price £	Material £	Labour hours	Labour £	Unit	Total rate £
S63 : SPRINKLERS						
Prefabricated black steel pipe; victualic joints; including all couplings and the like to BS 1387: 1985; fixing to backgrounds (Continued)						
Reducer						
65mm dia	18.87	23.40	0.48	9.36	nr	**32.76**
80mm dia	19.65	24.37	0.43	8.39	nr	**32.76**
100mm dia	24.38	30.23	0.46	8.97	nr	**39.21**
150mm dia	39.41	48.87	0.45	8.78	nr	**57.65**
Elbow; any degree						
65mm dia	10.82	13.42	0.56	10.92	nr	**24.34**
80mm dia	11.93	14.79	0.63	12.29	nr	**27.08**
100mm dia	15.10	18.73	0.71	13.85	nr	**32.58**
150mm dia	29.91	37.09	0.80	15.61	nr	**52.70**
Equal tee						
65mm dia	20.12	24.95	0.74	14.44	nr	**39.39**
80mm dia	22.38	27.75	0.83	16.19	nr	**43.94**
100mm dia	25.39	31.49	0.94	18.34	nr	**49.82**
150mm dia	51.85	64.30	1.05	20.48	nr	**84.78**
Y11 - PIPELINE ANCILLARIES						
SPRINKLER HEADS						
Sprinkler heads; brass body; frangible glass bulb; manufactured to standard operating temperature of 57-141 degrees Celsius; quick response; RTI<50						
conventional pattern; 15mm dia.	2.67	3.31	0.15	2.93	nr	**6.24**
sidewall pattern; 15mm dia.	3.80	4.71	0.15	2.93	nr	**7.64**
conventional pattern; 15mm dia.; satin chrome plated	3.12	3.87	0.15	2.93	nr	**6.80**
sidewall pattern; 15mm dia.; satin chrome plated	4.10	5.08	0.15	2.93	nr	**8.01**
Fully concealed; fusable link; 15mm dia	9.98	12.38	0.15	2.93	nr	**15.30**
VALVES						
Wet system alarm valves; including internal non-return valve; working pressure up to 12.5 bar; BS4504 PN16 flanged ends; bolted connections						
100mm dia.	863.90	1071.32	25.00	487.67	nr	**1558.99**
150mm dia.	1048.75	1300.55	25.00	487.67	nr	**1788.22**
Wet system by pass alarm valves; including internal non-return valve; working pressure up to 12.5 bar; BS4504 PN16 flanged ends; bolted connections						
100mm dia.	1523.50	1889.29	40.00	780.27	nr	**2669.56**
150mm dia.	1925.00	2387.19	25.00	487.67	nr	**2874.86**

Material Costs/Measured Work Prices - Mechanical Installations 161

S:PIPED SUPPLY SYSTEMS

Item	Net Price £	Material £	Labour hours	Labour £	Unit	Total rate £
Alternate system wet/dry alarm station; including butterfly valve, wet alarm valve, dry pipe differential pressure valve and pressure gauges; working pressure up to 10.5 bar; BS4505 PN16 flanged ends; bolted connections						
100mm dia.	1890.75	2344.72	40.00	780.27	nr	**3124.99**
150mm dia.	2203.00	2731.94	40.00	780.27	nr	**3512.21**
Alternate system wet/dry alarm station; including electrically supervised butterfly valve, water supply accelerator set, wet alarm valve, dry pipe differential pressure valve and pressure gauges; working pressure up to 10.5 bar; BS4505 PN16 flanged ends; bolted connections						
100mm dia.	2155.75	2673.35	45.00	877.81	nr	**3551.15**
150mm dia.	2468.00	3060.57	45.00	877.81	nr	**3938.37**
ALARM/GONGS						
Water operated motor alarm and gong; stainless steel and aluminum body and gong; screwed connections						
Connection to sprinkler system and drain pipework	261.25	323.98	6.00	117.04	nr	**441.02**

Material Costs/Measured Work Prices - Mechanical Installations

S:PIPED SUPPLY SYSTEMS

Item	Net Price £	Material £	Labour hours	Labour £	Unit	Total rate £
S65 : FIRE HYDRANTS						
EXTINGUISHERS						
Fire extinguishers; hand held; BS 5423; placed in position						
Water type; cartridge operated; for Class A fires						
Water type, 9 litres capacity; 55gm CO2 cartridge; Class A fires (fire rating 13A)	30.00	37.20	1.00	19.51	nr	**56.71**
Foam type, 9 litres capacity; 75 gm CO2 cartridge; Class A & B fires (fire rating 13A:183B)	42.00	52.08	1.00	19.51	nr	**71.59**
Dry powder type; cartridge operated; for Class A, B & C fires and electrical equipment fires						
Dry powder type, 1kg capacity; 12gm CO2 cartridge; Class A, B & C fires (fire rating 5A:34B)	24.00	29.76	1.00	19.51	nr	**49.27**
Dry powder type, 2kg capacity; 28gm CO2 cartridge; Class A, B & C fires (fire rating 13A:55B)	26.00	32.24	1.00	19.51	nr	**51.75**
Dry powder type, 4kg capacity; 90gm CO2 cartridge; Class A, B & C fires (fire rating 21A:183B)	37.00	45.88	1.00	19.51	nr	**65.39**
Dry powder type, 9kg capacity; 190gm CO2 cartridge; Class A, B & C fires (fire rating 43A:233B)	40.50	50.22	1.00	19.51	nr	**69.73**
Carbon dioxide type; for Class B fires and electrical equipment fires						
CO2 type with hose and horn, 2kg capacity, Class B fires (fire rating 34B)	35.40	43.90	1.00	19.51	nr	**63.41**
CO2 type with hose and horn, 5kg capacity, Class B fires (fire rating 55B)	58.60	72.67	1.00	19.51	nr	**92.18**
Glass fibre blanket, in GRP container						
1100 x 1100mm	14.00	17.36	0.50	9.75	nr	**27.11**
1200 x 1200mm	22.00	27.28	0.50	9.75	nr	**37.04**
1800 x 1200mm	28.00	34.72	0.50	9.75	nr	**44.48**
HYDRANTS						
Fire hydrants; bolted connections						
Underground hydrants, complete with frost plug to BS 750						
sluice valve pattern type 1	369.00	457.60	4.50	87.78	nr	**545.38**
screw down pattern type 2	369.00	457.60	4.50	87.78	nr	**545.38**
Stand pipe for underground hydrant screwed base; light alloy						
Single outlet	79.00	97.97	-	-	nr	**97.97**
Double outlet	115.00	142.61	-	-	nr	**142.61**

Material Costs/Measured Work Prices - Mechanical Installations

S:PIPED SUPPLY SYSTEMS

Item	Net Price £	Material £	Labour hours	Labour £	Unit	Total rate £
64mm diameter bronze/gunmetal outlet valves						
Oblique flanged landing valve	122.00	151.29	1.00	19.51	nr	**170.80**
Oblique screwed landing valve	122.00	151.29	1.00	19.51	nr	**170.80**
Cast iron surface box; fixing by others						
400 x 200 x 100mm	88.00	109.13	1.00	19.51	nr	**128.64**
500 x 200 x 150mm	116.00	143.85	1.00	19.51	nr	**163.36**
Frost Plug	21.00	26.04	0.25	4.88	nr	**30.92**

Material Costs/Measured Work Prices - Mechanical Installations

T:MECHANICAL/COOLING/HEATING SYSTEMS

Item	Net Price £	Material £	Labour hours	Labour £	Unit	Total rate £
T10 : GAS/OIL FIRED BOILERS						
Domestic water boilers; stove enamelled casing; electric controls; placing in position; assembling and connecting; electrical work elsewhere						
Gas fired; floor standing; connected to conventional flue						
30,000 to 40,000 Btu/Hr	370.95	450.45	8.59	151.19	nr	**601.64**
40,000 to 50,000 Btu/Hr	390.24	473.87	8.59	151.19	nr	**625.06**
50,000 to 60,000 Btu/Hr	415.65	504.73	8.88	156.30	nr	**661.02**
60,000 to 70,000 Btu/Hr	415.65	504.73	9.92	174.60	nr	**679.33**
70,000 to 80,000 Btu/Hr	535.34	650.06	10.66	187.62	nr	**837.69**
80,000 to 100,000 Btu/Hr	694.32	843.12	11.81	207.87	nr	**1050.98**
100,000 to 125,000 Btu/Hr	821.22	997.21	11.81	207.87	nr	**1205.07**
125,000 to 140,000 Btu/Hr	854.03	1037.05	12.68	223.18	nr	**1260.22**
Gas fired; wall hung; connected to conventional flue						
30,000 to 40,000 Btu/Hr	326.03	395.90	8.59	151.19	nr	**547.09**
40,000 to 50,000 Btu/Hr	420.92	511.13	8.59	151.19	nr	**662.32**
45,000 to 60,000 Btu/Hr	534.20	648.68	8.59	151.19	nr	**799.87**
Gas fired; floor standing; connected to balanced flue						
30,000 to 40,000 Btu/Hr	463.39	562.69	9.16	161.22	nr	**723.91**
40,000 to 50,000 Btu/Hr	482.13	585.45	10.95	192.73	nr	**778.18**
50,000 to 60,000 Btu/Hr	518.97	630.19	11.98	210.86	nr	**841.05**
60,000 to 70,000 Btu/Hr	611.55	742.61	12.78	224.94	nr	**967.55**
70,000 to 80,000 Btu/Hr	705.36	856.51	12.78	224.94	nr	**1081.45**
80,000 to 100,000 Btu/Hr	899.05	1091.72	15.45	271.93	nr	**1363.65**
100,000 to 125,000 Btu/Hr	1728.34	2098.73	17.65	310.65	nr	**2409.38**
Gas fired; wall hung; connected to balanced flue						
20,000 to 30,000 Btu/Hr	349.87	424.84	9.16	161.22	nr	**586.07**
30,000 to 40,000 Btu/Hr	400.37	486.17	9.16	161.22	nr	**647.39**
40,000 to 50,000 Btu/Hr	453.36	550.52	9.45	166.33	nr	**716.85**
50,000 to 60,000 Btu/Hr	543.60	660.09	9.74	171.43	nr	**831.52**
60,000 to 75,000 Btu/Hr	577.02	700.67	9.74	171.43	nr	**872.10**
Gas fired; wall hung; connected to fan flue (incl. flue kit)						
20,000 to 30,000 Btu/Hr	400.15	485.91	9.16	161.22	-	**647.13**
30,000 to 40,000 Btu/Hr	445.94	541.50	9.16	161.22	-	**702.72**
40,000 to 50,000 Btu/Hr	484.08	587.82	10.95	192.73	-	**780.54**
50,000 to 60,000 Btu/Hr	521.43	633.17	11.98	210.86	-	**844.03**
60,000 to 70,000 Btu/Hr	589.02	715.25	12.78	224.94	-	**940.19**
60,000 to 80,000 Btu/Hr	679.40	824.99	12.78	224.94	-	**1049.93**
80,000 to 100,000 Btu/Hr	879.80	1068.34	15.45	271.93	-	**1340.28**
100,000 to 120,000 Btu/Hr	1080.15	1311.62	17.65	310.65	-	**1622.28**

Material Costs/Measured Work Prices - Mechanical Installations

T:MECHANICAL/COOLING/HEATING SYSTEMS

Item	Net Price £	Material £	Labour hours	Labour £	Unit	Total rate £
Oil fired; floor standing; connected to conventional flue						
40,000 to 50,000 Btu/Hr	726.05	881.64	10.38	182.70	nr	**1064.34**
50,000 to 65,000 Btu/Hr	757.38	919.69	12.20	214.73	nr	**1134.42**
70,000 to 85,000 Btu/Hr	864.38	1049.62	14.30	251.69	nr	**1301.31**
88,000 to 110,000 Btu/Hr	947.68	1150.77	15.80	278.09	nr	**1428.86**
120,000 to 170,000 Btu/Hr	1069.96	1299.26	20.46	360.11	nr	**1659.37**
Fire place mounted natural gas fire and back boiler; cast iron water boiler; electric control box; fire output 3kW with wood surround						
10.50kW Rating (45,000 Btu/Hr)	201.16	244.27	8.88	156.30	nr	**400.56**
10.50kW Rating (57,000 Btu/Hr)	201.16	244.27	8.88	156.30	nr	**400.56**
Industrial cast iron sectional boilers; including controls, enamelled jacket and insulation; assembled on site and commissioned by supplier - costs included in material prices; electrical work elsewhere						
Gas fired; on/off type						
16-26 kW; 3 sections; 125mm dia. flue	1400.80	1700.99	8.00	140.81	nr	**1841.80**
26-33 kW; 4 sections; 125mm dia. flue	1493.50	1813.56	8.00	140.81	nr	**1954.36**
33-40 kW; 5 sections; 125mm dia. flue	1689.20	2051.20	8.00	140.81	nr	**2192.00**
35-50 kW; 3 sections; 153mm dia. flue	1854.00	2251.31	8.00	140.81	nr	**2392.12**
50-65 kW; 4 sections; 153mm dia. flue	1998.20	2426.41	8.00	140.81	nr	**2567.22**
65-80 kW; 5 sections; 153mm dia. flue	2173.30	2639.04	8.00	140.81	nr	**2779.84**
80-100 kW; 6 sections; 180mm dia. flue	2525.56	3066.79	8.00	140.81	nr	**3207.59**
100-120 kW; 7 sections; 180mm dia. flue	3193.00	3877.26	8.00	140.81	nr	**4018.07**
105-140 kW; 5 sections; 180mm dia. flue	3708.00	4502.62	8.00	140.81	nr	**4643.43**
140-180 kW; 6 sections; 180mm dia. flue	4223.00	5127.99	8.00	140.81	nr	**5268.80**
180-230 kW; 7 sections; 200mm dia. flue	4738.00	5753.35	8.00	140.81	nr	**5894.16**
230-280 kW; 8 sections; 200mm dia. flue	5047.00	6128.57	8.00	140.81	nr	**6269.38**
280-330 kW; 9 sections; 200mm dia. flue	5665.00	6879.01	8.00	140.81	nr	**7019.82**
Gas fired; high/low type						
105-140 kW; 5 sections; 180mm dia. flue	4429.00	5378.13	8.00	140.81	nr	**5518.94**
140-180 kW; 6 sections; 180mm dia. flue	4758.60	5778.37	8.00	140.81	nr	**5919.17**
180-230 kW; 7 sections; 200mm dia. flue	5397.20	6553.82	8.00	140.81	nr	**6694.63**
230-280 kW; 8 sections; 200mm dia. flue	5768.00	7004.08	8.00	140.81	nr	**7144.89**
280-330 kW; 9 sections; 200mm dia. flue	6180.00	7504.37	8.00	140.81	nr	**7645.18**
300-390 kW; 8 sections; 250mm dia. flue	7416.00	9005.25	12.00	211.21	nr	**9216.46**
390-450 kW; 9 sections; 250mm dia. flue	8167.90	9918.28	12.00	211.21	nr	**10129.49**
450-540 kW; 10 sections; 250mm dia. flue	8961.00	10881.34	12.00	211.21	nr	**11092.55**
540-600 kW; 11 sections; 300mm dia. flue	9167.00	11131.49	12.00	211.21	nr	**11342.70**
600-670 kW; 12 sections; 300mm dia. flue	11517.46	13985.65	12.00	211.21	nr	**14196.86**
670-720 kW; 13 sections; 300mm dia. flue	11723.46	14235.80	12.00	211.21	nr	**14447.01**
720-780 kW; 14 sections; 300mm dia. flue	11877.96	14423.41	12.00	211.21	nr	**14634.62**
754-812 kW; 14 sections; 400mm dia. flue	11980.96	14548.48	12.00	211.21	nr	**14759.69**
812-870 kW; 15 sections; 400mm dia. flue	15070.96	18300.67	12.00	211.21	nr	**18511.88**
870-928 kW; 16 sections; 400mm dia. flue	16306.96	19801.54	12.00	211.21	nr	**20012.75**
928-986 kW; 17 sections; 400mm dia. flue	16718.96	20301.83	12.00	211.21	nr	**20513.04**
986-1,044 kW; 18 sections; 400mm dia. flue	17439.96	21177.34	12.00	211.21	nr	**21388.55**

Material Costs/Measured Work Prices - Mechanical Installations

T:MECHANICAL/COOLING/HEATING SYSTEMS

Item	Net Price £	Material £	Labour hours	Labour £	Unit	Total rate £
T10 : GAS/OIL FIRED BOILERS						
Industrial cast iron sectional boilers; including controls, enamelled jacket and insulation; assembled on site and commissioned by supplier - costs included in material prices; electrical work elsewhere (Continued)						
1,044-1,102 kW; 19 sections; 400mm dia. flue	17851.96	21677.64	12.00	211.21	nr	**21888.84**
1,102-1,160 kW; 20 sections; 400mm dia. flue	18305.16	22227.96	12.00	211.21	nr	**22439.17**
1,160-1,218 kW; 21 sections; exceeding 400mm dia. flue	19293.96	23428.66	12.00	211.21	nr	**23639.87**
1,218-1,276 kW; 22 sections; exceeding 400mm dia. flue	20307.48	24659.37	12.00	211.21	nr	**24870.58**
1,276-1,334 kW; 23 sections; exceeding 400mm dia. flue	21018.18	25522.38	12.00	211.21	nr	**25733.59**
1,334-1,392 kW; 24 sections; exceeding 400mm dia. flue	21945.18	26648.03	12.00	211.21	nr	**26859.24**
1,392-1,450 kW; 25 sections; exceeding 400mm dia. flue	22563.18	27398.47	12.00	211.21	nr	**27609.68**
Oil fired; on/off type						
16-26 kW; 3 sections; 125mm dia. flue	1030.00	1250.73	8.00	140.81	nr	**1391.54**
26-33 kW; 4 sections; 125mm dia. flue	1133.00	1375.80	8.00	140.81	nr	**1516.61**
33-40 kW; 5 sections; 125mm dia. flue	1236.00	1500.87	8.00	140.81	nr	**1641.68**
35-50 kW; 3 sections; 153mm dia. flue	1452.30	1763.53	8.00	140.81	nr	**1904.33**
50-65 kW; 4 sections; 153mm dia. flue	1524.40	1851.08	8.00	140.81	nr	**1991.89**
65-80 kW; 5 sections; 153mm dia. flue	1586.20	1926.12	8.00	140.81	nr	**2066.93**
80-100 kW; 6 sections; 180mm dia. flue	2369.00	2876.68	8.00	140.81	nr	**3017.48**
100-120 kW; 7 sections; 180mm dia. flue	2678.00	3251.90	8.00	140.81	nr	**3392.70**
105-140 kW; 5 sections; 180mm dia. flue	3090.00	3752.19	8.00	140.81	nr	**3892.99**
140-180 kW; 6 sections; 180mm dia. flue	3502.00	4252.48	8.00	140.81	nr	**4393.29**
180-230 kW; 7 sections; 200mm dia. flue	4326.00	5253.06	8.00	140.81	nr	**5393.87**
230-280 kW; 8 sections; 200mm dia. flue	4738.00	5753.35	8.00	140.81	nr	**5894.16**
280-330 kW; 9 sections; 200mm dia. flue	5253.00	6378.72	8.00	140.81	nr	**6519.52**
Oil fired; high/low type						
105-140 kW; 5 sections; 180mm dia. flue	2482.30	3014.26	8.00	140.81	nr	**3155.06**
140-180 kW; 6 sections; 180mm dia. flue	2678.00	3251.90	8.00	140.81	nr	**3392.70**
180-230 kW; 7 sections; 200mm dia. flue	2987.00	3627.11	8.00	140.81	nr	**3767.92**
230-280 kW; 8 sections; 200mm dia. flue	3522.60	4277.49	8.00	140.81	nr	**4418.30**
280-330 kW; 9 sections; 200mm dia. flue	5562.00	6753.94	8.00	140.81	nr	**6894.74**
300-390 kW; 8 sections; 250mm dia. flue	6180.00	7504.37	12.00	211.21	nr	**7715.58**
390-450 kW; 9 sections; 250mm dia. flue	6643.50	8067.20	12.00	211.21	nr	**8278.41**
450-540 kW; 10 sections; 250mm dia. flue	7519.00	9130.32	12.00	211.21	nr	**9341.53**
540-600 kW; 11 sections; 300mm dia. flue	7900.10	9593.09	12.00	211.21	nr	**9804.30**
600-670 kW; 12 sections; 300mm dia. flue	8446.00	10255.98	12.00	211.21	nr	**10467.19**
670-720 kW; 13 sections; 300mm dia. flue	9579.00	11631.78	12.00	211.21	nr	**11842.99**
720-780 kW; 14 sections; 300mm dia. flue	9733.50	11819.39	12.00	211.21	nr	**12030.60**
754-812 kW; 14 sections; 400mm dia. flue	9867.40	11981.98	12.00	211.21	nr	**12193.19**
812-870 kW; 15 sections; 400mm dia. flue	13184.00	16009.33	12.00	211.21	nr	**16220.54**
870-928 kW; 16 sections; 400mm dia. flue	13688.70	16622.19	12.00	211.21	nr	**16833.40**

Material Costs/Measured Work Prices - Mechanical Installations

T:MECHANICAL/COOLING/HEATING SYSTEMS

Item	Net Price £	Material £	Labour hours	Labour £	Unit	Total rate £
928-986 kW; 17 sections; 400mm dia. flue	14162.50	17197.52	12.00	211.21	nr	**17408.73**
986-1,044 kW; 18 sections; 400mm dia. flue	14729.00	17885.42	12.00	211.21	nr	**18096.63**
1,044-1,102 kW; 19 sections; 400mm dia. flue	15450.00	18760.94	12.00	211.21	nr	**18972.14**
1,102-1,160 kW; 20 sections; 400 mm dia. flue	16315.20	19811.55	12.00	211.21	nr	**20022.76**
1,160-1,218 kW; 21 sections; exceeding 400mm dia. flue	17221.60	20912.19	12.00	211.21	nr	**21123.40**
1,218-1,276 kW; 22 sections; exceeding 400mm dia. flue	18076.50	21950.29	12.00	211.21	nr	**22161.50**
1,276-1,334 kW; 23 sections; exceeding 400mm dia. flue	18715.10	22725.75	12.00	211.21	nr	**22936.96**
1,334-1,392 kW; 24 sections; exceeding 400mm dia. flue	19812.05	24057.77	12.00	211.21	nr	**24268.98**
1,392-1,450 kW; 25 sections; exceeding 400mm dia. flue	20538.20	24939.54	12.00	211.21	nr	**25150.75**
Packaged water boilers; boiler mountings controls; enamelled casing; burner; insulation; all connections and commissioning						
Gas fired						
150 kW rating; on/off type	4359.69	5293.98	20.75	365.22	nr	**5659.19**
350 kW rating; high/low type	7073.90	8589.83	33.40	587.87	nr	**9177.70**
600 kW rating; high/low type	8935.68	10850.60	41.50	730.43	nr	**11581.03**
1500 kW rating; high/low type	15989.43	19415.97	51.88	913.13	nr	**20329.10**
3000 kW rating; high/low type	28305.15	34370.94	51.88	913.13	nr	**35284.07**
Oil fired						
150 kW rating; high/low type	3847.60	4672.14	20.75	365.22	nr	**5037.36**
350 kW rating; high/low type	5422.05	6584.00	33.40	587.87	nr	**7171.87**
600 kW rating; high/low type	7300.79	8865.35	41.50	730.43	nr	**9595.78**
1500 kW rating; high/low type	13393.97	16264.30	51.88	913.13	nr	**17177.43**
3000 kW rating; high/low type	25557.01	31033.88	51.88	913.13	nr	**31947.01**
Chimneys; applicable to domestic, medium sized industrial and commercial oil and gas appliances; stainless steel, twin wall, insulated; for use internally or externally						
Straight length; 120mm long; including one locking band						
127mm dia.	28.94	35.14	0.49	8.62	nr	**43.77**
152mm dia.	32.43	39.38	0.51	8.98	nr	**48.36**
175mm dia.	37.65	45.72	0.54	9.50	nr	**55.22**
203mm dia.	42.84	52.02	0.58	10.21	nr	**62.23**
254mm dia.	50.66	61.52	0.70	12.32	nr	**73.84**
304mm dia.	63.46	77.06	0.74	13.02	nr	**90.08**
355mm dia.	90.88	110.36	0.80	14.08	nr	**124.44**

168 **Material Costs/Measured Work Prices - Mechanical Installations**

T:MECHANICAL/COOLING/HEATING SYSTEMS

Item	Net Price £	Material £	Labour hours	Labour £	Unit	Total rate £
T10 : GAS/OIL FIRED BOILERS						
Chimneys; applicable to domestic, medium sized industrial and commercial oil and gas appliances; stainless steel, twin wall, insulated; for use internally or externally (Continued)						
Straight length; 300mm long; including one locking band						
127mm dia.	44.65	54.22	0.52	9.15	nr	**63.37**
152mm dia.	50.48	61.30	0.52	9.15	nr	**70.45**
178mm dia.	58.00	70.43	0.55	9.68	nr	**80.11**
203mm dia.	65.64	79.71	0.64	11.26	nr	**90.97**
254mm dia.	72.79	88.39	0.79	13.90	nr	**102.29**
304mm dia.	87.33	106.04	0.86	15.14	nr	**121.18**
355mm dia.	95.82	116.35	0.94	16.54	nr	**132.90**
400mm dia.	102.54	124.51	1.03	18.13	nr	**142.64**
450mm dia.	117.29	142.43	1.03	18.13	nr	**160.55**
500mm dia.	125.80	152.76	1.10	19.36	nr	**172.12**
550mm dia.	138.83	168.58	1.10	19.36	nr	**187.94**
600mm dia.	153.19	186.02	1.10	19.36	nr	**205.38**
Straight length; 500mm long; including one locking band						
127mm dia.	52.48	63.73	0.55	9.68	nr	**73.41**
152mm dia.	58.50	71.04	0.55	9.68	nr	**80.72**
178mm dia.	65.89	80.01	0.63	11.09	nr	**91.10**
203mm dia.	77.02	93.53	0.63	11.09	nr	**104.61**
254mm dia.	89.45	108.62	0.86	15.14	nr	**123.76**
304mm dia.	107.17	130.14	0.95	16.72	nr	**146.86**
355mm dia.	120.35	146.14	1.03	18.13	nr	**164.27**
400mm dia.	131.42	159.58	1.12	19.71	nr	**179.30**
450mm dia.	151.77	184.29	1.12	19.71	nr	**204.01**
500mm dia.	163.53	198.57	1.19	20.95	nr	**219.52**
550mm dia.	180.26	218.89	1.19	20.95	nr	**239.83**
600mm dia.	186.73	226.75	1.19	20.95	nr	**247.69**
Straight length; 1000mm long; including one locking band						
127mm dia.	93.81	113.91	0.62	10.91	nr	**124.83**
152mm dia.	104.55	126.96	0.68	11.97	nr	**138.92**
178mm dia.	117.64	142.85	0.74	13.02	nr	**155.87**
203mm dia.	138.56	168.25	0.80	14.08	nr	**182.33**
254mm dia.	157.03	190.68	0.87	15.31	nr	**205.99**
304mm dia.	181.26	220.10	1.06	18.66	nr	**238.76**
355mm dia.	208.00	252.57	1.16	20.42	nr	**272.99**
400mm dia.	222.75	270.49	1.26	22.18	nr	**292.66**
450mm dia.	235.04	285.41	1.26	22.18	nr	**307.59**
500mm dia.	255.06	309.72	1.33	23.41	nr	**333.13**
550mm dia.	280.52	340.64	1.33	23.41	nr	**364.04**
600mm dia.	294.70	357.85	1.33	23.41	nr	**381.26**

Material Costs/Measured Work Prices - Mechanical Installations

T:MECHANICAL/COOLING/HEATING SYSTEMS

Item	Net Price £	Material £	Labour hours	Labour £	Unit	Total rate £
Adjustable length; boiler removal; internal use only; including one locking band						
127mm dia.	43.00	52.21	0.52	9.15	nr	**61.37**
152mm dia.	48.60	59.02	0.55	9.68	nr	**68.70**
178mm dia.	55.06	66.86	0.59	10.38	nr	**77.24**
203mm dia.	63.15	76.68	0.64	11.26	nr	**87.95**
254mm dia.	93.64	113.71	0.79	13.90	nr	**127.61**
304mm dia.	112.15	136.18	0.86	15.14	nr	**151.32**
355mm dia.	125.77	152.72	0.99	17.42	nr	**170.15**
400mm dia.	220.08	267.24	0.91	16.02	nr	**283.26**
450mm dia.	235.58	286.06	0.91	16.02	nr	**302.08**
500mm dia.	256.94	312.00	0.99	17.42	nr	**329.43**
550mm dia.	280.30	340.37	0.99	17.42	nr	**357.79**
600mm dia.	293.43	356.31	0.99	17.42	nr	**373.74**
Inspection length; 500mm long; including one locking band						
127mm dia.	110.53	134.22	0.55	9.68	nr	**143.90**
152mm dia.	114.45	138.98	0.55	9.68	nr	**148.66**
178mm dia.	120.54	146.37	0.63	11.09	nr	**157.46**
203mm dia.	127.65	155.01	0.63	11.09	nr	**166.09**
254mm dia.	165.04	200.41	0.86	15.14	nr	**215.54**
304mm dia.	178.87	217.20	0.95	16.72	nr	**233.92**
355mm dia.	201.96	245.24	1.03	18.13	nr	**263.37**
400mm dia.	324.03	393.47	1.12	19.71	nr	**413.18**
450mm dia.	332.84	404.17	1.12	19.71	nr	**423.88**
500mm dia.	365.22	443.49	1.19	20.95	nr	**464.43**
550mm dia.	381.91	463.75	1.19	20.95	nr	**484.70**
600mm dia.	388.40	471.63	1.19	20.95	nr	**492.58**
Adapters						
127mm dia.	9.24	11.22	0.49	8.62	nr	**19.84**
152mm dia.	10.11	12.28	0.51	8.98	nr	**21.25**
178mm dia.	10.81	13.13	0.54	9.50	nr	**22.63**
203mm dia.	12.31	14.95	0.58	10.21	nr	**25.16**
254mm dia.	13.55	16.45	0.70	12.32	nr	**28.77**
304mm dia.	16.90	20.52	0.74	13.02	nr	**33.55**
355mm dia.	20.53	24.93	0.80	14.08	nr	**39.01**
400mm dia.	23.02	27.95	0.89	15.66	nr	**43.62**
450mm dia.	24.59	29.86	0.89	15.66	nr	**45.52**
500mm dia.	26.10	31.69	0.96	16.90	nr	**48.59**
550mm dia.	30.35	36.85	0.96	16.90	nr	**53.75**
600mm dia.	36.47	44.29	0.96	16.90	nr	**61.18**
Chimney fittings						
90 degree insulated tee; including two locking bands						
127mm dia.; including locking plug	107.46	130.49	1.89	33.27	nr	**163.75**
152mm dia.; including locking plug	124.03	150.61	2.04	35.91	nr	**186.52**
178mm dia.; including locking plug	135.55	164.60	2.39	42.07	nr	**206.66**
203mm dia.; including locking plug	158.72	192.73	2.56	45.06	nr	**237.79**
254mm dia.; including locking plug	160.56	194.97	2.95	51.92	nr	**246.89**
304mm dia.; including locking plug	199.95	242.80	3.41	60.02	nr	**302.82**

Material Costs/Measured Work Prices - Mechanical Installations

T:MECHANICAL/COOLING/HEATING SYSTEMS

Item	Net Price £	Material £	Labour hours	Labour £	Unit	Total rate £
T10 : GAS/OIL FIRED BOILERS						
Chimneys; applicable to domestic, medium sized industrial and commercial oil and gas appliances; stainless steel, twin wall, insulated; for use internally or externally (Continued)						
355mm dia.; including locking plug	254.91	309.54	3.77	66.36	nr	**375.89**
400mm dia.	335.90	407.88	4.25	74.80	nr	**482.69**
450mm dia.	350.10	425.13	4.76	83.78	nr	**508.91**
500mm dia.	396.17	481.07	5.12	90.12	nr	**571.19**
550mm dia.	425.85	517.11	5.61	98.74	nr	**615.85**
600mm dia.	443.10	538.06	5.98	105.25	nr	**643.31**
135 degree insulated tee; including two locking bands						
127mm dia.; including locking plug	139.37	169.24	1.89	33.27	nr	**202.50**
152mm dia; including locking plug.	150.69	182.98	2.04	35.91	nr	**218.89**
178mm dia.; including locking plug	164.76	200.07	2.39	42.07	nr	**242.13**
203mm dia.; including locking plug	209.10	253.91	2.56	45.06	nr	**298.97**
254mm dia.; including locking plug	237.59	288.51	2.95	51.92	nr	**340.43**
304mm dia.; including locking plug	179.01	217.37	3.41	60.02	nr	**277.39**
355mm dia.; including locking plug	352.24	427.73	3.77	66.36	nr	**494.08**
400mm dia.	463.68	563.05	4.25	74.80	nr	**637.85**
450mm dia.	501.95	609.52	4.76	83.78	nr	**693.30**
500mm dia.	584.78	710.10	5.12	90.12	nr	**800.21**
550mm dia.	599.98	728.56	5.61	98.74	nr	**827.30**
600mm dia.	633.21	768.91	5.98	105.25	nr	**874.16**
Wall sleeve; for 135 degree tee through wall						
127mm dia.	14.27	17.33	1.89	33.27	nr	**50.59**
152mm dia.	19.14	23.24	2.04	35.91	nr	**59.15**
178mm dia.	20.05	24.35	2.39	42.07	nr	**66.41**
203mm dia.	22.54	27.37	2.56	45.06	nr	**72.43**
254mm dia.	25.15	30.54	2.95	51.92	nr	**82.46**
304mm dia.	29.31	35.59	3.41	60.02	nr	**95.61**
355mm dia.	32.56	39.54	3.77	66.36	nr	**105.89**
15 degree insulated elbow; including two locking bands						
127mm dia.	74.58	90.56	1.57	27.63	nr	**118.20**
152mm dia.	82.88	100.64	1.79	31.51	nr	**132.15**
178mm dia.	88.64	107.64	2.05	36.08	nr	**143.72**
203mm dia.	93.97	114.11	2.33	41.01	nr	**155.12**
254mm dia.	96.78	117.52	2.45	43.12	nr	**160.64**
304mm dia.	122.36	148.58	3.43	60.37	nr	**208.95**
355mm dia.	163.69	198.77	4.71	82.90	nr	**281.67**

Material Costs/Measured Work Prices - Mechanical Installations 171

T:MECHANICAL/COOLING/HEATING SYSTEMS

Item	Net Price £	Material £	Labour hours	Labour £	Unit	Total rate £
30 degree insulated elbow; including two locking bands						
127mm dia.	74.58	90.56	1.44	25.35	nr	**115.91**
152mm dia.	82.88	100.64	1.62	28.51	nr	**129.15**
178mm dia.	88.64	107.64	1.89	33.27	nr	**140.90**
203mm dia.	93.97	114.11	2.17	38.19	nr	**152.30**
254mm dia.	96.78	117.52	2.16	38.02	nr	**155.54**
304mm dia.	122.36	148.58	2.74	48.23	nr	**196.81**
355mm dia.	163.38	198.39	3.17	55.79	nr	**254.19**
400mm dia.	163.69	198.77	3.53	62.13	nr	**260.90**
450mm dia.	189.71	230.36	3.88	68.29	nr	**298.66**
500mm dia.	198.89	241.51	4.24	74.63	nr	**316.14**
550mm dia.	213.45	259.19	4.61	81.14	nr	**340.33**
600mm dia.	232.75	282.63	4.96	87.30	nr	**369.93**
45 degree insulated elbow; including two locking bands						
127mm dia.	74.58	90.56	1.44	25.35	nr	**115.91**
152mm dia.	82.88	100.64	1.51	26.58	nr	**127.22**
178mm dia.	88.64	107.64	1.58	27.81	nr	**135.44**
203mm dia.	93.97	114.11	1.66	29.22	nr	**143.33**
254mm dia.	96.78	117.52	1.72	30.27	nr	**147.79**
304mm dia.	122.36	148.58	1.80	31.68	nr	**180.26**
355mm dia.	163.69	198.77	1.94	34.15	nr	**232.91**
400mm dia.	208.74	253.47	2.01	35.38	nr	**288.85**
450mm dia.	218.65	265.51	2.09	36.79	nr	**302.29**
500mm dia.	234.54	284.80	2.16	38.02	nr	**322.82**
550mm dia.	255.80	310.62	2.23	39.25	nr	**349.87**
600mm dia.	262.28	318.49	2.30	40.48	nr	**358.97**
Chimney supports						
Wall support, galvanised; including plate and brackets						
127mm dia.	46.63	56.62	2.24	39.43	nr	**96.05**
152mm dia.	51.27	62.26	2.44	42.95	nr	**105.20**
178mm dia.	56.60	68.73	2.52	44.35	nr	**113.08**
203mm dia.	58.93	71.56	2.77	48.75	nr	**120.31**
254mm dia.	70.28	85.34	2.98	52.45	nr	**137.79**
304mm dia.	79.47	96.50	3.46	60.90	nr	**157.40**
355mm dia.	106.37	129.17	4.08	71.81	nr	**200.98**
400mm dia.; including 300mm support length and collar	273.23	331.78	4.80	84.48	nr	**416.27**
450mm dia.; including 300mm support length and collar	291.31	353.74	5.62	98.92	nr	**452.65**
500mm dia.; including 300mm support length and collar	318.25	386.45	6.24	109.83	nr	**496.28**
550mm dia.; including 300mm support length and collar	348.19	422.81	6.97	122.68	nr	**545.48**
600mm dia.; including 300mm support length and collar	367.57	446.34	7.49	131.83	nr	**578.17**

Material Costs/Measured Work Prices - Mechanical Installations

T:MECHANICAL/COOLING/HEATING SYSTEMS

Item	Net Price £	Material £	Labour hours	Labour £	Unit	Total rate £
T10 : GAS/OIL FIRED BOILERS						
Chimneys; applicable to domestic, medium sized industrial and commercial oil and gas appliances; stainless steel, twin wall, insulated; for use internally or externally (Continued)						
Ceiling/floor support						
127mm dia.	14.43	17.52	1.86	32.74	nr	**50.26**
152mm dia.	16.05	19.49	2.14	37.67	nr	**57.16**
178mm dia.	19.21	23.33	1.93	33.97	nr	**57.30**
203mm dia.	28.93	35.13	2.74	48.23	nr	**83.36**
254mm dia.	33.00	40.07	3.21	56.50	nr	**96.57**
304mm dia.	37.72	45.80	3.68	64.77	nr	**110.57**
355mm dia.	44.95	54.58	4.28	75.33	nr	**129.91**
400mm dia.	228.09	276.97	4.86	85.54	nr	**362.51**
450mm dia.	240.13	291.59	5.46	96.10	nr	**387.69**
500mm dia.	253.49	307.81	6.04	106.31	nr	**414.12**
550mm dia.	277.50	336.97	6.65	117.05	nr	**454.01**
600mm dia.	282.24	342.72	7.24	127.43	nr	**470.15**
Ceiling/floor firestop spacer						
127mm dia.	2.81	3.41	0.66	11.62	nr	**15.03**
152mm dia.	3.13	3.80	0.69	12.14	nr	**15.95**
178mm dia.	3.53	4.29	0.70	12.32	nr	**16.61**
203mm dia.	4.19	5.09	0.87	15.31	nr	**20.40**
254mm dia.	4.32	5.25	0.91	16.02	nr	**21.26**
304mm dia.	5.20	6.31	0.95	16.72	nr	**23.04**
355mm dia.	9.47	11.50	0.99	17.42	nr	**28.92**
Wall band; internal or external use						
127mm dia.	17.00	20.64	1.03	18.13	nr	**38.77**
152mm dia.	17.75	21.55	1.07	18.83	nr	**40.39**
178mm dia.	18.45	22.40	1.11	19.54	nr	**41.94**
203mm dia.	19.48	23.65	1.18	20.77	nr	**44.42**
254mm dia.	20.27	24.61	1.30	22.88	nr	**47.49**
304mm dia.	21.90	26.59	1.45	25.52	nr	**52.11**
355mm dia.	23.31	28.31	1.65	29.04	nr	**57.35**
400mm dia.	28.90	35.09	1.85	32.56	nr	**67.65**
450mm dia.	30.80	37.40	2.39	42.07	nr	**79.47**
500mm dia.	37.02	44.95	2.25	39.60	nr	**84.56**
550mm dia.	38.78	47.09	2.45	43.12	nr	**90.21**
600mm dia.	40.99	49.77	2.66	46.82	nr	**96.59**

Material Costs/Measured Work Prices - Mechanical Installations 173

T:MECHANICAL/COOLING/HEATING SYSTEMS

Item	Net Price £	Material £	Labour hours	Labour £	Unit	Total rate £
Chimney flashings and terminals						
Insulated top stub; including one locking band						
127mm dia.	41.88	50.85	1.49	26.23	nr	**77.08**
152mm dia.	47.35	57.50	1.90	33.44	nr	**90.94**
178mm dia.	50.98	61.91	1.92	33.79	nr	**95.70**
203mm dia.	54.25	65.88	2.20	38.72	nr	**104.60**
254mm dia.	57.60	69.94	2.49	43.83	nr	**113.77**
304mm dia.	78.74	95.61	2.79	49.11	nr	**144.72**
355mm dia.	103.95	126.23	3.19	56.15	nr	**182.37**
400mm dia.	100.23	121.71	3.59	63.19	nr	**184.90**
450mm dia.	108.89	132.23	3.97	69.88	nr	**202.10**
500mm dia.	125.96	152.95	4.38	77.09	nr	**230.04**
550mm dia.	131.80	160.04	4.78	84.13	nr	**244.18**
600mm dia.	136.41	165.64	5.17	91.00	nr	**256.64**
Rain cap; including one locking band						
127mm dia.	22.38	27.18	1.49	26.23	nr	**53.40**
152mm dia.	23.39	28.40	1.54	27.11	nr	**55.51**
178mm dia.	25.76	31.28	1.72	30.27	nr	**61.55**
203mm dia.	30.81	37.41	2.00	35.20	nr	**72.61**
254mm dia.	40.51	49.19	2.49	43.83	nr	**93.02**
304mm dia.	54.61	66.31	2.80	49.28	nr	**115.60**
355mm dia.	73.17	88.85	3.19	56.15	nr	**145.00**
400mm dia.	72.99	88.63	3.45	60.72	nr	**149.35**
450mm dia.	79.38	96.39	3.97	69.88	nr	**166.27**
500mm dia.	85.84	104.24	4.38	77.09	nr	**181.33**
550mm dia.	92.11	111.85	4.78	84.13	nr	**195.98**
600mm dia.	98.47	119.57	5.17	91.00	nr	**210.57**
Round top; including one locking band						
127mm dia	42.80	51.97	1.49	26.23	nr	**78.20**
152mm dia	46.67	56.67	1.65	29.04	nr	**85.71**
178mm dia	53.21	64.61	1.92	33.79	nr	**98.41**
203mm dia	62.62	76.04	2.20	38.72	nr	**114.76**
254mm dia	74.03	89.89	2.49	43.83	nr	**133.72**
304mm dia	97.23	118.07	2.80	49.28	nr	**167.35**
355mm dia	129.72	157.52	3.19	56.15	nr	**213.67**
Coping cap; including one locking band						
127mm dia.	23.97	29.11	1.49	26.23	nr	**55.33**
152mm dia.	25.11	30.49	1.65	29.04	nr	**59.53**
178mm dia.	27.63	33.55	1.92	33.79	nr	**67.34**
203mm dia.	33.15	40.25	2.20	38.72	nr	**78.98**
254mm dia.	40.51	49.19	2.49	43.83	nr	**93.02**
304mm dia.	54.61	66.31	2.79	49.11	nr	**115.42**
355mm dia.	73.17	88.85	3.19	56.15	nr	**145.00**

174   *Material Costs/Measured Work Prices - Mechanical Installations*

T:MECHANICAL/COOLING/HEATING SYSTEMS

Item	Net Price £	Material £	Labour hours	Labour £	Unit	Total rate £
T10 : GAS/OIL FIRED BOILERS						
Chimneys; applicable to domestic, medium sized industrial and commercial oil and gas appliances; stainless steel, twin wall, insulated; for use internally or externally (Continued)						
Storm collar						
127mm dia.	4.55	5.53	0.52	9.15	nr	**14.68**
152mm dia.	4.86	5.90	0.55	9.68	nr	**15.58**
178mm dia.	5.39	6.55	0.57	10.03	nr	**16.58**
203mm dia.	5.64	6.85	0.66	11.62	nr	**18.47**
254mm dia.	7.07	8.59	0.66	11.62	nr	**20.20**
304mm dia.	7.37	8.95	0.72	12.67	nr	**21.62**
355mm dia.	7.87	9.56	0.77	13.55	nr	**23.11**
400mm dia.	20.88	25.35	0.82	14.43	nr	**39.79**
450mm dia.	22.98	27.90	0.87	15.31	nr	**43.22**
500mm dia.	25.06	30.43	0.92	16.19	nr	**46.62**
550mm dia.	27.15	32.97	0.98	17.25	nr	**50.22**
600mm dia.	29.24	35.51	1.03	18.13	nr	**53.63**
Flat flashing; including storm collar and sealant						
127mm dia.	25.52	30.99	1.49	26.23	nr	**57.21**
152mm dia.	26.37	32.02	1.65	29.04	nr	**61.06**
178mm dia.	27.67	33.60	1.92	33.79	nr	**67.39**
203mm dia.	30.30	36.79	2.20	38.72	nr	**75.52**
254mm dia.	41.41	50.28	2.49	43.83	nr	**94.11**
304mm dia.	49.48	60.08	2.80	49.28	nr	**109.37**
355mm dia.	77.97	94.68	3.20	56.32	nr	**151.00**
400mm dia.	108.77	132.08	3.59	63.19	nr	**195.27**
450mm dia.	124.95	151.73	3.97	69.88	nr	**221.60**
500mm dia.	135.37	164.38	4.38	77.09	nr	**241.47**
550mm dia.	143.73	174.53	4.78	84.13	nr	**258.66**
600mm dia.	148.92	180.83	5.17	91.00	nr	**271.83**
5 - 30 degree rigid adjustable flashing; including storm collar and sealant						
127mm dia.	43.47	52.79	1.49	26.23	nr	**79.01**
152mm dia.	45.80	55.61	1.65	29.04	nr	**84.66**
178mm dia.	48.71	59.15	1.92	33.79	nr	**92.94**
203mm dia.	51.21	62.18	2.20	38.72	nr	**100.91**
254mm dia.	53.91	65.46	2.49	43.83	nr	**109.29**
304mm dia.	66.72	81.02	2.80	49.28	nr	**130.30**
355mm dia.	75.88	92.14	3.19	56.15	nr	**148.29**
400mm dia.	244.63	297.05	3.59	63.19	nr	**360.24**
450mm dia.	285.49	346.67	3.97	69.88	nr	**416.55**
500mm dia.	304.78	370.09	4.38	77.09	nr	**447.19**
550mm dia.	318.93	387.28	4.77	83.96	nr	**471.23**
600mm dia.	347.00	421.36	5.17	91.00	nr	**512.36**

Material Costs/Measured Work Prices - Mechanical Installations   175

T:MECHANICAL/COOLING/HEATING SYSTEMS

Item	Net Price £	Material £	Labour hours	Labour £	Unit	Total rate £
Domestic and small commercial; twin walled gas vent system suitable for gas fired appliances; domestic gas boilers; small commercial boilers with internal or external flues						
152mm long						
100mm dia.	4.52	5.49	0.52	9.15	nr	**14.64**
125mm dia.	5.56	6.75	0.52	9.15	nr	**15.90**
150mm dia.	6.02	7.31	0.52	9.15	nr	**16.46**
305mm long						
100mm dia.	6.86	8.33	0.52	9.15	nr	**17.48**
125mm dia.	8.05	9.78	0.52	9.15	nr	**18.93**
150mm dia.	9.54	11.58	0.52	9.15	nr	**20.74**
457mm long						
100mm dia.	7.59	9.22	0.55	9.68	nr	**18.90**
125mm dia.	8.53	10.36	0.55	9.68	nr	**20.04**
150mm dia.	10.56	12.82	0.55	9.68	nr	**22.50**
914mm long						
100mm dia.	13.56	16.47	0.62	10.91	nr	**27.38**
125mm dia.	15.80	19.19	0.62	10.91	nr	**30.10**
150mm dia.	18.12	22.00	0.62	10.91	nr	**32.92**
1524mm long						
100mm dia.	19.59	23.79	0.82	14.43	nr	**38.22**
125mm dia.	24.10	29.26	0.84	14.78	nr	**44.05**
150mm dia.	25.88	31.43	0.84	14.78	nr	**46.21**
Adjustable length 305mm long						
100mm dia.	8.68	10.54	0.56	9.86	nr	**20.40**
125mm dia.	9.75	11.84	0.56	9.86	nr	**21.70**
150mm dia.	12.29	14.92	0.56	9.86	nr	**24.78**
Adjustable length 457mm long						
100mm dia.	11.70	14.21	0.56	9.86	nr	**24.06**
125mm dia.	14.19	17.23	0.56	9.86	nr	**27.09**
150mm dia.	15.79	19.17	0.56	9.86	nr	**29.03**
Adjustable elbow 0 - 90 deg						
100mm dia.	9.90	12.02	0.48	8.45	nr	**20.47**
125mm dia.	11.70	14.21	0.48	8.45	nr	**22.66**
150mm dia.	14.66	17.80	0.48	8.45	nr	**26.25**
Draughthood connector						
100mm dia.	3.06	3.72	0.48	8.45	nr	**12.16**
125mm dia.	3.44	4.18	0.48	8.45	nr	**12.63**
150mm dia.	3.74	4.54	0.48	8.45	nr	**12.99**
Adaptor						
100mm dia.	7.41	9.00	0.48	8.45	nr	**17.45**
125mm dia.	7.57	9.19	0.48	8.45	nr	**17.64**
150mm dia.	7.73	9.39	0.48	8.45	nr	**17.83**

176 **Material Costs/Measured Work Prices - Mechanical Installations**

T:MECHANICAL/COOLING/HEATING SYSTEMS

Item	Net Price £	Material £	Labour hours	Labour £	Unit	Total rate £
T10 : GAS/OIL FIRED BOILERS						
Domestic and small commercial; twin walled gas vent system suitable for gas fired appliances; domestic gas boilers; small commercial boilers with internal or external flues (Continued)						
Support plate						
100mm dia.	5.35	6.50	0.48	8.45	nr	**14.94**
125mm dia.	5.69	6.91	0.48	8.45	nr	**15.36**
150mm dia.	6.09	7.40	0.48	8.45	nr	**15.84**
Wall band						
100mm dia.	4.83	5.87	0.48	8.45	nr	**14.31**
125mm dia.	5.15	6.25	0.48	8.45	nr	**14.70**
150mm dia.	6.53	7.93	0.48	8.45	nr	**16.38**
Firestop						
100mm dia.	2.10	2.55	0.48	8.45	nr	**11.00**
125mm dia.	2.10	2.55	0.48	8.45	nr	**11.00**
150mm dia.	2.40	2.91	0.48	8.45	nr	**11.36**
Flat flashing						
125mm dia.	14.25	17.30	0.55	9.68	nr	**26.98**
150mm dia.	19.83	24.08	0.55	9.68	nr	**33.76**
Adjustable flashing 5-30 deg.						
100mm dia.	37.15	45.11	0.55	9.68	nr	**54.79**
125mm dia.	58.00	70.43	0.55	9.68	nr	**80.11**
Storm collar						
100mm dia.	2.96	3.59	0.55	9.68	nr	**13.27**
125mm dia.	3.03	3.68	0.55	9.68	nr	**13.36**
150mm dia.	3.11	3.78	0.55	9.68	nr	**13.46**
Gas vent terminal						
100mm dia.	11.30	13.72	0.55	9.68	nr	**23.40**
125mm dia.	12.42	15.08	0.55	9.68	nr	**24.76**
150mm dia.	15.93	19.34	0.55	9.68	nr	**29.02**
Twin wall galvanised steel flue box, 125mm dia.; (fitted where no chimney exists) for gas fire						
Free standing	71.83	87.22	2.15	37.84	nr	**125.06**
Recess	71.83	87.22	2.15	37.84	nr	**125.06**
Back boiler	52.92	64.26	2.40	42.24	nr	**106.50**

Material Costs/Measured Work Prices - Mechanical Installations

T:MECHANICAL/COOLING/HEATING SYSTEMS

Item	Net Price £	Material £	Labour hours	Labour £	Unit	Total rate £
T13 : PACKAGED STEAM GENERATORS						
Packaged steam boilers; boiler mountings centrifugal water feed pump; insulation; and sheet steel wrap around casing; plastic coated						
Gas fired						
293 kW rating	13400.30	16271.98	86.45	1521.59	nr	**17793.57**
1465 kW rating	28582.50	34707.73	148.22	2608.79	nr	**37316.52**
2930 kW rating	41045.50	49841.55	207.50	3652.17	nr	**53493.72**
Oil fired						
293 kW rating	12154.00	14758.60	86.45	1521.59	nr	**16280.19**
1465 kW rating	26440.10	32106.21	148.22	2608.79	nr	**34715.00**
2930 kW rating	40015.50	48590.82	207.50	3652.17	nr	**52242.99**

Material Costs/Measured Work Prices - Mechanical Installations

T:MECHANICAL/COOLING/HEATING SYSTEMS

Item	Net Price £	Material £	Labour hours	Labour £	Unit	Total rate £
T14 : HEAT PUMPS						
Split systems; ceiling void units; electrical work elsewhere						
Low profile heat pump units; four stage remote control; indoor						
Cooling 13.7kW, heating 15.5kW	1189.27	1444.13	16.60	292.17	nr	**1736.30**
Cooling 16.4kW, heating 18.8kW	1538.31	1867.97	16.60	292.17	nr	**2160.14**
Cooling 21.7kW, heating 24.9kW	1906.44	2314.99	16.60	292.17	nr	**2607.16**
Low profile heat pump units; four stage remote control; outdoor						
Cooling 13.7kW, heating 15.5kW	1457.68	1770.06	16.60	292.17	nr	**2062.23**
Cooling 16.4kW, heating 18.8kW	2781.68	3377.79	16.60	292.17	nr	**3669.97**
Cooling 21.7kW, heating 24.9kW	2914.29	3538.82	16.60	292.17	nr	**3831.00**

Material Costs/Measured Work Prices - Mechanical Installations 179

T:MECHANICAL/COOLING/HEATING SYSTEMS

Item	Net Price £	Material £	Labour hours	Labour £	Unit	Total rate £
T31 : LOW TEMPERATURE HOT WATER HEATING						
Y 10 - PIPELINES						
SCREWED STEEL						
Black steel pipes; screwed and socketed joints; BS 1387: 1985						
Varnished; light						
15mm dia.	1.12	1.36	0.52	9.15	m	**10.51**
20mm dia.	1.46	1.77	0.55	9.68	m	**11.45**
25mm dia.	2.00	2.43	0.60	10.56	m	**12.99**
32mm dia.	2.48	3.01	0.67	11.79	m	**14.80**
40mm dia.	3.07	3.73	0.75	13.20	m	**16.93**
50mm dia.	4.01	4.87	0.85	14.96	m	**19.83**
65mm dia.	5.95	7.23	0.93	16.37	m	**23.59**
80mm dia.	7.15	8.68	1.07	18.83	m	**27.52**
100mm dia.	10.48	12.73	1.46	25.70	m	**38.42**
Varnished; medium						
8mm dia.	0.88	1.07	0.51	8.98	m	**10.04**
10mm dia.	0.88	1.07	0.51	8.98	m	**10.04**
15mm dia.	0.98	1.19	0.52	9.15	m	**10.34**
20mm dia.	1.15	1.40	0.55	9.68	m	**11.08**
25mm dia.	1.65	2.00	0.60	10.56	m	**12.56**
32mm dia.	2.05	2.49	0.67	11.79	m	**14.28**
40mm dia.	2.38	2.89	0.75	13.20	m	**16.09**
50mm dia.	3.37	4.09	0.85	14.96	m	**19.05**
65mm dia.	4.65	5.65	0.93	16.37	m	**22.02**
80mm dia.	6.05	7.35	1.07	18.83	m	**26.18**
100mm dia.	8.68	10.54	1.46	25.70	m	**36.24**
125mm dia.	13.27	16.11	1.72	30.27	m	**46.39**
150mm dia.	15.76	19.14	1.96	34.50	m	**53.63**
Varnished; heavy						
15mm dia.	1.15	1.40	0.52	9.15	m	**10.55**
20mm dia.	1.38	1.68	0.55	9.68	m	**11.36**
25mm dia.	2.00	2.43	0.60	10.56	m	**12.99**
32mm dia.	2.49	3.02	0.67	11.79	m	**14.82**
40mm dia.	2.91	3.53	0.75	13.20	m	**16.73**
50mm dia.	4.06	4.93	0.85	14.96	m	**19.89**
65mm dia.	5.59	6.79	0.93	16.37	m	**23.16**
80mm dia.	7.14	8.67	1.07	18.83	m	**27.50**
100mm dia.	10.08	12.24	1.46	25.70	m	**37.94**
125mm dia.	14.05	17.06	1.72	30.27	m	**47.33**
150mm dia.	16.78	20.38	1.96	34.50	m	**54.87**

180   *Material Costs/Measured Work Prices - Mechanical Installations*

T:MECHANICAL/COOLING/HEATING SYSTEMS

Item	Net Price £	Material £	Labour hours	Labour £	Unit	Total rate £
T31 : LOW TEMPERATURE HOT WATER HEATING						
Y 10 - PIPELINES						
SCREWED STEEL (Continued)						
Extra Over black steel screwed pipes; black steel flanges, screwed and drilled; metric; BS 4504						
Screwed flanges; PN6						
15mm dia.	2.64	3.21	0.35	6.16	nr	9.37
20mm dia.	2.64	3.21	0.47	8.27	nr	11.48
25mm dia.	2.69	3.27	0.53	9.33	nr	12.59
32mm dia.	3.48	4.23	0.62	10.91	nr	15.14
40mm dia.	3.48	4.23	0.70	12.32	nr	16.55
50mm dia.	4.16	5.05	0.84	14.78	nr	19.84
65mm dia.	5.28	6.41	1.03	18.13	nr	24.54
80mm dia.	6.48	7.87	1.23	21.65	nr	29.52
100mm dia.	7.64	9.28	1.41	24.82	nr	34.09
125mm dia.	14.06	17.07	1.77	31.15	nr	48.23
150mm dia.	18.27	22.19	2.21	38.90	nr	61.08
Screwed flanges; PN16						
15mm dia.	2.56	3.11	0.35	6.16	nr	9.27
20mm dia.	2.58	3.13	0.47	8.27	nr	11.41
25mm dia.	2.63	3.19	0.53	9.33	nr	12.52
32mm dia.	3.60	4.37	0.62	10.91	nr	15.28
40mm dia.	3.60	4.37	0.70	12.32	nr	16.69
50mm dia.	4.26	5.17	0.84	14.78	nr	19.96
65mm dia.	5.18	6.29	1.03	18.13	nr	24.42
80mm dia.	6.78	8.23	1.23	21.65	nr	29.88
100mm dia.	7.82	9.50	1.41	24.82	nr	34.31
125mm dia.	18.14	22.03	1.77	31.15	nr	53.18
150mm dia.	18.87	22.91	2.21	38.90	nr	61.81
Extra Over black steel screwed pipes; black steel flanges, screwed and drilled; imperial; BS 10						
Screwed flanges; Table E						
1/2" dia.	3.07	3.73	0.35	6.16	nr	9.89
3/4" dia.	3.09	3.75	0.47	8.27	nr	12.02
1" dia.	3.14	3.81	0.53	9.33	nr	13.14
1 1/4" dia.	3.31	4.02	0.62	10.91	nr	14.93
1 1/2" dia.	3.31	4.02	0.70	12.32	nr	16.34
2" dia.	3.35	4.07	0.84	14.78	nr	18.85
2 1/2" dia.	4.13	5.02	1.03	18.13	nr	23.14
3" dia.	4.56	5.54	1.23	21.65	nr	27.19
4" dia.	6.49	7.88	1.41	24.82	nr	32.70
5" dia.	11.82	14.35	1.77	31.15	nr	45.51
6" dia.	12.54	15.23	2.21	38.90	nr	54.13

Material Costs/Measured Work Prices - Mechanical Installations 181

T:MECHANICAL/COOLING/HEATING SYSTEMS

Item	Net Price £	Material £	Labour hours	Labour £	Unit	Total rate £
Extra Over black steel screwed pipes; black steel flange connections						
Bolted connection between pair of flanges; including gasket, bolts, nuts and washers						
50mm dia.	5.16	6.27	0.53	9.33	nr	**15.59**
65mm dia.	5.98	7.26	0.53	9.33	nr	**16.59**
80mm dia.	8.64	10.49	0.53	9.33	nr	**19.82**
100mm dia.	9.63	11.69	0.61	10.74	nr	**22.43**
125mm dia.	17.49	21.24	0.61	10.74	nr	**31.97**
150mm dia.	18.56	22.54	0.90	15.84	nr	**38.38**
Extra Over black steel screwed pipes; black heavy steel tubular fittings; BS 1387						
Long screw connection with socket and backnut						
15mm dia.	2.77	3.36	0.63	11.09	nr	**14.45**
20mm dia.	3.29	4.00	0.84	14.78	nr	**18.78**
25mm dia.	4.70	5.71	0.95	16.72	nr	**22.43**
32mm dia.	5.97	7.25	1.11	19.54	nr	**26.79**
40mm dia.	7.03	8.54	1.28	22.53	nr	**31.07**
50mm dia.	10.20	12.39	1.53	26.93	nr	**39.32**
65mm dia.	20.76	25.21	1.87	32.91	nr	**58.12**
80mm dia.	28.02	34.02	2.21	38.90	nr	**72.92**
100mm dia.	44.59	54.15	3.05	53.68	nr	**107.83**
Running nipple						
15mm dia.	0.44	0.53	0.50	8.80	nr	**9.33**
20mm dia.	0.56	0.68	0.68	11.97	nr	**12.65**
25mm dia.	0.68	0.83	0.77	13.55	nr	**14.38**
32mm dia.	0.96	1.17	0.90	15.84	nr	**17.01**
40mm dia.	1.29	1.57	1.03	18.13	nr	**19.70**
50mm dia.	1.96	2.38	1.23	21.65	nr	**24.03**
65mm dia.	4.15	5.04	1.50	26.40	nr	**31.44**
80mm dia.	6.47	7.86	1.78	31.33	nr	**39.19**
100mm dia.	10.12	12.29	2.38	41.89	nr	**54.18**
Barrel nipple						
15mm dia.	0.39	0.47	0.50	8.80	nr	**9.27**
20mm dia.	0.44	0.53	0.68	11.97	nr	**12.50**
25mm dia.	0.63	0.77	0.77	13.55	nr	**14.32**
32mm dia.	0.79	0.96	0.90	15.84	nr	**16.80**
40mm dia.	0.98	1.19	1.03	18.13	nr	**19.32**
50mm dia.	1.39	1.69	1.23	21.65	nr	**23.34**
65mm dia.	2.52	3.06	1.50	26.40	nr	**29.46**
80mm dia.	3.51	4.26	1.78	31.33	nr	**35.59**
100mm dia.	6.36	7.72	2.38	41.89	nr	**49.61**
125mm dia.	11.80	14.33	2.87	50.51	nr	**64.84**
150mm dia.	18.60	22.59	3.39	59.67	nr	**82.25**

Material Costs/Measured Work Prices - Mechanical Installations

T:MECHANICAL/COOLING/HEATING SYSTEMS

Item	Net Price £	Material £	Labour hours	Labour £	Unit	Total rate £
T31 : LOW TEMPERATURE HOT WATER HEATING						
Y 10 - PIPELINES						
SCREWED STEEL (Continued)						
Close taper nipple						
15mm dia.	0.64	0.78	0.50	8.80	nr	**9.58**
20mm dia.	0.71	0.86	0.68	11.97	nr	**12.83**
25mm dia.	0.98	1.19	0.77	13.55	nr	**14.74**
32mm dia.	1.23	1.49	0.90	15.84	nr	**17.33**
40mm dia.	1.51	1.83	1.03	18.13	nr	**19.96**
50mm dia.	2.16	2.62	1.23	21.65	nr	**24.27**
65mm dia.	4.21	5.11	1.50	26.40	nr	**31.51**
80mm dia.	5.62	6.82	1.78	31.33	nr	**38.15**
100mm dia.	10.65	12.93	2.38	41.89	nr	**54.82**
125mm dia.	18.87	22.91	2.87	50.51	nr	**73.43**
150mm dia.	29.43	35.74	3.39	59.67	nr	**95.40**
90 degree bend with socket						
15mm dia.	1.62	1.97	0.64	11.26	nr	**13.23**
20mm dia.	2.11	2.56	0.85	14.96	nr	**17.52**
25mm dia.	2.39	2.90	0.97	17.07	nr	**19.98**
32mm dia.	3.67	4.46	1.12	19.71	nr	**24.17**
40mm dia.	4.15	5.04	1.29	22.70	nr	**27.74**
50mm dia.	7.00	8.50	1.55	27.28	nr	**35.78**
65mm dia.	12.74	15.47	1.89	33.27	nr	**48.74**
80mm dia.	19.66	23.87	2.24	39.43	nr	**63.30**
100mm dia.	34.57	41.98	3.09	54.39	nr	**96.36**
125mm dia.	130.18	158.08	3.39	59.67	nr	**217.74**
Extra Over black steel screwed pipes; black heavy steel fittings; BS 1740						
Plug						
15mm dia.	0.62	0.75	0.30	5.28	nr	**6.03**
20mm dia.	0.75	0.91	0.40	7.04	nr	**7.95**
25mm dia.	0.95	1.15	0.45	7.92	nr	**9.07**
32mm dia.	1.25	1.52	0.53	9.33	nr	**10.85**
40mm dia.	1.48	1.80	0.61	10.74	nr	**12.53**
50mm dia.	2.14	2.60	0.72	12.67	nr	**15.27**
65mm dia.	5.21	6.33	0.88	15.49	nr	**21.82**
80mm dia.	9.59	11.65	1.04	18.30	nr	**29.95**
100mm dia.	19.65	23.86	1.53	26.93	nr	**50.79**
150mm dia.	37.86	45.97	2.73	48.05	nr	**94.02**

Material Costs/Measured Work Prices - Mechanical Installations

T:MECHANICAL/COOLING/HEATING SYSTEMS

Item	Net Price £	Material £	Labour hours	Labour £	Unit	Total rate £
Socket						
15mm dia.	0.35	0.42	0.64	11.26	nr	**11.69**
20mm dia.	0.39	0.47	0.85	14.96	nr	**15.43**
25mm dia.	0.52	0.63	0.97	17.07	nr	**17.70**
32mm dia.	0.75	0.91	1.12	19.71	nr	**20.62**
40mm dia.	0.97	1.18	1.29	22.70	nr	**23.88**
50mm dia.	1.45	1.76	1.55	27.28	nr	**29.04**
65mm dia.	2.48	3.01	1.89	33.27	nr	**36.28**
80mm dia.	3.47	4.21	2.24	39.43	nr	**43.64**
100mm dia.	6.62	8.04	3.09	54.39	nr	**62.43**
125mm dia.	13.18	16.00	3.92	69.00	nr	**85.00**
150mm dia.	19.75	23.98	4.74	83.43	nr	**107.41**
Cone seat unions						
15mm dia.	3.30	4.01	0.64	11.26	nr	**15.27**
20mm dia.	4.43	5.38	0.85	14.96	nr	**20.34**
25mm dia.	6.19	7.52	0.97	17.07	nr	**24.59**
32mm dia.	12.22	14.84	1.12	19.71	nr	**34.55**
40mm dia.	14.03	17.04	1.29	22.70	nr	**39.74**
50mm dia.	20.78	25.23	1.55	27.28	nr	**52.51**
Elbow, male/female						
15mm dia.	2.46	2.99	0.64	11.26	nr	**14.25**
20mm dia.	3.34	4.06	0.85	14.96	nr	**19.02**
25mm dia.	5.12	6.22	0.97	17.07	nr	**23.29**
32mm dia.	10.15	12.33	1.12	19.71	nr	**32.04**
40mm dia.	10.15	12.33	1.29	22.70	nr	**35.03**
50mm dia.	17.71	21.51	1.55	27.28	nr	**48.79**
Elbow, female/female						
15mm dia.	2.20	2.67	0.64	11.26	nr	**13.94**
20mm dia.	2.68	3.25	0.85	14.96	nr	**18.21**
25mm dia.	3.65	4.43	0.97	17.07	nr	**21.51**
32mm dia.	7.27	8.83	1.12	19.71	nr	**28.54**
40mm dia.	7.27	8.83	1.29	22.70	nr	**31.53**
50mm dia.	9.38	11.39	1.55	27.28	nr	**38.67**
65mm dia.	21.15	25.68	1.89	33.27	nr	**58.95**
80mm dia.	31.09	37.75	2.24	39.43	nr	**77.18**
100mm dia.	58.65	71.22	3.09	54.39	nr	**125.61**
Equal tee						
15mm dia.	2.47	3.00	0.91	16.02	nr	**19.02**
20mm dia.	3.05	3.70	1.22	21.47	nr	**25.18**
25mm dia.	4.91	5.96	1.40	24.64	nr	**30.60**
32mm dia.	9.29	11.28	1.62	28.51	nr	**39.79**
40mm dia.	9.29	11.28	1.86	32.74	nr	**44.02**
50mm dia.	13.09	15.90	2.21	38.90	nr	**54.79**
65mm dia.	36.32	44.10	2.72	47.87	nr	**91.98**
80mm dia.	39.48	47.94	3.21	56.50	nr	**104.44**
100mm dia.	67.93	82.49	4.44	78.15	nr	**160.63**

Material Costs/Measured Work Prices - Mechanical Installations

T:MECHANICAL/COOLING/HEATING SYSTEMS

Item	Net Price £	Material £	Labour hours	Labour £	Unit	Total rate £
T31 : LOW TEMPERATURE HOT WATER HEATING						
Y 10 - PIPELINES						
SCREWED STEEL (Continued)						
Extra Over black steel screwed pipes; black malleable iron fittings; BS 143						
Cap						
15mm dia.	0.30	0.36	0.32	5.63	nr	**6.00**
20mm dia.	0.34	0.41	0.43	7.57	nr	**7.98**
25mm dia.	0.42	0.51	0.49	8.62	nr	**9.13**
32mm dia.	0.62	0.75	0.58	10.21	nr	**10.96**
40mm dia.	0.79	0.96	0.66	11.62	nr	**12.58**
50mm dia.	1.53	1.86	0.78	13.73	nr	**15.59**
65mm dia.	2.44	2.96	0.96	16.90	nr	**19.86**
80mm dia.	2.76	3.35	1.13	19.89	nr	**23.24**
100mm dia.	6.05	7.35	1.70	29.92	nr	**37.27**
Plain plug, hollow						
15mm dia.	0.24	0.29	0.28	4.93	nr	**5.22**
20mm dia.	0.30	0.36	0.38	6.69	nr	**7.05**
25mm dia.	0.36	0.44	0.44	7.74	nr	**8.18**
32mm dia.	0.52	0.63	0.51	8.98	nr	**9.61**
40mm dia.	0.85	1.03	0.59	10.38	nr	**11.42**
50mm dia.	1.18	1.43	0.70	12.32	nr	**13.75**
65mm dia.	1.79	2.17	0.85	14.96	nr	**17.13**
80mm dia.	2.68	3.25	1.00	17.60	nr	**20.86**
100mm dia.	4.93	5.99	1.44	25.35	nr	**31.33**
125mm dia.	12.89	15.65	1.98	34.85	nr	**50.50**
150mm dia.	15.46	18.77	2.53	44.53	nr	**63.30**
Plain plug, solid						
15mm dia.	0.62	0.75	0.28	4.93	nr	**5.68**
20mm dia.	0.62	0.75	0.38	6.69	nr	**7.44**
25mm dia.	0.92	1.12	0.44	7.74	nr	**8.86**
32mm dia.	1.13	1.37	0.51	8.98	nr	**10.35**
40mm dia.	1.48	1.80	0.59	10.38	nr	**12.18**
50mm dia.	1.96	2.38	0.70	12.32	nr	**14.70**
65mm dia.	2.02	2.45	0.85	14.96	nr	**17.41**
80mm dia.	3.01	3.65	1.00	17.60	nr	**21.26**
Elbow, male/female						
15mm dia.	0.38	0.46	0.64	11.26	nr	**11.73**
20mm dia.	0.52	0.63	0.85	14.96	nr	**15.59**
25mm dia.	0.85	1.03	0.97	17.07	nr	**18.11**
32mm dia.	1.40	1.70	1.12	19.71	nr	**21.41**
40mm dia.	2.08	2.53	1.29	22.70	nr	**25.23**
50mm dia.	2.68	3.25	1.55	27.28	nr	**30.54**
65mm dia.	5.16	6.27	1.89	33.27	nr	**39.53**
80mm dia.	7.05	8.56	2.24	39.43	nr	**47.99**
100mm dia.	12.32	14.96	3.09	54.39	nr	**69.35**

Material Costs/Measured Work Prices - Mechanical Installations

T:MECHANICAL/COOLING/HEATING SYSTEMS

Item	Net Price £	Material £	Labour hours	Labour £	Unit	Total rate £
Elbow						
15mm dia.	0.34	0.41	0.64	11.26	nr	**11.68**
20mm dia.	0.46	0.56	0.85	14.96	nr	**15.52**
25mm dia.	0.72	0.87	0.97	17.07	nr	**17.95**
32mm dia.	1.18	1.43	1.12	19.71	nr	**21.15**
40mm dia.	2.00	2.43	1.29	22.70	nr	**25.13**
50mm dia.	2.34	2.84	1.55	27.28	nr	**30.12**
65mm dia.	4.47	5.43	1.89	33.27	nr	**38.69**
80mm dia.	6.56	7.97	2.24	39.43	nr	**47.39**
100mm dia.	11.27	13.69	3.09	54.39	nr	**68.07**
125mm dia.	24.14	29.31	4.44	78.15	nr	**107.46**
150mm dia.	44.95	54.58	5.79	101.91	nr	**156.49**
45 degree elbow						
15mm dia.	0.72	0.87	0.64	11.26	nr	**12.14**
20mm dia.	0.90	1.09	0.85	14.96	nr	**16.05**
25mm dia.	1.32	1.60	0.97	17.07	nr	**18.68**
32mm dia.	2.42	2.94	1.12	19.71	nr	**22.65**
40mm dia.	2.98	3.62	1.29	22.70	nr	**26.32**
50mm dia.	4.08	4.95	1.55	27.28	nr	**32.24**
65mm dia.	5.31	6.45	1.89	33.27	nr	**39.71**
80mm dia.	7.99	9.70	2.24	39.43	nr	**49.13**
100mm dia.	15.43	18.74	3.09	54.39	nr	**73.12**
150mm dia.	41.31	50.16	5.79	101.91	nr	**152.07**
Bend, male/female						
15mm dia.	0.60	0.73	0.64	11.26	nr	**11.99**
20mm dia.	0.88	1.07	0.85	14.96	nr	**16.03**
25mm dia.	1.28	1.55	0.97	17.07	nr	**18.63**
32mm dia.	1.82	2.21	1.12	19.71	nr	**21.92**
40mm dia.	2.68	3.25	1.29	22.70	nr	**25.96**
50mm dia.	5.04	6.12	1.55	27.28	nr	**33.40**
65mm dia.	7.47	9.07	1.89	33.27	nr	**42.34**
80mm dia.	10.12	12.29	2.24	39.43	nr	**51.71**
100mm dia.	25.07	30.44	3.09	54.39	nr	**84.83**
Bend, male						
15mm dia.	1.22	1.48	0.64	11.26	nr	**12.75**
20mm dia.	1.36	1.65	0.85	14.96	nr	**16.61**
25mm dia.	2.00	2.43	0.97	17.07	nr	**19.50**
32mm dia.	3.91	4.75	1.12	19.71	nr	**24.46**
40mm dia.	5.48	6.65	1.29	22.70	nr	**29.36**
50mm dia.	7.33	8.90	1.55	27.28	nr	**36.18**
Bend, female						
15mm dia.	0.62	0.75	0.64	11.26	nr	**12.02**
20mm dia.	0.88	1.07	0.85	14.96	nr	**16.03**
25mm dia.	1.24	1.51	0.97	17.07	nr	**18.58**
32mm dia.	2.10	2.55	1.12	19.71	nr	**22.26**
40mm dia.	2.51	3.05	1.29	22.70	nr	**25.75**
50mm dia.	3.96	4.81	1.55	27.28	nr	**32.09**
65mm dia.	7.47	9.07	1.89	33.27	nr	**42.34**
80mm dia.	11.07	13.44	2.24	39.43	nr	**52.87**
100mm dia.	23.23	28.21	3.09	54.39	nr	**82.59**
125mm dia.	59.36	72.08	4.44	78.15	nr	**150.23**
150mm dia.	90.67	110.10	5.79	101.91	nr	**212.01**

T:MECHANICAL/COOLING/HEATING SYSTEMS

Item	Net Price £	Material £	Labour hours	Labour £	Unit	Total rate £
T31 : LOW TEMPERATURE HOT WATER HEATING						
Y 10 - PIPELINES						
SCREWED STEEL (Continued)						
Return bend						
15mm dia.	2.44	2.96	0.64	11.26	nr	**14.23**
20mm dia.	3.96	4.81	0.85	14.96	nr	**19.77**
25mm dia.	4.93	5.99	0.97	17.07	nr	**23.06**
32mm dia.	6.89	8.37	1.12	19.71	nr	**28.08**
40mm dia.	8.20	9.96	1.29	22.70	nr	**32.66**
50mm dia.	12.51	15.19	1.55	27.28	nr	**42.47**
Equal socket, parallel thread						
15mm dia.	0.32	0.39	0.64	11.26	nr	**11.65**
20mm dia.	0.38	0.46	0.85	14.96	nr	**15.42**
25mm dia.	0.52	0.63	0.97	17.07	nr	**17.70**
32mm dia.	0.88	1.07	1.12	19.71	nr	**20.78**
40mm dia.	1.18	1.43	1.29	22.70	nr	**24.14**
50mm dia.	1.79	2.17	1.55	27.28	nr	**29.45**
65mm dia.	2.89	3.51	1.89	33.27	nr	**36.77**
80mm dia.	3.99	4.85	2.24	39.43	nr	**44.27**
100mm dia.	6.77	8.22	3.09	54.39	nr	**62.61**
Concentric reducing socket						
20 x 15mm dia.	0.46	0.56	0.76	13.38	nr	**13.94**
25 x 15mm dia.	0.62	0.75	0.85	14.96	nr	**15.71**
25 x 20mm dia.	0.58	0.70	0.86	15.14	nr	**15.84**
32 x 25mm dia.	0.98	1.19	1.01	17.78	nr	**18.97**
40 x 25mm dia.	1.15	1.40	1.16	20.42	nr	**21.81**
40 x 32mm dia.	1.28	1.55	1.16	20.42	nr	**21.97**
50 x 25mm dia.	2.20	2.67	1.38	24.29	nr	**26.96**
50 x 40mm dia.	1.79	2.17	1.38	24.29	nr	**26.46**
65 x 50mm dia.	3.02	3.67	1.69	29.75	nr	**33.41**
80 x 50mm dia.	3.77	4.58	2.00	35.20	nr	**39.78**
100 x 50mm dia.	7.51	9.12	2.75	48.40	nr	**57.52**
100 x 80mm dia.	6.96	8.45	2.75	48.40	nr	**56.85**
150 x 100mm dia.	18.38	22.32	4.10	72.16	nr	**94.48**
Eccentric reducing socket						
20 x 15mm dia.	0.82	1.00	0.73	12.85	nr	**13.84**
25 x 15mm dia.	2.36	2.87	0.85	14.96	nr	**17.83**
25 x 20mm dia.	2.68	3.25	0.85	14.96	nr	**18.21**
32 x 25mm dia.	3.06	3.72	1.01	17.78	nr	**21.49**
40 x 25mm dia.	3.51	4.26	1.16	20.42	nr	**24.68**
40 x 32mm dia.	1.90	2.31	1.16	20.42	nr	**22.72**
50 x 25mm dia.	2.28	2.77	1.38	24.29	nr	**27.06**
50 x 40mm dia.	2.28	2.77	1.38	24.29	nr	**27.06**
65 x 50mm dia.	3.77	4.58	1.69	29.75	nr	**34.32**
80 x 50mm dia.	6.12	7.43	2.00	35.20	nr	**42.63**

Material Costs/Measured Work Prices - Mechanical Installations

T:MECHANICAL/COOLING/HEATING SYSTEMS

Item	Net Price £	Material £	Labour hours	Labour £	Unit	Total rate £
Hexagon bush						
20 x 15mm dia.	0.30	0.36	0.37	6.51	nr	**6.88**
25 x 15mm dia.	0.36	0.44	0.43	7.57	nr	**8.01**
25 x 20mm dia.	0.38	0.46	0.43	7.57	nr	**8.03**
32 x 25mm dia.	0.44	0.53	0.51	8.98	nr	**9.51**
40 x 25mm dia.	0.62	0.75	0.58	10.21	nr	**10.96**
40 x 32mm dia.	1.15	1.40	0.58	10.21	nr	**11.60**
50 x 25mm dia.	1.28	1.55	0.71	12.50	nr	**14.05**
50 x 40mm dia.	1.18	1.43	0.71	12.50	nr	**13.93**
65 x 50mm dia.	1.98	2.40	0.85	14.96	nr	**17.36**
80 x 50mm dia.	2.98	3.62	1.00	17.60	nr	**21.22**
100 x 50mm dia.	6.60	8.01	1.52	26.75	nr	**34.77**
100 x 80mm dia.	5.49	6.67	1.52	26.75	nr	**33.42**
150 x 100mm dia.	17.39	21.12	2.57	45.23	nr	**66.35**
Hexagon nipple						
15mm dia.	0.32	0.39	0.28	4.93	nr	**5.32**
20mm dia.	0.36	0.44	0.38	6.69	nr	**7.13**
25mm dia.	0.52	0.63	0.44	7.74	nr	**8.38**
32mm dia.	0.85	1.03	0.51	8.98	nr	**10.01**
40mm dia.	0.98	1.19	0.59	10.38	nr	**11.57**
50mm dia.	1.79	2.17	0.70	12.32	nr	**14.49**
65mm dia.	2.58	3.13	0.85	14.96	nr	**18.09**
80mm dia.	3.72	4.52	1.00	17.60	nr	**22.12**
100mm dia.	6.32	7.67	1.44	25.35	nr	**33.02**
150mm dia.	17.87	21.70	2.32	40.83	nr	**62.53**
Union, male/female						
15mm dia.	2.85	3.46	0.64	11.26	nr	**14.73**
20mm dia.	3.85	4.68	0.85	14.96	nr	**19.64**
25mm dia.	4.87	5.91	0.97	17.07	nr	**22.99**
32mm dia.	6.74	8.18	1.12	19.71	nr	**27.90**
40mm dia.	8.01	9.73	1.29	22.70	nr	**32.43**
50mm dia.	10.71	13.01	1.55	27.28	nr	**40.29**
65mm dia.	20.14	24.46	1.89	33.27	nr	**57.72**
80mm dia.	27.69	33.62	2.24	39.43	nr	**73.05**
Union, female						
15mm dia.	2.73	3.31	0.64	11.26	nr	**14.58**
20mm dia.	3.05	3.70	0.85	14.96	nr	**18.66**
25mm dia.	4.01	4.87	0.97	17.07	nr	**21.94**
32mm dia.	5.53	6.72	1.12	19.71	nr	**26.43**
40mm dia.	6.71	8.15	1.29	22.70	nr	**30.85**
50mm dia.	8.50	10.32	1.55	27.28	nr	**37.60**
65mm dia.	16.48	20.01	1.89	33.27	nr	**53.28**
80mm dia.	26.90	32.66	2.24	39.43	nr	**72.09**
100mm dia.	50.59	61.43	3.09	54.39	nr	**115.82**
Union elbow, male/female						
15mm dia.	4.88	5.93	0.55	9.68	nr	**15.61**
20mm dia.	6.59	8.00	0.85	14.96	nr	**22.96**
25mm dia.	7.95	9.65	0.97	17.07	nr	**26.73**

Material Costs/Measured Work Prices - Mechanical Installations

T:MECHANICAL/COOLING/HEATING SYSTEMS

Item	Net Price £	Material £	Labour hours	Labour £	Unit	Total rate £
T31 : LOW TEMPERATURE HOT WATER HEATING						
Y 10 - PIPELINES						
SCREWED STEEL (Continued)						
Twin elbow						
15mm dia.	1.62	1.97	0.91	16.02	nr	**17.98**
20mm dia.	1.79	2.17	1.22	21.47	nr	**23.65**
25mm dia.	2.89	3.51	1.39	24.47	nr	**27.97**
32mm dia.	5.12	6.22	1.62	28.51	nr	**34.73**
40mm dia.	6.49	7.88	1.86	32.74	nr	**40.62**
50mm dia.	8.33	10.12	2.21	38.90	nr	**49.01**
65mm dia.	13.04	15.83	2.72	47.87	nr	**63.71**
80mm dia.	20.41	24.78	3.21	56.50	nr	**81.28**
Equal tee						
15mm dia.	0.46	0.56	0.91	16.02	nr	**16.58**
20mm dia.	0.68	0.83	1.22	21.47	nr	**22.30**
25mm dia.	0.98	1.19	1.39	24.47	nr	**25.66**
32mm dia.	1.62	1.97	1.62	28.51	nr	**30.48**
40mm dia.	2.20	2.67	1.86	32.74	nr	**35.41**
50mm dia.	3.18	3.86	2.21	38.90	nr	**42.76**
65mm dia.	6.44	7.82	2.72	47.87	nr	**55.69**
80mm dia.	7.51	9.12	3.21	56.50	nr	**65.62**
100mm dia.	13.61	16.53	4.44	78.15	nr	**94.67**
125mm dia.	33.37	40.52	5.38	94.69	nr	**135.21**
150mm dia.	53.17	64.56	6.31	111.06	nr	**175.63**
Tee reducing on branch						
20 x 15mm dia.	0.62	0.75	1.22	21.47	nr	**22.23**
25 x 15mm dia.	0.85	1.03	1.39	24.47	nr	**25.50**
25 x 20mm dia.	0.90	1.09	1.39	24.47	nr	**25.56**
32 x 25mm dia.	1.58	1.92	1.62	28.51	nr	**30.43**
40 x 25mm dia.	2.08	2.53	1.86	32.74	nr	**35.26**
40 x 32mm dia.	2.72	3.30	1.86	32.74	nr	**36.04**
50 x 25mm dia.	2.77	3.36	2.21	38.90	nr	**42.26**
50 x 40mm dia.	3.82	4.64	2.21	38.90	nr	**43.54**
65 x 50mm dia.	5.72	6.95	2.72	47.87	nr	**54.82**
80 x 50mm dia.	7.72	9.37	3.21	56.50	nr	**65.87**
100 x 50mm dia.	11.27	13.69	4.44	78.15	nr	**91.83**
100 x 80mm dia.	17.39	21.12	4.44	78.15	nr	**99.26**
150 x 100mm dia.	39.16	47.55	6.31	111.06	nr	**158.61**

Material Costs/Measured Work Prices - Mechanical Installations   189

T:MECHANICAL/COOLING/HEATING SYSTEMS

Item	Net Price £	Material £	Labour hours	Labour £	Unit	Total rate £
Equal pitcher tee						
15mm dia.	1.28	1.55	0.91	16.02	nr	**17.57**
20mm dia.	1.58	1.92	1.22	21.47	nr	**23.39**
25mm dia.	2.36	2.87	1.39	24.47	nr	**27.33**
32mm dia.	3.23	3.92	1.62	28.51	nr	**32.44**
40mm dia.	5.00	6.07	1.86	32.74	nr	**38.81**
50mm dia.	7.01	8.51	2.21	38.90	nr	**47.41**
65mm dia.	9.66	11.73	2.72	47.87	nr	**59.60**
80mm dia.	13.28	16.13	3.21	56.50	nr	**72.62**
100mm dia.	29.88	36.28	4.44	78.15	nr	**114.43**
Cross						
15mm dia.	1.10	1.34	1.00	17.60	nr	**18.94**
20mm dia.	1.66	2.02	1.33	23.41	nr	**25.42**
25mm dia.	2.10	2.55	1.51	26.58	nr	**29.13**
32mm dia.	2.77	3.36	1.77	31.15	nr	**34.52**
40mm dia.	3.72	4.52	2.02	35.55	nr	**40.07**
50mm dia.	5.78	7.02	2.42	42.59	nr	**49.61**
65mm dia.	7.99	9.70	2.97	52.27	nr	**61.98**
80mm dia.	10.62	12.90	3.50	61.60	nr	**74.50**
100mm dia.	19.32	23.46	4.84	85.19	nr	**108.65**
PRESS FIT						
Press fit jointing system; operating temperature -20 C to +120 C; operating pressure 16 bar ; butyl rubber O ring mechanical joint						
Carbon steel						
Pipework						
15mm dia	2.04	2.48	0.46	8.10	m	**10.57**
22mm dia	2.91	3.53	0.48	8.45	m	**11.98**
28mm dia	3.74	4.54	0.52	9.15	m	**13.69**
35mm dia	4.78	5.80	0.56	9.86	m	**15.66**
42mm dia	5.99	7.27	0.58	10.21	m	**17.48**
54mm dia	7.63	9.27	0.66	11.62	m	**20.88**
Extra over for Carbon Steel pressfit fittings						
Coupling						
15mm dia	0.93	1.13	0.36	6.34	nr	**7.47**
22mm dia	1.13	1.37	0.36	6.34	nr	**7.71**
28mm dia	1.38	1.68	0.44	7.74	nr	**9.42**
35mm dia	2.32	2.82	0.44	7.74	nr	**10.56**
42mm dia	3.00	3.64	0.52	9.15	nr	**12.80**
54mm dia	3.55	4.31	0.60	10.56	nr	**14.87**

Material Costs/Measured Work Prices - Mechanical Installations

T:MECHANICAL/COOLING/HEATING SYSTEMS

Item	Net Price £	Material £	Labour hours	Labour £	Unit	Total rate £
T31 : LOW TEMPERATURE HOT WATER HEATING						
Y 10 - PIPELINES						
PRESS FIT						
Press fit jointing system; operating temperature -20 C to +120 C; operating pressure 16 bar ; butyl rubber O ring mechanical joint						
Carbon steel (Continued)						
Reducer						
22 x 15mm dia	0.82	1.00	0.36	6.34	nr	**7.33**
28 x 15mm dia	1.30	1.58	0.40	7.04	nr	**8.62**
28 x 22mm dia	0.96	1.17	0.40	7.04	nr	**8.21**
35 x 22mm dia	1.51	1.83	0.40	7.04	nr	**8.87**
35 x 28mm dia	1.31	1.59	0.44	7.74	nr	**9.34**
42 x 35mm dia	2.86	3.47	0.48	8.45	nr	**11.92**
54 x 22mm dia	10.06	12.22	0.48	8.45	nr	**20.66**
54 x 28mm dia	10.44	12.68	0.52	9.15	nr	**21.83**
54 x 42mm dia	3.36	4.08	0.56	9.86	nr	**13.94**
90 degree bend						
15mm dia	1.37	1.66	0.36	6.34	nr	**8.00**
22mm dia	1.76	2.14	0.36	6.34	nr	**8.47**
28mm dia	2.35	2.85	0.44	7.74	nr	**10.60**
35mm dia	-	5.74	0.44	7.74	nr	**13.49**
42mm dia	7.95	9.65	0.52	9.15	nr	**18.81**
54mm dia	8.89	10.80	0.60	10.56	nr	**21.36**
45 degree bend						
15mm dia	1.62	1.97	0.36	6.34	nr	**8.30**
22mm dia	1.77	2.15	0.36	6.34	nr	**8.49**
28mm dia	2.36	2.87	0.44	7.74	nr	**10.61**
35mm dia	4.63	5.62	0.44	7.74	nr	**13.37**
42mm dia	5.62	6.82	0.52	9.15	nr	**15.98**
54mm dia	5.93	7.20	0.60	10.56	nr	**17.76**
Equal tee						
15mm dia	2.51	3.05	0.54	9.50	nr	**12.55**
22mm dia	3.00	3.64	0.54	9.50	nr	**13.15**
28mm dia	3.65	4.43	0.66	11.62	nr	**16.05**
35mm dia	6.01	7.30	0.66	11.62	nr	**18.91**
42mm dia	8.60	10.44	0.78	13.73	nr	**24.17**
54mm dia	10.30	12.51	0.90	15.84	nr	**28.35**

Material Costs/Measured Work Prices - Mechanical Installations

T:MECHANICAL/COOLING/HEATING SYSTEMS

Item	Net Price £	Material £	Labour hours	Labour £	Unit	Total rate £
Reducing tee						
22 x 15mm dia	2.96	3.59	0.54	9.50	nr	**13.10**
28 x 15mm dia	3.81	4.63	0.62	10.91	nr	**15.54**
28 x 22mm dia	4.15	5.04	0.62	10.91	nr	**15.95**
35 x 15mm dia	5.63	6.84	0.62	10.91	nr	**17.75**
35 x 22mm dia	6.07	7.37	0.62	10.91	nr	**18.28**
35 x 28mm dia	6.15	7.47	0.62	10.91	nr	**18.38**
42 x 22mm dia	7.86	9.54	0.70	12.32	nr	**21.87**
42 x 28mm dia	8.15	9.90	0.70	12.32	nr	**22.22**
42 x 35mm dia	7.95	9.65	0.70	12.32	nr	**21.97**
54 x 22mm dia	9.36	11.37	0.82	14.43	nr	**25.80**
54 x 28mm dia	9.54	11.58	0.82	14.43	nr	**26.02**
54 x 35mm dia	9.83	11.94	0.82	14.43	nr	**26.37**
54 x 42mm dia	10.30	12.51	0.82	14.43	nr	**26.94**

Material Costs/Measured Work Prices - Mechanical Installations

T:MECHANICAL/COOLING/HEATING SYSTEMS

Item	Net Price £	Material £	Labour hours	Labour £	Unit	Total rate £
MECHANICAL GROOVED						
Mechanical grooved jointing system; working temprrature not exceeding 82 degrees C BS 5750; mechanical joints and painted finish						
Grooved Joints						
65 mm	5.38	6.53	0.58	10.21	m	**16.74**
80 mm	6.51	7.91	0.68	11.97	m	**19.87**
100 mm	8.88	10.78	0.79	13.90	m	**24.69**
125 mm	12.90	15.66	1.02	17.95	m	**33.62**
150mm	13.92	16.90	1.15	20.24	m	**37.14**
Extra over mechanical grooved system fittings						
Couplings						
65mm	7.98	9.69	0.41	7.22	nr	**16.91**
80mm	8.28	10.05	0.41	7.22	nr	**17.27**
100mm	10.31	12.52	0.66	11.62	nr	**24.14**
125mm	19.56	23.75	0.68	11.97	nr	**35.72**
150mm	16.08	19.53	0.80	14.08	nr	**33.61**
Concentric reducers						
80mm	26.86	32.62	0.59	10.38	nr	**43.00**
100mm	41.54	50.44	0.71	12.50	nr	**62.94**
125mm	55.72	67.66	0.85	14.96	nr	**82.62**
150mm	53.28	64.70	0.98	17.25	nr	**81.95**
Short radius elbow; 90 degree						
65mm	27.94	33.93	0.53	9.33	nr	**43.26**
80mm	28.79	34.96	0.61	10.74	nr	**45.70**
100mm	37.01	44.94	0.80	14.08	nr	**59.02**
125mm	66.06	80.22	0.90	15.84	nr	**96.06**
150mm	67.15	81.54	0.94	16.54	nr	**98.08**
Short radius elbow; 45 degree						
65mm	26.23	31.85	0.53	9.33	nr	**41.18**
80mm	28.10	34.12	0.61	10.74	nr	**44.86**
100mm	34.96	42.45	0.80	14.08	nr	**56.53**
125mm	63.30	76.87	0.90	15.84	nr	**92.71**
150mm	58.79	71.39	0.94	16.54	nr	**87.93**
Equal tee						
65mm	45.51	55.26	0.83	14.61	nr	**69.87**
80mm	47.69	57.91	0.93	16.37	nr	**74.28**
100mm	56.50	68.61	1.18	20.77	nr	**89.38**
125mm	126.34	153.41	1.37	24.11	nr	**177.53**
150mm	110.99	134.78	1.43	25.17	nr	**159.94**

Material Costs/Measured Work Prices - Mechanical Installations

T:MECHANICAL/COOLING/HEATING SYSTEMS

Item	Net Price £	Material £	Labour hours	Labour £	Unit	Total rate £
T31 : LOW TEMPERATURE HOT WATER HEATING						
Y 10 - PIPELINES						
BLACK WELDED STEEL						
Black steel pipes; butt welded joints; BS 1387: 1985; including protective painting						
Varnished; light						
15mm dia.	1.04	1.26	0.47	8.27	m	**9.54**
20mm dia.	1.36	1.65	0.50	8.80	m	**10.45**
25mm dia.	1.86	2.26	0.55	9.68	m	**11.94**
32mm dia.	2.30	2.79	0.63	11.09	m	**13.88**
40mm dia.	2.84	3.45	0.71	12.50	m	**15.95**
50mm dia.	3.70	4.49	0.81	14.26	m	**18.75**
65mm dia.	5.44	6.61	0.94	16.54	m	**23.15**
80mm dia.	6.49	7.88	1.06	18.66	m	**26.54**
100mm dia.	9.41	11.43	1.34	23.59	m	**35.01**
Varnished; medium						
8mm dia.	0.80	0.97	0.47	8.27	m	**9.24**
10mm dia.	0.79	0.96	0.47	8.27	m	**9.23**
15mm dia.	0.89	1.08	0.47	8.27	m	**9.35**
20mm dia.	1.04	1.26	0.50	8.80	m	**10.06**
25mm dia.	1.49	1.81	0.55	9.68	m	**11.49**
32mm dia.	1.85	2.25	0.63	11.09	m	**13.34**
40mm dia.	2.15	2.61	0.71	12.50	m	**15.11**
50mm dia.	3.03	3.68	0.81	14.26	m	**17.94**
65mm dia.	4.11	4.99	0.94	16.54	m	**21.54**
80mm dia.	5.34	6.48	1.06	18.66	m	**25.14**
100mm dia.	7.55	9.17	1.34	23.59	m	**32.75**
125mm dia.	11.15	13.54	1.57	27.63	m	**41.17**
150mm dia.	12.95	15.73	1.77	31.15	m	**46.88**
Varnished; heavy						
15mm dia.	1.05	1.27	0.47	8.27	m	**9.55**
20mm dia.	1.26	1.53	0.50	8.80	m	**10.33**
25mm dia.	1.82	2.21	0.55	9.68	m	**11.89**
32mm dia.	2.27	2.76	0.63	11.09	m	**13.84**
40mm dia.	2.65	3.22	0.71	12.50	m	**15.71**
50mm dia.	3.67	4.46	0.81	14.26	m	**18.71**
65mm dia.	5.00	6.07	0.94	16.54	m	**22.62**
80mm dia.	6.36	7.72	1.06	18.66	m	**26.38**
100mm dia.	8.87	10.77	1.34	23.59	m	**34.36**
125mm dia.	11.89	14.44	1.57	27.63	m	**42.07**
150mm dia.	13.90	16.88	1.77	31.15	m	**48.03**

Material Costs/Measured Work Prices - Mechanical Installations

T:MECHANICAL/COOLING/HEATING SYSTEMS

Item	Net Price £	Material £	Labour hours	Labour £	Unit	Total rate £
T31 : LOW TEMPERATURE HOT WATER HEATING						
Y 10 - PIPELINES						
BLACK WELDED STEEL						
Black steel pipes; butt welded joints; BS 1387: 1985; including protective painting (Continued)						
Extra Over black steel butt welded pipes; black steel flanges, welding and drilled; metric; BS 4504						
Welded flanges; PN6						
15mm dia.	1.66	2.02	0.59	10.38	nr	**12.40**
20mm dia.	1.77	2.15	0.69	12.14	nr	**14.29**
25mm dia.	1.97	2.39	0.84	14.78	nr	**17.18**
32mm dia.	2.30	2.79	1.00	17.60	nr	**20.39**
40mm dia.	2.45	2.98	1.11	19.54	nr	**22.51**
50mm dia.	2.64	3.21	1.37	24.11	nr	**27.32**
65mm dia.	3.09	3.75	1.54	27.11	nr	**30.86**
80mm dia.	4.08	4.95	1.67	29.39	nr	**34.35**
100mm dia.	4.44	5.39	2.22	39.07	nr	**44.47**
125mm dia.	6.83	8.29	2.61	45.94	nr	**54.23**
150mm dia.	7.62	9.25	2.99	52.63	nr	**61.88**
Welded flanges; PN16						
15mm dia.	1.69	2.05	0.59	10.38	nr	**12.44**
20mm dia.	1.84	2.23	0.69	12.14	nr	**14.38**
25mm dia.	2.00	2.43	0.84	14.78	nr	**17.21**
32mm dia.	2.70	3.28	1.00	17.60	nr	**20.88**
40mm dia.	2.80	3.40	1.11	19.54	nr	**22.94**
50mm dia.	3.15	3.83	1.37	24.11	nr	**27.94**
65mm dia.	3.66	4.44	1.54	27.11	nr	**31.55**
80mm dia.	4.90	5.95	1.67	29.39	nr	**35.34**
100mm dia.	5.31	6.45	2.22	39.07	nr	**45.52**
125mm dia.	7.46	9.06	2.61	45.94	nr	**55.00**
150mm dia.	9.12	11.07	2.99	52.63	nr	**63.70**
Blank flanges, slip on for welding; PN6						
15mm dia.	1.32	1.60	0.48	8.45	nr	**10.05**
20mm dia.	1.48	1.80	0.55	9.68	nr	**11.48**
25mm dia.	1.70	2.06	0.64	11.26	nr	**13.33**
32mm dia.	2.05	2.49	0.76	13.38	nr	**15.87**
40mm dia.	2.28	2.77	0.84	14.78	nr	**17.55**
50mm dia.	2.39	2.90	1.01	17.78	nr	**20.68**
65mm dia.	2.88	3.50	1.30	22.88	nr	**26.38**
80mm dia.	3.41	4.14	1.41	24.82	nr	**28.96**
100mm dia.	3.99	4.85	1.78	31.33	nr	**36.17**
125mm dia.	5.58	6.78	2.06	36.26	nr	**43.03**
150mm dia.	6.84	8.31	2.35	41.36	nr	**49.67**

Material Costs/Measured Work Prices - Mechanical Installations

T:MECHANICAL/COOLING/HEATING SYSTEMS

Item	Net Price £	Material £	Labour hours	Labour £	Unit	Total rate £
Blank flanges, slip on for welding; PN16						
15mm dia.	1.34	1.63	0.48	8.45	nr	**10.08**
20mm dia.	1.56	1.89	0.55	9.68	nr	**11.57**
25mm dia.	1.71	2.08	0.64	11.26	nr	**13.34**
32mm dia.	2.11	2.56	0.76	13.38	nr	**15.94**
40mm dia.	2.28	2.77	0.84	14.78	nr	**17.55**
50mm dia.	2.45	2.98	1.01	17.78	nr	**20.75**
65mm dia.	2.91	3.53	1.30	22.88	nr	**26.41**
80mm dia.	3.66	4.44	1.41	24.82	nr	**29.26**
100mm dia.	4.20	5.10	1.78	31.33	nr	**36.43**
125mm dia.	5.81	7.06	2.06	36.26	nr	**43.31**
150mm dia.	6.97	8.46	2.35	41.36	nr	**49.83**
Extra Over black steel butt welded pipes; black steel flanges, welding and drilled; imperial; BS 10						
Welded flanges; Table E						
1/2" dia.	2.36	2.87	0.59	10.38	nr	**13.25**
3/4" dia.	2.38	2.89	0.69	12.14	nr	**15.03**
1" dia.	2.43	2.95	0.84	14.78	nr	**17.74**
1 1/4" dia.	2.60	3.16	1.00	17.60	nr	**20.76**
1 1/2" dia.	2.60	3.16	1.11	19.54	nr	**22.69**
2" dia.	2.64	3.21	1.37	24.11	nr	**27.32**
2 1/2" dia.	3.22	3.91	1.54	27.11	nr	**31.02**
3" dia.	3.67	4.46	1.67	29.39	nr	**33.85**
4" dia.	5.12	6.22	2.22	39.07	nr	**45.29**
5" dia.	9.15	11.11	2.61	45.94	nr	**57.05**
6" dia.	9.86	11.97	2.99	52.63	nr	**64.60**
Blank flanges, slip on for welding; Table E						
1/2" dia.	1.32	1.60	0.48	8.45	nr	**10.05**
3/4" dia.	1.32	1.60	0.55	9.68	nr	**11.28**
1" dia.	1.32	1.60	0.64	11.26	nr	**12.87**
1 1/4" dia.	1.60	1.94	0.76	13.38	nr	**15.32**
1 1/2" dia.	1.81	2.20	0.84	14.78	nr	**16.98**
2" dia.	1.83	2.22	1.01	17.78	nr	**20.00**
2 1/2" dia.	2.55	3.10	1.30	22.88	nr	**25.98**
3" dia.	2.86	3.47	1.41	24.82	nr	**28.29**
4" dia.	3.70	4.49	1.78	31.33	nr	**35.82**
5" dia.	5.17	6.28	2.06	36.26	nr	**42.54**
6" dia.	7.13	8.66	2.35	41.36	nr	**50.02**
Extra Over black steel butt welded pipes; black steel flange connections						
Bolted connection between pair of flanges; including gasket, bolts, nuts and washers						
50mm dia.	5.16	6.27	0.50	8.80	nr	**15.07**
65mm dia.	5.98	7.26	0.50	8.80	nr	**16.06**
80mm dia.	8.64	10.49	0.50	8.80	nr	**19.29**
100mm dia.	9.63	11.69	0.50	8.80	nr	**20.49**
125mm dia.	17.49	21.24	0.50	8.80	nr	**30.04**
150mm dia.	18.56	22.54	0.88	15.49	nr	**38.03**

Material Costs/Measured Work Prices - Mechanical Installations

T:MECHANICAL/COOLING/HEATING SYSTEMS

Item	Net Price £	Material £	Labour hours	Labour £	Unit	Total rate £
T31 : LOW TEMPERATURE HOT WATER HEATING						
Y 10 - PIPELINES						
BLACK WELDED STEEL						
Black steel pipes; butt welded joints; BS 1387: 1985; including protective painting (Continued)						
Extra over fittings; BS 1965; butt welded						
Cap						
20mm dia.	5.96	7.24	0.38	6.69	nr	**13.93**
25mm dia.	6.25	7.59	0.47	8.27	nr	**15.86**
32mm dia.	6.09	7.40	0.59	10.38	nr	**17.78**
40mm dia.	5.28	6.41	0.70	12.32	nr	**18.73**
50mm dia.	4.33	5.26	0.99	17.42	nr	**22.68**
65mm dia.	5.20	6.31	1.35	23.76	nr	**30.08**
80mm dia.	7.99	9.70	1.66	29.22	nr	**38.92**
100mm dia.	7.57	9.19	2.23	39.25	nr	**48.44**
125mm dia.	19.79	24.03	3.03	53.33	nr	**77.36**
150mm dia.	18.10	21.98	3.79	66.71	nr	**88.69**
Concentric reducer						
20 x 15mm dia.	2.20	2.67	0.69	12.14	nr	**14.82**
25 x 15mm dia.	2.20	2.67	0.87	15.31	nr	**17.98**
25 x 20mm dia.	1.82	2.21	0.87	15.31	nr	**17.52**
32 x 25mm dia.	2.12	2.57	1.08	19.01	nr	**21.58**
40 x 25mm dia.	2.33	2.83	1.38	24.29	nr	**27.12**
40 x 32mm dia.	2.33	2.83	1.38	24.29	nr	**27.12**
50 x 25mm dia.	2.60	3.16	1.82	32.03	nr	**35.19**
50 x 40mm dia.	2.60	3.16	1.82	32.03	nr	**35.19**
65 x 50mm dia.	3.29	4.00	2.52	44.35	nr	**48.35**
80 x 50mm dia.	3.49	4.24	3.24	57.03	nr	**61.26**
100 x 50mm dia.	5.31	6.45	4.08	71.81	nr	**78.26**
100 x 80mm dia.	4.73	5.74	4.08	71.81	nr	**77.55**
125 x 80mm dia.	9.17	11.14	4.71	82.90	nr	**94.03**
150 x 100mm dia.	9.17	11.14	5.33	93.81	nr	**104.95**
Eccentric reducer						
20 x 15mm dia.	2.76	3.35	0.69	12.14	nr	**15.50**
25 x 15mm dia.	2.85	3.46	0.87	15.31	nr	**18.77**
25 x 20mm dia.	2.75	3.34	0.87	15.31	nr	**18.65**
32 x 25mm dia.	2.94	3.57	1.08	19.01	nr	**22.58**
40 x 25mm dia.	3.11	3.78	1.38	24.29	nr	**28.07**
40 x 32mm dia.	3.11	3.78	1.38	24.29	nr	**28.07**
50 x 25mm dia.	4.89	5.94	1.82	32.03	nr	**37.97**
50 x 40mm dia.	4.89	5.94	1.82	32.03	nr	**37.97**
65 x 50mm dia.	5.01	6.08	2.52	44.35	nr	**50.44**
80 x 50mm dia.	5.23	6.35	3.24	57.03	nr	**63.38**
100 x 50mm dia.	6.76	8.21	4.08	71.81	nr	**80.02**
100 x 80mm dia.	9.74	11.83	4.08	71.81	nr	**83.64**
125 x 80mm dia.	13.95	16.94	4.71	82.90	nr	**99.84**
150 x 100mm dia.	13.95	16.94	5.33	93.81	nr	**110.75**

Material Costs/Measured Work Prices - Mechanical Installations

T:MECHANICAL/COOLING/HEATING SYSTEMS

Item	Net Price £	Material £	Labour hours	Labour £	Unit	Total rate £
45 degree elbow, long radius						
15mm dia.	2.68	3.25	0.56	9.86	nr	**13.11**
20mm dia.	2.68	3.25	0.75	13.20	nr	**16.45**
25mm dia.	2.84	3.45	0.93	16.37	nr	**19.82**
32mm dia.	2.91	3.53	1.17	20.59	nr	**24.13**
40mm dia.	2.94	3.57	1.46	25.70	nr	**29.27**
50mm dia.	3.03	3.68	1.97	34.67	nr	**38.35**
65mm dia.	4.60	5.59	2.70	47.52	nr	**53.11**
80mm dia.	4.59	5.57	3.32	58.43	nr	**64.01**
100mm dia.	6.37	7.74	4.09	71.99	nr	**79.72**
125mm dia.	13.85	16.82	4.94	86.95	nr	**103.77**
150mm dia.	17.27	20.97	5.78	101.73	nr	**122.70**
90 degree elbow, long radius						
15mm dia.	2.68	3.25	0.56	9.86	nr	**13.11**
20mm dia.	2.68	3.25	0.75	13.20	nr	**16.45**
25mm dia.	2.84	3.45	0.93	16.37	nr	**19.82**
32mm dia.	2.91	3.53	1.17	20.59	nr	**24.13**
40mm dia.	2.94	3.57	1.46	25.70	nr	**29.27**
50mm dia.	3.03	3.68	1.97	34.67	nr	**38.35**
65mm dia.	4.60	5.59	2.70	47.52	nr	**53.11**
80mm dia.	4.59	5.57	3.32	58.43	nr	**64.01**
100mm dia.	6.37	7.74	4.09	71.99	nr	**79.72**
125mm dia.	13.85	16.82	4.94	86.95	nr	**103.77**
150mm dia.	17.27	20.97	5.78	101.73	nr	**122.70**
Equal tee						
15mm dia.	11.62	14.11	0.82	14.43	nr	**28.54**
20mm dia.	11.62	14.11	1.10	19.36	nr	**33.47**
25mm dia.	11.99	14.56	1.35	23.76	nr	**38.32**
32mm dia.	12.34	14.98	1.63	28.69	nr	**43.67**
40mm dia.	12.34	14.98	2.14	37.67	nr	**52.65**
50mm dia.	12.41	15.07	3.02	53.15	nr	**68.22**
65mm dia.	22.10	26.84	3.61	63.54	nr	**90.37**
80mm dia.	23.40	28.41	4.18	73.57	nr	**101.99**
100mm dia.	30.06	36.50	5.24	92.23	nr	**128.73**
125mm dia.	66.96	81.31	6.70	117.93	nr	**199.23**
150mm dia.	72.55	88.10	8.45	148.73	nr	**236.82**
Extra Over black steel butt welded pipes; labours						
Made bend						
15mm dia.	-	-	0.42	7.39	nr	**7.39**
20mm dia.	-	-	0.42	7.39	nr	**7.39**
25mm dia.	-	-	0.50	8.80	nr	**8.80**
32mm dia.	-	-	0.62	10.91	nr	**10.91**
40mm dia.	-	-	0.74	13.02	nr	**13.02**
50mm dia.	-	-	0.89	15.66	nr	**15.66**
65mm dia.	-	-	1.05	18.48	nr	**18.48**
80mm dia.	-	-	1.13	19.89	nr	**19.89**
100mm dia.	-	-	2.90	51.04	nr	**51.04**
125mm dia.	-	-	3.56	62.66	nr	**62.66**
150mm dia.	-	-	4.18	73.57	nr	**73.57**

Material Costs/Measured Work Prices - Mechanical Installations

T:MECHANICAL/COOLING/HEATING SYSTEMS

Item	Net Price £	Material £	Labour hours	Labour £	Unit	Total rate £
T31 : LOW TEMPERATURE HOT WATER HEATING						
Y 10 - PIPELINES						
BLACK WELDED STEEL						
Black steel pipes; butt welded joints; BS 1387: 1985; including protective painting (Continued)						
Splay cut end						
15mm dia.	-	-	0.14	2.46	nr	**2.46**
20mm dia.	-	-	0.16	2.82	nr	**2.82**
25mm dia.	-	-	0.18	3.17	nr	**3.17**
32mm dia.	-	-	0.25	4.40	nr	**4.40**
40mm dia.	-	-	0.27	4.75	nr	**4.75**
50mm dia.	-	-	0.31	5.46	nr	**5.46**
65mm dia.	-	-	0.35	6.16	nr	**6.16**
80mm dia.	-	-	0.40	7.04	nr	**7.04**
100mm dia.	-	-	0.48	8.45	nr	**8.45**
125mm dia.	-	-	0.56	9.86	nr	**9.86**
150mm dia.	-	-	0.64	11.26	nr	**11.26**
Screwed joint to fitting						
15mm dia.	-	-	0.30	5.28	nr	**5.28**
20mm dia.	-	-	0.40	7.04	nr	**7.04**
25mm dia.	-	-	0.46	8.10	nr	**8.10**
32mm dia.	-	-	0.53	9.33	nr	**9.33**
40mm dia.	-	-	0.61	10.74	nr	**10.74**
50mm dia.	-	-	0.73	12.85	nr	**12.85**
65mm dia.	-	-	0.89	15.66	nr	**15.66**
80mm dia.	-	-	1.05	18.48	nr	**18.48**
100mm dia.	-	-	1.46	25.70	nr	**25.70**
125mm dia.	-	-	2.10	36.96	nr	**36.96**
150mm dia.	-	-	2.73	48.05	nr	**48.05**
Straight butt weld						
15mm dia.	-	-	0.31	5.46	nr	**5.46**
20mm dia.	-	-	0.42	7.39	nr	**7.39**
25mm dia.	-	-	0.52	9.15	nr	**9.15**
32mm dia.	-	-	0.69	12.14	nr	**12.14**
40mm dia.	-	-	0.83	14.61	nr	**14.61**
50mm dia.	-	-	1.22	21.47	nr	**21.47**
65mm dia.	-	-	1.57	27.63	nr	**27.63**
80mm dia.	-	-	1.95	34.32	nr	**34.32**
100mm dia.	-	-	2.38	41.89	nr	**41.89**
125mm dia.	-	-	2.83	49.81	nr	**49.81**
150mm dia.	-	-	3.27	57.55	nr	**57.55**

Material Costs/Measured Work Prices - Mechanical Installations   199

T:MECHANICAL/COOLING/HEATING SYSTEMS

Item	Net Price £	Material £	Labour hours	Labour £	Unit	Total rate £
Branch weld						
15mm dia.	-	-	0.48	8.45	nr	**8.45**
20mm dia.	-	-	0.64	11.26	nr	**11.26**
25mm dia.	-	-	0.80	14.08	nr	**14.08**
32mm dia.	-	-	1.05	18.48	nr	**18.48**
40mm dia.	-	-	1.18	20.77	nr	**20.77**
50mm dia.	-	-	1.64	28.87	nr	**28.87**
65mm dia.	-	-	2.10	36.96	nr	**36.96**
80mm dia.	-	-	2.60	45.76	nr	**45.76**
100mm dia.	-	-	3.18	55.97	nr	**55.97**
125mm dia.	-	-	3.79	66.71	nr	**66.71**
150mm dia.	-	-	4.40	77.44	nr	**77.44**
Welded reducing joint						
15mm dia.	-	-	0.60	10.56	nr	**10.56**
20mm dia.	-	-	0.80	14.08	nr	**14.08**
25mm dia.	-	-	1.00	17.60	nr	**17.60**
32mm dia.	-	-	1.32	23.23	nr	**23.23**
40mm dia.	-	-	1.60	28.16	nr	**28.16**
50mm dia.	-	-	2.40	42.24	nr	**42.24**
65mm dia.	-	-	3.21	56.50	nr	**56.50**
80mm dia.	-	-	4.00	70.40	nr	**70.40**
100mm dia.	-	-	4.41	77.62	nr	**77.62**
125mm dia.	-	-	4.81	84.66	nr	**84.66**
150mm dia.	-	-	5.21	91.70	nr	**91.70**

CARBON WELDED STEEL

Hot finished seamless carbon steel pipe; BS 806 and BS 3601; wall thickness to BS 3600; butt welded joints; including protective painting

Item	Net Price £	Material £	Labour hours	Labour £	Unit	Total rate £
Pipework						
200mm dia	16.35	19.85	2.04	35.91	m	**55.76**
250mm dia	23.18	28.15	2.54	44.71	m	**72.85**
300mm dia	28.23	34.28	2.99	52.63	m	**86.91**
350mm dia	31.97	38.82	3.52	61.95	m	**100.78**
400mm dia	36.35	44.14	4.08	71.81	m	**115.95**

Extra over fittings; BS 1965 part 1; butt welded

Item	Net Price £	Material £	Labour hours	Labour £	Unit	Total rate £
Cap						
200mm dia	16.34	19.84	3.70	65.12	nr	**84.96**
250mm dia	34.71	42.15	4.73	83.25	nr	**125.40**
300mm dia	39.28	47.70	5.65	99.44	nr	**147.14**
350mm dia	39.53	48.00	6.68	117.57	nr	**165.57**
400mm dia	47.04	57.12	7.70	135.53	nr	**192.65**

T:MECHANICAL/COOLING/HEATING SYSTEMS

Item	Net Price £	Material £	Labour hours	Labour £	Unit	Total rate £
T31 : LOW TEMPERATURE HOT WATER HEATING						
Y 10 - PIPELINES						
BLACK WELDED STEEL						
Hot finished seamless carbon steel pipe; BS 806 and BS 3601; wall thickness to BS 3600; butt welded joints; including protective painting (Continued)						
Concentric reducer						
200mm x 150mm dia	13.90	16.88	7.27	127.96	nr	**144.84**
250mm x 150mm dia	23.76	28.85	9.05	159.29	nr	**188.14**
250mm x 200mm dia	16.44	19.96	9.10	160.17	nr	**180.13**
300mm x 150mm dia	59.49	72.24	10.75	189.21	nr	**261.45**
300mm x 200mm dia	33.90	41.16	10.75	189.21	nr	**230.37**
300mm x 250mm dia	30.14	36.60	11.15	196.25	nr	**232.85**
350mm x 200mm dia	56.90	69.09	12.50	220.01	nr	**289.10**
350mm x 250mm dia	52.73	64.03	12.70	223.53	nr	**287.56**
350mm x 300mm dia	50.33	61.12	13.00	228.81	nr	**289.93**
400mm x 250mm dia	106.16	128.91	14.46	254.51	nr	**383.42**
400mm x 300mm dia	88.49	107.45	14.51	255.39	nr	**362.84**
400mm x 350mm dia	84.47	102.57	15.16	266.83	nr	**369.40**
Eccentric reducer						
200mm x 150mm dia	32.95	40.01	7.27	127.96	nr	**167.97**
250mm x 150mm dia	36.62	44.47	9.05	159.29	nr	**203.75**
250mm x 200mm dia	30.77	37.36	9.10	160.17	nr	**197.53**
300mm x 150mm dia	68.27	82.90	10.75	189.21	nr	**272.11**
300mm x 200mm dia	65.08	79.03	10.75	189.21	nr	**268.24**
300mm x 250mm dia	52.38	63.60	11.15	196.25	nr	**259.85**
350mm x 200mm dia	94.56	114.82	12.50	220.01	nr	**334.83**
350mm x 250mm dia	79.53	96.57	12.70	223.53	nr	**320.10**
350mm x 300mm dia	75.92	92.19	13.00	228.81	nr	**321.00**
400mm x 250mm dia	159.53	193.72	14.46	254.51	nr	**448.22**
400mm x 300mm dia	132.33	160.69	14.51	255.39	nr	**416.08**
400mm x 350mm dia	126.31	153.38	15.16	266.83	nr	**420.21**
45 degree elbow						
200mm dia	18.81	22.84	7.75	136.41	nr	**159.25**
250mm dia	35.90	43.59	10.05	176.89	nr	**220.48**
300mm dia	55.52	67.42	12.20	214.73	nr	**282.15**
350mm dia	89.18	108.29	14.65	257.85	nr	**366.14**
400mm dia	113.90	138.31	17.12	301.33	nr	**439.63**
90 degree elbow						
200mm dia	23.30	28.29	7.75	136.41	nr	**164.70**
250mm dia	42.54	51.66	10.05	176.89	nr	**228.54**
300mm dia	65.68	79.76	12.20	214.73	nr	**294.49**
350mm dia	103.01	125.08	14.65	257.85	nr	**382.94**
400mm dia	131.57	159.77	17.12	301.33	nr	**461.09**

Material Costs/Measured Work Prices - Mechanical Installations   201

T:MECHANICAL/COOLING/HEATING SYSTEMS

Item	Net Price £	Material £	Labour hours	Labour £	Unit	Total rate £
Equal tee						
200mm dia	56.55	68.67	11.25	198.01	nr	**266.68**
250mm dia	90.21	109.54	14.53	255.74	nr	**365.28**
300mm dia	130.60	158.59	17.55	308.89	nr	**467.48**
350mm dia	153.38	186.25	20.98	369.26	nr	**555.51**
400mm dia	262.38	318.61	24.38	429.11	nr	**747.72**
Extra Over black steel butt welded pipes; labours						
Straight butt weld						
200mm dia	-	-	4.08	71.81	nr	**71.81**
250mm dia	-	-	5.20	91.52	nr	**91.52**
300mm dia	-	-	6.22	109.48	nr	**109.48**
350mm dia	-	-	7.33	129.01	nr	**129.01**
400mm dia	-	-	8.41	148.02	nr	**148.02**
Branch weld						
100mm dia.	6.38	7.75	3.46	60.90	nr	**68.65**
125mm dia.	14.23	17.28	4.23	74.45	nr	**91.73**
150mm dia.	20.37	24.74	5.00	88.00	nr	**112.74**
Extra Over black steel butt welded pipes; black steel flanges, welding and drilled; metric; BS 4504						
Welded flanges; PN16						
200mm dia	13.64	16.56	4.10	72.16	nr	**88.73**
250mm dia	25.55	31.03	5.33	93.81	nr	**124.84**
300mm dia	29.64	35.99	6.40	112.65	nr	**148.64**
350mm dia	33.74	40.97	7.43	130.77	nr	**171.74**
400mm dia	43.64	52.99	8.45	148.73	nr	**201.72**
Welded flanges; PN25						
200mm dia	24.97	30.32	4.10	72.16	nr	**102.48**
250mm dia	35.95	43.65	5.33	93.81	nr	**137.47**
300mm dia	45.94	55.78	6.40	112.65	nr	**168.43**
Blank flanges, slip on for welding; PN16						
200mm dia	19.55	23.74	2.70	47.52	nr	**71.26**
250mm dia	30.27	36.76	3.48	61.25	nr	**98.01**
300mm dia	42.89	52.08	4.20	73.92	nr	**126.00**
350mm dia	58.59	71.15	4.78	84.13	nr	**155.28**
400mm dia	75.43	91.59	5.35	94.16	nr	**185.76**
Blank flanges, slip on for welding; PN25						
200mm dia	38.96	47.31	2.70	47.52	nr	**94.83**
250mm dia	57.92	70.33	3.48	61.25	nr	**131.58**
300mm dia	79.89	97.01	4.20	73.92	nr	**170.93**

202   *Material Costs/Measured Work Prices - Mechanical Installations*

T:MECHANICAL/COOLING/HEATING SYSTEMS

Item	Net Price £	Material £	Labour hours	Labour £	Unit	Total rate £
T31 : LOW TEMPERATURE HOT WATER HEATING						
Y 10 - PIPELINES						
BLACK WELDED STEEL (Continued)						
Extra Over black steel butt welded pipes; black steel flange connections						
Bolted connection between pair of flanges; including gasket, bolts, nuts and washers						
200mm dia	21.83	26.51	3.83	67.41	nr	93.92
250mm dia	40.72	49.45	4.93	86.77	nr	136.22
300mm dia	48.86	59.33	5.90	103.84	nr	163.18
FIXINGS						
For steel pipes; black malleable iron						
Single pipe bracket, screw on, black malleable iron; screwed to wood						
15mm dia.	0.38	0.46	0.14	2.46	nr	2.93
20mm dia.	0.43	0.52	0.14	2.46	nr	2.99
25mm dia.	0.50	0.61	0.17	2.99	nr	3.60
32mm dia.	0.68	0.83	0.19	3.34	nr	4.17
40mm dia.	0.91	1.10	0.22	3.87	nr	4.98
50mm dia.	1.19	1.45	0.22	3.87	nr	5.32
65mm dia.	1.59	1.93	0.28	4.93	nr	6.86
80mm dia.	2.17	2.63	0.32	5.63	nr	8.27
100mm dia.	3.17	3.85	0.35	6.16	nr	10.01
Single pipe bracket, screw on, black malleable iron; plugged and screwed						
15mm dia.	0.38	0.46	0.25	4.40	nr	4.86
20mm dia.	0.43	0.52	0.25	4.40	nr	4.92
25mm dia.	0.50	0.61	0.30	5.28	nr	5.89
32mm dia.	0.68	0.83	0.32	5.63	nr	6.46
40mm dia.	0.91	1.10	0.32	5.63	nr	6.74
50mm dia.	1.19	1.45	0.32	5.63	nr	7.08
65mm dia.	1.59	1.93	0.35	6.16	nr	8.09
80mm dia.	2.17	2.63	0.42	7.39	nr	10.03
100mm dia.	3.17	3.85	0.42	7.39	nr	11.24
Single pipe bracket for building in, black malleable iron						
15mm dia.	0.93	1.13	0.10	1.76	nr	2.89
20mm dia.	0.93	1.13	0.11	1.94	nr	3.07
25mm dia.	0.93	1.13	0.12	2.11	nr	3.24
32mm dia.	1.05	1.27	0.14	2.46	nr	3.74
40mm dia.	1.05	1.27	0.15	2.64	nr	3.92
50mm dia.	1.09	1.32	0.16	2.82	nr	4.14

Material Costs/Measured Work Prices - Mechanical Installations

T:MECHANICAL/COOLING/HEATING SYSTEMS

Item	Net Price £	Material £	Labour hours	Labour £	Unit	Total rate £
Pipe ring, single socket, black malleable iron						
15mm dia.	0.43	0.52	0.10	1.76	nr	**2.28**
20mm dia.	0.50	0.61	0.11	1.94	nr	**2.54**
25mm dia.	0.55	0.67	0.12	2.11	nr	**2.78**
32mm dia.	0.57	0.69	0.14	2.46	nr	**3.16**
40mm dia.	0.72	0.87	0.15	2.64	nr	**3.51**
50mm dia.	0.93	1.13	0.16	2.82	nr	**3.95**
65mm dia.	1.34	1.63	0.30	5.28	nr	**6.91**
80mm dia.	1.61	1.96	0.35	6.16	nr	**8.12**
100mm dia.	2.44	2.96	0.40	7.04	nr	**10.00**
125mm dia.	4.91	5.96	0.60	10.56	nr	**16.53**
150mm dia.	5.50	6.68	0.77	13.55	nr	**20.23**
200mm dia	7.33	8.90	0.90	15.84	nr	**24.74**
250mm dia	9.16	11.12	1.10	19.36	nr	**30.48**
300mm dia	11.00	13.36	1.25	22.00	nr	**35.36**
350mm dia	12.83	15.58	1.50	26.40	nr	**41.98**
400mm dia	14.66	17.80	1.75	30.80	nr	**48.60**
Pipe ring, double socket, black malleable iron						
15mm dia.	0.53	0.64	0.10	1.76	nr	**2.40**
20mm dia.	0.59	0.72	0.11	1.94	nr	**2.65**
25mm dia.	0.66	0.80	0.12	2.11	nr	**2.91**
32mm dia.	0.77	0.94	0.14	2.46	nr	**3.40**
40mm dia.	0.91	1.10	0.15	2.64	nr	**3.75**
50mm dia.	1.02	1.24	0.16	2.82	nr	**4.05**
Screw on backplate, black malleable iron; screwed to wood						
15mm dia.	0.53	0.64	0.14	2.46	nr	**3.11**
Screw on backplate, black malleable iron; plugged and screwed						
15mm dia.	0.53	0.64	0.25	4.40	nr	**5.04**
For steel pipes; galvanised iron						
Single pipe bracket, screw on, galvanised iron; screwed to wood						
15mm dia.	0.53	0.64	0.14	2.46	nr	**3.11**
20mm dia.	0.58	0.70	0.14	2.46	nr	**3.17**
25mm dia.	0.67	0.81	0.17	2.99	nr	**3.81**
32mm dia.	0.92	1.12	0.19	3.34	nr	**4.46**
40mm dia.	1.22	1.48	0.22	3.87	nr	**5.35**
50mm dia.	1.62	1.97	0.22	3.87	nr	**5.84**
65mm dia.	2.14	2.60	0.28	4.93	nr	**7.53**
80mm dia.	2.93	3.56	0.32	5.63	nr	**9.19**
100mm dia.	4.26	5.17	0.35	6.16	nr	**11.33**

Material Costs/Measured Work Prices - Mechanical Installations

T:MECHANICAL/COOLING/HEATING SYSTEMS

Item	Net Price £	Material £	Labour hours	Labour £	Unit	Total rate £
T31 : LOW TEMPERATURE HOT WATER HEATING						
Y 10 - PIPELINES						
FIXINGS						
For steel pipes; galvanised iron (Continued)						
Single pipe bracket, screw on, galvanised iron; plugged and screwed						
15mm dia.	0.53	0.64	0.25	4.40	nr	5.04
20mm dia.	0.58	0.70	0.25	4.40	nr	5.10
25mm dia.	0.67	0.81	0.30	5.28	nr	6.09
32mm dia.	0.92	1.12	0.32	5.63	nr	6.75
40mm dia.	1.22	1.48	0.32	5.63	nr	7.11
50mm dia.	1.62	1.97	0.32	5.63	nr	7.60
65mm dia.	2.14	2.60	0.35	6.16	nr	8.76
80mm dia.	2.93	3.56	0.42	7.39	nr	10.95
100mm dia.	4.26	5.17	0.42	7.39	nr	12.57
Single pipe bracket for building in, galvanised iron						
15mm dia.	1.38	1.68	0.10	1.76	nr	3.44
20mm dia.	1.38	1.68	0.11	1.94	nr	3.61
25mm dia.	1.38	1.68	0.12	2.11	nr	3.79
32mm dia.	1.57	1.91	0.14	2.46	nr	4.37
40mm dia.	1.59	1.93	0.15	2.64	nr	4.57
50mm dia.	1.64	1.99	0.16	2.82	nr	4.81
Pipe ring, single socket, galvanised iron						
15mm dia.	0.58	0.70	0.10	1.76	nr	2.46
20mm dia.	0.67	0.81	0.11	1.94	nr	2.75
25mm dia.	0.73	0.89	0.12	2.11	nr	3.00
32mm dia.	0.76	0.92	0.15	2.64	nr	3.56
40mm dia.	0.98	1.19	0.15	2.64	nr	3.83
50mm dia.	1.26	1.53	0.16	2.82	nr	4.35
65mm dia.	1.80	2.19	0.30	5.28	nr	7.47
80mm dia.	2.16	2.62	0.35	6.16	nr	8.78
100mm dia.	3.30	4.01	0.40	7.04	nr	11.05
125mm dia.	6.61	8.03	0.60	10.56	nr	18.59
150mm dia.	7.41	9.00	0.77	13.55	nr	22.55
Pipe ring, double socket, galvanised iron						
15mm dia.	0.70	0.85	0.10	1.76	nr	2.61
20mm dia.	0.79	0.96	0.11	1.94	nr	2.90
25mm dia.	0.89	1.08	0.12	2.11	nr	3.19
32mm dia.	1.04	1.26	0.14	2.46	nr	3.73
40mm dia.	1.22	1.48	0.15	2.64	nr	4.12
50mm dia.	1.37	1.66	0.16	2.82	nr	4.48
Screw on backplate, galvanised iron; screwed to wood						
15mm dia.	0.71	0.86	0.14	2.46	nr	3.33

Material Costs/Measured Work Prices - Mechanical Installations

T:MECHANICAL/COOLING/HEATING SYSTEMS

Item	Net Price £	Material £	Labour hours	Labour £	Unit	Total rate £
Screw on backplate, galvanised iron; plugged and screwed						
15mm dia.	0.71	0.86	0.25	4.40	nr	**5.26**
Fabricated hangers and brackets (Note: It has been assumed there would be sufficient quantities required to gain the benefit of bulk purchase)						
Galvanised steel; including inserts, bolts, nuts, washers; fixed to backgrounds						
41 x 21mm	1.92	2.33	0.29	5.10	m	**7.44**
41 x 41mm	2.74	3.33	0.29	5.10	m	**8.43**
Threaded rods; metric thread; including nuts, washers etc						
10mm dia	0.92	1.12	0.18	3.17	m	**4.29**
12mm dia	1.02	1.24	0.18	3.17	m	**4.41**
Floor or ceiling cover plates						
Plastic						
15mm dia.	0.21	0.26	0.16	2.82	nr	**3.07**
20mm dia.	0.23	0.28	0.22	3.87	nr	**4.15**
25mm dia.	0.26	0.32	0.22	3.87	nr	**4.19**
32mm dia.	0.38	0.46	0.24	4.22	nr	**4.69**
40mm dia.	0.57	0.69	0.26	4.58	nr	**5.27**
50mm dia.	0.63	0.77	0.26	4.58	nr	**5.34**
Chromium plated						
15mm dia.	1.39	1.69	0.16	2.82	nr	**4.50**
20mm dia.	1.48	1.80	0.17	2.99	nr	**4.79**
25mm dia.	1.54	1.87	0.21	3.70	nr	**5.57**
32mm dia.	1.58	1.92	0.22	3.87	nr	**5.79**
40mm dia.	1.77	2.15	0.26	4.58	nr	**6.73**
50mm dia.	2.12	2.57	0.26	4.58	nr	**7.15**
Pipe roller and chair						
Roller and chair; black malleable						
Up to 50mm dia.	2.37	2.88	0.20	3.52	nr	**6.40**
65mm dia.	2.41	2.93	0.20	3.52	nr	**6.45**
80mm dia.	3.45	4.19	0.20	3.52	nr	**7.71**
100mm dia.	3.71	4.51	0.20	3.52	nr	**8.03**
125mm dia.	4.06	4.93	0.20	3.52	nr	**8.45**
150mm dia.	4.49	5.45	0.30	5.28	nr	**10.73**
175mm dia.	11.28	13.70	0.30	5.28	nr	**18.98**
200mm dia.	11.28	13.70	0.30	5.28	nr	**18.98**
250mm dia.	16.11	19.56	0.30	5.28	nr	**24.84**
300mm dia.	16.92	20.55	0.30	5.28	nr	**25.83**

Material Costs/Measured Work Prices - Mechanical Installations

T:MECHANICAL/COOLING/HEATING SYSTEMS

Item	Net Price £	Material £	Labour hours	Labour £	Unit	Total rate £
T31 : LOW TEMPERATURE HOT WATER HEATING						
Y 10 - PIPELINES						
FIXINGS						
Pipe roller and chair (Continued)						
Roller and chair; galvanised						
Up to 50mm dia.	3.30	4.01	0.20	3.52	nr	7.53
65mm dia.	3.62	4.40	0.20	3.52	nr	7.92
80mm dia.	5.17	6.28	0.20	3.52	nr	9.80
100mm dia.	5.56	6.75	0.20	3.52	nr	10.27
125mm dia.	6.08	7.38	0.20	3.52	nr	10.90
150mm dia.	6.75	8.20	0.30	5.28	nr	13.48
175mm dia.	16.92	20.55	0.30	5.28	nr	25.83
200mm dia.	16.92	20.55	0.30	5.28	nr	25.83
250mm dia.	24.17	29.35	0.30	5.28	nr	34.63
300mm dia.	25.38	30.82	0.30	5.28	nr	36.10
Roller bracket; black malleable						
25mm dia.	2.06	2.50	0.20	3.52	nr	6.02
32mm dia.	2.22	2.70	0.20	3.52	nr	6.22
40mm dia.	2.31	2.81	0.20	3.52	nr	6.33
50mm dia.	2.39	2.90	0.20	3.52	nr	6.42
65mm dia.	2.80	3.40	0.20	3.52	nr	6.92
80mm dia.	3.86	4.69	0.20	3.52	nr	8.21
100mm dia.	4.60	5.59	0.20	3.52	nr	9.11
125mm dia.	6.25	7.59	0.20	3.52	nr	11.11
150mm dia.	6.25	7.59	0.30	5.28	nr	12.87
175mm dia.	13.53	16.43	0.30	5.28	nr	21.71
200mm dia.	13.53	16.43	0.30	5.28	nr	21.71
250mm dia.	17.69	21.48	0.30	5.28	nr	26.76
300mm dia.	23.84	28.95	0.30	5.28	nr	34.23
350mm dia	41.72	50.66	0.30	5.28	nr	55.94
400mm dia	47.68	57.90	0.30	5.28	nr	63.18
Roller bracket; galvanised						
25mm dia.	3.09	3.75	0.20	3.52	nr	7.27
32mm dia.	3.34	4.06	0.20	3.52	nr	7.58
40mm dia.	3.46	4.20	0.20	3.52	nr	7.72
50mm dia.	3.57	4.34	0.20	3.52	nr	7.86
65mm dia.	4.19	5.09	0.20	3.52	nr	8.61
80mm dia.	5.80	7.04	0.20	3.52	nr	10.56
100mm dia.	6.91	8.39	0.20	3.52	nr	11.91
125mm dia.	9.38	11.39	0.20	3.52	nr	14.91
150mm dia.	9.38	11.39	0.30	5.28	nr	16.67
175mm dia.	20.30	24.65	0.30	5.28	nr	29.93
200mm dia.	20.30	24.65	0.30	5.28	nr	29.93
250mm dia.	26.53	32.22	0.30	5.28	nr	37.50
300mm dia.	35.76	43.42	0.30	5.28	nr	48.70

Material Costs/Measured Work Prices - Mechanical Installations

T:MECHANICAL/COOLING/HEATING SYSTEMS

Item	Net Price £	Material £	Labour hours	Labour £	Unit	Total rate £
T31 : LOW TEMPERATURE HOT WATER HEATING						
Y11 - PIPELINE - ANCILLAIRES						
EXPANSION JOINTS						
Axial movement bellows expansion joints; stainless steel						
Screwed ends for steel pipework; up to 6 bar G at 100 degrees Celsius						
15mm dia.	52.99	64.35	0.68	11.97	nr	**76.31**
20mm dia.	68.13	82.73	0.81	14.26	nr	**96.99**
25mm dia.	71.38	86.68	0.93	16.37	nr	**103.05**
32mm dia.	85.44	103.75	1.06	18.66	nr	**122.41**
40mm dia.	101.20	122.89	1.16	20.42	nr	**143.30**
50mm dia.	111.39	135.26	1.19	20.95	nr	**156.21**
Screwed ends for steel pipework; aluminium and steel outer sleeves; up to 16 bar G at 120 degrees Celsius						
20mm dia.	109.23	132.64	1.32	23.23	nr	**155.87**
25mm dia.	113.56	137.90	1.52	26.75	nr	**164.65**
32mm dia.	129.78	157.59	1.80	31.68	nr	**189.27**
40mm dia.	144.92	175.98	2.03	35.73	nr	**211.71**
50mm dia.	160.06	194.36	2.26	39.78	nr	**234.14**
Flanged ends for steel pipework; aluminium and steel outer sleeves; up to 16 bar G at 120 degrees Celsius						
20mm dia.	207.65	252.15	0.53	9.33	nr	**261.48**
25mm dia.	208.73	253.46	0.64	11.26	nr	**264.73**
32mm dia.	216.30	262.65	0.74	13.02	nr	**275.68**
40mm dia.	221.71	269.22	0.82	14.43	nr	**283.66**
50mm dia.	229.28	278.41	0.89	15.66	nr	**294.08**
Flanged ends for steel pipework; up to 16 bar G at 120 degrees Celsius						
65mm dia.	162.22	196.98	1.10	19.36	nr	**216.34**
80mm dia.	197.91	240.32	1.31	23.06	nr	**263.38**
100mm dia.	226.03	274.47	1.78	31.33	nr	**305.80**
150mm dia.	340.67	413.68	3.08	54.21	nr	**467.89**
Screwed ends for non-ferrous pipework; up to 6 bar G at 100 degrees Celsius						
20mm dia.	76.79	93.25	0.72	12.67	nr	**105.92**
25mm dia.	81.11	98.49	0.84	14.78	nr	**113.28**
32mm dia.	97.33	118.19	1.02	17.95	nr	**136.14**
40mm dia.	112.48	136.58	1.11	19.54	nr	**156.12**
50mm dia.	127.62	154.97	1.18	20.77	nr	**175.74**

Material Costs/Measured Work Prices - Mechanical Installations

T:MECHANICAL/COOLING/HEATING SYSTEMS

Item	Net Price £	Material £	Labour hours	Labour £	Unit	Total rate £
T31 : LOW TEMPERATURE HOT WATER HEATING						
Y11 - PIPELINE - ANCILLAIRES						
EXPANSION JOINTS (Continued)						
Flanged ends for steel, copper or non-ferrous pipework; up to 16 bar G at 120 degrees Celsius						
65mm dia.	214.14	260.03	0.87	15.31	nr	275.34
80mm dia.	249.83	303.37	0.95	16.72	nr	320.09
100mm dia.	289.84	351.95	1.15	20.24	nr	372.19
150mm dia.	429.36	521.37	1.36	23.94	nr	545.31
Angular movement bellows expansion joints; stainless steel						
Flanged ends for steel pipework; up to 16 bar G at 120 degrees Celsius						
50mm dia.	321.21	390.05	0.71	12.50	nr	402.54
65mm dia.	340.67	413.68	0.83	14.61	nr	428.28
80mm dia.	403.40	489.85	0.91	16.02	nr	505.87
100mm dia.	482.35	585.72	0.97	17.07	nr	602.79
125mm dia.	564.54	685.52	1.16	20.42	nr	705.94
150mm dia.	590.50	717.04	1.18	20.77	nr	737.81
Universal lateral movement bellows expansion joints; stainless steel						
Flanged ends for steel pipework; up to 16 bar G at 120 degrees Celsius						
50mm dia.	452.07	548.95	0.89	15.66	nr	564.61
65mm dia.	472.62	573.90	1.10	19.36	nr	593.26
80mm dia.	512.63	622.49	1.31	23.06	nr	645.54
100mm dia.	612.13	743.31	1.78	31.33	nr	774.64
125mm dia.	994.98	1208.20	3.06	53.86	nr	1262.06
150mm dia.	1124.76	1365.80	3.08	54.21	nr	1420.01
Universal movement expansion joints; reinforced neoprene flexible connector						
Spherical expansion joints; flanged to BS 10, Table E; up to 10 bar at 100 degrees Celsius						
40mm dia.	101.66	123.45	0.82	14.43	nr	137.88
50mm dia.	103.82	126.07	0.89	15.66	nr	141.73
65mm dia.	112.48	136.58	1.10	19.36	nr	155.95
80mm dia.	128.70	156.28	1.31	23.06	nr	179.34
100mm dia.	152.49	185.17	1.78	31.33	nr	216.50
150mm dia.	218.46	265.28	3.08	54.21	nr	319.49

Material Costs/Measured Work Prices - Mechanical Installations 209

T:MECHANICAL/COOLING/HEATING SYSTEMS

Item	Net Price £	Material £	Labour hours	Labour £	Unit	Total rate £
T31 : LOW TEMPERATURE HOT WATER HEATING						
Y11 - PIPELINE - ANCILLAIRES						
EXPANSION JOINTS (Continued)						
Universal movement expansion joints; reinforced neoprene flexible connector (Continued)						
Hose connector; BSP threaded union ends; up to 8 bar at 100 degrees Celsius						
20mm dia.	27.04	32.83	1.32	23.23	nr	**56.07**
25mm dia.	36.77	44.65	1.52	26.75	nr	**71.40**
32mm dia.	41.10	49.91	1.80	31.68	nr	**81.59**
40mm dia.	54.08	65.67	2.03	35.73	nr	**101.40**
50mm dia.	70.30	85.37	2.26	39.78	nr	**125.14**
VALVES						
Isolating valves						
Bronze wedge non-rising stem; BS 5154, series B, PN 32; wheelhead operated; working pressure saturated steam up to 14 bar; cold services up to 32 bar; screwed ends to steel						
15mm dia.	14.14	17.17	1.11	19.54	nr	**36.71**
20mm dia.	18.23	22.14	1.28	22.53	nr	**44.67**
25mm dia.	24.71	30.01	1.49	26.23	nr	**56.23**
32mm dia.	37.16	45.12	1.88	33.09	nr	**78.21**
40mm dia.	46.70	56.71	2.31	40.66	nr	**97.37**
50mm dia.	67.49	81.95	2.80	49.28	nr	**131.24**
Bronze wedge non-rising stem; BS 5154, series B, PN 20; wheelhead operated; working pressure saturated steam up to 9 bar; cold services up to 20 bar; screwed ends to steel						
15mm dia.	9.25	11.23	0.84	14.78	nr	**26.02**
20mm dia.	13.11	15.92	1.01	17.78	nr	**33.70**
25mm dia.	16.98	20.62	1.19	20.95	nr	**41.56**
32mm dia.	24.20	29.39	1.38	24.29	nr	**53.68**
40mm dia.	33.11	40.21	1.62	28.51	nr	**68.72**
50mm dia.	47.41	57.57	1.94	34.15	nr	**91.72**

Material Costs/Measured Work Prices - Mechanical Installations

T:MECHANICAL/COOLING/HEATING SYSTEMS

Item	Net Price £	Material £	Labour hours	Labour £	Unit	Total rate £
T31 : LOW TEMPERATURE HOT WATER HEATING						
Y11 - PIPELINE - ANCILLAIRES						
VALVES (Continued)						
Bronze wedge non-rising stem; BS 5154, PN 16; wheelhead operated; working pressure saturated steam up to 7 bar; cold services up to 16 bar; BS4504 flanged ends; bolted connections						
15mm dia.	37.69	45.77	1.24	21.82	nr	67.59
20mm dia.	48.72	59.16	1.31	23.06	nr	82.22
25mm dia.	63.89	77.58	1.43	25.17	nr	102.75
32mm dia.	83.23	101.07	1.53	26.93	nr	128.00
40mm dia.	99.71	121.08	1.63	28.69	nr	149.77
50mm dia.	139.06	168.86	1.71	30.10	nr	198.96
65mm dia.	212.14	257.60	1.88	33.09	nr	290.69
80mm dia.	300.01	364.30	2.03	35.73	nr	400.03
100mm dia.	532.08	646.10	2.81	49.46	nr	695.56
Cast iron, trim material bronze; non rising stem, inside screwed; BS 5150, PN6; wheelhead operated; working pressure up to 6 bar at 120 degrees C; BS4504 flanged ends; bolted connections						
40mm dia.	102.37	124.31	1.79	31.51	nr	155.81
50mm dia.	102.37	124.31	1.85	32.56	nr	156.87
65mm dia.	102.45	124.41	2.00	35.20	nr	159.61
80mm dia.	118.40	143.77	2.27	39.95	nr	183.73
100mm dia.	156.44	189.97	2.76	48.58	nr	238.54
125mm dia.	223.91	271.89	6.05	106.48	nr	378.38
150mm dia.	257.66	312.88	8.03	141.33	nr	454.21
200mm dia.	490.77	595.94	9.17	161.40	nr	757.34
250mm dia.	755.15	916.98	10.72	188.68	nr	1105.66
300mm dia.	895.67	1087.61	11.75	206.81	nr	1294.42
Cast iron, trim material bronze; non rising stem, inside screwed; BS 5150, PN10; wheelhead operated; working pressure saturated steam up to 8.4 bar, cold services up to 10 bar; BS4504 flanged ends; bolted connections						
50mm dia.	111.65	135.58	1.85	32.56	nr	168.14
65mm dia.	129.44	157.18	2.00	35.20	nr	192.38
80mm dia.	171.77	208.58	2.27	39.95	nr	248.53
100mm dia.	242.32	294.25	2.76	48.58	nr	342.83
125mm dia.	279.13	338.95	6.05	106.48	nr	445.43
150mm dia.	506.12	614.58	8.03	141.33	nr	755.92
200mm dia.	754.57	916.27	9.17	161.40	nr	1077.67
250mm dia.	960.08	1165.83	10.72	188.68	nr	1354.51
300mm dia.	1275.57	1548.92	11.75	206.81	nr	1755.73
350mm dia.	2143.62	2603.00	12.67	223.00	nr	2826.00

Material Costs/Measured Work Prices - Mechanical Installations

T:MECHANICAL/COOLING/HEATING SYSTEMS

Item	Net Price £	Material £	Labour hours	Labour £	Unit	Total rate £
Cast iron, trim material bronze; non rising stem, inside screwed; BS 5150, PN16; wheelhead operated; working pressure saturated steam up to 12.8 bar, cold services up to 16 bar; BS4504 flanged ends; bolted connections						
40mm dia.	105.45	128.05	1.79	31.51	nr	159.55
50mm dia.	105.45	128.05	1.85	32.56	nr	160.61
65mm dia.	125.47	152.36	2.00	35.20	nr	187.56
80mm dia.	139.69	169.63	2.27	39.95	nr	209.58
100mm dia.	185.95	225.80	2.76	48.58	nr	274.38
125mm dia.	259.99	315.71	6.05	106.48	nr	422.19
150mm dia.	306.73	372.46	8.03	141.33	nr	513.80
200mm dia.	563.57	684.34	9.17	161.40	nr	845.74
250mm dia.	885.24	1074.95	10.72	188.68	nr	1263.63
300mm dia.	996.16	1209.64	11.75	206.81	nr	1416.45
350mm dia.	1883.76	2287.45	12.67	223.00	nr	2510.45
400mm dia.	2363.28	2869.73	14.30	251.69	nr	3121.42
450mm dia.	3331.49	4045.43	15.09	265.60	nr	4311.02
Cast iron, trim material bronze; non rising stem; BS 5163 series A, PN16; wheelhead operated; working pressure cold services up to 16 bar; BS4504 flanged ends; bolted connections						
50mm dia.	196.01	238.01	1.85	32.56	nr	270.58
65mm dia.	208.62	253.33	2.00	35.20	nr	288.53
80mm dia.	215.04	261.12	2.27	39.95	nr	301.08
100mm dia.	278.30	337.94	2.76	48.58	nr	386.52
125mm dia.	370.89	450.37	6.05	106.48	nr	556.86
150mm dia.	451.23	547.93	8.03	141.33	nr	689.26
Ball valves						
Malleable iron body; lever operated stainless steel ball and stem; Class 125; cold working pressure up to 12 bar; flanged ends (BS 4504 16/11); bolted connections						
40mm dia.	101.37	123.10	1.54	27.11	nr	150.20
50mm dia.	127.54	154.87	1.64	28.87	nr	183.73
80mm dia.	213.29	259.00	1.92	33.79	nr	292.79
100mm dia.	394.24	478.73	2.80	49.28	nr	528.01
150mm dia.	536.32	651.26	12.05	212.09	nr	863.35
Malleable iron body; lever operated stainless steel ball and stem; cold working pressure up to 16 bar; screwed ends to steel						
20mm dia.	22.31	27.09	1.34	23.59	nr	50.68
25mm dia.	22.93	27.85	1.40	24.65	nr	52.50
32mm dia.	31.59	38.36	1.46	25.73	nr	64.09
40mm dia.	31.59	38.36	1.54	27.12	nr	65.47
50mm dia.	37.78	45.87	1.64	28.90	nr	74.77

Material Costs/Measured Work Prices - Mechanical Installations

T:MECHANICAL/COOLING/HEATING SYSTEMS

Item	Net Price £	Material £	Labour hours	Labour £	Unit	Total rate £
T31 : LOW TEMPERATURE HOT WATER HEATING						
Y11 - PIPELINE - ANCILLAIRES						
VALVES (Continued)						
Carbon steel body; lever operated stainless steel ball and stem; Class 150; cold working pressure up to 19 bar; screwed ends to steel						
15mm dia.	16.38	19.89	0.84	14.79	nr	34.68
20mm dia.	17.20	20.89	1.14	20.07	nr	40.96
25mm dia.	19.69	23.90	1.30	22.89	nr	46.79
Globe valves						
Bronze; renewable disc; BS 5154 series B, PN32; working pressure saturated steam up to 14 bar; cold services up to 32 bar; screwed ends to steel						
15mm dia.	9.93	12.06	0.77	13.55	nr	25.61
20mm dia.	13.51	16.41	1.03	18.13	nr	34.53
25mm dia.	20.65	25.08	1.19	20.95	nr	46.02
32mm dia.	29.15	35.40	1.38	24.29	nr	59.69
40mm dia.	36.25	44.02	1.62	28.51	nr	72.53
50mm dia.	57.33	69.62	1.61	28.34	nr	97.95
Bronze; needle valve; BS 5154, series B, PN32; working pressure saturated steam up to 14 bar; cold services up to 32 bar; BS4504 screwed ends to steel						
15mm dia.	14.42	17.51	1.07	18.83	nr	36.34
20mm dia.	24.41	29.64	1.18	20.77	nr	50.41
25mm dia.	34.40	41.77	1.27	22.35	nr	64.12
32mm dia.	71.77	87.15	1.35	23.76	nr	110.91
40mm dia.	113.06	137.29	1.47	25.87	nr	163.16
50mm dia.	143.20	173.89	1.61	28.34	nr	202.23
Bronze; renewable disc; BS 5154, series B, PN16; working pressure saturated steam up to 7 bar; cold services up to 16 bar; BS4504 flanged ends; bolted connections						
15mm dia.	33.42	40.58	1.16	20.42	nr	61.00
20mm dia.	40.63	49.34	1.26	22.18	nr	71.51
25mm dia.	59.31	72.02	1.38	24.29	nr	96.31
32mm dia.	77.01	93.51	1.47	25.87	nr	119.39
40mm dia.	94.71	115.01	1.56	27.46	nr	142.46
50mm dia.	138.29	167.93	1.71	30.10	nr	198.02

Material Costs/Measured Work Prices - Mechanical Installations

T:MECHANICAL/COOLING/HEATING SYSTEMS

Item	Net Price £	Material £	Labour hours	Labour £	Unit	Total rate £
Bronze; renewable disc; BS 2060, class 250; working pressure saturated steam up to 24 bar; cold services up to 38 bar; flanged ends (BS 10 table H); bolted connections						
15mm dia.	91.10	110.62	1.16	20.42	nr	**131.04**
20mm dia.	106.17	128.92	1.26	22.18	nr	**151.10**
25mm dia.	145.51	176.69	1.38	24.29	nr	**200.98**
32mm dia.	190.07	230.80	1.47	25.87	nr	**256.68**
40mm dia.	240.87	292.49	1.56	27.46	nr	**319.95**
50mm dia.	353.93	429.78	1.71	30.10	nr	**459.87**
65mm dia.	522.63	634.63	1.88	33.09	nr	**667.72**
80mm dia.	1313.25	1594.68	2.03	35.73	nr	**1630.41**
Check valves						
Bronze, swing pattern; BS 5154 series B, PN 25; working pressure saturated steam up to 10.5 bar, cold services up to 25 bar; screwed ends						
15mm dia.	10.92	13.26	0.77	13.55	nr	**26.81**
20mm dia.	12.99	15.77	1.03	18.13	nr	**33.90**
25mm dia.	17.98	21.83	1.19	20.95	nr	**42.78**
32mm dia.	26.90	32.66	1.38	24.29	nr	**56.95**
40mm dia.	33.48	40.65	1.62	28.51	nr	**69.17**
50mm dia.	52.65	63.93	1.94	34.15	nr	**98.08**
65mm dia.	104.52	126.92	2.45	43.12	nr	**170.04**
80mm dia.	156.78	190.38	2.83	49.81	nr	**240.19**
Bronze, horizontal lift pattern; BS 5154 series B, PN32; working pressure saturated steam up to 14 bar, cold services up to 32 bar; screwed connections to steel						
15mm dia.	14.20	17.24	0.96	16.90	nr	**34.14**
20mm dia.	19.89	24.15	1.07	18.83	nr	**42.99**
25mm dia.	26.54	32.23	1.17	20.59	nr	**52.82**
32mm dia.	38.39	46.62	1.33	23.41	nr	**70.03**
40mm dia.	48.68	59.11	1.41	24.82	nr	**83.93**
50mm dia.	75.81	92.06	1.55	27.28	nr	**119.34**
65mm dia.	256.10	310.99	1.80	31.68	nr	**342.67**
80mm dia.	385.09	467.62	1.99	35.03	nr	**502.65**
Bronze, oblique lift pattern; BS 5154 series B, PN32; working pressure saturated steam up to 14 bar, cold services up to 32 bar; screwed connections to steel						
15mm dia.	23.73	28.82	0.96	16.90	nr	**45.71**
20mm dia.	26.06	31.64	1.07	18.83	nr	**50.48**
25mm dia.	39.72	48.23	1.17	20.59	nr	**68.82**
32mm dia.	57.39	69.69	1.33	23.41	nr	**93.10**
40mm dia.	63.60	77.23	1.41	24.82	nr	**102.05**
50mm dia.	91.21	110.76	1.55	27.28	nr	**138.04**

Material Costs/Measured Work Prices - Mechanical Installations

T:MECHANICAL/COOLING/HEATING SYSTEMS

Item	Net Price £	Material £	Labour hours	Labour £	Unit	Total rate £
T31 : LOW TEMPERATURE HOT WATER HEATING						
Y11 - PIPELINE - ANCILLAIRES						
VALVES (Continued)						
Cast iron, swing pattern; BS 5153 PN6; working pressure for cold services up to 6 bar; BS 4504 flanged ends; bolted connections						
50mm dia.	116.54	141.51	1.86	32.74	nr	**174.25**
65mm dia.	132.23	160.57	2.00	35.20	nr	**195.77**
80mm dia.	148.84	180.74	2.56	45.06	nr	**225.79**
100mm dia.	190.66	231.52	2.76	48.58	nr	**280.10**
125mm dia.	284.46	345.42	6.05	106.48	nr	**451.90**
150mm dia.	319.81	388.35	8.11	142.74	nr	**531.09**
200mm dia.	707.28	858.85	9.26	162.98	nr	**1021.83**
250mm dia.	1076.30	1306.95	10.72	188.68	nr	**1495.63**
300mm dia.	1429.94	1736.38	11.75	206.81	nr	**1943.19**
Cast iron, horizontal lift pattern; BS 5153 PN16; working pressure for saturated steam 13 bar, cold services up to 16 bar; BS 4504 flanged ends; bolted connections						
50mm dia.	116.62	141.61	1.86	32.74	nr	**174.35**
65mm dia.	181.75	220.70	2.00	35.20	nr	**255.90**
80mm dia.	221.24	268.65	2.56	45.06	nr	**313.71**
100mm dia.	288.56	350.40	2.96	52.10	nr	**402.50**
125mm dia.	454.13	551.45	7.76	136.58	nr	**688.03**
150mm dia.	493.12	598.80	10.50	184.81	nr	**783.60**
Cast iron semi lugged butterfly valve; BS5155 PN16; working pressure 16 bar at 120 degrees Celsius; EPDM seating; flanged ends (BS 4504, Part 1, Table 16); bolted connections						
50mm dia	42.07	51.09	2.20	38.72	nr	**89.81**
65mm dia	42.07	51.09	2.31	40.66	nr	**91.74**
80mm dia	50.48	61.30	2.88	50.69	nr	**111.99**
100mm dia	75.97	92.25	3.11	54.74	nr	**146.99**
125mm dia	111.11	134.92	5.02	88.36	nr	**223.28**
150mm dia	128.93	156.56	6.98	122.85	nr	**279.41**
200mm dia	225.16	273.41	8.25	145.21	nr	**418.62**
250mm dia	449.51	545.84	10.47	184.28	nr	**730.12**
300mm dia	570.19	692.38	11.48	202.06	nr	**894.44**

T:MECHANICAL/COOLING/HEATING SYSTEMS

Item	Net Price £	Material £	Labour hours	Labour £	Unit	Total rate £
Cast iron fully lugged butterfly valve; BS5155 PN16; working pressure 16 bar at 120 degrees Celsius; EPDM seating; flanged ends (BS 4504, Part 1, Table 16); bolted connections						
50mm dia	65.97	80.11	2.20	38.72	nr	118.83
65mm dia	65.97	80.11	2.31	40.66	nr	120.77
80mm dia	78.09	94.82	2.88	50.69	nr	145.52
100mm dia	111.43	135.31	3.11	54.74	nr	190.05
125mm dia	164.55	199.81	5.02	88.36	nr	288.17
150mm dia	186.21	226.11	6.98	122.85	nr	348.97
200mm dia	297.07	360.73	8.25	145.21	nr	505.94
250mm dia	517.06	627.87	10.47	184.28	nr	812.15
300mm dia	642.24	779.87	11.48	202.06	nr	981.93
350mm dia	2447.05	2971.45	12.00	211.21	nr	3182.66
400mm dia	3298.29	4005.11	12.50	220.01	nr	4225.12
450mm dia	4218.85	5122.95	13.00	228.81	nr	5351.76
Commissioning valves						
Bronze commissioning set; metering station; double regulating valve; BS5154 PN20 Series B; working pressure 20 bar at 100 degrees Celsius; screwed ends to steel						
15mm dia.	34.44	41.82	1.08	19.01	nr	60.83
20mm dia.	56.53	68.64	1.46	25.70	nr	94.34
25mm dia.	68.67	83.39	1.68	29.57	nr	112.96
32mm dia.	90.48	109.87	1.95	34.32	nr	144.19
40mm dia.	120.21	145.97	2.27	39.95	nr	185.92
50mm dia.	173.78	211.02	2.73	48.05	nr	259.07
Cast iron commissioning set; metering station; double regulating valve; BS5152 PN16; woking pressure 16 bar at 120 degrees Celsius; flanged ends (BS 4504, Part 1, Table 16); bolted connections						
65mm dia.	319.57	388.05	1.80	31.68	nr	419.74
80mm dia.	388.99	472.35	2.56	45.06	nr	517.41
100mm dia.	511.93	621.64	2.30	40.48	nr	662.12
125mm dia.	782.19	949.81	2.44	42.95	nr	992.76
150mm dia.	981.55	1191.90	2.90	51.04	nr	1242.94
200mm dia	2573.10	3124.52	8.26	145.38	nr	3269.90
250mm dia	3709.35	4504.26	10.49	184.63	nr	4688.90
300mm dia	5844.82	7097.36	11.49	202.23	nr	7299.60

216

Material Costs/Measured Work Prices - Mechanical Installations

T:MECHANICAL/COOLING/HEATING SYSTEMS

Item	Net Price £	Material £	Labour hours	Labour £	Unit	Total rate £
T31 : LOW TEMPERATURE HOT WATER HEATING						
Y11 - PIPELINE - ANCILLAIRES						
VALVES (Continued)						
Cast iron orifice valve; complete with controlled test points; BS5152 PN16; working pressure 16 bar at 120 degrees Celsius; flanged ends (BS 4504, Part 1, Table 16); bolted connections						
65mm dia.	181.06	219.86	2.00	35.20	nr	**255.06**
80mm dia.	217.56	264.18	2.56	45.06	nr	**309.24**
100mm dia.	280.40	340.49	2.96	52.10	nr	**392.59**
125mm dia.	428.46	520.28	7.76	136.58	nr	**656.86**
150mm dia.	531.58	645.50	10.50	184.81	nr	**830.31**
200mm dia	638.73	775.61	8.26	145.38	nr	**920.99**
250mm dia	709.78	861.89	10.49	184.63	nr	**1046.52**
300mm dia	1098.61	1334.04	11.49	202.23	nr	**1536.28**
Cast iron, double regulating valve; LTHW, MTHW and chilled water; BS5152 PN16; working pressure 16 bar at 120 degrees Celsius; flanged ends (BS 4504, Part 1, Table 16); bolted connections						
65mm dia.	228.27	277.19	2.00	35.20	nr	**312.39**
80mm dia.	281.53	341.86	2.56	45.06	nr	**386.92**
100mm dia.	383.49	465.67	2.96	52.10	nr	**517.77**
125mm dia.	553.92	672.63	7.76	136.58	nr	**809.21**
150mm dia.	715.23	868.50	10.50	184.81	nr	**1053.31**
200mm dia	1934.37	2348.91	8.26	145.38	nr	**2494.29**
250mm dia	2999.58	3642.39	10.49	184.63	nr	**3827.02**
300mm dia	4746.22	5763.33	11.49	202.23	nr	**5965.57**
Bronze autoflow commissioning valve; PN25 ; working pressure 25 bar at 100 degrees Celsius; screwed ends to steel						
15mm dia	53.79	65.32	0.82	14.43	nr	**79.75**
20mm dia	55.69	67.62	1.08	19.01	nr	**86.63**
25mm dia	81.33	98.76	1.27	22.35	nr	**121.11**
32mm dia	106.00	128.72	1.50	26.40	nr	**155.12**
40mm dia	132.90	161.38	1.76	30.98	nr	**192.36**
50mm dia	167.71	203.65	2.13	37.49	nr	**241.14**
Ductile iron autoflow commissioning valves; PN16; working pressure 16 bar at 120 degrees Celsius; for ANSI 150 flanged ends						
65mm dia	402.95	489.30	2.31	40.66	nr	**529.96**
80mm dia	445.50	540.97	2.88	50.69	nr	**591.66**
100mm dia	694.72	843.60	3.11	54.74	nr	**898.34**
150mm dia	1172.38	1423.62	6.98	122.85	nr	**1546.47**
200mm dia	1692.01	2054.61	8.26	145.38	nr	**2199.99**
250mm dia	2349.96	2853.56	10.49	184.63	nr	**3038.19**
300mm dia	3001.24	3644.41	11.49	202.23	nr	**3846.64**

Material Costs/Measured Work Prices - Mechanical Installations

T:MECHANICAL/COOLING/HEATING SYSTEMS

Item	Net Price £	Material £	Labour hours	Labour £	Unit	Total rate £
T31 : LOW TEMPERATURE HOT WATER HEATING						
Y11 - PIPELINE - ANCILLAIRES						
VALVES (Continued)						
Strainers						
Bronze strainer; Y type; PN32 ; working pressure 25 bar at 100 degrees Celsius; screwed ends to steel						
15mm dia	11.91	14.46	0.82	14.43	nr	**28.90**
20mm dia	15.13	18.37	1.08	19.01	nr	**37.38**
25mm dia	21.33	25.90	1.27	22.35	nr	**48.25**
32mm dia	34.45	41.83	1.50	26.40	nr	**68.23**
40mm dia	46.79	56.82	1.76	30.98	nr	**87.79**
50mm dia	78.37	95.16	2.13	37.49	nr	**132.65**
Cast iron strainer; Y type; PN16; working pressure 16 bar at 120 degrees Celsius; BS 4504 flanged ends						
65mm dia	84.47	102.57	2.31	40.66	nr	**143.23**
80mm dia	97.82	118.78	2.88	50.69	nr	**169.47**
100mm dia	144.84	175.88	3.11	54.74	nr	**230.62**
125mm dia	299.02	363.10	5.02	88.36	nr	**451.46**
150mm dia	387.40	470.42	6.98	122.85	nr	**593.27**
200mm dia	601.48	730.38	8.26	145.38	nr	**875.76**
250mm dia	909.56	1104.48	10.49	184.63	nr	**1289.11**
300mm dia	1525.44	1852.34	11.49	202.23	nr	**2054.57**
Regulators						
Gunmetal; self-acting two port thermostatic regulator; single seat; water or water; normally closed or normally open; screwed ends; complete with sensing element; 2m long capillary tube						
15mm dia.	319.71	388.23	1.37	24.11	nr	**412.34**
20mm dia.	327.95	398.23	1.24	21.82	nr	**420.06**
25mm dia.	338.66	411.24	1.34	23.59	nr	**434.82**
Gunmetal; self-acting two port thermostatic regulator; double seat; water or steam; flanged ends (BS 4504 PN25); with sensing element; 2m long capillary tube; steel body						
65mm dia.	1078.62	1309.76	1.23	19.71	nr	**1329.48**
80mm dia.	1273.08	1545.90	1.62	25.96	nr	**1571.86**

218 **Material Costs/Measured Work Prices - Mechanical Installations**

T:MECHANICAL/COOLING/HEATING SYSTEMS

Item	Net Price £	Material £	Labour hours	Labour £	Unit	Total rate £
T31 : LOW TEMPERATURE HOT WATER HEATING						
Y11 - PIPELINE - ANCILLAIRES						
VALVES (Continued)						
Control valves; electrically operated (electrical work elsewhere)						
Cast iron; butterfly type; two position electrically controlled 240V motor and linkage mechanism; for low pressure hot water; max 6 bar 120 degree celsius; flanged ends; counter flanges						
25mm dia.	228.35	277.29	1.47	25.87	nr	303.16
32mm dia.	235.95	286.51	1.52	26.75	nr	313.27
40mm dia.	329.70	400.35	1.61	28.34	nr	428.69
50mm dia.	339.06	411.72	1.71	30.10	nr	441.82
65mm dia.	347.55	422.03	2.51	44.18	nr	466.21
80mm dia.	361.15	438.54	2.69	47.35	nr	485.89
100mm dia.	378.99	460.21	2.81	49.46	nr	509.67
125mm dia.	419.78	509.74	2.94	51.75	nr	561.49
150mm dia.	454.90	552.39	3.33	58.61	nr	611.00
200mm dia.	564.23	685.14	3.67	64.59	nr	749.74
Cast iron; three way 240V motorized; for low pressure hot water; max 6 bar 120 degree celsius; flanged ends and drilled (BS 10, Table F)						
25mm dia.	255.21	309.90	1.99	35.03	nr	344.93
40mm dia.	267.68	325.04	2.13	37.49	nr	362.53
50mm dia.	264.28	320.92	3.21	56.50	nr	377.41
65mm dia.	295.29	358.57	3.23	56.85	nr	415.42
80mm dia.	330.87	401.78	3.50	61.60	nr	463.38
Two port normally closed motorised valve; electric actuator; spring return; domestic usage						
22mm dia.	37.17	45.14	1.18	20.77	nr	65.90
28mm dia.	50.39	61.19	1.35	23.76	nr	84.95
Two port on/off motorised valve; electric actuator; spring return; domestic usage						
22mm dia.	37.17	45.14	1.18	20.77	nr	65.90
Three port motorised valve; electric actuator; spring return; domestic usage						
22mm dia.	54.25	65.88	1.18	20.77	nr	86.64
Safety and relief valves						
Bronze relief valve; spring type; side outlet; working pressure up to 20.7 bar at 120 degrees Celsius; screwed ends to steel						
15mm dia.	36.79	44.67	0.26	4.58	nr	49.25
20mm dia.	45.24	54.93	0.36	6.34	nr	61.27

Material Costs/Measured Work Prices - Mechanical Installations

219

T:MECHANICAL/COOLING/HEATING SYSTEMS

Item	Net Price £	Material £	Labour hours	Labour £	Unit	Total rate £
Bronze relief valve; spring type; side outlet; working pressure up to 17.2 bar at 120 degrees Celsius; screwed ends to steel						
25mm dia.	61.10	74.19	0.38	6.69	nr	**80.88**
32mm dia.	89.45	108.62	0.48	8.45	nr	**117.07**
Bronze relief valve; spring type; side outlet; working pressure up to 13.8 bar at 120 degrees Celsius; screwed ends to steel						
40mm dia.	107.35	130.36	0.64	11.26	nr	**141.62**
50mm dia.	149.88	182.00	0.76	13.38	nr	**195.38**
65mm dia.	237.97	288.97	0.94	16.54	nr	**305.51**
80mm dia.	312.22	379.13	1.10	19.36	nr	**398.49**
Cocks; screwed joints to steel						
Bronze gland cock; complete with malleable iron lever; working pressure cold services up to 10 bar; screwed ends						
15mm dia.	16.11	19.56	0.77	13.55	nr	**33.12**
20mm dia.	22.55	27.38	1.03	18.13	nr	**45.51**
25mm dia.	31.72	38.52	1.19	20.95	nr	**59.46**
32mm dia.	55.01	66.80	1.38	24.29	nr	**91.09**
40mm dia.	78.05	94.78	1.62	28.51	nr	**123.29**
50mm dia.	133.28	161.84	1.94	34.15	nr	**195.99**
Bronze three-way plug cock; complete with malleable iron lever; working pressure cold services up to 10 bar; screwed ends						
15mm dia.	35.21	42.76	0.77	13.55	nr	**56.31**
20mm dia.	37.96	46.09	1.03	18.13	nr	**64.22**
25mm dia.	51.06	62.00	1.19	20.95	nr	**82.95**
32mm dia.	70.56	85.68	1.38	24.29	nr	**109.97**
40mm dia.	86.61	105.17	1.62	28.51	nr	**133.68**
Air vents; including regulating, adjusting and testing						
Automatic air vent; pressures up to 7 bar; 93 degree celsius; screwed ends to steel						
15mm dia.	63.86	77.55	0.80	14.08	nr	**91.63**
Automatic air vent; pressures up to 7 bar; 93 degree celsius; lockhead isolating valve; screwed ends to steel						
15mm dia.	68.69	83.41	0.83	14.61	nr	**98.02**
Automatic air vent; pressures up to 17 bar; 204 degree celsius; flanged end (BS10, Table H); bolted connections to counter flange (measured separately)						
15mm dia.	287.85	349.54	0.83	14.61	nr	**364.14**

Material Costs/Measured Work Prices - Mechanical Installations

T:MECHANICAL/COOLING/HEATING SYSTEMS

Item	Net Price £	Material £	Labour hours	Labour £	Unit	Total rate £
T31 : LOW TEMPERATURE HOT WATER HEATING						
Y11 - PIPELINE - ANCILLAIRES						
VALVES (Continued)						
Radiator valves						
Bronze; wheelhead or lockshield; chromium plated finish; screwed joints to steel						
Straight						
15mm dia	21.56	26.18	0.59	10.38	nr	**36.56**
20mm dia	28.08	34.10	0.73	12.85	nr	**46.95**
25mm dia	35.09	42.61	0.85	14.96	nr	**57.57**
Angled						
15mm dia	15.34	18.63	0.59	10.38	nr	**29.01**
20mm dia	20.22	24.55	0.73	12.85	nr	**37.40**
25mm dia	26.07	31.66	0.85	14.96	nr	**46.62**
Bronze; wheelhead or lockshield; chromium plated finish; compression joints to copper						
Straight						
8mm dia	7.49	9.19	0.23	4.05	nr	**13.23**
10mm dia	7.49	9.19	0.23	4.05	nr	**13.23**
15mm dia	21.56	26.18	0.59	10.38	nr	**36.56**
20mm dia	28.08	34.10	0.73	12.85	nr	**46.95**
25mm dia	35.09	42.61	0.85	14.96	nr	**57.57**
Angled						
15mm dia	15.34	18.63	0.59	10.38	nr	**29.01**
20mm dia	20.22	24.55	0.73	12.85	nr	**37.40**
25mm dia	26.07	31.66	0.85	14.96	nr	**46.62**
Bronze; wheelhead or lockshield; chromium plated finish; compression joints to copper						
Twin entry						
8mm dia	21.89	26.85	0.23	4.05	nr	**30.90**
10mm dia	24.06	29.51	0.23	4.05	nr	**33.56**
Bronze; thermostatic head; chromium plated finish; compression joints to copper						
Straight						
15mm dia	19.59	23.79	0.59	10.38	nr	**34.17**
20mm dia	21.91	26.61	0.73	12.85	nr	**39.45**
Angled						
15mm dia	20.36	24.72	0.59	10.38	nr	**35.11**
20mm dia	26.11	31.71	0.73	12.85	nr	**44.55**

Material Costs/Measured Work Prices - Mechanical Installations　221

T:MECHANICAL/COOLING/HEATING SYSTEMS

Item	Net Price £	Material £	Labour hours	Labour £	Unit	Total rate £
T31 : LOW TEMPERATURE HOT WATER HEATING						
Y11 - PIPELINE - ANCILLAIRES						
GAUGES						
Thermometers and pressure gauges Dial thermometer; coated steel case and dial; glass window; brass pocket; BS 5235; pocket length 100mm; screwed end						
Back entry						
100mm dia face	36.67	44.53	0.81	14.26	nr	**58.79**
150mm dia face	42.14	51.17	0.81	14.26	nr	**65.43**
Bottom entry						
100mm dia face	36.67	44.53	0.81	14.26	nr	**58.79**
150mm dia face	42.14	51.17	0.81	14.26	nr	**65.43**
Dial pressure/altitude gauge; bronze bourdon tube type; coated steel case and dial; glass window BS 1780; screwed end						
100mm dia face	23.16	28.12	0.81	14.26	nr	**42.38**
150mm dia face	21.22	25.77	0.81	14.26	nr	**40.03**

222

Material Costs/Measured Work Prices - Mechanical Installations

T:MECHANICAL/COOLING/HEATING SYSTEMS

Item	Net Price £	Material £	Labour hours	Labour £	Unit	Total rate £
T31 : LOW TEMPERATURE HOT WATER HEATING						
EQUIPMENT						
Y20 - PUMPS						
Centrifugal heating pump; belt drive; 3 phase, 1450 rpm motor; max. pressure 400kN/m2; max. temperature 125 degree C; bed plate; coupling guard bolted connections; supply only mating flanges;includes fixing on prepared base; electrical work elsewhere						
Heating pumps						
40mm pump size; 4.0 litre/sec maximum delivery; 70 kPa maximum head; 0.55kW maximum motor rating	1012.28	1229.21	7.59	133.59	nr	**1362.80**
40mm pump size; 4.0 litre/sec maximum delivery; 130 kPa maximum head; 1.1kW maximum motor rating	1073.93	1304.07	8.09	142.39	nr	**1446.46**
50mm pump size; 8.5 litre/sec maximum delivery; 90 kPa maximum head; 1.1kW maximum motor rating	1315.10	1596.93	8.67	152.60	nr	**1749.52**
50mm pump size; 8.5 litre/sec maximum delivery; 190 kPa maximum head; 2.2kW maximum motor rating	1563.85	1898.98	11.20	197.13	nr	**2096.11**
50mm pump size; 8.5 litre/sec maximum delivery; 215 kPa maximum head; 3.0 kW maximum motor rating	1596.29	1938.37	11.70	205.93	nr	**2144.30**
65mm pump size; 14.0 litre/sec maximum delivery; 90 kPa maximum head; 1.50kW maximum motor rating	1358.36	1649.46	11.70	205.93	nr	**1855.39**
65mm pump size; 14.0 litre/sec maximum delivery; 160 kPa maximum head; 3.0 kW maximum motor rating	1596.29	1938.37	11.70	205.93	nr	**2144.30**
65mm pump size; 14.5 litre/sec maximum delivery; 210 kPa maximum head; 4.0 kW maximum motor rating	1622.25	1969.90	11.70	205.93	nr	**2175.83**
80mm pump size; 22.0 litre/sec maximum delivery; 130 kPa maximum head; 4.0 kW maximum motor rating	1528.16	1855.64	13.64	240.07	nr	**2095.72**
80mm pump size; 22.0 litre/sec maximum delivery; 200 kPa maximum head; 5.5 kW maximum motor rating	2374.97	2883.93	13.64	240.07	nr	**3124.00**
80mm pump size; 22.0 litre/sec maximum delivery; 250 kPa maximum head; 7.50kW maximum motor rating	2673.47	3246.39	13.64	240.07	nr	**3486.47**
100mm pump size; 30.0 litre/sec maximum delivery; 100 kPa maximum head; 4.0 kW maximum motor rating	1622.25	1969.90	19.15	337.06	nr	**2306.95**
100mm pump size; 36.0 litre/sec maximum delivery; 25 kPa maximum head; 11.0kW maximum motor rating	2235.46	2714.52	19.15	337.06	nr	**3051.57**

Material Costs/Measured Work Prices - Mechanical Installations

T:MECHANICAL/COOLING/HEATING SYSTEMS

Item	Net Price £	Material £	Labour hours	Labour £	Unit	Total rate £
Centrifugal heating pump; belt drive; 3 phase, 1450 rpm motor; max. pressure 400kN/m2; max. temperature 125 degree C; bed plate; coupling guard bolted connections; supply only mating flanges;includes fixing on prepared base; electrical work elsewhere (Continued)						
100mm pump size; 36.0 litre/sec maximum delivery; 550 kPa maximum head; 30.0 kW maximum motor rating	4973.82	6039.71	19.15	337.06	nr	**6376.76**
Extra for single phase motor						
0.37 kW	166.75	202.48	6.02	106.03	nr	**308.51**
0.55 kW	166.75	202.48	6.02	106.03	nr	**308.51**
0.75 kW	166.75	202.48	6.02	106.03	nr	**308.51**
1.10 kW	166.75	202.48	6.02	106.03	nr	**308.51**
Centrifugal heating pump; close coupled; 3 phase, 1450 rpm motor; max. pressure 400kN/m2; max. temperature 110 degree C; bed plate; coupling guard; bolted connections; supply only mating flanges; includes fixing on prepared base; electrical work elsewhere						
Heating pumps						
40mm pump size; 4.0 litre/sec maximum delivery; 4 kPa maximum head; 0.55kW maximum motor rating	493.16	598.84	7.31	128.66	nr	**727.51**
40mm pump size; 4.0 litre/sec maximum delivery; 23 kPa maximum head; 2.20kW maximum motor rating	528.85	642.18	7.31	128.66	nr	**770.84**
40mm pump size; 4.0 litre/sec maximum delivery; 75 kPa maximum head; 11 kW maximum motor rating	590.50	717.04	7.31	128.66	nr	**845.71**
50mm pump size; 7.0 litre/sec maximum delivery; 65 kPa maximum head; 0.75 kW maximum motor rating	572.11	694.71	8.01	140.98	nr	**835.70**
50mm pump size; 10.0 litre/sec maximum delivery; 33 kPa maximum head; 5.5kW maximum motor rating	531.02	644.82	8.01	140.98	nr	**785.80**
50mm pump size; 4.0 litre/sec maximum delivery; 120 kPa maximum head; 1.1 kW maximum motor rating	649.98	789.27	8.01	140.98	nr	**930.25**
80mm pump size; 16.0 litre/sec maximum delivery; 80 kPa maximum head; 2.2 kW maximum motor rating	829.51	1007.27	12.35	217.37	nr	**1224.64**
80mm pump size; 16.0 litre/sec maximum delivery; 120 kPa maximum head; 3.0 kW maximum motor rating	906.30	1100.52	12.35	217.37	nr	**1317.89**

Material Costs/Measured Work Prices - Mechanical Installations

T:MECHANICAL/COOLING/HEATING SYSTEMS

Item	Net Price £	Material £	Labour hours	Labour £	Unit	Total rate £
T31 : LOW TEMPERATURE HOT WATER HEATING						
EQUIPMENT						
Y20 - PUMPS (Continued)						
100mm pump size; 28.0 litre/sec maximum delivery; 40 kPa maximum head; 2.2 kW maximum motor rating	829.51	1007.27	17.86	314.35	nr	**1321.62**
100mm pump size; 28.0 litre/sec maximum delivery; 90 kPa maximum head; 4.0 kW maximum motor rating	979.84	1189.82	17.86	314.35	nr	**1504.17**
125mm pump size; 40.0 litre/sec maximum delivery; 50 kPa maximum head; 3.0 kW maximum motor rating	1152.88	1399.94	25.85	454.98	nr	**1854.92**
125mm pump size; 40.0 litre/sec maximum delivery; 120 kPa maximum head; 7.5 kW maximum motor rating	1917.50	2328.42	25.85	454.98	nr	**2783.40**
150mm pump size; 70.0 litre/sec maximum delivery; 75 kPa maximum head; 7.5 kW maximum motor rating	1246.97	1514.20	30.43	535.59	nr	**2049.79**
150mm pump size; 70.0 litre/sec maximum delivery; 120 kPa maximum head; 11.0 kW maximum motor rating	2029.98	2465.00	30.43	535.59	nr	**3000.60**
150mm pump size; 70.0 litre/sec maximum delivery; 150 kPa maximum head; 15.0 kW maximum motor rating	2108.93	2560.87	30.43	535.59	nr	**3096.47**
Glandless domestic heating pump; for low pressure domestic hot water heating systems; 240 volt; 50Hz electric motor; max working pressure 1000N/m2 and max temperature of 110 degree C; includes fixing; electrical work elsewhere						
Single speed						
20mm BSP unions	65.97	80.11	1.58	27.81	nr	**107.92**
Three speed; pump only						
40mm	65.97	80.11	1.58	27.81	nr	**107.92**

Material Costs/Measured Work Prices - Mechanical Installations   225

T:MECHANICAL/COOLING/HEATING SYSTEMS

Item	Net Price £	Material £	Labour hours	Labour £	Unit	Total rate £
Pipeline mounted circulator; for low and medium pressure heating and hot water services; silent running; 3 phase; 1450 rpm motor; max pressure 600 kN/m2; max temperature 110 degree C; bolted connections; supply only mating flanges; includes fixing; electrical elsewhere						
Circulator						
32mm pump size; 2.0 litre/sec maximum delivery; 17 kPa maximum head; 0.2kW maximum motor rating	217.38	263.96	6.44	113.35	nr	**377.31**
50mm pump size; 3.0 litre/sec maximum delivery; 20 kPa maximum head; 0.2kW maximum motor rating	170.88	207.50	6.86	120.74	nr	**328.24**
65mm pump size; 5.0 litre/sec maximum delivery; 30 kPa maximum head; 0.37 kW maximum motor rating	340.67	413.68	7.48	131.65	nr	**545.33**
65mm pump size; 8.0 litre/sec maximum delivery; 37 kPa maximum head; 0.75 kW maximum motor rating	436.93	530.56	7.48	131.65	nr	**662.22**
80mm pump size; 12.0 litre/sec maximum delivery; 42 kPa maximum head; 1.1 kW maximum motor rating	949.56	1153.05	8.01	140.98	nr	**1294.03**
100mm pump size; 25.0 litre/sec maximum delivery; 37 kPa maximum head; 2.2 kW maximum motor rating	829.51	1007.27	9.11	160.34	nr	**1167.62**
Dual pipeline mounted circulator; for low and medium pressure heating and hot water services; silent running; 3 phase; 1450 rpm motor; max pressure 600 kN/m2; max temperature 110 degree C; bolted connections; supply only mating flanges; includes fixing; electrical work elsewhere						
Circulator						
40mm pump size; 2.0 litre/sec maximum delivery; 17 kPa maximum head; 0.8kW maximum motor rating	680.26	826.04	7.88	138.69	nr	**964.73**
50mm pump size; 3.0 litre/sec maximum delivery; 20 kPa maximum head; 0.2 kW maximum motor rating	853.30	1036.16	8.01	140.98	nr	**1177.14**
65mm pump size; 5.0 litre/sec maximum delivery; 30 kPa maximum head; 0.37 kW maximum motor rating	1116.11	1355.29	9.20	161.93	nr	**1517.22**
65mm pump size; 8.0 litre/sec maximum delivery; 37 kPa maximum head; 0.75 kW maximum motor rating	1195.06	1451.16	9.20	161.93	nr	**1613.09**
80mm pump size; 12.0 litre/sec maximum delivery; 42 kPa maximum head; 1.1 kW maximum motor rating	1422.17	1726.94	9.45	166.33	nr	**1893.27**

T:MECHANICAL/COOLING/HEATING SYSTEMS

Item	Net Price £	Material £	Labour hours	Labour £	Unit	Total rate £
T31 : LOW TEMPERATURE HOT WATER HEATING						
EQUIPMENT						
Y20 - PUMPS (Continued)						
Glandless accelerator pumps; for low and medium pressure heating and hot water services; silent running; 3 phase; 1450 rpm motor; max pressure 600 kN/m2; max temperature 120 degree C; bolted connections; supply only mating flanges; includes fixing; electrical work elsewhere						
Accelerator pump						
40mm pump size; 4.0 litre/sec maximum delivery; 15 kPa maximum head; 0.35kW maximum motor rating	320.12	388.72	6.94	122.15	nr	**510.87**
50mm pump size; 6.0 litre/sec maximum delivery; 20 kPa maximum head; 0.45kW maximum motor rating	401.43	487.46	7.35	129.37	nr	**616.82**
80mm pump size; 13.0 litre/sec maximum delivery; 28 kPa maximum head; 0.58kW maximum motor rating	986.64	1198.08	7.76	136.58	nr	**1334.66**

NEW FROM SPON PRESS

Understanding the Building Regulations
2nd Edition
Simon Polley, BRCS, UK

This is a new edition of the highly successful introductory guide to current Building Regulations and Approval Documents. Including the major revisions to part B, it is an essential tool for those involved in design and construction and for those who require knowledge of building control.

Thoroughly revised and updated, it will provide all the information necessary to design and build to the building regulations. This is an essential tool for construction professionals requiring a 'pocket book' guide to the regulations.

Reviews of the first edition: *'...covers all the requirements of the Building Regulations as we know them today. It is clear and concise in its explanations...a good book...'* **Clerk of Works Journal**

> **Contents**: Preface. Acknowledgements. Introduction. The Building Regulations 2000. Approved Document to support Regulation 7: Materials and Workmanship. Approved Document A: Structure. Approved Document B: Fire Safety. Approved Document C: Site Preparation and Resistance to Moisture. Approved Document D: Toxic Substances. Approved Document E: Resistance to the Passage of Sound. Approved Document F: Ventilation. Approved Document G: Hygiene. Approved Document H: Drainage and Waste Disposal. Approved Document K: Protection from Falling Collision and Impact. Document L: Conservation of Fuel and Power. Approved Document M: Access and Facilities for Disabled People. Approved Document N: Glazing - Safety in Relation to Impact, Opening and Cleaning. Further Information. Index.

August 2001: 234x156: 208pp :illus.55 line figs. Pb: 0-419-24720-3: £16.99

To Order: Tel: +44 (0) 8700 768853, or +44 (0) 1264 343071 Fax: +44 (0) 1264 343005, or
Post: Spon Press Customer Services, ITPS Andover, Hants, SP10 5BE, UK Email: book.orders@tandf.co.uk.

Postage & Packing: UK: 5% of order value (min. charge £1, max. charge £10) for 3-5 days delivery. Option of next day delivery at an additional £6.50. Europe: 10% of order value (min. charge £2.95, max. charge £20) for delivery surface post. Option of airmail at an additional £6.50. ROW: 15% of order value (min.charge£6.50, max. charge £30) for airmail delivery.

For a complete listing of all our titles visit: www.sponpress.com

NEW FROM SPON PRESS

Understanding Active Noise Cancellation

Colin H. Hansen, University of Adelaide, Australia

Understanding Active Noise Cancellation provides a concise introduction to the fundamentals and applications of active control of vibration and sound for the non-expert. It is also a useful quick reference for the specialist engineer. The book emphasises the practical applications of technology, and complex control algorithms and structures are only discussed to the extent that they aid understanding. This is an ideal book for those seeking an overview of the key issues: fundamentals, control systems, transducers, applications and possible future directions.

Contents: 1. A little History. 2. Foundations of Active Control. 3. The Electronic Control System. 4. Active Noise Control Sources. 5. Reference and Error Sensing. 6. Applications of Active Noise Control. References. Appendices. Index.

June 2001: 234x156: 176pp
60 line drawings.
Pb: 0-415-23192-2: £24.99
Hb: 0-415-23191-4: £65.00

To Order: Tel: +44 (0) 8700 768853, or +44 (0) 1264 343071 Fax: +44 (0) 1264 343005, or
Post: Spon Press Customer Services, ITPS Andover, Hants, SP10 5BE, UK Email: book.orders@tandf.co.uk.

Postage & Packing: UK: 5% of order value (min. charge £1, max. charge £10) for 3-5 days delivery. Option of next day delivery at an additional £6.50. Europe: 10% of order value (min. charge £2.95, max. charge £20) for delivery surface post. Option of airmail at an additional £6.50. ROW: 15% of order value (min.charge£6.50, max. charge £30) for airmail delivery.

For a complete listing of all our titles visit: www.sponpress.com

Material Costs/Measured Work Prices - Mechanical Installations   227

T:MECHANICAL/COOLING/HEATING SYSTEMS

Item	Net Price £	Material £	Labour hours	Labour £	Unit	Total rate £
T31 : LOW TEMPERATURE HOT WATER HEATING						
EQUIPMENT (Continued)						
Y23 - CALORIFIERS						
Non-storage calorifiers; mild steel; heater battery duty 82.71 degrees C to BS 853, maximum test on shell 11.55 bar, tubes 26.25 bar						
Horizontal; primary water at 116 degree C (on) 90 degree C (off)						
88 kW capacity	429.75	521.85	5.00	88.00	nr	**609.85**
176 kW capacity	735.10	892.63	7.04	123.95	nr	**1016.58**
293 kW capacity	904.73	1098.61	9.01	158.57	nr	**1257.18**
586 kW capacity	1526.74	1853.92	22.22	391.13	nr	**2245.05**
879 kW capacity	1786.85	2169.77	28.57	502.88	nr	**2672.65**
1465 kW capacity	2488.03	3021.21	50.00	880.04	nr	**3901.25**
Vertical; primary water at 116 degree C (on) 93 degree C (off)						
88 kW capacity	441.06	535.58	5.00	88.00	nr	**623.58**
176 kW capacity	752.06	913.23	7.04	123.95	nr	**1037.18**
293 kW capacity	949.97	1153.55	9.01	158.57	nr	**1312.11**
586 kW capacity	1526.74	1853.92	22.22	391.13	nr	**2245.05**
879 kW capacity	1809.47	2197.24	28.57	502.88	nr	**2700.12**
1465 kW capacity	2544.57	3089.87	50.00	880.04	nr	**3969.91**

Material Costs/Measured Work Prices - Mechanical Installations

T:MECHANICAL/COOLING/HEATING SYSTEMS

Item	Net Price £	Material £	Labour hours	Labour £	Unit	Total rate £
T31 : LOW TEMPERATURE HOT WATER HEATING						
EQUIPMENT (Continued)						
Perimeter heating metal casing standard finish top outlet; punched louvre grill; including backplates						
Standard unit						
60 x 200mm	21.23	25.78	2.00	35.20	m	**60.98**
60 x 300mm	23.09	28.04	2.00	35.20	m	**63.24**
60 x 450mm	28.63	34.76	2.00	35.20	m	**69.96**
60 x 525mm	30.47	37.00	2.00	35.20	m	**72.20**
60 x 600mm	33.24	40.36	2.00	35.20	m	**75.56**
90 x 250mm	23.09	28.04	2.00	35.20	m	**63.24**
90 x 300mm	24.01	29.15	2.00	35.20	m	**64.35**
90 x 375mm	26.78	32.52	2.00	35.20	m	**67.72**
90 x 450mm	29.56	35.89	2.00	35.20	m	**71.09**
90 x 525mm	32.33	39.25	2.00	35.20	m	**74.45**
90 x 600mm	34.17	41.49	2.00	35.20	m	**76.69**
HEAT EMITTERS						
Perimeter heating metal casing standard finish sloping outlet; punched louvre grill; including backplates						
Sloping outlet						
60 x 200mm	21.23	25.78	2.00	35.20	m	**60.98**
60 x 300mm	23.09	28.04	2.00	35.20	m	**63.24**
60 x 450mm	28.63	34.76	2.00	35.20	m	**69.96**
60 x 525mm	30.47	37.00	2.00	35.20	m	**72.20**
60 x 600mm	33.24	40.36	2.00	35.20	m	**75.56**
90 x 250mm	23.09	28.04	2.00	35.20	m	**63.24**
90 x 300mm	24.01	29.15	2.00	35.20	m	**64.35**
90 x 375mm	26.78	32.52	2.00	35.20	m	**67.72**
90 x 450mm	29.56	35.89	2.00	35.20	m	**71.09**
90 x 525mm	32.33	39.25	2.00	35.20	m	**74.45**
90 x 600mm	34.17	41.49	2.00	35.20	m	**76.69**
Perimeter heating metal casing standard finish flat front outlet; punched louvregrill; including backplates.						
60 x 200mm	21.23	25.78	2.00	35.20	m	**60.98**
60 x 300mm	23.09	28.04	2.00	35.20	m	**63.24**
60 x 450mm	28.63	34.76	2.00	35.20	m	**69.96**
60 x 525mm	30.47	37.00	2.00	35.20	m	**72.20**
60 x 600mm	33.24	40.36	2.00	35.20	m	**75.56**
90 x 250mm	23.09	28.04	2.00	35.20	m	**63.24**
90 x 300mm	24.01	29.15	2.00	35.20	m	**64.35**
90 x 375mm	26.78	32.52	2.00	35.20	m	**67.72**
90 x 450mm	29.56	35.89	2.00	35.20	m	**71.09**
90 x 525mm	32.33	39.25	2.00	35.20	m	**74.45**
90 x 600mm	34.17	41.49	2.00	35.20	m	**76.69**

Material Costs/Measured Work Prices - Mechanical Installations

T:MECHANICAL/COOLING/HEATING SYSTEMS

Item	Net Price £	Material £	Labour hours	Labour £	Unit	Total rate £
Perimeter heating metal casing standard finish top outlet; punched louvre grill; including backplates						
Standard unit						
60 x 200mm	26.54	32.23	2.00	35.20	m	**67.43**
60 x 300mm	28.87	35.05	2.00	35.20	m	**70.25**
60 x 450mm	35.78	43.45	2.00	35.20	m	**78.65**
60 x 525mm	38.09	46.25	2.00	35.20	m	**81.45**
60 x 600mm	41.55	50.45	2.00	35.20	m	**85.66**
90 x 250mm	27.90	33.88	2.00	35.20	m	**69.08**
90 x 300mm	30.01	36.44	2.00	35.20	m	**71.64**
90 x 375mm	33.48	40.65	2.00	35.20	m	**75.85**
90 x 450mm	36.95	44.86	2.00	35.20	m	**80.07**
90 x 525mm	40.41	49.07	2.00	35.20	m	**84.27**
90 x 600mm	42.71	51.87	2.00	35.20	m	**87.07**
Extra over for dampers						
Damper	9.23	11.21	0.25	4.40	nr	**15.61**
Extra over for fittings						
60mm End caps	7.39	8.97	0.25	4.40	nr	**13.37**
90mm End caps	12.01	14.58	0.25	4.40	nr	**18.98**
60mm Corners	15.71	19.07	0.25	4.40	nr	**23.47**
90mm Corners	23.09	28.04	0.25	4.40	nr	**32.44**
Radient Strip Heaters						
Black 1.25" steel tube, aluminium radiant plates including insulation, sliding brackets, cover plates, end closures; weld or screwed BSP ends						
One tube						
1500mm long	52.05	63.20	3.11	54.74	nr	**117.94**
3000mm long	82.33	99.97	3.11	54.74	nr	**154.71**
4500mm long	111.67	135.60	3.11	54.74	nr	**190.34**
6000mm long	153.94	186.93	3.11	54.74	nr	**241.67**
Two tube						
1500mm long	97.17	117.99	4.15	73.04	nr	**191.04**
3000mm long	153.49	186.38	4.15	73.04	nr	**259.43**
4500mm long	209.60	254.52	4.15	73.04	nr	**327.56**
6000mm long	282.38	342.89	4.15	73.04	nr	**415.94**
Pressed steel panel type radiators; fixed with and including brackets; taking down once for decoration; refixing						
300mm high; single panel						
500mm length	13.28	16.13	2.03	35.73	nr	**51.86**
1000mm length	26.54	32.23	2.03	35.73	nr	**67.96**
1500mm length	34.77	42.22	2.03	35.73	nr	**77.95**
2000mm length	39.01	47.37	2.47	43.47	nr	**90.84**
2500mm length	43.23	52.49	2.97	52.27	nr	**104.77**
3000mm length	51.74	62.83	3.22	56.67	nr	**119.50**

230   *Material Costs/Measured Work Prices - Mechanical Installations*

T:MECHANICAL/COOLING/HEATING SYSTEMS

Item	Net Price £	Material £	Labour hours	Labour £	Unit	Total rate £
T31 : LOW TEMPERATURE HOT WATER HEATING						
EQUIPMENT						
Pressed steel panel type radiators; fixed with and including brackets; taking down once for decoration; refixing (Continued)						
300mm high; double panel; convector						
500mm length	25.53	31.00	2.13	37.49	nr	68.49
1000mm length	51.09	62.04	2.13	37.49	nr	99.53
1500mm length	76.62	93.04	2.13	37.49	nr	130.53
2000mm length	102.17	124.06	2.57	45.23	nr	169.30
2500mm length	127.70	155.07	3.07	54.03	nr	209.10
3000mm length	153.25	186.09	3.31	58.26	nr	244.35
450mm high; single panel						
500mm length	12.39	15.05	2.08	36.61	nr	51.65
1000mm length	24.79	30.10	2.08	36.61	nr	66.71
1600mm length	39.66	48.16	2.53	44.53	nr	92.69
2000mm length	49.58	60.20	2.97	52.27	nr	112.48
2400mm length	59.49	72.24	3.47	61.07	nr	133.31
3000mm length	74.37	90.31	3.82	67.24	nr	157.54
450mm high; double panel; convector						
500mm length	22.70	27.56	2.18	38.37	nr	65.93
1000mm length	45.40	55.13	2.18	38.37	nr	93.50
1600mm length	83.14	100.96	2.63	46.29	nr	147.25
2000mm length	139.77	169.72	3.06	53.86	nr	223.58
2400mm length	167.74	203.69	3.37	59.31	nr	263.00
3000mm length	209.67	254.60	3.92	69.00	nr	323.60
600mm high; single panel						
500mm length	16.60	20.16	2.18	38.37	nr	58.53
1000mm length	33.20	40.31	2.43	42.77	nr	83.08
1600mm length	53.12	64.50	3.13	55.09	nr	119.59
2000mm length	66.40	80.63	3.77	66.36	nr	146.98
2400mm length	79.69	96.77	4.07	71.64	nr	168.40
3000mm length	99.61	120.96	5.11	89.94	nr	210.90
600mm high; double panel; convector						
500mm length	28.58	34.70	2.28	40.13	nr	74.83
1000mm length	57.16	69.41	2.28	40.13	nr	109.54
1600mm length	104.70	127.14	3.23	56.85	nr	183.99
2000mm length	176.02	213.74	3.87	68.12	nr	281.86
2400mm length	211.22	256.48	4.17	73.40	nr	329.88
3000mm length	264.02	320.60	5.24	92.23	nr	412.83
700mm high; single panel						
500mm length	19.43	23.59	2.23	39.25	nr	62.84
1000mm length	38.84	47.16	2.83	49.81	nr	96.9
1600mm length	62.14	75.46	3.73	65.65	nr	141.1
2000mm length	77.68	94.33	4.46	78.50	nr	172.8
2400mm length	93.23	113.21	4.48	78.85	nr	192.0
3000mm length	116.52	141.49	5.24	92.23	nr	233.7

Material Costs/Measured Work Prices - Mechanical Installations   231

T:MECHANICAL/COOLING/HEATING SYSTEMS

Item	Net Price £	Material £	Labour hours	Labour £	Unit	Total rate £
700mm high; double panel; convector						
500mm length	37.15	45.11	2.33	41.01	nr	**86.12**
1000mm length	99.94	121.36	3.08	54.21	nr	**175.57**
1600mm length	159.90	194.17	3.83	67.41	nr	**261.58**
2000mm length	199.88	242.71	4.17	73.40	nr	**316.11**
2400mm length	239.86	291.26	4.37	76.92	nr	**368.18**
3000mm length	299.82	364.07	4.82	84.84	nr	**448.91**
Quality sheet steel cased units; extruded aluminium grilles for LPHW centrifugal fans; filter choice of 3 speeds; single phase thermostatic controls; 3/42" BSP connections; all 230mm deep complete with access locks						
Free standing flat top 695mm high medium speed rating E.A.T. at 18 degree						
695mm length 1 row 1.94 kW 75 l/sec 39C	504.85	613.04	2.73	48.05	nr	**661.09**
695mm length 2 row 2.64 kW 75 l/sec 47C	504.85	613.04	2.73	48.05	nr	**661.09**
895mm length 1 row 4.02 kW 150 l/sec 40C	568.88	690.79	2.73	48.05	nr	**738.84**
895mm length 2 row 5.62 kW 150 l/sec 49C	568.88	690.79	2.73	48.05	nr	**738.84**
1195mm length 1 row 6.58 kW 250 l/sec 40C	647.68	786.48	3.00	52.80	nr	**839.28**
1195mm length 2 row 9.27 kW 250 l/sec 48C	647.68	786.48	3.00	52.80	nr	**839.28**
1495mm length 1 row 9.04 kW 340 l/sec 40C	722.80	877.70	3.26	57.38	nr	**935.07**
1495mm length 2 row 12.73 kW 340 l/sec 49C	722.80	877.70	3.26	57.38	nr	**935.07**
Free standing flat top 695mm high medium speed rating E.A.T. at 18 degree						
695mm length 1 row 1.94 kW 75 l/sec 39C	523.32	635.47	2.73	48.05	nr	**683.52**
695mm length 2 row 2.64 kW 75 l/sec 47C	523.32	635.47	2.73	48.05	nr	**683.52**
895mm length 1 row 4.02 kW 150 l/sec 40C	587.36	713.23	2.73	48.05	nr	**761.28**
895mm length 2 row 5.62 kW 150 l/sec 49C	587.36	713.23	2.73	48.05	nr	**761.28**
1195mm length 1 row 6.58 kW 250 l/sec 40C	666.15	808.91	3.00	52.80	nr	**861.71**
1195mm length 2 row 9.27 kW 250 l/sec 48C	666.15	808.91	3.00	52.80	nr	**861.71**
1495mm length 1 row 9.04 kW 340 l/sec 40C	741.27	900.12	3.26	57.38	nr	**957.50**
1495mm length 2 row 12.73 kW 340 l/sec 49C	741.27	900.12	3.26	57.38	nr	**957.50**
Free standing flat top 695mm high medium speed rating c/w plinth						
695mm length 1 row 1.94 kW 75 l/sec 39C	530.09	643.69	2.73	48.05	nr	**691.74**
695mm length 2 row 2.64 kW 75 l/sec 47C	530.09	643.69	2.73	48.05	nr	**691.74**
895mm length 1 row 4.02 kW 150 l/sec 40C	597.32	725.33	2.73	48.05	nr	**773.38**
895mm length 2 row 5.62 kW 150 l/sec 49C	597.32	725.33	2.73	48.05	nr	**773.38**
1195mm length 1 row 6.58 kW 250 l/sec 40C	680.07	825.81	3.00	52.80	nr	**878.61**
1195mm length 2 row 9.27 kW 250 l/sec 48C	680.07	825.81	3.00	52.80	nr	**878.61**
1495mm length 1 row 9.04 kW 340 l/sec 40C	758.93	921.57	3.26	57.38	nr	**978.95**
1495mm length 2 row 12.73 kW 340 l/sec 49C	758.93	921.57	3.26	57.38	nr	**978.95**

Material Costs/Measured Work Prices - Mechanical Installations

T:MECHANICAL/COOLING/HEATING SYSTEMS

Item	Net Price £	Material £	Labour hours	Labour £	Unit	Total rate £
T31 : LOW TEMPERATURE HOT WATER HEATING						
EQUIPMENT						
Quality sheet steel cased units; extruded aluminium grilles for LPHW centrifugal fans; filter choice of 3 speeds; single phase thermostatic controls; 3/42" BSP connections; all 230mm deep complete with access locks (Continued)						
Free standing sloping top 695mm high medium speed rating c/w plinth						
695mm length 1 row 1.94 kW 75 l/sec 39C	548.56	666.12	2.73	48.05	nr	714.17
695mm length 2 row 2.64 kW 75 l/sec 47C	548.56	666.12	2.73	48.05	nr	714.17
895mm length 1 row 4.02 kW 150 l/sec 40C	615.79	747.75	2.73	48.05	nr	795.80
895mm length 2 row 5.62 kW 150 l/sec 49C	615.79	747.75	2.73	48.05	nr	795.80
1195mm length 1 row 6.58 kW 250 l/sec 40C	698.54	848.24	3.00	52.80	nr	901.04
1195mm length 2 row 9.27 kW 250 l/sec 48C	698.54	848.24	3.00	52.80	nr	901.04
1495mm length 1 row 9.04 kW 340 l/sec 40C	777.40	944.00	3.26	57.38	nr	1001.38
1495mm length 2 row 12.73 kW 340 l/sec 49C	777.40	944.00	3.26	57.38	nr	1001.38
Wall mounted reversed air floor high level sloping discharge						
695mm length 1 row 1.94 kW 75 l/sec 39C	554.10	672.84	2.73	48.05	nr	720.89
695mm length 2 row 2.64 kW 75 l/sec 47C	554.10	672.84	2.73	48.05	nr	720.89
895mm length 1 row 4.02 kW 150 l/sec 40C	570.12	692.30	2.73	48.05	nr	740.35
895mm length 2 row 5.62 kW 150 l/sec 49C	570.12	692.30	2.73	48.05	nr	740.35
1195mm length 1 row 6.58 kW 250 l/sec 40C	693.84	842.53	3.00	52.80	nr	895.33
1195mm length 2 row 9.27 kW 250 l/sec 48C	693.84	842.53	3.00	52.80	nr	895.33
1495mm length 1 row 9.04 kW 340 l/sec 40C	752.34	913.57	3.26	57.38	nr	970.95
1495mm length 2 row 12.73 kW 340 l/sec 49C	752.34	913.57	3.26	57.38	nr	970.95
Ceiling mounted sloping inlet/outlet 665mm width						
895mm length 1 row 4.02 kW 150 l/sec 40C	626.75	761.06	4.15	73.04	nr	834.11
895mm length 2 row 5.62 kW 150 l/sec 49C	626.75	761.06	4.15	73.04	nr	834.11
1195mm length 1 row 6.58 kW 250 l/sec 40C	704.33	855.27	4.15	73.04	nr	928.31
1195mm length 2 row 9.27 kW 250 l/sec 48C	704.33	855.27	4.15	73.04	nr	928.31
1495mm length 1 row 9.04 kW 340 l/sec 40C	773.28	938.99	4.15	73.04	nr	1012.04
1495mm length 2 row 12.73 kW 340 l/sec 49C	773.28	938.99	4.15	73.04	nr	1012.04
Free standing unit extended height 1700/1900/2100mm						
895mm length 1 row 4.02 kW 150 l/sec 40C	725.26	880.68	3.11	54.74	nr	935.42
895mm length 2 row 5.62 kW 150 l/sec 49C	725.26	880.68	3.11	54.74	nr	935.42
1195mm length 1 row 6.58 kW 250 l/sec 40C	847.16	1028.71	3.11	54.74	nr	1083.44
1195mm length 2 row 9.27 kW 250 l/sec 48C	847.16	1028.71	3.11	54.74	nr	1083.44
1495mm length 1 row 9.04 kW 340 l/sec 40C	932.13	1131.89	3.11	54.74	nr	1186.62
1495mm length 2 row 12.73 kW 340 l/sec 49C	932.13	1131.89	3.11	54.74	nr	1186.62

Material Costs/Measured Work Prices - Mechanical Installations

T:MECHANICAL/COOLING/HEATING SYSTEMS

Item	Net Price £	Material £	Labour hours	Labour £	Unit	Total rate £
PIPE FREEZING						
Freeze isolation of carbon steel or copper pipelines containing static water, either side of work location, freeze duration not exceeding 4 hours assuming that flow and return circuits are treated concurrently and activities undertaken during normal working hours						
4 freezes						
2" diameter	388.31	471.52	-	-	nr	**471.52**
2.5" diameter	388.31	471.52	-	-	nr	**471.52**
3" diameter	443.93	539.06	-	-	nr	**539.06**
4" diameter	499.55	606.60	-	-	nr	**606.60**
6" diameter	831.21	1009.34	-	-	nr	**1009.34**
8" diameter	1274.11	1547.15	-	-	nr	**1547.15**
42mm diameter	333.72	405.24	-	-	nr	**405.24**
54 mm diameter	388.31	471.52	-	-	nr	**471.52**

T:MECHANICAL/COOLING/HEATING SYSTEMS

Item	Net Price £	Material £	Labour hours	Labour £	Unit	Total rate £
T31 : LOW TEMPERATURE HOT WATER HEATING						
Y53 - CONTROL COMPONENTS - MECHANICAL						
Room thermostats; light and medium duty; installed and connected to provide system control						
Range 3 to 27 degree C; 240 Volt						
1 amp; on/off type	19.52	23.70	0.30	5.28	nr	**28.98**
Range 0 to +15 degree C; 240 Volt						
6 amp; frost thermostat	13.16	15.98	0.30	5.28	nr	**21.26**
Range 3 to 27 degree C; 250 Volt						
2 amp; changeover type; dead zone	32.64	39.64	0.30	5.28	nr	**44.92**
2 amp; changeover type	15.14	18.38	0.30	5.28	nr	**23.66**
2 amp; changeover type; concealed setting	19.10	23.19	0.30	5.28	nr	**28.47**
6 amp; on/off type	11.79	14.32	0.30	5.28	nr	**19.60**
6 amp; temperature set-back	24.39	29.62	0.30	5.28	nr	**34.90**
16 amp; on/off type	18.16	22.05	0.30	5.28	nr	**27.33**
16 amp; on/off type; concealed setting	19.80	24.04	0.30	5.28	nr	**29.32**
20 amp; on/off type; concealed setting; industrial non-ventilated cover	21.62	26.25	0.30	5.28	nr	**31.53**
20 amp; indicated "off" position	21.26	25.81	0.30	5.28	nr	**31.09**
20 amp; manual; double pole on/off and neon indicator	38.70	46.99	0.30	5.28	nr	**52.27**
20 amp; indicated "off" position	25.32	30.75	0.30	5.28	nr	**36.03**
Range 10 to 40 degree C; 240 Volt						
20 amp; changeover contacts	23.07	28.01	0.30	5.28	nr	**33.29**
2 amp; 'Heating-Cooling' switch	49.91	60.60	0.30	5.28	nr	**65.88**
Surface thermostats						
Cylinder thermostat						
6 amp; changeover type; with cable	12.59	15.67	0.25	4.40	nr	**20.07**
Electrical thermostats; installed and connected to provide system control						
Range 5 to 30 degree C; 230 Volt Standard Port Single Time						
10 amp with sensor	14.19	17.23	0.30	5.28	nr	**22.51**
Range 5 to 30 degree C; 230 Volt Standard Port Double Time						
10 amp with sensor	16.08	19.53	0.30	5.28	nr	**24.81**
10 amp with sensor and on/off switch	24.53	29.79	0.30	5.28	nr	**35.07**

Material Costs/Measured Work Prices - Mechanical Installations

235

T:MECHANICAL/COOLING/HEATING SYSTEMS

Item	Net Price £	Material £	Labour hours	Labour £	Unit	Total rate £
Radiator thermostats						
Angled valve body; thermostatic head; built in sensor						
15mm; liquid filled	11.80	14.33	0.84	14.79	nr	**29.12**
15mm; wax filled	11.80	14.33	0.84	14.79	nr	**29.12**
Immersion thermostats; stem type; domestic water boilers; fitted; electrical work elsewhere						
Temperature range 0 to 40 degree C						
Non standard; 280mm stem	7.18	8.72	0.25	4.40	nr	**13.12**
Temperature range 18 to 88 degree C						
13 amp; 178mm stem	4.77	5.79	0.25	4.40	nr	**10.19**
20 amp; 178mm stem	7.41	9.00	0.25	4.40	nr	**13.40**
Non standard; pocket clip; 280mm stem	6.87	8.35	0.25	4.40	nr	**12.75**
Temperature range 40 to 80 degree C						
13 amp; 178mm stem	2.61	3.17	0.25	4.40	nr	**7.57**
20 amp; 178mm stem	4.95	6.01	0.25	4.40	nr	**10.41**
Non standard; pocket clip; 280mm stem	7.70	9.35	0.25	4.40	nr	**13.75**
13 amp; 457mm stem	3.09	3.75	0.25	4.40	nr	**8.15**
20 amp; 457mm stem	5.42	6.58	0.25	4.40	nr	**10.98**
Temperature range 50 to 100 degree C						
Non standard; 1780mm stem	6.49	7.88	0.25	4.40	nr	**12.28**
Non standard; 280mm stem	6.76	8.21	0.25	4.40	nr	**12.61**
Pockets for thermostats						
For 178mm stem	8.48	10.30	0.25	4.40	nr	**14.70**
For 280mm stem	8.40	10.20	0.25	4.40	nr	**14.60**
Immersion thermostats; stem type; industrial installations; fitted; electrical work elsewhere						
Temperature range 5 to 105 degree C						
For 305mm stem	106.54	129.37	0.50	8.80	nr	**138.17**

Material Costs/Measured Work Prices - Mechanical Installations

T:MECHANICAL/COOLING/HEATING SYSTEMS

Item	Net Price £	Material £	Labour hours	Labour £	Unit	Total rate £
T31 : THERMAL INSULATION						
Y50 -THERMAL INSULATION						
For flexible closed cell insulation see Section S10 - Cold Water						
Mineral fibre sectional insulation; Bright Class O foil faced; Bright Class O foil taped joints; 19mm aluminium bands						
Concealed pipework						
20mm thick						
15mm diameter	2.13	2.76	0.15	2.50	m	**5.26**
20mm diameter	2.27	2.94	0.15	2.50	m	**5.44**
25mm diameter	2.44	3.16	0.15	2.50	m	**5.66**
32mm diameter	2.72	3.52	0.15	2.50	m	**6.03**
40mm diameter	2.91	3.77	0.15	2.50	m	**6.27**
50mm diameter	3.32	4.30	0.15	2.50	m	**6.80**
Extra over for fittings concealed insulation						
Flange/union						
15mm diameter	1.06	1.37	0.13	2.17	nr	**3.54**
20mm diameter	1.14	1.48	0.13	2.17	nr	**3.65**
25mm diameter	1.22	1.58	0.13	2.17	nr	**3.75**
32mm diameter	1.36	1.76	0.13	2.17	nr	**3.93**
40mm diameter	1.45	1.88	0.13	2.17	nr	**4.05**
50mm diameter	1.66	2.15	0.13	2.17	nr	**4.32**
Valves						
15mm diameter	2.13	2.76	0.15	2.50	nr	**5.26**
20mm diameter	2.44	3.16	0.15	2.50	nr	**5.66**
25mm diameter	2.44	3.16	0.15	2.50	nr	**5.66**
32mm diameter	2.72	3.52	0.15	2.50	nr	**6.03**
40mm diameter	2.91	3.77	0.15	2.50	nr	**6.27**
50mm diameter	3.32	4.30	0.15	2.50	nr	**6.80**
Pumps						
15mm diameter	4.25	5.50	0.45	7.51	nr	**13.02**
20mm diameter	4.54	5.88	0.45	7.51	nr	**13.39**
25mm diameter	4.88	6.32	0.45	7.51	nr	**13.83**
32mm diameter	5.43	7.03	0.45	7.51	nr	**14.54**
40mm diameter	5.82	7.54	0.45	7.51	nr	**15.05**
50mm diameter	6.65	8.61	0.45	7.51	nr	**16.12**
Expansion Bellows						
15mm diameter	4.25	5.50	0.22	3.67	nr	**9.18**
20mm diameter	4.54	5.88	0.22	3.67	nr	**9.55**
25mm diameter	4.88	6.32	0.22	3.67	nr	**9.99**
32mm diameter	5.43	7.03	0.22	3.67	nr	**10.70**
40mm diameter	5.82	7.54	0.22	3.67	nr	**11.21**
50mm diameter	6.65	8.61	0.22	3.67	nr	**12.28**

Material Costs/Measured Work Prices - Mechanical Installations 237

T:MECHANICAL/COOLING/HEATING SYSTEMS

Item	Net Price £	Material £	Labour hours	Labour £	Unit	Total rate £
25mm thick						
15mm diameter	2.34	3.03	0.15	2.50	m	**5.53**
20mm diameter	2.52	3.26	0.15	2.50	m	**5.77**
25mm diameter	2.82	3.65	0.15	2.50	m	**6.16**
32mm diameter	3.07	3.98	0.15	2.50	m	**6.48**
40mm diameter	3.29	4.26	0.15	2.50	m	**6.76**
50mm diameter	3.77	4.88	0.15	2.50	m	**7.39**
65mm diameter	4.30	5.57	0.15	2.50	m	**8.07**
80mm diameter	4.72	6.11	0.22	3.67	m	**9.78**
100mm diameter	6.22	8.05	0.22	3.67	m	**11.73**
125mm diameter	7.19	9.31	0.22	3.67	m	**12.98**
150mm diameter	8.57	11.10	0.22	3.67	m	**14.77**
200mm diameter	12.11	15.68	0.25	4.17	m	**19.86**
250mm diameter	14.50	18.78	0.25	4.17	m	**22.95**
300mm diameter	15.50	20.07	0.25	4.17	m	**24.25**
Extra over for fittings concealed insulation						
Flange/union						
15mm diameter	1.17	1.52	0.13	2.17	nr	**3.69**
20mm diameter	1.26	1.63	0.13	2.17	nr	**3.80**
25mm diameter	1.41	1.83	0.13	2.17	nr	**4.00**
32mm diameter	1.54	1.99	0.13	2.17	nr	**4.16**
40mm diameter	1.64	2.12	0.13	2.17	nr	**4.29**
50mm diameter	1.88	2.43	0.13	2.17	nr	**4.60**
65mm diameter	2.15	2.78	0.13	2.17	nr	**4.95**
80mm diameter	2.36	3.06	0.18	3.00	nr	**6.06**
100mm diameter	3.11	4.03	0.18	3.00	nr	**7.03**
125mm diameter	3.60	4.66	0.18	3.00	nr	**7.67**
150mm diameter	4.28	5.54	0.18	3.00	nr	**8.55**
200mm diameter	6.05	7.83	0.22	3.67	nr	**11.51**
250mm diameter	7.25	9.39	0.22	3.67	nr	**13.06**
300mm diameter	7.75	10.04	0.22	3.67	nr	**13.71**
Valves						
15mm diameter	2.34	3.03	0.15	2.50	nr	**5.53**
20mm diameter	2.52	3.26	0.15	2.50	nr	**5.77**
25mm diameter	2.82	3.65	0.15	2.50	nr	**6.16**
32mm diameter	3.07	3.98	0.15	2.50	nr	**6.48**
40mm diameter	3.29	4.26	0.15	2.50	nr	**6.76**
50mm diameter	3.77	4.88	0.15	2.50	nr	**7.39**
65mm diameter	4.30	5.57	0.15	2.50	nr	**8.07**
80mm diameter	4.72	6.11	0.20	3.34	nr	**9.45**
100mm diameter	6.22	8.05	0.20	3.34	nr	**11.39**
125mm diameter	7.19	9.31	0.20	3.34	nr	**12.65**
150mm diameter	8.57	11.10	0.20	3.34	nr	**14.44**
200mm diameter	12.11	15.68	0.25	4.17	nr	**19.86**
250mm diameter	14.50	18.78	0.25	4.17	nr	**22.95**
300mm diameter	15.50	20.07	0.25	4.17	nr	**24.25**

238 **Material Costs/Measured Work Prices - Mechanical Installations**

T:MECHANICAL/COOLING/HEATING SYSTEMS

Item	Net Price £	Material £	Labour hours	Labour £	Unit	Total rate £
T31 : THERMAL INSULATION						
Y50 -THERMAL INSULATION						
Mineral fibre sectional insulation; Bright Class O foil faced; Bright Class O foil taped joints; 19mm aluminium bands (Continued)						
Pumps						
15mm diameter	4.68	6.06	0.45	7.51	nr	**13.57**
20mm diameter	5.05	6.54	0.45	7.51	nr	**14.05**
25mm diameter	5.63	7.29	0.45	7.51	nr	**14.80**
32mm diameter	6.14	7.95	0.45	7.51	nr	**15.46**
40mm diameter	6.58	8.52	0.45	7.51	nr	**16.03**
50mm diameter	7.54	9.76	0.45	7.51	nr	**17.28**
65mm diameter	8.61	11.15	0.45	7.51	nr	**18.66**
80mm diameter	9.43	12.21	0.60	10.02	nr	**22.23**
100mm diameter	12.43	16.10	0.60	10.02	nr	**26.11**
125mm diameter	14.38	18.62	0.60	10.02	nr	**28.64**
150mm diameter	17.14	22.20	0.60	10.02	nr	**32.21**
200mm diameter	24.21	31.35	0.75	12.52	nr	**43.87**
250mm diameter	28.99	37.54	0.75	12.52	nr	**50.06**
300mm diameter	31.00	40.15	0.75	12.52	nr	**52.66**
Expansion Bellows						
15mm diameter	4.68	6.06	0.22	3.67	nr	**9.73**
20mm diameter	5.05	6.54	0.22	3.67	nr	**10.21**
25mm diameter	5.63	7.29	0.22	3.67	nr	**10.96**
32mm diameter	6.14	7.95	0.22	3.67	nr	**11.62**
40mm diameter	6.58	8.52	0.22	3.67	nr	**12.19**
50mm diameter	7.54	9.76	0.22	3.67	nr	**13.44**
65mm diameter	8.61	11.15	0.22	3.67	nr	**14.82**
80mm diameter	9.43	12.21	0.29	4.84	nr	**17.05**
100mm diameter	12.43	16.10	0.29	4.84	nr	**20.94**
125mm diameter	14.38	18.62	0.29	4.84	nr	**23.46**
150mm diameter	17.14	22.20	0.29	4.84	nr	**27.04**
200mm diameter	24.21	31.35	0.36	6.01	nr	**37.36**
250mm diameter	28.99	37.54	0.36	6.01	nr	**43.55**
300mm diameter	31.00	40.15	0.36	6.01	nr	**46.15**
30mm thick						
15mm diameter	3.04	3.94	0.15	2.50	m	**6.44**
20mm diameter	3.26	4.22	0.15	2.50	m	**6.73**
25mm diameter	3.45	4.47	0.15	2.50	m	**6.97**
32mm diameter	3.76	4.87	0.15	2.50	m	**7.37**
40mm diameter	3.98	5.15	0.15	2.50	m	**7.66**
50mm diameter	4.55	5.89	0.15	2.50	m	**8.40**
65mm diameter	5.15	6.67	0.15	2.50	m	**9.17**
80mm diameter	5.62	7.28	0.22	3.67	m	**10.95**
100mm diameter	7.27	9.41	0.22	3.67	m	**13.09**
125mm diameter	8.37	10.84	0.22	3.67	m	**14.51**
150mm diameter	9.83	12.73	0.22	3.67	m	**16.40**
200mm diameter	13.75	17.81	0.25	4.17	m	**21.98**
250mm diameter	16.35	21.17	0.25	4.17	m	**25.35**
300mm diameter	17.33	22.44	0.25	4.17	m	**26.62**
350mm diameter	19.04	24.66	0.25	4.17	m	**28.83**

Material Costs/Measured Work Prices - Mechanical Installations

T:MECHANICAL/COOLING/HEATING SYSTEMS

Item	Net Price £	Material £	Labour hours	Labour £	Unit	Total rate £
Extra over for fittings concealed insulation						
Flange/union						
15mm diameter	1.52	1.97	0.13	2.17	nr	**4.14**
20mm diameter	1.63	2.11	0.13	2.17	nr	**4.28**
25mm diameter	1.72	2.23	0.13	2.17	nr	**4.40**
32mm diameter	1.88	2.43	0.13	2.17	nr	**4.60**
40mm diameter	1.99	2.58	0.13	2.17	nr	**4.75**
50mm diameter	2.28	2.95	0.13	2.17	nr	**5.12**
65mm diameter	2.58	3.34	0.13	2.17	nr	**5.51**
80mm diameter	2.81	3.64	0.18	3.00	nr	**6.64**
100mm diameter	3.64	4.71	0.18	3.00	nr	**7.72**
125mm diameter	4.18	5.41	0.18	3.00	nr	**8.42**
150mm diameter	4.92	6.37	0.18	3.00	nr	**9.38**
200mm diameter	6.87	8.90	0.22	3.67	nr	**12.57**
250mm diameter	8.18	10.59	0.22	3.67	nr	**14.27**
300mm diameter	8.67	11.23	0.22	3.67	nr	**14.90**
350mm diameter	9.52	12.33	0.22	3.67	nr	**16.00**
Valves						
15mm diameter	3.26	4.22	0.15	2.50	nr	**6.73**
20mm diameter	3.26	4.22	0.15	2.50	nr	**6.73**
25mm diameter	3.45	4.47	0.15	2.50	nr	**6.97**
32mm diameter	3.76	4.87	0.15	2.50	nr	**7.37**
40mm diameter	3.98	5.15	0.15	2.50	nr	**7.66**
50mm diameter	4.55	5.89	0.15	2.50	nr	**8.40**
65mm diameter	5.15	6.67	0.15	2.50	nr	**9.17**
80mm diameter	5.62	7.28	0.20	3.34	nr	**10.62**
100mm diameter	7.27	9.41	0.20	3.34	nr	**12.75**
125mm diameter	8.37	10.84	0.20	3.34	nr	**14.18**
150mm diameter	9.83	12.73	0.20	3.34	nr	**16.07**
200mm diameter	13.75	17.81	0.25	4.17	nr	**21.98**
250mm diameter	16.35	21.17	0.25	4.17	nr	**25.35**
300mm diameter	17.33	22.44	0.25	4.17	nr	**26.62**
350mm diameter	19.04	24.66	0.25	4.17	nr	**28.83**
Pumps						
15mm diameter	6.09	7.89	0.45	7.51	nr	**15.40**
20mm diameter	6.51	8.43	0.45	7.51	nr	**15.94**
25mm diameter	6.89	8.92	0.45	7.51	nr	**16.43**
32mm diameter	7.53	9.75	0.45	7.51	nr	**17.26**
40mm diameter	7.96	10.31	0.45	7.51	nr	**17.82**
50mm diameter	9.11	11.80	0.45	7.51	nr	**19.31**
65mm diameter	10.30	13.34	0.45	7.51	nr	**20.85**
80mm diameter	11.24	14.56	0.60	10.02	nr	**24.57**
100mm diameter	14.55	18.84	0.60	10.02	nr	**28.86**
125mm diameter	16.73	21.67	0.60	10.02	nr	**31.68**
150mm diameter	19.66	25.46	0.60	10.02	nr	**35.48**
200mm diameter	27.49	35.60	0.75	12.52	nr	**48.12**
250mm diameter	32.70	42.35	0.75	12.52	nr	**54.87**
300mm diameter	34.67	44.90	0.75	12.52	nr	**57.42**
350mm diameter	38.07	49.30	0.75	12.52	nr	**61.82**

240   *Material Costs/Measured Work Prices - Mechanical Installations*

T:MECHANICAL/COOLING/HEATING SYSTEMS

Item	Net Price £	Material £	Labour hours	Labour £	Unit	Total rate £
T31 : THERMAL INSULATION						
Y50 -THERMAL INSULATION						
Mineral fibre sectional insulation; Bright Class O foil faced; Bright Class O foil taped joints; 19mm aluminium bands (Continued)						
Expansion Bellows						
15mm diameter	6.09	7.89	0.22	3.67	nr	**11.56**
20mm diameter	6.51	8.43	0.22	3.67	nr	**12.10**
25mm diameter	6.89	8.92	0.22	3.67	nr	**12.59**
32mm diameter	7.53	9.75	0.22	3.67	nr	**13.42**
40mm diameter	7.96	10.31	0.22	3.67	nr	**13.98**
50mm diameter	9.11	11.80	0.22	3.67	nr	**15.47**
65mm diameter	10.30	13.34	0.22	3.67	nr	**17.01**
80mm diameter	11.24	14.56	0.29	4.84	nr	**19.40**
100mm diameter	14.55	18.84	0.29	4.84	nr	**23.68**
125mm diameter	16.73	21.67	0.29	4.84	nr	**26.51**
150mm diameter	19.66	25.46	0.29	4.84	nr	**30.30**
200mm diameter	27.49	35.60	0.36	6.01	nr	**41.61**
250mm diameter	32.70	42.35	0.36	6.01	nr	**48.36**
300mm diameter	34.67	44.90	0.36	6.01	nr	**50.91**
350mm diameter	38.07	49.30	0.36	6.01	nr	**55.31**
40mm thick						
15mm diameter	3.90	5.05	0.15	2.50	m	**7.55**
20mm diameter	4.02	5.21	0.15	2.50	m	**7.71**
25mm diameter	4.32	5.59	0.15	2.50	m	**8.10**
32mm diameter	4.61	5.97	0.15	2.50	m	**8.47**
40mm diameter	4.84	6.27	0.15	2.50	m	**8.77**
50mm diameter	5.48	7.10	0.15	2.50	m	**9.60**
65mm diameter	6.15	7.96	0.15	2.50	m	**10.47**
80mm diameter	6.69	8.66	0.22	3.67	m	**12.34**
100mm diameter	8.68	11.24	0.22	3.67	m	**14.91**
125mm diameter	9.81	12.70	0.22	3.67	m	**16.38**
150mm diameter	11.45	14.83	0.22	3.67	m	**18.50**
200mm diameter	15.82	20.49	0.25	4.17	m	**24.66**
250mm diameter	18.54	24.01	0.25	4.17	m	**28.18**
300mm diameter	19.82	25.67	0.25	4.17	m	**29.84**
350mm diameter	21.92	28.39	0.25	4.17	m	**32.56**
400mm diameter	24.43	31.64	0.25	4.17	m	**35.81**
Extra over for fittings concealed insulation						
Flange/union						
15mm diameter	1.95	2.53	0.13	2.17	nr	**4.70**
20mm diameter	2.01	2.60	0.13	2.17	nr	**4.77**
25mm diameter	2.16	2.80	0.13	2.17	nr	**4.97**
32mm diameter	2.30	2.98	0.13	2.17	nr	**5.15**
40mm diameter	2.42	3.13	0.13	2.17	nr	**5.30**
50mm diameter	2.74	3.55	0.13	2.17	nr	**5.72**
65mm diameter	3.07	3.98	0.13	2.17	nr	**6.15**
80mm diameter	3.35	4.34	0.18	3.00	nr	**7.34**

Material Costs/Measured Work Prices - Mechanical Installations   241

T:MECHANICAL/COOLING/HEATING SYSTEMS

Item	Net Price £	Material £	Labour hours	Labour £	Unit	Total rate £
100mm diameter	4.34	5.62	0.18	3.00	nr	**8.63**
125mm diameter	4.90	6.35	0.18	3.00	nr	**9.35**
150mm diameter	5.73	7.42	0.18	3.00	nr	**10.43**
200mm diameter	7.91	10.24	0.22	3.67	nr	**13.92**
250mm diameter	9.27	12.00	0.22	3.67	nr	**15.68**
300mm diameter	9.91	12.83	0.22	3.67	nr	**16.51**
350mm diameter	10.96	14.19	0.22	3.67	nr	**17.87**
400mm diameter	12.22	15.82	0.22	3.67	nr	**19.50**
Valves						
15mm diameter	3.90	5.05	0.15	2.50	nr	**7.55**
20mm diameter	4.02	5.21	0.15	2.50	nr	**7.71**
25mm diameter	4.32	5.59	0.15	2.50	nr	**8.10**
32mm diameter	4.61	5.97	0.15	2.50	nr	**8.47**
40mm diameter	4.84	6.27	0.15	2.50	nr	**8.77**
50mm diameter	5.48	7.10	0.15	2.50	nr	**9.60**
65mm diameter	6.15	7.96	0.15	2.50	nr	**10.47**
80mm diameter	6.69	8.66	0.20	3.34	nr	**12.00**
100mm diameter	8.68	11.24	0.20	3.34	nr	**14.58**
125mm diameter	9.81	12.70	0.20	3.34	nr	**16.04**
150mm diameter	11.45	14.83	0.20	3.34	nr	**18.17**
200mm diameter	15.82	20.49	0.25	4.17	nr	**24.66**
250mm diameter	18.54	24.01	0.25	4.17	nr	**28.18**
300mm diameter	19.82	25.67	0.25	4.17	nr	**29.84**
350mm diameter	21.92	28.39	0.25	4.17	nr	**32.56**
400mm diameter	24.43	31.64	0.25	4.17	nr	**35.81**
Pumps						
15mm diameter	7.81	10.11	0.45	7.51	nr	**17.63**
20mm diameter	8.03	10.40	0.45	7.51	nr	**17.91**
25mm diameter	8.65	11.20	0.45	7.51	nr	**18.71**
32mm diameter	9.21	11.93	0.45	7.51	nr	**19.44**
40mm diameter	9.69	12.55	0.45	7.51	nr	**20.06**
50mm diameter	10.96	14.19	0.45	7.51	nr	**21.70**
65mm diameter	12.29	15.92	0.45	7.51	nr	**23.43**
80mm diameter	13.38	17.33	0.60	10.02	nr	**27.34**
100mm diameter	17.35	22.47	0.60	10.02	nr	**32.48**
125mm diameter	19.61	25.39	0.60	10.02	nr	**35.41**
150mm diameter	22.90	29.66	0.60	10.02	nr	**39.67**
200mm diameter	31.64	40.97	0.75	12.52	nr	**53.49**
250mm diameter	37.08	48.02	0.75	12.52	nr	**60.54**
300mm diameter	39.64	51.33	0.75	12.52	nr	**63.85**
350mm diameter	43.83	56.76	0.75	12.52	nr	**69.28**
400mm diameter	48.86	63.27	0.75	12.52	nr	**75.79**
Expansion Bellows						
15mm diameter	7.81	10.11	0.22	3.67	nr	**13.79**
20mm diameter	8.03	10.40	0.22	3.67	nr	**14.07**
25mm diameter	8.65	11.20	0.22	3.67	nr	**14.87**
32mm diameter	9.21	11.93	0.22	3.67	nr	**15.60**
40mm diameter	9.69	12.55	0.22	3.67	nr	**16.22**
50mm diameter	10.96	14.19	0.22	3.67	nr	**17.87**
65mm diameter	12.29	15.92	0.22	3.67	nr	**19.59**
80mm diameter	13.38	17.33	0.29	4.84	nr	**22.17**
100mm diameter	17.35	22.47	0.29	4.84	nr	**27.31**
125mm diameter	19.61	25.39	0.29	4.84	nr	**30.24**

Material Costs/Measured Work Prices - Mechanical Installations

T:MECHANICAL/COOLING/HEATING SYSTEMS

Item	Net Price £	Material £	Labour hours	Labour £	Unit	Total rate £
T31 : THERMAL INSULATION						
Y50 -THERMAL INSULATION						
Mineral fibre sectional insulation; Bright Class O foil faced; Bright Class O foil taped joints; 19mm aluminium bands (Continued)						
Expansion Bellows						
150mm diameter	22.90	29.66	0.29	4.84	nr	**34.50**
200mm diameter	31.64	40.97	0.36	6.01	nr	**46.98**
250mm diameter	37.08	48.02	0.36	6.01	nr	**54.03**
300mm diameter	39.64	51.33	0.36	6.01	nr	**57.34**
350mm diameter	21.92	28.39	0.36	6.01	nr	**34.40**
400mm diameter	48.86	63.27	0.36	6.01	nr	**69.28**
50mm thick						
15mm diameter	5.42	7.02	0.15	2.50	m	**9.52**
20mm diameter	5.70	7.38	0.15	2.50	m	**9.89**
25mm diameter	6.06	7.85	0.15	2.50	m	**10.35**
32mm diameter	6.33	8.20	0.15	2.50	m	**10.70**
40mm diameter	6.66	8.62	0.15	2.50	m	**11.13**
50mm diameter	7.47	9.67	0.15	2.50	m	**12.18**
65mm diameter	8.16	10.57	0.15	2.50	m	**13.07**
80mm diameter	8.74	11.32	0.22	3.67	m	**14.99**
100mm diameter	11.17	14.47	0.22	3.67	m	**18.14**
125mm diameter	12.54	16.24	0.22	3.67	m	**19.91**
150mm diameter	14.47	18.74	0.22	3.67	m	**22.41**
200mm diameter	19.76	25.59	0.25	4.17	m	**29.76**
250mm diameter	22.82	29.55	0.25	4.17	m	**33.73**
300mm diameter	24.18	31.31	0.25	4.17	m	**35.49**
350mm diameter	26.69	34.56	0.25	4.17	m	**38.74**
400mm diameter	29.60	38.33	0.25	4.17	m	**42.51**
Extra over for fittings concealed insulation						
Flange/union						
15mm diameter	2.71	3.51	0.13	2.17	nr	**5.68**
20mm diameter	2.85	3.69	0.13	2.17	nr	**5.86**
25mm diameter	3.03	3.92	0.13	2.17	nr	**6.09**
32mm diameter	3.17	4.11	0.13	2.17	nr	**6.28**
40mm diameter	3.33	4.31	0.13	2.17	nr	**6.48**
50mm diameter	3.73	4.83	0.13	2.17	nr	**7.00**
65mm diameter	4.08	5.28	0.13	2.17	nr	**7.45**
80mm diameter	4.37	5.66	0.18	3.00	nr	**8.66**
100mm diameter	5.59	7.24	0.18	3.00	nr	**10.24**
125mm diameter	6.27	8.12	0.18	3.00	nr	**11.12**
150mm diameter	7.24	9.38	0.18	3.00	nr	**12.38**
200mm diameter	9.88	12.79	0.22	3.67	nr	**16.47**
250mm diameter	11.41	14.78	0.22	3.67	nr	**18.45**
300mm diameter	12.09	15.66	0.22	3.67	nr	**19.33**
350mm diameter	13.35	17.29	0.22	3.67	nr	**20.96**
400mm diameter	14.80	19.17	0.22	3.67	nr	**22.84**

Material Costs/Measured Work Prices - Mechanical Installations 243

T:MECHANICAL/COOLING/HEATING SYSTEMS

Item	Net Price £	Material £	Labour hours	Labour £	Unit	Total rate £
Valves						
15mm diameter	5.42	7.02	0.15	2.50	nr	**9.52**
20mm diameter	5.70	7.38	0.15	2.50	nr	**9.89**
25mm diameter	6.06	7.85	0.15	2.50	nr	**10.35**
32mm diameter	6.33	8.20	0.15	2.50	nr	**10.70**
40mm diameter	6.66	8.62	0.15	2.50	nr	**11.13**
50mm diameter	7.47	9.67	0.15	2.50	nr	**12.18**
65mm diameter	8.16	10.57	0.15	2.50	nr	**13.07**
80mm diameter	8.74	11.32	0.20	3.34	nr	**14.66**
100mm diameter	11.17	14.47	0.20	3.34	nr	**17.80**
125mm diameter	12.54	16.24	0.20	3.34	nr	**19.58**
150mm diameter	14.47	18.74	0.20	3.34	nr	**22.08**
200mm diameter	19.76	25.59	0.25	4.17	nr	**29.76**
250mm diameter	22.82	29.55	0.25	4.17	nr	**33.73**
300mm diameter	24.18	31.31	0.25	4.17	nr	**35.49**
350mm diameter	26.69	34.56	0.25	4.17	nr	**38.74**
400mm diameter	29.60	38.33	0.25	4.17	nr	**42.51**
Pumps						
15mm diameter	10.83	14.02	0.45	7.51	nr	**21.54**
20mm diameter	11.40	14.76	0.45	7.51	nr	**22.27**
25mm diameter	12.12	15.70	0.45	7.51	nr	**23.21**
32mm diameter	12.67	16.41	0.45	7.51	nr	**23.92**
40mm diameter	13.32	17.25	0.45	7.51	nr	**24.76**
50mm diameter	14.93	19.33	0.45	7.51	nr	**26.85**
65mm diameter	16.32	21.13	0.45	7.51	nr	**28.65**
80mm diameter	17.48	22.64	0.60	10.02	nr	**32.65**
100mm diameter	22.34	28.93	0.60	10.02	nr	**38.95**
125mm diameter	25.07	32.47	0.60	10.02	nr	**42.48**
150mm diameter	28.95	37.49	0.60	10.02	nr	**47.51**
200mm diameter	39.53	51.19	0.75	12.52	nr	**63.71**
250mm diameter	45.63	59.09	0.75	12.52	nr	**71.61**
300mm diameter	48.37	62.64	0.75	12.52	nr	**75.16**
350mm diameter	53.38	69.13	0.75	12.52	nr	**81.65**
400mm diameter	59.19	76.65	0.75	12.52	nr	**89.17**
Expansion Bellows						
15mm diameter	10.83	14.02	0.22	3.67	nr	**17.70**
20mm diameter	11.40	14.76	0.22	3.67	nr	**18.44**
25mm diameter	12.12	15.70	0.22	3.67	nr	**19.37**
32mm diameter	12.67	16.41	0.22	3.67	nr	**20.08**
40mm diameter	13.32	17.25	0.22	3.67	nr	**20.92**
50mm diameter	14.93	19.33	0.22	3.67	nr	**23.01**
65mm diameter	16.32	21.13	0.22	3.67	nr	**24.81**
80mm diameter	17.48	22.64	0.29	4.84	nr	**27.48**
100mm diameter	22.34	28.93	0.29	4.84	nr	**33.77**
125mm diameter	25.07	32.47	0.29	4.84	nr	**37.31**
150mm diameter	28.95	37.49	0.29	4.84	nr	**42.33**
200mm diameter	39.53	51.19	0.36	6.01	nr	**57.20**
250mm diameter	45.63	59.09	0.36	6.01	nr	**65.10**
300mm diameter	48.37	62.64	0.36	6.01	nr	**68.65**
350mm diameter	53.38	69.13	0.36	6.01	nr	**75.14**
400mm diameter	59.19	76.65	0.36	6.01	nr	**82.66**

Material Costs/Measured Work Prices - Mechanical Installations

T:MECHANICAL/COOLING/HEATING SYSTEMS

Item	Net Price £	Material £	Labour hours	Labour £	Unit	Total rate £
T31 : THERMAL INSULATION						
Y50 -THERMAL INSULATION						
Mineral fibre sectional insulation; Bright Class O foil faced; Bright Class O foil taped joints; 22 swg plain/embossed aluminium cladding; pop rivited						
Plantroom pipework						
20mm thick						
15mm diameter	3.58	4.64	0.44	7.34	m	**11.98**
20mm diameter	3.79	4.91	0.44	7.34	m	**12.25**
25mm diameter	4.06	5.26	0.44	7.34	m	**12.60**
32mm diameter	4.40	5.70	0.44	7.34	m	**13.04**
40mm diameter	4.63	6.00	0.44	7.34	m	**13.34**
50mm diameter	5.16	6.68	0.44	7.34	m	**14.03**
Extra over for fittings plantroom insulation						
Flange/union						
15mm diameter	4.01	5.19	0.58	9.68	nr	**14.87**
20mm diameter	4.25	5.50	0.58	9.68	nr	**15.19**
25mm diameter	4.56	5.91	0.58	9.68	nr	**15.59**
32mm diameter	4.98	6.45	0.58	9.68	nr	**16.13**
40mm diameter	5.26	6.81	0.58	9.68	nr	**16.49**
50mm diameter	5.91	7.65	0.58	9.68	nr	**17.34**
Bends						
15mm diameter	1.97	2.55	0.44	7.34	nr	**9.90**
20mm diameter	2.09	2.71	0.44	7.34	nr	**10.05**
25mm diameter	2.24	2.90	0.44	7.34	nr	**10.25**
32mm diameter	2.42	3.13	0.44	7.34	nr	**10.48**
40mm diameter	2.54	3.29	0.44	7.34	nr	**10.63**
50mm diameter	2.84	3.68	0.44	7.34	nr	**11.02**
Tees						
15mm diameter	1.18	1.53	0.44	7.34	nr	**8.87**
20mm diameter	1.25	1.62	0.44	7.34	nr	**8.96**
25mm diameter	1.34	1.74	0.44	7.34	nr	**9.08**
32mm diameter	1.45	1.88	0.44	7.34	nr	**9.22**
40mm diameter	1.53	1.98	0.44	7.34	nr	**9.33**
50mm diameter	1.70	2.20	0.44	7.34	nr	**9.55**
Valves						
15mm diameter	2.13	2.76	0.78	13.02	nr	**15.78**
20mm diameter	2.44	3.16	0.78	13.02	nr	**16.18**
25mm diameter	2.44	3.16	0.78	13.02	nr	**16.18**
32mm diameter	2.72	3.52	0.78	13.02	nr	**16.54**
40mm diameter	2.91	3.77	0.78	13.02	nr	**16.79**
50mm diameter	3.32	4.30	0.78	13.02	nr	**17.32**

Material Costs/Measured Work Prices - Mechanical Installations

T:MECHANICAL/COOLING/HEATING SYSTEMS

Item	Net Price £	Material £	Labour hours	Labour £	Unit	Total rate £
Pumps						
15mm diameter	11.78	15.26	2.34	39.06	nr	**54.32**
20mm diameter	12.50	16.19	2.34	39.06	nr	**55.25**
25mm diameter	13.42	17.38	2.34	39.06	nr	**56.44**
32mm diameter	14.64	18.96	2.34	39.06	nr	**58.02**
40mm diameter	15.47	20.03	2.34	39.06	nr	**59.09**
50mm diameter	17.39	22.52	2.34	39.06	nr	**61.58**
Expansion Bellows						
15mm diameter	9.42	12.20	1.05	17.53	nr	**29.73**
20mm diameter	10.00	12.95	1.05	17.53	nr	**30.48**
25mm diameter	10.74	13.91	1.05	17.53	nr	**31.44**
32mm diameter	11.71	15.16	1.05	17.53	nr	**32.69**
40mm diameter	12.37	16.02	1.05	17.53	nr	**33.55**
50mm diameter	13.91	18.01	1.05	17.53	nr	**35.54**
25mm thick						
15mm diameter	3.93	5.09	0.44	7.34	m	**12.43**
20mm diameter	4.25	5.50	0.44	7.34	m	**12.85**
25mm diameter	4.67	6.05	0.44	7.34	m	**13.39**
32mm diameter	4.97	6.44	0.44	7.34	m	**13.78**
40mm diameter	5.34	6.92	0.44	7.34	m	**14.26**
50mm diameter	5.98	7.74	0.44	7.34	m	**15.09**
65mm diameter	6.76	8.75	0.44	7.34	m	**16.10**
80mm diameter	7.33	9.49	0.52	8.68	m	**18.17**
100mm diameter	9.10	11.78	0.52	8.68	m	**20.46**
125mm diameter	10.57	13.69	0.52	8.68	m	**22.37**
150mm diameter	12.30	15.93	0.52	8.68	m	**24.61**
200mm diameter	16.80	21.76	0.60	10.02	m	**31.77**
250mm diameter	19.96	25.85	0.60	10.02	m	**35.86**
300mm diameter	22.35	28.94	0.60	10.02	m	**38.96**
Extra over for fittings plantroom insulation						
Flange/union						
15mm diameter	4.40	5.70	0.58	9.68	nr	**15.38**
20mm diameter	4.76	6.16	0.58	9.68	nr	**15.85**
25mm diameter	5.25	6.80	0.58	9.68	nr	**16.48**
32mm diameter	5.63	7.29	0.58	9.68	nr	**16.97**
40mm diameter	6.05	7.83	0.58	9.68	nr	**17.52**
50mm diameter	6.82	8.83	0.58	9.68	nr	**18.51**
65mm diameter	7.73	10.01	0.58	9.68	nr	**19.69**
80mm diameter	8.41	10.89	0.67	11.18	nr	**22.07**
100mm diameter	10.64	13.78	0.67	11.18	nr	**24.96**
125mm diameter	12.36	16.01	0.67	11.18	nr	**27.19**
150mm diameter	14.48	18.75	0.67	11.18	nr	**29.94**
200mm diameter	20.01	25.91	0.87	14.52	nr	**40.44**
250mm diameter	23.84	30.87	0.87	14.52	nr	**45.40**
300mm diameter	26.34	34.11	0.87	14.52	nr	**48.63**

Material Costs/Measured Work Prices - Mechanical Installations

T:MECHANICAL/COOLING/HEATING SYSTEMS

Item	Net Price £	Material £	Labour hours	Labour £	Unit	Total rate £
T31 : THERMAL INSULATION						
Y50 -THERMAL INSULATION						
Mineral fibre sectional insulation; Bright Class O foil faced; Bright Class O foil taped joints; 22 swg plain/embossed aluminium cladding; pop rivited (Continued)						
Bends						
15mm diameter	2.16	2.80	0.44	7.34	nr	**10.14**
20mm diameter	2.34	3.03	0.44	7.34	nr	**10.38**
25mm diameter	2.57	3.33	0.44	7.34	nr	**10.67**
32mm diameter	2.74	3.55	0.44	7.34	nr	**10.89**
40mm diameter	2.94	3.81	0.44	7.34	nr	**11.15**
50mm diameter	3.29	4.26	0.44	7.34	nr	**11.61**
65mm diameter	3.72	4.82	0.44	7.34	nr	**12.16**
80mm diameter	4.03	5.22	0.52	8.68	nr	**13.90**
100mm diameter	5.00	6.47	0.52	8.68	nr	**15.16**
125mm diameter	5.82	7.54	0.52	8.68	nr	**16.22**
150mm diameter	6.76	8.75	0.52	8.68	nr	**17.43**
200mm diameter	9.24	11.97	0.60	10.02	nr	**21.98**
250mm diameter	10.98	14.22	0.60	10.02	nr	**24.23**
300mm diameter	12.29	15.92	0.60	10.02	nr	**25.93**
Tees						
15mm diameter	1.30	1.68	0.44	7.34	nr	**9.03**
20mm diameter	1.40	1.81	0.44	7.34	nr	**9.16**
25mm diameter	1.54	1.99	0.44	7.34	nr	**9.34**
32mm diameter	1.64	2.12	0.44	7.34	nr	**9.47**
40mm diameter	1.76	2.28	0.44	7.34	nr	**9.62**
50mm diameter	1.97	2.55	0.44	7.34	nr	**9.90**
65mm diameter	2.23	2.89	0.44	7.34	nr	**10.23**
80mm diameter	2.42	3.13	0.52	8.68	nr	**11.81**
100mm diameter	3.00	3.88	0.52	8.68	nr	**12.57**
125mm diameter	3.49	4.52	0.52	8.68	nr	**13.20**
150mm diameter	4.06	5.26	0.52	8.68	nr	**13.94**
200mm diameter	5.54	7.17	0.60	10.02	nr	**17.19**
250mm diameter	6.59	8.53	0.60	10.02	nr	**18.55**
300mm diameter	7.38	9.56	0.60	10.02	nr	**19.57**
Valves						
15mm diameter	6.99	9.05	0.78	13.02	nr	**22.07**
20mm diameter	7.56	9.79	0.78	13.02	nr	**22.81**
25mm diameter	8.34	10.80	0.78	13.02	nr	**23.82**
32mm diameter	8.95	11.59	0.78	13.02	nr	**24.61**
40mm diameter	9.60	12.43	0.78	13.02	nr	**25.45**
50mm diameter	10.82	14.01	0.78	13.02	nr	**27.03**
65mm diameter	12.28	15.90	0.78	13.02	nr	**28.92**
80mm diameter	13.36	17.30	0.92	15.36	nr	**32.66**
100mm diameter	16.90	21.89	0.92	15.36	nr	**37.24**
125mm diameter	19.63	25.42	0.92	15.36	nr	**40.78**
150mm diameter	23.00	29.79	0.92	15.36	nr	**45.14**
200mm diameter	31.78	41.16	1.12	18.70	nr	**59.85**
250mm diameter	37.87	49.04	1.12	18.70	nr	**67.74**
300mm diameter	41.83	54.17	1.12	18.70	nr	**72.87**

Material Costs/Measured Work Prices - Mechanical Installations 247

T:MECHANICAL/COOLING/HEATING SYSTEMS

Item	Net Price £	Material £	Labour hours	Labour £	Unit	Total rate £
Pumps						
15mm diameter	12.94	16.76	2.34	39.06	nr	**55.82**
20mm diameter	13.99	18.12	2.34	39.06	nr	**57.18**
25mm diameter	15.45	20.01	2.34	39.06	nr	**59.07**
32mm diameter	16.57	21.46	2.34	39.06	nr	**60.52**
40mm diameter	17.78	23.03	2.34	39.06	nr	**62.09**
50mm diameter	20.05	25.96	2.34	39.06	nr	**65.03**
65mm diameter	22.74	29.45	2.34	39.06	nr	**68.51**
80mm diameter	24.74	32.04	2.76	46.07	nr	**78.11**
100mm diameter	31.29	40.52	2.76	46.07	nr	**86.59**
125mm diameter	36.34	47.06	2.76	46.07	nr	**93.13**
150mm diameter	42.59	55.15	2.76	46.07	nr	**101.23**
200mm diameter	58.86	76.22	3.36	56.09	nr	**132.31**
250mm diameter	70.12	90.81	3.36	56.09	nr	**146.89**
300mm diameter	77.46	100.31	3.36	56.09	nr	**156.40**
Expansion Bellows						
15mm diameter	10.35	13.40	1.05	17.53	nr	**30.93**
20mm diameter	11.19	14.49	1.05	17.53	nr	**32.02**
25mm diameter	12.36	16.01	1.05	17.53	nr	**33.53**
32mm diameter	13.26	17.17	1.05	17.53	nr	**34.70**
40mm diameter	14.23	18.43	1.05	17.53	nr	**35.95**
50mm diameter	16.04	20.77	1.05	17.53	nr	**38.30**
65mm diameter	18.19	23.56	1.05	17.53	nr	**41.08**
80mm diameter	19.79	25.63	1.26	21.03	nr	**46.66**
100mm diameter	25.03	32.41	1.26	21.03	nr	**53.45**
125mm diameter	29.07	37.65	1.26	21.03	nr	**58.68**
150mm diameter	34.07	44.12	1.26	21.03	nr	**65.15**
200mm diameter	47.09	60.98	1.53	25.54	nr	**86.52**
250mm diameter	56.10	72.65	1.53	25.54	nr	**98.19**
300mm diameter	61.97	80.25	1.53	25.54	nr	**105.79**
30mm thick						
15mm diameter	4.85	6.28	0.44	7.34	m	**13.63**
20mm diameter	5.13	6.64	0.44	7.34	m	**13.99**
25mm diameter	5.43	7.03	0.44	7.34	m	**14.38**
32mm diameter	5.95	7.71	0.44	7.34	m	**15.05**
40mm diameter	6.20	8.03	0.44	7.34	m	**15.37**
50mm diameter	6.88	8.91	0.44	7.34	m	**16.25**
65mm diameter	7.76	10.05	0.44	7.34	m	**17.39**
80mm diameter	8.48	10.98	0.52	8.68	m	**19.66**
100mm diameter	10.42	13.49	0.52	8.68	m	**22.17**
125mm diameter	11.91	15.42	0.52	8.68	m	**24.10**
150mm diameter	13.88	17.97	0.52	8.68	m	**26.65**
200mm diameter	18.65	24.15	0.60	10.02	m	**34.17**
250mm diameter	22.07	28.58	0.60	10.02	m	**38.60**
300mm diameter	24.29	31.46	0.60	10.02	m	**41.47**
350mm diameter	27.05	35.03	0.60	10.02	m	**45.05**

Material Costs/Measured Work Prices - Mechanical Installations

T:MECHANICAL/COOLING/HEATING SYSTEMS

Item	Net Price £	Material £	Labour hours	Labour £	Unit	Total rate £
T31 : THERMAL INSULATION						
Y50 -THERMAL INSULATION						
Mineral fibre sectional insulation; Bright Class O foil faced; Bright Class O foil taped joints; 22 swg plain/embossed aluminium cladding; pop rivited (Continued)						
Extra over for fittings plantroom insulation						
Flange/union						
15mm diameter	5.52	7.15	0.58	9.68	nr	**16.83**
20mm diameter	5.86	7.59	0.58	9.68	nr	**17.27**
25mm diameter	6.21	8.04	0.58	9.68	nr	**17.72**
32mm diameter	6.80	8.81	0.58	9.68	nr	**18.49**
40mm diameter	7.11	9.21	0.58	9.68	nr	**18.89**
50mm diameter	7.97	10.32	0.58	9.68	nr	**20.00**
65mm diameter	9.00	11.65	0.58	9.68	nr	**21.34**
80mm diameter	9.83	12.73	0.67	11.18	nr	**23.91**
100mm diameter	12.28	15.90	0.67	11.18	nr	**27.09**
125mm diameter	14.06	18.21	0.67	11.18	nr	**29.39**
150mm diameter	16.44	21.29	0.67	11.18	nr	**32.47**
200mm diameter	22.39	29.00	0.87	14.52	nr	**43.52**
250mm diameter	26.55	34.38	0.87	14.52	nr	**48.90**
300mm diameter	28.88	37.40	0.87	14.52	nr	**51.92**
350mm diameter	32.04	41.49	0.87	14.52	nr	**56.01**
Bends						
15mm diameter	2.67	3.46	0.44	7.34	nr	**10.80**
20mm diameter	2.82	3.65	0.44	7.34	nr	**11.00**
25mm diameter	2.99	3.87	0.44	7.34	nr	**11.22**
32mm diameter	3.27	4.23	0.44	7.34	nr	**11.58**
40mm diameter	3.41	4.42	0.44	7.34	nr	**11.76**
50mm diameter	3.79	4.91	0.44	7.34	nr	**12.25**
65mm diameter	4.27	5.53	0.44	7.34	nr	**12.87**
80mm diameter	4.66	6.03	0.52	8.68	nr	**14.71**
100mm diameter	5.73	7.42	0.52	8.68	nr	**16.10**
125mm diameter	6.55	8.48	0.52	8.68	nr	**17.16**
150mm diameter	7.64	9.89	0.52	8.68	nr	**18.57**
200mm diameter	10.26	13.29	0.60	10.02	nr	**23.30**
250mm diameter	12.14	15.72	0.60	10.02	nr	**25.74**
300mm diameter	13.36	17.30	0.60	10.02	nr	**27.32**
350mm diameter	14.88	19.27	0.60	10.02	nr	**29.29**
Tees						
15mm diameter	1.60	2.07	0.44	7.34	nr	**9.42**
20mm diameter	1.69	2.19	0.44	7.34	nr	**9.53**
25mm diameter	1.79	2.32	0.44	7.34	nr	**9.66**
32mm diameter	1.96	2.54	0.44	7.34	nr	**9.88**
40mm diameter	2.05	2.65	0.44	7.34	nr	**10.00**
50mm diameter	2.27	2.94	0.44	7.34	nr	**10.28**
65mm diameter	2.56	3.32	0.44	7.34	nr	**10.66**
80mm diameter	2.80	3.63	0.52	8.68	nr	**12.31**

Material Costs/Measured Work Prices - Mechanical Installations

T:MECHANICAL/COOLING/HEATING SYSTEMS

Item	Net Price £	Material £	Labour hours	Labour £	Unit	Total rate £
100mm diameter	3.44	4.45	0.52	8.68	nr	**13.13**
125mm diameter	3.93	5.09	0.52	8.68	nr	**13.77**
150mm diameter	4.58	5.93	0.52	8.68	nr	**14.61**
200mm diameter	6.16	7.98	0.60	10.02	nr	**17.99**
250mm diameter	7.28	9.43	0.60	10.02	nr	**19.44**
300mm diameter	8.02	10.39	0.60	10.02	nr	**20.40**
350mm diameter	8.93	11.56	0.60	10.02	nr	**21.58**
Valves						
15mm diameter	8.77	11.36	0.78	13.02	nr	**24.38**
20mm diameter	9.30	12.04	0.78	13.02	nr	**25.06**
25mm diameter	9.86	12.77	0.78	13.02	nr	**25.79**
32mm diameter	10.80	13.99	0.78	13.02	nr	**27.01**
40mm diameter	11.30	14.63	0.78	13.02	nr	**27.65**
50mm diameter	12.65	16.38	0.78	13.02	nr	**29.40**
65mm diameter	14.30	18.52	0.78	13.02	nr	**31.54**
80mm diameter	15.61	20.21	0.92	15.36	nr	**35.57**
100mm diameter	19.50	25.25	0.92	15.36	nr	**40.61**
125mm diameter	22.33	28.92	0.92	15.36	nr	**44.27**
150mm diameter	26.11	33.81	0.92	15.36	nr	**49.17**
200mm diameter	35.56	46.05	1.12	18.70	nr	**64.75**
250mm diameter	42.16	54.60	1.12	18.70	nr	**73.29**
300mm diameter	45.87	59.40	1.12	18.70	nr	**78.10**
350mm diameter	50.89	65.90	1.12	18.70	nr	**84.60**
Pumps						
15mm diameter	16.24	21.03	2.34	39.06	nr	**60.09**
20mm diameter	17.23	22.31	2.34	39.06	nr	**61.37**
25mm diameter	18.25	23.63	2.34	39.06	nr	**62.69**
32mm diameter	20.00	25.90	2.34	39.06	nr	**64.96**
40mm diameter	20.92	27.09	2.34	39.06	nr	**66.15**
50mm diameter	23.43	30.34	2.34	39.06	nr	**69.40**
65mm diameter	26.47	34.28	2.34	39.06	nr	**73.34**
80mm diameter	28.91	37.44	2.76	46.07	nr	**83.51**
100mm diameter	36.11	46.76	2.76	46.07	nr	**92.83**
125mm diameter	41.36	53.56	2.76	46.07	nr	**99.63**
150mm diameter	48.36	62.63	2.76	46.07	nr	**108.70**
200mm diameter	65.85	85.28	3.36	56.09	nr	**141.36**
250mm diameter	78.08	101.11	3.36	56.09	nr	**157.20**
300mm diameter	84.95	110.01	3.36	56.09	nr	**166.10**
350mm diameter	94.23	122.03	3.36	56.09	nr	**178.12**
Expansion Bellows						
15mm diameter	12.99	16.82	1.05	17.53	nr	**34.35**
20mm diameter	13.78	17.85	1.05	17.53	nr	**35.37**
25mm diameter	14.60	18.91	1.05	17.53	nr	**36.43**
32mm diameter	16.00	20.72	1.05	17.53	nr	**38.25**
40mm diameter	16.74	21.68	1.05	17.53	nr	**39.21**
50mm diameter	18.74	24.27	1.05	17.53	nr	**41.80**
65mm diameter	21.18	27.43	1.05	17.53	nr	**44.96**
80mm diameter	23.13	29.95	1.26	21.03	nr	**50.99**
100mm diameter	28.89	37.41	1.26	21.03	nr	**58.45**
125mm diameter	33.09	42.85	1.26	21.03	nr	**63.88**
150mm diameter	38.69	50.10	1.26	21.03	nr	**71.14**
200mm diameter	52.68	68.22	1.53	25.54	nr	**93.76**

250   *Material Costs/Measured Work Prices - Mechanical Installations*

T:MECHANICAL/COOLING/HEATING SYSTEMS

Item	Net Price £	Material £	Labour hours	Labour £	Unit	Total rate £
T31 : THERMAL INSULATION						
Y50 -THERMAL INSULATION						
Mineral fibre sectional insulation; Bright Class O foil faced; Bright Class O foil taped joints; 22 swg plain/embossed aluminium cladding; pop rivited (Continued)						
Expansion Bellows						
250mm diameter	62.46	80.89	1.53	25.54	nr	**106.43**
300mm diameter	67.96	88.01	1.53	25.54	nr	**113.55**
350mm diameter	75.39	97.63	1.53	25.54	nr	**123.17**
40mm thick						
15mm diameter	5.93	7.68	0.44	7.34	m	**15.02**
20mm diameter	6.23	8.07	0.44	7.34	m	**15.41**
25mm diameter	6.63	8.59	0.44	7.34	m	**15.93**
32mm diameter	6.95	9.00	0.44	7.34	m	**16.34**
40mm diameter	7.39	9.57	0.44	7.34	m	**16.91**
50mm diameter	8.18	10.59	0.44	7.34	m	**17.94**
65mm diameter	9.12	11.81	0.44	7.34	m	**19.16**
80mm diameter	9.81	12.70	0.52	8.68	m	**21.38**
100mm diameter	14.37	18.61	0.52	8.68	m	**27.29**
125mm diameter	13.46	17.43	0.52	8.68	m	**26.11**
150mm diameter	15.70	20.33	0.52	8.68	m	**29.01**
200mm diameter	20.75	26.87	0.60	10.02	m	**36.89**
250mm diameter	24.32	31.49	0.60	10.02	m	**41.51**
300mm diameter	26.89	34.82	0.60	10.02	m	**44.84**
350mm diameter	30.05	38.91	0.60	10.02	m	**48.93**
400mm diameter	33.78	43.75	0.60	10.02	m	**53.76**
Extra over for fittings plantroom insulation						
Flange/union						
15mm diameter	6.85	8.87	0.58	9.68	nr	**18.55**
20mm diameter	7.15	9.26	0.58	9.68	nr	**18.94**
25mm diameter	7.65	9.91	0.58	9.68	nr	**19.59**
32mm diameter	8.05	10.42	0.58	9.68	nr	**20.11**
40mm diameter	8.54	11.06	0.58	9.68	nr	**20.74**
50mm diameter	9.52	12.33	0.58	9.68	nr	**22.01**
65mm diameter	10.63	13.77	0.58	9.68	nr	**23.45**
80mm diameter	11.47	14.85	0.67	11.18	nr	**26.04**
100mm diameter	14.37	18.61	0.67	11.18	nr	**29.79**
125mm diameter	16.08	20.82	0.67	11.18	nr	**32.01**
150mm diameter	18.77	24.31	0.67	11.18	nr	**35.49**
200mm diameter	25.18	32.61	0.87	14.52	nr	**47.13**
250mm diameter	29.52	38.23	0.87	14.52	nr	**52.75**
300mm diameter	32.30	41.83	0.87	14.52	nr	**56.35**
350mm diameter	36.00	46.62	0.87	14.52	nr	**61.14**
400mm diameter	40.37	52.28	0.87	14.52	nr	**66.80**

Material Costs/Measured Work Prices - Mechanical Installations

T:MECHANICAL/COOLING/HEATING SYSTEMS

Item	Net Price £	Material £	Labour hours	Labour £	Unit	Total rate £
Bends						
15mm diameter	3.26	4.22	0.44	7.34	nr	**11.57**
20mm diameter	3.42	4.43	0.44	7.34	nr	**11.77**
25mm diameter	3.65	4.73	0.44	7.34	nr	**12.07**
32mm diameter	3.82	4.95	0.44	7.34	nr	**12.29**
40mm diameter	4.07	5.27	0.44	7.34	nr	**12.62**
50mm diameter	4.50	5.83	0.44	7.34	nr	**13.17**
65mm diameter	5.02	6.50	0.44	7.34	nr	**13.85**
80mm diameter	5.39	6.98	0.52	8.68	nr	**15.66**
100mm diameter	6.65	8.61	0.52	8.68	nr	**17.29**
125mm diameter	7.40	9.58	0.52	8.68	nr	**18.26**
150mm diameter	8.63	11.18	0.52	8.68	nr	**19.86**
200mm diameter	11.41	14.78	0.60	10.02	nr	**24.79**
250mm diameter	13.28	17.20	0.60	10.02	nr	**27.21**
300mm diameter	14.79	19.15	0.60	10.02	nr	**29.17**
350mm diameter	16.53	21.41	0.60	10.02	nr	**31.42**
400mm diameter	18.58	24.06	0.60	10.02	nr	**34.08**
Tees						
15mm diameter	1.96	2.54	0.44	7.34	nr	**9.88**
20mm diameter	2.05	2.65	0.44	7.34	nr	**10.00**
25mm diameter	2.19	2.84	0.44	7.34	nr	**10.18**
32mm diameter	2.29	2.97	0.44	7.34	nr	**10.31**
40mm diameter	2.44	3.16	0.44	7.34	nr	**10.50**
50mm diameter	2.70	3.50	0.44	7.34	nr	**10.84**
65mm diameter	3.01	3.90	0.44	7.34	nr	**11.24**
80mm diameter	3.24	4.20	0.52	8.68	nr	**12.88**
100mm diameter	3.99	5.17	0.52	8.68	nr	**13.85**
125mm diameter	4.44	5.75	0.52	8.68	nr	**14.43**
150mm diameter	5.18	6.71	0.52	8.68	nr	**15.39**
200mm diameter	6.85	8.87	0.60	10.02	nr	**18.89**
250mm diameter	8.03	10.40	0.60	10.02	nr	**20.41**
300mm diameter	8.87	11.49	0.60	10.02	nr	**21.50**
350mm diameter	9.92	12.85	0.60	10.02	nr	**22.86**
400mm diameter	11.15	14.44	0.60	10.02	nr	**24.45**
Valves						
15mm diameter	5.93	7.68	0.78	13.02	nr	**20.70**
20mm diameter	11.36	14.71	0.78	13.02	nr	**27.73**
25mm diameter	12.14	15.72	0.78	13.02	nr	**28.74**
32mm diameter	12.78	16.55	0.78	13.02	nr	**29.57**
40mm diameter	13.56	17.56	0.78	13.02	nr	**30.58**
50mm diameter	15.11	19.57	0.78	13.02	nr	**32.59**
65mm diameter	16.88	21.86	0.78	13.02	nr	**34.88**
80mm diameter	18.22	23.59	0.92	15.36	nr	**38.95**
100mm diameter	22.83	29.56	0.92	15.36	nr	**44.92**
125mm diameter	25.53	33.06	0.92	15.36	nr	**48.42**
150mm diameter	29.81	38.60	0.92	15.36	nr	**53.96**
200mm diameter	39.99	51.79	1.12	18.70	nr	**70.48**
250mm diameter	46.88	60.71	1.12	18.70	nr	**79.41**
300mm diameter	51.30	66.43	1.12	18.70	nr	**85.13**
350mm diameter	57.18	74.05	1.12	18.70	nr	**92.74**
400mm diameter	64.12	83.04	1.12	18.70	nr	**101.73**

Material Costs/Measured Work Prices - Mechanical Installations

T:MECHANICAL/COOLING/HEATING SYSTEMS

Item	Net Price £	Material £	Labour hours	Labour £	Unit	Total rate £
T31 : THERMAL INSULATION						
Y50 -THERMAL INSULATION						
Mineral fibre sectional insulation; Bright Class O foil faced; Bright Class O foil taped joints; 22 swg plain/embossed aluminium cladding; pop rivited (Continued)						
Pumps						
15mm diameter	20.14	26.08	2.34	39.06	nr	**65.14**
20mm diameter	21.03	27.23	2.34	39.06	nr	**66.29**
25mm diameter	22.49	29.12	2.34	39.06	nr	**68.19**
32mm diameter	23.67	30.65	2.34	39.06	nr	**69.71**
40mm diameter	25.11	32.52	2.34	39.06	nr	**71.58**
50mm diameter	27.99	36.25	2.34	39.06	nr	**75.31**
65mm diameter	31.26	40.48	2.34	39.06	nr	**79.54**
80mm diameter	33.74	43.69	2.76	46.07	nr	**89.76**
100mm diameter	42.27	54.74	2.76	46.07	nr	**100.81**
125mm diameter	47.28	61.23	2.76	46.07	nr	**107.30**
150mm diameter	55.21	71.50	2.76	46.07	nr	**117.57**
200mm diameter	74.06	95.91	3.36	56.09	nr	**151.99**
250mm diameter	86.82	112.43	3.36	56.09	nr	**168.52**
300mm diameter	94.99	123.01	3.36	56.09	nr	**179.10**
350mm diameter	105.88	137.11	3.36	56.09	nr	**193.20**
400mm diameter	118.74	153.77	3.36	56.09	nr	**209.86**
Expansion Bellows						
15mm diameter	16.11	20.86	1.05	17.53	nr	**38.39**
20mm diameter	16.83	21.79	1.05	17.53	nr	**39.32**
25mm diameter	17.99	23.30	1.05	17.53	nr	**40.82**
32mm diameter	18.93	24.51	1.05	17.53	nr	**42.04**
40mm diameter	20.08	26.00	1.05	17.53	nr	**43.53**
50mm diameter	22.39	29.00	1.05	17.53	nr	**46.52**
65mm diameter	25.01	32.39	1.05	17.53	nr	**49.92**
80mm diameter	26.99	34.95	1.26	21.03	nr	**55.98**
100mm diameter	33.82	43.80	1.26	21.03	nr	**64.83**
125mm diameter	37.82	48.98	1.26	21.03	nr	**70.01**
150mm diameter	44.16	57.19	1.26	21.03	nr	**78.22**
200mm diameter	59.24	76.72	1.53	25.54	nr	**102.26**
250mm diameter	69.46	89.95	1.53	25.54	nr	**115.49**
300mm diameter	75.99	98.41	1.53	25.54	nr	**123.95**
350mm diameter	84.70	109.69	1.53	25.54	nr	**135.23**
400mm diameter	94.99	123.01	1.53	25.54	nr	**148.55**
50mm thick						
15mm diameter	7.74	10.02	0.44	7.34	m	**17.37**
20mm diameter	8.16	10.57	0.44	7.34	m	**17.91**
25mm diameter	8.64	11.19	0.44	7.34	m	**18.53**
32mm diameter	9.08	11.76	0.44	7.34	m	**19.10**
40mm diameter	9.50	12.30	0.44	7.34	m	**19.65**
50mm diameter	10.48	13.57	0.44	7.34	m	**20.92**
65mm diameter	11.27	14.59	0.44	7.34	m	**21.94**
80mm diameter	12.10	15.67	0.52	8.68	m	**24.35**

Material Costs/Measured Work Prices - Mechanical Installations   253

T:MECHANICAL/COOLING/HEATING SYSTEMS

Item	Net Price £	Material £	Labour hours	Labour £	Unit	Total rate £
100mm diameter	14.95	19.36	0.52	8.68	m	**28.04**
125mm diameter	16.55	21.43	0.52	8.68	m	**30.11**
150mm diameter	19.01	24.62	0.52	8.68	m	**33.30**
200mm diameter	24.92	32.27	0.60	10.02	m	**42.29**
250mm diameter	28.80	37.30	0.60	10.02	m	**47.31**
300mm diameter	31.35	40.60	0.60	10.02	m	**50.61**
350mm diameter	34.94	45.25	0.60	10.02	m	**55.26**
400mm diameter	39.08	50.61	0.60	10.02	m	**60.62**
Extra over for fittings plantroom insulation						
Flange/union						
15mm diameter	9.12	11.81	0.58	9.68	nr	**21.49**
20mm diameter	9.62	12.46	0.58	9.68	nr	**22.14**
25mm diameter	10.19	13.20	0.58	9.68	nr	**22.88**
32mm diameter	10.70	13.86	0.58	9.68	nr	**23.54**
40mm diameter	11.21	14.52	0.58	9.68	nr	**24.20**
50mm diameter	12.44	16.11	0.58	9.68	nr	**25.79**
65mm diameter	13.44	17.40	0.58	9.68	nr	**27.09**
80mm diameter	14.41	18.66	0.67	11.18	nr	**29.84**
100mm diameter	18.02	23.34	0.67	11.18	nr	**34.52**
125mm diameter	20.04	25.95	0.67	11.18	nr	**37.14**
150mm diameter	23.06	29.86	0.67	11.18	nr	**41.05**
200mm diameter	30.65	39.69	0.87	14.52	nr	**54.21**
250mm diameter	35.42	45.87	0.87	14.52	nr	**60.39**
300mm diameter	38.22	49.49	0.87	14.52	nr	**64.02**
350mm diameter	42.49	55.02	0.87	14.52	nr	**69.55**
400mm diameter	47.41	61.40	0.87	14.52	nr	**75.92**
Bend						
15mm diameter	4.26	5.52	0.44	7.34	nr	**12.86**
20mm diameter	4.49	5.81	0.44	7.34	nr	**13.16**
25mm diameter	4.75	6.15	0.44	7.34	nr	**13.50**
32mm diameter	5.00	6.47	0.44	7.34	nr	**13.82**
40mm diameter	5.22	6.76	0.44	7.34	nr	**14.10**
50mm diameter	5.77	7.47	0.44	7.34	nr	**14.82**
65mm diameter	6.20	8.03	0.44	7.34	nr	**15.37**
80mm diameter	6.65	8.61	0.52	8.68	nr	**17.29**
100mm diameter	8.22	10.64	0.52	8.68	nr	**19.33**
125mm diameter	9.10	11.78	0.52	8.68	nr	**20.46**
150mm diameter	10.46	13.55	0.52	8.68	nr	**22.23**
200mm diameter	13.71	17.75	0.60	10.02	nr	**27.77**
250mm diameter	15.84	20.51	0.60	10.02	nr	**30.53**
300mm diameter	17.24	22.33	0.60	10.02	nr	**32.34**
350mm diameter	19.22	24.89	0.60	10.02	nr	**34.91**
400mm diameter	21.49	27.83	0.60	10.02	nr	**37.85**

Material Costs/Measured Work Prices - Mechanical Installations

T:MECHANICAL/COOLING/HEATING SYSTEMS

Item	Net Price £	Material £	Labour hours	Labour £	Unit	Total rate £
T31 : THERMAL INSULATION						
Y50 -THERMAL INSULATION						
Mineral fibre sectional insulation; Bright Class O foil faced; Bright Class O foil taped joints; 22 swg plain/embossed aluminium cladding; pop rivited (Continued)						
Tee						
15mm diameter	2.56	3.32	0.44	7.34	nr	**10.66**
20mm diameter	2.69	3.48	0.44	7.34	nr	**10.83**
25mm diameter	2.85	3.69	0.44	7.34	nr	**11.04**
32mm diameter	3.00	3.88	0.44	7.34	nr	**11.23**
40mm diameter	3.13	4.05	0.44	7.34	nr	**11.40**
50mm diameter	3.46	4.48	0.44	7.34	nr	**11.83**
65mm diameter	3.72	4.82	0.44	7.34	nr	**12.16**
80mm diameter	3.99	5.17	0.52	8.68	nr	**13.85**
100mm diameter	4.93	6.38	0.52	8.68	nr	**15.06**
125mm diameter	5.46	7.07	0.52	8.68	nr	**15.75**
150mm diameter	6.27	8.12	0.52	8.68	nr	**16.80**
200mm diameter	8.22	10.64	0.60	10.02	nr	**20.66**
250mm diameter	9.51	12.32	0.60	10.02	nr	**22.33**
300mm diameter	10.35	13.40	0.60	10.02	nr	**23.42**
350mm diameter	11.53	14.93	0.60	10.02	nr	**24.95**
400mm diameter	12.90	16.71	0.60	10.02	nr	**26.72**
Valves						
15mm diameter	14.49	18.76	0.78	13.02	nr	**31.78**
20mm diameter	15.27	19.77	0.78	13.02	nr	**32.79**
25mm diameter	16.18	20.95	0.78	13.02	nr	**33.97**
32mm diameter	16.99	22.00	0.78	13.02	nr	**35.02**
40mm diameter	17.80	23.05	0.78	13.02	nr	**36.07**
50mm diameter	19.75	25.58	0.78	13.02	nr	**38.60**
65mm diameter	21.34	27.64	0.78	13.02	nr	**40.66**
80mm diameter	22.89	29.64	0.92	15.36	nr	**45.00**
100mm diameter	28.62	37.06	0.92	15.36	nr	**52.42**
125mm diameter	31.82	41.21	0.92	15.36	nr	**56.56**
150mm diameter	36.62	47.42	0.92	15.36	nr	**62.78**
200mm diameter	48.69	63.05	1.12	18.70	nr	**81.75**
250mm diameter	56.26	72.86	1.12	18.70	nr	**91.55**
300mm diameter	60.71	78.62	1.12	18.70	nr	**97.32**
350mm diameter	67.48	87.39	1.12	18.70	nr	**106.08**
400mm diameter	75.29	97.50	1.12	18.70	nr	**116.20**
Pumps						
15mm diameter	26.83	34.74	2.34	39.06	nr	**73.81**
20mm diameter	28.28	36.62	2.34	39.06	nr	**75.68**
25mm diameter	29.96	38.80	2.34	39.06	nr	**77.86**
32mm diameter	31.47	40.75	2.34	39.06	nr	**79.81**
40mm diameter	32.97	42.70	2.34	39.06	nr	**81.76**
50mm diameter	36.57	47.36	2.34	39.06	nr	**86.42**
65mm diameter	29.52	38.23	2.34	39.06	nr	**77.29**
80mm diameter	42.40	54.91	2.76	46.07	nr	**100.98**

Material Costs/Measured Work Prices - Mechanical Installations

T:MECHANICAL/COOLING/HEATING SYSTEMS

Item	Net Price £	Material £	Labour hours	Labour £	Unit	Total rate £
100mm diameter	52.99	68.62	2.76	46.07	nr	**114.69**
125mm diameter	58.93	76.31	2.76	46.07	nr	**122.39**
150mm diameter	67.82	87.83	2.76	46.07	nr	**133.90**
200mm diameter	90.16	116.76	3.36	56.09	nr	**172.84**
250mm diameter	104.18	134.91	3.36	56.09	nr	**191.00**
300mm diameter	112.42	145.58	3.36	56.09	nr	**201.67**
350mm diameter	124.97	161.84	3.36	56.09	nr	**217.92**
400mm diameter	139.43	180.56	3.36	56.09	nr	**236.65**
Expansion Bellows						
15mm diameter	21.46	27.79	1.05	17.53	nr	**45.32**
20mm diameter	22.62	29.29	1.05	17.53	nr	**46.82**
25mm diameter	23.97	31.04	1.05	17.53	nr	**48.57**
32mm diameter	25.17	32.60	1.05	17.53	nr	**50.12**
40mm diameter	26.37	34.15	1.05	17.53	nr	**51.68**
50mm diameter	29.26	37.89	1.05	17.53	nr	**55.42**
65mm diameter	31.61	40.94	1.05	17.53	nr	**58.46**
80mm diameter	33.92	43.93	1.26	21.03	nr	**64.96**
100mm diameter	42.39	54.90	1.26	21.03	nr	**75.93**
125mm diameter	47.14	61.05	1.26	21.03	nr	**82.08**
150mm diameter	54.26	70.27	1.26	21.03	nr	**91.30**
200mm diameter	72.13	93.41	1.53	25.54	nr	**118.95**
250mm diameter	83.34	107.93	1.53	25.54	nr	**133.47**
300mm diameter	89.94	116.47	1.53	25.54	nr	**142.01**
350mm diameter	99.97	129.46	1.53	25.54	nr	**155.00**
400mm diameter	111.54	144.44	1.53	25.54	nr	**169.98**

Mineral fibre sectional insulation; Bright Class O foil faced; Bright Class O foil taped joints; 0.8mm Polyisobutylene sheeting; welded joints

External pipework

	Net Price £	Material £	Labour hours	Labour £	Unit	Total rate £
20mm thick						
15mm diameter	3.30	4.27	0.30	5.01	m	**9.28**
20mm diameter	3.52	4.56	0.30	5.01	m	**9.57**
25mm diameter	3.79	4.91	0.30	5.01	m	**9.92**
32mm diameter	4.17	5.40	0.30	5.01	m	**10.41**
40mm diameter	4.44	5.75	0.30	5.01	m	**10.76**
50mm diameter	5.02	6.50	0.30	5.01	m	**11.51**

Extra over for fittings external insulation

	Net Price £	Material £	Labour hours	Labour £	Unit	Total rate £
Flange/union						
15mm diameter	5.06	6.55	0.75	12.52	nr	**19.07**
20mm diameter	5.37	6.95	0.75	12.52	nr	**19.47**
25mm diameter	5.77	7.47	0.75	12.52	nr	**19.99**
32mm diameter	6.28	8.13	0.75	12.52	nr	**20.65**
40mm diameter	6.63	8.59	0.75	12.52	nr	**21.11**
50mm diameter	7.43	9.62	0.75	12.52	nr	**22.14**

Material Costs/Measured Work Prices - Mechanical Installations

T:MECHANICAL/COOLING/HEATING SYSTEMS

Item	Net Price £	Material £	Labour hours	Labour £	Unit	Total rate £
T31 : THERMAL INSULATION						
Y50 -THERMAL INSULATION						
Mineral fibre sectional insulation; Bright Class O foil faced; Bright Class O foil taped joints; 0.8mm Polyisobutylene sheeting; welded joints (Continued)						
Bends						
15mm diameter	0.83	1.07	0.30	5.01	nr	**6.08**
20mm diameter	0.88	1.14	0.30	5.01	nr	**6.15**
25mm diameter	0.95	1.23	0.30	5.01	nr	**6.24**
32mm diameter	1.04	1.35	0.30	5.01	nr	**6.35**
40mm diameter	1.11	1.44	0.30	5.01	nr	**6.45**
50mm diameter	1.25	1.62	0.30	5.01	nr	**6.63**
Tees						
15mm diameter	0.83	1.07	0.30	5.01	nr	**6.08**
20mm diameter	0.88	1.14	0.30	5.01	nr	**6.15**
25mm diameter	0.95	1.23	0.30	5.01	nr	**6.24**
32mm diameter	1.04	1.35	0.30	5.01	nr	**6.35**
40mm diameter	1.11	1.44	0.30	5.01	nr	**6.45**
50mm diameter	1.25	1.62	0.30	5.01	nr	**6.63**
Valves						
15mm diameter	8.03	10.40	1.03	17.19	nr	**27.59**
20mm diameter	8.53	11.05	1.03	17.19	nr	**28.24**
25mm diameter	9.16	11.86	1.03	17.19	nr	**29.06**
32mm diameter	9.97	12.91	1.03	17.19	nr	**30.10**
40mm diameter	10.53	13.64	1.03	17.19	nr	**30.83**
50mm diameter	11.80	15.28	1.03	17.19	nr	**32.47**
Pumps						
15mm diameter	14.88	19.27	3.10	51.75	nr	**71.02**
20mm diameter	15.80	20.46	3.10	51.75	nr	**72.21**
25mm diameter	16.97	21.98	3.10	51.75	nr	**73.72**
32mm diameter	18.47	23.92	3.10	51.75	nr	**75.67**
40mm diameter	19.51	25.27	3.10	51.75	nr	**77.01**
50mm diameter	21.85	28.30	3.10	51.75	nr	**80.04**
Expansion Bellows						
15mm diameter	11.90	15.41	1.42	23.70	nr	**39.11**
20mm diameter	12.64	16.37	1.42	23.70	nr	**40.07**
25mm diameter	13.58	17.59	1.42	23.70	nr	**41.29**
32mm diameter	14.78	19.14	1.42	23.70	nr	**42.84**
40mm diameter	15.60	20.20	1.42	23.70	nr	**43.91**
50mm diameter	17.48	22.64	1.42	23.70	nr	**46.34**
25mm thick						
15mm diameter	3.64	4.71	0.30	5.01	m	**9.72**
20mm diameter	3.91	5.06	0.30	5.01	m	**10.07**
25mm diameter	4.30	5.57	0.30	5.01	m	**10.58**
32mm diameter	4.66	6.03	0.30	5.01	m	**11.04**

T:MECHANICAL/COOLING/HEATING SYSTEMS

Item	Net Price £	Material £	Labour hours	Labour £	Unit	Total rate £
40mm diameter	4.95	6.41	0.30	5.01	m	**11.42**
50mm diameter	5.59	7.24	0.30	5.01	m	**12.25**
65mm diameter	6.34	8.21	0.30	5.01	m	**13.22**
80mm diameter	6.93	8.97	0.40	6.68	m	**15.65**
100mm diameter	8.76	11.34	0.40	6.68	m	**18.02**
125mm diameter	10.08	13.05	0.40	6.68	m	**19.73**
150mm diameter	11.83	15.32	0.40	6.68	m	**22.00**
200mm diameter	16.05	20.78	0.50	8.35	m	**29.13**
250mm diameter	19.16	24.81	0.50	8.35	m	**33.16**
300mm diameter	20.84	26.99	0.50	8.35	m	**35.33**
Extra over for fittings external insulation						
Flange/union						
15mm diameter	5.57	7.21	0.75	12.52	nr	**19.73**
20mm diameter	6.00	7.77	0.75	12.52	nr	**20.29**
25mm diameter	6.58	8.52	0.75	12.52	nr	**21.04**
32mm diameter	7.05	9.13	0.75	12.52	nr	**21.65**
40mm diameter	7.54	9.76	0.75	12.52	nr	**22.28**
50mm diameter	8.45	10.94	0.75	12.52	nr	**23.46**
65mm diameter	9.56	12.38	0.75	12.52	nr	**24.90**
80mm diameter	10.40	13.47	0.89	14.86	nr	**28.32**
100mm diameter	12.92	16.73	0.89	14.86	nr	**31.59**
125mm diameter	14.96	19.37	0.89	14.86	nr	**34.23**
150mm diameter	17.41	22.55	0.89	14.86	nr	**37.40**
200mm diameter	23.56	30.51	1.15	19.20	nr	**49.71**
250mm diameter	28.04	36.31	1.15	19.20	nr	**55.51**
300mm diameter	31.15	40.34	1.15	19.20	nr	**59.54**
Bends						
15mm diameter	0.91	1.18	0.30	5.01	nr	**6.19**
20mm diameter	0.98	1.27	0.30	5.01	nr	**6.28**
25mm diameter	1.07	1.39	0.30	5.01	nr	**6.39**
32mm diameter	1.16	1.50	0.30	5.01	nr	**6.51**
40mm diameter	1.24	1.61	0.30	5.01	nr	**6.61**
50mm diameter	1.40	1.81	0.30	5.01	nr	**6.82**
65mm diameter	1.59	2.06	0.30	5.01	nr	**7.07**
80mm diameter	1.73	2.24	0.40	6.68	nr	**8.92**
100mm diameter	2.19	2.84	0.40	6.68	nr	**9.51**
125mm diameter	2.52	3.26	0.40	6.68	nr	**9.94**
150mm diameter	2.96	3.83	0.40	6.68	nr	**10.51**
200mm diameter	4.01	5.19	0.50	8.35	nr	**13.54**
250mm diameter	4.79	6.20	0.50	8.35	nr	**14.55**
300mm diameter	5.21	6.75	0.50	8.35	nr	**15.09**
Tees						
15mm diameter	0.91	1.18	0.30	5.01	nr	**6.19**
20mm diameter	0.98	1.27	0.30	5.01	nr	**6.28**
25mm diameter	1.07	1.39	0.30	5.01	nr	**6.39**
32mm diameter	1.16	1.50	0.30	5.01	nr	**6.51**
40mm diameter	1.24	1.61	0.30	5.01	nr	**6.61**
50mm diameter	1.40	1.81	0.30	5.01	nr	**6.82**
65mm diameter	1.59	2.06	0.30	5.01	nr	**7.07**
80mm diameter	1.73	2.24	0.40	6.68	nr	**8.92**
100mm diameter	2.19	2.84	0.40	6.68	nr	**9.51**

Material Costs/Measured Work Prices - Mechanical Installations

T:MECHANICAL/COOLING/HEATING SYSTEMS

Item	Net Price £	Material £	Labour hours	Labour £	Unit	Total rate £
T31 : THERMAL INSULATION						
Y50 -THERMAL INSULATION						
Mineral fibre sectional insulation; Bright Class O foil faced; Bright Class O foil taped joints; 0.8mm Polyisobutylene sheeting; welded joints (Continued)						
125mm diameter	2.52	3.26	0.40	6.68	nr	**9.94**
150mm diameter	2.96	3.83	0.40	6.68	nr	**10.51**
200mm diameter	4.01	5.19	0.50	8.35	nr	**13.54**
250mm diameter	4.79	6.20	0.50	8.35	nr	**14.55**
300mm diameter	5.21	6.75	0.50	8.35	nr	**15.09**
Valves						
15mm diameter	8.84	11.45	1.03	17.19	nr	**28.64**
20mm diameter	9.53	12.34	1.03	17.19	nr	**29.53**
25mm diameter	10.45	13.53	1.03	17.19	nr	**30.73**
32mm diameter	11.20	14.50	1.03	17.19	nr	**31.70**
40mm diameter	11.97	15.50	1.03	17.19	nr	**32.69**
50mm diameter	13.43	17.39	1.03	17.19	nr	**34.59**
65mm diameter	15.19	19.67	1.03	17.19	nr	**36.86**
80mm diameter	16.51	21.38	1.25	20.87	nr	**42.25**
100mm diameter	20.53	26.59	1.25	20.87	nr	**47.45**
125mm diameter	23.75	30.76	1.25	20.87	nr	**51.62**
150mm diameter	27.66	35.82	1.25	20.87	nr	**56.69**
200mm diameter	37.42	48.46	1.55	25.87	nr	**74.33**
250mm diameter	44.53	57.67	1.55	25.87	nr	**83.54**
300mm diameter	49.47	64.06	1.55	25.87	nr	**89.94**
Pumps						
15mm diameter	16.38	21.21	3.10	51.75	nr	**72.96**
20mm diameter	17.65	22.86	3.10	51.75	nr	**74.60**
25mm diameter	19.35	25.06	3.10	51.75	nr	**76.81**
32mm diameter	20.75	26.87	3.10	51.75	nr	**78.62**
40mm diameter	22.17	28.71	3.10	51.75	nr	**80.46**
50mm diameter	24.87	32.21	3.10	51.75	nr	**83.95**
65mm diameter	28.13	36.43	3.10	51.75	nr	**88.18**
80mm diameter	30.58	39.60	3.75	62.60	nr	**102.20**
100mm diameter	38.01	49.22	3.75	62.60	nr	**111.82**
125mm diameter	43.99	56.97	3.75	62.60	nr	**119.56**
150mm diameter	51.22	66.33	3.75	62.60	nr	**128.93**
200mm diameter	69.29	89.73	4.65	77.62	nr	**167.35**
250mm diameter	44.53	57.67	4.65	77.62	nr	**135.29**
300mm diameter	49.47	64.06	4.65	77.62	nr	**141.68**
Expansion Bellows						
15mm diameter	13.10	16.96	1.42	23.70	nr	**40.67**
20mm diameter	14.12	18.29	1.42	23.70	nr	**41.99**
25mm diameter	15.48	20.05	1.42	23.70	nr	**43.75**
32mm diameter	16.60	21.50	1.42	23.70	nr	**45.20**
40mm diameter	17.74	22.97	1.42	23.70	nr	**46.68**
50mm diameter	19.89	25.76	1.42	23.70	nr	**49.46**

Material Costs/Measured Work Prices - Mechanical Installations

T:MECHANICAL/COOLING/HEATING SYSTEMS

Item	Net Price £	Material £	Labour hours	Labour £	Unit	Total rate £
65mm diameter	22.50	29.14	1.42	23.70	nr	**52.84**
80mm diameter	24.46	31.68	1.75	29.21	nr	**60.89**
100mm diameter	30.41	39.38	1.75	29.21	nr	**68.59**
125mm diameter	35.19	45.57	1.75	29.21	nr	**74.78**
150mm diameter	40.97	53.06	1.75	29.21	nr	**82.27**
200mm diameter	55.43	71.78	2.17	36.22	nr	**108.00**
250mm diameter	65.97	85.43	2.17	36.22	nr	**121.65**
300mm diameter	73.28	94.90	3.17	52.92	nr	**147.81**
30mm thick						
15mm diameter	4.48	5.80	0.30	5.01	m	**10.81**
20mm diameter	4.77	6.18	0.30	5.01	m	**11.18**
25mm diameter	5.06	6.55	0.30	5.01	m	**11.56**
32mm diameter	5.48	7.10	0.30	5.01	m	**12.10**
40mm diameter	5.78	7.49	0.30	5.01	m	**12.49**
50mm diameter	6.51	8.43	0.30	5.01	m	**13.44**
65mm diameter	7.32	9.48	0.30	5.01	m	**14.49**
80mm diameter	7.96	10.31	0.40	6.68	m	**16.99**
100mm diameter	9.95	12.89	0.40	6.68	m	**19.56**
125mm diameter	11.39	14.75	0.40	6.68	m	**21.43**
150mm diameter	13.23	17.13	0.40	6.68	m	**23.81**
200mm diameter	17.82	23.08	0.50	8.35	m	**31.42**
250mm diameter	21.14	27.38	0.50	8.35	m	**35.72**
300mm diameter	22.80	29.53	0.50	8.35	m	**37.87**
350mm diameter	24.93	32.28	0.50	8.35	m	**40.63**
Extra over for fittings external insulation						
Flange/union						
15mm diameter	6.81	8.82	0.75	12.52	nr	**21.34**
20mm diameter	7.22	9.35	0.75	12.52	nr	**21.87**
25mm diameter	7.65	9.91	0.75	12.52	nr	**22.43**
32mm diameter	8.34	10.80	0.75	12.52	nr	**23.32**
40mm diameter	8.72	11.29	0.75	12.52	nr	**23.81**
50mm diameter	9.73	12.60	0.75	12.52	nr	**25.12**
65mm diameter	10.95	14.18	0.75	12.52	nr	**26.70**
80mm diameter	11.93	15.45	0.89	14.86	nr	**30.31**
100mm diameter	14.69	19.02	0.89	14.86	nr	**33.88**
125mm diameter	16.78	21.73	0.89	14.86	nr	**36.59**
150mm diameter	19.50	25.25	0.89	14.86	nr	**40.11**
200mm diameter	26.06	33.75	1.15	19.20	nr	**52.94**
250mm diameter	30.86	39.96	1.15	19.20	nr	**59.16**
300mm diameter	33.81	43.78	1.15	19.20	nr	**62.98**
350mm diameter	37.35	48.37	1.15	19.20	nr	**67.56**
Bends						
15mm diameter	1.12	1.45	0.30	5.01	nr	**6.46**
20mm diameter	1.19	1.54	0.30	5.01	nr	**6.55**
25mm diameter	1.27	1.64	0.30	5.01	nr	**6.65**
32mm diameter	1.37	1.77	0.30	5.01	nr	**6.78**
40mm diameter	1.44	1.86	0.30	5.01	nr	**6.87**
50mm diameter	1.63	2.11	0.30	5.01	nr	**7.12**
65mm diameter	1.83	2.37	0.30	5.01	nr	**7.38**
80mm diameter	1.99	2.58	0.40	6.68	nr	**9.25**

Material Costs/Measured Work Prices - Mechanical Installations

T:MECHANICAL/COOLING/HEATING SYSTEMS

Item	Net Price £	Material £	Labour hours	Labour £	Unit	Total rate £
T31 : THERMAL INSULATION						
Y50 -THERMAL INSULATION						
Mineral fibre sectional insulation; Bright Class O foil faced; Bright Class O foil taped joints; 0.8mm Polyisobutylene sheeting; welded joints (Continued)						
Bends						
100mm diameter	2.49	3.22	0.40	6.68	nr	9.90
125mm diameter	2.85	3.69	0.40	6.68	nr	10.37
150mm diameter	3.31	4.29	0.40	6.68	nr	10.96
200mm diameter	4.46	5.78	0.50	8.35	nr	14.12
250mm diameter	5.29	6.85	0.50	8.35	nr	15.20
300mm diameter	5.70	7.38	0.50	8.35	nr	15.73
350mm diameter	6.23	8.07	0.50	8.35	nr	16.41
Tees						
15mm diameter	1.12	1.45	0.30	5.01	nr	6.46
20mm diameter	1.19	1.54	0.30	5.01	nr	6.55
25mm diameter	1.27	1.64	0.30	5.01	nr	6.65
32mm diameter	1.37	1.77	0.30	5.01	nr	6.78
40mm diameter	1.44	1.86	0.30	5.01	nr	6.87
50mm diameter	1.63	2.11	0.30	5.01	nr	7.12
65mm diameter	1.83	2.37	0.30	5.01	nr	7.38
80mm diameter	1.99	2.58	0.40	6.68	nr	9.25
100mm diameter	2.49	3.22	0.40	6.68	nr	9.90
125mm diameter	2.85	3.69	0.40	6.68	nr	10.37
150mm diameter	3.31	4.29	0.40	6.68	nr	10.96
200mm diameter	4.46	5.78	0.50	8.35	nr	14.12
250mm diameter	5.29	6.85	0.50	8.35	nr	15.20
300mm diameter	5.70	7.38	0.50	8.35	nr	15.73
350mm diameter	6.23	8.07	0.50	8.35	nr	16.41
Valves						
15mm diameter	10.82	14.01	1.03	17.19	nr	31.21
20mm diameter	11.46	14.84	1.03	17.19	nr	32.03
25mm diameter	12.15	15.73	1.03	17.19	nr	32.93
32mm diameter	13.24	17.15	1.03	17.19	nr	34.34
40mm diameter	13.86	17.95	1.03	17.19	nr	35.14
50mm diameter	15.45	20.01	1.03	17.19	nr	37.20
65mm diameter	17.39	22.52	1.03	17.19	nr	39.71
80mm diameter	18.95	24.54	1.25	20.87	nr	45.41
100mm diameter	23.32	30.20	1.25	20.87	nr	51.07
125mm diameter	26.65	34.51	1.25	20.87	nr	55.38
150mm diameter	30.96	40.09	1.25	20.87	nr	60.96
200mm diameter	41.39	53.60	1.55	25.87	nr	79.47
250mm diameter	49.01	63.47	1.55	25.87	nr	89.34
300mm diameter	53.70	69.54	1.55	25.87	nr	95.41
350mm diameter	59.32	76.82	1.55	25.87	nr	102.69

Material Costs/Measured Work Prices - Mechanical Installations 261

T:MECHANICAL/COOLING/HEATING SYSTEMS

Item	Net Price £	Material £	Labour hours	Labour £	Unit	Total rate £
Pumps						
15mm diameter	20.03	25.94	3.10	51.75	nr	**77.69**
20mm diameter	21.23	27.49	3.10	51.75	nr	**79.24**
25mm diameter	22.50	29.14	3.10	51.75	nr	**80.88**
32mm diameter	24.53	31.77	3.10	51.75	nr	**83.51**
40mm diameter	25.66	33.23	3.10	51.75	nr	**84.98**
50mm diameter	28.60	37.04	3.10	51.75	nr	**88.78**
65mm diameter	32.21	41.71	3.10	51.75	nr	**93.46**
80mm diameter	35.10	45.45	3.75	62.60	nr	**108.05**
100mm diameter	43.19	55.93	3.75	62.60	nr	**118.53**
125mm diameter	49.35	63.91	3.75	62.60	nr	**126.51**
150mm diameter	57.34	74.26	3.75	62.60	nr	**136.85**
200mm diameter	76.65	99.26	4.65	77.62	nr	**176.88**
250mm diameter	90.77	117.55	4.65	77.62	nr	**195.17**
300mm diameter	99.44	128.77	4.65	77.62	nr	**206.40**
350mm diameter	109.85	142.26	4.65	77.62	nr	**219.88**
Expansion Bellows						
15mm diameter	16.02	20.75	1.42	23.70	nr	**44.45**
20mm diameter	16.98	21.99	1.42	23.70	nr	**45.69**
25mm diameter	18.00	23.31	1.42	23.70	nr	**47.01**
32mm diameter	19.62	25.41	1.42	23.70	nr	**49.11**
40mm diameter	20.53	26.59	1.42	23.70	nr	**50.29**
50mm diameter	22.88	29.63	1.42	23.70	nr	**53.33**
65mm diameter	25.77	33.37	1.42	23.70	nr	**57.08**
80mm diameter	28.08	36.36	1.75	29.21	nr	**65.58**
100mm diameter	34.55	44.74	1.75	29.21	nr	**73.95**
125mm diameter	39.48	51.13	1.75	29.21	nr	**80.34**
150mm diameter	45.87	59.40	1.75	29.21	nr	**88.61**
200mm diameter	61.32	79.41	2.17	36.22	nr	**115.63**
250mm diameter	72.61	94.03	2.17	36.22	nr	**130.25**
300mm diameter	79.55	103.02	2.17	36.22	nr	**139.24**
350mm diameter	87.88	113.80	2.17	36.22	nr	**150.03**
40mm thick						
15mm diameter	5.61	7.26	0.30	5.01	m	**12.27**
20mm diameter	5.80	7.51	0.30	5.01	m	**12.52**
25mm diameter	6.20	8.03	0.30	5.01	m	**13.04**
32mm diameter	6.59	8.53	0.30	5.01	m	**13.54**
40mm diameter	6.91	8.95	0.30	5.01	m	**13.96**
50mm diameter	7.70	9.97	0.30	5.01	m	**14.98**
65mm diameter	8.58	11.11	0.30	5.01	m	**16.12**
80mm diameter	9.30	12.04	0.40	6.68	m	**18.72**
100mm diameter	11.62	15.05	0.40	6.68	m	**21.72**
125mm diameter	13.09	16.95	0.40	6.68	m	**23.63**
150mm diameter	15.11	19.57	0.40	6.68	m	**26.24**
200mm diameter	20.16	26.11	0.50	8.35	m	**34.45**
250mm diameter	23.60	30.56	0.50	8.35	m	**38.91**
300mm diameter	25.56	33.10	0.50	8.35	m	**41.45**
350mm diameter	28.08	36.36	0.50	8.35	m	**44.71**
400mm diameter	31.26	40.48	0.50	8.35	m	**48.83**

T:MECHANICAL/COOLING/HEATING SYSTEMS

Item	Net Price £	Material £	Labour hours	Labour £	Unit	Total rate £
T31 : THERMAL INSULATION						
Y50 -THERMAL INSULATION						
Mineral fibre sectional insulation; Bright Class O foil faced; Bright Class O foil taped joints; 0.8mm Polyisobutylene sheeting; welded joints (Continued)						
Extra over for fittings external insulation						
Flange/union						
15mm diameter	8.38	10.85	0.75	12.52	nr	**23.37**
20mm diameter	8.75	11.33	0.75	12.52	nr	**23.85**
25mm diameter	9.33	12.08	0.75	12.52	nr	**24.60**
32mm diameter	9.83	12.73	0.75	12.52	nr	**25.25**
40mm diameter	10.39	13.46	0.75	12.52	nr	**25.97**
50mm diameter	11.51	14.91	0.75	12.52	nr	**27.42**
65mm diameter	12.82	16.60	0.75	12.52	nr	**29.12**
80mm diameter	13.82	17.90	0.89	14.86	nr	**32.75**
100mm diameter	17.02	22.04	0.89	14.86	nr	**36.90**
125mm diameter	19.03	24.64	0.89	14.86	nr	**39.50**
150mm diameter	22.06	28.57	0.89	14.86	nr	**43.42**
200mm diameter	29.08	37.66	1.15	19.20	nr	**56.86**
250mm diameter	34.08	44.13	1.15	19.20	nr	**63.33**
300mm diameter	37.46	48.51	1.15	19.20	nr	**67.71**
350mm diameter	41.55	53.81	1.15	19.20	nr	**73.00**
400mm diameter	46.53	60.26	1.15	19.20	nr	**79.45**
Bends						
15mm diameter	1.40	1.81	0.30	5.01	nr	**6.82**
20mm diameter	1.45	1.88	0.30	5.01	nr	**6.89**
25mm diameter	1.55	2.01	0.30	5.01	nr	**7.01**
32mm diameter	1.65	2.14	0.30	5.01	nr	**7.14**
40mm diameter	1.73	2.24	0.30	5.01	nr	**7.25**
50mm diameter	1.93	2.50	0.30	5.01	nr	**7.51**
65mm diameter	2.15	2.78	0.30	5.01	nr	**7.79**
80mm diameter	2.33	3.02	0.40	6.68	nr	**9.69**
100mm diameter	2.91	3.77	0.40	6.68	nr	**10.45**
125mm diameter	3.27	4.23	0.40	6.68	nr	**10.91**
150mm diameter	3.78	4.90	0.40	6.68	nr	**11.57**
200mm diameter	5.04	6.53	0.50	8.35	nr	**14.87**
250mm diameter	5.90	7.64	0.50	8.35	nr	**15.99**
300mm diameter	6.39	8.28	0.50	8.35	nr	**16.62**
350mm diameter	7.02	9.09	0.50	8.35	nr	**17.44**
400mm diameter	7.82	10.13	0.50	8.35	nr	**18.47**
Tees						
15mm diameter	1.40	1.81	0.30	5.01	nr	**6.82**
20mm diameter	1.45	1.88	0.30	5.01	nr	**6.89**
25mm diameter	1.55	2.01	0.30	5.01	nr	**7.01**
32mm diameter	1.65	2.14	0.30	5.01	nr	**7.14**
40mm diameter	1.73	2.24	0.30	5.01	nr	**7.25**
50mm diameter	1.93	2.50	0.30	5.01	nr	**7.51**
65mm diameter	2.15	2.78	0.30	5.01	nr	**7.79**
80mm diameter	2.33	3.02	0.40	6.68	nr	**9.69**

Material Costs/Measured Work Prices - Mechanical Installations

T:MECHANICAL/COOLING/HEATING SYSTEMS

Item	Net Price £	Material £	Labour hours	Labour £	Unit	Total rate £
100mm diameter	2.91	3.77	0.40	6.68	nr	**10.45**
125mm diameter	3.27	4.23	0.40	6.68	nr	**10.91**
150mm diameter	3.78	4.90	0.40	6.68	nr	**11.57**
200mm diameter	5.04	6.53	0.50	8.35	nr	**14.87**
250mm diameter	5.90	7.64	0.50	8.35	nr	**15.99**
300mm diameter	6.39	8.28	0.50	8.35	nr	**16.62**
350mm diameter	7.02	9.09	0.50	8.35	nr	**17.44**
400mm diameter	7.82	10.13	0.50	8.35	nr	**18.47**
Valves						
15mm diameter	13.30	17.22	1.03	17.19	nr	**34.42**
20mm diameter	13.90	18.00	1.03	17.19	nr	**35.19**
25mm diameter	14.82	19.19	1.03	17.19	nr	**36.39**
32mm diameter	15.61	20.21	1.03	17.19	nr	**37.41**
40mm diameter	16.50	21.37	1.03	17.19	nr	**38.56**
50mm diameter	18.28	23.67	1.03	17.19	nr	**40.87**
65mm diameter	20.36	26.37	1.03	17.19	nr	**43.56**
80mm diameter	21.94	28.41	1.25	20.87	nr	**49.28**
100mm diameter	27.03	35.00	1.25	20.87	nr	**55.87**
125mm diameter	30.23	39.15	1.25	20.87	nr	**60.01**
150mm diameter	35.04	45.38	1.25	20.87	nr	**66.24**
200mm diameter	46.19	59.82	1.55	25.87	nr	**85.69**
250mm diameter	54.12	70.09	1.55	25.87	nr	**95.96**
300mm diameter	59.50	77.05	1.55	25.87	nr	**102.93**
350mm diameter	65.99	85.46	1.55	25.87	nr	**111.33**
400mm diameter	73.89	95.69	1.55	25.87	nr	**121.56**
Pumps						
15mm diameter	24.64	31.91	3.10	51.75	nr	**83.66**
20mm diameter	25.74	33.33	3.10	51.75	nr	**85.08**
25mm diameter	27.44	35.53	3.10	51.75	nr	**87.28**
32mm diameter	28.90	37.43	3.10	51.75	nr	**89.17**
40mm diameter	30.55	39.56	3.10	51.75	nr	**91.31**
50mm diameter	33.86	43.85	3.10	51.75	nr	**95.60**
65mm diameter	37.70	48.82	3.10	51.75	nr	**100.57**
80mm diameter	40.64	52.63	3.75	62.60	nr	**115.23**
100mm diameter	50.06	64.83	3.75	62.60	nr	**127.42**
125mm diameter	55.98	72.49	3.75	62.60	nr	**135.09**
150mm diameter	64.90	84.05	3.75	62.60	nr	**146.64**
200mm diameter	85.54	110.77	4.65	77.62	nr	**188.39**
250mm diameter	100.22	129.78	4.65	77.62	nr	**207.41**
300mm diameter	110.19	142.70	4.65	77.62	nr	**220.32**
350mm diameter	122.21	158.26	4.65	77.62	nr	**235.88**
400mm diameter	136.84	177.21	4.65	77.62	nr	**254.83**
Expansion Bellows						
15mm diameter	19.71	25.52	1.42	23.70	nr	**49.23**
20mm diameter	20.59	26.66	1.42	23.70	nr	**50.37**
25mm diameter	21.95	28.43	1.42	23.70	nr	**52.13**
32mm diameter	23.12	29.94	1.42	23.70	nr	**53.64**
40mm diameter	24.44	31.65	1.42	23.70	nr	**55.35**
50mm diameter	27.09	35.08	1.42	23.70	nr	**58.79**
65mm diameter	30.16	39.06	1.42	23.70	nr	**62.76**
80mm diameter	32.51	42.10	1.75	29.21	nr	**71.31**

Material Costs/Measured Work Prices - Mechanical Installations

T:MECHANICAL/COOLING/HEATING SYSTEMS

Item	Net Price £	Material £	Labour hours	Labour £	Unit	Total rate £
T31 : THERMAL INSULATION						
Y50 -THERMAL INSULATION						
Mineral fibre sectional insulation; Bright Class O foil faced; Bright Class O foil taped joints; 0.8mm Polyisobutylene sheeting; welded joints (Continued)						
Expansion Bellows						
100mm diameter	40.05	51.86	1.75	29.21	nr	**81.08**
125mm diameter	44.78	57.99	1.75	29.21	nr	**87.20**
150mm diameter	51.92	67.24	1.75	29.21	nr	**96.45**
200mm diameter	68.43	88.62	2.17	36.22	nr	**124.84**
250mm diameter	80.18	103.83	2.17	36.22	nr	**140.06**
300mm diameter	88.15	114.15	2.17	36.22	nr	**150.38**
350mm diameter	97.77	126.61	2.17	36.22	nr	**162.84**
400mm diameter	109.47	141.76	2.17	36.22	nr	**177.99**
50mm thick						
15mm diameter	7.39	9.57	0.30	5.01	m	**14.58**
20mm diameter	7.75	10.04	0.30	5.01	m	**15.04**
25mm diameter	8.20	10.62	0.30	5.01	m	**15.63**
32mm diameter	8.85	11.46	0.30	5.01	m	**16.47**
40mm diameter	8.99	11.64	0.30	5.01	m	**16.65**
50mm diameter	9.96	12.90	0.30	5.01	m	**17.91**
65mm diameter	10.86	14.06	0.30	5.01	m	**19.07**
80mm diameter	11.62	15.05	0.40	6.68	m	**21.72**
100mm diameter	14.38	18.62	0.40	6.68	m	**25.30**
125mm diameter	16.09	20.84	0.40	6.68	m	**27.51**
150mm diameter	18.40	23.83	0.40	6.68	m	**30.50**
200mm diameter	24.37	31.56	0.50	8.35	m	**39.91**
250mm diameter	28.14	36.44	0.50	8.35	m	**44.79**
300mm diameter	30.18	39.08	0.50	8.35	m	**47.43**
350mm diameter	33.12	42.89	0.50	8.35	m	**51.24**
400mm diameter	36.69	47.51	0.50	8.35	m	**55.86**
Extra over for fittings external insulation						
Flange/union						
15mm diameter	10.89	14.10	0.75	12.52	nr	**26.62**
20mm diameter	11.46	14.84	0.75	12.52	nr	**27.36**
25mm diameter	12.11	15.68	0.75	12.52	nr	**28.20**
32mm diameter	12.72	16.47	0.75	12.52	nr	**28.99**
40mm diameter	13.30	17.22	0.75	12.52	nr	**29.74**
50mm diameter	14.67	19.00	0.75	12.52	nr	**31.52**
65mm diameter	15.86	20.54	0.75	12.52	nr	**33.06**
80mm diameter	17.00	22.02	0.89	14.86	nr	**36.87**
100mm diameter	20.90	27.07	0.89	14.86	nr	**41.92**
125mm diameter	23.23	30.08	0.89	14.86	nr	**44.94**
150mm diameter	26.60	34.45	0.89	14.86	nr	**49.30**
200mm diameter	34.80	45.07	1.15	19.20	nr	**64.26**
250mm diameter	40.22	52.08	1.15	19.20	nr	**71.28**

Material Costs/Measured Work Prices - Mechanical Installations 265

T:MECHANICAL/COOLING/HEATING SYSTEMS

Item	Net Price £	Material £	Labour hours	Labour £	Unit	Total rate £
300mm diameter	43.63	56.50	1.15	19.20	nr	**75.70**
350mm diameter	48.28	62.52	1.15	19.20	nr	**81.72**
400mm diameter	53.80	69.67	1.15	19.20	nr	**88.87**
Bend						
15mm diameter	1.85	2.40	0.30	5.01	nr	**7.40**
20mm diameter	1.94	2.51	0.30	5.01	nr	**7.52**
25mm diameter	2.05	2.65	0.30	5.01	nr	**7.66**
32mm diameter	2.15	2.78	0.30	5.01	nr	**7.79**
40mm diameter	2.25	2.91	0.30	5.01	nr	**7.92**
50mm diameter	2.49	3.22	0.30	5.01	nr	**8.23**
65mm diameter	2.72	3.52	0.30	5.01	nr	**8.53**
80mm diameter	2.90	3.76	0.40	6.68	nr	**10.43**
100mm diameter	3.59	4.65	0.40	6.68	nr	**11.33**
125mm diameter	4.02	5.21	0.40	6.68	nr	**11.88**
150mm diameter	4.60	5.96	0.40	6.68	nr	**12.63**
200mm diameter	6.09	7.89	0.50	8.35	nr	**16.23**
250mm diameter	7.04	9.12	0.50	8.35	nr	**17.46**
300mm diameter	7.55	9.78	0.50	8.35	nr	**18.12**
350mm diameter	8.28	10.72	0.50	8.35	nr	**19.07**
400mm diameter	9.17	11.88	0.50	8.35	nr	**20.22**
Tee						
15mm diameter	1.85	2.40	0.30	5.01	nr	**7.40**
20mm diameter	1.94	2.51	0.30	5.01	nr	**7.52**
25mm diameter	2.05	2.65	0.30	5.01	nr	**7.66**
32mm diameter	2.15	2.78	0.30	5.01	nr	**7.79**
40mm diameter	2.25	2.91	0.30	5.01	nr	**7.92**
50mm diameter	2.49	3.22	0.30	5.01	nr	**8.23**
65mm diameter	2.72	3.52	0.30	5.01	nr	**8.53**
80mm diameter	2.90	3.76	0.40	6.68	nr	**10.43**
100mm diameter	3.59	4.65	0.40	6.68	nr	**11.33**
125mm diameter	4.02	5.21	0.40	6.68	nr	**11.88**
150mm diameter	4.60	5.96	0.40	6.68	nr	**12.63**
200mm diameter	6.09	7.89	0.50	8.35	nr	**16.23**
250mm diameter	7.04	9.12	0.50	8.35	nr	**17.46**
300mm diameter	7.55	9.78	0.50	8.35	nr	**18.12**
350mm diameter	8.28	10.72	0.50	8.35	nr	**19.07**
400mm diameter	9.17	11.88	0.50	8.35	nr	**20.22**
Valves						
15mm diameter	17.30	22.40	1.03	17.19	nr	**39.60**
20mm diameter	18.20	23.57	1.03	17.19	nr	**40.76**
25mm diameter	19.24	24.92	1.03	17.19	nr	**42.11**
32mm diameter	20.20	26.16	1.03	17.19	nr	**43.35**
40mm diameter	21.13	27.36	1.03	17.19	nr	**44.56**
50mm diameter	23.30	30.17	1.03	17.19	nr	**47.37**
65mm diameter	25.20	32.63	1.03	17.19	nr	**49.83**
80mm diameter	27.00	34.97	1.25	20.87	nr	**55.83**
100mm diameter	33.20	42.99	1.25	20.87	nr	**63.86**
125mm diameter	36.90	47.79	1.25	20.87	nr	**68.65**
150mm diameter	42.24	54.70	1.25	20.87	nr	**75.57**
200mm diameter	55.27	71.57	1.55	25.87	nr	**97.45**
250mm diameter	63.87	82.71	1.55	25.87	nr	**108.59**

Material Costs/Measured Work Prices - Mechanical Installations

T:MECHANICAL/COOLING/HEATING SYSTEMS

Item	Net Price £	Material £	Labour hours	Labour £	Unit	Total rate £
T31 : THERMAL INSULATION						
Y50 -THERMAL INSULATION						
Mineral fibre sectional insulation; Bright Class O foil faced; Bright Class O foil taped joints; 0.8mm Polyisobutylene sheeting; welded joints (Continued)						
Valves						
300mm diameter	69.29	89.73	1.55	25.87	nr	**115.60**
350mm diameter	76.68	99.30	1.55	25.87	nr	**125.17**
400mm diameter	85.44	110.64	1.55	25.87	nr	**136.52**
Pumps						
15mm diameter	32.04	41.49	3.10	51.75	nr	**93.24**
20mm diameter	33.70	43.64	3.10	51.75	nr	**95.39**
25mm diameter	35.62	46.13	3.10	51.75	nr	**97.88**
32mm diameter	37.41	48.45	3.10	51.75	nr	**100.19**
40mm diameter	39.13	50.67	3.10	51.75	nr	**102.42**
50mm diameter	43.16	55.89	3.10	51.75	nr	**107.64**
65mm diameter	46.66	60.42	3.10	51.75	nr	**112.17**
80mm diameter	50.01	64.76	3.75	62.60	nr	**127.36**
100mm diameter	61.47	79.60	3.75	62.60	nr	**142.20**
125mm diameter	68.33	88.49	3.75	62.60	nr	**151.08**
150mm diameter	78.22	101.29	3.75	62.60	nr	**163.89**
200mm diameter	102.35	132.54	4.65	77.62	nr	**210.16**
250mm diameter	118.28	153.17	4.65	77.62	nr	**230.79**
300mm diameter	128.32	166.17	4.65	77.62	nr	**243.79**
350mm diameter	142.00	183.89	4.65	77.62	nr	**261.51**
400mm diameter	158.23	204.91	4.65	77.62	nr	**282.53**
Expansion Bellows						
15mm diameter	25.63	33.19	1.42	23.70	nr	**56.89**
20mm diameter	26.96	34.91	1.42	23.70	nr	**58.62**
25mm diameter	28.50	36.91	1.42	23.70	nr	**60.61**
32mm diameter	29.93	38.76	1.42	23.70	nr	**62.46**
40mm diameter	31.30	40.53	1.42	23.70	nr	**64.24**
50mm diameter	34.52	44.70	1.42	23.70	nr	**68.41**
65mm diameter	37.33	48.34	1.42	23.70	nr	**72.05**
80mm diameter	40.01	51.81	1.75	29.21	nr	**81.03**
100mm diameter	49.18	63.69	1.75	29.21	nr	**92.90**
125mm diameter	54.67	70.80	1.75	29.21	nr	**100.01**
150mm diameter	62.58	81.04	1.75	29.21	nr	**110.25**
200mm diameter	81.88	106.03	2.17	36.22	nr	**142.26**
250mm diameter	94.62	122.53	2.17	36.22	nr	**158.76**
300mm diameter	102.66	132.94	2.17	36.22	nr	**169.17**
350mm diameter	113.60	147.11	2.17	36.22	nr	**183.33**
400mm diameter	126.58	163.92	2.17	36.22	nr	**200.14**

Material Costs/Measured Work Prices - Mechanical Installations 267

T:MECHANICAL/COOLING/HEATING SYSTEMS

Item	Net Price £	Material £	Labour hours	Labour £	Unit	Total rate £
T33 : STEAM HEATING						
Y10 - PIPELINES						
For prices for pipework refer to Section						
T31 - Low Temperature Hot Water						
Heating						
Y11 - PIPELINE ANCILLARIES						
Steam traps and accessories						
Cast iron; inverted bucket type; steam trap						
pressure range up to 17 bar at 210 degree						
celsius; screwed ends						
1/2" dia.	79.78	96.88	0.85	14.96	nr	**111.84**
3/4" dia.	117.12	142.22	1.13	19.89	nr	**162.11**
1" dia.	183.32	222.61	1.35	23.76	nr	**246.37**
11/2" dia.	339.49	412.24	1.80	31.68	nr	**443.92**
2" dia.	523.66	635.88	2.18	38.37	nr	**674.25**
Cast iron; inverted bucket type; steam trap						
pressure range up to 17 bar at 210 degree						
celsius; flanged ends (BS 10 table H); bolted						
connections						
15mm dia.	191.81	232.92	1.15	20.24	nr	**253.16**
20mm dia.	223.21	271.05	1.25	22.00	nr	**293.05**
25mm dia.	342.03	415.33	1.33	23.41	nr	**438.74**
40mm dia.	528.75	642.06	1.46	25.70	nr	**667.76**
50mm dia.	645.03	783.26	1.60	28.16	nr	**811.42**
Steam traps and strainers						
Stainless steel; thermodynamic trap with pressure						
range up to 42 bar; temperature range to 400						
degree celsius; screwed ends to steel						
15mm dia.	56.44	68.54	0.84	14.79	nr	**83.33**
20mm dia.	82.75	100.48	1.14	20.07	nr	**120.55**
Stainless steel; thermodynamic trap with pressure						
range up to 24 bar; temperature range to 288						
degree celsius; flanged ends (BS 10, Table H);						
bolted connections						
15mm dia.	214.73	260.74	1.24	21.84	nr	**282.58**
20mm dia.	219.82	266.93	1.34	23.59	nr	**290.52**
25mm dia.	237.64	288.57	1.40	24.65	nr	**313.22**
Malleable iron pipeline strainer; max steam						
working pressure 14 bar and temperature range						
to 230 degree celsius; screwed ends to steel						
1/2" dia.	8.19	9.95	0.84	14.79	nr	**24.74**
3/4" dia.	10.95	13.30	1.14	20.07	nr	**33.37**
1" dia.	16.17	19.63	1.30	22.89	nr	**42.52**
11/2" dia.	26.78	32.52	1.50	26.43	nr	**58.95**
2" dia.	47.96	58.23	1.74	30.66	nr	**88.90**

Material Costs/Measured Work Prices - Mechanical Installations

T:MECHANICAL/COOLING/HEATING SYSTEMS

Item	Net Price £	Material £	Labour hours	Labour £	Unit	Total rate £
T33 : STEAM HEATING						
Y11 - PIPELINE ANCILLARIES (Continued)						
Bronze pipeline strainer; max steam working pressure 25 bar; flanged ends (BS 19, Table H); bolted connections						
15mm dia.	97.60	118.52	1.24	21.84	nr	**140.36**
20mm dia.	117.97	143.25	1.34	23.59	nr	**166.85**
25mm dia.	135.80	164.90	1.40	24.65	nr	**189.55**
32mm dia.	210.48	255.59	1.46	25.73	nr	**281.32**
40mm dia.	238.49	289.60	1.54	27.12	nr	**316.72**
50mm dia.	367.50	446.25	1.64	28.90	nr	**475.15**
65mm dia.	406.54	493.66	2.50	44.00	nr	**537.66**
80mm dia.	506.69	615.27	2.91	51.17	nr	**666.43**
100mm dia.	877.58	1065.64	3.51	61.76	nr	**1127.40**
Balanced pressure thermostatic steam trap and strainer; max working pressure up to 13 bar; screwed ends to steel						
1/2" dia.	36.37	44.17	1.26	22.19	nr	**66.36**
3/4" dia.	39.30	47.72	1.71	30.14	nr	**77.86**
Bimetallic thermostatic steam trap and strainer; max working pressure up to 21 bar; flanged ends						
15mm	118.82	144.28	1.24	21.84	nr	**166.12**
20mm	118.82	144.28	1.34	23.59	nr	**167.88**
Sight glasses						
Pressed brass; straight; single window; screwed ends to steel						
15mm dia.	24.20	29.39	0.84	14.79	nr	**44.18**
20mm dia.	26.90	32.66	1.14	20.07	nr	**52.73**
25mm dia.	33.62	40.82	1.30	22.89	nr	**63.71**
Gunmetal; straight; double window; screwed ends to steel						
15mm dia.	39.04	47.40	0.84	14.79	nr	**62.19**
20mm dia.	42.86	52.05	1.14	20.07	nr	**72.12**
25mm dia.	52.95	64.30	1.30	22.89	nr	**87.19**
32mm dia.	87.41	106.14	1.35	23.79	nr	**129.93**
40mm dia.	87.41	106.14	1.74	30.66	nr	**136.81**
50mm dia.	105.90	128.59	2.08	36.67	nr	**165.26**
SG Iron flanged; BS 4504, PN 25						
15mm dia.	79.85	96.96	1.00	17.60	nr	**114.56**
20mm dia.	94.13	114.31	1.25	22.00	nr	**136.31**
25mm dia.	119.35	144.92	1.50	26.43	nr	**171.35**
32mm dia.	131.96	160.23	1.70	29.93	nr	**190.17**
40mm dia.	173.14	210.24	2.00	35.20	nr	**245.44**
50mm dia.	208.44	253.11	2.30	40.55	nr	**293.66**

Material Costs/Measured Work Prices - Mechanical Installations

T:MECHANICAL/COOLING/HEATING SYSTEMS

Item	Net Price £	Material £	Labour hours	Labour £	Unit	Total rate £
Check valve and sight glass; gun metal; screwed						
15mm dia.	39.54	48.02	0.84	14.79	nr	**62.81**
20mm dia.	41.65	50.57	1.14	20.07	nr	**70.64**
25mm dia.	70.60	85.73	1.30	22.89	nr	**108.62**
Pressure reducing valves						
Pressure reducing valve for steam; maximum range of 17 bar and 232 degree celsius; screwed ends to steel						
15mm dia.	425.39	516.55	0.87	15.31	nr	**531.86**
20mm dia.	449.08	545.32	0.91	16.02	nr	**561.33**
25mm dia.	507.79	616.61	1.35	23.76	nr	**640.37**
Pressure reducing valve for steam; maximum range of 17 bar and 232 degree celsius; flanged ends (BS 10, Table H or BS 4504, Tables 25 and 16)						
25mm dia.	507.79	616.61	1.70	29.92	nr	**646.53**
32mm dia.	577.83	701.66	1.87	32.91	nr	**734.57**
40mm dia.	690.10	837.99	2.12	37.31	nr	**875.30**
50mm dia.	796.19	966.81	2.57	45.23	nr	**1012.05**
Safety and relief valves						
Bronze safety valve; 'pop' type; side outlet; including easing lever; working pressure saturated steam up to 20.7 bar; screwed ends to steel						
15mm dia.	70.45	85.55	0.32	5.63	nr	**91.18**
20mm dia.	86.65	105.22	0.40	7.04	nr	**112.26**
Bronze safety valve; 'pop' type; side outlet; including easing lever; working pressure saturated steam up to 17.2 bar; screwed ends to steel						
25mm dia.	113.32	137.60	0.47	8.27	nr	**145.88**
32mm dia.	152.20	184.82	0.56	9.86	nr	**194.67**
Bronze safety valve; 'pop' type; side outlet; including easing lever; working pressure saturated steam up to 13.8 bar; screwed ends to steel						
40mm dia.	211.75	257.13	0.64	11.26	nr	**268.39**
50mm dia.	280.04	340.05	0.76	13.38	nr	**353.43**
65mm dia.	408.86	496.48	0.94	16.54	nr	**513.02**
80mm dia.	458.64	556.93	1.10	19.36	nr	**576.29**

270   *Material Costs/Measured Work Prices - Mechanical Installations*

T:MECHANICAL/COOLING/HEATING SYSTEMS

Item	Net Price £	Material £	Labour hours	Labour £	Unit	Total rate £
T33 : STEAM HEATING						
EQUIPMENT						
Y23 - CALORIFIERS						
Non-storage calorifiers; mild steel; heater battery duty 82.71 degrees C to BS 853, maximum test on shell 11.55 bar, tubes 26.25 bar						
Horizontal; steam at 3.2 bar						
88 kW capacity	395.82	480.64	8.00	140.81	nr	**621.45**
176 kW capacity	593.73	720.97	12.05	212.06	nr	**933.02**
293 kW capacity	622.01	755.31	14.08	247.90	nr	**1003.21**
586 kW capacity	814.27	988.77	37.04	651.88	nr	**1640.65**
879 kW capacity	1012.17	1229.08	40.00	704.03	nr	**1933.11**
1465 kW capacity	1611.56	1956.92	45.45	800.04	nr	**2756.95**
Horizontal; steam at 4.8 bar						
88 kW capacity	395.82	480.64	5.00	88.00	nr	**568.65**
176 kW capacity	593.73	720.97	8.00	140.81	nr	**861.77**
293 kW capacity	616.35	748.43	12.05	212.06	nr	**960.49**
586 kW capacity	802.96	975.03	22.22	391.13	nr	**1366.16**
879 kW capacity	1000.86	1215.34	28.57	502.88	nr	**1718.22**
1465 kW capacity	1611.56	1956.92	40.00	704.03	nr	**2660.95**
Vertical; steam at 3.2 bar						
88 kW capacity	407.13	494.38	8.00	140.81	nr	**635.18**
176 kW capacity	610.70	741.57	12.05	212.06	nr	**953.63**
293 kW capacity	644.63	782.77	14.08	247.90	nr	**1030.67**
586 kW capacity	809.62	983.12	37.04	651.88	nr	**1635.00**
879 kW capacity	1017.83	1235.95	40.00	704.03	nr	**1939.98**
1465 kW capacity	1639.83	1991.25	45.45	800.04	nr	**2791.28**
Vertical; steam at 4.8 bar						
88 kW capacity	407.13	494.38	5.00	88.00	nr	**582.38**
176 kW capacity	610.70	741.57	12.05	212.06	nr	**953.63**
293 kW capacity	644.63	782.77	12.05	212.06	nr	**994.83**
586 kW capacity	792.11	961.86	22.22	391.13	nr	**1352.99**
879 kW capacity	1017.83	1235.95	28.57	502.88	nr	**1738.83**
1465 kW capacity	1639.83	1991.25	40.00	704.03	nr	**2695.28**

Material Costs/Measured Work Prices - Mechanical Installations

T:MECHANICAL/COOLING/HEATING SYSTEMS

Item	Net Price £	Material £	Labour hours	Labour £	Unit	Total rate £
T42 : LOCAL HEATING UNITS						
Unit heater; horizontal or vertical discharge; recirculating type for industrial and commercial user for heights up to 3m, normal speed; EAT 15C; fixed to existing suspension rods; complete with enclosures; includes connections or hot water services; electrical work included elsewhere						
Low pressure hot water						
7.5 kW 265 l/sec	283.45	344.19	6.53	114.93	nr	**459.13**
15.4 kW 575 l/sec	343.12	416.65	7.54	132.71	nr	**549.36**
26.9 kW 1040 l/sec	464.95	564.59	8.65	152.25	nr	**716.84**
48.0 kW 1620 l/sec	612.88	744.22	9.35	164.57	nr	**908.79**
Steam, 2 Bar						
9.2 kW 265 l/sec	405.27	492.12	6.53	114.93	nr	**607.05**
18.8 kW 575 l/sec	438.84	532.88	6.82	120.04	nr	**652.92**
34.8 kW 1040 l/sec	504.73	612.89	6.82	120.04	nr	**732.93**
51.6 kW 1625 l/sec	683.76	830.29	7.10	124.97	nr	**955.26**

Material Costs/Measured Work Prices - Mechanical Installations

T:MECHANICAL/COOLING/HEATING SYSTEMS

Item	Net Price £	Material £	Labour hours	Labour £	Unit	Total rate £
T61 : CHILLED WATER						
Y10 - PIPELINES						
For prices for pipework refer to Section T31 - Low Temperature Hot Water Heating						
Y11 - PIPELINE ANCILLARIES						
For prices for ancillaries refer to Section T31 - Low Temperature Hot Water Heating						
Y24 - TRACE HEATING						
Trace Heating; for freeze protection or temperature maintainance of pipework; to BS 6351; including fixing to parent structures by plastic pull ties.						
Straight laid						
15mm	17.97	21.82	0.27	4.75	m	**26.57**
25mm	17.97	21.82	0.27	4.75	m	**26.57**
28mm	17.97	21.82	0.27	4.75	m	**26.57**
32mm	17.97	21.82	0.30	5.28	m	**27.10**
35mm	17.97	21.82	0.31	5.46	m	**27.28**
50mm	17.97	21.82	0.34	5.98	m	**27.81**
100mm	17.97	21.82	0.40	7.04	m	**28.86**
150mm	17.97	21.82	0.40	7.04	m	**28.86**
Helically Wound						
15mm	27.56	33.47	1.00	17.60	m	**51.07**
25mm	27.56	33.47	1.00	17.60	m	**51.07**
28mm	27.56	33.47	1.00	17.60	m	**51.07**
32mm	27.56	33.47	1.00	17.60	m	**51.07**
35mm	27.56	33.47	1.00	17.60	m	**51.07**
50mm	27.56	33.47	1.00	17.60	m	**51.07**
100mm	27.56	33.47	1.00	17.60	m	**51.07**
150mm	27.56	33.47	1.00	17.60	m	**51.07**
Accessories for Trace Heating; Weatherproof; polycarbonate enclosure to IP Standards; fully installed.						
Connection Junction Box						
100*100*75mm	29.96	36.38	1.40	24.64	nr	**61.02**
Single Air Thermostat						
150*150*75mm	41.94	50.93	1.42	24.99	nr	**75.92**

Material Costs/Measured Work Prices - Mechanical Installations

T:MECHANICAL/COOLING/HEATING SYSTEMS

Item	Net Price £	Material £	Labour hours	Labour £	Unit	Total rate £
Single Capillary Thermostat						
150*150*75mm	59.91	72.75	1.46	25.70	nr	**98.45**
Twin Capillary Thermostat						
150*150*75mm	77.89	94.58	1.46	25.70	nr	**120.28**
EQUIPMENT						
Y44 - CHILLED BEAMS						
Exposed below ceiling; cooling 0.42kW/m						
Passive	120.94	146.86	3.53	62.13	m	**208.99**
Ventilated/Active	157.70	191.50	3.53	62.13	m	**253.63**
Flush with ceiling; cooling 0.42kW/m						
Passive	120.94	146.86	3.53	62.13	m	**208.99**
Ventilated/Active	157.70	191.50	3.53	62.13	m	**253.63**
LEAK DETECTION						
Leak detection system consisting of a central control module connected by a leader cable to water sensing cables						
Control Modules						
Alarm Only	266.87	324.06	4.00	64.11	nr	**388.17**
Alarm and location	1805.85	2192.84	8.00	128.21	nr	**2321.06**
Cables						
Sensing - 3m length	79.06	96.00	4.00	64.11	nr	**160.11**
Sensing - 7.5m length	109.24	132.65	4.00	64.11	nr	**196.76**
Sensing - 15m length	195.89	237.87	8.00	128.21	nr	**366.08**
Leader - 3.5m length	34.51	41.91	2.00	32.05	nr	**73.96**
End terminal						
End terminal	14.27	17.33	0.05	0.80	nr	**18.13**

Material Costs/Measured Work Prices - Mechanical Installations

T:MECHANICAL/COOLING/HEATING SYSTEMS

Item	Net Price £	Material £	Labour hours	Labour £	Unit	Total rate £
T70 : LOCAL COOLING UNITS						
One piece package; ceiling void unit; electrical work elsewhere						
Low profile; heat pump units; four stage remote control; indoor or outdoor						
Cooling 13.7kW, heating 15.5kW	5691.73	6911.47	40.00	704.03	nr	**7615.50**
Cooling 16.4kW, heating 18.8kW	5831.77	7081.52	40.00	704.03	nr	**7785.55**
Cooling 21.7kW, heating 24.9kW	7237.46	8788.45	40.00	704.03	nr	**9492.48**
Low profile; cooling only units; four stage remote control; indoor or outdoor						
Cooling 7.20kW	1726.37	1899.01	40.00	704.03	nr	**2603.04**
Cooling 10.6kW	2003.59	2203.94	40.00	704.03	nr	**2907.98**
Cooling 12.7kW	2168.06	2384.86	40.00	704.03	nr	**3088.90**
Cooling 20.50kW	3324.49	3656.94	40.00	704.03	nr	**4360.97**
Cooling 25.00kW	3687.92	4056.71	40.00	704.03	nr	**4760.74**
Split systems; ceiling void units; electrical work elsewhere						
Low profile cooling only units; four stage remote control; indoor						
Cooling 2.4kW	1154.63	1402.07	16.60	292.17	nr	**1694.24**
Cooling 3.5kW	1493.50	1813.56	16.60	292.17	nr	**2105.73**
Cooling 5.1kW	1850.91	2247.56	16.60	292.17	nr	**2539.73**
Cooling 6.5kW	2111.50	2563.99	16.60	292.17	nr	**2856.17**
Low profile cooling only units; four stage remote control; outdoor						
Cooling 13.7kW	1155.32	1402.91	16.60	292.17	nr	**1695.08**
Cooling 16.4kW	2586.47	3140.75	16.60	292.17	nr	**3432.92**
Cooling 21.7kW	2800.78	3400.99	16.60	292.17	nr	**3693.16**
Chilled water cassette						
Standard model with remote thermostat / fan speed controller; suitable for 2,3 or 4 way blow; including fascia						
Cooling duty 3.25kW	465.74	565.55	4.15	73.04	nr	**638.59**
Cooling duty 3.75kW	610.02	740.75	4.15	73.04	nr	**813.79**
Cooling duty 5.80kW	725.66	881.17	4.15	73.04	nr	**954.21**
Cooling duty 7.6kW	943.14	1145.25	4.15	73.04	nr	**1218.30**

Material Costs/Measured Work Prices - Mechanical Installations

U:VENTILATION/AIR CONDITIONING SYSTEMS

Item	Net Price £	Material £	Labour hours	Labour £	Unit	Total rate £
U10 : DUCTWORK : CIRCULAR						
Y30 - AIR DUCTLINES						
Galvanised sheet metal DW144 class B spirally wound circular section ductwork; including all necessary stiffeners, joints, couplers in the running length and duct supports						
Straight duct						
80mm dia.	3.45	4.57	0.15	3.27	m	**7.84**
100mm dia.	3.68	4.88	0.87	18.99	m	**23.86**
150mm dia.	4.94	6.55	0.87	18.99	m	**25.53**
200mm dia.	6.27	8.31	0.87	18.99	m	**27.30**
250mm dia.	7.87	10.43	1.21	26.41	m	**36.84**
300mm dia.	9.36	12.40	1.21	26.41	m	**38.81**
355mm dia.	12.59	16.68	1.21	26.41	m	**43.09**
400mm dia.	14.06	18.63	1.21	26.41	m	**45.04**
450mm dia.	15.49	20.52	1.21	26.41	m	**46.93**
500mm dia.	16.90	22.39	1.21	26.41	m	**48.80**
630mm dia.	22.28	29.52	1.39	30.34	m	**59.86**
710mm dia.	25.58	33.89	1.39	30.34	m	**64.23**
800mm dia.	28.71	38.04	1.44	31.43	m	**69.47**
900mm dia.	35.96	47.65	1.46	31.86	m	**79.51**
1000mm dia.	39.63	52.51	1.65	36.01	m	**88.52**
1120mm dia.	82.13	108.82	2.43	53.03	m	**161.86**
1250mm dia.	89.62	118.75	2.43	53.03	m	**171.78**
1400mm dia.	101.16	134.04	2.77	60.45	m	**194.49**
1600mm dia.	113.55	150.45	3.06	66.78	m	**217.24**
Extra over fittings; Circular duct class B						
End cap						
80mm dia.	1.90	2.52	0.15	3.27	nr	**5.79**
100mm dia.	1.91	2.53	0.15	3.27	nr	**5.80**
150mm dia.	2.58	3.42	0.15	3.27	nr	**6.69**
200mm dia.	2.79	3.70	0.20	4.36	nr	**8.06**
250mm dia.	3.99	5.29	0.29	6.33	nr	**11.62**
300mm dia.	4.37	5.79	0.29	6.33	nr	**12.12**
355mm dia.	6.27	8.31	0.44	9.60	nr	**17.91**
400mm dia.	6.32	8.37	0.44	9.60	nr	**17.98**
450mm dia.	6.81	9.02	0.44	9.60	nr	**18.63**
500mm dia.	7.16	9.49	0.44	9.60	nr	**19.09**
630mm dia.	8.17	10.83	0.58	12.66	nr	**23.48**
710mm dia.	10.17	13.48	0.69	15.06	nr	**28.53**
800mm dia.	14.48	19.19	0.81	17.68	nr	**36.86**
900mm dia.	16.47	21.82	0.92	20.08	nr	**41.90**
1000mm dia.	19.79	26.22	1.04	22.70	nr	**48.92**
1120mm dia.	30.58	40.52	1.16	25.32	nr	**65.84**
1250mm dia.	35.70	47.30	1.16	25.32	nr	**72.62**
1400mm dia.	49.97	66.21	1.16	25.32	nr	**91.53**
1600mm dia.	61.52	81.51	1.16	25.32	nr	**106.83**

Material Costs/Measured Work Prices - Mechanical Installations

U:VENTILATION/AIR CONDITIONING SYSTEMS

Item	Net Price £	Material £	Labour hours	Labour £	Unit	Total rate £
U10 : DUCTWORK : CIRCULAR						
Y30 - AIR DUCTLINES						
Galvanised sheet metal DW144 class B spirally wound circular section ductwork; including all necessary stiffeners, joints, couplers in the running length and duct supports (Continued)						
Reducer						
100mm dia.	5.26	6.97	0.29	6.33	nr	13.30
150mm dia.	6.38	8.45	0.29	6.33	nr	14.78
200mm dia.	7.11	9.42	0.44	9.60	nr	19.02
250mm dia.	8.02	10.63	0.58	12.66	nr	23.28
300mm dia.	9.79	12.97	0.58	12.66	nr	25.63
355mm dia.	11.59	15.36	0.87	18.99	nr	34.34
400mm dia.	13.17	17.45	0.87	18.99	nr	36.44
450mm dia.	14.12	18.71	0.87	18.99	nr	37.70
500mm dia.	15.61	20.68	0.87	18.99	nr	39.67
630mm dia.	23.57	31.23	0.87	18.99	nr	50.22
710mm dia.	24.22	32.09	0.96	20.95	nr	53.04
800mm dia.	32.48	43.04	1.06	23.13	nr	66.17
900mm dia.	33.55	44.45	1.16	25.32	nr	69.77
1000mm dia.	35.46	46.98	1.25	27.28	nr	74.27
1120mm dia.	73.31	97.14	3.47	75.73	nr	172.87
1250mm dia.	80.83	107.10	3.47	75.73	nr	182.83
1400mm dia.	110.00	145.75	4.05	88.39	nr	234.14
1600mm dia.	114.45	151.65	4.62	100.83	nr	252.48
90 degree segmented radius bend						
80mm dia.	2.66	3.52	0.29	6.33	nr	9.85
100mm dia.	4.83	6.40	0.29	6.33	nr	12.73
150mm dia.	4.83	6.40	0.29	6.33	nr	12.73
200mm dia.	6.71	8.89	0.44	9.60	nr	18.49
250mm dia.	9.57	12.68	0.58	12.66	nr	25.34
300mm dia.	10.31	13.66	0.58	12.66	nr	26.32
355mm dia.	10.87	14.40	0.87	18.99	nr	33.39
400mm dia.	13.21	17.50	0.87	18.99	nr	36.49
450mm dia.	14.20	18.82	0.87	18.99	nr	37.80
500mm dia.	15.33	20.31	0.87	18.99	nr	39.30
630mm dia.	20.31	26.91	0.87	18.99	nr	45.90
710mm dia.	28.28	37.47	0.96	20.95	nr	58.42
800mm dia.	31.00	41.08	1.06	23.13	nr	64.21
900mm dia.	36.65	48.56	1.16	25.32	nr	73.88
1000mm dia.	49.12	65.08	1.25	27.28	nr	92.36
1120mm dia.	156.27	207.06	3.47	75.73	nr	282.79
1250mm dia.	197.22	261.32	3.47	75.73	nr	337.05
1400mm dia.	262.15	347.35	4.05	88.39	nr	435.74
1600mm dia.	255.09	337.99	4.62	100.83	nr	438.82
45 degree radius bend						
80mm dia.	2.66	3.52	0.29	6.33	nr	9.85
100mm dia.	2.73	3.62	0.29	6.33	nr	9.95
150mm dia.	3.84	5.09	0.29	6.33	nr	11.42
200mm dia.	5.15	6.82	0.40	8.73	nr	15.55

Material Costs/Measured Work Prices - Mechanical Installations

U:VENTILATION/AIR CONDITIONING SYSTEMS

Item	Net Price £	Material £	Labour hours	Labour £	Unit	Total rate £
250mm dia.	7.69	10.19	0.58	12.66	nr	**22.85**
300mm dia.	8.84	11.71	0.58	12.66	nr	**24.37**
355mm dia.	9.22	12.22	0.87	18.99	nr	**31.20**
400mm dia.	11.35	15.04	0.87	18.99	nr	**34.03**
450mm dia.	11.69	15.49	0.87	18.99	nr	**34.48**
500mm dia.	12.41	16.44	0.87	18.99	nr	**35.43**
630mm dia.	16.42	21.76	0.87	18.99	nr	**40.74**
710mm dia.	22.06	29.23	0.96	20.95	nr	**50.18**
800mm dia.	26.32	34.87	1.06	23.13	nr	**58.01**
900mm dia.	30.08	39.86	1.16	25.32	nr	**65.17**
1000mm dia.	41.29	54.71	1.25	27.28	nr	**81.99**
1120mm dia.	138.78	183.88	3.47	75.73	nr	**259.61**
1250mm dia.	151.59	200.86	3.47	75.73	nr	**276.59**
1400mm dia.	216.03	286.24	4.05	88.39	nr	**374.63**
1600mm dia.	255.09	337.99	4.62	100.83	nr	**438.82**
90 degree equal twin bend						
80mm dia.	13.99	18.54	0.58	12.66	nr	**31.20**
100mm dia.	14.31	18.96	0.58	12.66	nr	**31.62**
150mm dia.	19.53	25.88	0.58	12.66	nr	**38.54**
200mm dia.	25.23	33.43	0.87	18.99	nr	**52.42**
250mm dia.	35.59	47.16	1.16	25.32	nr	**72.47**
300mm dia.	38.00	50.35	1.16	25.32	nr	**75.67**
355mm dia.	43.52	57.66	1.73	37.76	nr	**95.42**
400mm dia.	46.43	61.52	1.73	37.76	nr	**99.28**
450mm dia.	49.69	65.84	1.73	37.76	nr	**103.60**
500mm dia.	54.12	71.71	1.73	37.76	nr	**109.47**
630mm dia.	79.02	104.70	1.73	37.76	nr	**142.46**
710mm dia.	105.20	139.39	1.82	39.72	nr	**179.11**
800mm dia.	128.92	170.82	1.93	42.12	nr	**212.94**
900mm dia.	153.46	203.33	2.02	44.09	nr	**247.42**
1000mm dia.	202.53	268.35	2.11	46.05	nr	**314.40**
1120mm dia.	347.54	460.49	4.62	100.83	nr	**561.32**
1250mm dia.	382.77	507.17	4.62	100.83	nr	**608.00**
1400mm dia.	647.73	858.24	4.62	100.83	nr	**959.07**
1600mm dia.	650.02	861.28	4.62	100.83	nr	**962.11**
Conical branch						
80mm dia.	14.77	19.57	0.58	12.66	nr	**32.23**
100mm dia.	14.73	19.52	0.58	12.66	nr	**32.18**
150mm dia.	16.64	22.05	0.58	12.66	nr	**34.71**
200mm dia.	16.66	22.07	0.87	18.99	nr	**41.06**
250mm dia.	19.89	26.35	1.16	25.32	nr	**51.67**
300mm dia.	21.87	28.98	1.16	25.32	nr	**54.29**
355mm dia.	25.27	33.48	1.73	37.76	nr	**71.24**
400mm dia.	24.73	32.77	1.73	37.76	nr	**70.52**
450mm dia.	29.28	38.80	1.73	37.76	nr	**76.55**
500mm dia.	30.51	40.43	1.73	37.76	nr	**78.18**
630mm dia.	43.84	58.09	1.73	37.76	nr	**95.84**
710mm dia.	53.73	71.19	1.82	39.72	nr	**110.91**
800mm dia.	65.62	86.95	1.93	42.12	nr	**129.07**
900mm dia.	71.85	95.20	2.02	44.09	nr	**139.29**
1000mm dia.	81.73	108.29	2.11	46.05	nr	**154.34**
1120mm dia.	135.84	179.99	4.62	100.83	nr	**280.82**

Material Costs/Measured Work Prices - Mechanical Installations

U:VENTILATION/AIR CONDITIONING SYSTEMS

Item	Net Price £	Material £	Labour hours	Labour £	Unit	Total rate £
U10 : DUCTWORK : CIRCULAR						
Y30 - AIR DUCTLINES						
Galvanised sheet metal DW144 class B spirally wound circular section ductwork; including all necessary stiffeners, joints, couplers in the running length and duct supports (Continued)						
1250mm dia.	135.84	179.99	5.20	113.49	nr	**293.48**
1400mm dia .	164.97	218.59	5.20	113.49	nr	**332.07**
1600mm dia.	194.75	258.04	5.20	113.49	nr	**371.53**
45 degree branch						
80mm dia.	11.01	14.59	0.58	12.66	nr	**27.25**
100mm dia.	10.98	14.55	0.58	12.66	nr	**27.21**
150mm dia.	14.36	19.03	0.58	12.66	nr	**31.69**
200mm dia.	15.04	19.93	1.16	25.32	nr	**45.24**
250mm dia.	15.77	20.90	1.16	25.32	nr	**46.21**
300mm dia.	16.80	22.26	1.16	25.32	nr	**47.58**
355mm dia.	21.56	28.57	1.73	37.76	nr	**66.32**
400mm dia.	21.38	28.33	1.73	37.76	nr	**66.09**
450mm dia.	22.69	30.06	1.73	37.76	nr	**67.82**
500mm dia.	24.14	31.99	1.73	37.76	nr	**69.74**
630mm dia.	32.43	42.97	1.73	37.76	nr	**80.73**
710mm dia.	38.37	50.84	1.82	39.72	nr	**90.56**
800mm dia.	53.33	70.66	1.93	42.12	nr	**112.78**
900mm dia.	61.47	81.45	2.31	50.41	nr	**131.86**
1000mm dia.	67.89	89.95	2.31	50.41	nr	**140.37**
1120mm dia.	102.43	135.72	4.62	100.83	nr	**236.55**
1250mm dia.	110.75	146.74	4.62	100.83	nr	**247.57**
1400mm dia.	127.45	168.87	4.62	100.83	nr	**269.70**
1600mm dia.	157.89	209.20	4.62	100.83	nr	**310.03**
Galvanised sheet metal DW144 class C spirally wound circular section ductwork; including all necessary stiffeners, joints, couplers in the running length and duct supports						
Straight duct						
80mm dia.	3.45	4.57	0.87	18.99	m	**23.56**
100mm dia.	3.68	4.88	0.87	18.99	m	**23.86**
150mm dia.	4.94	6.55	0.87	18.99	m	**25.53**
200mm dia.	6.27	8.31	0.87	18.99	m	**27.30**
250mm dia.	7.87	10.43	1.21	26.41	m	**36.84**
300mm dia.	9.36	12.40	1.21	26.41	m	**38.81**
355mm dia.	12.59	16.68	1.21	26.41	m	**43.09**
400mm dia.	14.06	18.63	1.21	26.41	m	**45.04**
450mm dia.	15.49	20.52	1.21	26.41	m	**46.93**
500mm dia.	16.90	22.39	1.21	26.41	m	**48.80**
630mm dia.	22.28	29.52	1.39	30.34	m	**59.86**
710mm dia.	25.58	33.89	1.39	30.34	m	**64.23**
800mm dia.	28.71	38.04	1.44	31.43	m	**69.47**
900mm dia.	35.96	47.65	1.46	31.86	m	**79.51**

Material Costs/Measured Work Prices - Mechanical Installations 279

U:VENTILATION/AIR CONDITIONING SYSTEMS

Item	Net Price £	Material £	Labour hours	Labour £	Unit	Total rate £
1000mm dia.	39.63	52.51	1.65	36.01	m	**88.52**
1120mm dia.	82.13	108.82	2.43	53.03	m	**161.86**
1250mm dia.	89.62	118.75	2.43	53.03	m	**171.78**
1400mm dia.	101.16	134.04	2.77	60.45	m	**194.49**
1600mm dia.	113.55	150.45	3.06	66.78	m	**217.24**
Extra over fittings; Circular duct class C						
End cap						
80mm dia.	1.90	2.52	0.15	3.27	nr	**5.79**
100mm dia.	1.91	2.53	0.15	3.27	nr	**5.80**
150mm dia.	2.58	3.42	0.15	3.27	nr	**6.69**
200mm dia.	2.79	3.70	0.20	4.36	nr	**8.06**
250mm dia.	3.99	5.29	0.29	6.33	nr	**11.62**
300mm dia.	4.37	5.79	0.29	6.33	nr	**12.12**
355mm dia.	6.27	8.31	0.44	9.60	nr	**17.91**
400mm dia.	6.32	8.37	0.44	9.60	nr	**17.98**
450mm dia.	6.81	9.02	0.44	9.60	nr	**18.63**
500mm dia.	7.16	9.49	0.44	9.60	nr	**19.09**
630mm dia.	8.17	10.83	0.58	12.66	nr	**23.48**
710mm dia.	10.17	13.48	0.69	15.06	nr	**28.53**
800mm dia.	14.48	19.19	0.81	17.68	nr	**36.86**
900mm dia.	16.47	21.82	0.92	20.08	nr	**41.90**
1000mm dia.	19.79	26.22	1.04	22.70	nr	**48.92**
1120mm dia.	30.58	40.52	1.16	25.32	nr	**65.84**
1250mm dia.	35.70	47.30	1.16	25.32	nr	**72.62**
1400mm dia.	49.97	66.21	1.16	25.32	nr	**91.53**
1600mm dia.	61.52	81.51	1.16	25.32	nr	**106.83**
Reducer						
100mm dia.	5.26	6.97	0.29	6.33	nr	**13.30**
150mm dia.	6.38	8.45	0.29	6.33	nr	**14.78**
200mm dia.	7.11	9.42	0.44	9.60	nr	**19.02**
250mm dia.	8.02	10.63	0.58	12.66	nr	**23.28**
300mm dia.	9.79	12.97	0.58	12.66	nr	**25.63**
355mm dia.	11.59	15.36	0.87	18.99	nr	**34.34**
400mm dia.	13.17	17.45	0.87	18.99	nr	**36.44**
450mm dia.	14.12	18.71	0.87	18.99	nr	**37.70**
500mm dia.	15.61	20.68	0.87	18.99	nr	**39.67**
630mm dia.	23.57	31.23	0.87	18.99	nr	**50.22**
710mm dia.	24.22	32.09	0.96	20.95	nr	**53.04**
800mm dia.	32.48	43.04	1.06	23.13	nr	**66.17**
900mm dia.	33.55	44.45	1.16	25.32	nr	**69.77**
1000mm dia.	35.46	46.98	1.25	27.28	nr	**74.27**
1120mm dia.	73.31	97.14	3.47	75.73	nr	**172.87**
1250mm dia.	80.83	107.10	3.47	75.73	nr	**182.83**
1400mm dia.	110.00	145.75	4.05	88.39	nr	**234.14**
1600mm dia.	114.45	151.65	4.62	100.83	nr	**252.48**
90 degree segmented radius bend						
80mm dia.	2.66	3.52	0.29	6.33	nr	**9.85**
100mm dia.	2.79	3.70	0.29	6.33	nr	**10.03**
150mm dia.	4.83	6.40	0.29	6.33	nr	**12.73**
200mm dia.	6.71	8.89	0.44	9.60	nr	**18.49**

Material Costs/Measured Work Prices - Mechanical Installations

U:VENTILATION/AIR CONDITIONING SYSTEMS

Item	Net Price £	Material £	Labour hours	Labour £	Unit	Total rate £
U10 : DUCTWORK : CIRCULAR						
Y30 - AIR DUCTLINES						
Galvanised sheet metal DW144 class C spirally wound circular section ductwork; including all necessary stiffeners, joints, couplers in the running length and duct supports (Continued)						
90 degree segmented radius bend						
250mm dia.	9.57	12.68	0.58	12.66	nr	**25.34**
300mm dia.	10.31	13.66	0.58	12.66	nr	**26.32**
355mm dia.	10.87	14.40	0.87	18.99	nr	**33.39**
400mm dia.	13.21	17.50	0.87	18.99	nr	**36.49**
450mm dia.	14.20	18.82	0.87	18.99	nr	**37.80**
500mm dia.	15.33	20.31	0.87	18.99	nr	**39.30**
630mm dia.	20.31	26.91	0.87	18.99	nr	**45.90**
710mm dia.	28.28	37.47	0.96	20.95	nr	**58.42**
800mm dia.	31.00	41.08	1.06	23.13	nr	**64.21**
900mm dia.	36.65	48.56	1.16	25.32	nr	**73.88**
1000mm dia.	49.12	65.08	1.25	27.28	nr	**92.36**
1120mm dia.	156.27	207.06	3.47	75.73	nr	**282.79**
1250mm dia.	197.22	261.32	3.47	75.73	nr	**337.05**
1400mm dia.	262.15	347.35	4.05	88.39	nr	**435.74**
1600mm dia.	255.09	337.99	4.62	100.83	nr	**438.82**
45 degree radius bend						
80mm dia.	2.66	3.52	0.29	6.33	nr	**9.85**
100mm dia.	2.73	3.62	0.29	6.33	nr	**9.95**
150mm dia.	3.84	5.09	0.29	6.33	nr	**11.42**
200mm dia.	5.15	6.82	0.44	9.60	nr	**16.43**
250mm dia.	7.69	10.19	0.58	12.66	nr	**22.85**
300mm dia.	8.84	11.71	0.58	12.66	nr	**24.37**
355mm dia.	9.22	12.22	0.87	18.99	nr	**31.20**
400mm dia.	11.35	15.04	0.87	18.99	nr	**34.03**
450mm dia.	11.69	15.49	0.87	18.99	nr	**34.48**
500mm dia.	12.41	16.44	0.87	18.99	nr	**35.43**
630mm dia.	16.42	21.76	0.87	18.99	nr	**40.74**
710mm dia.	22.06	29.23	0.96	20.95	nr	**50.18**
800mm dia.	26.32	34.87	1.06	23.13	nr	**58.01**
900mm dia.	30.08	39.86	1.16	25.32	nr	**65.17**
1000mm dia.	41.29	54.71	1.25	27.28	nr	**81.99**
1120mm dia.	138.78	183.88	3.47	75.73	nr	**259.61**
1250mm dia.	151.59	200.86	3.47	75.73	nr	**276.59**
1400mm dia.	202.96	268.92	5.20	113.49	nr	**382.41**
1600mm dia.	216.03	286.24	4.62	100.83	nr	**387.07**
90 degree equal twin bend						
80mm dia.	13.99	18.54	0.58	12.66	nr	**31.20**
100mm dia.	14.31	18.96	0.58	12.66	nr	**31.62**
150mm dia.	19.53	25.88	0.58	12.66	nr	**38.54**
200mm dia.	25.23	33.43	0.87	18.99	nr	**52.42**
250mm dia.	35.59	47.16	1.16	25.32	nr	**72.47**
300mm dia.	38.00	50.35	1.16	25.32	nr	**75.67**

Material Costs/Measured Work Prices - Mechanical Installations   281

U:VENTILATION/AIR CONDITIONING SYSTEMS

Item	Net Price £	Material £	Labour hours	Labour £	Unit	Total rate £
355mm dia.	43.52	57.66	1.73	37.76	nr	**95.42**
400mm dia.	46.43	61.52	1.73	37.76	nr	**99.28**
450mm dia.	49.69	65.84	1.73	37.76	nr	**103.60**
500mm dia.	54.12	71.71	1.73	37.76	nr	**109.47**
630mm dia.	79.02	104.70	1.73	37.76	nr	**142.46**
710mm dia.	105.20	139.39	1.82	39.72	nr	**179.11**
800mm dia.	128.92	170.82	2.02	44.09	nr	**214.90**
900mm dia.	153.46	203.33	2.02	44.09	nr	**247.42**
1000mm dia.	202.53	268.35	2.11	46.05	nr	**314.40**
1120mm dia.	347.54	460.49	4.62	100.83	nr	**561.32**
1250mm dia.	382.77	507.17	4.62	100.83	nr	**608.00**
1400mm dia.	647.73	858.24	4.62	100.83	nr	**959.07**
1600mm dia.	651.85	863.70	4.62	100.83	nr	**964.53**
Conical branch						
80mm dia.	14.77	19.57	0.58	12.66	nr	**32.23**
100mm dia.	14.73	19.52	0.58	12.66	nr	**32.18**
150mm dia.	15.64	20.72	0.58	12.66	nr	**33.38**
200mm dia.	16.66	22.07	0.87	18.99	nr	**41.06**
250mm dia.	19.89	26.35	1.16	25.32	nr	**51.67**
300mm dia.	21.87	28.98	1.16	25.32	nr	**54.29**
355mm dia.	25.27	33.48	1.73	37.76	nr	**71.24**
400mm dia.	24.73	32.77	1.73	37.76	nr	**70.52**
450mm dia.	29.28	38.80	1.73	37.76	nr	**76.55**
500mm dia.	30.51	40.43	1.73	37.76	nr	**78.18**
630mm dia.	43.84	58.09	1.73	37.76	nr	**95.84**
710mm dia.	53.73	71.19	1.82	39.72	nr	**110.91**
800mm dia.	65.62	86.95	1.93	42.12	nr	**129.07**
900mm dia.	71.85	95.20	2.02	44.09	nr	**139.29**
1000mm dia.	81.73	108.29	2.11	46.05	nr	**154.34**
1120mm dia.	119.88	158.84	4.62	100.83	nr	**259.67**
1250mm dia.	135.84	179.99	4.62	100.83	nr	**280.82**
1400mm dia.	153.94	203.97	4.62	100.83	nr	**304.80**
1600mm dia.	172.41	228.44	4.62	100.83	nr	**329.27**
45 degree branch						
80mm dia.	11.01	14.59	0.58	12.66	nr	**27.25**
100mm dia.	10.98	14.55	0.58	12.66	nr	**27.21**
150mm dia.	14.36	19.03	0.58	12.66	nr	**31.69**
200mm dia.	15.04	19.93	0.87	18.99	nr	**38.92**
250mm dia.	15.77	20.90	1.16	25.32	nr	**46.21**
300mm dia.	16.80	22.26	1.16	25.32	nr	**47.58**
355mm dia.	21.56	28.57	1.73	37.76	nr	**66.32**
400mm dia.	21.38	28.33	1.73	37.76	nr	**66.09**
450mm dia.	22.69	30.06	1.73	37.76	nr	**67.82**
500mm dia.	24.14	31.99	1.73	37.76	nr	**69.74**
630mm dia.	32.43	42.97	1.73	37.76	nr	**80.73**
710mm dia.	38.37	50.84	1.82	39.72	nr	**90.56**
800mm dia.	53.33	70.66	2.13	46.49	nr	**117.15**
900mm dia.	61.47	81.45	2.31	50.41	nr	**131.86**
1000mm dia.	67.89	89.95	2.31	50.41	nr	**140.37**
1120mm dia.	102.43	135.72	4.62	100.83	nr	**236.55**
1250mm dia.	110.75	146.74	4.62	100.83	nr	**247.57**
1400mm dia.	122.15	161.85	4.62	100.83	nr	**262.68**
1600mm dia.	142.58	188.92	4.62	100.83	nr	**289.75**

Material Costs/Measured Work Prices - Mechanical Installations

U:VENTILATION/AIR CONDITIONING SYSTEMS

Item	Net Price £	Material £	Labour hours	Labour £	Unit	Total rate £
U10 : DUCTWORK : FLAT OVAL						
Y30 - AIR DUCTLINES						
Galvanised sheet metal DW144 class B spirally wound circular section ductwork; including all necessary stiffeners, joints, couplers in the running length and duct supports						
Straight duct						
345 x 102mm	7.86	10.41	2.71	59.14	m	69.56
427 x 102mm	8.70	11.53	2.99	65.26	m	76.78
508 x 102mm	9.53	12.63	3.14	68.53	m	81.16
559 x 152mm	14.09	18.67	3.43	74.86	m	93.53
531 x 203mm	14.08	18.66	3.43	74.86	m	93.51
851 x 203mm	20.42	27.06	5.72	124.84	m	151.89
582 x 254mm	14.93	19.78	3.62	79.01	m	98.79
1303 x 254mm	53.14	70.41	8.13	177.43	m	247.84
632 x 305mm	16.43	21.77	3.93	85.77	m	107.54
1275 x 305mm	42.04	55.70	8.13	177.43	m	233.14
765 x 356mm	18.23	24.15	5.72	124.84	m	148.99
1247 x 356mm	42.14	55.84	8.13	177.43	m	233.27
1727 x 356mm	61.78	81.86	10.41	227.19	m	309.05
737 x 406mm	18.05	23.92	5.72	124.84	m	148.75
818 x 406mm	21.64	28.67	6.21	135.53	m	164.20
978 x 406mm	25.43	33.69	6.92	151.03	m	184.72
1379 x 406mm	46.72	61.90	8.75	190.97	m	252.87
1699 x 406mm	61.93	82.06	10.41	227.19	m	309.25
709 x 457mm	18.09	23.97	5.72	124.84	m	148.81
1671 x 457mm	62.10	82.28	10.31	225.01	m	307.29
678 x 508mm	18.14	24.04	5.72	124.84	m	148.87
Extra over fittings; Flat Oval duct class C						
End cap						
345 x 102mm	16.94	22.45	0.20	4.36	nr	26.81
427 x 102mm	18.23	24.15	0.20	4.36	nr	28.52
508 x 102mm	19.37	25.67	0.20	4.36	nr	30.03
559 x 152mm	23.11	30.62	0.29	6.33	nr	36.95
531 x 203mm	23.63	31.31	0.29	6.33	nr	37.64
851 x 203mm	28.44	37.68	0.44	9.60	nr	47.29
582 x 254mm	25.99	34.44	0.44	9.60	nr	44.04
1303 x 254mm	50.70	67.18	0.44	9.60	nr	76.78
632 x 305mm	28.55	37.83	0.69	15.06	nr	52.89
1275 x 305mm	48.16	63.81	0.44	9.60	nr	73.41
765 x 356mm	40.91	54.21	0.69	15.06	nr	69.26
1727 x 356mm	50.83	67.35	0.69	15.06	nr	82.41
737 x 406mm	32.60	43.20	1.04	22.70	nr	65.89
818 x 406mm	34.16	45.26	0.44	9.60	nr	54.86
978 x 406mm	36.81	48.77	0.44	9.60	nr	58.38
1379 x 406mm	53.57	70.98	0.44	9.60	nr	80.58
1699 x 406mm	63.58	84.24	0.69	15.06	nr	99.30
709 x 457mm	33.56	44.47	1.04	22.70	nr	67.16
1671 x 457mm	64.56	85.54	0.44	9.60	nr	95.14
678 x 508mm	30.28	40.12	1.04	22.70	nr	62.82

Material Costs/Measured Work Prices - Mechanical Installations

U:VENTILATION/AIR CONDITIONING SYSTEMS

Item	Net Price £	Material £	Labour hours	Labour £	Unit	Total rate £
Reducer						
345 x 102mm	21.06	27.90	0.95	20.73	nr	48.64
427 x 102mm	21.30	28.22	1.06	23.13	nr	51.36
508 x 102mm	21.53	28.53	1.13	24.66	nr	53.19
559 x 152mm	24.64	32.65	1.26	27.50	nr	60.15
531 x 203mm	30.27	40.11	1.26	27.50	nr	67.61
851 x 203mm	30.01	39.76	1.16	25.32	nr	65.08
582 x 254mm	32.57	43.16	1.34	29.25	nr	72.40
1303 x 254mm	53.44	70.81	1.16	25.32	nr	96.12
632 x 305mm	43.87	58.13	0.70	15.28	nr	73.40
1275 x 305mm	56.77	75.22	1.16	25.32	nr	100.54
765 x 356mm	49.85	66.05	1.16	25.32	nr	91.37
1247 x 356mm	58.67	77.74	1.16	25.32	nr	103.05
1727 x 356mm	58.36	77.33	1.25	27.28	nr	104.61
737 x 406mm	53.78	71.26	1.16	25.32	nr	96.58
818 x 406mm	53.74	71.21	1.27	27.72	nr	98.92
978 x 406mm	55.77	73.90	1.44	31.43	nr	105.32
1379 x 406mm	59.97	79.46	1.44	31.43	nr	110.89
1699 x 406mm	58.98	78.15	1.44	31.43	nr	109.58
709 x 457mm	55.37	73.37	1.16	25.32	nr	98.68
1671 x 457mm	60.40	80.03	1.44	31.43	nr	111.46
678 x 508mm	57.65	76.39	1.16	25.32	nr	101.70
90 degree radius bend						
345 x 102mm	42.54	56.37	0.29	6.33	nr	62.69
427 x 102mm	41.59	55.11	0.58	12.66	nr	67.77
508 x 102mm	40.85	54.13	0.58	12.66	nr	66.78
559 x 152mm	46.54	61.67	0.58	12.66	nr	74.32
531 x 203mm	55.21	73.15	0.87	18.99	nr	92.14
851 x 203mm	69.42	91.98	0.87	18.99	nr	110.97
582 x 254mm	54.79	72.60	0.87	18.99	nr	91.58
1303 x 254mm	104.02	137.83	0.96	20.95	nr	158.78
632 x 305mm	57.56	76.27	0.87	18.99	nr	95.25
1275 x 305mm	135.25	179.21	0.96	20.95	nr	200.16
765 x 356mm	66.61	88.26	0.87	18.99	nr	107.25
1247 x 356mm	136.53	180.90	0.96	20.95	nr	201.85
1727 x 356mm	153.95	203.98	1.25	27.28	nr	231.26
737 x 406mm	71.60	94.87	0.96	20.95	nr	115.82
818 x 406mm	83.66	110.85	0.87	18.99	nr	129.84
978 x 406mm	77.60	102.82	0.96	20.95	nr	123.77
90 degree radius bend						
1379 x 406mm	128.43	170.17	1.16	25.32	nr	195.49
1699 x 406mm	158.50	210.01	1.25	27.28	nr	237.29
709 x 457mm	79.86	105.81	0.87	18.99	nr	124.80
1671 x 457mm	198.51	263.03	1.25	27.28	nr	290.31
678 x 508mm	80.73	106.97	0.87	18.99	nr	125.95
45 degree radius bend						
345 x 102mm	27.77	36.80	1.03	22.48	nr	59.27
427 x 102mm	27.44	36.36	0.95	20.73	nr	57.09
508 x 102mm	27.14	35.96	0.82	17.90	nr	53.86
559 x 152mm	31.81	42.15	0.79	17.24	nr	59.39
531 x 203mm	38.75	51.34	0.85	18.55	nr	69.89
851 x 203mm	46.84	62.06	0.18	3.93	nr	65.99

Material Costs/Measured Work Prices - Mechanical Installations

U:VENTILATION/AIR CONDITIONING SYSTEMS

Item	Net Price £	Material £	Labour hours	Labour £	Unit	Total rate £
U10 : DUCTWORK : FLAT OVAL						
Y30 - AIR DUCTLINES						
Galvanised sheet metal DW144 class B spirally wound circular section ductwork; including all necessary stiffeners, joints, couplers in the running length and duct supports (Continued)						
45 degree radius bend						
582 x 254mm	39.24	51.99	0.76	16.59	nr	**68.58**
1303 x 254mm	85.59	113.41	1.16	25.32	nr	**138.72**
632 x 305mm	43.01	56.99	0.58	12.66	nr	**69.65**
1275 x 305mm	100.14	132.69	1.16	25.32	nr	**158.00**
765 x 356mm	46.15	61.15	0.87	18.99	nr	**80.14**
1247 x 356mm	100.75	133.49	1.16	25.32	nr	**158.81**
1727 x 356mm	94.21	124.83	1.26	27.50	nr	**152.33**
737 x 406mm	48.39	64.12	0.69	15.06	nr	**79.18**
818 x 406mm	56.84	75.31	0.40	8.73	nr	**84.04**
978 x 406mm	54.13	71.72	0.87	18.99	nr	**90.71**
1379 x 406mm	97.32	128.95	1.16	25.32	nr	**154.27**
1699 x 406mm	96.28	127.57	1.27	27.72	nr	**155.29**
709 x 457mm	54.19	71.80	0.81	17.68	nr	**89.48**
1671 x 457mm	114.87	152.20	1.26	27.50	nr	**179.70**
678 x 508mm	53.90	71.42	0.92	20.08	nr	**91.50**
90 degree hard bend with turning vanes						
345 x 102mm	55.30	73.27	0.55	12.00	nr	**85.28**
427 x 102mm	55.13	73.05	1.16	25.32	nr	**98.36**
508 x 102mm	54.94	72.80	1.16	25.32	nr	**98.11**
559 x 152mm	56.21	74.48	1.16	25.32	nr	**99.79**
531 x 203mm	57.99	76.84	1.73	37.76	nr	**114.59**
851 x 203mm	78.92	104.57	1.73	37.76	nr	**142.33**
582 x 254mm	64.37	85.29	1.73	37.76	nr	**123.05**
1303 x 254mm	96.01	127.21	1.82	39.72	nr	**166.93**
632 x 305mm	74.12	98.21	1.73	37.76	nr	**135.97**
1275 x 305mm	102.87	136.30	1.82	39.72	nr	**176.02**
765 x 356mm	83.96	111.25	1.73	37.76	nr	**149.00**
1247 x 356mm	103.63	137.31	1.82	39.72	nr	**177.03**
1727 x 356mm	159.32	211.10	1.82	39.72	nr	**250.82**
737 x 406mm	88.29	116.98	1.73	37.76	nr	**154.74**
818 x 406mm	81.58	108.09	1.73	37.76	nr	**145.85**
978 x 406mm	100.56	133.24	1.73	37.76	nr	**171.00**
1379 x 406mm	129.95	172.18	1.82	39.72	nr	**211.90**
1699 x 406mm	163.44	216.56	2.11	46.05	nr	**262.61**
709 x 457mm	98.82	130.94	1.73	37.76	nr	**168.69**
1671 x 457mm	173.28	229.60	2.11	46.05	nr	**275.65**
678 x 508mm	101.91	135.03	1.82	39.72	nr	**174.75**
90 degree branch						
345 x 102mm	34.66	45.92	0.58	12.66	nr	**58.58**
427 x 102mm	36.33	48.14	0.58	12.66	nr	**60.80**
508 x 102mm	37.69	49.94	1.16	25.32	nr	**75.26**

Material Costs/Measured Work Prices - Mechanical Installations

U:VENTILATION/AIR CONDITIONING SYSTEMS

Item	Net Price £	Material £	Labour hours	Labour £	Unit	Total rate £
559 x 152mm	48.24	63.92	1.16	25.32	nr	**89.23**
531 x 203mm	51.76	68.58	1.16	25.32	nr	**93.90**
851 x 203mm	65.93	87.36	1.73	37.76	nr	**125.11**
582 x 254mm	56.27	74.56	1.73	37.76	nr	**112.31**
1303 x 254mm	101.66	134.70	1.82	39.72	nr	**174.42**
632 x 305mm	62.09	82.27	1.73	37.76	nr	**120.03**
1275 x 305mm	101.94	135.07	1.82	39.72	nr	**174.79**
765 x 356mm	67.27	89.13	1.73	37.76	nr	**126.89**
1247 x 356mm	104.87	138.95	1.82	39.72	nr	**178.67**
1727 x 356mm	130.59	173.03	2.11	46.05	nr	**219.08**
737 x 406mm	74.40	98.58	1.73	37.76	nr	**136.34**
818 x 406mm	80.45	106.60	1.73	37.76	nr	**144.35**
978 x 406mm	86.81	115.02	0.96	20.95	nr	**135.97**
1379 x 406mm	118.24	156.67	1.93	42.12	nr	**198.79**
1699 x 406mm	134.78	178.58	2.11	46.05	nr	**224.63**
709 x 457mm	80.53	106.70	1.73	37.76	nr	**144.46**
1671 x 457mm	133.06	176.30	2.11	46.05	nr	**222.35**
678 x 508mm	84.08	111.41	1.73	37.76	nr	**149.16**
45 degree branch						
345 x 102mm	40.26	53.34	0.58	12.66	nr	**66.00**
427 x 102mm	42.86	56.79	0.58	12.66	nr	**69.45**
508 x 102mm	45.19	59.88	1.16	25.32	nr	**85.19**
599 x 152mm	31.61	41.88	1.73	37.76	nr	**79.64**
531 x 203mm	60.32	79.92	1.73	37.76	nr	**117.68**
851 x 203mm	81.58	108.09	1.73	37.76	nr	**145.85**
582 x 254mm	65.45	86.72	1.73	37.76	nr	**124.48**
1303 x 254mm	131.13	173.75	0.96	20.95	nr	**194.70**
632 x 305mm	72.04	95.45	1.73	37.76	nr	**133.21**
1275 x 305mm	129.52	171.61	1.82	39.72	nr	**211.33**
765 x 356mm	78.50	104.01	1.73	37.76	nr	**141.77**
1247 x 356mm	130.90	173.44	1.82	39.72	nr	**213.16**
1727 x 356mm	142.76	189.16	1.82	39.72	nr	**228.88**
737 x 406mm	86.15	114.15	1.73	37.76	nr	**151.91**
818 x 406mm	94.77	125.57	1.73	37.76	nr	**163.33**
978 x 406mm	104.66	138.67	1.73	37.76	nr	**176.43**
1379 x 406mm	148.95	197.36	1.93	42.12	nr	**239.48**
1699 x 406mm	178.14	236.04	2.19	47.80	nr	**283.83**
709 x 457mm	92.39	122.42	1.73	37.76	nr	**160.17**
1671 x 457mm	173.35	229.69	2.11	46.05	nr	**275.74**
678 x 508mm	95.01	125.89	1.73	37.76	nr	**163.64**

**For rates for access doors see U10 :
DUCTWORK : RECTANGULAR**

Material Costs/Measured Work Prices - Mechanical Installations

U:VENTILATION/AIR CONDITIONING SYSTEMS

Item	Net Price £	Material £	Labour hours	Labour £	Unit	Total rate £
U10 : DUCTWORK : RECTANGULAR - CLASS B						
Y30 - AIR DUCTLINES						
Galvanised sheet metal DW144 class B rectangular section ductwork; including all necessary stiffeners, joints, couplers in the running length and duct supports						
Ductwork up to 400mm longest side						
Sum of two sides 200mm	13.46	17.83	1.19	25.97	m	**43.81**
Sum of two sides 250mm	14.07	18.64	1.19	25.97	m	**44.61**
Sum of two sides 300mm	14.61	19.36	1.19	25.97	m	**45.33**
Sum of two sides 350mm	15.98	21.17	1.19	25.97	m	**47.14**
Sum of two sides 400mm	16.53	21.90	1.16	25.32	m	**47.22**
Sum of two sides 450mm	17.09	22.64	1.16	25.32	m	**47.96**
Sum of two sides 500mm	17.73	23.49	1.16	25.32	m	**48.81**
Sum of two sides 550mm	18.29	24.23	1.27	27.72	m	**51.95**
Sum of two sides 600mm	18.85	24.98	1.27	27.72	m	**52.69**
Sum of two sides 650mm	19.40	25.70	1.27	27.72	m	**53.42**
Sum of two sides 700mm	19.96	26.45	1.27	27.72	m	**54.16**
Sum of two sides 750mm	20.52	27.19	1.27	27.72	m	**54.91**
Sum of two sides 800mm	21.07	27.92	1.27	27.72	m	**55.63**
Extra Over fittings; Rectangular ductwork class B; upto 400mm longest side						
End Cap						
Sum of two sides 200mm	8.94	11.85	0.38	8.29	nr	**20.14**
Sum of two sides 250mm	9.31	12.34	0.38	8.29	nr	**20.63**
Sum of two sides 300mm	9.68	12.83	0.38	8.29	nr	**21.12**
Sum of two sides 350mm	10.05	13.32	0.38	8.29	nr	**21.61**
Sum of two sides 400mm	10.42	13.81	0.38	8.29	nr	**22.10**
Sum of two sides 450mm	10.79	14.30	0.38	8.29	nr	**22.59**
Sum of two sides 500mm	11.17	14.80	0.38	8.29	nr	**23.09**
Sum of two sides 550mm	11.53	15.28	0.38	8.29	nr	**23.57**
Sum of two sides 600mm	11.91	15.78	0.38	8.29	nr	**24.07**
Sum of two sides 650mm	12.28	16.27	0.38	8.29	nr	**24.56**
Sum of two sides 700mm	12.65	16.76	0.38	8.29	nr	**25.05**
Sum of two sides 750mm	13.02	17.25	0.38	8.29	nr	**25.54**
Sum of two sides 800mm	13.39	17.74	0.38	8.29	nr	**26.04**
Reducer						
Sum of two sides 200mm	11.81	15.65	1.40	30.55	nr	**46.20**
Sum of two sides 250mm	12.47	16.52	1.40	30.55	nr	**47.08**
Sum of two sides 300mm	13.12	17.38	1.40	30.55	nr	**47.94**
Sum of two sides 350mm	16.95	22.46	1.42	30.99	nr	**53.45**
Sum of two sides 400mm	17.62	23.35	1.42	30.99	nr	**54.34**
Sum of two sides 450mm	18.29	24.23	1.42	30.99	nr	**55.23**
Sum of two sides 500mm	18.95	25.11	1.42	30.99	nr	**56.10**
Sum of two sides 550mm	19.63	26.01	1.69	36.88	nr	**62.89**
Sum of two sides 600mm	20.29	26.88	1.69	36.88	nr	**63.77**

Material Costs/Measured Work Prices - Mechanical Installations

U:VENTILATION/AIR CONDITIONING SYSTEMS

Item	Net Price £	Material £	Labour hours	Labour £	Unit	Total rate £
Sum of two sides 650mm	20.96	27.77	1.69	36.88	nr	**64.66**
Sum of two sides 700mm	21.63	28.66	1.69	36.88	nr	**65.54**
Sum of two sides 750mm	22.30	29.55	1.92	41.90	nr	**71.45**
Sum of two sides 800mm	22.96	30.42	1.92	41.90	nr	**72.33**
Offset						
Sum of two sides 200mm	24.43	32.37	1.63	35.57	nr	**67.94**
Sum of two sides 250mm	25.19	33.38	1.63	35.57	nr	**68.95**
Sum of two sides 300mm	25.94	34.37	1.63	35.57	nr	**69.94**
Sum of two sides 350mm	29.74	39.41	1.65	36.01	nr	**75.42**
Sum of two sides 400mm	30.26	40.09	1.65	36.01	nr	**76.11**
Sum of two sides 450mm	31.03	41.11	1.65	36.01	nr	**77.13**
Sum of two sides 500mm	31.79	42.12	1.65	36.01	nr	**78.13**
Sum of two sides 550mm	32.30	42.80	1.92	41.90	nr	**84.70**
Sum of two sides 600mm	32.80	43.46	1.92	41.90	nr	**85.36**
Sum of two sides 650mm	33.56	44.47	1.92	41.90	nr	**86.37**
Sum of two sides 700mm	34.06	45.13	1.92	41.90	nr	**87.03**
Sum of two sides 750mm	34.56	45.79	1.92	41.90	nr	**87.70**
Sum of two sides 800mm	35.04	46.43	1.92	41.90	nr	**88.33**
Square to round						
Sum of two sides 200mm	20.05	26.57	1.63	35.57	nr	**62.14**
Sum of two sides 250mm	20.90	27.69	1.63	35.57	nr	**63.27**
Sum of two sides 300mm	21.76	28.83	1.63	35.57	nr	**64.41**
Sum of two sides 350mm	23.85	31.60	1.65	36.01	nr	**67.61**
Sum of two sides 400mm	24.68	32.70	1.65	36.01	nr	**68.71**
Sum of two sides 450mm	25.51	33.80	1.65	36.01	nr	**69.81**
Sum of two sides 500mm	26.34	34.90	1.65	36.01	nr	**70.91**
Sum of two sides 550mm	27.17	36.00	1.92	41.90	nr	**77.90**
Sum of two sides 600mm	27.99	37.09	1.92	41.90	nr	**78.99**
Sum of two sides 650mm	28.82	38.19	1.92	41.90	nr	**80.09**
Sum of two sides 700mm	29.65	39.29	1.92	41.90	nr	**81.19**
Sum of two sides 750mm	30.48	40.39	1.92	41.90	nr	**82.29**
Sum of two sides 800mm	31.30	41.47	1.92	41.90	nr	**83.38**
90 degree radius bend						
Sum of two sides 200mm	13.96	18.50	1.22	26.63	nr	**45.12**
Sum of two sides 250mm	14.18	18.79	1.22	26.63	nr	**45.41**
Sum of two sides 300mm	14.38	19.05	1.22	26.63	nr	**45.68**
Sum of two sides 350mm	17.66	23.40	1.25	27.28	nr	**50.68**
Sum of two sides 400mm	18.09	23.97	1.25	27.28	nr	**51.25**
Sum of two sides 450mm	18.28	24.22	1.25	27.28	nr	**51.50**
Sum of two sides 500mm	18.46	24.46	1.25	27.28	nr	**51.74**
Sum of two sides 550mm	18.88	25.02	1.33	29.03	nr	**54.04**
Sum of two sides 600mm	19.31	25.59	1.33	29.03	nr	**54.61**
Sum of two sides 650mm	19.48	25.81	1.33	29.03	nr	**54.84**
Sum of two sides 700mm	19.91	26.38	1.33	29.03	nr	**55.41**
Sum of two sides 750mm	20.33	26.94	1.40	30.55	nr	**57.49**
Sum of two sides 800mm	20.76	27.51	1.40	30.55	nr	**58.06**
45 degree radius bend						
Sum of two sides 200mm	14.84	19.66	0.89	19.42	nr	**39.09**
Sum of two sides 250mm	15.27	20.23	1.12	24.44	nr	**44.68**
Sum of two sides 300mm	15.70	20.80	1.12	24.44	nr	**45.25**

288   *Material Costs/Measured Work Prices - Mechanical Installations*

U:VENTILATION/AIR CONDITIONING SYSTEMS

Item	Net Price £	Material £	Labour hours	Labour £	Unit	Total rate £
U10 : DUCTWORK : RECTANGULAR - CLASS B						
Y30 - AIR DUCTLINES						
Galvanised sheet metal DW144 class B rectangular section ductwork; including all necessary stiffeners, joints, couplers in the running length and duct supports (Continued)						
45 degree radius bend						
Sum of two sides 350mm	19.33	25.61	1.12	24.44	nr	**50.06**
Sum of two sides 400mm	19.89	26.35	1.10	24.01	nr	**50.36**
Sum of two sides 450mm	20.31	26.91	1.10	24.01	nr	**50.92**
Sum of two sides 500mm	20.74	27.48	1.10	24.01	nr	**51.49**
Sum of two sides 550mm	21.29	28.21	1.16	25.32	nr	**53.53**
Sum of two sides 600mm	21.84	28.94	1.16	25.32	nr	**54.25**
Sum of two sides 650mm	22.26	29.49	1.16	25.32	nr	**54.81**
Sum of two sides 700mm	22.81	30.22	1.16	25.32	nr	**55.54**
Sum of two sides 750mm	23.36	30.95	1.22	26.63	nr	**57.58**
Sum of two sides 800mm	23.91	31.68	1.22	26.63	nr	**58.31**
90 degree mitre bend						
Sum of two sides 200mm	21.93	29.06	1.29	28.15	nr	**57.21**
Sum of two sides 250mm	23.15	30.67	1.29	28.15	nr	**58.83**
Sum of two sides 300mm	24.37	32.29	1.29	28.15	nr	**60.44**
Sum of two sides 350mm	28.66	37.97	1.29	28.15	nr	**66.13**
Sum of two sides 400mm	30.19	40.00	1.29	28.15	nr	**68.16**
Sum of two sides 450mm	31.40	41.60	1.29	28.15	nr	**69.76**
Sum of two sides 500mm	32.60	43.20	1.29	28.15	nr	**71.35**
Sum of two sides 550mm	34.18	45.29	1.39	30.34	nr	**75.62**
Sum of two sides 600mm	35.75	47.37	1.39	30.34	nr	**77.70**
Sum of two sides 650mm	36.99	49.01	1.39	30.34	nr	**79.35**
Sum of two sides 700mm	38.59	51.13	1.39	30.34	nr	**81.47**
Sum of two sides 750mm	40.19	53.25	1.46	31.86	nr	**85.12**
Sum of two sides 800mm	41.79	55.37	1.46	31.86	nr	**87.24**
Branch						
Sum of two sides 200mm	16.13	21.37	0.92	20.08	nr	**41.45**
Sum of two sides 250mm	16.96	22.47	0.92	20.08	nr	**42.55**
Sum of two sides 300mm	17.79	23.57	0.92	20.08	nr	**43.65**
Sum of two sides 350mm	20.11	26.65	0.95	20.73	nr	**47.38**
Sum of two sides 400mm	20.91	27.71	0.95	20.73	nr	**48.44**
Sum of two sides 450mm	21.72	28.78	0.95	20.73	nr	**49.51**
Sum of two sides 500mm	22.56	29.89	0.95	20.73	nr	**50.63**
Sum of two sides 550mm	23.38	30.98	1.03	22.48	nr	**53.46**
Sum of two sides 600mm	24.18	32.04	1.03	22.48	nr	**54.52**
Sum of two sides 650mm	25.00	33.13	1.03	22.48	nr	**55.60**
Sum of two sides 700mm	25.81	34.20	1.03	22.48	nr	**56.68**
Sum of two sides 750mm	26.62	35.27	1.03	22.48	nr	**57.75**
Sum of two sides 800mm	27.43	36.34	1.03	22.48	nr	**58.82**
Grille neck						
Sum of two sides 200mm	17.65	23.39	1.10	24.01	nr	**47.39**
Sum of two sides 250mm	18.55	24.58	1.10	24.01	nr	**48.59**
Sum of two sides 300mm	19.46	25.78	1.10	24.01	nr	**49.79**

Material Costs/Measured Work Prices - Mechanical Installations

U:VENTILATION/AIR CONDITIONING SYSTEMS

Item	Net Price £	Material £	Labour hours	Labour £	Unit	Total rate £
Sum of two sides 350mm	20.36	26.98	1.16	25.32	nr	**52.29**
Sum of two sides 400mm	21.26	28.17	1.16	25.32	nr	**53.49**
Sum of two sides 450mm	22.16	29.36	1.16	25.32	nr	**54.68**
Sum of two sides 500mm	23.07	30.57	1.16	25.32	nr	**55.88**
Sum of two sides 550mm	23.97	31.76	1.18	25.75	nr	**57.51**
Sum of two sides 600mm	24.87	32.95	1.18	25.75	nr	**58.71**
Sum of two sides 650mm	25.78	34.16	1.18	25.75	nr	**59.91**
Sum of two sides 700mm	26.69	35.36	1.18	25.75	nr	**61.12**
Sum of two sides 750mm	27.58	36.54	1.18	25.75	nr	**62.30**
Sum of two sides 800mm	28.48	37.74	1.18	25.75	nr	**63.49**
Ductwork 401 to 600mm longest side						
Sum of two sides 600mm	21.34	28.28	1.27	27.72	m	**55.99**
Sum of two sides 650mm	22.10	29.28	1.27	27.72	m	**57.00**
Sum of two sides 700mm	22.94	30.40	1.27	27.72	m	**58.11**
Sum of two sides 750mm	23.70	31.40	1.27	27.72	m	**59.12**
Sum of two sides 800mm	24.47	32.42	1.27	27.72	m	**60.14**
Sum of two sides 850mm	25.23	33.43	1.27	27.72	m	**61.15**
Sum of two sides 900mm	26.00	34.45	1.27	27.72	m	**62.17**
Sum of two sides 950mm	26.76	35.46	1.37	29.90	m	**65.36**
Sum of two sides 1000mm	27.53	36.48	1.37	29.90	m	**66.38**
Sum of two sides 1050mm	28.46	37.71	1.37	29.90	m	**67.61**
Sum of two sides 1100mm	29.22	38.72	1.37	29.90	m	**68.62**
Sum of two sides 1150mm	29.98	39.72	1.37	29.90	m	**69.62**
Sum of two sides 1200mm	30.75	40.74	1.37	29.90	m	**70.64**
Extra over fittings; Ductwork 401 to 600mm longest side						
End Cap						
Sum of two sides 600mm	12.32	16.32	0.38	8.29	nr	**24.62**
Sum of two sides 650mm	12.82	16.99	0.38	8.29	nr	**25.28**
Sum of two sides 700mm	13.33	17.66	0.38	8.29	nr	**25.96**
Sum of two sides 750mm	13.83	18.32	0.38	8.29	nr	**26.62**
Sum of two sides 800mm	14.33	18.99	0.38	8.29	nr	**27.28**
Sum of two sides 850mm	14.33	18.99	0.58	12.66	nr	**31.65**
Sum of two sides 900mm	15.34	20.33	0.58	12.66	nr	**32.98**
Sum of two sides 950mm	15.84	20.99	0.58	12.66	nr	**33.65**
Sum of two sides 1000mm	16.34	21.65	0.58	12.66	nr	**34.31**
Sum of two sides 1050mm	16.84	22.31	0.58	12.66	nr	**34.97**
Sum of two sides 1100mm	17.35	22.99	0.58	12.66	nr	**35.65**
Sum of two sides 1150mm	17.85	23.65	0.58	12.66	nr	**36.31**
Sum of two sides 1200mm	18.36	24.33	0.58	12.66	nr	**36.99**
Reducer						
Sum of two sides 600mm	19.68	26.08	1.69	36.88	nr	**62.96**
Sum of two sides 650mm	20.24	26.82	1.69	36.88	nr	**63.70**
Sum of two sides 700mm	20.79	27.55	1.69	36.88	nr	**64.43**
Sum of two sides 750mm	21.35	28.29	1.92	41.90	nr	**70.19**
Sum of two sides 800mm	21.90	29.02	1.92	41.90	nr	**70.92**
Sum of two sides 850mm	22.46	29.76	1.92	41.90	nr	**71.66**
Sum of two sides 900mm	23.01	30.49	1.92	41.90	nr	**72.39**
Sum of two sides 950mm	23.57	31.23	1.95	42.56	nr	**73.79**
Sum of two sides 1000mm	24.12	31.96	2.18	47.58	nr	**79.54**

Material Costs/Measured Work Prices - Mechanical Installations

U:VENTILATION/AIR CONDITIONING SYSTEMS

Item	Net Price £	Material £	Labour hours	Labour £	Unit	Total rate £
U10 : DUCTWORK : RECTANGULAR - CLASS B						
Y30 - AIR DUCTLINES						
Galvanised sheet metal DW144 class B rectangular section ductwork; including all necessary stiffeners, joints, couplers in the running length and duct supports (Continued)						
Sum of two sides 1050mm	24.85	32.93	2.18	47.58	nr	**80.50**
Sum of two sides 1100mm	25.41	33.67	2.18	47.58	nr	**81.25**
Sum of two sides 1150mm	25.96	34.40	2.18	47.58	nr	**81.97**
Sum of two sides 1200mm	26.52	35.14	2.18	47.58	nr	**82.72**
Offset						
Sum of two sides 600mm	32.26	42.74	1.92	41.90	nr	**84.65**
Sum of two sides 650mm	33.01	43.74	1.92	41.90	nr	**85.64**
Sum of two sides 700mm	33.76	44.73	1.92	41.90	nr	**86.64**
Sum of two sides 750mm	34.25	45.38	1.92	41.90	nr	**87.28**
Sum of two sides 800mm	34.73	46.02	1.92	41.90	nr	**87.92**
Sum of two sides 850mm	35.20	46.64	1.92	41.90	nr	**88.54**
Sum of two sides 900mm	35.67	47.26	1.92	41.90	nr	**89.17**
Sum of two sides 950mm	36.13	47.87	2.18	47.58	nr	**95.45**
Sum of two sides 1000mm	36.87	48.85	2.18	47.58	nr	**96.43**
Sum of two sides 1050mm	37.45	49.62	0.58	12.66	nr	**62.28**
Sum of two sides 1100mm	37.90	50.22	2.18	47.58	nr	**97.80**
Sum of two sides 1150mm	38.34	50.80	2.18	47.58	nr	**98.38**
Sum of two sides 1200mm	38.77	51.37	2.18	47.58	nr	**98.95**
Square to round						
Sum of two sides 600mm	27.55	36.50	1.33	29.03	nr	**65.53**
Sum of two sides 650mm	28.38	37.60	1.33	29.03	nr	**66.63**
Sum of two sides 700mm	29.20	38.69	1.33	29.03	nr	**67.72**
Sum of two sides 750mm	30.03	39.79	1.40	30.55	nr	**70.34**
Sum of two sides 800mm	30.85	40.88	1.40	30.55	nr	**71.43**
Sum of two sides 850mm	31.67	41.96	1.40	30.55	nr	**72.52**
Sum of two sides 900mm	32.50	43.06	1.40	30.55	nr	**73.62**
Sum of two sides 950mm	33.33	44.16	1.82	39.72	nr	**83.88**
Sum of two sides 1000mm	34.15	45.25	1.82	39.72	nr	**84.97**
Sum of two sides 1050mm	35.02	46.40	1.82	39.72	nr	**86.12**
Sum of two sides 1100mm	35.85	47.50	1.82	39.72	nr	**87.22**
Sum of two sides 1150mm	36.67	48.59	1.82	39.72	nr	**88.31**
Sum of two sides 1200mm	37.50	49.69	1.82	39.72	nr	**89.41**
90 degree radius bend						
Sum of two sides 600mm	17.61	23.33	1.16	25.32	nr	**48.65**
Sum of two sides 650mm	17.93	23.76	1.16	25.32	nr	**49.07**
Sum of two sides 700mm	18.24	24.17	1.16	25.32	nr	**49.48**
Sum of two sides 750mm	-	-	1.22	26.63	nr	**26.63**
Sum of two sides 800mm	19.40	25.70	1.22	26.63	nr	**52.33**
Sum of two sides 850mm	19.98	26.47	1.22	26.63	nr	**53.10**
Sum of two sides 900mm	20.55	27.23	1.22	26.63	nr	**53.85**
Sum of two sides 950mm	21.13	28.00	1.40	30.55	nr	**58.55**

Material Costs/Measured Work Prices - Mechanical Installations

U:VENTILATION/AIR CONDITIONING SYSTEMS

Item	Net Price £	Material £	Labour hours	Labour £	Unit	Total rate £
Sum of two sides 1000mm	21.43	28.39	1.40	30.55	nr	**58.95**
Sum of two sides 1050mm	22.13	29.32	1.40	30.55	nr	**59.88**
Sum of two sides 1100mm	22.71	30.09	1.40	30.55	nr	**60.65**
Sum of two sides 1150mm	23.29	30.86	1.40	30.55	nr	**61.41**
Sum of two sides 1200mm	23.86	31.61	1.40	30.55	nr	**62.17**
45 degree bend						
Sum of two sides 600mm	21.17	28.05	1.16	25.32	nr	**53.37**
Sum of two sides 650mm	21.58	28.59	1.16	25.32	nr	**53.91**
Sum of two sides 700mm	21.98	29.12	1.39	30.34	nr	**59.46**
Sum of two sides 750mm	22.52	29.84	1.46	31.86	nr	**61.70**
Sum of two sides 800mm	23.05	30.54	1.46	31.86	nr	**62.41**
Sum of two sides 850mm	23.59	31.26	1.46	31.86	nr	**63.12**
Sum of two sides 900mm	24.13	31.97	1.46	31.86	nr	**63.84**
Sum of two sides 950mm	24.66	32.67	1.88	41.03	nr	**73.70**
Sum of two sides 1000mm	25.06	33.20	1.88	41.03	nr	**74.23**
Sum of two sides 1050mm	25.77	34.15	1.88	41.03	nr	**75.18**
Sum of two sides 1100mm	26.31	34.86	1.88	41.03	nr	**75.89**
Sum of two sides 1150mm	26.84	35.56	1.88	41.03	nr	**76.59**
Sum of two sides 1200mm	27.37	36.27	1.88	41.03	nr	**77.30**
90 degree mitire bend						
Sum of two sides 600mm	36.88	48.87	1.39	30.34	nr	**79.20**
Sum of two sides 650mm	37.78	50.06	1.39	30.34	nr	**80.39**
Sum of two sides 700mm	38.67	51.24	2.16	47.14	nr	**98.38**
Sum of two sides 750mm	40.05	53.07	2.26	49.32	nr	**102.39**
Sum of two sides 800mm	41.44	54.91	2.26	49.32	nr	**104.23**
Sum of two sides 850mm	42.83	56.75	2.26	49.32	nr	**106.07**
Sum of two sides 900mm	44.22	58.59	2.26	49.32	nr	**107.92**
Sum of two sides 950mm	45.60	60.42	3.01	65.69	nr	**126.11**
Sum of two sides 1000mm	46.60	61.74	3.01	65.69	nr	**127.44**
Sum of two sides 1050mm	48.12	63.76	3.01	65.69	nr	**129.45**
Sum of two sides 1100mm	49.53	65.63	3.01	65.69	nr	**131.32**
Sum of two sides 1150mm	50.94	67.50	3.01	65.69	nr	**133.19**
Sum of two sides 1200mm	52.35	69.36	3.01	65.69	nr	**135.06**
Branch						
Sum of two sides 600mm	24.85	32.93	1.03	22.48	nr	**55.41**
Sum of two sides 650mm	25.69	34.04	1.03	22.48	nr	**56.52**
Sum of two sides 700mm	26.52	35.14	1.03	22.48	nr	**57.62**
Sum of two sides 750mm	27.35	36.24	1.03	22.48	nr	**58.72**
Sum of two sides 800mm	28.18	37.34	1.03	22.48	nr	**59.82**
Sum of two sides 850mm	29.01	38.44	1.03	22.48	nr	**60.92**
Sum of two sides 900mm	29.84	39.54	1.03	22.48	nr	**62.02**
Sum of two sides 950mm	30.68	40.65	1.29	28.15	nr	**68.80**
Sum of two sides 1000mm	31.51	41.75	1.29	28.15	nr	**69.90**
Sum of two sides 1050mm	34.13	45.22	1.29	28.15	nr	**73.38**
Sum of two sides 1100mm	33.29	44.11	1.29	28.15	nr	**72.26**
Sum of two sides 1150mm	34.13	45.22	1.29	28.15	nr	**73.38**
Sum of two sides 1200mm	34.96	46.32	1.29	28.15	nr	**74.48**
Grille neck						
Sum of two sides 600mm	25.57	33.88	1.18	25.75	nr	**59.63**
Sum of two sides 650mm	26.49	35.10	1.18	25.75	nr	**60.85**
Sum of two sides 700mm	27.42	36.33	1.18	25.75	nr	**62.08**
Sum of two sides 750mm	28.34	37.55	1.18	25.75	nr	**63.30**

U:VENTILATION/AIR CONDITIONING SYSTEMS

Item	Net Price £	Material £	Labour hours	Labour £	Unit	Total rate £
U10 : DUCTWORK : RECTANGULAR - CLASS B						
Y30 - AIR DUCTLINES						
Galvanised sheet metal DW144 class B rectangular section ductwork; including all necessary stiffeners, joints, couplers in the running length and duct supports (Continued)						
Grille neck						
Sum of two sides 800mm	29.27	38.78	1.18	25.75	nr	**64.54**
Sum of two sides 850mm	30.19	40.00	1.18	25.75	nr	**65.75**
Sum of two sides 900mm	31.12	41.23	1.18	25.75	nr	**66.99**
Sum of two sides 950mm	32.04	42.45	1.44	31.43	nr	**73.88**
Sum of two sides 1000mm	32.97	43.69	1.44	31.43	nr	**75.11**
Sum of two sides 1050mm	33.89	44.90	1.44	31.43	nr	**76.33**
Sum of two sides 1100mm	34.82	46.14	1.44	31.43	nr	**77.56**
Sum of two sides 1150mm	35.74	47.36	1.44	31.43	nr	**78.78**
Sum of two sides 1200mm	36.67	48.59	1.44	31.43	nr	**80.02**
Ductwork 601 to 800mm longest side						
Sum of two sides 900mm	29.96	30.97	1.27	27.72	m	**67.41**
Sum of two sides 950mm	30.67	40.64	1.37	29.90	m	**70.54**
Sum of two sides 1000mm	31.36	41.55	1.37	29.90	m	**71.45**
Sum of two sides 1050mm	32.07	42.49	1.37	29.90	m	**72.39**
Sum of two sides 1100mm	32.77	43.42	1.37	29.90	m	**73.32**
Sum of two sides 1150mm	33.47	44.35	1.37	29.90	m	**74.25**
Sum of two sides 1200mm	34.33	45.49	1.37	29.90	m	**75.39**
Sum of two sides 1250mm	35.04	46.43	1.37	29.90	m	**76.33**
Sum of two sides 1300mm	35.74	47.36	1.40	30.55	m	**77.91**
Sum of two sides 1350mm	36.44	48.28	1.40	30.55	m	**78.84**
Sum of two sides 1400mm	37.14	49.21	1.40	30.55	m	**79.76**
Sum of two sides 1450mm	37.84	50.14	1.40	30.55	m	**80.69**
Sum of two sides 1500mm	38.55	51.08	1.48	32.30	m	**83.38**
Sum of two sides 1550mm	39.25	52.01	1.48	32.30	m	**84.31**
Sum of two sides 1600mm	39.95	52.93	1.55	33.83	m	**86.76**
Extra over fittings: Ductwork 601 to 800mm longest side						
End Cap						
Sum of two sides 900mm	15.83	20.97	0.58	12.66	nr	**33.63**
Sum of two sides 950mm	16.32	21.62	0.58	12.66	nr	**34.28**
Sum of two sides 1000mm	16.81	22.27	0.58	12.66	nr	**34.93**
Sum of two sides 1050mm	17.29	22.91	0.58	12.66	nr	**35.57**
Sum of two sides 1100mm	17.79	23.57	0.58	12.66	nr	**36.23**
Sum of two sides 1150mm	18.28	24.22	0.58	12.66	nr	**36.88**
Sum of two sides 1200mm	18.77	24.87	0.58	12.66	nr	**37.53**
Sum of two sides 1250mm	19.25	25.51	0.58	12.66	nr	**38.16**
Sum of two sides 1300mm	19.74	26.16	0.58	12.66	nr	**38.81**
Sum of two sides 1350mm	20.23	26.80	0.58	12.66	nr	**39.46**
Sum of two sides 1400mm	20.72	27.45	0.58	12.66	nr	**40.11**
Sum of two sides 1450mm	21.78	28.86	0.58	12.66	nr	**41.52**

Material Costs/Measured Work Prices - Mechanical Installations

U:VENTILATION/AIR CONDITIONING SYSTEMS

Item	Net Price £	Material £	Labour hours	Labour £	Unit	Total rate £
Sum of two sides 1500mm	22.92	30.37	0.58	12.66	nr	**43.03**
Sum of two sides 1550mm	24.06	31.88	0.58	12.66	nr	**44.54**
Sum of two sides 1600mm	25.20	33.39	0.58	12.66	nr	**46.05**
Reducer						
Sum of two sides 900mm	22.25	29.48	1.92	41.90	nr	**71.38**
Sum of two sides 950mm	22.77	30.17	1.95	42.56	nr	**72.73**
Sum of two sides 1000mm	23.28	30.85	2.18	47.58	nr	**78.42**
Sum of two sides 1050mm	23.80	31.54	2.18	47.58	nr	**79.11**
Sum of two sides 1100mm	24.31	32.21	2.18	47.58	nr	**79.79**
Sum of two sides 1150mm	24.83	32.90	2.18	47.58	nr	**80.48**
Sum of two sides 1200mm	25.52	33.81	2.18	47.58	nr	**81.39**
Sum of two sides 1250mm	26.04	34.50	2.18	47.58	nr	**82.08**
Sum of two sides 1300mm	26.55	35.18	2.30	50.20	nr	**85.38**
Sum of two sides 1350mm	27.07	35.87	2.30	50.20	nr	**86.06**
Sum of two sides 1400mm	27.59	36.56	2.30	50.20	nr	**86.75**
Sum of two sides 1450mm	29.25	38.76	2.30	50.20	nr	**88.95**
Sum of two sides 1500mm	31.06	41.15	2.47	53.91	nr	**95.06**
Sum of two sides 1550mm	32.89	43.58	2.47	53.91	nr	**97.49**
Sum of two sides 1600mm	34.70	45.98	2.47	53.91	nr	**99.88**
Offset						
Sum of two sides 900mm	36.28	48.07	1.92	41.90	nr	**89.97**
Sum of two sides 950mm	36.46	48.37	2.18	47.58	nr	**95.89**
Sum of two sides 1000mm	36.62	48.52	2.18	47.58	nr	**96.10**
Sum of two sides 1050mm	36.78	48.73	2.18	47.58	nr	**96.31**
Sum of two sides 1100mm	36.94	48.95	2.18	47.58	nr	**96.52**
Sum of two sides 1150mm	37.09	49.14	2.18	47.58	nr	**96.72**
Sum of two sides 1200mm	37.35	49.49	2.18	47.58	nr	**97.07**
Sum of two sides 1250mm	37.48	49.66	2.18	47.58	nr	**97.24**
Sum of two sides 1300mm	37.60	49.82	2.30	50.20	nr	**100.02**
Sum of two sides 1350mm	37.71	49.97	2.30	50.20	nr	**100.16**
Sum of two sides 1400mm	38.23	50.65	2.30	50.20	nr	**100.85**
Sum of two sides 1450mm	40.05	53.07	2.30	50.20	nr	**103.26**
Sum of two sides 1500mm	42.11	55.80	2.47	53.91	nr	**109.70**
Sum of two sides 1550mm	44.15	58.50	2.47	53.91	nr	**112.41**
Sum of two sides 1600mm	46.19	61.20	2.47	53.91	nr	**115.11**
Square to round						
Sum of two sides 900mm	29.11	38.57	1.40	30.55	nr	**69.13**
Sum of two sides 950mm	30.07	39.84	1.82	39.72	nr	**79.56**
Sum of two sides 1000mm	31.02	41.10	1.82	39.72	nr	**80.82**
Sum of two sides 1050mm	31.97	42.36	1.82	39.72	nr	**82.08**
Sum of two sides 1100mm	32.93	43.63	1.82	39.72	nr	**83.35**
Sum of two sides 1150mm	33.88	44.89	1.82	39.72	nr	**84.61**
Sum of two sides 1200mm	34.89	46.23	1.82	39.72	nr	**85.95**
Sum of two sides 1250mm	35.84	47.49	1.82	39.72	nr	**87.21**
Sum of two sides 1300mm	36.79	48.75	2.15	46.92	nr	**95.67**
Sum of two sides 1350mm	37.75	50.02	2.15	46.92	nr	**96.94**
Sum of two sides 1400mm	38.70	51.28	2.15	46.92	nr	**98.20**
Sum of two sides 1450mm	41.08	54.43	2.15	46.92	nr	**101.35**
Sum of two sides 1500mm	43.66	57.85	2.38	51.94	nr	**109.79**
Sum of two sides 1550mm	46.25	61.28	2.38	51.94	nr	**113.22**
Sum of two sides 1600mm	48.82	64.69	2.38	51.94	nr	**116.63**

U:VENTILATION/AIR CONDITIONING SYSTEMS

Item	Net Price £	Material £	Labour hours	Labour £	Unit	Total rate £
U10 : DUCTWORK : RECTANGULAR - CLASS B						
Y30 - AIR DUCTLINES						
Galvanised sheet metal DW144 class B rectangular section ductwork; including all necessary stiffeners, joints, couplers in the running length and duct supports (Continued)						
90 degree radius bend						
Sum of two sides 900mm	19.39	25.69	1.22	26.63	nr	**52.32**
Sum of two sides 950mm	20.05	26.57	1.40	30.55	nr	**57.12**
Sum of two sides 1000mm	20.71	27.44	1.40	30.55	nr	**57.12**
Sum of two sides 1050mm	21.37	28.32	1.40	30.55	nr	**58.87**
Sum of two sides 1100mm	22.04	29.20	1.40	30.55	nr	**59.76**
Sum of two sides 1150mm	22.70	30.08	1.40	30.55	nr	**60.63**
Sum of two sides 1200mm	23.46	31.08	1.40	30.55	nr	**61.64**
Sum of two sides 1250mm	24.12	31.96	1.40	30.55	nr	**62.51**
Sum of two sides 1300mm	24.78	32.83	1.91	41.69	nr	**74.52**
Sum of two sides 1350mm	25.45	33.72	1.91	41.69	nr	**75.41**
Sum of two sides 1400mm	25.70	34.05	1.91	41.69	nr	**75.74**
Sum of two sides 1450mm	27.50	36.44	1.91	41.69	nr	**78.12**
Sum of two sides 1500mm	29.46	39.03	2.11	46.05	nr	**85.08**
Sum of two sides 1550mm	31.41	41.62	2.11	46.05	nr	**87.67**
Sum of two sides 1600mm	33.37	44.22	2.11	46.05	nr	**90.27**
45 degree bend						
Sum of two sides 900mm	23.57	31.23	1.22	26.63	nr	**57.86**
Sum of two sides 950mm	24.13	31.97	1.40	30.55	nr	**62.53**
Sum of two sides 1000mm	24.69	32.71	1.40	30.55	nr	**63.27**
Sum of two sides 1050mm	25.25	33.46	1.40	30.55	nr	**64.01**
Sum of two sides 1100mm	25.81	34.20	1.88	41.03	nr	**75.23**
Sum of two sides 1150mm	26.37	34.94	1.88	41.03	nr	**75.97**
Sum of two sides 1200mm	27.10	35.91	1.88	41.03	nr	**76.94**
Sum of two sides 1250mm	27.66	36.65	1.88	41.03	nr	**77.68**
Sum of two sides 1300mm	28.23	37.40	2.26	49.32	nr	**86.73**
Sum of two sides 1350mm	28.79	38.15	2.26	49.32	nr	**87.47**
Sum of two sides 1400mm	29.14	38.61	2.26	49.32	nr	**87.93**
Sum of two sides 1450mm	30.84	40.86	2.26	49.32	nr	**90.19**
Sum of two sides 1500mm	32.70	43.33	2.49	54.34	nr	**97.67**
Sum of two sides 1550mm	34.57	45.81	2.49	54.34	nr	**100.15**
Sum of two sides 1600mm	36.41	48.24	2.49	54.34	nr	**102.59**
90 degree mitre bend						
Sum of two sides 900mm	43.93	58.21	1.22	26.63	nr	**84.83**
Sum of two sides 950mm	45.45	60.22	1.40	30.55	nr	**90.78**
Sum of two sides 1000mm	46.98	62.25	1.40	30.55	nr	**92.80**
Sum of two sides 1050mm	48.51	64.28	1.40	30.55	nr	**94.83**
Sum of two sides 1100mm	50.03	66.29	3.01	65.69	nr	**131.98**
Sum of two sides 1150mm	51.46	68.18	3.01	65.69	nr	**133.88**
Sum of two sides 1200mm	53.17	70.45	3.01	65.69	nr	**136.14**
Sum of two sides 1250mm	54.70	72.48	3.01	65.69	nr	**138.17**
Sum of two sides 1300mm	56.23	74.50	3.67	80.10	nr	**154.60**

Material Costs/Measured Work Prices - Mechanical Installations

U:VENTILATION/AIR CONDITIONING SYSTEMS

Item	Net Price £	Material £	Labour hours	Labour £	Unit	Total rate £
Sum of two sides 1350mm	57.75	76.52	3.67	80.10	nr	**156.62**
Sum of two sides 1400mm	58.73	77.82	3.67	80.10	nr	**157.91**
Sum of two sides 1450mm	61.99	82.14	3.67	80.10	nr	**162.23**
Sum of two sides 1500mm	65.49	86.77	4.07	88.83	nr	**175.60**
Sum of two sides 1550mm	34.57	45.81	4.07	88.83	nr	**134.63**
Sum of two sides 1600mm	72.86	96.54	4.07	88.83	nr	**185.37**
Branch						
Sum of two sides 900mm	30.43	40.32	1.22	26.63	nr	**66.95**
Sum of two sides 950mm	31.30	41.47	1.40	30.55	nr	**72.03**
Sum of two sides 1000mm	32.18	42.64	1.40	30.55	nr	**73.19**
Sum of two sides 1050mm	33.05	43.79	1.40	30.55	nr	**74.35**
Sum of two sides 1100mm	33.92	44.94	1.29	28.15	nr	**73.10**
Sum of two sides 1150mm	34.79	46.10	1.29	28.15	nr	**74.25**
Sum of two sides 1200mm	35.79	47.42	1.29	28.15	nr	**75.58**
Sum of two sides 1250mm	36.66	48.57	1.29	28.15	nr	**76.73**
Sum of two sides 1300mm	37.54	49.74	1.39	30.34	nr	**80.08**
Sum of two sides 1350mm	38.41	50.89	1.39	30.34	nr	**81.23**
Sum of two sides 1400mm	39.28	52.05	1.39	30.34	nr	**82.38**
Sum of two sides 1450mm	41.68	55.23	1.39	30.34	nr	**85.56**
Sum of two sides 1500mm	44.29	58.68	1.64	35.79	nr	**94.48**
Sum of two sides 1550mm	46.89	62.13	1.64	35.79	nr	**97.92**
Sum of two sides 1600mm	49.50	65.59	1.64	35.79	nr	**101.38**
Grille neck						
Sum of two sides 900mm	30.61	40.56	1.22	26.63	nr	**67.18**
Sum of two sides 950mm	31.54	41.79	1.40	30.55	nr	**72.34**
Sum of two sides 1000mm	32.47	43.02	1.40	30.55	nr	**73.58**
Sum of two sides 1050mm	33.40	44.26	1.40	30.55	nr	**74.81**
Sum of two sides 1100mm	34.32	45.47	1.44	31.43	nr	**76.90**
Sum of two sides 1150mm	35.25	46.71	1.44	31.43	nr	**78.13**
Sum of two sides 1200mm	36.18	47.94	1.44	31.43	nr	**79.37**
Sum of two sides 1250mm	37.11	49.17	1.44	31.43	nr	**80.60**
Sum of two sides 1300mm	38.04	50.40	1.69	36.88	nr	**87.29**
Sum of two sides 1350mm	38.97	51.64	1.69	36.88	nr	**88.52**
Sum of two sides 1400mm	39.90	52.87	1.69	36.88	nr	**89.75**
Sum of two sides 1450mm	42.54	56.37	1.69	36.88	nr	**93.25**
Sum of two sides 1500mm	45.42	60.18	1.79	39.07	nr	**99.25**
Sum of two sides 1550mm	48.31	64.01	1.79	39.07	nr	**103.08**
Sum of two sides 1600mm	51.19	67.83	1.79	39.07	nr	**106.89**
Ductwork 801 to 1000mm longest side						
Sum of two sides 1100mm	40.28	53.37	1.37	29.90	m	**83.27**
Sum of two sides 1150mm	41.34	54.78	1.37	29.90	m	**84.68**
Sum of two sides 1200mm	42.40	56.18	1.37	29.90	m	**86.08**
Sum of two sides 1250mm	43.46	57.58	1.37	29.90	m	**87.48**
Sum of two sides 1300mm	44.52	58.99	1.40	30.55	m	**89.54**
Sum of two sides 1350mm	45.58	60.39	0.71	15.59	m	**75.98**
Sum of two sides 1400mm	46.80	62.01	1.40	30.55	m	**92.56**
Sum of two sides 1450mm	47.86	63.41	1.40	30.55	m	**93.97**
Sum of two sides 1500mm	48.92	64.82	1.48	32.30	m	**97.12**
Sum of two sides 1550mm	49.98	66.22	1.48	32.30	m	**98.52**
Sum of two sides 1600mm	51.04	67.63	1.55	33.83	m	**101.46**
Sum of two sides 1650mm	52.03	68.94	1.55	33.83	m	**102.77**

Material Costs/Measured Work Prices - Mechanical Installations

U:VENTILATION/AIR CONDITIONING SYSTEMS

Item	Net Price £	Material £	Labour hours	Labour £	Unit	Total rate £
U10 : DUCTWORK : RECTANGULAR - CLASS B						
Y30 - AIR DUCTLINES						
Galvanised sheet metal DW144 class B rectangular section ductwork; including all necessary stiffeners, joints, couplers in the running length and duct supports (Continued)						
Extra over fittings; Ductwork 801 to 1000mm longest side						
End Cap						
Sum of two sides 1700mm	53.16	70.44	1.55	33.83	m	**104.27**
Sum of two sides 1750mm	54.22	71.84	1.55	33.83	m	**105.67**
Sum of two sides 1800mm	55.44	73.46	1.61	35.14	m	**108.60**
Sum of two sides 1850mm	56.50	74.86	1.61	35.14	m	**110.00**
Sum of two sides 1900mm	57.56	76.27	1.61	35.14	m	**111.40**
Sum of two sides 1950mm	58.62	77.67	1.61	35.14	m	**112.81**
Sum of two sides 2000mm	59.68	79.08	1.61	35.14	m	**114.21**
Extra over fittings; Ductwork 801 to 1000mm longest side						
End Cap						
Sum of two sides 1100mm	19.28	25.55	1.44	31.43	nr	**56.97**
Sum of two sides 1150mm	19.98	26.47	1.44	31.43	nr	**57.90**
Sum of two sides 1200mm	20.68	27.40	1.44	31.43	nr	**58.83**
Sum of two sides 1250mm	21.38	28.33	1.44	31.43	nr	**59.76**
Sum of two sides 1300mm	22.09	29.27	1.44	31.43	nr	**60.70**
Sum of two sides 1350mm	22.79	30.20	1.44	31.43	nr	**61.62**
Sum of two sides 1400mm	23.49	31.12	1.44	31.43	nr	**62.55**
Sum of two sides 1450mm	24.77	32.82	1.44	31.43	nr	**64.25**
Sum of two sides 1500mm	26.12	34.61	1.44	31.43	nr	**66.04**
Sum of two sides 1550mm	27.48	36.41	1.44	31.43	nr	**67.84**
Sum of two sides 1600mm	28.83	38.20	1.44	31.43	nr	**69.63**
Sum of two sides 1650mm	30.18	39.99	1.44	31.43	nr	**71.42**
Sum of two sides 1700mm	31.54	41.79	1.44	31.43	nr	**73.22**
Sum of two sides 1750mm	32.89	43.58	1.44	31.43	nr	**75.01**
Sum of two sides 1800mm	34.24	45.37	1.44	31.43	nr	**76.80**
Sum of two sides 1850mm	35.60	47.17	1.44	31.43	nr	**78.60**
Sum of two sides 1900mm	36.95	48.96	1.44	31.43	nr	**80.39**
Sum of two sides 1950mm	38.31	50.76	1.44	31.43	nr	**82.19**
Sum of two sides 2000mm	39.66	52.55	1.44	31.43	nr	**83.98**
Reducer						
Sum of two sides 1100mm	18.57	24.61	1.44	31.43	nr	**56.03**
Sum of two sides 1150mm	19.05	25.24	1.44	31.43	nr	**56.67**
Sum of two sides 1200mm	19.53	25.88	1.44	31.43	nr	**57.30**
Sum of two sides 1250mm	20.02	26.53	1.44	31.43	nr	**57.95**
Sum of two sides 1300mm	20.50	27.16	1.69	36.88	nr	**64.05**
Sum of two sides 1350mm	20.98	27.80	1.69	36.88	nr	**64.68**
Sum of two sides 1400mm	21.61	28.63	1.69	36.88	nr	**65.52**

Material Costs/Measured Work Prices - Mechanical Installations

U:VENTILATION/AIR CONDITIONING SYSTEMS

Item	Net Price £	Material £	Labour hours	Labour £	Unit	Total rate £
Sum of two sides 1450mm	23.24	30.79	1.69	36.88	nr	**67.68**
Sum of two sides 1500mm	25.03	33.16	2.47	53.91	nr	**87.07**
Sum of two sides 1550mm	26.81	35.52	2.47	53.91	nr	**89.43**
Sum of two sides 1600mm	28.60	37.90	2.47	53.91	nr	**91.80**
Sum of two sides 1650mm	30.38	40.25	2.47	53.91	nr	**94.16**
Sum of two sides 1700mm	32.17	42.63	2.47	53.91	nr	**96.53**
Sum of two sides 1750mm	33.95	44.98	2.47	53.91	nr	**98.89**
Sum of two sides 1800mm	35.88	47.54	2.59	56.53	nr	**104.07**
Sum of two sides 1850mm	37.66	49.90	2.59	56.53	nr	**106.43**
Sum of two sides 1900mm	39.45	52.27	2.71	59.14	nr	**111.42**
Sum of two sides 1950mm	41.24	54.64	2.71	59.14	nr	**113.79**
Sum of two sides 2000mm	43.02	57.00	2.71	59.14	nr	**116.15**
Offset						
Sum of two sides 1100mm	39.54	52.39	1.44	31.43	nr	**83.82**
Sum of two sides 1150mm	39.73	52.64	1.44	31.43	nr	**84.07**
Sum of two sides 1200mm	39.90	52.87	1.44	31.43	nr	**84.29**
Sum of two sides 1250mm	39.90	52.87	1.44	31.43	nr	**84.29**
Sum of two sides 1300mm	40.19	53.25	1.69	36.88	nr	**90.14**
Sum of two sides 1350mm	40.32	53.42	1.69	36.88	nr	**90.31**
Sum of two sides 1400mm	40.54	53.72	1.69	36.88	nr	**90.60**
Sum of two sides 1450mm	42.34	56.10	1.69	36.88	nr	**92.98**
Sum of two sides 1500mm	44.35	58.76	2.47	53.91	nr	**112.67**
Sum of two sides 1550mm	46.35	61.41	2.47	53.91	nr	**115.32**
Sum of two sides 1600mm	48.33	64.04	2.47	53.91	nr	**117.94**
Sum of two sides 1650mm	49.64	65.77	2.47	53.91	nr	**119.68**
Sum of two sides 1700mm	52.24	69.22	2.59	56.53	nr	**125.74**
Sum of two sides 1750mm	54.17	71.78	2.59	56.53	nr	**128.30**
Sum of two sides 1800mm	56.15	74.40	2.61	56.96	nr	**131.36**
Sum of two sides 1850mm	58.73	77.82	2.61	56.96	nr	**134.78**
Sum of two sides 1900mm	60.61	80.31	2.71	59.14	nr	**139.45**
Sum of two sides 1950mm	62.48	82.79	2.71	59.14	nr	**141.93**
Sum of two sides 2000mm	64.32	85.22	2.71	59.14	nr	**144.37**
Square to round						
Sum of two sides 1100mm	26.70	35.38	1.44	31.43	nr	**66.80**
Sum of two sides 1150mm	28.02	37.13	1.44	31.43	nr	**68.55**
Sum of two sides 1200mm	29.35	38.89	1.44	31.43	nr	**70.32**
Sum of two sides 1250mm	30.68	40.65	1.44	31.43	nr	**72.08**
Sum of two sides 1300mm	32.00	42.40	1.69	36.88	nr	**79.28**
Sum of two sides 1350mm	33.33	44.16	1.69	36.88	nr	**81.05**
Sum of two sides 1400mm	34.68	45.95	1.69	36.88	nr	**82.83**
Sum of two sides 1450mm	37.44	49.61	1.69	36.88	nr	**86.49**
Sum of two sides 1500mm	40.39	53.52	2.38	51.94	nr	**105.46**
Sum of two sides 1550mm	43.34	57.43	2.38	51.94	nr	**109.37**
Sum of two sides 1600mm	46.30	61.35	2.38	51.94	nr	**113.29**
Sum of two sides 1650mm	49.25	65.26	2.38	51.94	nr	**117.20**
Sum of two sides 1700mm	52.21	69.18	2.55	55.65	nr	**124.83**
Sum of two sides 1750mm	55.16	73.09	2.55	55.65	nr	**128.74**
Sum of two sides 1800mm	58.14	77.04	2.55	55.65	nr	**132.69**
Sum of two sides 1850mm	61.09	80.94	2.55	55.65	nr	**136.60**
Sum of two sides 1900mm	64.05	84.87	2.83	61.76	nr	**146.63**
Sum of two sides 1950mm	67.00	88.78	2.83	61.76	nr	**150.54**
Sum of two sides 2000mm	69.95	92.68	2.83	61.76	nr	**154.45**

Material Costs/Measured Work Prices - Mechanical Installations

U:VENTILATION/AIR CONDITIONING SYSTEMS

Item	Net Price £	Material £	Labour hours	Labour £	Unit	Total rate £
U10 : DUCTWORK : RECTANGULAR - CLASS B						
Y30 - AIR DUCTLINES						
Galvanised sheet metal DW144 class B rectangular section ductwork; including all necessary stiffeners, joints, couplers in the running length and duct supports (Continued)						
90 degree radius bend						
Sum of two sides 1100mm	19.29	25.56	1.44	31.43	nr	56.99
Sum of two sides 1150mm	19.79	26.22	1.44	31.43	nr	57.65
Sum of two sides 1200mm	20.28	26.87	1.44	31.43	nr	58.30
Sum of two sides 1250mm	21.23	28.13	1.44	31.43	nr	59.56
Sum of two sides 1300mm	21.27	28.18	1.69	36.88	nr	65.07
Sum of two sides 1350mm	21.77	28.85	1.69	36.88	nr	65.73
Sum of two sides 1400mm	22.35	29.61	1.69	36.88	nr	66.50
Sum of two sides 1450mm	23.98	31.77	1.69	36.88	nr	68.66
Sum of two sides 1500mm	25.78	34.16	2.11	46.05	nr	80.21
Sum of two sides 1550mm	27.58	36.54	2.11	46.05	nr	82.59
Sum of two sides 1600mm	29.37	38.92	2.11	46.05	nr	84.97
Sum of two sides 1650mm	31.82	42.16	2.11	46.05	nr	88.21
Sum of two sides 1700mm	32.97	43.69	2.26	49.32	nr	93.01
Sum of two sides 1750mm	34.77	46.07	2.26	49.32	nr	95.39
Sum of two sides 1800mm	36.65	48.56	2.26	49.32	nr	97.88
Sum of two sides 1850mm	37.75	50.02	2.26	49.32	nr	99.34
Sum of two sides 1900mm	39.54	52.39	2.48	54.13	nr	106.52
Sum of two sides 1950mm	41.33	54.76	2.48	54.13	nr	108.89
Sum of two sides 2000mm	43.12	57.13	2.48	54.13	nr	111.26
45 degree bend						
Sum of two sides 1100mm	24.00	31.80	1.44	31.43	nr	63.23
Sum of two sides 1150mm	24.28	32.17	1.44	31.43	nr	63.60
Sum of two sides 1200mm	24.67	32.69	1.44	31.43	nr	64.12
Sum of two sides 1250mm	25.06	33.20	1.44	31.43	nr	64.63
Sum of two sides 1300mm	25.44	33.71	1.69	36.88	nr	70.59
Sum of two sides 1350mm	25.83	34.22	1.69	36.88	nr	71.11
Sum of two sides 1400mm	26.38	34.95	1.69	36.88	nr	71.84
Sum of two sides 1450mm	27.91	36.98	1.69	36.88	nr	73.86
Sum of two sides 1500mm	29.60	39.22	2.49	54.34	nr	93.56
Sum of two sides 1550mm	31.29	41.46	2.49	54.34	nr	95.80
Sum of two sides 1600mm	32.98	43.70	2.49	54.34	nr	98.04
Sum of two sides 1650mm	35.00	46.38	2.49	54.34	nr	100.72
Sum of two sides 1700mm	36.36	48.18	2.67	58.27	nr	106.45
Sum of two sides 1750mm	38.05	50.42	2.67	58.27	nr	108.69
Sum of two sides 1800mm	39.90	52.87	2.67	58.27	nr	111.14
Sum of two sides 1850mm	41.25	54.66	2.67	58.27	nr	112.93
Sum of two sides 1900mm	42.93	56.88	3.06	66.78	nr	123.67
Sum of two sides 1950mm	44.62	59.12	3.06	66.78	nr	125.90
Sum of two sides 2000mm	46.30	61.35	3.06	66.78	nr	128.13

Material Costs/Measured Work Prices - Mechanical Installations 299

U:VENTILATION/AIR CONDITIONING SYSTEMS

Item	Net Price £	Material £	Labour hours	Labour £	Unit	Total rate £
90 degree mitre bend						
Sum of two sides 1100mm	41.08	54.43	1.44	31.43	nr	**85.86**
Sum of two sides 1150mm	43.48	57.61	1.44	31.43	nr	**89.04**
Sum of two sides 1200mm	45.89	60.80	1.44	31.43	nr	**92.23**
Sum of two sides 1250mm	48.29	63.98	1.44	31.43	nr	**95.41**
Sum of two sides 1300mm	50.70	67.18	1.69	36.88	nr	**104.06**
Sum of two sides 1350mm	53.10	70.36	1.69	36.88	nr	**107.24**
Sum of two sides 1400mm	55.57	73.63	1.69	36.88	nr	**110.51**
Sum of two sides 1450mm	56.69	75.11	1.69	36.88	nr	**112.00**
Sum of two sides 1500mm	64.05	84.87	4.07	88.83	nr	**173.69**
Sum of two sides 1550mm	68.40	90.63	4.07	88.83	nr	**179.46**
Sum of two sides 1600mm	72.76	96.41	4.07	88.83	nr	**185.23**
Sum of two sides 1650mm	78.00	103.35	4.07	88.83	nr	**192.18**
Sum of two sides 1700mm	81.48	107.96	2.80	61.11	nr	**169.07**
Sum of two sides 1750mm	85.84	113.74	2.80	61.11	nr	**174.85**
Sum of two sides 1800mm	90.26	119.59	2.67	58.27	nr	**177.87**
Sum of two sides 1850mm	93.73	124.19	2.67	58.27	nr	**182.46**
Sum of two sides 1900mm	98.10	129.98	2.95	64.38	nr	**194.37**
Sum of two sides 1950mm	102.47	135.77	2.95	64.38	nr	**200.16**
Sum of two sides 2000mm	106.85	141.58	2.95	64.38	nr	**205.96**
Branch						
Sum of two sides 1100mm	34.92	46.27	1.44	31.43	nr	**77.70**
Sum of two sides 1150mm	35.83	47.47	1.44	31.43	nr	**78.90**
Sum of two sides 1200mm	36.73	48.67	1.44	31.43	nr	**80.09**
Sum of two sides 1250mm	37.64	49.87	1.44	31.43	nr	**81.30**
Sum of two sides 1300mm	38.55	51.08	1.64	35.79	nr	**86.87**
Sum of two sides 1350mm	39.46	52.28	1.64	35.79	nr	**88.08**
Sum of two sides 1400mm	40.49	53.65	1.64	35.79	nr	**89.44**
Sum of two sides 1450mm	42.92	56.87	1.64	35.79	nr	**92.66**
Sum of two sides 1500mm	45.57	60.38	1.64	35.79	nr	**96.17**
Sum of two sides 1550mm	48.21	63.88	1.64	35.79	nr	**99.67**
Sum of two sides 1600mm	50.86	67.39	1.64	35.79	nr	**103.18**
Sum of two sides 1650mm	53.50	70.89	1.64	35.79	nr	**106.68**
Sum of two sides 1700mm	56.15	74.40	1.69	36.88	nr	**111.28**
Sum of two sides 1750mm	58.79	77.90	1.69	36.88	nr	**114.78**
Sum of two sides 1800mm	61.55	81.55	1.69	36.88	nr	**118.44**
Sum of two sides 1850mm	64.20	85.06	1.69	36.88	nr	**121.95**
Sum of two sides 1900mm	66.85	88.58	1.85	40.38	nr	**128.95**
Sum of two sides 1950mm	69.49	92.07	1.85	40.38	nr	**132.45**
Sum of two sides 2000mm	72.13	95.57	1.85	40.38	nr	**135.95**
Grille neck						
Sum of two sides 1100mm	35.24	46.69	1.44	31.43	nr	**78.12**
Sum of two sides 1150mm	36.21	47.98	1.44	31.43	nr	**79.41**
Sum of two sides 1200mm	37.17	49.25	1.44	31.43	nr	**80.68**
Sum of two sides 1250mm	38.14	50.54	1.44	31.43	nr	**81.96**
Sum of two sides 1300mm	39.10	51.81	1.69	36.88	nr	**88.69**
Sum of two sides 1350mm	40.07	53.09	1.69	36.88	nr	**89.98**
Sum of two sides 1400mm	41.03	54.36	1.69	36.88	nr	**91.25**
Sum of two sides 1450mm	43.72	57.93	1.69	36.88	nr	**94.81**
Sum of two sides 1500mm	46.63	61.78	1.79	39.07	nr	**100.85**
Sum of two sides 1550mm	49.55	65.65	1.79	39.07	nr	**104.72**
Sum of two sides 1600mm	52.47	69.52	1.79	39.07	nr	**108.59**

Material Costs/Measured Work Prices - Mechanical Installations

U:VENTILATION/AIR CONDITIONING SYSTEMS

Item	Net Price £	Material £	Labour hours	Labour £	Unit	Total rate £
U10 : DUCTWORK : RECTANGULAR - CLASS B						
Y30 - AIR DUCTLINES						
Galvanised sheet metal DW144 class B rectangular section ductwork; including all necessary stiffeners, joints, couplers in the running length and duct supports (Continued)						
Grille neck						
Sum of two sides 1650mm	55.39	73.39	1.79	39.07	nr	**112.46**
Sum of two sides 1700mm	58.31	77.26	1.86	40.59	nr	**117.85**
Sum of two sides 1750mm	61.22	81.12	1.86	40.59	nr	**121.71**
Sum of two sides 1800mm	64.15	85.00	2.02	44.09	nr	**129.08**
Sum of two sides 1850mm	67.06	88.85	2.02	44.09	nr	**132.94**
Sum of two sides 1900mm	69.98	92.72	2.02	44.09	nr	**136.81**
Sum of two sides 1950mm	72.90	96.59	2.02	44.09	nr	**140.68**
Sum of two sides 2000mm	75.82	100.46	2.02	44.09	nr	**144.55**
Ductwork 1001 to 1250mm longest side						
Sum of two sides 1300mm	51.32	68.00	1.40	30.55	m	**98.55**
Sum of two sides 1350mm	52.50	69.56	1.40	30.55	m	**100.12**
Sum of two sides 1400mm	53.87	71.38	1.40	30.55	m	**101.93**
Sum of two sides 1450mm	55.14	73.06	1.40	30.55	m	**103.61**
Sum of two sides 1500mm	56.42	74.76	1.48	32.30	m	**107.06**
Sum of two sides 1550mm	57.69	76.44	1.48	32.30	m	**108.74**
Sum of two sides 1600mm	59.13	78.35	1.55	33.83	m	**112.18**
Sum of two sides 1650mm	60.40	80.03	1.55	33.83	m	**113.86**
Sum of two sides 1700mm	61.68	81.73	1.55	33.83	m	**115.55**
Sum of two sides 1750mm	62.95	83.41	1.55	33.83	m	**117.24**
Sum of two sides 1800mm	64.22	85.09	1.61	35.14	m	**120.23**
Sum of two sides 1850mm	65.49	86.77	1.61	35.14	m	**121.91**
Sum of two sides 1900mm	66.77	88.47	1.61	35.14	m	**123.61**
Sum of two sides 1950mm	68.12	90.26	1.61	35.14	m	**125.40**
Sum of two sides 2000mm	69.48	92.06	1.61	35.14	m	**127.20**
Sum of two sides 2050mm	70.75	93.74	1.61	35.14	m	**128.88**
Sum of two sides 2100mm	72.03	95.44	2.17	47.36	m	**142.80**
Sum of two sides 2150mm	73.30	97.12	2.17	47.36	m	**144.48**
Sum of two sides 2200mm	74.57	98.81	2.19	47.80	m	**146.60**
Sum of two sides 2250mm	75.85	100.50	2.19	47.80	m	**148.30**
Sum of two sides 2300mm	77.12	102.18	2.19	47.80	m	**149.98**
Sum of two sides 2350mm	78.39	103.87	2.19	47.80	m	**151.66**
Sum of two sides 2400mm	79.67	105.56	2.38	51.94	m	**157.51**
Sum of two sides 2450mm	80.94	107.25	2.38	51.94	m	**159.19**
Sum of two sides 2500mm	82.22	108.94	2.38	51.94	m	**160.88**

Material Costs/Measured Work Prices - Mechanical Installations

U:VENTILATION/AIR CONDITIONING SYSTEMS

Item	Net Price £	Material £	Labour hours	Labour £	Unit	Total rate £
Extra over fittings; Ductwork 1001 to 1250mm longest side						
End Cap						
Sum of two sides 1300mm	22.25	29.48	1.69	36.88	nr	66.36
Sum of two sides 1350mm	22.96	30.42	1.69	36.88	nr	67.31
Sum of two sides 1400mm	23.67	31.36	1.69	36.88	nr	68.25
Sum of two sides 1450mm	24.95	33.06	1.69	36.88	nr	69.94
Sum of two sides 1500mm	26.31	34.86	1.69	36.88	nr	71.74
Sum of two sides 1550mm	27.67	36.66	1.69	36.88	nr	73.55
Sum of two sides 1600mm	29.02	38.45	1.69	36.88	nr	75.34
Sum of two sides 1650mm	30.38	40.25	1.69	36.88	nr	77.14
Sum of two sides 1700mm	31.75	42.07	1.69	36.88	nr	78.95
Sum of two sides 1750mm	33.11	43.87	1.69	36.88	nr	80.75
Sum of two sides 1800mm	34.47	45.67	1.69	36.88	nr	82.56
Sum of two sides 1850mm	35.83	47.47	1.69	36.88	nr	84.36
Sum of two sides 1900mm	37.19	49.28	1.69	36.88	nr	86.16
Sum of two sides 1950mm	38.55	51.08	1.69	36.88	nr	87.96
Sum of two sides 2000mm	39.91	52.88	1.69	36.88	nr	89.76
Sum of two sides 2050mm	41.26	54.67	1.69	36.88	nr	91.55
Sum of two sides 2100mm	42.62	56.47	1.69	36.88	nr	93.36
Sum of two sides 2150mm	43.99	58.29	1.69	36.88	nr	95.17
Sum of two sides 2200mm	45.35	60.09	1.69	36.88	nr	96.97
Sum of two sides 2250mm	46.71	61.89	1.69	36.88	nr	98.77
Sum of two sides 2300mm	48.07	63.69	1.69	36.88	nr	100.58
Sum of two sides 2350mm	49.43	65.49	1.69	36.88	nr	102.38
Sum of two sides 2400mm	50.79	67.30	1.69	36.88	nr	104.18
Sum of two sides 2450mm	52.15	69.10	1.69	36.88	nr	105.98
Sum of two sides 2500mm	53.51	70.90	1.69	36.88	nr	107.78
Reducer						
Sum of two sides 1300mm	17.98	23.82	1.69	36.88	nr	60.71
Sum of two sides 1350mm	18.37	24.34	1.69	36.88	nr	61.22
Sum of two sides 1400mm	18.76	24.86	1.69	36.88	nr	61.74
Sum of two sides 1450mm	20.30	26.90	1.69	36.88	nr	63.78
Sum of two sides 1500mm	21.99	29.14	2.47	53.91	nr	83.04
Sum of two sides 1550mm	23.69	31.39	2.47	53.91	nr	85.30
Sum of two sides 1600mm	25.53	33.83	2.47	53.91	nr	87.73
Sum of two sides 1650mm	27.22	36.07	2.47	53.91	nr	89.97
Sum of two sides 1700mm	28.91	38.31	2.47	53.91	nr	92.21
Sum of two sides 1750mm	30.61	40.56	2.47	53.91	nr	94.47
Sum of two sides 1800mm	32.30	42.80	2.59	56.53	nr	99.32
Sum of two sides 1850mm	33.99	45.04	2.59	56.53	nr	101.56
Sum of two sides 1900mm	35.68	47.28	2.71	59.14	nr	106.42
Sum of two sides 1950mm	37.53	49.73	2.71	59.14	nr	108.87
Sum of two sides 2000mm	39.22	51.97	2.59	56.53	nr	108.49
Sum of two sides 2050mm	40.91	54.21	2.71	59.14	nr	113.35
Sum of two sides 2100mm	42.61	56.46	2.92	63.73	nr	120.19
Sum of two sides 2150mm	44.30	58.70	2.92	63.73	nr	122.43
Sum of two sides 2200mm	45.99	60.94	2.92	63.73	nr	124.66
Sum of two sides 2250mm	47.69	63.19	2.92	63.73	nr	126.92
Sum of two sides 2300mm	49.38	65.43	2.92	63.73	nr	129.16
Sum of two sides 2350mm	51.08	67.68	2.92	63.73	nr	131.41
Sum of two sides 2400mm	52.77	69.92	3.12	68.09	nr	138.01
Sum of two sides 2450mm	54.46	72.16	3.12	68.09	nr	140.25
Sum of two sides 2500mm	56.15	74.40	3.12	68.09	nr	142.49

302   *Material Costs/Measured Work Prices - Mechanical Installations*

U:VENTILATION/AIR CONDITIONING SYSTEMS

Item	Net Price £	Material £	Labour hours	Labour £	Unit	Total rate £
U10 : DUCTWORK : RECTANGULAR - CLASS B						
Y30 - AIR DUCTLINES						
Galvanised sheet metal DW144 class B rectangular section ductwork; including all necessary stiffeners, joints, couplers in the running length and duct supports (Continued)						
Offset						
Sum of two sides 1300mm	51.66	68.45	1.69	36.88	nr	**105.33**
Sum of two sides 1350mm	51.19	67.83	1.69	36.88	nr	**104.71**
Sum of two sides 1400mm	52.26	69.24	1.69	36.88	nr	**106.13**
Sum of two sides 1450mm	54.23	71.85	1.69	36.88	nr	**108.74**
Sum of two sides 1500mm	56.41	74.74	2.47	53.91	nr	**128.65**
Sum of two sides 1550mm	58.56	77.59	2.47	53.91	nr	**131.50**
Sum of two sides 1600mm	60.81	80.57	2.47	53.91	nr	**134.48**
Sum of two sides 1650mm	62.89	83.33	2.47	53.91	nr	**137.24**
Sum of two sides 1700mm	64.95	86.06	2.59	56.53	nr	**142.58**
Sum of two sides 1750mm	66.98	88.75	2.59	56.53	nr	**145.27**
Sum of two sides 1800mm	68.98	91.40	2.61	56.96	nr	**148.36**
Sum of two sides 1850mm	70.94	94.00	2.61	56.96	nr	**150.96**
Sum of two sides 1900mm	72.89	96.58	2.71	59.14	nr	**155.72**
Sum of two sides 1950mm	73.90	97.92	2.71	59.14	nr	**157.06**
Sum of two sides 2000mm	76.75	101.69	2.71	59.14	nr	**160.84**
Sum of two sides 2050mm	78.59	104.13	2.71	59.14	nr	**163.28**
Sum of two sides 2100mm	80.41	106.54	2.92	63.73	nr	**170.27**
Sum of two sides 2150mm	82.20	108.92	2.92	63.73	nr	**172.64**
Sum of two sides 2200mm	83.96	111.25	3.26	71.15	nr	**182.40**
Sum of two sides 2250mm	86.75	114.94	3.26	71.15	nr	**186.09**
Sum of two sides 2300mm	89.05	117.99	3.26	71.15	nr	**189.14**
Sum of two sides 2350mm	91.84	121.69	3.26	71.15	nr	**192.84**
Sum of two sides 2400mm	94.62	125.37	3.48	75.95	nr	**201.32**
Sum of two sides 2450mm	97.40	129.06	3.48	75.95	nr	**205.00**
Sum of two sides 2500mm	100.19	132.75	3.48	75.95	nr	**208.70**
Square to round						
Sum of two sides 1300mm	29.30	38.82	1.69	36.88	nr	**75.71**
Sum of two sides 1350mm	30.54	40.47	1.69	36.88	nr	**77.35**
Sum of two sides 1400mm	31.77	42.10	1.69	36.88	nr	**78.98**
Sum of two sides 1450mm	34.44	45.63	1.69	36.88	nr	**82.52**
Sum of two sides 1500mm	37.30	49.42	2.38	51.94	nr	**101.37**
Sum of two sides 1550mm	40.16	53.21	2.38	51.94	nr	**105.15**
Sum of two sides 1600mm	43.05	57.04	2.38	51.94	nr	**108.98**
Sum of two sides 1650mm	45.91	60.83	2.38	51.94	nr	**112.77**
Sum of two sides 1700mm	48.78	64.63	2.55	55.65	nr	**120.29**
Sum of two sides 1750mm	51.63	68.41	2.55	55.65	nr	**124.06**
Sum of two sides 1800mm	54.50	72.21	2.55	55.65	nr	**127.87**
Sum of two sides 1850mm	57.36	76.00	2.55	55.65	nr	**131.65**
Sum of two sides 1900mm	60.23	79.80	2.83	61.76	nr	**141.57**
Sum of two sides 1950mm	63.11	83.62	2.83	61.76	nr	**145.38**
Sum of two sides 2000mm	65.97	87.41	2.83	61.76	nr	**149.17**
Sum of two sides 2050mm	68.84	91.21	2.83	61.76	nr	**152.98**

Material Costs/Measured Work Prices - Mechanical Installations   303

U:VENTILATION/AIR CONDITIONING SYSTEMS

Item	Net Price £	Material £	Labour hours	Labour £	Unit	Total rate £
Sum of two sides 2100mm	71.70	95.00	3.85	84.02	nr	**179.03**
Sum of two sides 2150mm	74.56	98.79	3.85	84.02	nr	**182.82**
Sum of two sides 2200mm	77.42	102.58	4.18	91.23	nr	**193.81**
Sum of two sides 2250mm	80.29	106.38	4.18	91.23	nr	**197.61**
Sum of two sides 2300mm	83.15	110.17	4.22	92.10	nr	**202.27**
Sum of two sides 2350mm	86.01	113.96	4.22	92.10	nr	**206.06**
Sum of two sides 2400mm	88.87	117.75	4.68	102.14	nr	**219.89**
Sum of two sides 2450mm	91.73	121.54	4.68	102.14	nr	**223.68**
Sum of two sides 2500mm	94.59	125.33	4.70	102.58	nr	**227.91**
90 degree radius bend						
Sum of two sides 1300mm	13.72	18.18	1.69	36.88	nr	**55.06**
Sum of two sides 1350mm	14.80	19.61	1.69	36.88	nr	**56.49**
Sum of two sides 1400mm	14.29	18.93	1.69	36.88	nr	**55.82**
Sum of two sides 1450mm	15.72	20.83	1.69	36.88	nr	**57.71**
Sum of two sides 1500mm	17.30	22.92	2.11	46.05	nr	**68.97**
Sum of two sides 1550mm	18.89	25.03	2.11	46.05	nr	**71.08**
Sum of two sides 1600mm	20.53	27.20	2.11	46.05	nr	**73.25**
Sum of two sides 1650mm	22.12	29.31	2.11	46.05	nr	**75.36**
Sum of two sides 1700mm	23.70	31.40	2.19	47.80	nr	**79.20**
Sum of two sides 1750mm	25.29	33.51	2.19	47.80	nr	**81.31**
Sum of two sides 1800mm	26.88	35.62	2.19	47.80	nr	**83.41**
Sum of two sides 1850mm	28.46	37.71	2.19	47.80	nr	**85.51**
Sum of two sides 1900mm	30.05	39.82	2.48	54.13	nr	**93.94**
Sum of two sides 1950mm	32.67	43.29	2.26	49.32	nr	**92.61**
Sum of two sides 2000mm	33.28	44.10	2.26	49.32	nr	**93.42**
Sum of two sides 2050mm	34.86	46.19	2.26	49.32	nr	**95.51**
Sum of two sides 2100mm	36.45	48.30	2.48	54.13	nr	**102.42**
Sum of two sides 2150mm	38.03	50.39	2.48	54.13	nr	**104.51**
Sum of two sides 2200mm	39.62	52.50	2.48	54.13	nr	**106.62**
Sum of two sides 2250mm	40.14	53.19	2.48	54.13	nr	**107.31**
Sum of two sides 2300mm	41.71	55.27	2.48	54.13	nr	**109.39**
Sum of two sides 2350mm	43.29	57.36	2.48	54.13	nr	**111.48**
Sum of two sides 2400mm	44.86	59.44	3.90	85.12	nr	**144.56**
Sum of two sides 2450mm	46.43	61.52	3.90	85.12	nr	**146.64**
Sum of two sides 2500mm	47.98	63.57	3.90	85.12	nr	**148.69**
45 degree bend						
Sum of two sides 1300mm	21.61	28.63	1.69	36.88	nr	**65.52**
Sum of two sides 1350mm	22.28	29.52	1.69	36.88	nr	**66.40**
Sum of two sides 1400mm	22.16	29.36	1.69	36.88	nr	**66.25**
Sum of two sides 1450mm	23.58	31.24	1.69	36.88	nr	**68.13**
Sum of two sides 1500mm	25.16	33.34	2.49	54.34	nr	**87.68**
Sum of two sides 1550mm	26.73	35.42	2.49	54.34	nr	**89.76**
Sum of two sides 1600mm	28.46	37.71	2.49	54.34	nr	**92.05**
Sum of two sides 1650mm	30.03	39.79	2.49	54.34	nr	**94.13**
Sum of two sides 1700mm	31.61	41.88	2.67	58.27	nr	**100.15**
Sum of two sides 1750mm	33.19	43.98	2.67	58.27	nr	**102.25**
Sum of two sides 1800mm	34.77	46.07	2.67	58.27	nr	**104.34**
Sum of two sides 1850mm	36.34	48.15	2.67	58.27	nr	**106.42**
Sum of two sides 1900mm	37.92	50.24	3.06	66.78	nr	**117.03**
Sum of two sides 1950mm	40.13	53.17	3.06	66.78	nr	**119.96**
Sum of two sides 2000mm	41.22	54.62	3.06	66.78	nr	**121.40**
Sum of two sides 2050mm	42.80	56.71	3.06	66.78	nr	**123.49**

Material Costs/Measured Work Prices - Mechanical Installations

U:VENTILATION/AIR CONDITIONING SYSTEMS

Item	Net Price £	Material £	Labour hours	Labour £	Unit	Total rate £
U10 : DUCTWORK : RECTANGULAR - CLASS B						
Y30 - AIR DUCTLINES						
Galvanised sheet metal DW144 class B rectangular section ductwork; including all necessary stiffeners, joints, couplers in the running length and duct supports (Continued)						
45 degree bend						
Sum of two sides 2100mm	44.37	58.79	4.05	88.39	nr	147.18
Sum of two sides 2150mm	45.95	60.88	4.05	88.39	nr	149.27
Sum of two sides 2200mm	47.53	62.98	4.05	88.39	nr	151.37
Sum of two sides 2250mm	48.57	64.36	4.15	90.57	nr	154.93
Sum of two sides 2300mm	50.14	66.44	4.39	95.81	nr	162.25
Sum of two sides 2350mm	51.71	68.52	4.39	95.81	nr	164.33
Sum of two sides 2400mm	53.28	70.60	4.85	105.85	nr	176.45
Sum of two sides 2450mm	54.85	72.68	4.85	105.85	nr	178.53
Sum of two sides 2500mm	56.42	74.76	4.85	105.85	nr	180.61
90 degree mitre bend						
Sum of two sides 1300mm	55.39	73.39	1.69	36.88	nr	110.28
Sum of two sides 1350mm	59.65	79.04	1.69	36.88	nr	115.92
Sum of two sides 1400mm	61.08	80.93	1.69	36.88	nr	117.81
Sum of two sides 1450mm	65.64	86.97	1.69	36.88	nr	123.86
Sum of two sides 1500mm	70.44	93.33	2.80	61.11	nr	154.44
Sum of two sides 1550mm	75.23	99.68	2.80	61.11	nr	160.79
Sum of two sides 1600mm	80.06	106.08	2.80	61.11	nr	167.19
Sum of two sides 1650mm	84.86	112.44	2.80	61.11	nr	173.55
Sum of two sides 1700mm	89.66	118.80	2.95	64.38	nr	183.18
Sum of two sides 1750mm	94.45	125.15	2.95	64.38	nr	189.53
Sum of two sides 1800mm	99.25	131.51	2.95	64.38	nr	195.89
Sum of two sides 1850mm	104.05	137.87	2.95	64.38	nr	202.25
Sum of two sides 1900mm	108.85	144.23	4.05	88.39	nr	232.62
Sum of two sides 1950mm	115.01	152.39	4.05	88.39	nr	240.78
Sum of two sides 2000mm	118.47	156.97	4.05	88.39	nr	245.36
Sum of two sides 2050mm	123.27	163.33	4.05	88.39	nr	251.72
Sum of two sides 2100mm	128.06	169.68	4.07	88.83	nr	258.51
Sum of two sides 2150mm	132.86	176.04	4.07	88.83	nr	264.87
Sum of two sides 2200mm	137.66	182.40	4.07	88.83	nr	271.23
Sum of two sides 2250mm	141.11	186.97	4.07	88.83	nr	275.80
Sum of two sides 2300mm	145.91	193.33	4.39	95.81	nr	289.14
Sum of two sides 2350mm	150.72	199.70	4.39	95.81	nr	295.51
Sum of two sides 2400mm	155.52	206.06	4.85	105.85	nr	311.91
Sum of two sides 2450mm	160.33	212.44	4.85	105.85	nr	318.29
Sum of two sides 2500mm	165.13	218.80	4.85	105.85	nr	324.65
Branch						
Sum of two sides 1300mm	39.22	51.97	1.44	31.43	nr	83.39
Sum of two sides 1350mm	40.14	53.19	1.44	31.43	nr	84.61
Sum of two sides 1400mm	41.07	54.42	1.44	31.43	nr	85.85
Sum of two sides 1450mm	43.52	57.66	1.44	31.43	nr	89.09
Sum of two sides 1500mm	46.18	61.19	1.64	35.79	nr	96.98

Material Costs/Measured Work Prices - Mechanical Installations

U:VENTILATION/AIR CONDITIONING SYSTEMS

Item	Net Price £	Material £	Labour hours	Labour £	Unit	Total rate £
Sum of two sides 1550mm	48.84	64.71	1.64	35.79	nr	**100.51**
Sum of two sides 1600mm	51.63	68.41	1.64	35.79	nr	**104.20**
Sum of two sides 1650mm	54.29	71.93	1.64	35.79	nr	**107.73**
Sum of two sides 1700mm	56.96	75.47	1.64	35.79	nr	**111.26**
Sum of two sides 1750mm	59.62	79.00	1.64	35.79	nr	**114.79**
Sum of two sides 1800mm	62.28	82.52	1.64	35.79	nr	**118.31**
Sum of two sides 1850mm	64.94	86.05	1.64	35.79	nr	**121.84**
Sum of two sides 1900mm	67.61	89.58	1.69	36.88	nr	**126.47**
Sum of two sides 1950mm	70.39	93.27	1.69	36.88	nr	**130.15**
Sum of two sides 2000mm	73.05	96.79	1.69	36.88	nr	**133.67**
Sum of two sides 2050mm	75.71	100.32	1.69	36.88	nr	**137.20**
Sum of two sides 2100mm	78.38	103.85	1.85	40.38	nr	**144.23**
Sum of two sides 2150mm	81.04	107.38	1.85	40.38	nr	**147.75**
Sum of two sides 2200mm	83.70	110.90	1.85	40.38	nr	**151.28**
Sum of two sides 2250mm	86.36	114.43	1.85	40.38	nr	**154.80**
Sum of two sides 2300mm	89.03	117.96	2.61	56.96	nr	**174.93**
Sum of two sides 2350mm	91.69	121.49	2.61	56.96	nr	**178.45**
Sum of two sides 2400mm	94.35	125.01	2.61	56.96	nr	**181.98**
Sum of two sides 2450mm	97.02	128.55	2.61	56.96	nr	**185.51**
Sum of two sides 2500mm	99.69	132.09	2.61	56.96	nr	**189.05**
Grille neck						
Sum of two sides 1300mm	39.91	52.88	1.79	39.07	nr	**91.95**
Sum of two sides 1350mm	40.90	54.19	1.79	39.07	nr	**93.26**
Sum of two sides 1400mm	41.90	55.52	1.79	39.07	nr	**94.58**
Sum of two sides 1450mm	44.61	59.11	1.79	39.07	nr	**98.17**
Sum of two sides 1500mm	47.56	63.02	1.79	39.07	nr	**102.08**
Sum of two sides 1550mm	50.51	66.93	1.79	39.07	nr	**105.99**
Sum of two sides 1600mm	53.46	70.83	1.79	39.07	nr	**109.90**
Sum of two sides 1650mm	56.41	74.74	1.79	39.07	nr	**113.81**
Sum of two sides 1700mm	59.35	78.64	1.86	40.59	nr	**119.23**
Sum of two sides 1750mm	62.30	82.55	1.86	40.59	nr	**123.14**
Sum of two sides 1800mm	65.26	86.47	2.02	44.09	nr	**130.56**
Sum of two sides 1850mm	68.21	90.38	2.02	44.09	nr	**134.46**
Sum of two sides 1900mm	71.15	94.27	2.02	44.09	nr	**138.36**
Sum of two sides 1950mm	74.10	98.18	2.02	44.09	nr	**142.27**
Sum of two sides 2000mm	77.05	102.09	2.02	44.09	nr	**146.18**
Sum of two sides 2050mm	80.00	106.00	2.02	44.09	nr	**150.09**
Sum of two sides 2100mm	82.95	109.91	2.61	56.96	nr	**166.87**
Sum of two sides 2150mm	85.90	113.82	2.61	56.96	nr	**170.78**
Sum of two sides 2200mm	88.85	117.73	2.61	56.96	nr	**174.69**
Sum of two sides 2250mm	91.80	121.64	2.61	56.96	nr	**178.60**
Sum of two sides 2300mm	94.75	125.54	2.61	56.96	nr	**182.51**
Sum of two sides 2350mm	97.70	129.45	2.61	56.96	nr	**186.41**
Sum of two sides 2400mm	100.65	133.36	2.88	62.85	nr	**196.22**
Sum of two sides 2450mm	103.60	137.27	2.88	62.85	nr	**200.12**
Sum of two sides 2500mm	106.55	141.18	2.88	62.85	nr	**204.03**
Ductwork 1251 to 1600mm longest side						
Sum of two sides 1700mm	68.98	91.40	1.55	33.83	m	**125.23**
Sum of two sides 1750mm	70.34	93.20	1.55	33.83	m	**127.03**
Sum of two sides 1800mm	71.71	95.02	1.61	35.14	m	**130.15**
Sum of two sides 1850mm	73.10	96.86	1.61	35.14	m	**132.00**
Sum of two sides 1900mm	74.50	98.71	1.61	35.14	m	**133.85**

Material Costs/Measured Work Prices - Mechanical Installations

U:VENTILATION/AIR CONDITIONING SYSTEMS

Item	Net Price £	Material £	Labour hours	Labour £	Unit	Total rate £
U10 : DUCTWORK : RECTANGULAR - CLASS B						
Y30 - AIR DUCTLINES						
Galvanised sheet metal DW144 class B rectangular section ductwork; including all necessary stiffeners, joints, couplers in the running length and duct supports (Continued)						
Ductwork 1251 to 1600mm longest side						
Sum of two sides 1950mm	75.81	100.45	1.61	35.14	m	**135.59**
Sum of two sides 2000mm	77.13	102.20	1.61	35.14	m	**137.33**
Sum of two sides 2050mm	78.44	103.93	1.61	35.14	m	**139.07**
Sum of two sides 2100mm	79.75	105.67	2.17	47.36	m	**153.03**
Sum of two sides 2150mm	81.07	107.42	2.17	47.36	m	**154.78**
Sum of two sides 2200mm	82.39	109.17	2.19	47.80	m	**156.96**
Sum of two sides 2250mm	83.78	111.01	2.19	47.80	m	**158.80**
Sum of two sides 2300mm	85.18	112.86	2.19	47.80	m	**160.66**
Sum of two sides 2350mm	86.54	114.67	2.19	47.80	m	**162.46**
Sum of two sides 2400mm	87.90	116.47	2.38	51.94	m	**168.41**
Sum of two sides 2450mm	89.21	118.20	2.38	51.94	m	**170.15**
Sum of two sides 2500mm	90.53	119.95	2.38	51.94	m	**171.89**
Sum of two sides 2550mm	91.84	121.69	2.38	51.94	m	**173.63**
Sum of two sides 2600mm	93.15	123.42	2.64	57.62	m	**181.04**
Sum of two sides 2650mm	94.47	125.17	2.64	57.62	m	**182.79**
Sum of two sides 2700mm	95.79	126.92	2.66	58.05	m	**184.98**
Sum of two sides 2750mm	97.15	128.72	2.66	58.05	m	**186.78**
Sum of two sides 2800mm	98.51	130.53	2.95	64.38	m	**194.91**
Sum of two sides 2850mm	99.86	132.31	2.95	64.38	m	**196.70**
Sum of two sides 2900mm	101.21	134.10	2.96	64.60	m	**198.70**
Sum of two sides 2950mm	102.57	135.91	2.96	64.60	m	**200.51**
Sum of two sides 3000mm	103.93	137.71	3.15	68.75	m	**206.45**
Sum of two sides 3050mm	105.24	139.44	3.15	68.75	m	**208.19**
Sum of two sides 3100mm	106.56	141.19	3.15	68.75	m	**209.94**
Sum of two sides 3150mm	107.87	142.93	3.15	68.75	m	**211.68**
Sum of two sides 3200mm	109.19	144.68	3.18	69.40	m	**214.08**
Extra over fittings; Ductwork 1251 to 1600mm longest side						
End Cap						
Sum of two sides 1700mm	25.87	34.28	0.58	12.66	nr	**46.94**
Sum of two sides 1750mm	27.70	36.70	0.58	12.66	nr	**49.36**
Sum of two sides 1800mm	29.54	39.14	0.58	12.66	nr	**51.80**
Sum of two sides 1850mm	31.38	41.58	0.58	12.66	nr	**54.24**
Sum of two sides 1900mm	33.22	44.02	0.58	12.66	nr	**56.67**
Sum of two sides 1950mm	35.06	46.45	0.58	12.66	nr	**59.11**
Sum of two sides 2000mm	36.90	48.89	0.58	12.66	nr	**61.55**
Sum of two sides 2050mm	38.74	51.33	0.58	12.66	nr	**63.99**
Sum of two sides 2100mm	40.58	53.77	0.87	18.99	nr	**72.76**
Sum of two sides 2150mm	42.42	56.21	0.87	18.99	nr	**75.19**
Sum of two sides 2200mm	44.26	58.64	0.87	18.99	nr	**77.63**
Sum of two sides 2250mm	46.10	61.08	0.87	18.99	nr	**80.07**

Material Costs/Measured Work Prices - Mechanical Installations

U:VENTILATION/AIR CONDITIONING SYSTEMS

Item	Net Price £	Material £	Labour hours	Labour £	Unit	Total rate £
Sum of two sides 2300mm	47.94	63.52	0.87	18.99	nr	**82.51**
Sum of two sides 2350mm	49.78	65.96	0.87	18.99	nr	**84.95**
Sum of two sides 2400mm	51.62	68.40	0.87	18.99	nr	**87.38**
Sum of two sides 2450mm	53.46	70.83	0.87	18.99	nr	**89.82**
Sum of two sides 2500mm	55.30	73.27	0.87	18.99	nr	**92.26**
Sum of two sides 2550mm	57.14	75.71	0.87	18.99	nr	**94.70**
Sum of two sides 2600mm	58.98	78.15	0.87	18.99	nr	**97.14**
Sum of two sides 2650mm	60.82	80.59	0.87	18.99	nr	**99.57**
Sum of two sides 2700mm	62.66	83.02	0.87	18.99	nr	**102.01**
Sum of two sides 2750mm	64.50	85.46	0.87	18.99	nr	**104.45**
Sum of two sides 2800mm	66.34	87.90	1.16	25.32	nr	**113.22**
Sum of two sides 2850mm	68.18	90.34	1.16	25.32	nr	**115.66**
Sum of two sides 2900mm	70.02	92.78	1.16	25.32	nr	**118.09**
Sum of two sides 2950mm	71.86	95.21	1.16	25.32	nr	**120.53**
Sum of two sides 3000mm	76.65	101.56	1.73	37.76	nr	**139.32**
Sum of two sides 3050mm	75.31	99.79	1.73	37.76	nr	**137.54**
Sum of two sides 3100mm	76.92	101.92	1.80	39.28	nr	**141.20**
Sum of two sides 3150mm	78.53	104.05	1.80	39.28	nr	**143.34**
Sum of two sides 3200mm	80.15	106.20	1.80	39.28	nr	**145.48**
Reducer						
Sum of two sides 1700mm	31.14	41.26	2.47	53.91	nr	**95.17**
Sum of two sides 1750mm	32.88	43.57	2.47	53.91	nr	**97.47**
Sum of two sides 1800mm	34.61	45.86	2.59	56.53	nr	**102.38**
Sum of two sides 1850mm	36.41	48.24	2.59	56.53	nr	**104.77**
Sum of two sides 1900mm	38.20	50.62	2.71	59.14	nr	**109.76**
Sum of two sides 1950mm	39.94	52.92	2.71	59.14	nr	**112.07**
Sum of two sides 2000mm	41.68	55.23	2.71	59.14	nr	**114.37**
Sum of two sides 2050mm	43.42	57.53	2.71	59.14	nr	**116.68**
Sum of two sides 2100mm	45.15	59.82	2.92	63.73	nr	**123.55**
Sum of two sides 2150mm	46.89	62.13	2.92	63.73	nr	**125.86**
Sum of two sides 2200mm	48.63	64.43	2.92	63.73	nr	**128.16**
Sum of two sides 2250mm	50.42	66.81	2.92	63.73	nr	**130.53**
Sum of two sides 2300mm	52.22	69.19	2.92	63.73	nr	**132.92**
Sum of two sides 2350mm	53.95	71.48	2.92	63.73	nr	**135.21**
Sum of two sides 2400mm	55.69	73.79	3.12	68.09	nr	**141.88**
Sum of two sides 2450mm	57.43	76.09	3.12	68.09	nr	**144.19**
Sum of two sides 2500mm	59.17	78.40	3.12	68.09	nr	**146.49**
Sum of two sides 2550mm	60.90	80.69	3.12	68.09	nr	**148.79**
Sum of two sides 2600mm	62.14	82.34	3.12	68.09	nr	**150.43**
Sum of two sides 2650mm	64.38	85.30	3.12	68.09	nr	**153.40**
Sum of two sides 2700mm	66.11	87.60	3.12	68.09	nr	**155.69**
Sum of two sides 2750mm	67.85	89.90	3.12	68.09	nr	**157.99**
Sum of two sides 2800mm	69.59	92.21	3.95	86.21	nr	**178.41**
Sum of two sides 2850mm	71.38	94.58	3.95	86.21	nr	**180.79**
Sum of two sides 2900mm	73.18	96.96	3.97	86.64	nr	**183.61**
Sum of two sides 2950mm	74.92	99.27	3.97	86.64	nr	**185.91**
Sum of two sides 3000mm	76.65	101.56	4.52	98.65	nr	**200.21**
Sum of two sides 3050mm	77.93	103.26	4.52	98.65	nr	**201.90**
Sum of two sides 3100mm	79.21	104.95	4.52	98.65	nr	**203.60**
Sum of two sides 3150mm	80.49	106.65	4.52	98.65	nr	**205.30**
Sum of two sides 3200mm	81.78	108.36	4.52	98.65	nr	**207.01**

Material Costs/Measured Work Prices - Mechanical Installations

U:VENTILATION/AIR CONDITIONING SYSTEMS

Item	Net Price £	Material £	Labour hours	Labour £	Unit	Total rate £
U10 : DUCTWORK : RECTANGULAR - CLASS B						
Y30 - AIR DUCTLINES						
Galvanised sheet metal DW144 class B rectangular section ductwork; including all necessary stiffeners, joints, couplers in the running length and duct supports (Continued)						
Offset						
Sum of two sides 1700mm	75.82	100.46	2.59	56.53	nr	156.99
Sum of two sides 1750mm	78.87	104.50	2.59	56.53	nr	161.03
Sum of two sides 1800mm	81.92	108.54	2.61	56.96	nr	165.51
Sum of two sides 1850mm	83.97	111.26	2.61	56.96	nr	168.22
Sum of two sides 1900mm	86.02	113.98	2.71	59.14	nr	173.12
Sum of two sides 1950mm	87.94	116.52	2.71	59.14	nr	175.67
Sum of two sides 2000mm	89.87	119.08	2.71	59.14	nr	178.22
Sum of two sides 2050mm	91.72	121.53	2.71	59.14	nr	180.67
Sum of two sides 2100mm	93.58	123.99	2.92	63.73	nr	187.72
Sum of two sides 2150mm	95.38	126.38	2.92	63.73	nr	190.11
Sum of two sides 2200mm	97.17	128.75	3.26	71.15	nr	199.90
Sum of two sides 2250mm	98.94	131.10	3.26	71.15	nr	202.24
Sum of two sides 2300mm	100.70	133.43	3.26	71.15	nr	204.58
Sum of two sides 2350mm	103.60	137.27	3.26	71.15	nr	208.42
Sum of two sides 2400mm	106.49	141.10	3.47	75.73	nr	216.83
Sum of two sides 2450mm	108.13	143.27	3.47	75.73	nr	219.00
Sum of two sides 2500mm	109.76	145.43	3.48	75.95	nr	221.38
Sum of two sides 2550mm	111.68	147.98	3.48	75.95	nr	223.93
Sum of two sides 2600mm	113.60	150.52	3.49	76.17	nr	226.69
Sum of two sides 2650mm	116.44	154.28	3.49	76.17	nr	230.45
Sum of two sides 2700mm	119.28	158.05	3.50	76.39	nr	234.43
Sum of two sides 2750mm	122.13	161.82	3.50	76.39	nr	238.21
Sum of two sides 2800mm	124.97	165.59	4.34	94.72	nr	260.30
Sum of two sides 2850mm	127.83	169.37	4.34	94.72	nr	264.09
Sum of two sides 2900mm	130.70	173.18	4.76	103.89	nr	277.06
Sum of two sides 2950mm	133.54	176.94	4.76	103.89	nr	280.83
Sum of two sides 3000mm	136.38	180.70	5.32	116.11	nr	296.81
Sum of two sides 3050mm	138.54	183.57	5.32	116.11	nr	299.67
Sum of two sides 3100mm	140.70	186.43	5.35	116.76	nr	303.19
Sum of two sides 3150mm	142.86	189.29	5.35	116.76	nr	306.05
Sum of two sides 3200mm	145.02	192.15	5.35	116.76	nr	308.91
Square to round						
Sum of two sides 1700mm	42.65	56.51	2.55	55.65	nr	112.16
Sum of two sides 1750mm	45.51	60.30	2.55	55.65	nr	115.95
Sum of two sides 1800mm	48.38	64.10	2.55	55.65	nr	119.76
Sum of two sides 1850mm	51.24	67.89	2.55	55.65	nr	123.55
Sum of two sides 1900mm	54.10	71.68	2.83	61.76	nr	133.45
Sum of two sides 1950mm	56.97	75.49	2.83	61.76	nr	137.25
Sum of two sides 2000mm	59.83	79.27	2.83	61.76	nr	141.04
Sum of two sides 2050mm	62.70	83.08	2.83	61.76	nr	144.84
Sum of two sides 2100mm	65.57	86.88	3.85	84.02	nr	170.90
Sum of two sides 2150mm	68.43	90.67	3.85	84.02	nr	174.69

Material Costs/Measured Work Prices - Mechanical Installations

U:VENTILATION/AIR CONDITIONING SYSTEMS

Item	Net Price £	Material £	Labour hours	Labour £	Unit	Total rate £
Sum of two sides 2200mm	71.30	94.47	4.18	91.23	nr	**185.70**
Sum of two sides 2250mm	74.16	98.26	4.18	91.23	nr	**189.49**
Sum of two sides 2300mm	77.02	102.05	4.22	92.10	nr	**194.15**
Sum of two sides 2350mm	79.89	105.85	4.22	92.10	nr	**197.95**
Sum of two sides 2400mm	82.76	109.66	4.68	102.14	nr	**211.80**
Sum of two sides 2450mm	85.62	113.45	4.68	102.14	nr	**215.59**
Sum of two sides 2500mm	88.48	117.24	4.70	102.58	nr	**219.81**
Sum of two sides 2550mm	91.35	121.04	4.70	102.58	nr	**223.61**
Sum of two sides 2600mm	94.21	124.83	4.70	102.58	nr	**227.40**
Sum of two sides 2650mm	97.08	128.63	4.70	102.58	nr	**231.21**
Sum of two sides 2700mm	99.95	132.43	4.71	102.79	nr	**235.23**
Sum of two sides 2750mm	102.81	136.22	4.71	102.79	nr	**239.02**
Sum of two sides 2800mm	105.68	140.03	8.19	178.74	nr	**318.77**
Sum of two sides 2850mm	108.54	143.82	8.19	178.74	nr	**322.56**
Sum of two sides 2900mm	111.40	147.60	8.62	188.13	nr	**335.73**
Sum of two sides 2950mm	114.27	151.41	8.62	188.13	nr	**339.54**
Sum of two sides 3000mm	117.14	155.21	8.75	190.97	nr	**346.18**
Sum of two sides 3050mm	119.43	158.24	8.75	190.97	nr	**349.21**
Sum of two sides 3100mm	121.73	161.29	8.75	190.97	nr	**352.26**
Sum of two sides 3150mm	124.02	164.33	8.75	190.97	nr	**355.29**
Sum of two sides 3200mm	126.32	167.37	8.75	190.97	nr	**358.34**
90 degree radius bend						
Sum of two sides 1700mm	72.42	95.96	2.19	47.80	nr	**143.75**
Sum of two sides 1750mm	75.02	99.40	2.19	47.80	nr	**147.20**
Sum of two sides 1800mm	77.63	102.86	2.19	47.80	nr	**150.66**
Sum of two sides 1850mm	81.37	107.82	2.19	47.80	nr	**155.61**
Sum of two sides 1900mm	85.11	112.77	2.26	49.32	nr	**162.09**
Sum of two sides 1950mm	88.72	117.55	2.26	49.32	nr	**166.88**
Sum of two sides 2000mm	92.33	122.34	2.26	49.32	nr	**171.66**
Sum of two sides 2050mm	95.93	127.11	2.26	49.32	nr	**176.43**
Sum of two sides 2100mm	99.54	131.89	2.48	54.13	nr	**186.02**
Sum of two sides 2150mm	103.14	136.66	2.48	54.13	nr	**190.79**
Sum of two sides 2200mm	106.75	141.44	2.48	54.13	nr	**195.57**
Sum of two sides 2250mm	110.49	146.40	2.48	54.13	nr	**200.52**
Sum of two sides 2300mm	114.24	151.37	2.48	54.13	nr	**205.49**
Sum of two sides 2350mm	116.61	154.51	2.48	54.13	nr	**208.63**
Sum of two sides 2400mm	118.98	157.65	3.90	85.12	nr	**242.76**
Sum of two sides 2450mm	122.56	162.39	3.90	85.12	nr	**247.51**
Sum of two sides 2500mm	126.13	167.12	3.90	85.12	nr	**252.24**
Sum of two sides 2550mm	129.70	171.85	3.90	85.12	nr	**256.97**
Sum of two sides 2600mm	133.28	176.60	4.26	92.97	nr	**269.57**
Sum of two sides 2650mm	136.86	181.34	4.26	92.97	nr	**274.31**
Sum of two sides 2700mm	140.43	186.07	4.55	99.30	nr	**285.37**
Sum of two sides 2750mm	142.65	189.01	4.55	99.30	nr	**288.31**
Sum of two sides 2800mm	144.86	191.94	4.55	99.30	nr	**291.24**
Sum of two sides 2850mm	149.93	198.66	4.55	99.30	nr	**297.96**
Sum of two sides 2900mm	154.99	205.36	6.87	149.94	nr	**355.30**
Sum of two sides 2950mm	157.14	208.21	6.87	149.94	nr	**358.15**
Sum of two sides 3000mm	159.28	211.05	7.00	152.77	nr	**363.82**
Sum of two sides 3050mm	161.91	214.53	7.00	152.77	nr	**367.30**
Sum of two sides 3100mm	164.54	218.02	7.00	152.77	nr	**370.79**
Sum of two sides 3150mm	167.17	221.50	7.00	152.77	nr	**374.27**
Sum of two sides 3200mm	169.81	225.00	7.00	152.77	nr	**377.77**

310   *Material Costs/Measured Work Prices - Mechanical Installations*

U:VENTILATION/AIR CONDITIONING SYSTEMS

Item	Net Price £	Material £	Labour hours	Labour £	Unit	Total rate £
U10 : DUCTWORK : RECTANGULAR - CLASS B						
Y30 - AIR DUCTLINES						
Galvanised sheet metal DW144 class B rectangular section ductwork; including all necessary stiffeners, joints, couplers in the running length and duct supports (Continued)						
45 degree bend						
Sum of two sides 1700mm	35.27	46.73	2.67	58.27	nr	105.00
Sum of two sides 1750mm	36.45	48.30	2.67	58.27	nr	106.57
Sum of two sides 1800mm	37.63	49.86	2.67	58.27	nr	108.13
Sum of two sides 1850mm	39.37	52.17	2.67	58.27	nr	110.44
Sum of two sides 1900mm	41.11	54.47	3.06	66.78	nr	121.25
Sum of two sides 1950mm	42.79	56.70	3.06	66.78	nr	123.48
Sum of two sides 2000mm	44.47	58.92	3.06	66.78	nr	125.71
Sum of two sides 2050mm	46.15	61.15	3.06	66.78	nr	127.93
Sum of two sides 2100mm	47.82	63.36	4.05	88.39	nr	151.75
Sum of two sides 2150mm	49.50	65.59	4.05	88.39	nr	153.98
Sum of two sides 2200mm	51.18	67.81	4.05	88.39	nr	156.20
Sum of two sides 2250mm	52.93	70.13	4.05	88.39	nr	158.52
Sum of two sides 2300mm	54.67	72.44	4.39	95.81	nr	168.25
Sum of two sides 2350mm	55.43	73.44	4.39	95.81	nr	169.25
Sum of two sides 2400mm	56.79	75.25	4.85	105.85	nr	181.10
Sum of two sides 2450mm	58.45	77.45	4.85	105.85	nr	183.30
Sum of two sides 2500mm	60.11	79.65	4.85	105.85	nr	185.50
Sum of two sides 2550mm	61.78	81.86	4.85	105.85	nr	187.71
Sum of two sides 2600mm	63.44	84.06	4.87	106.29	nr	190.34
Sum of two sides 2650mm	65.10	86.26	4.87	106.29	nr	192.54
Sum of two sides 2700mm	66.77	88.47	4.87	106.29	nr	194.76
Sum of two sides 2750mm	67.75	89.77	4.87	106.29	nr	196.05
Sum of two sides 2800mm	68.73	91.07	8.81	192.27	nr	283.34
Sum of two sides 2850mm	71.13	94.25	8.81	192.27	nr	286.52
Sum of two sides 2900mm	73.54	97.44	8.81	192.27	nr	289.72
Sum of two sides 2950mm	74.49	98.70	8.81	192.27	nr	290.97
Sum of two sides 3000mm	75.43	99.94	9.31	203.19	nr	303.13
Sum of two sides 3050mm	76.62	101.52	9.31	203.19	nr	304.71
Sum of two sides 3100mm	77.81	103.10	9.31	203.19	nr	306.29
Sum of two sides 3150mm	79.00	104.67	9.31	203.19	nr	307.86
Sum of two sides 3200mm	80.20	106.27	9.39	204.93	nr	311.20
90 degree mitre bend						
Sum of two sides 1700mm	84.89	112.48	2.67	58.27	nr	170.75
Sum of two sides 1750mm	87.83	116.37	2.67	58.27	nr	174.65
Sum of two sides 1800mm	90.79	120.30	2.80	61.11	nr	181.41
Sum of two sides 1850mm	95.46	126.48	2.80	61.11	nr	187.59
Sum of two sides 1900mm	100.14	132.69	2.95	64.38	nr	197.07
Sum of two sides 1950mm	104.83	138.90	2.95	64.38	nr	203.28
Sum of two sides 2000mm	109.52	145.11	2.95	64.38	nr	209.50
Sum of two sides 2050mm	114.21	151.33	2.95	64.38	nr	215.71
Sum of two sides 2100mm	118.89	157.53	4.05	88.39	nr	245.92
Sum of two sides 2150mm	123.58	163.74	4.05	88.39	nr	252.13

Material Costs/Measured Work Prices - Mechanical Installations

U:VENTILATION/AIR CONDITIONING SYSTEMS

Item	Net Price £	Material £	Labour hours	Labour £	Unit	Total rate £
Sum of two sides 2200mm	128.27	169.96	4.05	88.39	nr	**258.35**
Sum of two sides 2250mm	132.95	176.16	4.05	88.39	nr	**264.55**
Sum of two sides 2300mm	137.63	182.36	4.39	95.81	nr	**278.17**
Sum of two sides 2350mm	140.54	186.22	4.39	95.81	nr	**282.03**
Sum of two sides 2400mm	143.45	190.07	4.85	105.85	nr	**295.92**
Sum of two sides 2450mm	148.15	196.30	4.85	105.85	nr	**302.15**
Sum of two sides 2500mm	152.84	202.51	4.85	105.85	nr	**308.36**
Sum of two sides 2550mm	157.54	208.74	4.85	105.85	nr	**314.59**
Sum of two sides 2600mm	162.24	214.97	4.87	106.29	nr	**321.25**
Sum of two sides 2650mm	166.94	221.20	4.87	106.29	nr	**327.48**
Sum of two sides 2700mm	171.65	227.44	4.87	106.29	nr	**333.72**
Sum of two sides 2750mm	174.51	231.23	4.87	106.29	nr	**337.51**
Sum of two sides 2800mm	177.37	235.02	8.81	192.27	nr	**427.29**
Sum of two sides 2850mm	184.65	244.66	8.81	192.27	nr	**436.94**
Sum of two sides 2900mm	191.93	254.31	14.81	323.22	nr	**577.53**
Sum of two sides 2950mm	194.86	258.19	14.81	323.22	nr	**581.41**
Sum of two sides 3000mm	197.79	262.07	15.20	331.73	nr	**593.81**
Sum of two sides 3050mm	201.81	267.40	15.20	331.73	nr	**599.13**
Sum of two sides 3100mm	205.84	272.74	15.60	340.46	nr	**613.20**
Sum of two sides 3150mm	209.86	278.06	15.60	340.46	nr	**618.53**
Sum of two sides 3200mm	213.89	283.40	15.60	340.46	nr	**623.87**
Branch						
Sum of two sides 1700mm	61.24	81.14	1.69	36.88	nr	**118.03**
Sum of two sides 1750mm	63.97	84.76	1.69	36.88	nr	**121.64**
Sum of two sides 1800mm	66.70	88.38	1.69	36.88	nr	**125.26**
Sum of two sides 1850mm	69.49	92.07	1.69	36.88	nr	**128.96**
Sum of two sides 1900mm	72.28	95.77	1.85	40.38	nr	**136.15**
Sum of two sides 1950mm	75.01	99.39	1.85	40.38	nr	**139.76**
Sum of two sides 2000mm	77.74	103.01	1.85	40.38	nr	**143.38**
Sum of two sides 2050mm	80.46	106.61	1.85	40.38	nr	**146.99**
Sum of two sides 2100mm	83.19	110.23	2.61	56.96	nr	**167.19**
Sum of two sides 2150mm	85.92	113.84	2.61	56.96	nr	**170.81**
Sum of two sides 2200mm	88.65	117.46	2.61	56.96	nr	**174.42**
Sum of two sides 2250mm	91.44	121.16	2.61	56.96	nr	**178.12**
Sum of two sides 2300mm	94.23	124.85	2.61	56.96	nr	**181.82**
Sum of two sides 2350mm	96.96	128.47	2.16	47.14	nr	**175.61**
Sum of two sides 2400mm	102.26	135.49	2.88	62.85	nr	**198.35**
Sum of two sides 2450mm	102.41	135.69	2.88	62.85	nr	**198.55**
Sum of two sides 2500mm	105.14	139.31	2.88	62.85	nr	**202.17**
Sum of two sides 2550mm	107.87	142.93	2.88	62.85	nr	**205.78**
Sum of two sides 2600mm	110.60	146.54	2.88	62.85	nr	**209.40**
Sum of two sides 2650mm	113.33	150.16	2.88	62.85	nr	**213.02**
Sum of two sides 2700mm	116.06	153.78	2.88	62.85	nr	**216.63**
Sum of two sides 2750mm	118.78	157.38	2.88	62.85	nr	**220.24**
Sum of two sides 2800mm	121.51	161.00	3.94	85.99	nr	**246.99**
Sum of two sides 2850mm	124.30	164.70	3.94	85.99	nr	**250.69**
Sum of two sides 2900mm	127.10	168.41	3.94	85.99	nr	**254.40**
Sum of two sides 2950mm	129.82	172.01	3.94	85.99	nr	**258.00**
Sum of two sides 3000mm	132.55	175.63	4.83	105.41	nr	**281.04**
Sum of two sides 3050mm	134.67	178.44	4.83	105.41	nr	**283.85**
Sum of two sides 3100mm	136.79	181.25	4.83	105.41	nr	**286.66**
Sum of two sides 3150mm	138.91	184.06	4.83	105.41	nr	**289.47**
Sum of two sides 3200mm	141.03	186.86	4.83	105.41	nr	**292.28**

312   *Material Costs/Measured Work Prices - Mechanical Installations*

U:VENTILATION/AIR CONDITIONING SYSTEMS

Item	Net Price £	Material £	Labour hours	Labour £	Unit	Total rate £
U10 : DUCTWORK : RECTANGULAR - CLASS B						
Y30 - AIR DUCTLINES						
Galvanised sheet metal DW144 class B rectangular section ductwork; including all necessary stiffeners, joints, couplers in the running length and duct supports (Continued)						
Grille neck						
Sum of two sides 1700mm	60.21	79.78	1.86	40.59	nr	**120.37**
Sum of two sides 1750mm	63.21	83.75	1.86	40.59	nr	**124.35**
Sum of two sides 1800mm	66.22	87.74	2.02	44.09	nr	**131.83**
Sum of two sides 1850mm	69.22	91.72	2.02	44.09	nr	**135.80**
Sum of two sides 1900mm	72.22	95.69	2.02	44.09	nr	**139.78**
Sum of two sides 1950mm	75.23	99.68	2.02	44.09	nr	**143.77**
Sum of two sides 2000mm	78.24	103.67	2.02	44.09	nr	**147.75**
Sum of two sides 2050mm	81.23	107.63	2.02	44.09	nr	**151.72**
Sum of two sides 2100mm	84.24	111.62	2.61	56.96	nr	**168.58**
Sum of two sides 2150mm	87.24	115.59	2.61	56.96	nr	**172.56**
Sum of two sides 2200mm	90.25	119.58	2.61	56.96	nr	**176.54**
Sum of two sides 2250mm	93.25	123.56	2.61	56.96	nr	**180.52**
Sum of two sides 2300mm	96.25	127.53	2.61	56.96	nr	**184.49**
Sum of two sides 2350mm	99.26	131.52	2.61	56.96	nr	**188.48**
Sum of two sides 2400mm	102.26	135.49	2.88	62.85	nr	**198.35**
Sum of two sides 2450mm	105.26	139.47	2.88	62.85	nr	**202.32**
Sum of two sides 2500mm	108.27	143.46	2.88	62.85	nr	**206.31**
Sum of two sides 2550mm	111.27	147.43	2.88	62.85	nr	**210.29**
Sum of two sides 2600mm	114.27	151.41	2.88	62.85	nr	**214.26**
Sum of two sides 2650mm	117.28	155.40	2.88	62.85	nr	**218.25**
Sum of two sides 2700mm	120.28	159.37	2.88	62.85	nr	**222.23**
Sum of two sides 2750mm	123.28	163.35	2.88	62.85	nr	**226.20**
Sum of two sides 2800mm	126.29	167.33	3.94	85.99	nr	**253.32**
Sum of two sides 2850mm	129.29	171.31	3.94	85.99	nr	**257.30**
Sum of two sides 2900mm	132.30	175.30	4.12	89.92	nr	**265.21**
Sum of two sides 2950mm	135.30	179.27	4.12	89.92	nr	**269.19**
Sum of two sides 3000mm	138.30	183.25	5.00	109.12	nr	**292.37**
Sum of two sides 3050mm	140.62	186.32	5.00	109.12	nr	**295.44**
Sum of two sides 3100mm	142.94	189.40	5.00	109.12	nr	**298.52**
Sum of two sides 3150mm	145.26	192.47	5.00	109.12	nr	**301.59**
Sum of two sides 3200mm	147.58	195.54	5.00	109.12	nr	**304.67**
Ductwork 1601 to 2000mm longest side						
Sum of two sides 2100mm	90.39	119.77	2.17	47.36	m	**167.13**
Sum of two sides 2150mm	91.97	121.86	2.17	47.36	m	**169.22**
Sum of two sides 2200mm	93.15	123.42	2.17	47.36	m	**170.78**
Sum of two sides 2250mm	94.34	125.00	2.17	47.36	m	**172.36**
Sum of two sides 2300mm	95.92	127.09	2.19	47.80	m	**174.89**
Sum of two sides 2350mm	97.43	129.09	2.19	47.80	m	**176.89**
Sum of two sides 2400mm	98.94	131.10	2.38	51.94	m	**183.04**
Sum of two sides 2450mm	100.52	133.19	2.38	51.94	m	**185.13**
Sum of two sides 2500mm	102.10	135.28	2.38	51.94	m	**187.22**
Sum of two sides 2550mm	103.60	137.27	2.38	51.94	m	**189.21**

Material Costs/Measured Work Prices - Mechanical Installations

U:VENTILATION/AIR CONDITIONING SYSTEMS

Item	Net Price £	Material £	Labour hours	Labour £	Unit	Total rate £
Sum of two sides 2600mm	105.10	139.26	2.64	57.62	m	**196.87**
Sum of two sides 2650mm	107.30	142.17	2.64	57.62	m	**199.79**
Sum of two sides 2700mm	109.51	145.10	2.66	58.05	m	**203.15**
Sum of two sides 2750mm	111.79	148.12	2.66	58.05	m	**206.18**
Sum of two sides 2800mm	114.08	151.16	2.95	64.38	m	**215.54**
Sum of two sides 2850mm	114.88	152.22	2.95	64.38	m	**216.60**
Sum of two sides 2900mm	115.67	153.26	2.96	64.60	m	**217.86**
Sum of two sides 2950mm	116.47	154.32	2.96	64.60	m	**218.92**
Sum of two sides 3000mm	117.26	155.37	2.96	64.60	m	**219.97**
Sum of two sides 3050mm	119.79	158.72	2.96	64.60	m	**223.32**
Sum of two sides 3100mm	122.32	162.07	2.96	64.60	m	**226.67**
Sum of two sides 3150mm	124.84	165.41	2.96	64.60	m	**230.01**
Sum of two sides 3200mm	127.37	168.77	3.15	68.75	m	**237.51**
Sum of two sides 3250mm	128.91	170.81	3.15	68.75	m	**239.55**
Sum of two sides 3300mm	130.45	172.85	3.15	68.75	m	**241.59**
Sum of two sides 3350mm	131.99	174.89	3.15	68.75	m	**243.63**
Sum of two sides 3400mm	133.53	176.93	3.15	68.75	m	**245.67**
Sum of two sides 3450mm	135.03	178.91	3.15	68.75	m	**247.66**
Sum of two sides 3500mm	136.53	180.90	3.15	68.75	m	**249.65**
Sum of two sides 3550mm	138.03	182.89	3.15	68.75	m	**251.64**
Sum of two sides 3600mm	139.53	184.88	3.18	69.40	m	**254.28**
Sum of two sides 3650mm	141.07	186.92	3.18	69.40	m	**256.32**
Sum of two sides 3700mm	142.61	188.96	3.18	69.40	m	**258.36**
Sum of two sides 3750mm	144.15	191.00	3.18	69.40	m	**260.40**
Sum of two sides 3800mm	145.69	193.04	3.18	69.40	m	**262.44**
Sum of two sides 3850mm	147.19	195.03	3.18	69.40	m	**264.43**
Sum of two sides 3900mm	148.69	197.01	3.18	69.40	m	**266.42**
Sum of two sides 3950mm	150.19	199.00	3.18	69.40	m	**268.40**
Sum of two sides 4000mm	151.69	200.99	3.18	69.40	m	**270.39**
Extra over fittings; Ductwork 1601 to 2000mm longest side						
End Cap						
Sum of two sides 2100mm	46.70	61.88	0.87	18.99	nr	**80.86**
Sum of two sides 2150mm	48.56	64.34	0.87	18.99	nr	**83.33**
Sum of two sides 2200mm	50.43	66.82	0.87	18.99	nr	**85.81**
Sum of two sides 2250mm	52.30	69.30	0.87	18.99	nr	**88.28**
Sum of two sides 2300mm	54.16	71.76	0.87	18.99	nr	**90.75**
Sum of two sides 2350mm	56.03	74.24	0.87	18.99	nr	**93.23**
Sum of two sides 2400mm	57.90	76.72	0.87	18.99	nr	**95.70**
Sum of two sides 2450mm	59.76	79.18	0.87	18.99	nr	**98.17**
Sum of two sides 2500mm	61.63	81.66	0.87	18.99	nr	**100.65**
Sum of two sides 2550mm	63.50	84.14	0.87	18.99	nr	**103.12**
Sum of two sides 2600mm	65.37	86.62	0.87	18.99	nr	**105.60**
Sum of two sides 2650mm	67.24	89.09	0.87	18.99	nr	**108.08**
Sum of two sides 2700mm	69.10	91.56	0.87	18.99	nr	**110.54**
Sum of two sides 2750mm	70.97	94.04	0.87	18.99	nr	**113.02**
Sum of two sides 2800mm	72.84	96.51	1.16	25.32	nr	**121.83**
Sum of two sides 2850mm	74.71	98.99	1.16	25.32	nr	**124.31**
Sum of two sides 2900mm	76.57	101.46	1.16	25.32	nr	**126.77**
Sum of two sides 2950mm	78.44	103.93	1.16	25.32	nr	**129.25**
Sum of two sides 3000mm	80.31	106.41	1.73	37.76	nr	**144.17**
Sum of two sides 3050mm	81.95	108.58	1.73	37.76	nr	**146.34**

U:VENTILATION/AIR CONDITIONING SYSTEMS

Item	Net Price £	Material £	Labour hours	Labour £	Unit	Total rate £
U10 : DUCTWORK : RECTANGULAR - CLASS B						
Y30 - AIR DUCTLINES						
Galvanised sheet metal DW144 class B rectangular section ductwork; including all necessary stiffeners, joints, couplers in the running length and duct supports (Continued)						
Sum of two sides 3100mm	83.59	110.76	1.80	39.28	nr	**150.04**
Sum of two sides 3150mm	85.23	112.93	1.80	39.28	nr	**152.21**
Sum of two sides 3200mm	86.86	115.09	1.80	39.28	nr	**154.37**
Sum of two sides 3250mm	88.47	117.22	1.80	39.28	nr	**156.51**
Sum of two sides 3300mm	90.08	119.36	1.80	39.28	nr	**158.64**
Sum of two sides 3350mm	91.68	121.48	1.80	39.28	nr	**160.76**
Sum of two sides 3400mm	93.29	123.61	1.80	39.28	nr	**162.89**
Sum of two sides 3450mm	94.89	125.73	1.80	39.28	nr	**165.01**
Sum of two sides 3500mm	96.50	127.86	1.80	39.28	nr	**167.15**
Sum of two sides 3550mm	98.10	129.98	1.80	39.28	nr	**169.27**
Sum of two sides 3600mm	99.70	132.10	1.80	39.28	nr	**171.39**
Sum of two sides 3650mm	101.31	134.24	1.80	39.28	nr	**173.52**
Sum of two sides 3700mm	102.91	136.36	1.80	39.28	nr	**175.64**
Sum of two sides 3750mm	104.52	138.49	1.80	39.28	nr	**177.77**
Sum of two sides 3800mm	106.12	140.61	1.80	39.28	nr	**179.89**
Sum of two sides 3850mm	107.73	142.74	1.80	39.28	nr	**182.03**
Sum of two sides 3900mm	109.33	144.86	1.80	39.28	nr	**184.15**
Sum of two sides 3950mm	110.94	147.00	1.80	39.28	nr	**186.28**
Sum of two sides 4000mm	112.55	149.13	1.80	39.28	nr	**188.41**
Reducer						
Sum of two sides 2100mm	32.40	42.93	2.61	56.96	nr	**99.89**
Sum of two sides 2150mm	34.38	45.55	2.61	56.96	nr	**102.52**
Sum of two sides 2200mm	36.36	48.18	2.61	56.96	nr	**105.14**
Sum of two sides 2250mm	38.35	50.81	2.61	56.96	nr	**107.78**
Sum of two sides 2300mm	40.33	53.44	2.61	56.96	nr	**110.40**
Sum of two sides 2350mm	42.20	55.91	2.61	56.96	nr	**112.88**
Sum of two sides 2400mm	44.07	58.39	2.88	62.85	nr	**121.25**
Sum of two sides 2450mm	46.05	61.02	2.88	62.85	nr	**123.87**
Sum of two sides 2500mm	48.04	63.65	3.12	68.09	nr	**131.75**
Sum of two sides 2550mm	48.04	63.65	3.12	68.09	nr	**131.75**
Sum of two sides 2600mm	51.88	68.74	3.12	68.09	nr	**136.83**
Sum of two sides 2650mm	53.81	71.30	3.12	68.09	nr	**139.39**
Sum of two sides 2700mm	55.74	73.86	3.12	68.09	nr	**141.95**
Sum of two sides 2750mm	57.66	76.40	3.12	68.09	nr	**144.49**
Sum of two sides 2800mm	59.59	78.96	3.95	86.21	nr	**165.16**
Sum of two sides 2850mm	61.54	81.54	3.95	86.21	nr	**167.75**
Sum of two sides 2900mm	63.50	84.14	3.97	86.64	nr	**170.78**
Sum of two sides 2950mm	65.45	86.72	3.97	86.64	nr	**173.36**
Sum of two sides 3000mm	67.41	89.32	4.52	98.65	nr	**187.97**
Sum of two sides 3050mm	68.05	90.17	4.52	98.65	nr	**188.81**
Sum of two sides 3100mm	68.70	91.03	4.52	98.65	nr	**189.67**
Sum of two sides 3150mm	69.35	91.89	4.52	98.65	nr	**190.54**
Sum of two sides 3200mm	69.99	92.74	4.52	98.65	nr	**191.38**

Material Costs/Measured Work Prices - Mechanical Installations

U:VENTILATION/AIR CONDITIONING SYSTEMS

Item	Net Price £	Material £	Labour hours	Labour £	Unit	Total rate £
Sum of two sides 3250mm	71.42	94.63	4.52	98.65	nr	193.28
Sum of two sides 3300mm	72.85	96.53	4.52	98.65	nr	195.17
Sum of two sides 3350mm	74.28	98.42	4.52	98.65	nr	197.07
Sum of two sides 3400mm	75.71	100.32	4.52	98.65	nr	198.96
Sum of two sides 3450mm	77.11	102.17	4.52	98.65	nr	200.82
Sum of two sides 3500mm	78.51	104.03	4.52	98.65	nr	202.67
Sum of two sides 3550mm	79.91	105.88	4.52	98.65	nr	204.53
Sum of two sides 3600mm	81.31	107.74	4.52	98.65	nr	206.38
Sum of two sides 3650mm	82.74	109.63	4.52	98.65	nr	208.28
Sum of two sides 3700mm	84.17	111.53	4.52	98.65	nr	210.17
Sum of two sides 3750mm	85.60	113.42	4.52	98.65	nr	212.07
Sum of two sides 3800mm	87.03	115.31	4.52	98.65	nr	213.96
Sum of two sides 3850mm	88.43	117.17	4.52	98.65	nr	215.82
Sum of two sides 3900mm	89.83	119.02	4.52	98.65	nr	217.67
Sum of two sides 3950mm	91.23	120.88	4.52	98.65	nr	219.53
Sum of two sides 4000mm	92.63	122.73	4.52	98.65	nr	221.38
Offset						
Sum of two sides 2100mm	98.69	130.76	2.61	56.96	nr	187.73
Sum of two sides 2150mm	100.70	133.43	2.61	56.96	nr	190.39
Sum of two sides 2200mm	102.71	136.09	2.61	56.96	nr	193.05
Sum of two sides 2250mm	104.71	138.74	2.61	56.96	nr	195.70
Sum of two sides 2300mm	106.72	141.40	2.61	56.96	nr	198.37
Sum of two sides 2350mm	111.45	147.67	2.61	56.96	nr	204.63
Sum of two sides 2400mm	116.18	153.94	2.88	62.85	nr	216.79
Sum of two sides 2450mm	118.13	156.52	2.88	62.85	nr	219.38
Sum of two sides 2500mm	120.07	159.09	3.48	75.95	nr	235.04
Sum of two sides 2550mm	121.88	161.49	3.48	75.95	nr	237.44
Sum of two sides 2600mm	123.69	163.89	3.49	76.17	nr	240.06
Sum of two sides 2650mm	125.40	166.16	3.49	76.17	nr	242.32
Sum of two sides 2700mm	127.12	168.43	3.50	76.39	nr	244.82
Sum of two sides 2750mm	128.83	170.70	3.50	76.39	nr	247.09
Sum of two sides 2800mm	130.54	172.97	4.34	94.72	nr	267.68
Sum of two sides 2850mm	132.14	175.09	4.34	94.72	nr	269.80
Sum of two sides 2900mm	133.73	177.19	4.76	103.89	nr	281.08
Sum of two sides 2950mm	135.32	179.30	4.76	103.89	nr	283.18
Sum of two sides 3000mm	136.91	181.41	5.32	116.11	nr	297.51
Sum of two sides 3050mm	137.48	182.16	5.32	116.11	nr	298.27
Sum of two sides 3100mm	138.05	182.92	5.35	116.76	nr	299.68
Sum of two sides 3150mm	138.62	183.67	5.35	116.76	nr	300.43
Sum of two sides 3200mm	139.19	184.43	5.35	116.76	nr	301.19
Sum of two sides 3250mm	71.42	94.63	5.35	116.76	nr	211.39
Sum of two sides 3300mm	143.93	190.71	5.35	116.76	nr	307.47
Sum of two sides 3350mm	146.30	193.85	5.35	116.76	nr	310.61
Sum of two sides 3400mm	148.67	196.99	5.35	116.76	nr	313.75
Sum of two sides 3450mm	151.03	200.11	5.35	116.76	nr	316.88
Sum of two sides 3500mm	153.39	203.24	5.35	116.76	nr	320.00
Sum of two sides 3550mm	155.75	206.37	5.35	116.76	nr	323.13
Sum of two sides 3600mm	158.11	209.50	5.35	116.76	nr	326.26
Sum of two sides 3650mm	160.47	212.62	5.35	116.76	nr	329.38
Sum of two sides 3700mm	162.84	215.76	5.35	116.76	nr	332.52
Sum of two sides 3750mm	165.21	218.90	5.35	116.76	nr	335.66
Sum of two sides 3800mm	167.58	222.04	5.35	116.76	nr	338.81
Sum of two sides 3850mm	169.94	225.17	5.35	116.76	nr	341.93

Material Costs/Measured Work Prices - Mechanical Installations

U:VENTILATION/AIR CONDITIONING SYSTEMS

Item	Net Price £	Material £	Labour hours	Labour £	Unit	Total rate £
U10 : DUCTWORK : RECTANGULAR - CLASS B						
Y30 - AIR DUCTLINES						
Galvanised sheet metal DW144 class B rectangular section ductwork; including all necessary stiffeners, joints, couplers in the running length and duct supports (Continued)						
Offset						
Sum of two sides 3900mm	172.30	228.30	5.35	116.76	nr	345.06
Sum of two sides 3950mm	174.66	231.42	5.35	116.76	nr	348.19
Sum of two sides 4000mm	177.02	234.55	5.35	116.76	nr	351.31
Square to round						
Sum of two sides 2100mm	72.28	95.77	2.61	56.96	nr	152.73
Sum of two sides 2150mm	75.22	99.67	2.61	56.96	nr	156.63
Sum of two sides 2200mm	78.17	103.58	2.61	56.96	nr	160.54
Sum of two sides 2250mm	81.11	107.47	2.61	56.96	nr	164.43
Sum of two sides 2300mm	84.05	111.37	2.61	56.96	nr	168.33
Sum of two sides 2350mm	87.00	115.28	2.61	56.96	nr	172.24
Sum of two sides 2400mm	89.95	119.18	2.88	62.85	nr	182.04
Sum of two sides 2450mm	92.90	123.09	2.88	62.85	nr	185.95
Sum of two sides 2500mm	95.84	126.99	4.70	102.58	nr	229.56
Sum of two sides 2550mm	98.79	130.90	4.70	102.58	nr	233.47
Sum of two sides 2600mm	101.73	134.79	4.70	102.58	nr	237.37
Sum of two sides 2650mm	104.68	138.70	4.70	102.58	nr	241.28
Sum of two sides 2700mm	107.62	142.60	4.71	102.79	nr	245.39
Sum of two sides 2750mm	110.57	146.51	4.71	102.79	nr	249.30
Sum of two sides 2800mm	113.52	150.41	8.19	178.74	nr	329.16
Sum of two sides 2850mm	116.46	154.31	8.19	178.74	nr	333.05
Sum of two sides 2900mm	119.41	158.22	8.19	178.74	nr	336.96
Sum of two sides 2950mm	122.35	162.11	8.19	178.74	nr	340.86
Sum of two sides 3000mm	125.30	166.02	8.19	178.74	nr	344.77
Sum of two sides 3050mm	126.74	167.93	8.19	178.74	nr	346.67
Sum of two sides 3100mm	128.17	169.83	8.19	178.74	nr	348.57
Sum of two sides 3150mm	129.61	171.73	8.19	178.74	nr	350.48
Sum of two sides 3200mm	131.05	173.64	8.19	178.74	nr	352.38
Sum of two sides 3250mm	133.21	176.50	8.19	178.74	nr	355.25
Sum of two sides 3300mm	135.37	179.37	8.19	178.74	nr	358.11
Sum of two sides 3350mm	137.52	182.21	8.62	188.13	nr	370.34
Sum of two sides 3400mm	139.68	185.08	8.62	188.13	nr	373.20
Sum of two sides 3450mm	141.84	187.94	8.62	188.13	nr	376.07
Sum of two sides 3500mm	144.00	190.80	8.62	188.13	nr	378.93
Sum of two sides 3550mm	146.16	193.66	8.62	188.13	nr	381.79
Sum of two sides 3600mm	148.31	196.51	8.62	188.13	nr	384.64
Sum of two sides 3650mm	150.47	199.37	8.62	188.13	nr	387.50
Sum of two sides 3700mm	152.63	202.23	8.62	188.13	nr	390.36
Sum of two sides 3750mm	154.78	205.08	8.75	190.97	nr	396.05
Sum of two sides 3800mm	156.94	207.95	8.75	190.97	nr	398.91
Sum of two sides 3850mm	159.10	210.81	8.75	190.97	nr	401.77
Sum of two sides 3900mm	161.26	213.67	8.75	190.97	nr	404.63
Sum of two sides 3950mm	163.41	216.52	8.75	190.97	nr	407.48
Sum of two sides 4000mm	165.57	219.38	8.75	190.97	nr	410.35

Material Costs/Measured Work Prices - Mechanical Installations 317

U:VENTILATION/AIR CONDITIONING SYSTEMS

Item	Net Price £	Material £	Labour hours	Labour £	Unit	Total rate £
90 degree radius bend						
Sum of two sides 2100mm	138.25	183.18	2.61	56.96	nr	**240.14**
Sum of two sides 2150mm	142.37	188.64	2.16	47.14	nr	**235.78**
Sum of two sides 2200mm	149.48	198.06	2.61	56.96	nr	**255.02**
Sum of two sides 2250mm	155.09	205.49	2.16	47.14	nr	**252.64**
Sum of two sides 2300mm	160.71	212.94	2.61	56.96	nr	**269.90**
Sum of two sides 2350mm	162.95	215.91	2.16	47.14	nr	**263.05**
Sum of two sides 2400mm	165.19	218.88	2.88	62.85	nr	**281.73**
Sum of two sides 2450mm	170.72	226.20	2.88	62.85	nr	**289.06**
Sum of two sides 2500mm	176.26	233.54	3.90	85.12	nr	**318.66**
Sum of two sides 2550mm	181.57	240.58	3.90	85.12	nr	**325.70**
Sum of two sides 2600mm	186.88	247.62	4.26	92.97	nr	**340.59**
Sum of two sides 2650mm	192.19	254.65	4.26	92.97	nr	**347.62**
Sum of two sides 2700mm	197.51	261.70	4.55	99.30	nr	**361.00**
Sum of two sides 2750mm	202.82	268.74	4.55	99.30	nr	**368.04**
Sum of two sides 2800mm	208.14	275.79	4.55	99.30	nr	**375.09**
Sum of two sides 2850mm	213.56	282.97	4.55	99.30	nr	**382.27**
Sum of two sides 2900mm	218.98	290.15	6.87	149.94	nr	**440.08**
Sum of two sides 2950mm	224.41	297.34	6.87	149.94	nr	**447.28**
Sum of two sides 3000mm	229.83	304.52	6.87	149.94	nr	**454.46**
Sum of two sides 3050mm	231.89	307.25	6.87	149.94	nr	**457.19**
Sum of two sides 3100mm	233.96	310.00	6.87	149.94	nr	**459.93**
Sum of two sides 3150mm	236.02	312.73	6.87	149.94	nr	**462.66**
Sum of two sides 3200mm	238.09	315.47	6.87	149.94	nr	**465.40**
Sum of two sides 3250mm	241.93	320.56	6.87	149.94	nr	**470.49**
Sum of two sides 3300mm	245.78	325.66	6.87	149.94	nr	**475.59**
Sum of two sides 3350mm	249.63	330.76	6.87	149.94	nr	**480.69**
Sum of two sides 3400mm	253.48	335.86	7.00	152.77	nr	**488.63**
Sum of two sides 3450mm	257.21	340.80	7.00	152.77	nr	**493.58**
Sum of two sides 3500mm	260.95	345.76	7.00	152.77	nr	**498.53**
Sum of two sides 3550mm	264.69	350.71	7.00	152.77	nr	**503.49**
Sum of two sides 3600mm	268.43	355.67	7.00	152.77	nr	**508.44**
Sum of two sides 3650mm	272.27	360.76	7.00	152.77	nr	**513.53**
Sum of two sides 3700mm	276.12	365.86	7.00	152.77	nr	**518.63**
Sum of two sides 3750mm	274.97	364.34	7.00	152.77	nr	**517.11**
Sum of two sides 3800mm	283.82	376.06	7.00	152.77	nr	**528.83**
Sum of two sides 3850mm	287.55	381.00	7.00	152.77	nr	**533.78**
Sum of two sides 3900mm	291.29	385.96	7.00	152.77	nr	**538.73**
Sum of two sides 3950mm	295.02	390.90	7.00	152.77	nr	**543.67**
Sum of two sides 4000mm	298.76	395.86	7.00	152.77	nr	**548.63**
45 degree bend						
Sum of two sides 2100mm	96.58	127.97	2.61	56.96	nr	**184.93**
Sum of two sides 2150mm	100.46	133.11	2.61	56.96	nr	**190.07**
Sum of two sides 2200mm	104.35	138.26	2.61	56.96	nr	**195.23**
Sum of two sides 2250mm	108.23	143.40	2.61	56.96	nr	**200.37**
Sum of two sides 2300mm	112.11	148.55	2.61	56.96	nr	**205.51**
Sum of two sides 2350mm	114.19	151.30	2.61	56.96	nr	**208.26**
Sum of two sides 2400mm	116.27	154.06	2.88	62.85	nr	**216.91**
Sum of two sides 2450mm	120.11	159.15	2.88	62.85	nr	**222.00**
Sum of two sides 2500mm	123.95	164.23	4.85	105.85	nr	**270.08**
Sum of two sides 2550mm	127.62	169.10	4.85	105.85	nr	**274.95**
Sum of two sides 2600mm	131.30	173.97	4.87	106.29	nr	**280.26**
Sum of two sides 2650mm	134.97	178.84	4.87	106.29	nr	**285.12**

Material Costs/Measured Work Prices - Mechanical Installations

U:VENTILATION/AIR CONDITIONING SYSTEMS

Item	Net Price £	Material £	Labour hours	Labour £	Unit	Total rate £
U10 : DUCTWORK : RECTANGULAR - CLASS B						
Y30 - AIR DUCTLINES						
Galvanised sheet metal DW144 class B rectangular section ductwork; including all necessary stiffeners, joints, couplers in the running length and duct supports (Continued)						
45 degree bend						
Sum of two sides 2700mm	138.65	183.71	4.87	106.29	nr	290.00
Sum of two sides 2750mm	142.32	188.57	4.87	106.29	nr	294.86
Sum of two sides 2800mm	145.99	193.44	8.81	192.27	nr	385.71
Sum of two sides 2850mm	149.75	198.42	8.81	192.27	nr	390.69
Sum of two sides 2900mm	153.51	203.40	8.81	192.27	nr	395.68
Sum of two sides 2950mm	157.26	208.37	8.81	192.27	nr	400.64
Sum of two sides 3000mm	161.02	213.35	9.31	203.19	nr	416.54
Sum of two sides 3050mm	162.79	215.70	9.31	203.19	nr	418.88
Sum of two sides 3100mm	164.56	218.04	9.31	203.19	nr	421.23
Sum of two sides 3150mm	166.33	220.39	9.31	203.19	nr	423.57
Sum of two sides 3200mm	168.10	222.73	9.31	203.19	nr	425.92
Sum of two sides 3250mm	170.81	226.32	9.31	203.19	nr	429.51
Sum of two sides 3300mm	173.51	229.90	9.31	203.19	nr	433.09
Sum of two sides 3350mm	176.22	233.49	9.31	203.19	nr	436.68
Sum of two sides 3400mm	178.93	237.08	9.31	203.19	nr	440.27
Sum of two sides 3450mm	181.55	240.55	9.39	204.93	nr	445.49
Sum of two sides 3500mm	184.17	244.03	9.39	204.93	nr	448.96
Sum of two sides 3550mm	186.80	247.51	9.39	204.93	nr	452.44
Sum of two sides 3600mm	189.42	250.98	9.39	204.93	nr	455.91
Sum of two sides 3650mm	192.12	254.56	9.39	204.93	nr	459.49
Sum of two sides 3700mm	194.83	258.15	9.39	204.93	nr	463.08
Sum of two sides 3750mm	197.54	261.74	9.39	204.93	nr	466.67
Sum of two sides 3800mm	200.24	265.32	9.39	204.93	nr	470.25
Sum of two sides 3850mm	202.87	268.80	9.39	204.93	nr	473.74
Sum of two sides 3900mm	205.49	272.27	9.39	204.93	nr	477.21
Sum of two sides 3950mm	208.11	275.75	9.39	204.93	nr	480.68
Sum of two sides 4000mm	210.73	279.22	9.39	204.93	nr	484.15
90 degree mitre bend						
Sum of two sides 2100mm	191.30	253.47	2.61	56.96	nr	310.43
Sum of two sides 2150mm	200.94	266.25	2.61	56.96	nr	323.21
Sum of two sides 2200mm	210.57	279.01	2.61	56.96	nr	335.97
Sum of two sides 2250mm	220.21	291.78	2.61	56.96	nr	348.74
Sum of two sides 2300mm	229.84	304.54	2.61	56.96	nr	361.50
Sum of two sides 2350mm	234.52	310.74	2.61	56.96	nr	367.70
Sum of two sides 2400mm	239.20	316.94	2.88	62.85	nr	379.79
Sum of two sides 2450mm	248.84	329.71	2.88	62.85	nr	392.57
Sum of two sides 2500mm	258.48	342.49	4.85	105.85	nr	448.34
Sum of two sides 2550mm	268.17	355.33	4.85	105.85	nr	461.17
Sum of two sides 2600mm	277.87	368.18	4.87	106.29	nr	474.46
Sum of two sides 2650mm	287.57	381.03	4.87	106.29	nr	487.32
Sum of two sides 2700mm	297.27	393.88	4.87	106.29	nr	500.17
Sum of two sides 2750mm	306.96	406.72	4.87	106.29	nr	513.01

Material Costs/Measured Work Prices - Mechanical Installations

U:VENTILATION/AIR CONDITIONING SYSTEMS

Item	Net Price £	Material £	Labour hours	Labour £	Unit	Total rate £
Sum of two sides 2800mm	316.66	419.57	8.81	192.27	nr	611.85
Sum of two sides 2850mm	326.33	432.39	8.81	192.27	nr	624.66
Sum of two sides 2900mm	336.00	445.20	14.81	323.22	nr	768.42
Sum of two sides 2950mm	345.67	458.01	14.81	323.22	nr	781.24
Sum of two sides 3000mm	355.34	470.83	15.20	331.73	nr	802.56
Sum of two sides 3050mm	361.40	478.86	15.20	331.73	nr	810.59
Sum of two sides 3100mm	367.47	486.90	15.20	331.73	nr	818.63
Sum of two sides 3150mm	373.53	494.93	15.20	331.73	nr	826.66
Sum of two sides 3200mm	379.60	502.97	15.20	331.73	nr	834.70
Sum of two sides 3250mm	388.20	514.37	15.20	331.73	nr	846.10
Sum of two sides 3300mm	396.80	525.76	15.20	331.73	nr	857.49
Sum of two sides 3350mm	405.40	537.15	15.20	331.73	nr	868.89
Sum of two sides 3400mm	414.00	548.55	15.20	331.73	nr	880.28
Sum of two sides 3450mm	422.12	559.31	15.20	331.73	nr	891.04
Sum of two sides 3500mm	430.24	570.07	15.60	340.46	nr	910.53
Sum of two sides 3550mm	438.36	580.83	15.60	340.46	nr	921.29
Sum of two sides 3600mm	446.48	591.59	15.60	340.46	nr	932.05
Sum of two sides 3650mm	454.57	602.31	15.60	340.46	nr	942.77
Sum of two sides 3700mm	462.67	613.04	15.60	340.46	nr	953.50
Sum of two sides 3750mm	470.76	623.76	15.60	340.46	nr	964.22
Sum of two sides 3800mm	478.85	634.48	15.60	340.46	nr	974.94
Sum of two sides 3850mm	486.97	645.24	15.60	340.46	nr	985.70
Sum of two sides 3900mm	495.09	655.99	15.60	340.46	nr	996.46
Sum of two sides 3950mm	503.21	666.75	15.60	340.46	nr	1007.22
Sum of two sides 4000mm	511.34	677.53	15.60	340.46	nr	1017.99
Branch						
Sum of two sides 2100mm	78.85	104.48	2.61	56.96	nr	161.44
Sum of two sides 2150mm	81.70	108.25	2.61	56.96	nr	165.21
Sum of two sides 2200mm	84.55	112.03	2.61	56.96	nr	168.99
Sum of two sides 2250mm	87.40	115.81	2.61	56.96	nr	172.77
Sum of two sides 2300mm	90.25	119.58	2.61	56.96	nr	176.54
Sum of two sides 2350mm	92.97	123.19	2.61	56.96	nr	180.15
Sum of two sides 2400mm	95.70	126.80	2.88	62.85	nr	189.66
Sum of two sides 2450mm	98.55	130.58	2.88	62.85	nr	193.43
Sum of two sides 2500mm	101.39	134.34	2.88	62.85	nr	197.20
Sum of two sides 2550mm	104.18	138.04	2.88	62.85	nr	200.89
Sum of two sides 2600mm	106.97	141.74	2.88	62.85	nr	204.59
Sum of two sides 2650mm	109.76	145.43	2.88	62.85	nr	208.29
Sum of two sides 2700mm	112.55	149.13	2.88	62.85	nr	211.98
Sum of two sides 2750mm	115.33	152.81	2.88	62.85	nr	215.67
Sum of two sides 2800mm	118.12	156.51	3.94	85.99	nr	242.50
Sum of two sides 2850mm	120.94	160.25	3.94	85.99	nr	246.23
Sum of two sides 2900mm	123.76	163.98	3.94	85.99	nr	249.97
Sum of two sides 2950mm	126.57	167.71	3.94	85.99	nr	253.69
Sum of two sides 3000mm	129.39	171.44	3.94	85.99	nr	257.43
Sum of two sides 3050mm	131.57	174.33	3.94	85.99	nr	260.32
Sum of two sides 3100mm	133.75	177.22	3.94	85.99	nr	263.21
Sum of two sides 3150mm	135.93	180.11	3.94	85.99	nr	266.10
Sum of two sides 3200mm	138.11	183.00	3.94	85.99	nr	268.98
Sum of two sides 3250mm	140.22	185.79	3.94	85.99	nr	271.78
Sum of two sides 3300mm	142.34	188.60	3.94	85.99	nr	274.59
Sum of two sides 3350mm	144.46	191.41	4.83	105.41	nr	296.82
Sum of two sides 3400mm	146.58	194.22	4.83	105.41	nr	299.63

Material Costs/Measured Work Prices - Mechanical Installations

U:VENTILATION/AIR CONDITIONING SYSTEMS

Item	Net Price £	Material £	Labour hours	Labour £	Unit	Total rate £
U10 : DUCTWORK : RECTANGULAR - CLASS B						
Y30 - AIR DUCTLINES						
Galvanised sheet metal DW144 class B rectangular section ductwork; including all necessary stiffeners, joints, couplers in the running length and duct supports (Continued)						
Branch						
Sum of two sides 3450mm	148.66	196.97	4.83	105.41	nr	302.39
Sum of two sides 3500mm	150.75	199.74	4.83	105.41	nr	305.16
Sum of two sides 3550mm	152.84	202.51	4.83	105.41	nr	307.93
Sum of two sides 3600mm	154.93	205.28	4.83	105.41	nr	310.70
Sum of two sides 3650mm	157.04	208.08	4.83	105.41	nr	313.49
Sum of two sides 3700mm	159.16	210.89	4.83	105.41	nr	316.30
Sum of two sides 3750mm	161.28	213.70	4.83	105.41	nr	319.11
Sum of two sides 3800mm	163.40	216.50	4.83	105.41	nr	321.92
Sum of two sides 3850mm	165.48	219.26	4.83	105.41	nr	324.67
Sum of two sides 3900mm	167.57	222.03	4.83	105.41	nr	327.44
Sum of two sides 3950mm	169.66	224.80	4.83	105.41	nr	330.21
Sum of two sides 4000mm	171.75	227.57	4.83	105.41	nr	332.98
Grille neck						
Sum of two sides 2100mm	82.70	109.58	2.61	56.96	nr	166.54
Sum of two sides 2150mm	85.77	113.65	2.61	56.96	nr	170.61
Sum of two sides 2200mm	88.84	117.71	2.61	56.96	nr	174.68
Sum of two sides 2250mm	91.91	121.78	2.61	56.96	nr	178.74
Sum of two sides 2300mm	94.97	125.84	2.61	56.96	nr	182.80
Sum of two sides 2350mm	98.04	129.90	2.61	56.96	nr	186.87
Sum of two sides 2400mm	101.11	133.97	2.88	62.85	nr	196.83
Sum of two sides 2450mm	104.18	138.04	2.88	62.85	nr	200.89
Sum of two sides 2500mm	107.25	142.11	2.88	62.85	nr	204.96
Sum of two sides 2550mm	110.32	146.17	2.88	62.85	nr	209.03
Sum of two sides 2600mm	113.38	150.23	2.88	62.85	nr	213.08
Sum of two sides 2650mm	116.45	154.30	2.88	62.85	nr	217.15
Sum of two sides 2700mm	119.52	158.36	2.88	62.85	nr	221.22
Sum of two sides 2750mm	122.59	162.43	2.88	62.85	nr	225.29
Sum of two sides 2800mm	125.66	166.50	3.94	85.99	nr	252.49
Sum of two sides 2850mm	128.72	170.55	3.94	85.99	nr	256.54
Sum of two sides 2900mm	131.79	174.62	4.12	89.92	nr	264.54
Sum of two sides 2950mm	134.86	178.69	4.12	89.92	nr	268.61
Sum of two sides 3000mm	137.93	182.76	4.12	89.92	nr	272.67
Sum of two sides 3050mm	140.31	185.91	4.12	89.92	nr	275.83
Sum of two sides 3100mm	142.70	189.08	4.12	89.92	nr	278.99
Sum of two sides 3150mm	145.08	192.23	4.12	89.92	nr	282.15
Sum of two sides 3200mm	147.46	195.38	4.12	89.92	nr	285.30
Sum of two sides 3250mm	149.75	198.42	4.12	89.92	nr	288.34
Sum of two sides 3300mm	152.03	201.44	4.12	89.92	nr	291.36
Sum of two sides 3350mm	154.31	204.46	5.00	109.12	nr	313.58
Sum of two sides 3400mm	156.59	207.48	5.00	109.12	nr	316.60
Sum of two sides 3450mm	158.87	210.50	5.00	109.12	nr	319.63
Sum of two sides 3500mm	161.15	213.52	5.00	109.12	nr	322.65

Material Costs/Measured Work Prices - Mechanical Installations

U:VENTILATION/AIR CONDITIONING SYSTEMS

Item	Net Price £	Material £	Labour hours	Labour £	Unit	Total rate £
Sum of two sides 3550mm	163.43	216.54	5.00	109.12	nr	325.67
Sum of two sides 3600mm	165.71	219.57	5.00	109.12	nr	328.69
Sum of two sides 3650mm	167.99	222.59	5.00	109.12	nr	331.71
Sum of two sides 3700mm	170.27	225.61	5.00	109.12	nr	334.73
Sum of two sides 3750mm	172.55	228.63	5.00	109.12	nr	337.75
Sum of two sides 3800mm	174.83	231.65	5.00	109.12	nr	340.77
Sum of two sides 3850mm	177.11	234.67	5.00	109.12	nr	343.79
Sum of two sides 3900mm	179.39	237.69	5.00	109.12	nr	346.81
Sum of two sides 3950mm	181.67	240.71	5.00	109.12	nr	349.84
Sum of two sides 4000mm	183.95	243.73	5.00	109.12	nr	352.86
Ductwork 2001 to 2500mm longest side						
Sum of two sides 2500mm	107.24	142.09	2.38	51.94	m	194.04
Sum of two sides 2550mm	110.31	146.16	2.38	51.94	m	198.10
Sum of two sides 2600mm	113.38	150.23	2.64	57.62	m	207.85
Sum of two sides 2650mm	116.45	154.30	2.64	57.62	m	211.91
Sum of two sides 2700mm	119.52	158.36	2.66	58.05	m	216.42
Sum of two sides 2750mm	122.59	162.43	2.66	58.05	m	220.49
Sum of two sides 2800mm	125.66	166.50	2.95	64.38	m	230.88
Sum of two sides 2850mm	128.72	170.55	2.95	64.38	m	234.94
Sum of two sides 2900mm	131.79	174.62	2.96	64.60	m	239.22
Sum of two sides 2950mm	134.86	178.69	2.96	64.60	m	243.29
Sum of two sides 3000mm	137.93	182.76	3.15	68.75	m	251.50
Sum of two sides 3050mm	140.31	185.91	3.15	68.75	m	254.66
Sum of two sides 3100mm	142.70	189.08	3.15	68.75	m	257.82
Sum of two sides 3150mm	145.08	192.23	3.15	68.75	m	260.98
Sum of two sides 3200mm	147.46	195.38	3.15	68.75	m	264.13
Sum of two sides 3250mm	149.75	198.42	3.15	68.75	m	267.17
Sum of two sides 3300mm	152.03	201.44	3.15	68.75	m	270.19
Sum of two sides 3350mm	154.31	204.46	3.15	68.75	m	273.21
Sum of two sides 3400mm	156.59	207.48	3.15	68.75	m	276.23
Sum of two sides 3450mm	158.87	210.50	3.15	68.75	m	279.25
Sum of two sides 3500mm	161.15	213.52	2.66	58.05	m	271.58
Sum of two sides 3550mm	163.43	216.54	3.18	69.40	m	285.95
Sum of two sides 3600mm	165.71	219.57	3.18	69.40	m	288.97
Sum of two sides 3650mm	169.02	223.95	3.18	69.40	m	293.35
Sum of two sides 3700mm	172.33	228.34	3.18	69.40	m	297.74
Sum of two sides 3750mm	173.58	229.99	3.18	69.40	m	299.40
Sum of two sides 3800mm	174.83	231.65	3.18	69.40	m	301.05
Sum of two sides 3850mm	177.11	234.67	3.18	69.40	m	304.07
Sum of two sides 3900mm	179.39	237.69	3.18	69.40	m	307.09
Sum of two sides 3950mm	181.67	240.71	3.18	69.40	m	310.11
Sum of two sides 4000mm	183.95	243.73	3.18	69.40	m	313.14
Extra over fittings; Ductwork 2001 to 2500mm longest side						
End Cap						
Sum of two sides 2500mm	61.64	81.67	0.87	18.99	nr	100.66
Sum of two sides 2550mm	63.50	84.14	0.87	18.99	nr	103.12
Sum of two sides 2600mm	65.37	86.62	0.87	18.99	nr	105.60
Sum of two sides 2650mm	67.24	89.09	0.87	18.99	nr	108.08
Sum of two sides 2700mm	69.10	91.56	0.87	18.99	nr	110.54
Sum of two sides 2750mm	70.97	94.04	0.87	18.99	nr	113.02

Material Costs/Measured Work Prices - Mechanical Installations

U:VENTILATION/AIR CONDITIONING SYSTEMS

Item	Net Price £	Material £	Labour hours	Labour £	Unit	Total rate £
U10 : DUCTWORK : RECTANGULAR - CLASS B						
Y30 - AIR DUCTLINES						
Galvanised sheet metal DW144 class B rectangular section ductwork; including all necessary stiffeners, joints, couplers in the running length and duct supports (Continued)						
End Cap						
Sum of two sides 2800mm	72.84	96.51	1.16	25.32	nr	121.83
Sum of two sides 2850mm	74.71	98.99	1.16	25.32	nr	124.31
Sum of two sides 2900mm	76.57	101.46	1.16	25.32	nr	126.77
Sum of two sides 2950mm	78.44	103.93	1.16	25.32	nr	129.25
Sum of two sides 3000mm	80.31	106.41	1.73	37.76	nr	144.17
Sum of two sides 3050mm	81.95	108.58	1.73	37.76	nr	146.34
Sum of two sides 3100mm	83.59	110.76	1.73	37.76	nr	148.51
Sum of two sides 3150mm	85.23	112.93	1.73	37.76	nr	150.69
Sum of two sides 3200mm	86.86	115.09	1.73	37.76	nr	152.85
Sum of two sides 3250mm	88.47	117.22	1.73	37.76	nr	154.98
Sum of two sides 3300mm	90.08	119.36	1.73	37.76	nr	157.11
Sum of two sides 3350mm	91.68	121.48	1.73	37.76	nr	159.23
Sum of two sides 3400mm	93.29	123.61	1.80	39.28	nr	162.89
Sum of two sides 3450mm	94.89	125.73	1.80	39.28	nr	165.01
Sum of two sides 3500mm	96.50	127.86	1.80	39.28	nr	167.15
Sum of two sides 3550mm	98.10	129.98	1.80	39.28	nr	169.27
Sum of two sides 3600mm	99.70	132.10	1.80	39.28	nr	171.39
Sum of two sides 3650mm	101.64	134.67	1.80	39.28	nr	173.96
Sum of two sides 3700mm	103.58	137.24	1.80	39.28	nr	176.53
Sum of two sides 3750mm	103.58	137.24	1.80	39.28	nr	176.53
Sum of two sides 3800mm	106.12	140.61	1.80	39.28	nr	179.89
Sum of two sides 3850mm	107.73	142.74	1.80	39.28	nr	182.03
Sum of two sides 3900mm	109.33	144.86	1.80	39.28	nr	184.15
Sum of two sides 3950mm	110.94	147.00	1.80	39.28	nr	186.28
Sum of two sides 4000mm	112.55	149.13	1.80	39.28	nr	188.41
Reducer						
Sum of two sides 2500mm	41.50	54.99	3.12	68.09	nr	123.08
Sum of two sides 2550mm	43.43	57.54	3.12	68.09	nr	125.64
Sum of two sides 2600mm	45.35	60.09	3.12	68.09	nr	128.18
Sum of two sides 2650mm	47.28	62.65	3.12	68.09	nr	130.74
Sum of two sides 2700mm	49.21	65.20	3.12	68.09	nr	133.30
Sum of two sides 2750mm	51.13	67.75	3.12	68.09	nr	135.84
Sum of two sides 2800mm	53.06	70.30	3.95	86.21	nr	156.5
Sum of two sides 2850mm	55.01	72.89	3.95	86.21	nr	159.1
Sum of two sides 2900mm	56.97	75.49	3.97	86.64	nr	162.13
Sum of two sides 2950mm	58.92	78.07	3.97	86.64	nr	164.7
Sum of two sides 3000mm	60.87	80.65	3.97	86.64	nr	167.3
Sum of two sides 3050mm	62.34	82.60	3.97	86.64	nr	169.2
Sum of two sides 3100mm	63.81	84.55	3.97	86.64	nr	171.19
Sum of two sides 3150mm	65.28	86.50	3.97	86.64	nr	173.14
Sum of two sides 3200mm	66.75	88.44	3.97	86.64	nr	175.08
Sum of two sides 3250mm	68.18	90.34	3.97	86.64	nr	176.9

Material Costs/Measured Work Prices - Mechanical Installations

U:VENTILATION/AIR CONDITIONING SYSTEMS

Item	Net Price £	Material £	Labour hours	Labour £	Unit	Total rate £
Sum of two sides 3300mm	69.61	92.23	3.97	86.64	nr	178.88
Sum of two sides 3350mm	71.04	94.13	3.97	86.64	nr	180.77
Sum of two sides 3400mm	72.47	96.02	4.52	98.65	nr	194.67
Sum of two sides 3450mm	73.09	96.84	4.52	98.65	nr	195.49
Sum of two sides 3500mm	73.71	97.67	4.52	98.65	nr	196.31
Sum of two sides 3550mm	74.33	98.49	4.52	98.65	nr	197.13
Sum of two sides 3600mm	74.95	99.31	4.52	98.65	nr	197.96
Sum of two sides 3650mm	77.01	102.04	4.52	98.65	nr	200.69
Sum of two sides 3700mm	79.07	104.77	4.52	98.65	nr	203.41
Sum of two sides 3750mm	79.87	105.83	4.52	98.65	nr	204.47
Sum of two sides 3800mm	80.66	106.87	4.52	98.65	nr	205.52
Sum of two sides 3850mm	82.06	108.73	4.52	98.65	nr	207.38
Sum of two sides 3900mm	83.46	110.58	4.52	98.65	nr	209.23
Sum of two sides 3950mm	84.86	112.44	4.52	98.65	nr	211.09
Sum of two sides 4000mm	86.26	114.29	4.52	98.65	nr	212.94
Offset						
Sum of two sides 2500mm	117.19	155.28	3.48	75.95	nr	231.23
Sum of two sides 2550mm	120.56	159.74	3.48	75.95	nr	235.69
Sum of two sides 2600mm	123.94	164.22	3.49	76.17	nr	240.39
Sum of two sides 2650mm	127.30	168.67	3.49	76.17	nr	244.84
Sum of two sides 2700mm	130.67	173.14	3.50	76.39	nr	249.52
Sum of two sides 2750mm	134.04	177.60	3.50	76.39	nr	253.99
Sum of two sides 2800mm	137.41	182.07	4.34	94.72	nr	276.79
Sum of two sides 2850mm	138.81	183.92	4.34	94.72	nr	278.64
Sum of two sides 2900mm	140.21	185.78	4.76	103.89	nr	289.66
Sum of two sides 2950mm	141.61	187.63	4.76	103.89	nr	291.52
Sum of two sides 3000mm	143.01	189.49	5.32	116.11	nr	305.60
Sum of two sides 3050mm	143.56	190.22	5.32	116.11	nr	306.32
Sum of two sides 3100mm	144.11	190.95	5.35	116.76	nr	307.71
Sum of two sides 3150mm	144.66	191.67	5.32	116.11	nr	307.78
Sum of two sides 3200mm	145.21	192.40	5.35	116.76	nr	309.16
Sum of two sides 3250mm	145.54	192.84	5.32	116.11	nr	308.95
Sum of two sides 3300mm	145.87	193.28	5.32	116.11	nr	309.38
Sum of two sides 3350mm	146.20	193.72	5.32	116.11	nr	309.82
Sum of two sides 3400mm	146.53	194.15	5.32	116.11	nr	310.26
Sum of two sides 3450mm	146.84	194.56	5.32	116.11	nr	310.67
Sum of two sides 3500mm	147.14	194.96	5.32	116.11	nr	311.07
Sum of two sides 3550mm	147.45	195.37	5.32	116.11	nr	311.48
Sum of two sides 3600mm	147.76	195.78	5.35	116.76	nr	312.54
Sum of two sides 3650mm	151.17	200.30	5.35	116.76	nr	317.06
Sum of two sides 3700mm	154.58	204.82	5.35	116.76	nr	321.58
Sum of two sides 3750mm	155.91	206.58	5.35	116.76	nr	323.34
Sum of two sides 3800mm	157.23	208.33	5.35	116.76	nr	325.09
Sum of two sides 3850mm	159.59	211.46	5.35	116.76	nr	328.22
Sum of two sides 3900mm	161.95	214.58	5.35	116.76	nr	331.35
Sum of two sides 3950mm	164.31	217.71	5.35	116.76	nr	334.47
Sum of two sides 4000mm	166.67	220.84	5.35	116.76	nr	337.60
Square to round						
Sum of two sides 2500mm	89.13	118.10	4.70	102.58	nr	220.67
Sum of two sides 2550mm	92.08	122.01	4.70	102.58	nr	224.58
Sum of two sides 2600mm	95.02	125.90	4.70	102.58	nr	228.48
Sum of two sides 2650mm	97.02	128.55	4.70	102.58	nr	231.13

324 *Material Costs/Measured Work Prices - Mechanical Installations*

U:VENTILATION/AIR CONDITIONING SYSTEMS

Item	Net Price £	Material £	Labour hours	Labour £	Unit	Total rate £
U10 : DUCTWORK : RECTANGULAR - CLASS B						
Y30 - AIR DUCTLINES						
Galvanised sheet metal DW144 class B rectangular section ductwork; including all necessary stiffeners, joints, couplers in the running length and duct supports (Continued)						
Square to Round						
Sum of two sides 2700mm	100.91	133.71	4.71	102.79	nr	**236.50**
Sum of two sides 2750mm	103.86	137.61	4.71	102.79	nr	**240.41**
Sum of two sides 2800mm	106.81	141.52	8.19	178.74	nr	**320.27**
Sum of two sides 2850mm	109.75	145.42	8.19	178.74	nr	**324.16**
Sum of two sides 2900mm	112.70	149.33	8.19	178.74	nr	**328.07**
Sum of two sides 2950mm	115.64	153.22	8.19	178.74	nr	**331.97**
Sum of two sides 3000mm	118.59	157.13	8.19	178.74	nr	**335.88**
Sum of two sides 3050mm	120.85	160.13	8.19	178.74	nr	**338.87**
Sum of two sides 3100mm	123.11	163.12	8.19	178.74	nr	**341.86**
Sum of two sides 3150mm	125.37	166.12	8.62	188.13	nr	**354.24**
Sum of two sides 3200mm	127.63	169.11	8.62	188.13	nr	**357.24**
Sum of two sides 3250mm	129.79	171.97	8.62	188.13	nr	**360.10**
Sum of two sides 3300mm	131.94	174.82	8.62	188.13	nr	**362.95**
Sum of two sides 3350mm	134.10	177.68	8.62	188.13	nr	**365.81**
Sum of two sides 3400mm	136.26	180.54	8.62	188.13	nr	**368.67**
Sum of two sides 3450mm	137.59	182.31	8.62	188.13	nr	**370.43**
Sum of two sides 3500mm	138.92	184.07	8.62	188.13	nr	**372.20**
Sum of two sides 3550mm	140.26	185.84	8.75	190.97	nr	**376.81**
Sum of two sides 3600mm	141.59	187.61	8.75	190.97	nr	**378.57**
Sum of two sides 3650mm	144.57	191.56	8.75	190.97	nr	**382.52**
Sum of two sides 3700mm	147.55	195.50	8.75	190.97	nr	**386.47**
Sum of two sides 3750mm	148.88	197.27	8.75	190.97	nr	**388.23**
Sum of two sides 3800mm	150.22	199.04	8.75	190.97	nr	**390.01**
Sum of two sides 3850mm	152.37	201.89	8.75	190.97	nr	**392.86**
Sum of two sides 3900mm	154.53	204.75	8.75	190.97	nr	**395.72**
Sum of two sides 3950mm	156.69	207.61	8.75	190.97	nr	**398.58**
Sum of two sides 4000mm	158.85	210.48	8.75	190.97	nr	**401.44**
90 degree radius bend						
Sum of two sides 2500mm	158.13	209.52	3.90	85.12	nr	**294.64**
Sum of two sides 2550mm	161.45	213.92	3.90	85.12	nr	**299.04**
Sum of two sides 2600mm	164.77	218.32	4.26	92.97	nr	**311.29**
Sum of two sides 2650mm	168.08	222.71	4.26	92.97	nr	**315.68**
Sum of two sides 2700mm	171.40	227.10	4.55	99.30	nr	**326.41**
Sum of two sides 2750mm	174.72	231.50	4.55	99.30	nr	**330.81**
Sum of two sides 2800mm	178.04	235.90	4.55	99.30	nr	**335.20**
Sum of two sides 2850mm	183.32	242.90	4.55	99.30	nr	**342.20**
Sum of two sides 2900mm	188.60	249.90	6.87	149.94	nr	**399.83**
Sum of two sides 2950mm	193.87	256.88	6.87	149.94	nr	**406.81**
Sum of two sides 3000mm	199.15	263.87	6.87	149.94	nr	**413.81**
Sum of two sides 3050mm	202.96	268.92	6.87	149.94	nr	**418.86**
Sum of two sides 3100mm	206.77	273.97	6.87	149.94	nr	**423.91**
Sum of two sides 3150mm	210.58	279.02	6.87	149.94	nr	**428.95**

Material Costs/Measured Work Prices - Mechanical Installations

U:VENTILATION/AIR CONDITIONING SYSTEMS

Item	Net Price £	Material £	Labour hours	Labour £	Unit	Total rate £
Sum of two sides 3200mm	214.40	284.08	6.87	149.94	nr	**434.01**
Sum of two sides 3250mm	218.10	288.98	6.87	149.94	nr	**438.92**
Sum of two sides 3300mm	221.80	293.88	6.87	149.94	nr	**443.82**
Sum of two sides 3350mm	225.50	298.79	6.87	149.94	nr	**448.72**
Sum of two sides 3400mm	229.21	303.70	6.87	149.94	nr	**453.64**
Sum of two sides 3450mm	230.75	305.74	7.00	152.77	nr	**458.52**
Sum of two sides 3500mm	232.30	307.80	7.00	152.77	nr	**460.57**
Sum of two sides 3550mm	233.84	309.84	7.00	152.77	nr	**462.61**
Sum of two sides 3600mm	235.39	311.89	7.00	152.77	nr	**464.66**
Sum of two sides 3650mm	238.64	316.20	7.00	152.77	nr	**468.97**
Sum of two sides 3700mm	241.89	320.50	7.00	152.77	nr	**473.28**
Sum of two sides 3750mm	241.04	319.38	7.00	152.77	nr	**472.15**
Sum of two sides 3800mm	250.20	331.51	7.00	152.77	nr	**484.29**
Sum of two sides 3850mm	253.80	336.29	7.00	152.77	nr	**489.06**
Sum of two sides 3900mm	257.41	341.07	7.00	152.77	nr	**493.84**
Sum of two sides 3950mm	261.01	345.84	7.00	152.77	nr	**498.61**
Sum of two sides 4000mm	264.62	350.62	7.00	152.77	nr	**503.39**
45 degree bend						
Sum of two sides 2500mm	114.63	151.88	4.85	105.85	nr	**257.73**
Sum of two sides 2550mm	117.31	155.44	4.85	105.85	nr	**261.29**
Sum of two sides 2600mm	119.99	158.99	4.87	106.29	nr	**265.27**
Sum of two sides 2650mm	122.66	162.52	4.87	106.29	nr	**268.81**
Sum of two sides 2700mm	125.34	166.08	4.87	106.29	nr	**272.36**
Sum of two sides 2750mm	130.69	173.16	4.87	106.29	nr	**279.45**
Sum of two sides 2800mm	130.69	173.16	8.81	192.27	nr	**365.44**
Sum of two sides 2850mm	134.38	178.05	8.81	192.27	nr	**370.33**
Sum of two sides 2900mm	138.06	182.93	8.81	192.27	nr	**375.20**
Sum of two sides 2950mm	141.75	187.82	8.81	192.27	nr	**380.09**
Sum of two sides 3000mm	145.43	192.69	9.31	203.19	nr	**395.88**
Sum of two sides 3050mm	148.13	196.27	9.31	203.19	nr	**399.46**
Sum of two sides 3100mm	150.82	199.84	9.31	203.19	nr	**403.02**
Sum of two sides 3150mm	153.52	203.41	9.31	203.19	nr	**406.60**
Sum of two sides 3200mm	156.21	206.98	9.31	203.19	nr	**410.17**
Sum of two sides 3250mm	158.84	210.46	9.31	203.19	nr	**413.65**
Sum of two sides 3300mm	161.48	213.96	9.31	203.19	nr	**417.15**
Sum of two sides 3350mm	164.11	217.45	9.31	203.19	nr	**420.63**
Sum of two sides 3400mm	166.75	220.94	9.31	203.19	nr	**424.13**
Sum of two sides 3450mm	168.26	222.94	9.31	203.19	nr	**426.13**
Sum of two sides 3500mm	169.78	224.96	9.31	203.19	nr	**428.15**
Sum of two sides 3550mm	171.30	226.97	9.31	203.19	nr	**430.16**
Sum of two sides 3600mm	172.82	228.99	9.39	204.93	nr	**433.92**
Sum of two sides 3650mm	175.66	232.75	9.39	204.93	nr	**437.68**
Sum of two sides 3700mm	178.50	236.51	9.39	204.93	nr	**441.45**
Sum of two sides 3750mm	180.93	239.73	9.39	204.93	nr	**444.67**
Sum of two sides 3800mm	183.35	242.94	9.39	204.93	nr	**447.87**
Sum of two sides 3850mm	185.91	246.33	9.39	204.93	nr	**451.26**
Sum of two sides 3900mm	188.47	249.72	9.39	204.93	nr	**454.66**
Sum of two sides 3950mm	191.03	253.11	9.39	204.93	nr	**458.05**
Sum of two sides 4000mm	193.58	256.49	9.39	204.93	nr	**461.43**
90 degree mitre bend						
Sum of two sides 2500mm	233.92	309.94	4.85	105.85	nr	**415.79**
Sum of two sides 2550mm	240.30	318.40	4.85	105.85	nr	**424.25**

Material Costs/Measured Work Prices - Mechanical Installations

U:VENTILATION/AIR CONDITIONING SYSTEMS

Item	Net Price £	Material £	Labour hours	Labour £	Unit	Total rate £
U10 : DUCTWORK : RECTANGULAR - CLASS B						
Y30 - AIR DUCTLINES						
Galvanised sheet metal DW144 class B rectangular section ductwork; including all necessary stiffeners, joints, couplers in the running length and duct supports (Continued)						
90 degree mitre bend						
Sum of two sides 2600mm	246.68	326.85	4.87	106.29	nr	433.14
Sum of two sides 2650mm	253.06	335.30	4.87	106.29	nr	441.59
Sum of two sides 2700mm	259.44	343.76	4.87	106.29	nr	450.04
Sum of two sides 2750mm	265.82	352.21	4.87	106.29	nr	458.50
Sum of two sides 2800mm	272.21	360.68	8.81	192.27	nr	552.95
Sum of two sides 2850mm	281.90	373.52	8.81	192.27	nr	565.79
Sum of two sides 2900mm	291.59	386.36	14.81	323.22	nr	709.58
Sum of two sides 2950mm	301.28	399.20	14.81	323.22	nr	722.42
Sum of two sides 3000mm	310.97	412.04	14.81	323.22	nr	735.26
Sum of two sides 3050mm	319.34	423.13	14.81	323.22	nr	746.35
Sum of two sides 3100mm	327.70	434.20	15.20	331.73	nr	765.94
Sum of two sides 3150mm	336.07	445.29	15.20	331.73	nr	777.03
Sum of two sides 3200mm	344.44	456.38	15.20	331.73	nr	788.12
Sum of two sides 3250mm	352.55	467.13	15.20	331.73	nr	798.86
Sum of two sides 3300mm	360.78	478.03	15.20	331.73	nr	809.77
Sum of two sides 3350mm	368.78	488.63	15.20	331.73	nr	820.37
Sum of two sides 3400mm	376.90	499.39	15.20	331.73	nr	831.13
Sum of two sides 3450mm	382.32	506.57	15.20	331.73	nr	838.31
Sum of two sides 3500mm	387.74	513.76	15.20	331.73	nr	845.49
Sum of two sides 3550mm	393.16	520.94	15.60	340.46	nr	861.40
Sum of two sides 3600mm	398.59	528.13	15.60	340.46	nr	868.60
Sum of two sides 3650mm	405.57	537.38	15.60	340.46	nr	877.84
Sum of two sides 3700mm	412.56	546.64	15.60	340.46	nr	887.11
Sum of two sides 3750mm	423.02	560.50	15.60	340.46	nr	900.97
Sum of two sides 3800mm	433.48	574.36	15.60	340.46	nr	914.82
Sum of two sides 3850mm	441.64	585.17	15.60	340.46	nr	925.64
Sum of two sides 3900mm	449.80	595.99	15.60	340.46	nr	936.45
Sum of two sides 3950mm	457.96	606.80	15.60	340.46	nr	947.26
Sum of two sides 4000mm	466.12	617.61	15.60	340.46	nr	958.07
Branch						
Sum of two sides 2500mm	101.58	134.59	2.88	62.85	nr	197.45
Sum of two sides 2550mm	104.37	138.29	2.88	62.85	nr	201.15
Sum of two sides 2600mm	107.15	141.97	2.88	62.85	nr	204.83
Sum of two sides 2650mm	109.94	145.67	2.88	62.85	nr	208.53
Sum of two sides 2700mm	112.73	149.37	2.88	62.85	nr	212.22
Sum of two sides 2750mm	115.51	153.05	2.88	62.85	nr	215.91
Sum of two sides 2800mm	118.30	156.75	3.94	85.99	nr	242.74
Sum of two sides 2850mm	121.12	160.48	3.94	85.99	nr	246.47
Sum of two sides 2900mm	123.94	164.22	3.94	85.99	nr	250.21
Sum of two sides 2950mm	126.75	167.94	3.94	85.99	nr	253.9
Sum of two sides 3000mm	129.57	171.68	3.94	85.99	nr	257.67
Sum of two sides 3050mm	131.75	174.57	3.94	85.99	nr	260.56

Material Costs/Measured Work Prices - Mechanical Installations

U:VENTILATION/AIR CONDITIONING SYSTEMS

Item	Net Price £	Material £	Labour hours	Labour £	Unit	Total rate £
Sum of two sides 3100mm	133.93	177.46	3.94	85.99	nr	263.45
Sum of two sides 3150mm	136.11	180.35	3.94	85.99	nr	266.33
Sum of two sides 3200mm	138.29	183.23	3.94	85.99	nr	269.22
Sum of two sides 3250mm	140.40	186.03	3.94	85.99	nr	272.02
Sum of two sides 3300mm	142.52	188.84	3.94	85.99	nr	274.83
Sum of two sides 3350mm	144.64	191.65	3.94	85.99	nr	277.64
Sum of two sides 3400mm	146.76	194.46	3.94	85.99	nr	280.45
Sum of two sides 3450mm	148.89	197.28	4.83	105.41	nr	302.69
Sum of two sides 3500mm	151.02	200.10	4.83	105.41	nr	305.51
Sum of two sides 3550mm	153.15	202.92	4.83	105.41	nr	308.34
Sum of two sides 3600mm	155.29	205.76	4.83	105.41	nr	311.17
Sum of two sides 3650mm	158.43	209.92	4.83	105.41	nr	315.33
Sum of two sides 3700mm	161.57	214.08	4.83	105.41	nr	319.49
Sum of two sides 3750mm	162.67	215.54	4.83	105.41	nr	320.95
Sum of two sides 3800mm	163.76	216.98	4.83	105.41	nr	322.39
Sum of two sides 3850mm	165.84	219.74	4.83	105.41	nr	325.15
Sum of two sides 3900mm	167.93	222.51	4.83	105.41	nr	327.92
Sum of two sides 3950mm	170.02	225.28	4.83	105.41	nr	330.69
Sum of two sides 4000mm	172.11	228.05	4.83	105.41	nr	333.46
Grille neck						
Sum of two sides 2500mm	107.24	142.09	2.88	62.85	nr	204.95
Sum of two sides 2550mm	110.31	146.16	2.88	62.85	nr	209.02
Sum of two sides 2600mm	113.38	150.23	2.88	62.85	nr	213.08
Sum of two sides 2650mm	116.45	154.30	2.88	62.85	nr	217.15
Sum of two sides 2700mm	119.52	158.36	2.88	62.85	nr	221.22
Sum of two sides 2750mm	122.59	162.43	2.88	62.85	nr	225.29
Sum of two sides 2800mm	125.66	166.50	3.94	85.99	nr	252.49
Sum of two sides 2850mm	128.72	170.55	3.94	85.99	nr	256.54
Sum of two sides 2900mm	131.79	174.62	3.94	85.99	nr	260.61
Sum of two sides 2950mm	134.86	178.69	3.94	85.99	nr	264.68
Sum of two sides 3000mm	137.93	182.76	3.94	85.99	nr	268.75
Sum of two sides 3050mm	140.31	185.91	4.12	89.92	nr	275.83
Sum of two sides 3100mm	142.70	189.08	4.12	89.92	nr	278.99
Sum of two sides 3150mm	145.08	192.23	4.12	89.92	nr	282.15
Sum of two sides 3200mm	147.46	195.38	4.12	89.92	nr	285.30
Sum of two sides 3250mm	149.75	198.42	4.12	89.92	nr	288.34
Sum of two sides 3300mm	152.03	201.44	4.12	89.92	nr	291.36
Sum of two sides 3350mm	154.31	204.46	4.12	89.92	nr	294.38
Sum of two sides 3400mm	156.59	207.48	4.12	89.92	nr	297.40
Sum of two sides 3450mm	158.87	210.50	5.00	109.12	nr	319.63
Sum of two sides 3500mm	161.15	213.52	5.00	109.12	nr	322.65
Sum of two sides 3550mm	163.43	216.54	5.00	109.12	nr	325.67
Sum of two sides 3600mm	165.71	219.57	5.00	109.12	nr	328.69
Sum of two sides 3650mm	169.02	223.95	5.00	109.12	nr	333.07
Sum of two sides 3700mm	172.33	228.34	5.00	109.12	nr	337.46
Sum of two sides 3750mm	173.58	229.99	5.00	109.12	nr	339.12
Sum of two sides 3800mm	147.83	195.87	5.00	109.12	nr	305.00
Sum of two sides 3850mm	177.11	234.67	5.00	109.12	nr	343.79
Sum of two sides 3900mm	179.39	237.69	5.00	109.12	nr	346.81
Sum of two sides 3950mm	181.67	240.71	5.00	109.12	nr	349.84
Sum of two sides 4000mm	183.95	243.73	5.00	109.12	nr	352.86

Material Costs/Measured Work Prices - Mechanical Installations

U:VENTILATION/AIR CONDITIONING SYSTEMS

Item	Net Price £	Material £	Labour hours	Labour £	Unit	Total rate £
U10 : DUCTWORK : RECTANGULAR - CLASS B						
Y30 - AIR DUCTLINES						
Galvanised sheet metal DW144 class B rectangular section ductwork; including all necessary stiffeners, joints, couplers in the running length and duct supports (Continued)						
Ductwork 2501 to 4000mm longest side						
Sum of two sides 3000mm	198.02	262.38	2.38	51.94	m	**314.32**
Sum of two sides 3050mm	200.83	266.10	2.38	51.94	m	**318.04**
Sum of two sides 3100mm	203.65	269.84	2.38	51.94	m	**321.78**
Sum of two sides 3150mm	206.47	273.57	2.38	51.94	m	**325.52**
Sum of two sides 3200mm	209.29	277.31	2.38	51.94	m	**329.25**
Sum of two sides 3250mm	211.90	280.77	2.38	51.94	m	**332.71**
Sum of two sides 3300mm	214.52	284.24	2.38	51.94	m	**336.18**
Sum of two sides 3350mm	217.14	287.71	2.64	57.62	m	**345.33**
Sum of two sides 3400mm	219.76	291.18	2.64	57.62	m	**348.80**
Sum of two sides 3450mm	222.37	294.64	2.64	57.62	m	**352.26**
Sum of two sides 3500mm	224.99	298.11	2.66	58.05	m	**356.17**
Sum of two sides 3550mm	227.61	301.58	2.66	58.05	m	**359.64**
Sum of two sides 3600mm	230.23	305.05	2.95	64.38	m	**369.44**
Sum of two sides 3650mm	233.05	308.79	2.95	64.38	m	**373.17**
Sum of two sides 3700mm	235.87	312.53	2.96	64.60	m	**377.13**
Sum of two sides 3750mm	238.68	316.25	2.96	64.60	m	**380.85**
Sum of two sides 3800mm	241.50	319.99	3.15	68.75	m	**388.74**
Sum of two sides 3850mm	244.96	324.57	3.15	68.75	m	**393.32**
Sum of two sides 3900mm	248.42	329.16	3.15	68.75	m	**397.90**
Sum of two sides 3950mm	251.88	333.74	3.15	68.75	m	**402.49**
Sum of two sides 4000mm	255.35	338.34	3.15	68.75	m	**407.09**
Extra over fittings; Ductwork 2501 to 4000mm longest side						
End Cap						
Sum of two sides 3000mm	80.69	106.91	1.73	37.76	nr	**144.67**
Sum of two sides 3050mm	82.34	109.10	1.73	37.76	nr	**146.86**
Sum of two sides 3100mm	83.98	111.27	1.73	37.76	nr	**149.03**
Sum of two sides 3150mm	85.63	113.46	1.73	37.76	nr	**151.22**
Sum of two sides 3200mm	87.27	115.63	1.73	37.76	nr	**153.39**
Sum of two sides 3250mm	88.89	117.78	1.73	37.76	nr	**155.54**
Sum of two sides 3300mm	90.50	119.91	1.73	37.76	nr	**157.67**
Sum of two sides 3350mm	92.11	122.05	1.73	37.76	nr	**159.80**
Sum of two sides 3400mm	93.72	124.18	1.73	37.76	nr	**161.94**
Sum of two sides 3450mm	95.33	126.31	1.73	37.76	nr	**164.07**
Sum of two sides 3500mm	96.94	128.45	1.73	37.76	nr	**166.20**
Sum of two sides 3550mm	98.55	130.58	1.73	37.76	nr	**168.34**
Sum of two sides 3600mm	100.16	132.71	1.73	37.76	nr	**170.47**
Sum of two sides 3650mm	101.78	134.86	1.73	37.76	nr	**172.62**
Sum of two sides 3700mm	103.39	136.99	1.73	37.76	nr	**174.75**
Sum of two sides 3750mm	105.00	139.13	1.73	37.76	nr	**176.88**
Sum of two sides 3800mm	106.61	141.26	1.73	37.76	nr	**179.01**

Material Costs/Measured Work Prices - Mechanical Installations 329

U:VENTILATION/AIR CONDITIONING SYSTEMS

Item	Net Price £	Material £	Labour hours	Labour £	Unit	Total rate £
Sum of two sides 3850mm	108.22	143.39	1.80	39.28	nr	**182.68**
Sum of two sides 3900mm	109.83	145.52	1.80	39.28	nr	**184.81**
Sum of two sides 3950mm	111.44	147.66	1.80	39.28	nr	**186.94**
Sum of two sides 4000mm	113.06	149.80	1.80	39.28	nr	**189.09**
Reducer						
Sum of two sides 3000mm	68.53	90.80	3.12	68.09	nr	**158.90**
Sum of two sides 3050mm	70.19	93.00	3.12	68.09	nr	**161.09**
Sum of two sides 3100mm	71.84	95.19	3.12	68.09	nr	**163.28**
Sum of two sides 3150mm	73.50	97.39	3.12	68.09	nr	**165.48**
Sum of two sides 3200mm	75.15	99.57	3.12	68.09	nr	**167.67**
Sum of two sides 3250mm	76.74	101.68	3.12	68.09	nr	**169.77**
Sum of two sides 3300mm	78.33	103.79	3.12	68.09	nr	**171.88**
Sum of two sides 3350mm	79.92	105.89	3.12	68.09	nr	**173.99**
Sum of two sides 3400mm	81.50	107.99	3.12	68.09	nr	**176.08**
Sum of two sides 3450mm	83.09	110.09	3.12	68.09	nr	**178.19**
Sum of two sides 3500mm	84.68	112.20	3.12	68.09	nr	**180.29**
Sum of two sides 3550mm	86.26	114.29	3.12	68.09	nr	**182.39**
Sum of two sides 3600mm	87.85	116.40	3.95	86.21	nr	**202.61**
Sum of two sides 3650mm	89.44	118.51	3.95	86.21	nr	**204.72**
Sum of two sides 3700mm	91.02	120.60	3.97	86.64	nr	**207.25**
Sum of two sides 3750mm	92.61	122.71	3.97	86.64	nr	**209.35**
Sum of two sides 3800mm	94.20	124.81	4.52	98.65	nr	**223.46**
Sum of two sides 3850mm	94.95	125.81	4.52	98.65	nr	**224.46**
Sum of two sides 3900mm	95.71	126.82	4.52	98.65	nr	**225.46**
Sum of two sides 3950mm	96.46	127.81	4.52	98.65	nr	**226.46**
Sum of two sides 4000mm	97.21	128.80	4.52	98.65	nr	**227.45**
Offset						
Sum of two sides 3000mm	101.75	134.82	3.48	75.95	nr	**210.77**
Sum of two sides 3050mm	103.84	137.59	3.48	75.95	nr	**213.54**
Sum of two sides 3100mm	105.94	140.37	3.48	75.95	nr	**216.32**
Sum of two sides 3150mm	108.03	143.14	3.48	75.95	nr	**219.09**
Sum of two sides 3200mm	110.12	145.91	3.48	75.95	nr	**221.86**
Sum of two sides 3250mm	112.21	148.68	3.48	75.95	nr	**224.63**
Sum of two sides 3300mm	114.31	151.46	3.48	75.95	nr	**227.41**
Sum of two sides 3350mm	116.40	154.23	3.49	76.17	nr	**230.40**
Sum of two sides 3400mm	118.49	157.00	3.49	76.17	nr	**233.17**
Sum of two sides 3450mm	145.70	193.05	3.49	76.17	nr	**269.22**
Sum of two sides 3500mm	122.68	162.55	3.50	76.39	nr	**238.94**
Sum of two sides 3550mm	124.77	165.32	3.50	76.39	nr	**241.71**
Sum of two sides 3600mm	126.87	168.10	3.50	76.39	nr	**244.49**
Sum of two sides 3650mm	128.96	170.87	3.50	76.39	nr	**247.26**
Sum of two sides 3700mm	131.05	173.64	4.76	103.89	nr	**277.53**
Sum of two sides 3750mm	133.15	176.42	4.76	103.89	nr	**280.31**
Sum of two sides 3800mm	135.24	179.19	5.32	116.11	nr	**295.30**
Sum of two sides 3850mm	137.33	181.96	5.32	116.11	nr	**298.07**
Sum of two sides 3900mm	139.43	184.74	5.35	116.76	nr	**301.51**
Sum of two sides 3950mm	141.52	187.51	5.32	116.11	nr	**303.62**
Sum of two sides 4000mm	143.61	190.28	5.35	116.76	nr	**307.04**
Square to round						
Sum of two sides 3000mm	118.63	157.18	4.70	102.58	nr	**259.76**
Sum of two sides 3050mm	120.90	160.19	4.70	102.58	nr	**262.77**

Material Costs/Measured Work Prices - Mechanical Installations

U:VENTILATION/AIR CONDITIONING SYSTEMS

Item	Net Price £	Material £	Labour hours	Labour £	Unit	Total rate £
U10 : DUCTWORK : RECTANGULAR - CLASS B						
Y30 - AIR DUCTLINES						
Galvanised sheet metal DW144 class B rectangular section ductwork; including all necessary stiffeners, joints, couplers in the running length and duct supports (Continued)						
Square to round						
Sum of two sides 3100mm	123.17	163.20	4.70	102.58	nr	265.78
Sum of two sides 3150mm	125.44	166.21	4.70	102.58	nr	268.78
Sum of two sides 3200mm	127.71	169.22	4.70	102.58	nr	271.79
Sum of two sides 3250mm	129.86	172.06	4.70	102.58	nr	274.64
Sum of two sides 3300mm	132.01	174.91	4.70	102.58	nr	277.49
Sum of two sides 3350mm	134.16	177.76	4.71	102.79	nr	280.56
Sum of two sides 3400mm	136.31	180.61	4.71	102.79	nr	283.40
Sum of two sides 3450mm	138.46	183.46	4.71	102.79	nr	286.25
Sum of two sides 3500mm	140.61	186.31	4.71	102.79	nr	289.10
Sum of two sides 3550mm	142.76	189.16	8.19	178.74	nr	367.90
Sum of two sides 3600mm	144.91	192.01	8.19	178.74	nr	370.75
Sum of two sides 3650mm	147.06	194.85	8.19	178.74	nr	373.60
Sum of two sides 3700mm	149.21	197.70	8.62	188.13	nr	385.83
Sum of two sides 3750mm	151.36	200.55	8.62	188.13	nr	388.68
Sum of two sides 3800mm	153.51	203.40	8.62	188.13	nr	391.53
Sum of two sides 3850mm	154.82	205.14	8.62	188.13	nr	393.26
Sum of two sides 3900mm	156.14	206.89	8.62	188.13	nr	395.01
Sum of two sides 3950mm	157.46	208.63	8.62	188.13	nr	396.76
Sum of two sides 4000mm	158.77	210.37	8.75	190.97	nr	401.34
90 degree radius bend						
Sum of two sides 3000mm	369.97	490.21	3.90	85.12	nr	575.33
Sum of two sides 3050mm	372.36	493.38	3.90	85.12	nr	578.49
Sum of two sides 3100mm	374.76	496.56	3.90	85.12	nr	581.67
Sum of two sides 3150mm	377.15	499.72	3.90	85.12	nr	584.84
Sum of two sides 3200mm	379.55	502.90	4.26	92.97	nr	595.88
Sum of two sides 3250mm	385.99	511.44	4.26	92.97	nr	604.41
Sum of two sides 3300mm	392.42	519.96	4.26	92.97	nr	612.93
Sum of two sides 3350mm	398.86	528.49	4.26	92.97	nr	621.46
Sum of two sides 3400mm	405.29	537.01	4.26	92.97	nr	629.98
Sum of two sides 3450mm	411.73	545.54	4.55	99.30	nr	644.84
Sum of two sides 3500mm	418.16	554.06	4.55	99.30	nr	653.36
Sum of two sides 3550mm	424.60	562.60	4.55	99.30	nr	661.90
Sum of two sides 3600mm	431.03	571.11	4.55	99.30	nr	670.42
Sum of two sides 3650mm	432.23	572.70	6.87	149.94	nr	722.64
Sum of two sides 3700mm	433.42	574.28	6.87	149.94	nr	724.22
Sum of two sides 3750mm	434.62	575.87	6.87	149.94	nr	725.81
Sum of two sides 3800mm	435.81	577.45	6.87	149.94	nr	727.38
Sum of two sides 3850mm	444.93	589.53	6.87	149.94	nr	739.47
Sum of two sides 3900mm	454.04	601.60	6.87	149.94	nr	751.54
Sum of two sides 3950mm	463.16	613.69	7.00	152.77	nr	766.46
Sum of two sides 4000mm	472.28	625.77	7.00	152.77	nr	778.54

Material Costs/Measured Work Prices - Mechanical Installations   331

U:VENTILATION/AIR CONDITIONING SYSTEMS

Item	Net Price £	Material £	Labour hours	Labour £	Unit	Total rate £
45 degree bend						
Sum of two sides 3000mm	180.30	238.90	4.85	105.85	nr	**344.75**
Sum of two sides 3050mm	181.48	240.46	4.85	105.85	nr	**346.31**
Sum of two sides 3100mm	182.67	242.04	4.85	105.85	nr	**347.89**
Sum of two sides 3150mm	183.85	243.60	4.85	105.85	nr	**349.45**
Sum of two sides 3200mm	185.04	245.18	4.85	105.85	nr	**351.03**
Sum of two sides 3250mm	188.24	249.42	4.87	106.29	nr	**355.70**
Sum of two sides 3300mm	191.44	253.66	4.87	106.29	nr	**359.94**
Sum of two sides 3350mm	194.65	257.91	4.87	106.29	nr	**364.20**
Sum of two sides 3400mm	197.85	262.15	4.87	106.29	nr	**368.44**
Sum of two sides 3450mm	201.06	266.40	4.87	106.29	nr	**372.69**
Sum of two sides 3500mm	204.26	270.64	4.87	106.29	nr	**376.93**
Sum of two sides 3550mm	207.46	274.88	8.81	192.27	nr	**467.16**
Sum of two sides 3600mm	210.67	279.14	8.81	192.27	nr	**471.41**
Sum of two sides 3650mm	211.25	279.91	8.81	192.27	nr	**472.18**
Sum of two sides 3700mm	211.84	280.69	8.81	192.27	nr	**472.96**
Sum of two sides 3750mm	212.42	281.46	8.81	192.27	nr	**473.73**
Sum of two sides 3800mm	213.00	282.23	9.31	203.19	nr	**485.41**
Sum of two sides 3850mm	217.50	288.19	9.31	203.19	nr	**491.37**
Sum of two sides 3900mm	221.99	294.14	9.31	203.19	nr	**497.32**
Sum of two sides 3950mm	226.48	300.09	9.31	203.19	nr	**503.27**
Sum of two sides 4000mm	230.97	306.04	9.31	203.19	nr	**509.22**
90 degree mitre bend						
Sum of two sides 3000mm	340.08	450.61	4.85	105.85	nr	**556.46**
Sum of two sides 3050mm	342.27	453.51	4.85	105.85	nr	**559.36**
Sum of two sides 3100mm	344.46	456.41	4.85	105.85	nr	**562.26**
Sum of two sides 3150mm	346.65	459.31	4.85	105.85	nr	**565.16**
Sum of two sides 3200mm	348.84	462.21	4.87	106.29	nr	**568.50**
Sum of two sides 3250mm	357.40	473.56	4.87	106.29	nr	**579.84**
Sum of two sides 3300mm	365.96	484.90	4.87	106.29	nr	**591.18**
Sum of two sides 3350mm	374.52	496.24	4.87	106.29	nr	**602.52**
Sum of two sides 3400mm	383.09	507.59	4.87	106.29	nr	**613.88**
Sum of two sides 3450mm	391.65	518.94	8.81	192.27	nr	**711.21**
Sum of two sides 3500mm	400.21	530.28	8.81	192.27	nr	**722.55**
Sum of two sides 3550mm	408.77	541.62	8.81	192.27	nr	**733.89**
Sum of two sides 3600mm	417.34	552.98	8.81	192.27	nr	**745.25**
Sum of two sides 3650mm	418.24	554.17	14.81	323.22	nr	**877.39**
Sum of two sides 3700mm	419.15	555.37	14.81	323.22	nr	**878.60**
Sum of two sides 3750mm	420.06	556.58	14.81	323.22	nr	**879.80**
Sum of two sides 3800mm	420.97	557.79	14.81	323.22	nr	**881.01**
Sum of two sides 3850mm	434.06	575.13	14.81	323.22	nr	**898.35**
Sum of two sides 3900mm	447.15	592.47	14.81	323.22	nr	**915.70**
Sum of two sides 3950mm	460.24	609.82	15.20	331.73	nr	**941.55**
Sum of two sides 4000mm	473.33	627.16	15.20	331.73	nr	**958.90**
Branch						
Sum of two sides 3000mm	158.09	209.47	2.88	62.85	nr	**272.32**
Sum of two sides 3050mm	160.80	213.06	2.88	62.85	nr	**275.91**
Sum of two sides 3100mm	344.46	456.41	2.88	62.85	nr	**519.26**
Sum of two sides 3150mm	166.23	220.25	2.88	62.85	nr	**283.11**
Sum of two sides 3200mm	168.94	223.85	2.88	62.85	nr	**286.70**
Sum of two sides 3250mm	171.56	227.32	2.88	62.85	nr	**290.17**
Sum of two sides 3300mm	174.19	230.80	2.88	62.85	nr	**293.66**

Material Costs/Measured Work Prices - Mechanical Installations

U:VENTILATION/AIR CONDITIONING SYSTEMS

Item	Net Price £	Material £	Labour hours	Labour £	Unit	Total rate £
U10 : DUCTWORK : RECTANGULAR - CLASS B						
Y30 - AIR DUCTLINES						
Galvanised sheet metal DW144 class B rectangular section ductwork; including all necessary stiffeners, joints, couplers in the running length and duct supports (Continued)						
Branch						
Sum of two sides 3350mm	176.81	234.27	2.88	62.85	nr	**297.13**
Sum of two sides 3400mm	179.43	237.74	2.88	62.85	nr	**300.60**
Sum of two sides 3450mm	182.05	241.22	3.94	85.99	nr	**327.21**
Sum of two sides 3500mm	184.67	244.69	3.94	85.99	nr	**330.68**
Sum of two sides 3550mm	187.30	248.17	3.94	85.99	nr	**334.16**
Sum of two sides 3600mm	189.92	251.64	3.94	85.99	nr	**337.63**
Sum of two sides 3650mm	192.54	255.12	3.94	85.99	nr	**341.10**
Sum of two sides 3700mm	195.16	258.59	3.94	85.99	nr	**344.58**
Sum of two sides 3750mm	197.78	262.06	3.94	85.99	nr	**348.05**
Sum of two sides 3800mm	200.40	265.53	4.83	105.41	nr	**370.94**
Sum of two sides 3850mm	203.02	269.00	4.83	105.41	nr	**374.41**
Sum of two sides 3900mm	205.64	272.47	4.83	105.41	nr	**377.89**
Sum of two sides 3950mm	208.27	275.96	4.83	105.41	nr	**381.37**
Sum of two sides 4000mm	210.89	279.43	4.83	105.41	nr	**384.84**
Grille neck						
Sum of two sides 3000mm	139.86	185.31	2.88	62.85	nr	**248.17**
Sum of two sides 3050mm	142.27	188.51	2.88	62.85	nr	**251.36**
Sum of two sides 3100mm	144.68	191.70	2.88	62.85	nr	**254.56**
Sum of two sides 3150mm	147.10	194.91	2.88	62.85	nr	**257.76**
Sum of two sides 3200mm	149.51	198.10	2.88	62.85	nr	**260.96**
Sum of two sides 3250mm	151.83	201.17	2.88	62.85	nr	**264.03**
Sum of two sides 3300mm	154.14	204.24	2.88	62.85	nr	**267.09**
Sum of two sides 3350mm	156.45	207.30	3.94	85.99	nr	**293.29**
Sum of two sides 3400mm	158.76	210.36	3.94	85.99	nr	**296.35**
Sum of two sides 3450mm	161.08	213.43	3.94	85.99	nr	**299.42**
Sum of two sides 3500mm	163.39	216.49	3.94	85.99	nr	**302.48**
Sum of two sides 3550mm	165.70	219.55	3.94	85.99	nr	**305.54**
Sum of two sides 3600mm	168.01	222.61	3.94	85.99	nr	**308.60**
Sum of two sides 3650mm	170.33	225.69	4.12	89.92	nr	**315.60**
Sum of two sides 3700mm	172.64	228.75	4.12	89.92	nr	**318.67**
Sum of two sides 3750mm	174.95	231.81	4.12	89.92	nr	**321.73**
Sum of two sides 3800mm	177.27	234.88	4.12	89.92	nr	**324.80**
Sum of two sides 3850mm	179.58	237.94	4.12	89.92	nr	**327.86**
Sum of two sides 3900mm	181.89	241.00	4.12	89.92	nr	**330.92**
Sum of two sides 3950mm	184.20	244.06	4.12	89.92	nr	**333.98**
Sum of two sides 4000mm	186.52	247.14	5.00	109.12	nr	**356.26**

Material Costs/Measured Work Prices - Mechanical Installations

U:VENTILATION/AIR CONDITIONING SYSTEMS

Item	Net Price £	Material £	Labour hours	Labour £	Unit	Total rate £
U10 : DUCTWORK : RECTANGULAR - CLASS B						
Y30 - ANCILLARIES						
Access doors, hollow steel construction; 25mm mineral wool insulation; removeable; fixed with cams; including sub-frame and integral sealing gaskets						
Rectangular duct						
150 x 150mm	13.44	17.81	1.25	27.28	nr	**45.09**
200 x 200mm	14.32	18.97	1.25	27.28	nr	**46.25**
300 x 150mm	13.95	18.48	1.25	27.28	nr	**45.76**
300 x 300mm	14.91	19.76	1.25	27.28	nr	**47.04**
400 x 400mm	15.36	20.35	1.35	29.49	nr	**49.85**
450 x 300mm	15.29	20.26	1.50	32.77	nr	**53.03**
450 x 450mm	15.74	20.86	1.50	32.77	nr	**53.63**
Access doors, hollow steel construction; 25mm mineral wool insulation;hinged; locked with cams; including sub-frame and integral sealing gaskets						
Rectangular duct						
150 x 150mm	16.00	21.20	1.25	27.28	nr	**48.48**
200 x 200mm	16.50	21.86	1.25	27.28	nr	**49.14**
300 x 150mm	16.75	22.19	1.25	27.28	nr	**49.47**
300 x 300mm	14.91	19.76	1.25	27.28	nr	**47.04**
400 x 400mm	19.08	25.28	1.35	29.49	nr	**54.77**
450 x 300mm	20.14	26.69	1.50	32.77	nr	**59.46**
450 x 450mm	15.74	20.86	1.50	32.77	nr	**53.63**

Material Costs/Measured Work Prices - Mechanical Installations

U:VENTILATION/AIR CONDITIONING SYSTEMS

Item	Net Price £	Material £	Labour hours	Labour £	Unit	Total rate £
U10 : DUCTWORK : RECTANGULAR - CLASS C						
Y30 - AIR DUCTLINES						
Galvanised sheet metal DW144 class C rectangular section ductwork; including all necessary stiffeners, joints, couplers in the running length and duct supports						
Ductwork up to 400mm longest side						
Sum of two sides 200mm	14.05	18.62	1.19	25.97	m	**44.59**
Sum of two sides 250mm	14.80	19.61	1.16	25.32	m	**44.93**
Sum of two sides 300mm	15.48	20.51	1.16	25.32	m	**45.83**
Sum of two sides 350mm	16.99	22.51	1.17	25.53	m	**48.05**
Sum of two sides 400mm	17.99	23.84	1.17	25.53	m	**49.37**
Sum of two sides 450mm	18.39	24.37	1.17	25.53	m	**49.90**
Sum of two sides 500mm	18.80	24.91	1.17	25.53	m	**50.44**
Sum of two sides 550mm	19.58	25.94	1.19	25.97	m	**51.91**
Sum of two sides 600mm	20.28	26.87	1.19	25.97	m	**52.84**
Sum of two sides 650mm	20.98	27.80	1.19	25.97	m	**53.77**
Sum of two sides 700mm	21.96	29.10	1.19	25.97	m	**55.07**
Sum of two sides 750mm	22.65	30.01	1.19	25.97	m	**55.98**
Sum of two sides 800mm	23.35	30.94	1.19	25.97	m	**56.91**
Extra over fittings; Ductwork up to 400mm longest side						
End Cap						
Sum of two sides 200mm	8.96	11.87	0.38	8.29	nr	**20.17**
Sum of two sides 250mm	9.34	12.38	0.38	8.29	nr	**20.67**
Sum of two sides 300mm	9.73	12.89	0.38	8.29	nr	**21.19**
Sum of two sides 350mm	10.12	13.41	0.38	8.29	nr	**21.70**
Sum of two sides 400mm	10.51	13.93	0.38	8.29	nr	**22.22**
Sum of two sides 450mm	10.90	14.44	0.38	8.29	nr	**22.74**
Sum of two sides 500mm	11.28	14.95	0.38	8.29	nr	**23.24**
Sum of two sides 550mm	11.67	15.46	0.38	8.29	nr	**23.76**
Sum of two sides 600mm	12.06	15.98	0.38	8.29	nr	**24.27**
Sum of two sides 650mm	12.45	16.50	0.38	8.29	nr	**24.79**
Sum of two sides 700mm	12.84	17.01	0.38	8.29	nr	**25.31**
Sum of two sides 750mm	13.22	17.52	0.38	8.29	nr	**25.81**
Sum of two sides 800mm	13.61	18.03	0.38	8.29	nr	**26.33**
Reducer						
Sum of two sides 200mm	11.71	15.52	1.40	30.55	nr	**46.07**
Sum of two sides 250mm	12.34	16.35	1.40	30.55	nr	**46.90**
Sum of two sides 300mm	12.97	17.19	1.40	30.55	nr	**47.74**
Sum of two sides 350mm	16.78	22.23	1.42	30.99	nr	**53.22**
Sum of two sides 400mm	17.42	23.08	1.42	30.99	nr	**54.07**
Sum of two sides 450mm	18.07	23.94	1.42	30.99	nr	**54.93**
Sum of two sides 500mm	18.71	24.79	1.42	30.99	nr	**55.78**
Sum of two sides 550mm	19.36	25.65	1.69	36.88	nr	**62.54**
Sum of two sides 600mm	20.01	26.51	1.69	36.88	nr	**63.40**
Sum of two sides 650mm	20.98	27.80	1.69	36.88	nr	**64.68**

Material Costs/Measured Work Prices - Mechanical Installations

U:VENTILATION/AIR CONDITIONING SYSTEMS

Item	Net Price £	Material £	Labour hours	Labour £	Unit	Total rate £
Sum of two sides 700mm	21.30	28.22	1.69	36.88	nr	**65.11**
Sum of two sides 750mm	21.95	29.08	1.92	41.90	nr	**70.99**
Sum of two sides 800mm	22.59	29.93	1.92	41.90	nr	**71.83**
Offset						
Sum of two sides 200mm	23.01	30.49	1.63	35.57	nr	**66.06**
Sum of two sides 250mm	23.81	31.55	1.63	35.57	nr	**67.12**
Sum of two sides 300mm	24.60	32.59	1.63	35.57	nr	**68.17**
Sum of two sides 350mm	28.44	37.68	1.65	36.01	nr	**73.69**
Sum of two sides 400mm	29.01	38.44	1.65	36.01	nr	**74.45**
Sum of two sides 450mm	29.82	39.51	1.65	36.01	nr	**75.52**
Sum of two sides 500mm	30.62	40.57	1.65	36.01	nr	**76.58**
Sum of two sides 550mm	31.18	41.31	1.92	41.90	nr	**83.22**
Sum of two sides 600mm	31.72	42.03	1.92	41.90	nr	**83.93**
Sum of two sides 650mm	32.53	43.10	1.92	41.90	nr	**85.01**
Sum of two sides 700mm	33.07	43.82	1.92	41.90	nr	**85.72**
Sum of two sides 750mm	33.60	44.52	1.92	41.90	nr	**86.42**
Sum of two sides 800mm	34.13	45.22	1.92	41.90	nr	**87.13**
Square to round						
Sum of two sides 200mm	19.94	26.42	1.22	26.63	nr	**53.05**
Sum of two sides 250mm	20.70	27.43	1.22	26.63	nr	**54.05**
Sum of two sides 300mm	21.46	28.43	1.22	26.63	nr	**55.06**
Sum of two sides 350mm	23.45	31.07	1.25	27.28	nr	**58.35**
Sum of two sides 400mm	24.18	32.04	1.25	27.28	nr	**59.32**
Sum of two sides 450mm	24.91	33.01	1.25	27.28	nr	**60.29**
Sum of two sides 500mm	25.64	33.97	1.25	27.28	nr	**61.25**
Sum of two sides 550mm	26.37	34.94	1.33	29.03	nr	**63.97**
Sum of two sides 600mm	27.10	35.91	1.33	29.03	nr	**64.93**
Sum of two sides 650mm	27.83	36.87	1.33	29.03	nr	**65.90**
Sum of two sides 700mm	28.56	37.84	1.33	29.03	nr	**66.87**
Sum of two sides 750mm	29.28	38.80	1.40	30.55	nr	**69.35**
Sum of two sides 800mm	30.01	39.76	1.40	30.55	nr	**70.32**
90 degree radius bend						
Sum of two sides 200mm	12.62	16.72	1.10	24.01	nr	**40.73**
Sum of two sides 250mm	12.85	17.03	1.10	24.01	nr	**41.03**
Sum of two sides 300mm	13.07	17.32	1.10	24.01	nr	**41.32**
Sum of two sides 350mm	16.36	21.68	1.12	24.44	nr	**46.12**
Sum of two sides 400mm	16.81	22.27	1.12	24.44	nr	**46.72**
Sum of two sides 450mm	17.02	22.55	1.12	24.44	nr	**47.00**
Sum of two sides 500mm	17.21	22.80	1.12	24.44	nr	**47.25**
Sum of two sides 550mm	17.65	23.39	1.16	25.32	nr	**48.70**
Sum of two sides 600mm	18.10	23.98	1.16	25.32	nr	**49.30**
Sum of two sides 650mm	18.29	24.23	1.16	25.32	nr	**49.55**
Sum of two sides 700mm	18.73	24.82	1.16	25.32	nr	**50.13**
Sum of two sides 750mm	19.17	25.40	1.16	25.32	nr	**50.72**
Sum of two sides 800mm	19.62	26.00	1.22	26.63	nr	**52.62**
45 degree radius bend						
Sum of two sides 200mm	14.13	18.72	1.29	28.15	nr	**46.88**
Sum of two sides 250mm	14.58	19.32	1.29	28.15	nr	**47.47**
Sum of two sides 300mm	15.03	19.91	1.29	28.15	nr	**48.07**
Sum of two sides 350mm	18.68	24.75	1.29	28.15	nr	**52.90**

Material Costs/Measured Work Prices - Mechanical Installations

U:VENTILATION/AIR CONDITIONING SYSTEMS

Item	Net Price £	Material £	Labour hours	Labour £	Unit	Total rate £
U10 : DUCTWORK : RECTANGULAR - CLASS C						
Y30 - AIR DUCTLINES						
Galvanised sheet metal DW144 class C rectangular section ductwork; including all necessary stiffeners, joints, couplers in the running length and duct supports (Continued)						
45 degree radius bend						
Sum of two sides 400mm	19.26	25.52	1.29	28.15	nr	**53.67**
Sum of two sides 450mm	19.71	26.12	1.29	28.15	nr	**54.27**
Sum of two sides 500mm	20.15	26.70	1.29	28.15	nr	**54.85**
Sum of two sides 550mm	20.73	27.47	1.39	30.34	nr	**57.80**
Sum of two sides 600mm	21.30	28.22	1.39	30.34	nr	**58.56**
Sum of two sides 650mm	21.74	28.81	1.39	30.34	nr	**59.14**
Sum of two sides 700mm	22.31	29.56	1.39	30.34	nr	**59.90**
Sum of two sides 750mm	22.88	30.32	1.46	31.86	nr	**62.18**
Sum of two sides 800mm	23.45	31.07	1.46	31.86	nr	**62.94**
90 degree mitire bend						
Sum of two sides 200mm	21.82	28.91	2.04	44.52	nr	**73.43**
Sum of two sides 250mm	22.78	30.18	2.04	44.52	nr	**74.71**
Sum of two sides 300mm	23.73	31.44	2.04	44.52	nr	**75.96**
Sum of two sides 350mm	27.75	36.77	2.09	45.61	nr	**82.38**
Sum of two sides 400mm	29.01	38.44	2.09	45.61	nr	**84.05**
Sum of two sides 450mm	29.95	39.68	2.09	45.61	nr	**85.30**
Sum of two sides 500mm	30.89	40.93	2.09	45.61	nr	**86.54**
Sum of two sides 550mm	32.20	42.66	2.15	46.92	nr	**89.59**
Sum of two sides 600mm	33.50	44.39	2.15	46.92	nr	**91.31**
Sum of two sides 650mm	34.48	45.69	2.15	46.92	nr	**92.61**
Sum of two sides 700mm	35.81	47.45	2.15	46.92	nr	**94.37**
Sum of two sides 750mm	37.14	49.21	2.26	49.32	nr	**98.53**
Sum of two sides 800mm	38.47	50.97	2.26	49.32	nr	**100.30**
Branch						
Sum of two sides 200mm	16.12	21.36	0.92	20.08	nr	**41.44**
Sum of two sides 250mm	16.95	22.46	0.92	20.08	nr	**42.54**
Sum of two sides 300mm	17.77	23.55	0.92	20.08	nr	**43.62**
Sum of two sides 350mm	20.10	26.63	0.95	20.73	nr	**47.37**
Sum of two sides 400mm	20.91	27.71	0.95	20.73	nr	**48.44**
Sum of two sides 450mm	21.71	28.77	0.95	20.73	nr	**49.50**
Sum of two sides 500mm	22.56	29.89	0.95	20.73	nr	**50.63**
Sum of two sides 550mm	23.38	30.98	1.03	22.48	nr	**53.46**
Sum of two sides 600mm	24.19	32.05	1.03	22.48	nr	**54.53**
Sum of two sides 650mm	25.00	33.13	1.03	22.48	nr	**55.60**
Sum of two sides 700mm	25.82	34.21	1.03	22.48	nr	**56.69**
Sum of two sides 750mm	26.63	35.28	1.03	22.48	nr	**57.76**
Sum of two sides 800mm	27.44	36.36	1.03	22.48	nr	**58.84**

Material Costs/Measured Work Prices - Mechanical Installations 337

U:VENTILATION/AIR CONDITIONING SYSTEMS

Item	Net Price £	Material £	Labour hours	Labour £	Unit	Total rate £
Grille neck						
Sum of two sides 200mm	17.72	23.48	1.10	24.01	nr	**47.49**
Sum of two sides 250mm	18.63	24.68	1.10	24.01	nr	**48.69**
Sum of two sides 300mm	19.54	25.89	1.10	24.01	nr	**49.90**
Sum of two sides 350mm	20.46	27.11	1.16	25.32	nr	**52.43**
Sum of two sides 400mm	21.37	28.32	1.16	25.32	nr	**53.63**
Sum of two sides 450mm	22.28	29.52	1.16	25.32	nr	**54.84**
Sum of two sides 500mm	23.19	30.73	1.16	25.32	nr	**56.04**
Sum of two sides 550mm	24.11	31.95	1.18	25.75	nr	**57.70**
Sum of two sides 600mm	25.02	33.15	1.18	25.75	nr	**58.90**
Sum of two sides 650mm	25.93	34.36	1.18	25.75	nr	**60.11**
Sum of two sides 700mm	26.84	35.56	1.18	25.75	nr	**61.32**
Sum of two sides 750mm	27.75	36.77	1.18	25.75	nr	**62.52**
Sum of two sides 800mm	28.67	37.99	1.18	25.75	nr	**63.74**
Ductwork 401 to 600mm longest side						
Sum of two sides 600mm	21.85	28.95	1.17	25.53	m	**54.49**
Sum of two sides 650mm	22.56	29.89	1.17	25.53	m	**55.43**
Sum of two sides 700mm	23.64	31.32	1.17	25.53	m	**56.86**
Sum of two sides 750mm	24.35	32.26	1.17	25.53	m	**57.80**
Sum of two sides 800mm	25.06	33.20	1.17	25.53	m	**58.74**
Sum of two sides 850mm	25.77	34.15	1.26	27.50	m	**61.64**
Sum of two sides 900mm	26.46	35.06	1.27	27.72	m	**62.78**
Sum of two sides 950mm	27.19	36.03	1.48	32.30	m	**68.33**
Sum of two sides 1000mm	27.89	36.95	1.48	32.30	m	**69.25**
Sum of two sides 1050mm	28.80	38.16	1.49	32.52	m	**70.68**
Sum of two sides 1100mm	29.51	39.10	1.49	32.52	m	**71.62**
Sum of two sides 1150mm	30.23	40.05	1.49	32.52	m	**72.57**
Sum of two sides 1200mm	30.94	41.00	1.49	32.52	m	**73.51**
Extra over fittings: Ductwork 401 to 600mm longest side						
End Cap						
Sum of two sides 600mm	11.78	15.61	0.38	8.29	nr	**23.90**
Sum of two sides 650mm	12.17	16.13	0.38	8.29	nr	**24.42**
Sum of two sides 700mm	12.84	17.01	0.38	8.29	nr	**25.31**
Sum of two sides 750mm	13.22	17.52	0.38	8.29	nr	**25.81**
Sum of two sides 800mm	13.61	18.03	0.38	8.29	nr	**26.33**
Sum of two sides 850mm	14.00	18.55	0.38	8.29	nr	**26.84**
Sum of two sides 900mm	14.38	19.05	0.38	8.29	nr	**27.35**
Sum of two sides 950mm	14.77	19.57	0.38	8.29	nr	**27.86**
Sum of two sides 1000mm	15.16	20.09	0.38	8.29	nr	**28.38**
Sum of two sides 1050mm	15.54	20.59	0.38	8.29	nr	**28.88**
Sum of two sides 1100mm	15.93	21.11	0.38	8.29	nr	**29.40**
Sum of two sides 1150mm	16.32	21.62	0.38	8.29	nr	**29.92**
Sum of two sides 1200mm	16.71	22.14	0.38	8.29	nr	**30.43**
Reducer						
Sum of two sides 600mm	18.03	23.89	1.69	36.88	nr	**60.77**
Sum of two sides 650mm	18.66	24.72	1.69	36.88	nr	**61.61**
Sum of two sides 700mm	19.86	26.31	1.69	36.88	nr	**63.20**
Sum of two sides 750mm	20.49	27.15	1.92	41.90	nr	**69.05**
Sum of two sides 800mm	21.12	27.98	1.92	41.90	nr	**69.89**

338　　　*Material Costs/Measured Work Prices - Mechanical Installations*

U:VENTILATION/AIR CONDITIONING SYSTEMS

Item	Net Price £	Material £	Labour hours	Labour £	Unit	Total rate £
U10 : DUCTWORK : RECTANGULAR - CLASS C						
Y30 - AIR DUCTLINES						
Galvanised sheet metal DW144 class C rectangular section ductwork; including all necessary stiffeners, joints, couplers in the running length and duct supports (Continued)						
Reducer						
Sum of two sides 850mm	21.75	28.82	1.92	41.90	nr	**70.72**
Sum of two sides 900mm	22.38	29.65	1.92	41.90	nr	**71.56**
Sum of two sides 950mm	23.01	30.49	2.18	47.58	nr	**78.07**
Sum of two sides 1000mm	23.64	31.32	2.18	47.58	nr	**78.90**
Sum of two sides 1050mm	24.43	32.37	2.18	47.58	nr	**79.95**
Sum of two sides 1100mm	25.06	33.20	2.18	47.58	nr	**80.78**
Sum of two sides 1150mm	25.70	34.05	2.18	47.58	nr	**81.63**
Sum of two sides 1200mm	26.33	34.89	2.18	47.58	nr	**82.46**
Offset						
Sum of two sides 600mm	31.20	41.34	1.92	41.90	nr	**83.24**
Sum of two sides 650mm	31.69	41.99	1.92	41.90	nr	**83.89**
Sum of two sides 700mm	33.30	44.12	1.92	41.90	nr	**86.03**
Sum of two sides 750mm	33.77	44.75	1.92	41.90	nr	**86.65**
Sum of two sides 800mm	34.24	45.37	1.92	41.90	nr	**87.27**
Sum of two sides 850mm	34.70	45.98	1.92	41.90	nr	**87.88**
Sum of two sides 900mm	35.16	46.59	1.92	41.90	nr	**88.49**
Sum of two sides 950mm	35.62	47.20	2.18	47.58	nr	**94.77**
Sum of two sides 1000mm	36.40	48.23	2.18	47.58	nr	**95.81**
Sum of two sides 1050mm	36.95	48.96	2.18	47.58	nr	**96.54**
Sum of two sides 1100mm	37.38	49.53	2.18	47.58	nr	**97.11**
Sum of two sides 1150mm	37.80	50.09	2.18	47.58	nr	**97.66**
Sum of two sides 1200mm	38.22	50.64	2.18	47.58	nr	**98.22**
Square to round						
Sum of two sides 600mm	24.92	33.02	1.33	29.03	nr	**62.05**
Sum of two sides 650mm	25.63	33.96	1.33	29.03	nr	**62.99**
Sum of two sides 700mm	26.99	35.76	1.33	29.03	nr	**64.79**
Sum of two sides 750mm	27.70	36.70	1.40	30.55	nr	**67.26**
Sum of two sides 800mm	28.42	37.66	1.40	30.55	nr	**68.21**
Sum of two sides 850mm	29.13	38.60	1.40	30.55	nr	**69.15**
Sum of two sides 900mm	29.84	39.54	1.40	30.55	nr	**70.09**
Sum of two sides 950mm	30.56	40.49	1.82	39.72	nr	**80.21**
Sum of two sides 1000mm	31.27	41.43	1.82	39.72	nr	**81.15**
Sum of two sides 1050mm	32.02	42.43	1.82	39.72	nr	**82.15**
Sum of two sides 1100mm	32.74	43.38	1.82	39.72	nr	**83.10**
Sum of two sides 1150mm	33.45	44.32	1.82	39.72	nr	**84.04**
Sum of two sides 1200mm	34.17	45.28	1.82	39.72	nr	**85.00**
90 degree radius bend						
Sum of two sides 600mm	16.36	21.68	1.16	25.32	nr	**46.99**
Sum of two sides 650mm	16.80	22.26	1.16	25.32	nr	**47.58**

Material Costs/Measured Work Prices - Mechanical Installations

U:VENTILATION/AIR CONDITIONING SYSTEMS

Item	Net Price £	Material £	Labour hours	Labour £	Unit	Total rate £
Sum of two sides 700mm	17.13	22.70	1.16	25.32	nr	**48.01**
Sum of two sides 750mm	17.56	23.27	1.16	25.32	nr	**48.58**
Sum of two sides 800mm	17.99	23.84	1.22	26.63	nr	**50.46**
Sum of two sides 850mm	18.42	24.41	1.22	26.63	nr	**51.03**
Sum of two sides 900mm	18.84	24.96	1.22	26.63	nr	**51.59**
Sum of two sides 950mm	30.56	40.49	1.40	30.55	nr	**71.05**
Sum of two sides 1000mm	19.36	25.65	1.40	30.55	nr	**56.21**
Sum of two sides 1050mm	19.89	26.35	1.40	30.55	nr	**56.91**
Sum of two sides 1100mm	20.31	26.91	1.40	30.55	nr	**57.47**
Sum of two sides 1150mm	33.45	44.32	1.40	30.55	nr	**74.88**
Sum of two sides 1200mm	21.17	28.05	1.40	30.55	nr	**58.60**
45 degree bend						
Sum of two sides 600mm	20.18	26.74	1.39	30.34	nr	**57.07**
Sum of two sides 650mm	20.75	27.49	1.39	30.34	nr	**57.83**
Sum of two sides 700mm	21.56	28.57	1.39	30.34	nr	**58.90**
Sum of two sides 750mm	22.12	29.31	1.46	31.86	nr	**61.17**
Sum of two sides 800mm	22.68	30.05	1.46	31.86	nr	**61.91**
Sum of two sides 850mm	23.24	30.79	1.46	31.86	nr	**62.66**
Sum of two sides 900mm	23.81	31.55	1.46	31.86	nr	**63.41**
Sum of two sides 950mm	24.37	32.29	1.88	41.03	nr	**73.32**
Sum of two sides 1000mm	24.76	32.81	1.88	41.03	nr	**73.84**
Sum of two sides 1050mm	25.49	33.77	1.88	41.03	nr	**74.80**
Sum of two sides 1100mm	26.05	34.52	1.88	41.03	nr	**75.55**
Sum of two sides 1150mm	26.61	35.26	1.88	41.03	nr	**76.29**
Sum of two sides 1200mm	27.17	36.00	1.88	41.03	nr	**77.03**
90 degree mitre bend						
Sum of two sides 600mm	30.98	41.05	2.15	46.92	nr	**87.97**
Sum of two sides 650mm	32.32	42.82	2.15	46.92	nr	**89.75**
Sum of two sides 700mm	33.67	44.61	2.15	46.92	nr	**91.54**
Sum of two sides 750mm	35.06	46.45	2.26	49.32	nr	**95.78**
Sum of two sides 800mm	36.45	48.30	2.26	49.32	nr	**97.62**
Sum of two sides 850mm	37.84	50.14	2.26	49.32	nr	**99.46**
Sum of two sides 900mm	39.23	51.98	2.26	49.32	nr	**101.30**
Sum of two sides 950mm	40.62	53.82	3.03	66.13	nr	**119.95**
Sum of two sides 1000mm	41.55	55.05	3.03	66.13	nr	**121.18**
Sum of two sides 1050mm	42.88	56.82	3.03	66.13	nr	**122.94**
Sum of two sides 1100mm	44.46	58.91	3.03	66.13	nr	**125.04**
Sum of two sides 1150mm	45.87	60.78	3.03	66.13	nr	**126.91**
Sum of two sides 1200mm	47.25	62.61	3.03	66.13	nr	**128.73**
Branch						
Sum of two sides 600mm	23.63	31.31	1.03	22.48	nr	**53.79**
Sum of two sides 650mm	24.44	32.38	1.03	22.48	nr	**54.86**
Sum of two sides 700mm	26.03	34.49	1.03	22.48	nr	**56.97**
Sum of two sides 750mm	26.85	35.58	1.03	22.48	nr	**58.06**
Sum of two sides 800mm	27.67	36.66	1.03	22.48	nr	**59.14**
Sum of two sides 850mm	28.48	37.74	1.03	22.48	nr	**60.22**
Sum of two sides 900mm	29.30	38.82	1.03	22.48	nr	**61.30**
Sum of two sides 950mm	30.12	39.91	1.29	28.15	nr	**68.06**
Sum of two sides 1000mm	30.94	41.00	1.29	28.15	nr	**69.15**
Sum of two sides 1050mm	31.88	42.24	1.29	28.15	nr	**70.39**
Sum of two sides 1100mm	32.70	43.33	1.29	28.15	nr	**71.48**
Sum of two sides 1150mm	33.52	44.41	1.29	28.15	nr	**72.57**
Sum of two sides 1200mm	34.34	45.50	1.29	28.15	nr	**73.65**

Material Costs/Measured Work Prices - Mechanical Installations

U:VENTILATION/AIR CONDITIONING SYSTEMS

Item	Net Price £	Material £	Labour hours	Labour £	Unit	Total rate £
U10 : DUCTWORK : RECTANGULAR - CLASS C						
Y30 - AIR DUCTLINES						
Galvanised sheet metal DW144 class C rectangular section ductwork; including all necessary stiffeners, joints, couplers in the running length and duct supports (Continued)						
Grille neck						
Sum of two sides 600mm	24.15	32.00	1.18	25.75	nr	**57.75**
Sum of two sides 650mm	25.07	33.22	1.18	25.75	nr	**58.97**
Sum of two sides 700mm	26.84	35.56	1.18	25.75	nr	**61.32**
Sum of two sides 750mm	27.75	36.77	1.18	25.75	nr	**62.52**
Sum of two sides 800mm	28.67	37.99	1.18	25.75	nr	**63.74**
Sum of two sides 850mm	29.58	39.19	1.18	25.75	nr	**64.95**
Sum of two sides 900mm	30.49	40.40	1.18	25.75	nr	**66.15**
Sum of two sides 950mm	31.40	41.60	1.44	31.43	nr	**73.03**
Sum of two sides 1000mm	32.32	42.82	1.44	31.43	nr	**74.25**
Sum of two sides 1050mm	33.23	44.03	1.44	31.43	nr	**75.46**
Sum of two sides 1100mm	34.14	45.24	1.44	31.43	nr	**76.66**
Sum of two sides 1150mm	35.05	46.44	1.44	31.43	nr	**77.87**
Sum of two sides 1200mm	35.97	47.66	1.44	31.43	nr	**79.09**
Ductwork 601 to 800mm longest side						
Sum of two sides 900mm	31.19	41.33	1.27	27.72	m	**69.04**
Sum of two sides 950mm	32.16	42.61	1.48	32.30	m	**74.91**
Sum of two sides 1000mm	33.05	43.79	1.48	32.30	m	**76.09**
Sum of two sides 1050mm	33.94	44.97	1.48	32.30	m	**77.27**
Sum of two sides 1100mm	34.83	46.15	1.49	32.52	m	**78.67**
Sum of two sides 1150mm	35.72	47.33	1.49	32.52	m	**79.85**
Sum of two sides 1200mm	36.62	48.52	1.49	32.52	m	**81.04**
Sum of two sides 1250mm	37.68	49.93	1.49	32.52	m	**82.44**
Sum of two sides 1300mm	38.57	51.11	1.51	32.96	m	**84.06**
Sum of two sides 1350mm	39.46	52.28	1.51	32.96	m	**85.24**
Sum of two sides 1400mm	40.35	53.46	1.55	33.83	m	**87.29**
Sum of two sides 1450mm	41.66	55.20	1.55	33.83	m	**89.03**
Sum of two sides 1500mm	42.55	56.38	1.61	35.14	m	**91.52**
Sum of two sides 1550mm	43.44	57.56	1.61	35.14	m	**92.70**
Sum of two sides 1600mm	44.34	58.75	1.62	35.36	m	**94.11**
Extra over fittings; Ductwork 601 to 800mm longest side						
End Cap						
Sum of two sides 900mm	15.78	20.91	0.38	8.29	nr	**29.20**
Sum of two sides 950mm	16.26	21.54	0.38	8.29	nr	**29.84**
Sum of two sides 1000mm	16.74	22.18	0.38	8.29	nr	**30.47**
Sum of two sides 1050mm	17.22	22.82	0.38	8.29	nr	**31.11**
Sum of two sides 1100mm	17.70	23.45	0.38	8.29	nr	**31.75**
Sum of two sides 1150mm	18.18	24.09	0.38	8.29	nr	**32.38**
Sum of two sides 1200mm	18.77	18.77	0.38	8.29	nr	**27.06**

Material Costs/Measured Work Prices - Mechanical Installations

U:VENTILATION/AIR CONDITIONING SYSTEMS

Item	Net Price £	Material £	Labour hours	Labour £	Unit	Total rate £
Sum of two sides 1250mm	19.25	19.25	0.38	8.29	nr	**27.54**
Sum of two sides 1300mm	19.74	19.74	0.38	8.29	nr	**28.03**
Sum of two sides 1350mm	20.23	20.23	0.38	8.29	nr	**28.52**
Sum of two sides 1400mm	20.59	27.28	0.38	8.29	nr	**35.58**
Sum of two sides 1450mm	21.92	29.04	0.38	8.29	nr	**37.34**
Sum of two sides 1500mm	23.05	30.54	0.38	8.29	nr	**38.83**
Sum of two sides 1550mm	24.18	32.04	0.38	8.29	nr	**40.33**
Sum of two sides 1600mm	25.32	33.55	0.38	8.29	nr	**41.84**
Reducer						
Sum of two sides 900mm	21.79	28.87	1.92	41.90	nr	**70.77**
Sum of two sides 950mm	22.21	29.43	2.18	47.58	nr	**77.01**
Sum of two sides 1000mm	22.64	30.00	2.18	47.58	nr	**77.58**
Sum of two sides 1050mm	23.06	30.55	2.18	47.58	nr	**78.13**
Sum of two sides 1100mm	23.48	31.11	2.18	47.58	nr	**78.69**
Sum of two sides 1150mm	23.90	31.67	2.18	47.58	nr	**79.25**
Sum of two sides 1200mm	24.33	32.24	2.18	47.58	nr	**79.81**
Sum of two sides 1250mm	24.93	33.03	2.18	47.58	nr	**80.61**
Sum of two sides 1300mm	25.35	33.59	2.30	50.20	nr	**83.79**
Sum of two sides 1350mm	25.77	34.15	2.30	50.20	nr	**84.34**
Sum of two sides 1400mm	26.19	34.70	2.30	50.20	nr	**84.90**
Sum of two sides 1450mm	28.26	37.44	2.30	50.20	nr	**87.64**
Sum of two sides 1500mm	29.98	39.72	2.47	53.91	nr	**93.63**
Sum of two sides 1550mm	31.71	42.02	2.47	53.91	nr	**95.92**
Sum of two sides 1600mm	33.44	44.31	2.47	53.91	nr	**98.21**
Offset						
Sum of two sides 900mm	36.15	47.90	1.92	41.90	nr	**89.80**
Sum of two sides 950mm	36.95	48.96	2.18	47.58	nr	**96.54**
Sum of two sides 1000mm	37.25	49.36	2.18	47.58	nr	**96.93**
Sum of two sides 1050mm	37.53	49.73	2.18	47.58	nr	**97.30**
Sum of two sides 1100mm	37.80	50.09	2.18	47.58	nr	**97.66**
Sum of two sides 1150mm	38.05	50.42	2.18	47.58	nr	**97.99**
Sum of two sides 1200mm	38.29	50.73	2.18	47.58	nr	**98.31**
Sum of two sides 1250mm	38.64	51.20	2.18	47.58	nr	**98.78**
Sum of two sides 1300mm	38.83	51.45	2.47	53.91	nr	**105.36**
Sum of two sides 1350mm	39.02	51.70	2.47	53.91	nr	**105.61**
Sum of two sides 1400mm	39.18	51.91	2.47	53.91	nr	**105.82**
Sum of two sides 1450mm	41.72	55.28	2.47	53.91	nr	**109.19**
Sum of two sides 1500mm	43.80	58.03	2.47	53.91	nr	**111.94**
Sum of two sides 1550mm	45.87	60.78	2.47	53.91	nr	**114.68**
Sum of two sides 1600mm	47.92	63.49	2.47	53.91	nr	**117.40**
Square to round						
Sum of two sides 900mm	23.95	31.73	1.40	30.55	nr	**62.29**
Sum of two sides 950mm	29.62	39.25	1.82	39.72	nr	**78.97**
Sum of two sides 1000mm	30.29	40.13	1.82	39.72	nr	**79.86**
Sum of two sides 1050mm	30.96	41.02	1.82	39.72	nr	**80.74**
Sum of two sides 1100mm	31.63	41.91	1.82	39.72	nr	**81.63**
Sum of two sides 1150mm	32.29	42.78	1.82	39.72	nr	**82.51**
Sum of two sides 1200mm	32.96	43.67	1.82	39.72	nr	**83.39**
Sum of two sides 1250mm	33.68	44.63	1.82	39.72	nr	**84.35**
Sum of two sides 1300mm	34.35	45.51	2.32	50.63	nr	**96.15**
Sum of two sides 1350mm	35.02	46.40	2.32	50.63	nr	**97.03**

Material Costs/Measured Work Prices - Mechanical Installations

U:VENTILATION/AIR CONDITIONING SYSTEMS

Item	Net Price £	Material £	Labour hours	Labour £	Unit	Total rate £
U10 : DUCTWORK : RECTANGULAR - CLASS C						
Y30 - AIR DUCTLINES						
Galvanised sheet metal DW144 class C rectangular section ductwork; including all necessary stiffeners, joints, couplers in the running length and duct supports (Continued)						
Square to round						
Sum of two sides 1400mm	35.68	47.28	2.32	50.63	nr	**97.91**
Sum of two sides 1450mm	38.37	50.84	2.56	55.87	nr	**106.71**
Sum of two sides 1500mm	40.66	53.87	2.56	55.87	nr	**109.75**
Sum of two sides 1550mm	42.96	56.92	2.58	56.31	nr	**113.23**
Sum of two sides 1600mm	45.26	59.97	2.58	56.31	nr	**116.28**
90 degree radius bend						
Sum of two sides 900mm	16.63	22.03	1.22	26.63	nr	**48.66**
Sum of two sides 950mm	16.79	22.25	1.40	30.55	nr	**52.80**
Sum of two sides 1000mm	17.44	23.11	1.40	30.55	nr	**53.66**
Sum of two sides 1050mm	18.08	23.96	1.40	30.55	nr	**54.51**
Sum of two sides 1100mm	18.73	24.82	1.40	30.55	nr	**55.37**
Sum of two sides 1150mm	19.38	25.68	1.40	30.55	nr	**56.23**
Sum of two sides 1200mm	20.02	26.53	1.40	30.55	nr	**57.08**
Sum of two sides 1250mm	20.78	27.53	1.40	30.55	nr	**58.09**
Sum of two sides 1300mm	21.42	28.38	1.91	41.69	nr	**70.07**
Sum of two sides 1350mm	22.07	29.24	1.91	41.69	nr	**70.93**
Sum of two sides 1400mm	22.72	30.10	1.91	41.69	nr	**71.79**
Sum of two sides 1450mm	24.35	32.26	1.91	41.69	nr	**73.95**
Sum of two sides 1500mm	26.29	34.83	2.11	46.05	nr	**80.88**
Sum of two sides 1550mm	28.23	37.40	2.11	46.05	nr	**83.45**
Sum of two sides 1600mm	30.18	39.99	2.11	46.05	nr	**86.04**
45 degree bend						
Sum of two sides 900mm	22.60	29.95	1.46	31.86	nr	**61.81**
Sum of two sides 950mm	22.88	30.32	1.88	41.03	nr	**71.35**
Sum of two sides 1000mm	23.41	31.02	1.88	41.03	nr	**72.05**
Sum of two sides 1050mm	23.94	31.72	1.88	41.03	nr	**72.75**
Sum of two sides 1100mm	24.47	32.42	1.88	41.03	nr	**73.45**
Sum of two sides 1150mm	25.00	33.13	1.88	41.03	nr	**74.16**
Sum of two sides 1200mm	25.53	33.83	1.88	41.03	nr	**74.86**
Sum of two sides 1250mm	26.23	34.75	1.88	41.03	nr	**75.78**
Sum of two sides 1300mm	26.76	35.46	2.26	49.32	nr	**84.78**
Sum of two sides 1350mm	27.28	36.15	2.26	49.32	nr	**85.47**
Sum of two sides 1400mm	27.81	36.85	2.44	53.25	nr	**90.10**
Sum of two sides 1450mm	29.69	39.34	2.44	53.25	nr	**92.59**
Sum of two sides 1500mm	31.52	41.76	2.44	53.25	nr	**95.02**
Sum of two sides 1550mm	33.34	44.18	2.68	58.49	nr	**102.67**
Sum of two sides 1600mm	35.17	46.60	2.68	58.49	nr	**105.09**

Material Costs/Measured Work Prices - Mechanical Installations 343

U:VENTILATION/AIR CONDITIONING SYSTEMS

Item	Net Price £	Material £	Labour hours	Labour £	Unit	Total rate £
90 degree mitre bend						
Sum of two sides 900mm	40.51	53.68	2.26	49.32	nr	**103.00**
Sum of two sides 950mm	40.78	54.03	3.03	66.13	nr	**120.16**
Sum of two sides 1000mm	41.92	55.54	3.03	66.13	nr	**121.67**
Sum of two sides 1050mm	43.06	57.05	3.03	66.13	nr	**123.18**
Sum of two sides 1100mm	44.20	58.56	3.03	66.13	nr	**124.69**
Sum of two sides 1150mm	45.34	60.08	3.03	66.13	nr	**126.20**
Sum of two sides 1200mm	46.48	61.59	3.03	66.13	nr	**127.71**
Sum of two sides 1250mm	47.71	63.22	3.03	66.13	nr	**129.34**
Sum of two sides 1300mm	48.85	64.73	3.85	84.02	nr	**148.75**
Sum of two sides 1350mm	49.99	66.24	3.85	84.02	nr	**150.26**
Sum of two sides 1400mm	51.13	67.75	3.85	84.02	nr	**151.77**
Sum of two sides 1450mm	54.28	71.92	3.85	84.02	nr	**155.95**
Sum of two sides 1500mm	57.39	76.04	4.25	92.75	nr	**168.80**
Sum of two sides 1550mm	60.50	80.16	4.25	92.75	nr	**172.92**
Sum of two sides 1600mm	63.59	84.26	4.26	92.97	nr	**177.23**
Branch						
Sum of two sides 900mm	29.87	39.58	1.03	22.48	nr	**62.06**
Sum of two sides 950mm	30.72	40.70	1.29	28.15	nr	**68.86**
Sum of two sides 1000mm	31.56	41.82	1.29	28.15	nr	**69.97**
Sum of two sides 1050mm	32.40	42.93	1.29	28.15	nr	**71.08**
Sum of two sides 1100mm	33.25	44.06	1.29	28.15	nr	**72.21**
Sum of two sides 1150mm	34.09	45.17	1.29	28.15	nr	**73.32**
Sum of two sides 1200mm	34.93	46.28	1.29	28.15	nr	**74.44**
Sum of two sides 1250mm	35.90	47.57	1.29	28.15	nr	**75.72**
Sum of two sides 1300mm	36.75	48.69	1.39	30.34	nr	**79.03**
Sum of two sides 1350mm	37.59	49.81	1.39	30.34	nr	**80.14**
Sum of two sides 1400mm	38.43	50.92	1.39	30.34	nr	**81.26**
Sum of two sides 1450mm	41.58	55.09	1.39	30.34	nr	**85.43**
Sum of two sides 1500mm	44.16	58.51	1.64	35.79	nr	**94.30**
Sum of two sides 1550mm	46.74	61.93	1.64	35.79	nr	**97.72**
Sum of two sides 1600mm	49.32	65.35	1.64	35.79	nr	**101.14**
Grille neck						
Sum of two sides 900mm	30.20	40.02	1.18	25.75	nr	**65.77**
Sum of two sides 950mm	31.11	41.22	1.44	31.43	nr	**72.65**
Sum of two sides 1000mm	32.02	42.43	1.44	31.43	nr	**73.85**
Sum of two sides 1050mm	32.93	43.63	1.44	31.43	nr	**75.06**
Sum of two sides 1100mm	33.84	44.84	1.44	31.43	nr	**76.27**
Sum of two sides 1150mm	34.75	46.04	1.44	31.43	nr	**77.47**
Sum of two sides 1200mm	35.66	47.25	1.44	31.43	nr	**78.68**
Sum of two sides 1250mm	36.57	48.46	1.44	31.43	nr	**79.88**
Sum of two sides 1300mm	37.48	49.66	1.69	36.88	nr	**86.54**
Sum of two sides 1350mm	38.39	50.87	1.69	36.88	nr	**87.75**
Sum of two sides 1400mm	39.30	52.07	1.69	36.88	nr	**88.96**
Sum of two sides 1450mm	42.79	56.70	1.69	36.88	nr	**93.58**
Sum of two sides 1500mm	45.65	60.49	1.79	39.07	nr	**99.55**
Sum of two sides 1550mm	48.52	64.29	1.79	39.07	nr	**103.36**
Sum of two sides 1600mm	51.39	68.09	1.79	39.07	nr	**107.16**

Material Costs/Measured Work Prices - Mechanical Installations

U:VENTILATION/AIR CONDITIONING SYSTEMS

Item	Net Price £	Material £	Labour hours	Labour £	Unit	Total rate £
U10 : DUCTWORK : RECTANGULAR - CLASS C						
Y30 - AIR DUCTLINES						
Galvanised sheet metal DW144 class C rectangular section ductwork; including all necessary stiffeners, joints, couplers in the running length and duct supports (Continued)						
Ductwork 801 to 1000mm longest side						
Sum of two sides 1100mm	40.12	53.16	1.49	32.52	m	**85.68**
Sum of two sides 1150mm	41.25	54.66	1.49	32.52	m	**87.18**
Sum of two sides 1200mm	42.30	56.05	1.49	32.52	m	**88.57**
Sum of two sides 1250mm	43.36	57.45	1.49	32.52	m	**89.97**
Sum of two sides 1300mm	44.41	58.84	1.51	32.96	m	**91.80**
Sum of two sides 1350mm	45.47	60.25	1.51	32.96	m	**93.20**
Sum of two sides 1400mm	46.52	61.64	1.55	33.83	m	**95.47**
Sum of two sides 1450mm	47.74	63.26	1.55	33.83	m	**97.08**
Sum of two sides 1500mm	48.80	64.66	1.61	35.14	m	**99.80**
Sum of two sides 1550mm	49.85	66.05	1.61	35.14	m	**101.19**
Sum of two sides 1600mm	50.91	67.46	1.62	35.36	m	**102.81**
Sum of two sides 1650mm	52.48	69.54	1.62	35.36	m	**104.89**
Sum of two sides 1700mm	53.61	71.03	1.74	37.97	m	**109.01**
Sum of two sides 1750mm	54.67	72.44	1.74	37.97	m	**110.41**
Sum of two sides 1800mm	55.89	74.05	1.76	38.41	m	**112.47**
Sum of two sides 1850mm	56.94	75.45	1.76	38.41	m	**113.86**
Sum of two sides 1900mm	58.00	76.85	1.81	39.50	m	**116.35**
Sum of two sides 1950mm	59.05	78.24	1.81	39.50	m	**117.74**
Sum of two sides 2000mm	60.11	79.65	1.82	39.72	m	**119.37**
Extra over fittings; Ductwork 801 to 1000mm longest side						
End Cap						
Sum of two sides 1100mm	17.25	22.86	0.38	8.29	nr	**31.15**
Sum of two sides 1150mm	17.76	23.53	0.38	8.29	nr	**31.83**
Sum of two sides 1200mm	18.28	24.22	0.38	8.29	nr	**32.51**
Sum of two sides 1250mm	18.80	24.91	0.38	8.29	nr	**33.20**
Sum of two sides 1300mm	19.31	25.59	0.38	8.29	nr	**33.88**
Sum of two sides 1350mm	19.82	26.26	0.38	8.29	nr	**34.55**
Sum of two sides 1400mm	20.34	26.95	0.38	8.29	nr	**35.24**
Sum of two sides 1450mm	20.85	27.63	0.38	8.29	nr	**35.92**
Sum of two sides 1500mm	21.94	29.07	0.38	8.29	nr	**37.36**
Sum of two sides 1550mm	23.11	30.62	0.38	8.29	nr	**38.91**
Sum of two sides 1600mm	24.27	32.16	0.38	8.29	nr	**40.45**
Sum of two sides 1650mm	26.37	34.94	0.38	8.29	nr	**43.23**
Sum of two sides 1700mm	27.53	36.48	0.58	12.66	nr	**49.14**
Sum of two sides 1750mm	28.70	38.03	0.58	12.66	nr	**50.69**
Sum of two sides 1800mm	29.86	39.56	0.58	12.66	nr	**52.22**
Sum of two sides 1850mm	31.03	41.11	0.58	12.66	nr	**53.77**
Sum of two sides 1900mm	32.20	42.66	0.58	12.66	nr	**55.32**
Sum of two sides 1950mm	33.36	44.20	0.58	12.66	nr	**56.86**
Sum of two sides 2000mm	34.53	45.75	0.58	12.66	nr	**58.41**

Material Costs/Measured Work Prices - Mechanical Installations   345

U:VENTILATION/AIR CONDITIONING SYSTEMS

Item	Net Price £	Material £	Labour hours	Labour £	Unit	Total rate £
Reducer						
Sum of two sides 1100mm	18.10	23.98	2.18	47.58	nr	**71.56**
Sum of two sides 1150mm	18.37	24.34	2.18	47.58	nr	**71.92**
Sum of two sides 1200mm	18.65	24.71	2.18	47.58	nr	**72.29**
Sum of two sides 1250mm	18.93	25.08	2.18	47.58	nr	**72.66**
Sum of two sides 1300mm	19.20	25.44	2.30	50.20	nr	**75.64**
Sum of two sides 1350mm	19.48	25.81	2.30	50.20	nr	**76.01**
Sum of two sides 1400mm	19.76	26.18	2.30	50.20	nr	**76.38**
Sum of two sides 1450mm	20.19	26.75	2.30	50.20	nr	**76.95**
Sum of two sides 1500mm	21.61	28.63	2.47	53.91	nr	**82.54**
Sum of two sides 1550mm	23.19	30.73	2.47	53.91	nr	**84.63**
Sum of two sides 1600mm	24.77	32.82	2.47	53.91	nr	**86.73**
Sum of two sides 1650mm	28.05	37.17	2.47	53.91	nr	**91.07**
Sum of two sides 1700mm	29.63	39.26	2.59	56.53	nr	**95.79**
Sum of two sides 1750mm	31.21	41.35	2.59	56.53	nr	**97.88**
Sum of two sides 1800mm	32.94	43.65	2.59	56.53	nr	**100.17**
Sum of two sides 1850mm	34.52	45.74	2.59	56.53	nr	**102.26**
Sum of two sides 1900mm	36.10	47.83	2.71	59.14	nr	**106.98**
Sum of two sides 1950mm	37.67	49.91	2.71	59.14	nr	**109.06**
Sum of two sides 2000mm	39.26	52.02	2.71	59.14	nr	**111.16**
Offset						
Sum of two sides 1100mm	34.85	46.18	2.18	47.58	nr	**93.75**
Sum of two sides 1150mm	35.61	47.18	2.18	47.58	nr	**94.76**
Sum of two sides 1200mm	35.72	47.33	2.18	47.58	nr	**94.91**
Sum of two sides 1250mm	35.80	47.44	2.18	47.58	nr	**95.01**
Sum of two sides 1300mm	35.85	47.50	2.47	53.91	nr	**101.41**
Sum of two sides 1350mm	35.88	47.54	2.47	53.91	nr	**101.45**
Sum of two sides 1400mm	35.87	47.53	2.47	53.91	nr	**101.43**
Sum of two sides 1450mm	35.95	47.63	2.47	53.91	nr	**101.54**
Sum of two sides 1500mm	37.60	49.82	2.47	53.91	nr	**103.73**
Sum of two sides 1550mm	39.45	52.27	2.47	53.91	nr	**106.18**
Sum of two sides 1600mm	41.27	54.68	2.47	53.91	nr	**108.59**
Sum of two sides 1650mm	43.94	58.22	2.47	53.91	nr	**112.13**
Sum of two sides 1700mm	46.49	61.60	2.61	56.96	nr	**118.56**
Sum of two sides 1750mm	48.19	63.85	2.61	56.96	nr	**120.81**
Sum of two sides 1800mm	49.94	66.17	2.61	56.96	nr	**123.13**
Sum of two sides 1850mm	52.45	69.50	2.61	56.96	nr	**126.46**
Sum of two sides 1900mm	54.08	71.66	2.71	59.14	nr	**130.80**
Sum of two sides 1950mm	55.68	73.78	2.71	59.14	nr	**132.92**
Sum of two sides 2000mm	57.25	75.86	2.71	59.14	nr	**135.00**
Square to round						
Sum of two sides 1100mm	26.16	34.66	1.82	39.72	nr	**74.38**
Sum of two sides 1150mm	26.57	35.21	1.82	39.72	nr	**74.93**
Sum of two sides 1200mm	26.98	35.75	1.82	39.72	nr	**75.47**
Sum of two sides 1250mm	27.39	36.29	1.82	39.72	nr	**76.01**
Sum of two sides 1300mm	27.80	36.84	2.32	50.63	nr	**87.47**
Sum of two sides 1350mm	28.21	37.38	2.32	50.63	nr	**88.01**
Sum of two sides 1400mm	28.63	37.93	2.32	50.63	nr	**88.57**
Sum of two sides 1450mm	29.07	38.52	2.32	50.63	nr	**89.15**
Sum of two sides 1500mm	30.91	40.96	2.56	55.87	nr	**96.83**
Sum of two sides 1550mm	32.94	43.65	2.56	55.87	nr	**99.52**
Sum of two sides 1600mm	34.98	46.35	2.58	56.31	nr	**102.66**

346   *Material Costs/Measured Work Prices - Mechanical Installations*

U:VENTILATION/AIR CONDITIONING SYSTEMS

Item	Net Price £	Material £	Labour hours	Labour £	Unit	Total rate £
U10 : DUCTWORK : RECTANGULAR - CLASS C						
Y30 - AIR DUCTLINES						
Galvanised sheet metal DW144 class C rectangular section ductwork; including all necessary stiffeners, joints, couplers in the running length and duct supports (Continued)						
Square to round						
Sum of two sides 1650mm	39.12	51.83	2.58	56.31	nr	108.14
Sum of two sides 1700mm	41.16	54.54	2.84	61.98	nr	116.52
Sum of two sides 1750mm	43.20	57.24	2.84	61.98	nr	119.22
Sum of two sides 1800mm	45.26	59.97	2.84	61.98	nr	121.95
Sum of two sides 1850mm	47.30	62.67	2.84	61.98	nr	124.65
Sum of two sides 1900mm	49.34	65.38	3.13	68.31	nr	133.69
Sum of two sides 1950mm	51.38	68.08	3.13	68.31	nr	136.39
Sum of two sides 2000mm	53.42	70.78	3.13	68.31	nr	139.09
90 degree radius bend						
Sum of two sides 1100mm	18.87	25.00	1.40	30.55	nr	55.56
Sum of two sides 1150mm	19.38	25.68	1.40	30.55	nr	56.23
Sum of two sides 1200mm	19.62	26.00	1.40	30.55	nr	56.55
Sum of two sides 1250mm	19.88	26.34	1.40	30.55	nr	56.90
Sum of two sides 1300mm	20.39	27.02	1.91	41.69	nr	68.70
Sum of two sides 1350mm	20.89	27.68	1.91	41.69	nr	69.36
Sum of two sides 1400mm	21.40	28.36	1.91	41.69	nr	70.04
Sum of two sides 1450mm	22.07	29.24	1.91	41.69	nr	70.93
Sum of two sides 1500mm	24.85	32.93	2.11	46.05	nr	78.98
Sum of two sides 1550mm	27.96	37.05	2.11	46.05	nr	83.10
Sum of two sides 1600mm	31.07	41.17	2.11	46.05	nr	87.22
Sum of two sides 1650mm	38.90	51.54	2.11	46.05	nr	97.59
Sum of two sides 1700mm	40.40	53.53	2.55	55.65	nr	109.18
Sum of two sides 1750mm	43.51	57.65	2.55	55.65	nr	113.30
Sum of two sides 1800mm	46.78	61.98	2.55	55.65	nr	117.64
Sum of two sides 1850mm	48.14	63.79	2.55	55.65	nr	119.44
Sum of two sides 1900mm	51.22	67.87	2.80	61.11	nr	128.98
Sum of two sides 1950mm	54.30	71.95	2.80	61.11	nr	133.06
Sum of two sides 2000mm	57.38	76.03	2.80	61.11	nr	137.14
45 degree bend						
Sum of two sides 1100mm	21.23	28.13	1.88	41.03	nr	69.16
Sum of two sides 1150mm	21.38	28.33	1.88	41.03	nr	69.36
Sum of two sides 1200mm	21.84	28.94	1.88	41.03	nr	69.97
Sum of two sides 1250mm	22.30	29.55	1.88	41.03	nr	70.58
Sum of two sides 1300mm	22.76	30.16	2.26	49.32	nr	79.48
Sum of two sides 1350mm	23.22	30.77	2.26	49.32	nr	80.09
Sum of two sides 1400mm	23.69	31.39	2.44	53.25	nr	84.64
Sum of two sides 1450mm	24.30	32.20	2.44	53.25	nr	85.45
Sum of two sides 1500mm	25.91	34.33	2.68	58.49	nr	92.82
Sum of two sides 1550mm	27.67	36.66	2.68	58.49	nr	95.15
Sum of two sides 1600mm	29.43	38.99	2.69	58.71	nr	97.70

Material Costs/Measured Work Prices - Mechanical Installations

U:VENTILATION/AIR CONDITIONING SYSTEMS

Item	Net Price £	Material £	Labour hours	Labour £	Unit	Total rate £
Sum of two sides 1650mm	33.32	44.15	2.69	58.71	nr	**102.86**
Sum of two sides 1700mm	34.68	45.95	2.96	64.60	nr	**110.55**
Sum of two sides 1750mm	36.45	48.30	2.96	64.60	nr	**112.90**
Sum of two sides 1800mm	38.36	50.83	2.96	64.60	nr	**115.43**
Sum of two sides 1850mm	39.69	52.59	2.96	64.60	nr	**117.19**
Sum of two sides 1900mm	41.45	54.92	3.26	71.15	nr	**126.07**
Sum of two sides 1950mm	43.20	57.24	3.26	71.15	nr	**128.39**
Sum of two sides 2000mm	44.96	59.57	3.26	71.15	nr	**130.72**
90 degree mitre bend						
Sum of two sides 1100mm	40.48	53.64	3.03	66.13	nr	**119.76**
Sum of two sides 1150mm	40.72	53.95	3.03	66.13	nr	**120.08**
Sum of two sides 1200mm	42.12	55.81	3.03	66.13	nr	**121.94**
Sum of two sides 1250mm	43.51	57.65	3.03	66.13	nr	**123.78**
Sum of two sides 1300mm	44.91	59.51	3.85	84.02	nr	**143.53**
Sum of two sides 1350mm	46.30	61.35	3.85	84.02	nr	**145.37**
Sum of two sides 1400mm	47.70	63.20	3.85	84.02	nr	**147.23**
Sum of two sides 1450mm	49.15	65.12	3.85	84.02	nr	**149.15**
Sum of two sides 1500mm	52.26	69.24	4.25	92.75	nr	**162.00**
Sum of two sides 1550mm	55.61	73.68	4.25	92.75	nr	**166.44**
Sum of two sides 1600mm	58.96	78.12	4.26	92.97	nr	**171.09**
Sum of two sides 1650mm	66.28	87.82	4.26	92.97	nr	**180.79**
Sum of two sides 1700mm	68.55	90.83	4.68	102.14	nr	**192.97**
Sum of two sides 1750mm	71.89	95.25	4.68	102.14	nr	**197.39**
Sum of two sides 1800mm	75.30	99.77	4.68	102.14	nr	**201.91**
Sum of two sides 1850mm	77.54	102.74	4.68	102.14	nr	**204.88**
Sum of two sides 1900mm	80.89	107.18	4.87	106.29	nr	**213.46**
Sum of two sides 1950mm	84.25	111.63	4.87	106.29	nr	**217.92**
Sum of two sides 2000mm	87.65	116.14	4.87	106.29	nr	**222.42**
Branch						
Sum of two sides 1100mm	33.13	43.90	1.29	28.15	nr	**72.05**
Sum of two sides 1150mm	33.97	45.01	1.29	28.15	nr	**73.16**
Sum of two sides 1200mm	34.83	46.15	1.29	28.15	nr	**74.30**
Sum of two sides 1250mm	35.68	47.28	1.29	28.15	nr	**75.43**
Sum of two sides 1300mm	36.53	48.40	1.39	30.34	nr	**78.74**
Sum of two sides 1350mm	37.38	49.53	1.39	30.34	nr	**79.86**
Sum of two sides 1400mm	38.24	50.67	1.39	30.34	nr	**81.00**
Sum of two sides 1450mm	39.21	51.95	1.39	30.34	nr	**82.29**
Sum of two sides 1500mm	41.58	55.09	1.64	35.79	nr	**90.89**
Sum of two sides 1550mm	44.17	58.53	1.64	35.79	nr	**94.32**
Sum of two sides 1600mm	46.76	61.96	1.64	35.79	nr	**97.75**
Sum of two sides 1650mm	51.88	68.74	1.64	35.79	nr	**104.53**
Sum of two sides 1700mm	54.47	72.17	1.69	36.88	nr	**109.06**
Sum of two sides 1750mm	57.05	75.59	1.69	36.88	nr	**112.47**
Sum of two sides 1800mm	59.76	79.18	1.69	36.88	nr	**116.07**
Sum of two sides 1850mm	62.35	82.61	1.69	36.88	nr	**119.50**
Sum of two sides 1900mm	64.64	85.65	1.85	40.38	nr	**126.02**
Sum of two sides 1950mm	67.52	89.46	1.85	40.38	nr	**129.84**
Sum of two sides 2000mm	70.11	92.90	1.85	40.38	nr	**133.27**

Material Costs/Measured Work Prices - Mechanical Installations

U:VENTILATION/AIR CONDITIONING SYSTEMS

Item	Net Price £	Material £	Labour hours	Labour £	Unit	Total rate £
U10 : DUCTWORK : RECTANGULAR - CLASS C						
Y30 - AIR DUCTLINES						
Galvanised sheet metal DW144 class C rectangular section ductwork; including all necessary stiffeners, joints, couplers in the running length and duct supports (Continued)						
Grille neck						
Sum of two sides 1100mm	32.96	43.67	1.44	31.43	nr	75.10
Sum of two sides 1150mm	33.88	44.89	1.44	31.43	nr	76.32
Sum of two sides 1200mm	34.81	46.12	1.44	31.43	nr	77.55
Sum of two sides 1250mm	35.74	47.36	1.44	31.43	nr	78.78
Sum of two sides 1300mm	36.66	48.57	1.69	36.88	nr	85.46
Sum of two sides 1350mm	37.59	49.81	1.69	36.88	nr	86.69
Sum of two sides 1400mm	38.52	51.04	1.69	36.88	nr	87.92
Sum of two sides 1450mm	39.45	52.27	1.69	36.88	nr	89.15
Sum of two sides 1500mm	42.09	55.77	1.79	39.07	nr	94.84
Sum of two sides 1550mm	44.97	59.59	1.79	39.07	nr	98.65
Sum of two sides 1600mm	47.85	63.40	1.79	39.07	nr	102.47
Sum of two sides 1650mm	53.54	70.94	1.79	39.07	nr	110.01
Sum of two sides 1700mm	56.43	74.77	1.86	40.59	nr	115.36
Sum of two sides 1750mm	59.30	78.57	1.86	40.59	nr	119.17
Sum of two sides 1800mm	62.19	82.40	1.86	40.59	nr	123.00
Sum of two sides 1850mm	65.06	86.20	1.86	40.59	nr	126.80
Sum of two sides 1900mm	67.95	90.03	2.02	44.09	nr	134.12
Sum of two sides 1950mm	70.83	93.85	2.02	44.09	nr	137.94
Sum of two sides 2000mm	73.71	97.67	2.02	44.09	nr	141.75
Ductwork 1001 to 1250mm longest side						
Sum of two sides 1300mm	52.19	69.15	1.51	32.96	m	102.11
Sum of two sides 1350mm	53.45	70.82	1.51	32.96	m	103.78
Sum of two sides 1400mm	54.81	72.62	1.55	33.83	m	106.45
Sum of two sides 1450mm	56.07	74.29	1.55	33.83	m	108.12
Sum of two sides 1500mm	57.34	75.98	1.61	35.14	m	111.11
Sum of two sides 1550mm	58.59	77.63	1.61	35.14	m	112.77
Sum of two sides 1600mm	59.86	79.31	1.62	35.36	m	114.67
Sum of two sides 1650mm	61.12	80.98	1.62	35.36	m	116.34
Sum of two sides 1700mm	62.55	82.88	1.74	37.97	m	120.85
Sum of two sides 1750mm	63.82	84.56	1.74	37.97	m	122.54
Sum of two sides 1800mm	65.08	86.23	1.76	38.41	m	124.64
Sum of two sides 1850mm	66.34	87.90	1.76	38.41	m	126.31
Sum of two sides 1900mm	67.61	89.58	1.81	39.50	m	129.09
Sum of two sides 1950mm	68.87	91.25	1.81	39.50	m	130.76
Sum of two sides 2000mm	70.14	92.94	1.82	39.72	m	132.66
Sum of two sides 2050mm	71.47	94.70	1.82	39.72	m	134.42
Sum of two sides 2100mm	72.82	96.49	2.53	55.22	m	151.70
Sum of two sides 2150mm	74.09	98.17	2.53	55.22	m	153.39
Sum of two sides 2200mm	75.35	99.84	2.55	55.65	m	155.49
Sum of two sides 2250mm	76.62	101.52	2.55	55.65	m	157.17
Sum of two sides 2300mm	77.88	103.19	2.56	55.87	m	159.06

Material Costs/Measured Work Prices - Mechanical Installations

U:VENTILATION/AIR CONDITIONING SYSTEMS

Item	Net Price £	Material £	Labour hours	Labour £	Unit	Total rate £
Sum of two sides 2350mm	79.15	104.87	2.56	55.87	m	**160.74**
Sum of two sides 2400mm	80.41	106.54	2.76	60.24	m	**166.78**
Sum of two sides 2450mm	81.67	108.21	2.76	60.24	m	**168.45**
Sum of two sides 2500mm	82.94	109.90	2.77	60.45	m	**170.35**
Extra over fittings; Ductwork 1001 to 1250mm longest side						
End Cap						
Sum of two sides 1300mm	20.65	27.36	0.38	8.29	nr	**35.65**
Sum of two sides 1350mm	21.32	28.25	0.38	8.29	nr	**36.54**
Sum of two sides 1400mm	22.00	29.15	0.38	8.29	nr	**37.44**
Sum of two sides 1450mm	22.67	30.04	0.38	8.29	nr	**38.33**
Sum of two sides 1500mm	23.35	30.94	0.38	8.29	nr	**39.23**
Sum of two sides 1550mm	24.59	32.58	0.38	8.29	nr	**40.88**
Sum of two sides 1600mm	25.92	34.34	0.38	8.29	nr	**42.64**
Sum of two sides 1650mm	27.24	36.09	0.38	8.29	nr	**44.39**
Sum of two sides 1700mm	28.57	37.86	0.58	12.66	nr	**50.51**
Sum of two sides 1750mm	29.89	39.60	0.58	12.66	nr	**52.26**
Sum of two sides 1800mm	31.22	41.37	0.58	12.66	nr	**54.02**
Sum of two sides 1850mm	32.55	43.13	0.58	12.66	nr	**55.79**
Sum of two sides 1900mm	33.87	44.88	0.58	12.66	nr	**57.54**
Sum of two sides 1950mm	35.20	46.64	0.58	12.66	nr	**59.30**
Sum of two sides 2000mm	36.53	48.40	0.58	12.66	nr	**61.06**
Sum of two sides 2050mm	37.85	50.15	0.58	12.66	nr	**62.81**
Sum of two sides 2100mm	39.18	51.91	0.87	18.99	nr	**70.90**
Sum of two sides 2150mm	40.50	53.66	0.87	18.99	nr	**72.65**
Sum of two sides 2200mm	41.83	55.42	0.87	18.99	nr	**74.41**
Sum of two sides 2250mm	43.16	57.19	0.87	18.99	nr	**76.17**
Sum of two sides 2300mm	44.48	58.94	0.87	18.99	nr	**77.92**
Sum of two sides 2350mm	45.81	60.70	0.87	18.99	nr	**79.69**
Sum of two sides 2400mm	47.13	62.45	0.87	18.99	nr	**81.43**
Sum of two sides 2450mm	48.46	64.21	0.87	18.99	nr	**83.20**
Sum of two sides 2500mm	49.79	65.97	0.87	18.99	nr	**84.96**
Reducer						
Sum of two sides 1300mm	16.88	22.37	2.30	50.20	nr	**72.56**
Sum of two sides 1350mm	17.31	22.94	2.30	50.20	nr	**73.13**
Sum of two sides 1400mm	17.73	23.49	2.30	50.20	nr	**73.69**
Sum of two sides 1450mm	18.14	24.04	2.30	50.20	nr	**74.23**
Sum of two sides 1500mm	18.56	24.59	2.47	53.91	nr	**78.50**
Sum of two sides 1550mm	20.12	26.66	2.47	53.91	nr	**80.57**
Sum of two sides 1600mm	21.85	28.95	2.47	53.91	nr	**82.86**
Sum of two sides 1650mm	23.57	31.23	2.47	53.91	nr	**85.14**
Sum of two sides 1700mm	25.44	33.71	2.59	56.53	nr	**90.23**
Sum of two sides 1750mm	27.16	35.99	2.59	56.53	nr	**92.51**
Sum of two sides 1800mm	28.88	38.27	2.59	56.53	nr	**94.79**
Sum of two sides 1850mm	30.61	40.56	2.59	56.53	nr	**97.08**
Sum of two sides 1900mm	32.33	42.84	2.71	59.14	nr	**101.98**
Sum of two sides 1950mm	34.05	45.12	2.71	59.14	nr	**104.26**
Sum of two sides 2000mm	35.78	47.41	2.71	59.14	nr	**106.55**
Sum of two sides 2050mm	37.65	49.89	2.71	59.14	nr	**109.03**
Sum of two sides 2100mm	39.37	52.17	2.92	63.73	nr	**115.89**
Sum of two sides 2150mm	41.09	54.44	2.92	63.73	nr	**118.17**

350   *Material Costs/Measured Work Prices - Mechanical Installations*

U:VENTILATION/AIR CONDITIONING SYSTEMS

Item	Net Price £	Material £	Labour hours	Labour £	Unit	Total rate £
U10 : DUCTWORK : RECTANGULAR - CLASS C						
Y30 - AIR DUCTLINES						
Galvanised sheet metal DW144 class C rectangular section ductwork; including all necessary stiffeners, joints, couplers in the running length and duct supports (Continued)						
Reducer						
Sum of two sides 2200mm	42.81	56.72	2.92	63.73	nr	**120.45**
Sum of two sides 2250mm	44.53	59.00	2.92	63.73	nr	**122.73**
Sum of two sides 2300mm	46.26	61.29	2.92	63.73	nr	**125.02**
Sum of two sides 2350mm	47.98	63.57	2.92	63.73	nr	**127.30**
Sum of two sides 2400mm	49.70	65.85	3.12	68.09	nr	**133.95**
Sum of two sides 2450mm	51.42	68.13	3.12	68.09	nr	**136.22**
Sum of two sides 2500mm	53.15	70.42	3.12	68.09	nr	**138.52**
Offset						
Sum of two sides 1300mm	47.79	63.32	2.47	53.91	nr	**117.23**
Sum of two sides 1350mm	48.29	63.98	2.47	53.91	nr	**117.89**
Sum of two sides 1400mm	49.56	65.67	2.47	53.91	nr	**119.57**
Sum of two sides 1450mm	50.51	66.93	2.47	53.91	nr	**120.83**
Sum of two sides 1500mm	50.93	67.48	2.47	53.91	nr	**121.39**
Sum of two sides 1550mm	53.02	70.25	2.47	53.91	nr	**124.16**
Sum of two sides 1600mm	55.33	73.31	2.47	53.91	nr	**127.22**
Sum of two sides 1650mm	57.61	76.33	2.47	53.91	nr	**130.24**
Sum of two sides 1700mm	59.97	79.46	2.61	56.96	nr	**136.42**
Sum of two sides 1750mm	62.18	82.39	2.61	56.96	nr	**139.35**
Sum of two sides 1800mm	64.36	85.28	2.61	56.96	nr	**142.24**
Sum of two sides 1850mm	66.50	88.11	2.61	56.96	nr	**145.07**
Sum of two sides 1900mm	68.61	90.91	2.71	59.14	nr	**150.05**
Sum of two sides 1950mm	70.70	93.68	2.71	59.14	nr	**152.82**
Sum of two sides 2000mm	72.75	96.39	2.71	59.14	nr	**155.54**
Sum of two sides 2050mm	73.82	97.81	2.71	59.14	nr	**156.96**
Sum of two sides 2100mm	76.82	101.79	2.92	63.73	nr	**165.51**
Sum of two sides 2150mm	78.77	104.37	2.92	63.73	nr	**168.10**
Sum of two sides 2200mm	80.69	106.91	3.26	71.15	nr	**178.06**
Sum of two sides 2250mm	82.57	109.41	3.26	71.15	nr	**180.55**
Sum of two sides 2300mm	84.43	111.87	3.26	71.15	nr	**183.02**
Sum of two sides 2350mm	87.37	115.77	3.26	71.15	nr	**186.91**
Sum of two sides 2400mm	89.78	118.96	3.47	75.73	nr	**194.69**
Sum of two sides 2450mm	92.71	122.84	3.47	75.73	nr	**·198.57**
Sum of two sides 2500mm	95.65	126.74	3.48	75.95	nr	**202.69**
Square to round						
Sum of two sides 1300mm	24.81	32.87	2.32	50.63	nr	**83.51**
Sum of two sides 1350mm	25.37	33.62	2.32	50.63	nr	**84.25**
Sum of two sides 1400mm	25.93	34.36	2.32	50.63	nr	**84.99**
Sum of two sides 1450mm	26.48	35.09	2.32	50.63	nr	**85.72**
Sum of two sides 1500mm	27.04	35.83	2.56	55.87	nr	**91.70**
Sum of two sides 1550mm	29.03	38.46	2.56	55.87	nr	**94.34**

Material Costs/Measured Work Prices - Mechanical Installations

U:VENTILATION/AIR CONDITIONING SYSTEMS

Item	Net Price £	Material £	Labour hours	Labour £	Unit	Total rate £
Sum of two sides 1600mm	31.22	41.37	2.58	56.31	nr	**97.67**
Sum of two sides 1650mm	33.40	44.26	2.58	56.31	nr	**100.56**
Sum of two sides 1700mm	35.61	47.18	2.84	61.98	nr	**109.17**
Sum of two sides 1750mm	37.80	50.09	2.84	61.98	nr	**112.07**
Sum of two sides 1800mm	39.98	52.97	2.84	61.98	nr	**114.96**
Sum of two sides 1850mm	42.17	55.88	2.84	61.98	nr	**117.86**
Sum of two sides 1900mm	60.23	79.80	3.13	68.31	nr	**148.12**
Sum of two sides 1950mm	46.54	61.67	3.13	68.31	nr	**129.98**
Sum of two sides 2000mm	48.73	64.57	3.13	68.31	nr	**132.88**
Sum of two sides 2050mm	50.94	67.50	3.13	68.31	nr	**135.81**
Sum of two sides 2100mm	53.13	70.40	4.26	92.97	nr	**163.37**
Sum of two sides 2150mm	55.32	73.30	4.26	92.97	nr	**166.27**
Sum of two sides 2200mm	57.50	76.19	4.27	93.19	nr	**169.38**
Sum of two sides 2250mm	59.69	79.09	4.27	93.19	nr	**172.28**
Sum of two sides 2300mm	61.88	81.99	4.30	93.85	nr	**175.84**
Sum of two sides 2350mm	64.06	84.88	4.30	93.85	nr	**178.73**
Sum of two sides 2400mm	66.25	87.78	4.77	104.10	nr	**191.88**
Sum of two sides 2450mm	68.43	90.67	4.77	104.10	nr	**194.77**
Sum of two sides 2500mm	70.62	93.57	4.79	104.54	nr	**198.11**
90 degree radius bend						
Sum of two sides 1300mm	21.51	28.50	1.91	41.69	nr	**70.19**
Sum of two sides 1350mm	22.10	29.28	1.91	41.69	nr	**70.97**
Sum of two sides 1400mm	21.89	29.00	1.91	41.69	nr	**70.69**
Sum of two sides 1450mm	22.46	29.76	1.91	41.69	nr	**71.44**
Sum of two sides 1500mm	23.04	30.53	2.11	46.05	nr	**76.58**
Sum of two sides 1550mm	24.72	32.75	2.11	46.05	nr	**78.80**
Sum of two sides 1600mm	26.60	35.24	2.11	46.05	nr	**81.29**
Sum of two sides 1650mm	28.48	37.74	2.11	46.05	nr	**83.79**
Sum of two sides 1700mm	-	40.27	2.55	55.65	nr	**95.92**
Sum of two sides 1750mm	32.29	42.78	2.55	55.65	nr	**98.44**
Sum of two sides 1800mm	34.16	45.26	2.55	55.65	nr	**100.91**
Sum of two sides 1850mm	36.04	47.75	2.55	55.65	nr	**103.41**
Sum of two sides 1900mm	37.92	50.24	2.80	61.11	nr	**111.35**
Sum of two sides 1950mm	39.79	52.72	2.80	61.11	nr	**113.83**
Sum of two sides 2000mm	41.67	55.21	2.80	61.11	nr	**116.32**
Sum of two sides 2050mm	44.62	59.12	2.80	61.11	nr	**120.23**
Sum of two sides 2100mm	45.48	60.26	2.80	61.11	nr	**121.37**
Sum of two sides 2150mm	47.35	62.74	2.80	61.11	nr	**123.85**
Sum of two sides 2200mm	49.23	65.23	2.80	61.11	nr	**126.34**
Sum of two sides 2250mm	51.11	67.72	4.35	94.94	nr	**162.66**
Sum of two sides 2300mm	52.98	70.20	4.35	94.94	nr	**165.14**
Sum of two sides 2350mm	53.74	71.21	4.35	94.94	nr	**166.14**
Sum of two sides 2400mm	55.61	73.68	4.35	94.94	nr	**168.62**
Sum of two sides 2450mm	57.47	76.15	4.35	94.94	nr	**171.08**
Sum of two sides 2500mm	59.33	78.61	4.35	94.94	nr	**173.55**
45 degree bend						
Sum of two sides 1300mm	20.72	27.45	2.26	49.32	nr	**76.78**
Sum of two sides 1350mm	21.20	28.09	2.26	49.32	nr	**77.41**
Sum of two sides 1400mm	21.27	28.18	2.44	53.25	nr	**81.43**
Sum of two sides 1450mm	21.73	28.79	2.44	53.25	nr	**82.04**
Sum of two sides 1500mm	22.20	29.41	2.68	58.49	nr	**87.90**
Sum of two sides 1550mm	23.80	31.54	2.68	58.49	nr	**90.02**

352   **Material Costs/Measured Work Prices - Mechanical Installations**

U:VENTILATION/AIR CONDITIONING SYSTEMS

Item	Net Price £	Material £	Labour hours	Labour £	Unit	Total rate £
U10 : DUCTWORK : RECTANGULAR - CLASS C						
Y30 - AIR DUCTLINES						
Galvanised sheet metal DW144 class C rectangular section ductwork; including all necessary stiffeners, joints, couplers in the running length and duct supports (Continued)						
45 degree bend						
Sum of two sides 1600mm	25.57	33.88	2.69	58.71	nr	92.59
Sum of two sides 1650mm	27.34	36.23	2.69	58.71	nr	94.93
Sum of two sides 1700mm	29.24	38.74	2.96	64.60	nr	103.34
Sum of two sides 1750mm	31.01	41.09	2.96	64.60	nr	105.69
Sum of two sides 1800mm	32.78	43.43	2.96	64.60	nr	108.03
Sum of two sides 1850mm	34.54	45.77	2.96	64.60	nr	110.37
Sum of two sides 1900mm	36.31	48.11	3.26	71.15	nr	119.26
Sum of two sides 1950mm	38.08	50.46	3.26	71.15	nr	121.60
Sum of two sides 2000mm	39.84	52.79	3.26	71.15	nr	123.94
Sum of two sides 2050mm	42.26	55.99	3.26	71.15	nr	127.14
Sum of two sides 2100mm	43.51	57.65	7.50	163.68	nr	221.34
Sum of two sides 2150mm	45.28	60.00	7.50	163.68	nr	223.68
Sum of two sides 2200mm	47.05	62.34	7.50	163.68	nr	226.03
Sum of two sides 2250mm	48.81	64.67	7.50	163.68	nr	228.36
Sum of two sides 2300mm	50.58	67.02	7.55	164.78	nr	231.79
Sum of two sides 2350mm	51.78	68.61	7.55	164.78	nr	233.38
Sum of two sides 2400mm	53.54	70.94	8.13	177.43	nr	248.37
Sum of two sides 2450mm	55.30	73.27	8.13	177.43	nr	250.71
Sum of two sides 2500mm	57.06	75.60	8.30	181.14	nr	256.75
90 degree mitre bend						
Sum of two sides 1300mm	43.26	57.32	3.85	84.02	nr	141.34
Sum of two sides 1350mm	46.83	62.05	3.85	84.02	nr	146.07
Sum of two sides 1400mm	48.94	64.85	3.85	84.02	nr	148.87
Sum of two sides 1450mm	52.52	69.59	3.85	84.02	nr	153.61
Sum of two sides 1500mm	56.10	74.33	4.25	92.75	nr	167.09
Sum of two sides 1550mm	61.40	81.36	4.25	92.75	nr	174.11
Sum of two sides 1600mm	66.93	88.68	4.26	92.97	nr	181.66
Sum of two sides 1650mm	72.46	96.01	4.26	92.97	nr	188.98
Sum of two sides 1700mm	89.66	118.80	4.68	102.14	nr	220.94
Sum of two sides 1750mm	94.45	125.15	4.68	102.14	nr	227.29
Sum of two sides 1800mm	99.25	131.51	4.68	102.14	nr	233.65
Sum of two sides 1850mm	104.05	137.87	4.68	102.14	nr	240.01
Sum of two sides 1900mm	108.85	144.23	4.87	106.29	nr	250.51
Sum of two sides 1950mm	105.69	140.04	4.87	106.29	nr	246.32
Sum of two sides 2000mm	111.22	147.37	4.87	106.29	nr	253.65
Sum of two sides 2050mm	118.18	156.59	4.87	106.29	nr	262.87
Sum of two sides 2100mm	122.31	162.06	7.50	163.68	nr	325.75
Sum of two sides 2150mm	127.84	169.39	7.50	163.68	nr	333.07
Sum of two sides 2200mm	133.38	176.73	7.50	163.68	nr	340.41
Sum of two sides 2250mm	138.91	184.06	7.50	163.68	nr	347.74
Sum of two sides 2300mm	144.44	191.38	7.55	164.78	nr	356.16

Material Costs/Measured Work Prices - Mechanical Installations

U:VENTILATION/AIR CONDITIONING SYSTEMS

Item	Net Price £	Material £	Labour hours	Labour £	Unit	Total rate £
Sum of two sides 2350mm	148.55	196.83	7.55	164.78	nr	361.60
Sum of two sides 2400mm	154.09	204.17	8.13	177.43	nr	381.60
Sum of two sides 2450mm	159.63	211.51	8.13	177.43	nr	388.94
Sum of two sides 2500mm	165.18	218.86	8.30	181.14	nr	400.01
Branch						
Sum of two sides 1300mm	37.42	49.58	1.39	30.34	nr	79.92
Sum of two sides 1350mm	38.38	50.85	1.39	30.34	nr	81.19
Sum of two sides 1400mm	39.34	52.13	1.39	30.34	nr	82.46
Sum of two sides 1450mm	40.30	53.40	1.39	30.34	nr	83.73
Sum of two sides 1500mm	41.26	54.67	1.64	35.79	nr	90.46
Sum of two sides 1550mm	43.99	58.29	1.64	35.79	nr	94.08
Sum of two sides 1600mm	46.96	62.22	1.64	35.79	nr	98.01
Sum of two sides 1650mm	49.94	66.17	1.64	35.79	nr	101.96
Sum of two sides 1700mm	53.05	70.29	1.69	36.88	nr	107.17
Sum of two sides 1750mm	56.03	74.24	1.69	36.88	nr	111.12
Sum of two sides 1800mm	59.00	78.17	1.69	36.88	nr	115.06
Sum of two sides 1850mm	61.97	82.11	1.69	36.88	nr	118.99
Sum of two sides 1900mm	64.95	86.06	1.85	40.38	nr	126.43
Sum of two sides 1950mm	67.92	89.99	1.85	40.38	nr	130.37
Sum of two sides 2000mm	70.89	93.93	1.85	40.38	nr	134.30
Sum of two sides 2050mm	74.01	98.06	1.85	40.38	nr	138.44
Sum of two sides 2100mm	76.99	102.01	2.61	56.96	nr	158.97
Sum of two sides 2150mm	79.96	105.95	2.61	56.96	nr	162.91
Sum of two sides 2200mm	82.93	109.88	2.61	56.96	nr	166.84
Sum of two sides 2250mm	85.90	113.82	2.61	56.96	nr	170.78
Sum of two sides 2300mm	88.87	117.75	2.61	56.96	nr	174.72
Sum of two sides 2350mm	91.85	121.70	2.61	56.96	nr	178.66
Sum of two sides 2400mm	94.82	125.64	2.88	62.85	nr	188.49
Sum of two sides 2450mm	97.80	129.59	2.88	62.85	nr	192.44
Sum of two sides 2500mm	100.77	133.52	2.88	62.85	nr	196.38
Grille neck						
Sum of two sides 1300mm	35.15	46.57	1.69	36.88	nr	83.46
Sum of two sides 1350mm	36.17	47.93	1.69	36.88	nr	84.81
Sum of two sides 1400mm	37.19	49.28	1.69	36.88	nr	86.16
Sum of two sides 1450mm	38.20	50.62	1.69	36.88	nr	87.50
Sum of two sides 1500mm	39.22	51.97	1.79	39.07	nr	91.03
Sum of two sides 1550mm	41.95	55.58	1.79	39.07	nr	94.65
Sum of two sides 1600mm	44.92	59.52	1.79	39.07	nr	98.59
Sum of two sides 1650mm	47.88	63.44	1.79	39.07	nr	102.51
Sum of two sides 1700mm	50.85	67.38	1.86	40.59	nr	107.97
Sum of two sides 1750mm	53.82	71.31	1.86	40.59	nr	111.91
Sum of two sides 1800mm	56.79	75.25	1.86	40.59	nr	115.84
Sum of two sides 1850mm	59.76	79.18	1.86	40.59	nr	119.78
Sum of two sides 1900mm	62.73	83.12	2.02	44.09	nr	127.20
Sum of two sides 1950mm	66.20	87.72	2.02	44.09	nr	131.80
Sum of two sides 2000mm	68.66	90.97	2.02	44.09	nr	135.06
Sum of two sides 2050mm	71.63	94.91	2.02	44.09	nr	139.00
Sum of two sides 2100mm	74.60	98.84	2.61	56.96	nr	155.81
Sum of two sides 2150mm	77.57	102.78	2.80	61.11	nr	163.89
Sum of two sides 2200mm	80.54	106.72	2.80	61.11	nr	167.82
Sum of two sides 2250mm	83.51	110.65	2.80	61.11	nr	171.76
Sum of two sides 2300mm	86.48	114.59	2.80	61.11	nr	175.69

354   *Material Costs/Measured Work Prices - Mechanical Installations*

U:VENTILATION/AIR CONDITIONING SYSTEMS

Item	Net Price £	Material £	Labour hours	Labour £	Unit	Total rate £
U10 : DUCTWORK : RECTANGULAR - CLASS C						
Y30 - AIR DUCTLINES						
Galvanised sheet metal DW144 class C rectangular section ductwork; including all necessary stiffeners, joints, couplers in the running length and duct supports (Continued)						
Grille neck						
Sum of two sides 2350mm	89.44	118.51	2.80	61.11	nr	179.62
Sum of two sides 2400mm	92.41	122.44	3.06	66.78	nr	189.23
Sum of two sides 2450mm	95.38	126.38	3.06	66.78	nr	193.16
Sum of two sides 2500mm	98.35	130.31	3.06	66.78	nr	197.10
Ductwork 1251 to 1600mm longest side						
Sum of two sides 1700mm	70.56	93.49	1.74	37.97	m	131.47
Sum of two sides 1750mm	72.04	95.45	1.74	37.97	m	133.43
Sum of two sides 1800mm	73.53	97.43	1.76	38.41	m	135.84
Sum of two sides 1850mm	75.05	99.44	1.76	38.41	m	137.85
Sum of two sides 1900mm	76.57	101.46	1.81	39.50	m	140.96
Sum of two sides 1950mm	78.01	103.36	1.81	39.50	m	142.87
Sum of two sides 2000mm	79.45	105.27	1.82	39.72	m	144.99
Sum of two sides 2050mm	80.90	107.19	1.82	39.72	m	146.91
Sum of two sides 2100mm	82.34	109.10	2.53	55.22	m	164.32
Sum of two sides 2150mm	83.78	111.01	2.53	55.22	m	166.22
Sum of two sides 2200mm	85.22	112.92	2.55	55.65	m	168.57
Sum of two sides 2250mm	86.74	114.93	2.55	55.65	m	170.58
Sum of two sides 2300mm	88.26	116.94	2.56	55.87	m	172.82
Sum of two sides 2350mm	89.74	118.91	2.56	55.87	m	174.78
Sum of two sides 2400mm	91.23	120.88	2.76	60.24	m	181.12
Sum of two sides 2450mm	92.67	122.79	2.76	60.24	m	183.02
Sum of two sides 2500mm	94.11	124.70	2.77	60.45	m	185.15
Sum of two sides 2550mm	95.55	126.60	2.77	60.45	m	187.06
Sum of two sides 2600mm	96.99	128.51	2.97	64.82	m	193.33
Sum of two sides 2650mm	98.43	130.42	2.97	64.82	m	195.24
Sum of two sides 2700mm	99.87	132.33	2.99	65.26	m	197.58
Sum of two sides 2750mm	101.36	134.30	2.99	65.26	m	199.56
Sum of two sides 2800mm	102.84	136.26	3.30	72.02	m	208.28
Sum of two sides 2850mm	104.32	138.22	3.30	72.02	m	210.25
Sum of two sides 2900mm	105.80	140.19	3.31	72.24	m	212.42
Sum of two sides 2950mm	107.28	142.15	3.31	72.24	m	214.39
Sum of two sides 3000mm	108.77	144.12	3.53	77.04	m	221.16
Sum of two sides 3050mm	110.21	146.03	3.53	77.04	m	223.07
Sum of two sides 3100mm	111.65	147.94	3.55	77.48	m	225.41
Sum of two sides 3150mm	113.09	149.84	3.55	77.48	m	227.32
Sum of two sides 3200mm	114.53	151.75	3.56	77.70	m	229.45

Material Costs/Measured Work Prices - Mechanical Installations

U:VENTILATION/AIR CONDITIONING SYSTEMS

Item	Net Price £	Material £	Labour hours	Labour £	Unit	Total rate £
Extra over fittings; Ductwork 1251 to 1600mm longest side						
End Cap						
Sum of two sides 1700mm	34.79	46.10	0.58	12.66	nr	**58.76**
Sum of two sides 1750mm	36.27	48.06	0.58	12.66	nr	**60.72**
Sum of two sides 1800mm	37.76	50.03	0.58	12.66	nr	**62.69**
Sum of two sides 1850mm	39.25	52.01	0.58	12.66	nr	**64.66**
Sum of two sides 1900mm	40.74	53.98	0.58	12.66	nr	**66.64**
Sum of two sides 1950mm	42.23	55.95	0.58	12.66	nr	**68.61**
Sum of two sides 2000mm	43.72	57.93	0.58	12.66	nr	**70.59**
Sum of two sides 2050mm	45.21	59.90	0.58	12.66	nr	**72.56**
Sum of two sides 2100mm	46.70	61.88	0.87	18.99	nr	**80.86**
Sum of two sides 2150mm	48.19	63.85	0.87	18.99	nr	**82.84**
Sum of two sides 2200mm	49.68	65.83	0.87	18.99	nr	**84.81**
Sum of two sides 2250mm	51.17	67.80	0.87	18.99	nr	**86.79**
Sum of two sides 2300mm	52.66	69.77	0.87	18.99	nr	**88.76**
Sum of two sides 2350mm	54.15	71.75	0.87	18.99	nr	**90.74**
Sum of two sides 2400mm	55.64	73.72	0.87	18.99	nr	**92.71**
Sum of two sides 2450mm	57.13	75.70	0.87	18.99	nr	**94.68**
Sum of two sides 2500mm	58.62	77.67	0.87	18.99	nr	**96.66**
Sum of two sides 2550mm	60.11	79.65	0.87	18.99	nr	**98.63**
Sum of two sides 2600mm	61.60	81.62	0.87	18.99	nr	**100.61**
Sum of two sides 2650mm	63.09	83.59	0.87	18.99	nr	**102.58**
Sum of two sides 2700mm	64.58	85.57	0.87	18.99	nr	**104.56**
Sum of two sides 2750mm	66.07	87.54	0.87	18.99	nr	**106.53**
Sum of two sides 2800mm	67.56	89.52	1.16	25.32	nr	**114.83**
Sum of two sides 2850mm	69.05	91.49	1.16	25.32	nr	**116.81**
Sum of two sides 2900mm	70.54	93.47	1.16	25.32	nr	**118.78**
Sum of two sides 2950mm	72.03	95.44	1.16	25.32	nr	**120.76**
Sum of two sides 3000mm	73.77	97.75	1.73	37.76	nr	**135.50**
Sum of two sides 3050mm	74.78	99.08	1.73	37.76	nr	**136.84**
Sum of two sides 3100mm	76.04	100.75	1.73	37.76	nr	**138.51**
Sum of two sides 3150mm	77.30	102.42	1.73	37.76	nr	**140.18**
Sum of two sides 3200mm	78.56	104.09	1.73	37.76	nr	**141.85**
Reducer						
Sum of two sides 1700mm	22.00	29.15	2.59	56.53	nr	**85.68**
Sum of two sides 1750mm	23.97	31.76	2.59	56.53	nr	**88.29**
Sum of two sides 1800mm	25.95	34.38	2.59	56.53	nr	**90.91**
Sum of two sides 1850mm	27.99	37.09	2.59	56.53	nr	**93.61**
Sum of two sides 1900mm	30.02	39.78	2.71	59.14	nr	**98.92**
Sum of two sides 1950mm	32.00	42.40	2.71	59.14	nr	**101.54**
Sum of two sides 2000mm	43.72	57.93	2.71	59.14	nr	**117.07**
Sum of two sides 2050mm	45.21	59.90	2.71	59.14	nr	**119.05**
Sum of two sides 2100mm	37.93	50.26	2.92	63.73	nr	**113.98**
Sum of two sides 2150mm	39.91	52.88	2.92	63.73	nr	**116.61**
Sum of two sides 2200mm	41.89	55.50	2.92	63.73	nr	**119.23**
Sum of two sides 2250mm	43.93	58.21	2.92	63.73	nr	**121.94**
Sum of two sides 2300mm	45.96	60.90	2.92	63.73	nr	**124.62**
Sum of two sides 2350mm	47.94	63.52	2.92	63.73	nr	**127.25**
Sum of two sides 2400mm	49.92	66.14	3.12	68.09	nr	**134.24**
Sum of two sides 2450mm	51.90	68.77	3.12	68.09	nr	**136.86**
Sum of two sides 2500mm	53.87	71.38	3.12	68.09	nr	**139.47**

U:VENTILATION/AIR CONDITIONING SYSTEMS

Item	Net Price £	Material £	Labour hours	Labour £	Unit	Total rate £
U10 : DUCTWORK : RECTANGULAR - CLASS C						
Y30 - AIR DUCTLINES						
Galvanised sheet metal DW144 class C rectangular section ductwork; including all necessary stiffeners, joints, couplers in the running length and duct supports (Continued)						
Reducer						
Sum of two sides 2550mm	55.85	74.00	3.12	68.09	nr	142.09
Sum of two sides 2600mm	57.83	76.62	3.16	68.97	nr	145.59
Sum of two sides 2650mm	63.09	83.59	3.16	68.97	nr	152.56
Sum of two sides 2700mm	61.79	81.87	3.16	68.97	nr	150.84
Sum of two sides 2750mm	63.77	84.50	3.16	68.97	nr	153.46
Sum of two sides 2800mm	65.74	87.11	4.00	87.30	nr	174.40
Sum of two sides 2850mm	67.78	89.81	4.00	87.30	nr	177.11
Sum of two sides 2900mm	69.82	92.51	4.01	87.52	nr	180.03
Sum of two sides 2950mm	71.79	95.12	4.01	87.52	nr	182.64
Sum of two sides 3000mm	73.77	97.75	4.56	99.52	nr	197.27
Sum of two sides 3050mm	75.29	99.76	4.56	99.52	nr	199.28
Sum of two sides 3100mm	76.81	101.77	4.56	99.52	nr	201.29
Sum of two sides 3150mm	78.33	103.79	4.56	99.52	nr	203.31
Sum of two sides 3200mm	79.85	105.80	4.56	99.52	nr	205.32
Offset						
Sum of two sides 1700mm	66.55	88.18	2.61	56.96	nr	145.14
Sum of two sides 1750mm	70.04	92.80	2.61	56.96	nr	149.77
Sum of two sides 1800mm	73.54	97.44	2.61	56.96	nr	154.40
Sum of two sides 1850mm	76.00	100.70	2.61	56.96	nr	157.66
Sum of two sides 1900mm	78.47	103.97	2.71	59.14	nr	163.12
Sum of two sides 1950mm	80.80	107.06	2.71	59.14	nr	166.20
Sum of two sides 2000mm	83.13	110.15	2.71	59.14	nr	169.29
Sum of two sides 2050mm	85.40	113.16	2.71	59.14	nr	172.30
Sum of two sides 2100mm	87.66	116.15	2.92	63.73	nr	179.88
Sum of two sides 2150mm	89.86	119.06	2.92	63.73	nr	182.79
Sum of two sides 2200mm	92.06	121.98	3.26	71.15	nr	193.13
Sum of two sides 2250mm	94.23	124.85	3.26	71.15	nr	196.00
Sum of two sides 2300mm	96.39	127.72	3.26	71.15	nr	198.87
Sum of two sides 2350mm	99.73	132.14	3.26	71.15	nr	203.29
Sum of two sides 2400mm	103.06	136.55	3.47	75.73	nr	212.29
Sum of two sides 2450mm	105.09	139.24	3.47	75.73	nr	214.98
Sum of two sides 2500mm	107.12	141.93	3.48	75.95	nr	217.88
Sum of two sides 2550mm	109.44	145.01	3.48	75.95	nr	220.96
Sum of two sides 2600mm	111.76	148.08	3.49	76.17	nr	224.25
Sum of two sides 2650mm	115.04	152.43	3.49	76.17	nr	228.60
Sum of two sides 2700mm	118.32	156.77	3.50	76.39	nr	233.16
Sum of two sides 2750mm	121.59	161.11	3.50	76.39	nr	237.49
Sum of two sides 2800mm	124.87	165.45	4.33	94.50	nr	259.95
Sum of two sides 2850mm	128.16	169.81	4.33	94.50	nr	264.31
Sum of two sides 2900mm	131.46	174.18	4.74	103.45	nr	277.63
Sum of two sides 2950mm	134.74	178.53	4.74	103.45	nr	281.98

Material Costs/Measured Work Prices - Mechanical Installations

U:VENTILATION/AIR CONDITIONING SYSTEMS

Item	Net Price £	Material £	Labour hours	Labour £	Unit	Total rate £
Sum of two sides 3000mm	138.02	182.88	5.31	115.89	nr	298.77
Sum of two sides 3050mm	140.61	186.31	5.31	115.89	nr	302.20
Sum of two sides 3100mm	143.20	189.74	5.34	116.54	nr	306.28
Sum of two sides 3150mm	145.79	193.17	5.34	116.54	nr	309.72
Sum of two sides 3200mm	148.38	196.60	5.35	116.76	nr	313.37
Square to round						
Sum of two sides 1700mm	32.70	43.33	2.84	61.98	nr	105.31
Sum of two sides 1750mm	35.41	46.92	2.84	61.98	nr	108.90
Sum of two sides 1800mm	38.13	50.52	2.84	61.98	nr	112.50
Sum of two sides 1850mm	40.84	54.11	2.84	61.98	nr	116.09
Sum of two sides 1900mm	43.55	57.70	3.13	68.31	nr	126.01
Sum of two sides 1950mm	46.27	61.31	3.13	68.31	nr	129.62
Sum of two sides 2000mm	48.98	64.90	3.13	68.31	nr	133.21
Sum of two sides 2050mm	51.70	68.50	3.13	68.31	nr	136.81
Sum of two sides 2100mm	54.42	72.11	4.26	92.97	nr	165.08
Sum of two sides 2150mm	57.13	75.70	4.26	92.97	nr	168.67
Sum of two sides 2200mm	59.85	79.30	4.27	93.19	nr	172.49
Sum of two sides 2250mm	62.56	82.89	4.27	93.19	nr	176.08
Sum of two sides 2300mm	65.27	86.48	4.30	93.85	nr	180.33
Sum of two sides 2350mm	67.98	90.07	4.30	93.85	nr	183.92
Sum of two sides 2400mm	70.70	93.68	4.77	104.10	nr	197.78
Sum of two sides 2450mm	73.42	97.28	4.77	104.10	nr	201.38
Sum of two sides 2500mm	76.13	100.87	4.79	104.54	nr	205.41
Sum of two sides 2550mm	78.85	104.48	4.79	104.54	nr	209.02
Sum of two sides 2600mm	81.56	108.07	4.95	108.03	nr	216.10
Sum of two sides 2650mm	84.28	111.67	4.95	108.03	nr	219.70
Sum of two sides 2700mm	87.00	115.28	4.95	108.03	nr	223.31
Sum of two sides 2750mm	89.71	118.87	4.95	108.03	nr	226.90
Sum of two sides 2800mm	92.43	122.47	8.49	185.29	nr	307.76
Sum of two sides 2850mm	95.14	126.06	8.49	185.29	nr	311.35
Sum of two sides 2900mm	97.85	129.65	8.88	193.80	nr	323.45
Sum of two sides 2950mm	100.56	133.24	8.88	193.80	nr	327.04
Sum of two sides 3000mm	103.28	136.85	9.02	196.86	nr	333.70
Sum of two sides 3050mm	105.42	139.68	9.02	196.86	nr	336.54
Sum of two sides 3100mm	107.57	142.53	9.02	196.86	nr	339.39
Sum of two sides 3150mm	109.71	145.37	9.02	196.86	nr	342.22
Sum of two sides 3200mm	111.86	148.21	9.09	198.39	nr	346.60
90 degree radius bend						
Sum of two sides 1700mm	72.77	96.42	2.55	55.65	nr	152.07
Sum of two sides 1750mm	75.30	99.77	2.55	55.65	nr	155.43
Sum of two sides 1800mm	77.83	103.12	2.55	55.65	nr	158.78
Sum of two sides 1850mm	81.53	108.03	2.55	55.65	nr	163.68
Sum of two sides 1900mm	85.22	112.92	2.80	61.11	nr	174.03
Sum of two sides 1950mm	88.78	117.63	2.80	61.11	nr	178.74
Sum of two sides 2000mm	92.35	122.36	2.80	61.11	nr	183.47
Sum of two sides 2050mm	95.91	127.08	2.80	61.11	nr	188.19
Sum of two sides 2100mm	99.47	131.80	2.61	56.96	nr	188.76
Sum of two sides 2150mm	103.03	136.51	2.61	56.96	nr	193.48
Sum of two sides 2200mm	106.59	141.23	2.62	57.18	nr	198.41
Sum of two sides 2250mm	110.28	146.12	2.62	57.18	nr	203.30
Sum of two sides 2300mm	113.98	151.02	2.63	57.40	nr	208.42
Sum of two sides 2350mm	116.27	154.06	2.63	57.40	nr	211.46

Material Costs/Measured Work Prices - Mechanical Installations

U:VENTILATION/AIR CONDITIONING SYSTEMS

Item	Net Price £	Material £	Labour hours	Labour £	Unit	Total rate £
U10 : DUCTWORK : RECTANGULAR - CLASS C						
Y30 - AIR DUCTLINES						
Galvanised sheet metal DW144 class C rectangular section ductwork; including all necessary stiffeners, joints, couplers in the running length and duct supports (Continued)						
90 degree radius bend						
Sum of two sides 2400mm	118.56	157.09	4.34	94.72	nr	**251.81**
Sum of two sides 2450mm	122.08	161.76	4.34	94.72	nr	**256.47**
Sum of two sides 2500mm	125.61	166.43	4.35	94.94	nr	**261.37**
Sum of two sides 2550mm	129.14	171.11	4.35	94.94	nr	**266.05**
Sum of two sides 2600mm	132.67	175.79	4.53	98.87	nr	**274.65**
Sum of two sides 2650mm	136.20	180.47	4.53	98.87	nr	**279.33**
Sum of two sides 2700mm	139.73	185.14	4.53	98.87	nr	**284.01**
Sum of two sides 2750mm	141.85	187.95	4.53	98.87	nr	**286.82**
Sum of two sides 2800mm	143.97	190.76	7.13	155.61	nr	**346.37**
Sum of two sides 2850mm	149.03	197.46	7.13	155.61	nr	**353.07**
Sum of two sides 2900mm	154.09	204.17	7.17	156.48	nr	**360.65**
Sum of two sides 2950mm	156.14	206.89	7.17	156.48	nr	**363.37**
Sum of two sides 3000mm	158.20	209.62	7.26	158.45	nr	**368.06**
Sum of two sides 3050mm	160.78	213.03	7.26	158.45	nr	**371.48**
Sum of two sides 3100mm	163.36	216.45	7.26	158.45	nr	**374.90**
Sum of two sides 3150mm	165.94	219.87	7.26	158.45	nr	**378.32**
Sum of two sides 3200mm	168.52	223.29	7.31	159.54	nr	**382.83**
45 degree bend						
Sum of two sides 1700mm	29.77	39.45	2.96	64.60	nr	**104.05**
Sum of two sides 1750mm	31.14	41.26	2.96	64.60	nr	**105.86**
Sum of two sides 1800mm	32.51	43.08	2.96	64.60	nr	**107.68**
Sum of two sides 1850mm	34.45	45.65	2.96	64.60	nr	**110.25**
Sum of two sides 1900mm	36.40	48.23	3.26	71.15	nr	**119.38**
Sum of two sides 1950mm	38.28	50.72	3.26	71.15	nr	**121.87**
Sum of two sides 2000mm	40.16	53.21	3.26	71.15	nr	**124.36**
Sum of two sides 2050mm	42.04	55.70	3.26	71.15	nr	**126.85**
Sum of two sides 2100mm	43.93	58.21	7.50	163.68	nr	**221.89**
Sum of two sides 2150mm	45.81	60.70	7.50	163.68	nr	**224.38**
Sum of two sides 2200mm	47.69	63.19	7.50	163.68	nr	**226.87**
Sum of two sides 2250mm	49.64	65.77	7.50	163.68	nr	**229.46**
Sum of two sides 2300mm	51.58	68.34	7.55	164.78	nr	**233.12**
Sum of two sides 2350mm	52.83	70.00	7.55	164.78	nr	**234.78**
Sum of two sides 2400mm	54.07	71.64	8.13	177.43	nr	**249.08**
Sum of two sides 2450mm	55.94	74.12	8.13	177.43	nr	**251.55**
Sum of two sides 2500mm	57.80	76.58	8.30	181.14	nr	**257.73**
Sum of two sides 2550mm	59.67	79.06	8.30	181.14	nr	**260.21**
Sum of two sides 2600mm	61.53	81.53	8.56	186.82	nr	**268.35**
Sum of two sides 2650mm	63.40	84.00	8.56	186.82	nr	**270.82**
Sum of two sides 2700mm	65.27	86.48	8.62	188.13	nr	**274.61**
Sum of two sides 2750mm	66.43	88.02	8.62	188.13	nr	**276.15**
Sum of two sides 2800mm	67.59	89.56	9.09	198.39	nr	**287.94**

Material Costs/Measured Work Prices - Mechanical Installations 359

U:VENTILATION/AIR CONDITIONING SYSTEMS

Item	Net Price £	Material £	Labour hours	Labour £	Unit	Total rate £
Sum of two sides 2850mm	70.22	93.04	9.09	198.39	nr	**291.43**
Sum of two sides 2900mm	72.85	96.53	9.09	198.39	nr	**294.91**
Sum of two sides 2950mm	73.97	98.01	9.09	198.39	nr	**296.40**
Sum of two sides 3000mm	75.10	99.51	9.62	209.95	nr	**309.46**
Sum of two sides 3050mm	76.49	101.35	9.62	209.95	nr	**311.30**
Sum of two sides 3100mm	77.89	103.20	9.62	209.95	nr	**313.16**
Sum of two sides 3150mm	79.28	105.05	9.62	209.95	nr	**315.00**
Sum of two sides 3200mm	80.67	106.89	9.62	209.95	nr	**316.84**
90 degree mitre bend						
Sum of two sides 1700mm	77.81	103.10	4.68	102.14	nr	**205.24**
Sum of two sides 1750mm	81.02	107.35	4.68	102.14	nr	**209.49**
Sum of two sides 1800mm	85.74	113.61	4.68	102.14	nr	**215.74**
Sum of two sides 1850mm	89.22	118.22	4.68	102.14	nr	**220.36**
Sum of two sides 1900mm	94.20	124.81	4.87	106.29	nr	**231.10**
Sum of two sides 1950mm	99.19	131.43	4.87	106.29	nr	**237.71**
Sum of two sides 2000mm	104.19	138.05	4.87	106.29	nr	**244.34**
Sum of two sides 2050mm	109.18	144.66	4.87	106.29	nr	**250.95**
Sum of two sides 2100mm	114.17	151.28	7.50	163.68	nr	**314.96**
Sum of two sides 2150mm	119.16	157.89	7.50	163.68	nr	**321.57**
Sum of two sides 2200mm	124.15	164.50	7.50	163.68	nr	**328.18**
Sum of two sides 2250mm	129.13	171.10	7.50	163.68	nr	**334.78**
Sum of two sides 2300mm	134.11	177.70	7.55	164.78	nr	**342.47**
Sum of two sides 2350mm	137.28	181.90	7.55	164.78	nr	**346.67**
Sum of two sides 2400mm	140.44	186.08	8.13	177.43	nr	**363.52**
Sum of two sides 2450mm	145.44	192.71	8.13	177.43	nr	**370.14**
Sum of two sides 2500mm	150.44	199.33	8.30	181.14	nr	**380.48**
Sum of two sides 2550mm	155.44	205.96	8.30	181.14	nr	**387.10**
Sum of two sides 2600mm	160.44	212.58	8.56	186.82	nr	**399.40**
Sum of two sides 2650mm	165.44	219.21	8.56	186.82	nr	**406.03**
Sum of two sides 2700mm	170.45	225.85	8.62	188.13	nr	**413.97**
Sum of two sides 2750mm	173.55	229.95	8.62	188.13	nr	**418.08**
Sum of two sides 2800mm	176.66	234.07	15.20	331.73	nr	**565.81**
Sum of two sides 2850mm	184.30	244.20	15.20	331.73	nr	**575.93**
Sum of two sides 2900mm	191.93	254.31	15.20	331.73	nr	**586.04**
Sum of two sides 2950mm	195.10	258.51	15.20	331.73	nr	**590.24**
Sum of two sides 3000mm	198.27	262.71	15.60	340.46	nr	**603.17**
Sum of two sides 3050mm	202.60	268.44	15.60	340.46	nr	**608.91**
Sum of two sides 3100mm	206.92	274.17	16.04	350.07	nr	**624.24**
Sum of two sides 3150mm	211.25	279.91	16.04	350.07	nr	**629.97**
Sum of two sides 3200mm	215.58	285.64	16.04	350.07	nr	**635.71**
Branch						
Sum of two sides 1700mm	59.36	78.65	1.69	36.88	nr	**115.54**
Sum of two sides 1750mm	62.25	82.48	1.69	36.88	nr	**119.36**
Sum of two sides 1800mm	65.14	86.31	1.69	36.88	nr	**123.19**
Sum of two sides 1850mm	68.10	90.23	1.69	36.88	nr	**127.12**
Sum of two sides 1900mm	71.05	94.14	1.85	40.38	nr	**134.52**
Sum of two sides 1950mm	73.95	97.98	1.85	40.38	nr	**138.36**
Sum of two sides 2000mm	76.84	101.81	1.85	40.38	nr	**142.19**
Sum of two sides 2050mm	79.73	105.64	1.85	40.38	nr	**146.02**
Sum of two sides 2100mm	82.62	109.47	2.61	56.96	nr	**166.43**
Sum of two sides 2150mm	85.52	113.31	2.61	56.96	nr	**170.28**
Sum of two sides 2200mm	88.41	117.14	2.61	56.96	nr	**174.11**

360   *Material Costs/Measured Work Prices - Mechanical Installations*

U:VENTILATION/AIR CONDITIONING SYSTEMS

Item	Net Price £	Material £	Labour hours	Labour £	Unit	Total rate £
U10 : DUCTWORK : RECTANGULAR - CLASS C						
Y30 - AIR DUCTLINES						
Galvanised sheet metal DW144 class C rectangular section ductwork; including all necessary stiffeners, joints, couplers in the running length and duct supports (Continued)						
Branch						
Sum of two sides 2250mm	91.36	121.05	2.61	56.96	nr	**178.01**
Sum of two sides 2300mm	94.32	124.97	2.61	56.96	nr	**181.94**
Sum of two sides 2350mm	97.21	128.80	2.61	56.96	nr	**185.77**
Sum of two sides 2400mm	100.10	132.63	2.88	62.85	nr	**195.49**
Sum of two sides 2450mm	103.00	136.47	2.88	62.85	nr	**199.33**
Sum of two sides 2500mm	105.89	140.30	2.88	62.85	nr	**203.16**
Sum of two sides 2550mm	108.78	144.13	2.88	62.85	nr	**206.99**
Sum of two sides 2600mm	111.68	147.98	2.88	62.85	nr	**210.83**
Sum of two sides 2650mm	114.57	151.81	2.88	62.85	nr	**214.66**
Sum of two sides 2700mm	117.46	155.63	2.88	62.85	nr	**218.49**
Sum of two sides 2750mm	120.35	159.46	2.88	62.85	nr	**222.32**
Sum of two sides 2800mm	123.24	163.29	3.94	85.99	nr	**249.28**
Sum of two sides 2850mm	126.20	167.22	3.94	85.99	nr	**253.20**
Sum of two sides 2900mm	129.16	171.14	3.94	85.99	nr	**257.13**
Sum of two sides 2950mm	132.05	174.97	3.94	85.99	nr	**260.96**
Sum of two sides 3000mm	134.94	178.80	4.83	105.41	nr	**284.21**
Sum of two sides 3050mm	137.22	181.82	4.83	105.41	nr	**287.23**
Sum of two sides 3100mm	139.51	184.85	4.83	105.41	nr	**290.26**
Sum of two sides 3150mm	141.79	187.87	4.83	105.41	nr	**293.28**
Sum of two sides 3200mm	144.08	190.91	4.83	105.41	nr	**296.32**
Grille neck						
Sum of two sides 1700mm	58.33	77.29	1.86	40.59	nr	**117.88**
Sum of two sides 1750mm	61.29	81.21	1.86	40.59	nr	**121.80**
Sum of two sides 1800mm	64.26	85.14	1.86	40.59	nr	**125.74**
Sum of two sides 1850mm	67.23	89.08	1.86	40.59	nr	**129.67**
Sum of two sides 1900mm	70.20	93.02	2.02	44.09	nr	**137.10**
Sum of two sides 1950mm	73.17	96.95	2.02	44.09	nr	**141.04**
Sum of two sides 2000mm	76.13	100.87	2.02	44.09	nr	**144.96**
Sum of two sides 2050mm	79.10	104.81	2.02	44.09	nr	**148.89**
Sum of two sides 2100mm	82.07	108.74	2.80	61.11	nr	**169.85**
Sum of two sides 2150mm	85.52	113.31	2.80	61.11	nr	**174.42**
Sum of two sides 2200mm	88.00	116.60	2.80	61.11	nr	**177.71**
Sum of two sides 2250mm	90.97	120.54	2.80	61.11	nr	**181.64**
Sum of two sides 2300mm	93.94	124.47	2.80	61.11	nr	**185.58**
Sum of two sides 2350mm	96.90	128.39	2.80	61.11	nr	**189.50**
Sum of two sides 2400mm	99.87	132.33	3.06	66.78	nr	**199.11**
Sum of two sides 2450mm	102.84	136.26	3.06	66.78	nr	**203.05**
Sum of two sides 2500mm	105.81	140.20	3.06	66.78	nr	**206.98**
Sum of two sides 2550mm	108.77	144.12	3.06	66.78	nr	**210.90**
Sum of two sides 2600mm	111.74	148.06	3.08	67.22	nr	**215.28**
Sum of two sides 2650mm	114.10	151.18	3.08	67.22	nr	**218.40**

Material Costs/Measured Work Prices - Mechanical Installations

U:VENTILATION/AIR CONDITIONING SYSTEMS

Item	Net Price £	Material £	Labour hours	Labour £	Unit	Total rate £
Sum of two sides 2700mm	117.68	155.93	3.08	67.22	nr	223.15
Sum of two sides 2750mm	120.64	159.85	3.08	67.22	nr	227.07
Sum of two sides 2800mm	123.61	163.78	4.13	90.14	nr	253.92
Sum of two sides 2850mm	126.58	167.72	4.13	90.14	nr	257.85
Sum of two sides 2900mm	129.55	171.65	4.13	90.14	nr	261.79
Sum of two sides 2950mm	132.51	175.58	4.13	90.14	nr	265.71
Sum of two sides 3000mm	135.48	179.51	5.02	109.56	nr	289.07
Sum of two sides 3050mm	137.76	182.53	5.02	109.56	nr	292.09
Sum of two sides 3100mm	140.04	185.55	5.02	109.56	nr	295.11
Sum of two sides 3150mm	142.33	188.59	5.02	109.56	nr	298.15
Sum of two sides 3200mm	144.61	191.61	5.02	109.56	nr	301.17
Ductwork 1601 to 2000mm longest side						
Sum of two sides 2100mm	96.83	128.30	2.53	55.22	m	183.52
Sum of two sides 2150mm	98.52	130.54	2.53	55.22	m	185.76
Sum of two sides 2200mm	100.20	132.76	2.55	55.65	m	188.42
Sum of two sides 2250mm	101.89	135.00	2.55	55.65	m	190.66
Sum of two sides 2300mm	103.57	137.23	2.55	55.65	m	192.88
Sum of two sides 2350mm	105.15	139.32	2.56	55.87	m	195.19
Sum of two sides 2400mm	106.74	141.43	2.56	55.87	m	197.30
Sum of two sides 2450mm	108.42	143.66	2.76	60.24	m	203.89
Sum of two sides 2500mm	110.11	145.90	2.76	60.24	m	206.13
Sum of two sides 2550mm	111.70	148.00	2.77	60.45	m	208.46
Sum of two sides 2600mm	113.29	150.11	2.77	60.45	m	210.56
Sum of two sides 2650mm	114.88	152.22	2.97	64.82	m	217.04
Sum of two sides 2700mm	116.47	154.32	2.97	64.82	m	219.14
Sum of two sides 2750mm	118.05	156.42	2.99	65.26	m	221.67
Sum of two sides 2800mm	119.64	158.52	2.99	65.26	m	223.78
Sum of two sides 2850mm	121.28	160.70	3.30	72.02	m	232.72
Sum of two sides 2900mm	122.92	162.87	3.30	72.02	m	234.89
Sum of two sides 2950mm	124.55	165.03	3.31	72.24	m	237.27
Sum of two sides 3000mm	126.19	167.20	3.31	72.24	m	239.44
Sum of two sides 3050mm	129.01	170.94	3.53	77.04	m	247.98
Sum of two sides 3100mm	131.83	174.67	3.53	77.04	m	251.72
Sum of two sides 3150mm	134.66	178.42	3.53	77.04	m	255.47
Sum of two sides 3200mm	137.48	182.16	3.53	77.04	m	259.20
Sum of two sides 3250mm	139.12	184.33	3.53	77.04	m	261.37
Sum of two sides 3300mm	140.75	186.49	3.53	77.04	m	263.53
Sum of two sides 3350mm	142.39	188.67	3.53	77.04	m	265.71
Sum of two sides 3400mm	144.03	190.84	3.55	77.48	m	268.32
Sum of two sides 3450mm	145.62	192.95	3.55	77.48	m	270.42
Sum of two sides 3500mm	147.21	195.05	3.55	77.48	m	272.53
Sum of two sides 3550mm	148.80	197.16	3.55	77.48	m	274.64
Sum of two sides 3600mm	150.38	199.25	3.55	77.48	m	276.73
Sum of two sides 3650mm	152.02	201.43	3.55	77.48	m	278.90
Sum of two sides 3700mm	153.66	203.60	3.56	77.70	m	281.30
Sum of two sides 3750mm	155.29	205.76	3.56	77.70	m	283.45
Sum of two sides 3800mm	156.93	207.93	3.56	77.70	m	285.63
Sum of two sides 3850mm	158.52	210.04	3.56	77.70	m	287.73
Sum of two sides 3900mm	160.11	212.15	3.56	77.70	m	289.84
Sum of two sides 3950mm	161.70	214.25	3.56	77.70	m	291.95
Sum of two sides 4000mm	163.29	216.36	3.56	77.70	m	294.05

U:VENTILATION/AIR CONDITIONING SYSTEMS

Item	Net Price £	Material £	Labour hours	Labour £	Unit	Total rate £
U10 : DUCTWORK : RECTANGULAR - CLASS C						
Y30 - AIR DUCTLINES						
Galvanised sheet metal DW144 class C rectangular section ductwork; including all necessary stiffeners, joints, couplers in the running length and duct supports (Continued)						
Extra over fittings; Ductwork 1601 to 2000mm longest side						
End Cap						
Sum of two sides 2100mm	46.96	62.22	0.87	18.99	nr	**81.21**
Sum of two sides 2150mm	48.46	64.21	0.87	18.99	nr	**83.20**
Sum of two sides 2200mm	49.96	66.20	0.87	18.99	nr	**85.18**
Sum of two sides 2250mm	51.46	68.18	0.87	18.99	nr	**87.17**
Sum of two sides 2300mm	52.96	70.17	0.87	18.99	nr	**89.16**
Sum of two sides 2350mm	54.45	72.15	0.87	18.99	nr	**91.13**
Sum of two sides 2400mm	55.95	74.13	0.87	18.99	nr	**93.12**
Sum of two sides 2450mm	57.44	76.11	0.87	18.99	nr	**95.10**
Sum of two sides 2500mm	58.94	78.10	0.87	18.99	nr	**97.08**
Sum of two sides 2550mm	60.43	80.07	0.87	18.99	nr	**99.06**
Sum of two sides 2600mm	61.93	82.06	0.87	18.99	nr	**101.04**
Sum of two sides 2650mm	63.43	84.04	0.87	18.99	nr	**103.03**
Sum of two sides 2700mm	64.92	86.02	0.87	18.99	nr	**105.01**
Sum of two sides 2750mm	66.42	88.01	0.87	18.99	nr	**106.99**
Sum of two sides 2800mm	67.92	89.99	1.16	25.32	nr	**115.31**
Sum of two sides 2850mm	69.41	91.97	1.16	25.32	nr	**117.28**
Sum of two sides 2900mm	70.91	93.96	1.16	25.32	nr	**119.27**
Sum of two sides 2950mm	72.40	95.93	1.16	25.32	nr	**121.25**
Sum of two sides 3000mm	73.90	97.92	1.73	37.76	nr	**135.67**
Sum of two sides 3050mm	75.17	99.60	1.73	37.76	nr	**137.36**
Sum of two sides 3100mm	76.44	101.28	1.73	37.76	nr	**139.04**
Sum of two sides 3150mm	77.70	102.95	1.73	37.76	nr	**140.71**
Sum of two sides 3200mm	78.97	104.64	1.73	37.76	nr	**142.39**
Sum of two sides 3250mm	80.21	106.28	1.73	37.76	nr	**144.03**
Sum of two sides 3300mm	81.44	107.91	1.73	37.76	nr	**145.66**
Sum of two sides 3350mm	82.67	109.54	1.73	37.76	nr	**147.29**
Sum of two sides 3400mm	82.91	109.86	1.73	37.76	nr	**147.61**
Sum of two sides 3450mm	85.14	112.81	1.73	37.76	nr	**150.57**
Sum of two sides 3500mm	86.38	114.45	1.73	37.76	nr	**152.21**
Sum of two sides 3550mm	87.61	116.08	1.73	37.76	nr	**153.84**
Sum of two sides 3600mm	88.84	117.71	1.73	37.76	nr	**155.47**
Sum of two sides 3650mm	90.08	119.36	1.73	37.76	nr	**157.11**
Sum of two sides 3700mm	91.31	120.99	1.73	37.76	nr	**158.74**
Sum of two sides 3750mm	92.54	122.62	1.73	37.76	nr	**160.37**
Sum of two sides 3800mm	93.78	124.26	1.73	37.76	nr	**162.02**
Sum of two sides 3850mm	95.01	125.89	1.73	37.76	nr	**163.64**
Sum of two sides 3900mm	96.24	127.52	1.73	37.76	nr	**165.27**
Sum of two sides 3950mm	97.48	129.16	1.73	37.76	nr	**166.92**
Sum of two sides 4000mm	98.71	130.79	1.73	37.76	nr	**168.55**

Material Costs/Measured Work Prices - Mechanical Installations   363

U:VENTILATION/AIR CONDITIONING SYSTEMS

Item	Net Price £	Material £	Labour hours	Labour £	Unit	Total rate £
Reducer						
Sum of two sides 2100mm	29.25	38.76	2.92	63.73	nr	**102.48**
Sum of two sides 2150mm	31.21	41.35	2.92	63.73	nr	**105.08**
Sum of two sides 2200mm	33.17	43.95	2.92	63.73	nr	**107.68**
Sum of two sides 2250mm	35.13	46.55	2.92	63.73	nr	**110.28**
Sum of two sides 2300mm	37.08	49.13	2.92	63.73	nr	**112.86**
Sum of two sides 2350mm	38.95	51.61	2.92	63.73	nr	**115.34**
Sum of two sides 2400mm	40.82	54.09	3.12	68.09	nr	**122.18**
Sum of two sides 2450mm	42.77	56.67	3.12	68.09	nr	**124.76**
Sum of two sides 2500mm	44.73	59.27	3.12	68.09	nr	**127.36**
Sum of two sides 2550mm	46.64	61.80	3.12	68.09	nr	**129.89**
Sum of two sides 2600mm	48.54	64.32	3.16	68.97	nr	**133.28**
Sum of two sides 2650mm	50.46	66.86	3.16	68.97	nr	**135.83**
Sum of two sides 2700mm	52.37	69.39	3.16	68.97	nr	**138.36**
Sum of two sides 2750mm	54.28	71.92	3.16	68.97	nr	**140.89**
Sum of two sides 2800mm	56.19	74.45	4.00	87.30	nr	**161.75**
Sum of two sides 2850mm	58.12	77.01	4.00	87.30	nr	**164.31**
Sum of two sides 2900mm	60.06	79.58	4.01	87.52	nr	**167.10**
Sum of two sides 2950mm	61.99	82.14	4.01	87.52	nr	**169.65**
Sum of two sides 3000mm	63.92	84.69	4.01	87.52	nr	**172.21**
Sum of two sides 3050mm	64.39	85.32	4.01	87.52	nr	**172.83**
Sum of two sides 3100mm	64.86	85.94	4.01	87.52	nr	**173.46**
Sum of two sides 3150mm	65.33	86.56	4.01	87.52	nr	**174.08**
Sum of two sides 3200mm	65.79	87.17	4.56	99.52	nr	**186.69**
Sum of two sides 3250mm	67.20	89.04	4.56	99.52	nr	**188.56**
Sum of two sides 3300mm	68.61	90.91	4.56	99.52	nr	**190.43**
Sum of two sides 3350mm	70.01	92.76	4.56	99.52	nr	**192.28**
Sum of two sides 3400mm	71.42	94.63	4.56	99.52	nr	**194.15**
Sum of two sides 3450mm	72.81	96.47	4.56	99.52	nr	**195.99**
Sum of two sides 3500mm	74.19	98.30	4.56	99.52	nr	**197.82**
Sum of two sides 3550mm	75.58	100.14	4.56	99.52	nr	**199.66**
Sum of two sides 3600mm	76.96	101.97	4.56	99.52	nr	**201.49**
Sum of two sides 3650mm	78.37	103.84	4.56	99.52	nr	**203.36**
Sum of two sides 3700mm	79.78	105.71	4.56	99.52	nr	**205.23**
Sum of two sides 3750mm	81.19	107.58	4.56	99.52	nr	**207.10**
Sum of two sides 3800mm	82.60	109.44	4.56	99.52	nr	**208.97**
Sum of two sides 3850mm	83.98	111.27	4.56	99.52	nr	**210.79**
Sum of two sides 3900mm	85.37	113.12	4.56	99.52	nr	**212.64**
Sum of two sides 3950mm	86.75	114.94	4.56	99.52	nr	**214.46**
Sum of two sides 4000mm	88.14	116.79	4.56	99.52	nr	**216.31**
Offset						
Sum of two sides 2100mm	90.31	119.66	2.92	63.73	nr	**183.39**
Sum of two sides 2150mm	92.09	122.02	2.92	63.73	nr	**185.75**
Sum of two sides 2200mm	93.87	124.38	3.26	71.15	nr	**195.53**
Sum of two sides 2250mm	95.65	126.74	3.26	71.15	nr	**197.88**
Sum of two sides 2300mm	97.43	129.09	3.26	71.15	nr	**200.24**
Sum of two sides 2350mm	102.52	135.84	3.26	71.15	nr	**206.99**
Sum of two sides 2400mm	107.62	142.60	3.47	75.73	nr	**218.33**
Sum of two sides 2450mm	109.32	144.85	3.47	75.73	nr	**220.58**
Sum of two sides 2500mm	111.02	147.10	3.48	75.95	nr	**223.05**
Sum of two sides 2550mm	112.59	149.18	3.48	75.95	nr	**225.13**
Sum of two sides 2600mm	114.16	151.26	3.49	76.17	nr	**227.43**
Sum of two sides 2650mm	115.61	153.18	3.49	76.17	nr	**229.35**

Material Costs/Measured Work Prices - Mechanical Installations

U:VENTILATION/AIR CONDITIONING SYSTEMS

Item	Net Price £	Material £	Labour hours	Labour £	Unit	Total rate £
U10 : DUCTWORK : RECTANGULAR - CLASS C						
Y30 - AIR DUCTLINES						
Galvanised sheet metal DW144 class C rectangular section ductwork; including all necessary stiffeners, joints, couplers in the running length and duct supports (Continued)						
Offset						
Sum of two sides 2700mm	117.06	155.10	3.50	76.39	nr	**231.49**
Sum of two sides 2750mm	118.51	157.03	3.50	76.39	nr	**233.41**
Sum of two sides 2800mm	119.96	158.95	4.33	94.50	nr	**253.45**
Sum of two sides 2850mm	121.25	160.66	4.33	94.50	nr	**255.16**
Sum of two sides 2900mm	122.55	162.38	4.74	103.45	nr	**265.83**
Sum of two sides 2950mm	123.84	164.09	4.74	103.45	nr	**267.54**
Sum of two sides 3000mm	125.14	165.81	5.31	115.89	nr	**281.70**
Sum of two sides 3050mm	125.35	166.09	5.31	115.89	nr	**281.98**
Sum of two sides 3100mm	125.57	166.38	5.34	116.54	nr	**282.92**
Sum of two sides 3150mm	125.78	166.66	5.34	116.54	nr	**283.20**
Sum of two sides 3200mm	126.00	166.95	5.35	116.76	nr	**283.71**
Sum of two sides 3250mm	128.38	170.10	5.35	116.76	nr	**286.87**
Sum of two sides 3300mm	130.76	173.26	5.35	116.76	nr	**290.02**
Sum of two sides 3350mm	133.15	176.42	5.35	116.76	nr	**293.19**
Sum of two sides 3400mm	135.53	179.58	5.35	116.76	nr	**296.34**
Sum of two sides 3450mm	137.91	182.73	5.35	116.76	nr	**299.49**
Sum of two sides 3500mm	140.30	185.90	5.35	116.76	nr	**302.66**
Sum of two sides 3550mm	142.68	189.05	5.35	116.76	nr	**305.81**
Sum of two sides 3600mm	145.07	192.22	5.35	116.76	nr	**308.98**
Sum of two sides 3650mm	147.45	195.37	5.35	116.76	nr	**312.13**
Sum of two sides 3700mm	149.83	198.52	5.35	116.76	nr	**315.29**
Sum of two sides 3750mm	152.22	201.69	5.35	116.76	nr	**318.45**
Sum of two sides 3800mm	154.60	204.84	5.35	116.76	nr	**321.61**
Sum of two sides 3850mm	156.98	208.00	5.35	116.76	nr	**324.76**
Sum of two sides 3900mm	159.37	211.17	5.35	116.76	nr	**327.93**
Sum of two sides 3950mm	161.75	214.32	5.35	116.76	nr	**331.08**
Sum of two sides 4000mm	164.14	217.49	5.35	116.76	nr	**334.25**
Square to round						
Sum of two sides 2100mm	57.13	75.70	4.26	92.97	nr	**168.67**
Sum of two sides 2150mm	60.22	79.79	4.26	92.97	nr	**172.76**
Sum of two sides 2200mm	63.30	83.87	4.27	93.19	nr	**177.06**
Sum of two sides 2250mm	66.39	87.97	4.27	93.19	nr	**181.16**
Sum of two sides 2300mm	69.47	92.05	4.30	93.85	nr	**185.89**
Sum of two sides 2350mm	72.58	96.17	4.30	93.85	nr	**190.01**
Sum of two sides 2400mm	127.75	169.27	4.77	104.10	nr	**273.37**
Sum of two sides 2450mm	78.78	104.38	4.77	104.10	nr	**208.49**
Sum of two sides 2500mm	81.47	107.95	4.79	104.54	nr	**212.49**
Sum of two sides 2550mm	84.97	112.59	4.79	104.54	nr	**217.13**
Sum of two sides 2600mm	88.06	116.68	4.95	108.03	nr	**224.71**
Sum of two sides 2650mm	91.16	120.79	4.95	108.03	nr	**228.82**
Sum of two sides 2700mm	94.26	124.89	4.95	108.03	nr	**232.93**

Material Costs/Measured Work Prices - Mechanical Installations

U:VENTILATION/AIR CONDITIONING SYSTEMS

Item	Net Price £	Material £	Labour hours	Labour £	Unit	Total rate £
Sum of two sides 2750mm	97.36	129.00	4.95	108.03	nr	**237.03**
Sum of two sides 2800mm	100.46	133.11	8.49	185.29	nr	**318.40**
Sum of two sides 2850mm	103.55	137.20	8.49	185.29	nr	**322.49**
Sum of two sides 2900mm	106.64	141.30	8.88	193.80	nr	**335.10**
Sum of two sides 2950mm	109.73	145.39	8.88	193.80	nr	**339.19**
Sum of two sides 3000mm	112.82	149.49	9.02	196.86	nr	**346.34**
Sum of two sides 3050mm	114.25	151.38	9.02	196.86	nr	**348.24**
Sum of two sides 3100mm	115.68	153.28	9.02	196.86	nr	**350.13**
Sum of two sides 3150mm	117.11	155.17	9.02	196.86	nr	**352.03**
Sum of two sides 3200mm	118.54	157.07	9.09	198.39	nr	**355.45**
Sum of two sides 3250mm	120.84	160.11	9.09	198.39	nr	**358.50**
Sum of two sides 3300mm	123.14	163.16	9.09	198.39	nr	**361.55**
Sum of two sides 3350mm	125.44	166.21	9.09	198.39	nr	**364.59**
Sum of two sides 3400mm	127.75	169.27	9.09	198.39	nr	**367.65**
Sum of two sides 3450mm	130.06	172.33	9.09	198.39	nr	**370.72**
Sum of two sides 3500mm	132.37	175.39	9.09	198.39	nr	**373.78**
Sum of two sides 3550mm	134.68	178.45	9.09	198.39	nr	**376.84**
Sum of two sides 3600mm	136.99	181.51	9.09	198.39	nr	**379.90**
Sum of two sides 3650mm	139.30	184.57	9.09	198.39	nr	**382.96**
Sum of two sides 3700mm	141.60	187.62	9.09	198.39	nr	**386.01**
Sum of two sides 3750mm	143.90	190.67	9.09	198.39	nr	**389.05**
Sum of two sides 3800mm	146.21	193.73	9.09	198.39	nr	**392.11**
Sum of two sides 3850mm	148.52	196.79	9.09	198.39	nr	**395.17**
Sum of two sides 3900mm	150.83	199.85	9.09	198.39	nr	**398.24**
Sum of two sides 3950mm	153.14	202.91	9.09	198.39	nr	**401.30**
Sum of two sides 4000mm	155.45	205.97	9.09	198.39	nr	**404.36**
90 degree radius bend						
Sum of two sides 2100mm	130.87	173.40	2.61	56.96	nr	**230.36**
Sum of two sides 2150mm	136.32	180.62	2.61	56.96	nr	**237.59**
Sum of two sides 2200mm	141.77	187.85	2.62	57.18	nr	**245.03**
Sum of two sides 2250mm	147.22	195.07	2.62	57.18	nr	**252.25**
Sum of two sides 2300mm	152.67	202.29	2.63	57.40	nr	**259.69**
Sum of two sides 2350mm	154.22	204.34	2.63	57.40	nr	**261.74**
Sum of two sides 2400mm	155.78	206.41	4.34	94.72	nr	**301.13**
Sum of two sides 2450mm	161.13	213.50	4.34	94.72	nr	**308.22**
Sum of two sides 2500mm	166.49	220.60	4.35	94.94	nr	**315.54**
Sum of two sides 2550mm	171.65	227.44	4.35	94.94	nr	**322.37**
Sum of two sides 2600mm	176.82	234.29	4.53	98.87	nr	**333.15**
Sum of two sides 2650mm	181.98	241.12	4.53	98.87	nr	**339.99**
Sum of two sides 2700mm	187.15	247.97	4.53	98.87	nr	**346.84**
Sum of two sides 2750mm	192.31	254.81	4.53	98.87	nr	**353.68**
Sum of two sides 2800mm	197.48	261.66	7.13	155.61	nr	**417.27**
Sum of two sides 2850mm	202.73	268.62	7.13	155.61	nr	**424.23**
Sum of two sides 2900mm	207.99	275.59	7.17	156.48	nr	**432.07**
Sum of two sides 2950mm	213.25	282.56	7.17	156.48	nr	**439.04**
Sum of two sides 3000mm	218.51	289.53	7.26	158.45	nr	**447.97**
Sum of two sides 3050mm	220.05	291.57	7.26	158.45	nr	**450.01**
Sum of two sides 3100mm	221.59	293.61	7.26	158.45	nr	**452.05**
Sum of two sides 3150mm	223.13	295.65	7.26	158.45	nr	**454.09**
Sum of two sides 3200mm	224.67	297.69	7.31	159.54	nr	**457.23**
Sum of two sides 3250mm	228.35	302.56	7.31	159.54	nr	**462.10**
Sum of two sides 3300mm	232.04	307.45	7.31	159.54	nr	**466.99**
Sum of two sides 3350mm	235.72	312.33	7.31	159.54	nr	**471.87**

Material Costs/Measured Work Prices - Mechanical Installations

U:VENTILATION/AIR CONDITIONING SYSTEMS

Item	Net Price £	Material £	Labour hours	Labour £	Unit	Total rate £
U10 : DUCTWORK : RECTANGULAR - CLASS C						
Y30 - AIR DUCTLINES						
Galvanised sheet metal DW144 class C rectangular section ductwork; including all necessary stiffeners, joints, couplers in the running length and duct supports (Continued)						
90 degree radius bend						
Sum of two sides 3400mm	239.41	317.22	7.31	159.54	nr	**476.76**
Sum of two sides 3450mm	242.99	321.96	7.31	159.54	nr	**481.50**
Sum of two sides 3500mm	246.58	326.72	7.31	159.54	nr	**486.26**
Sum of two sides 3550mm	250.17	331.48	7.31	159.54	nr	**491.01**
Sum of two sides 3600mm	253.76	336.23	7.31	159.54	nr	**495.77**
Sum of two sides 3650mm	257.44	341.11	7.31	159.54	nr	**500.65**
Sum of two sides 3700mm	261.13	346.00	7.31	159.54	nr	**505.54**
Sum of two sides 3750mm	264.81	350.87	7.31	159.54	nr	**510.41**
Sum of two sides 3800mm	268.50	355.76	7.31	159.54	nr	**515.30**
Sum of two sides 3850mm	272.08	360.51	7.31	159.54	nr	**520.04**
Sum of two sides 3900mm	275.67	365.26	7.31	159.54	nr	**524.80**
Sum of two sides 3950mm	279.26	370.02	7.31	159.54	nr	**529.56**
Sum of two sides 4000mm	282.85	374.78	7.31	159.54	nr	**534.31**
45 degree bend						
Sum of two sides 2100mm	88.51	117.28	7.50	163.68	nr	**280.96**
Sum of two sides 2150mm	92.36	122.38	7.50	163.68	nr	**286.06**
Sum of two sides 2200mm	96.22	127.49	7.50	163.68	nr	**291.18**
Sum of two sides 2250mm	100.08	132.61	7.50	163.68	nr	**296.29**
Sum of two sides 2300mm	103.94	137.72	7.55	164.78	nr	**302.50**
Sum of two sides 2350mm	105.73	140.09	7.55	164.78	nr	**304.87**
Sum of two sides 2400mm	107.53	142.48	8.13	177.43	nr	**319.91**
Sum of two sides 2450mm	111.33	147.51	8.13	177.43	nr	**324.95**
Sum of two sides 2500mm	115.14	152.56	8.30	181.14	nr	**333.70**
Sum of two sides 2550mm	118.80	157.41	8.30	181.14	nr	**338.55**
Sum of two sides 2600mm	122.46	162.26	8.56	186.82	nr	**349.08**
Sum of two sides 2650mm	126.11	167.10	8.56	186.82	nr	**353.91**
Sum of two sides 2700mm	129.77	171.95	8.62	188.13	nr	**360.07**
Sum of two sides 2750mm	133.43	176.79	8.62	188.13	nr	**364.92**
Sum of two sides 2800mm	137.09	181.64	9.09	198.39	nr	**380.03**
Sum of two sides 2850mm	140.82	186.59	9.09	198.39	nr	**384.97**
Sum of two sides 2900mm	144.55	191.53	9.09	198.39	nr	**389.91**
Sum of two sides 2950mm	148.29	196.48	9.09	198.39	nr	**394.87**
Sum of two sides 3000mm	152.02	201.43	9.62	209.95	nr	**411.38**
Sum of two sides 3050mm	153.58	203.49	9.62	209.95	nr	**413.45**
Sum of two sides 3100mm	155.13	205.55	9.62	209.95	nr	**415.50**
Sum of two sides 3150mm	156.69	207.61	9.62	209.95	nr	**417.57**
Sum of two sides 3200mm	158.24	209.67	9.62	209.95	nr	**419.62**
Sum of two sides 3250mm	160.92	213.22	9.62	209.95	nr	**423.17**
Sum of two sides 3300mm	163.61	216.78	9.62	209.95	nr	**426.74**
Sum of two sides 3350mm	166.29	220.33	9.62	209.95	nr	**430.29**
Sum of two sides 3400mm	168.97	223.89	9.62	209.95	nr	**433.84**

Material Costs/Measured Work Prices - Mechanical Installations

U:VENTILATION/AIR CONDITIONING SYSTEMS

Item	Net Price £	Material £	Labour hours	Labour £	Unit	Total rate £
Sum of two sides 3450mm	171.58	227.34	9.62	209.95	nr	**437.30**
Sum of two sides 3500mm	174.19	230.80	9.62	209.95	nr	**440.75**
Sum of two sides 3550mm	176.79	234.25	9.62	209.95	nr	**444.20**
Sum of two sides 3600mm	179.40	237.71	9.62	209.95	nr	**447.66**
Sum of two sides 3650mm	182.08	241.26	9.62	209.95	nr	**451.21**
Sum of two sides 3700mm	184.76	244.81	9.62	209.95	nr	**454.76**
Sum of two sides 3750mm	187.44	248.36	9.62	209.95	nr	**458.31**
Sum of two sides 3800mm	190.13	251.92	9.62	209.95	nr	**461.87**
Sum of two sides 3850mm	192.73	255.37	9.62	209.95	nr	**465.32**
Sum of two sides 3900mm	195.34	258.83	9.62	209.95	nr	**468.78**
Sum of two sides 3950mm	197.94	262.27	9.62	209.95	nr	**472.22**
Sum of two sides 4000mm	200.55	265.73	9.62	209.95	nr	**475.68**
90 degree mitre bend						
Sum of two sides 2100mm	149.24	197.74	7.50	163.68	nr	**361.43**
Sum of two sides 2150mm	155.91	206.58	7.50	163.68	nr	**370.27**
Sum of two sides 2200mm	162.57	215.41	7.50	163.68	nr	**379.09**
Sum of two sides 2250mm	169.24	224.24	7.50	163.68	nr	**387.93**
Sum of two sides 2300mm	175.90	233.07	7.55	164.78	nr	**397.84**
Sum of two sides 2350mm	178.67	236.74	7.55	164.78	nr	**401.51**
Sum of two sides 2400mm	181.44	240.41	8.13	177.43	nr	**417.84**
Sum of two sides 2450mm	188.01	249.11	8.13	177.43	nr	**426.55**
Sum of two sides 2500mm	194.58	257.82	8.30	181.14	nr	**438.96**
Sum of two sides 2550mm	200.96	266.27	8.30	181.14	nr	**447.42**
Sum of two sides 2600mm	207.34	274.73	8.56	186.82	nr	**461.54**
Sum of two sides 2650mm	213.72	283.18	8.56	186.82	nr	**470.00**
Sum of two sides 2700mm	220.10	291.63	8.62	188.13	nr	**479.76**
Sum of two sides 2750mm	226.48	300.09	8.62	188.13	nr	**488.21**
Sum of two sides 2800mm	232.86	308.54	15.20	331.73	nr	**640.27**
Sum of two sides 2850mm	239.33	317.11	15.20	331.73	nr	**648.85**
Sum of two sides 2900mm	245.80	325.69	15.20	331.73	nr	**657.42**
Sum of two sides 2950mm	252.28	334.27	15.20	331.73	nr	**666.00**
Sum of two sides 3000mm	258.75	342.84	15.60	340.46	nr	**683.31**
Sum of two sides 3050mm	261.51	346.50	15.60	340.46	nr	**686.96**
Sum of two sides 3100mm	264.26	350.14	16.04	350.07	nr	**700.21**
Sum of two sides 3150mm	267.02	353.80	16.04	350.07	nr	**703.87**
Sum of two sides 3200mm	269.77	357.45	16.04	350.07	nr	**707.51**
Sum of two sides 3250mm	274.67	363.94	16.04	350.07	nr	**714.00**
Sum of two sides 3300mm	279.57	370.43	16.04	350.07	nr	**720.50**
Sum of two sides 3350mm	284.47	376.92	16.04	350.07	nr	**726.99**
Sum of two sides 3400mm	289.37	383.42	16.04	350.07	nr	**733.48**
Sum of two sides 3450mm	294.17	389.78	16.04	350.07	nr	**739.84**
Sum of two sides 3500mm	298.97	396.14	16.04	350.07	nr	**746.20**
Sum of two sides 3550mm	303.78	402.51	16.04	350.07	nr	**752.58**
Sum of two sides 3600mm	308.58	408.87	16.04	350.07	nr	**758.94**
Sum of two sides 3650mm	313.48	415.36	16.04	350.07	nr	**765.43**
Sum of two sides 3700mm	318.38	421.85	16.04	350.07	nr	**771.92**
Sum of two sides 3750mm	323.28	428.35	16.04	350.07	nr	**778.41**
Sum of two sides 3800mm	328.18	434.84	16.04	350.07	nr	**784.91**
Sum of two sides 3850mm	332.98	441.20	16.04	350.07	nr	**791.27**
Sum of two sides 3900mm	337.78	447.56	16.04	350.07	nr	**797.63**
Sum of two sides 3950mm	342.59	453.93	16.04	350.07	nr	**804.00**
Sum of two sides 4000mm	347.46	460.38	16.04	350.07	nr	**810.45**

Material Costs/Measured Work Prices - Mechanical Installations

U:VENTILATION/AIR CONDITIONING SYSTEMS

Item	Net Price £	Material £	Labour hours	Labour £	Unit	Total rate £
U10 : DUCTWORK : RECTANGULAR - CLASS C						
Y30 - AIR DUCTLINES						
Galvanised sheet metal DW144 class C rectangular section ductwork; including all necessary stiffeners, joints, couplers in the running length and duct supports (Continued)						
Branch						
Sum of two sides 2100mm	83.67	110.86	2.61	56.96	nr	**167.82**
Sum of two sides 2150mm	86.64	114.80	2.61	56.96	nr	**171.76**
Sum of two sides 2200mm	89.62	118.75	2.61	56.96	nr	**175.71**
Sum of two sides 2250mm	92.59	122.68	2.61	56.96	nr	**179.64**
Sum of two sides 2300mm	95.57	126.63	2.61	56.96	nr	**183.59**
Sum of two sides 2350mm	98.42	130.41	2.61	56.96	nr	**187.37**
Sum of two sides 2400mm	101.26	134.17	2.88	62.85	nr	**197.02**
Sum of two sides 2450mm	104.24	138.12	2.88	62.85	nr	**200.97**
Sum of two sides 2500mm	107.21	142.05	2.88	62.85	nr	**204.91**
Sum of two sides 2550mm	110.12	145.91	2.88	62.85	nr	**208.76**
Sum of two sides 2600mm	113.04	149.78	2.88	62.85	nr	**212.63**
Sum of two sides 2650mm	115.95	153.63	2.88	62.85	nr	**216.49**
Sum of two sides 2700mm	118.86	157.49	2.88	62.85	nr	**220.34**
Sum of two sides 2750mm	121.77	161.35	2.88	62.85	nr	**224.20**
Sum of two sides 2800mm	124.68	165.20	3.94	85.99	nr	**251.19**
Sum of two sides 2850mm	127.63	169.11	3.94	85.99	nr	**255.10**
Sum of two sides 2900mm	130.57	173.01	3.94	85.99	nr	**258.99**
Sum of two sides 2950mm	133.51	176.90	3.94	85.99	nr	**262.89**
Sum of two sides 3000mm	136.46	180.81	4.83	105.41	nr	**286.22**
Sum of two sides 3050mm	138.76	183.86	4.83	105.41	nr	**289.27**
Sum of two sides 3100mm	141.06	186.90	4.83	105.41	nr	**292.32**
Sum of two sides 3150mm	143.36	189.95	4.83	105.41	nr	**295.36**
Sum of two sides 3200mm	145.67	193.01	4.83	105.41	nr	**298.43**
Sum of two sides 3250mm	147.91	195.98	4.83	105.41	nr	**301.39**
Sum of two sides 3300mm	150.15	198.95	4.83	105.41	nr	**304.36**
Sum of two sides 3350mm	152.40	201.93	4.83	105.41	nr	**307.34**
Sum of two sides 3400mm	154.64	204.90	4.83	105.41	nr	**310.31**
Sum of two sides 3450mm	156.85	207.83	4.83	105.41	nr	**313.24**
Sum of two sides 3500mm	159.06	210.75	4.83	105.41	nr	**316.17**
Sum of two sides 3550mm	161.27	213.68	4.83	105.41	nr	**319.10**
Sum of two sides 3600mm	163.48	216.61	4.83	105.41	nr	**322.02**
Sum of two sides 3650mm	165.72	219.58	4.83	105.41	nr	**324.99**
Sum of two sides 3700mm	167.97	222.56	4.83	105.41	nr	**327.97**
Sum of two sides 3750mm	170.21	225.53	4.83	105.41	nr	**330.94**
Sum of two sides 3800mm	172.45	228.50	4.83	105.41	nr	**333.91**
Sum of two sides 3850mm	174.66	231.42	4.83	105.41	nr	**336.84**
Sum of two sides 3900mm	176.88	234.37	4.83	105.41	nr	**339.78**
Sum of two sides 3950mm	179.09	237.29	4.83	105.41	nr	**342.71**
Sum of two sides 4000mm	181.30	240.22	4.83	105.41	nr	**345.64**

Material Costs/Measured Work Prices - Mechanical Installations   369

U:VENTILATION/AIR CONDITIONING SYSTEMS

Item	Net Price £	Material £	Labour hours	Labour £	Unit	Total rate £
Grille neck						
Sum of two sides 2100mm	83.42	110.53	2.80	61.11	nr	**171.64**
Sum of two sides 2150mm	86.41	114.49	2.80	61.11	nr	**175.60**
Sum of two sides 2200mm	89.41	118.47	2.80	61.11	nr	**179.58**
Sum of two sides 2250mm	92.41	122.44	2.80	61.11	nr	**183.55**
Sum of two sides 2300mm	95.41	126.42	2.80	61.11	nr	**187.53**
Sum of two sides 2350mm	98.41	130.39	2.80	61.11	nr	**191.50**
Sum of two sides 2400mm	101.41	134.37	3.06	66.78	nr	**201.15**
Sum of two sides 2450mm	104.41	138.34	3.06	66.78	nr	**205.13**
Sum of two sides 2500mm	107.41	142.32	3.06	66.78	nr	**209.10**
Sum of two sides 2550mm	110.12	145.91	3.06	66.78	nr	**212.69**
Sum of two sides 2600mm	113.41	150.27	3.08	67.22	nr	**217.49**
Sum of two sides 2650mm	116.41	154.24	3.08	67.22	nr	**221.46**
Sum of two sides 2700mm	118.86	157.49	3.08	67.22	nr	**224.71**
Sum of two sides 2750mm	122.40	162.18	3.08	67.22	nr	**229.40**
Sum of two sides 2800mm	124.68	165.20	4.13	90.14	nr	**255.34**
Sum of two sides 2850mm	128.40	170.13	4.13	90.14	nr	**260.27**
Sum of two sides 2900mm	130.57	173.01	4.13	90.14	nr	**263.14**
Sum of two sides 2950mm	133.51	176.90	4.13	90.14	nr	**267.04**
Sum of two sides 3000mm	137.40	182.06	5.02	109.56	nr	**291.61**
Sum of two sides 3050mm	139.72	185.13	5.02	109.56	nr	**294.69**
Sum of two sides 3100mm	142.03	188.19	5.02	109.56	nr	**297.75**
Sum of two sides 3150mm	144.35	191.26	5.02	109.56	nr	**300.82**
Sum of two sides 3200mm	146.66	194.32	5.02	109.56	nr	**303.88**
Sum of two sides 3250mm	148.87	197.25	5.02	109.56	nr	**306.81**
Sum of two sides 3300mm	151.08	200.18	5.02	109.56	nr	**309.74**
Sum of two sides 3350mm	153.29	203.11	5.02	109.56	nr	**312.67**
Sum of two sides 3400mm	155.50	206.04	5.02	109.56	nr	**315.60**
Sum of two sides 3450mm	157.72	208.98	5.02	109.56	nr	**318.54**
Sum of two sides 3500mm	159.93	211.91	5.02	109.56	nr	**321.47**
Sum of two sides 3550mm	162.14	214.84	5.02	109.56	nr	**324.39**
Sum of two sides 3600mm	164.35	217.76	5.02	109.56	nr	**327.32**
Sum of two sides 3650mm	166.56	220.69	5.02	109.56	nr	**330.25**
Sum of two sides 3700mm	168.77	223.62	5.02	109.56	nr	**333.18**
Sum of two sides 3750mm	170.99	226.56	5.02	109.56	nr	**336.12**
Sum of two sides 3800mm	173.20	229.49	5.02	109.56	nr	**339.05**
Sum of two sides 3850mm	175.41	232.42	5.02	109.56	nr	**341.98**
Sum of two sides 3900mm	177.62	235.35	5.02	109.56	nr	**344.91**
Sum of two sides 3950mm	179.83	238.27	5.02	109.56	nr	**347.83**
Sum of two sides 4000mm	182.05	241.22	5.02	109.56	nr	**350.78**
Ductwork 2001 to 2500mm longest side						
Sum of two sides 2500mm	165.59	219.41	2.77	60.45	m	**279.86**
Sum of two sides 2550mm	168.21	222.88	2.77	60.45	m	**283.33**
Sum of two sides 2600mm	170.82	226.34	2.97	64.82	m	**291.16**
Sum of two sides 2650mm	173.44	229.81	2.97	64.82	m	**294.63**
Sum of two sides 2700mm	176.05	233.27	2.99	65.26	m	**298.52**
Sum of two sides 2750mm	178.67	236.74	2.99	65.26	m	**301.99**
Sum of two sides 2800mm	181.28	240.20	3.30	72.02	m	**312.22**
Sum of two sides 2850mm	183.73	243.44	3.30	72.02	m	**315.46**
Sum of two sides 2900mm	186.19	246.70	3.31	72.24	m	**318.94**
Sum of two sides 2950mm	188.64	249.95	3.31	72.24	m	**322.19**
Sum of two sides 3000mm	191.09	253.19	3.53	77.04	m	**330.24**
Sum of two sides 3050mm	193.54	256.44	3.53	77.04	m	**333.48**

Material Costs/Measured Work Prices - Mechanical Installations

U:VENTILATION/AIR CONDITIONING SYSTEMS

Item	Net Price £	Material £	Labour hours	Labour £	Unit	Total rate £
U10 : DUCTWORK : RECTANGULAR - CLASS C						
Y30 - AIR DUCTLINES						
Galvanised sheet metal DW144 class C rectangular section ductwork; including all necessary stiffeners, joints, couplers in the running length and duct supports (Continued)						
Ductwork 2001 to 2500mm longest side						
Sum of two sides 3100mm	195.99	259.69	3.55	77.48	m	337.16
Sum of two sides 3150mm	198.45	262.95	3.55	77.48	m	340.42
Sum of two sides 3200mm	200.90	266.19	3.56	77.70	m	343.89
Sum of two sides 3250mm	203.35	269.44	3.56	77.70	m	347.13
Sum of two sides 3300mm	205.80	272.69	3.56	77.70	m	350.38
Sum of two sides 3350mm	208.25	275.93	3.56	77.70	m	353.63
Sum of two sides 3400mm	210.71	279.19	3.56	77.70	m	356.89
Sum of two sides 3450mm	214.41	284.09	3.56	77.70	m	361.79
Sum of two sides 3500mm	218.12	289.01	3.56	77.70	m	366.70
Sum of two sides 3550mm	221.82	293.91	3.56	77.70	m	371.61
Sum of two sides 3600mm	225.53	298.83	3.56	77.70	m	376.52
Sum of two sides 3650mm	227.98	302.07	3.56	77.70	m	379.77
Sum of two sides 3700mm	230.43	305.32	3.56	77.70	m	383.02
Sum of two sides 3750mm	232.88	308.57	3.56	77.70	m	386.26
Sum of two sides 3800mm	235.34	311.83	3.56	77.70	m	389.52
Sum of two sides 3850mm	237.79	315.07	3.56	77.70	m	392.77
Sum of two sides 3900mm	240.24	318.32	3.56	77.70	m	396.01
Sum of two sides 3950mm	242.69	321.56	3.56	77.70	m	399.26
Sum of two sides 4000mm	245.15	324.82	3.56	77.70	m	402.52
Extra over fittings; Ductwork 2001 to 2500mm longest side						
End Cap						
Sum of two sides 2500mm	50.99	67.56	0.87	18.99	nr	86.55
Sum of two sides 2550mm	60.43	80.07	0.87	18.99	nr	99.06
Sum of two sides 2600mm	61.93	82.06	0.87	18.99	nr	101.04
Sum of two sides 2650mm	63.43	84.04	0.87	18.99	nr	103.03
Sum of two sides 2700mm	64.92	86.02	0.87	18.99	nr	105.01
Sum of two sides 2750mm	66.42	88.01	0.87	18.99	nr	106.99
Sum of two sides 2800mm	67.92	89.99	1.16	25.32	nr	115.31
Sum of two sides 2850mm	69.41	91.97	1.16	25.32	nr	117.28
Sum of two sides 2900mm	70.91	93.96	1.16	25.32	nr	119.27
Sum of two sides 2950mm	72.40	95.93	1.16	25.32	nr	121.25
Sum of two sides 3000mm	73.90	97.92	1.73	37.76	nr	135.67
Sum of two sides 3050mm	75.17	99.60	1.73	37.76	nr	137.36
Sum of two sides 3100mm	76.44	101.28	1.73	37.76	nr	139.04
Sum of two sides 3150mm	77.70	102.95	1.73	37.76	nr	140.71
Sum of two sides 3200mm	78.97	104.64	1.73	37.76	nr	142.39
Sum of two sides 3250mm	80.21	106.28	1.73	37.76	nr	144.03
Sum of two sides 3300mm	81.44	107.91	1.73	37.76	nr	145.66
Sum of two sides 3350mm	82.67	109.54	1.73	37.76	nr	147.29

Material Costs/Measured Work Prices - Mechanical Installations

U:VENTILATION/AIR CONDITIONING SYSTEMS

Item	Net Price £	Material £	Labour hours	Labour £	Unit	Total rate £
Sum of two sides 3400mm	83.91	111.18	1.73	37.76	nr	**148.94**
Sum of two sides 3450mm	85.14	112.81	1.73	37.76	nr	**150.57**
Sum of two sides 3500mm	86.38	114.45	1.73	37.76	nr	**152.21**
Sum of two sides 3550mm	87.61	116.08	1.73	37.76	nr	**153.84**
Sum of two sides 3600mm	88.84	117.71	1.73	37.76	nr	**155.47**
Sum of two sides 3650mm	90.08	119.36	1.73	37.76	nr	**157.11**
Sum of two sides 3700mm	91.31	120.99	1.73	37.76	nr	**158.74**
Sum of two sides 3750mm	92.54	122.62	1.73	37.76	nr	**160.37**
Sum of two sides 3800mm	93.78	124.26	1.73	37.76	nr	**162.02**
Sum of two sides 3850mm	95.01	125.89	1.73	37.76	nr	**163.64**
Sum of two sides 3900mm	96.24	127.52	1.73	37.76	nr	**165.27**
Sum of two sides 3950mm	97.48	129.16	1.73	37.76	nr	**166.92**
Sum of two sides 4000mm	98.71	130.79	1.73	37.76	nr	**168.55**
Reducer						
Sum of two sides 2500mm	50.99	67.56	3.12	68.09	nr	**135.65**
Sum of two sides 2550mm	53.25	70.56	3.12	68.09	nr	**138.65**
Sum of two sides 2600mm	55.50	73.54	3.16	68.97	nr	**142.50**
Sum of two sides 2650mm	57.75	76.52	3.16	68.97	nr	**145.48**
Sum of two sides 2700mm	60.00	79.50	3.16	68.97	nr	**148.47**
Sum of two sides 2750mm	62.25	82.48	3.16	68.97	nr	**151.45**
Sum of two sides 2800mm	64.50	85.46	4.00	87.30	nr	**172.76**
Sum of two sides 2850mm	66.75	88.44	4.00	87.30	nr	**175.74**
Sum of two sides 2900mm	69.00	91.42	4.01	87.52	nr	**178.94**
Sum of two sides 2950mm	71.25	94.41	4.01	87.52	nr	**181.92**
Sum of two sides 3000mm	73.50	97.39	4.56	99.52	nr	**196.91**
Sum of two sides 3050mm	75.30	99.77	4.56	99.52	nr	**199.29**
Sum of two sides 3100mm	77.09	102.14	4.56	99.52	nr	**201.66**
Sum of two sides 3150mm	78.88	104.52	4.56	99.52	nr	**204.04**
Sum of two sides 3200mm	80.68	106.90	4.56	99.52	nr	**206.42**
Sum of two sides 3250mm	82.40	109.18	4.56	99.52	nr	**208.70**
Sum of two sides 3300mm	84.13	111.47	4.56	99.52	nr	**210.99**
Sum of two sides 3350mm	85.86	113.76	4.56	99.52	nr	**213.28**
Sum of two sides 3400mm	87.58	116.04	4.56	99.52	nr	**215.56**
Sum of two sides 3450mm	88.30	117.00	4.56	99.52	nr	**216.52**
Sum of two sides 3500mm	89.03	117.96	4.56	99.52	nr	**217.48**
Sum of two sides 3550mm	89.75	118.92	4.56	99.52	nr	**218.44**
Sum of two sides 3600mm	90.47	119.87	4.56	99.52	nr	**219.39**
Sum of two sides 3650mm	92.20	122.17	4.56	99.52	nr	**221.69**
Sum of two sides 3700mm	93.92	124.44	4.56	99.52	nr	**223.96**
Sum of two sides 3750mm	95.65	126.74	4.56	99.52	nr	**226.26**
Sum of two sides 3800mm	97.37	129.02	4.56	99.52	nr	**228.54**
Sum of two sides 3850mm	99.10	131.31	4.56	99.52	nr	**230.83**
Sum of two sides 3900mm	100.82	133.59	4.56	99.52	nr	**233.11**
Sum of two sides 3950mm	102.55	135.88	4.56	99.52	nr	**235.40**
Sum of two sides 4000mm	104.27	138.16	4.56	99.52	nr	**237.68**
Offset						
Sum of two sides 2500mm	125.77	166.65	3.48	75.95	nr	**242.59**
Sum of two sides 2550mm	129.64	171.77	3.48	75.95	nr	**247.72**
Sum of two sides 2600mm	133.51	176.90	3.49	76.17	nr	**253.07**
Sum of two sides 2650mm	137.37	182.02	3.49	76.17	nr	**258.18**
Sum of two sides 2700mm	141.24	187.14	3.50	76.39	nr	**263.53**
Sum of two sides 2750mm	145.11	192.27	3.50	76.39	nr	**268.66**

372   *Material Costs/Measured Work Prices - Mechanical Installations*

U:VENTILATION/AIR CONDITIONING SYSTEMS

Item	Net Price £	Material £	Labour hours	Labour £	Unit	Total rate £
U10 : DUCTWORK : RECTANGULAR - CLASS C						
Y30 - AIR DUCTLINES						
Galvanised sheet metal DW144 class C rectangular section ductwork; including all necessary stiffeners, joints, couplers in the running length and duct supports (Continued)						
Offset						
Sum of two sides 2800mm	148.98	197.40	4.33	94.50	nr	**291.90**
Sum of two sides 2850mm	152.85	202.53	4.33	94.50	nr	**297.03**
Sum of two sides 2900mm	156.72	207.65	4.74	103.45	nr	**311.10**
Sum of two sides 2950mm	160.59	212.78	4.74	103.45	nr	**316.23**
Sum of two sides 3000mm	164.46	217.91	5.31	115.89	nr	**333.80**
Sum of two sides 3050mm	168.33	223.04	5.31	115.89	nr	**338.93**
Sum of two sides 3100mm	172.20	228.16	5.34	116.54	nr	**344.71**
Sum of two sides 3150mm	176.07	233.29	5.34	116.54	nr	**349.84**
Sum of two sides 3200mm	179.94	238.42	5.35	116.76	nr	**355.18**
Sum of two sides 3250mm	183.81	243.55	5.35	116.76	nr	**360.31**
Sum of two sides 3300mm	187.68	248.68	5.35	116.76	nr	**365.44**
Sum of two sides 3350mm	191.55	253.80	5.35	116.76	nr	**370.57**
Sum of two sides 3400mm	195.42	258.93	5.35	116.76	nr	**375.69**
Sum of two sides 3450mm	199.29	264.06	5.35	116.76	nr	**380.82**
Sum of two sides 3500mm	203.16	269.19	5.35	116.76	nr	**385.95**
Sum of two sides 3550mm	207.03	274.31	5.35	116.76	nr	**391.08**
Sum of two sides 3600mm	210.90	279.44	5.35	116.76	nr	**396.20**
Sum of two sides 3650mm	214.77	284.57	5.35	116.76	nr	**401.33**
Sum of two sides 3700mm	218.64	289.70	5.35	116.76	nr	**406.46**
Sum of two sides 3750mm	222.51	294.83	5.35	116.76	nr	**411.59**
Sum of two sides 3800mm	226.38	299.95	5.35	116.76	nr	**416.72**
Sum of two sides 3850mm	230.25	305.08	5.35	116.76	nr	**421.84**
Sum of two sides 3900mm	234.12	310.21	5.35	116.76	nr	**426.97**
Sum of two sides 3950mm	237.99	315.34	5.35	116.76	nr	**432.10**
Sum of two sides 4000mm	241.86	320.46	5.35	116.76	nr	**437.23**
Square to round						
Sum of two sides 2500mm	65.61	86.93	4.79	104.54	nr	**191.47**
Sum of two sides 2550mm	68.54	90.82	4.79	104.54	nr	**195.36**
Sum of two sides 2600mm	71.47	94.70	4.95	108.03	nr	**202.73**
Sum of two sides 2650mm	74.40	98.58	4.95	108.03	nr	**206.61**
Sum of two sides 2700mm	77.32	102.45	4.95	108.03	nr	**210.48**
Sum of two sides 2750mm	80.24	106.32	4.95	108.03	nr	**214.35**
Sum of two sides 2800mm	83.17	110.20	8.49	185.29	nr	**295.49**
Sum of two sides 2850mm	86.09	114.07	8.49	185.29	nr	**299.36**
Sum of two sides 2900mm	89.02	117.95	8.88	193.80	nr	**311.75**
Sum of two sides 2950mm	91.94	121.82	8.88	193.80	nr	**315.62**
Sum of two sides 3000mm	94.86	125.69	9.02	196.86	nr	**322.55**
Sum of two sides 3050mm	97.10	128.66	9.02	196.86	nr	**325.52**
Sum of two sides 3100mm	99.34	131.63	9.02	196.86	nr	**328.48**
Sum of two sides 3150mm	101.58	134.59	9.02	196.86	nr	**331.45**
Sum of two sides 3200mm	103.82	137.56	9.09	198.39	nr	**335.95**

Material Costs/Measured Work Prices - Mechanical Installations

U:VENTILATION/AIR CONDITIONING SYSTEMS

Item	Net Price £	Material £	Labour hours	Labour £	Unit	Total rate £
Sum of two sides 3250mm	105.95	140.38	9.09	198.39	nr	**338.77**
Sum of two sides 3300mm	108.09	143.22	9.09	198.39	nr	**341.60**
Sum of two sides 3350mm	110.23	146.05	9.09	198.39	nr	**344.44**
Sum of two sides 3400mm	112.36	148.88	9.09	198.39	nr	**347.26**
Sum of two sides 3450mm	113.49	150.37	9.09	198.39	nr	**348.76**
Sum of two sides 3500mm	114.63	151.88	9.09	198.39	nr	**350.27**
Sum of two sides 3550mm	115.76	153.38	9.09	198.39	nr	**351.77**
Sum of two sides 3600mm	116.89	154.88	9.09	198.39	nr	**353.26**
Sum of two sides 3650mm	119.03	157.71	9.09	198.39	nr	**356.10**
Sum of two sides 3700mm	121.17	160.55	9.09	198.39	nr	**358.94**
Sum of two sides 3750mm	123.30	163.37	9.09	198.39	nr	**361.76**
Sum of two sides 3800mm	125.44	166.21	9.09	198.39	nr	**364.59**
Sum of two sides 3850mm	127.57	169.03	9.09	198.39	nr	**367.42**
Sum of two sides 3900mm	129.71	171.87	9.09	198.39	nr	**370.25**
Sum of two sides 3950mm	131.84	174.69	9.09	198.39	nr	**373.07**
Sum of two sides 4000mm	133.98	177.52	9.09	198.39	nr	**375.91**
90 degree radius bend						
Sum of two sides 2500mm	206.28	273.32	4.35	94.94	nr	**368.26**
Sum of two sides 2550mm	208.73	276.57	4.35	94.94	nr	**371.50**
Sum of two sides 2600mm	211.18	279.81	4.53	98.87	nr	**378.68**
Sum of two sides 2650mm	213.62	283.05	4.53	98.87	nr	**381.91**
Sum of two sides 2700mm	216.07	286.29	4.53	98.87	nr	**385.16**
Sum of two sides 2750mm	218.52	289.54	4.53	98.87	nr	**388.40**
Sum of two sides 2800mm	220.97	292.79	7.13	155.61	nr	**448.39**
Sum of two sides 2850mm	227.21	301.05	7.13	155.61	nr	**456.66**
Sum of two sides 2900mm	233.46	309.33	7.17	156.48	nr	**465.82**
Sum of two sides 2950mm	239.71	317.62	7.17	156.48	nr	**474.10**
Sum of two sides 3000mm	245.96	325.90	7.26	158.45	nr	**484.34**
Sum of two sides 3050mm	250.83	332.35	7.26	158.45	nr	**490.80**
Sum of two sides 3100mm	255.71	338.82	7.26	158.45	nr	**497.26**
Sum of two sides 3150mm	260.59	345.28	7.26	158.45	nr	**503.73**
Sum of two sides 3200mm	265.47	351.75	7.31	159.54	nr	**511.29**
Sum of two sides 3250mm	270.14	357.94	7.31	159.54	nr	**517.47**
Sum of two sides 3300mm	274.81	364.12	7.31	159.54	nr	**523.66**
Sum of two sides 3350mm	279.48	370.31	7.31	159.54	nr	**529.85**
Sum of two sides 3400mm	284.16	376.51	7.31	159.54	nr	**536.05**
Sum of two sides 3450mm	286.14	379.14	7.31	159.54	nr	**538.67**
Sum of two sides 3500mm	288.13	381.77	7.31	159.54	nr	**541.31**
Sum of two sides 3550mm	290.12	384.41	7.31	159.54	nr	**543.95**
Sum of two sides 3600mm	292.11	387.05	7.31	159.54	nr	**546.58**
Sum of two sides 3650mm	296.78	393.23	7.31	159.54	nr	**552.77**
Sum of two sides 3700mm	301.45	399.42	7.31	159.54	nr	**558.96**
Sum of two sides 3750mm	306.12	405.61	7.31	159.54	nr	**565.15**
Sum of two sides 3800mm	310.79	411.80	7.31	159.54	nr	**571.33**
Sum of two sides 3850mm	315.46	417.98	7.31	159.54	nr	**577.52**
Sum of two sides 3900mm	320.14	424.19	7.31	159.54	nr	**583.72**
Sum of two sides 3950mm	324.81	430.37	7.31	159.54	nr	**589.91**
Sum of two sides 4000mm	329.48	436.56	7.31	159.54	nr	**596.10**
45 degree bend						
Sum of two sides 2500mm	155.13	205.55	8.30	181.14	nr	**386.69**
Sum of two sides 2550mm	157.90	209.22	8.30	181.14	nr	**390.36**
Sum of two sides 2600mm	160.68	212.90	8.56	186.82	nr	**399.72**

374

Material Costs/Measured Work Prices - Mechanical Installations

U:VENTILATION/AIR CONDITIONING SYSTEMS

Item	Net Price £	Material £	Labour hours	Labour £	Unit	Total rate £
U10 : DUCTWORK : RECTANGULAR - CLASS C						
Y30 - AIR DUCTLINES						
Galvanised sheet metal DW144 class C rectangular section ductwork; including all necessary stiffeners, joints, couplers in the running length and duct supports (Continued)						
45 degree bend						
Sum of two sides 2650mm	163.45	216.57	8.56	186.82	nr	**403.39**
Sum of two sides 2700mm	166.22	220.24	8.62	188.13	nr	**408.37**
Sum of two sides 2750mm	168.99	223.91	8.62	188.13	nr	**412.04**
Sum of two sides 2800mm	171.76	227.58	9.09	198.39	nr	**425.97**
Sum of two sides 2850mm	176.43	233.77	9.09	198.39	nr	**432.16**
Sum of two sides 2900mm	181.10	239.96	9.09	198.39	nr	**438.34**
Sum of two sides 2950mm	185.77	246.15	9.09	198.39	nr	**444.53**
Sum of two sides 3000mm	190.44	252.33	9.62	209.95	nr	**462.29**
Sum of two sides 3050mm	194.20	257.31	9.62	209.95	nr	**467.27**
Sum of two sides 3100mm	197.96	262.30	9.62	209.95	nr	**472.25**
Sum of two sides 3150mm	201.72	267.28	9.62	209.95	nr	**477.23**
Sum of two sides 3200mm	205.48	272.26	9.62	209.95	nr	**482.21**
Sum of two sides 3250mm	209.10	277.06	9.62	209.95	nr	**487.01**
Sum of two sides 3300mm	212.72	281.85	9.62	209.95	nr	**491.81**
Sum of two sides 3350mm	216.34	286.65	9.62	209.95	nr	**496.60**
Sum of two sides 3400mm	219.96	291.45	9.62	209.95	nr	**501.40**
Sum of two sides 3450mm	222.17	294.38	9.62	209.95	nr	**504.33**
Sum of two sides 3500mm	224.39	297.32	9.62	209.95	nr	**507.27**
Sum of two sides 3550mm	226.60	300.25	9.62	209.95	nr	**510.20**
Sum of two sides 3600mm	228.82	303.19	9.62	209.95	nr	**513.14**
Sum of two sides 3650mm	228.82	303.19	9.62	209.95	nr	**513.14**
Sum of two sides 3700mm	236.06	312.78	9.62	209.95	nr	**522.73**
Sum of two sides 3750mm	239.68	317.58	9.62	209.95	nr	**527.53**
Sum of two sides 3800mm	243.30	322.37	9.62	209.95	nr	**532.33**
Sum of two sides 3850mm	246.92	327.17	9.62	209.95	nr	**537.12**
Sum of two sides 3900mm	250.54	331.97	9.62	209.95	nr	**541.92**
Sum of two sides 3950mm	254.16	336.76	9.62	209.95	nr	**546.71**
Sum of two sides 4000mm	257.78	341.56	9.62	209.95	nr	**551.51**
90 degree mitre bend						
Sum of two sides 2500mm	149.00	197.43	8.30	181.14	nr	**378.57**
Sum of two sides 2550mm	150.75	199.74	8.30	181.14	nr	**380.89**
Sum of two sides 2600mm	152.50	202.06	8.56	186.82	nr	**388.88**
Sum of two sides 2650mm	154.25	204.38	8.56	186.82	nr	**391.20**
Sum of two sides 2700mm	156.00	206.70	8.62	188.13	nr	**394.83**
Sum of two sides 2750mm	157.75	209.02	8.62	188.13	nr	**397.15**
Sum of two sides 2800mm	159.50	211.34	15.20	331.73	nr	**543.07**
Sum of two sides 2850mm	166.13	220.12	15.20	331.73	nr	**551.86**
Sum of two sides 2900mm	172.75	228.89	15.20	331.73	nr	**560.63**
Sum of two sides 2950mm	179.38	237.68	15.20	331.73	nr	**569.41**
Sum of two sides 3000mm	186.00	246.45	15.20	331.73	nr	**578.18**
Sum of two sides 3050mm	191.57	253.83	15.20	331.73	nr	**585.56**

Material Costs/Measured Work Prices - Mechanical Installations

U:VENTILATION/AIR CONDITIONING SYSTEMS

Item	Net Price £	Material £	Labour hours	Labour £	Unit	Total rate £
Sum of two sides 3100mm	197.14	261.21	16.04	350.07	nr	**611.28**
Sum of two sides 3150mm	202.71	268.59	16.04	350.07	nr	**618.66**
Sum of two sides 3200mm	208.29	275.98	16.04	350.07	nr	**626.05**
Sum of two sides 3250mm	213.65	283.09	16.04	350.07	nr	**633.15**
Sum of two sides 3300mm	219.02	290.20	16.04	350.07	nr	**640.27**
Sum of two sides 3350mm	224.38	297.30	16.04	350.07	nr	**647.37**
Sum of two sides 3400mm	229.75	304.42	16.04	350.07	nr	**654.49**
Sum of two sides 3450mm	231.85	307.20	16.04	350.07	nr	**657.27**
Sum of two sides 3500mm	233.96	310.00	16.04	350.07	nr	**660.06**
Sum of two sides 3550mm	236.06	312.78	16.04	350.07	nr	**662.85**
Sum of two sides 3600mm	238.17	315.58	16.04	350.07	nr	**665.64**
Sum of two sides 3650mm	244.14	323.49	16.04	350.07	nr	**673.55**
Sum of two sides 3700mm	250.11	331.40	16.04	350.07	nr	**681.46**
Sum of two sides 3750mm	256.08	339.31	16.04	350.07	nr	**689.37**
Sum of two sides 3800mm	262.06	347.23	16.04	350.07	nr	**697.30**
Sum of two sides 3850mm	267.42	354.33	16.04	350.07	nr	**704.40**
Sum of two sides 3900mm	272.79	361.45	16.04	350.07	nr	**711.51**
Sum of two sides 3950mm	278.15	368.55	16.04	350.07	nr	**718.62**
Sum of two sides 4000mm	283.59	375.76	16.04	350.07	nr	**725.82**
Branch						
Sum of two sides 2500mm	129.73	171.89	2.88	62.85	nr	**234.75**
Sum of two sides 2550mm	133.15	176.42	2.88	62.85	nr	**239.28**
Sum of two sides 2600mm	136.58	180.97	2.88	62.85	nr	**243.82**
Sum of two sides 2650mm	140.01	185.51	2.88	62.85	nr	**248.37**
Sum of two sides 2700mm	143.43	190.04	2.88	62.85	nr	**252.90**
Sum of two sides 2750mm	146.86	194.59	2.88	62.85	nr	**257.44**
Sum of two sides 2800mm	150.29	199.13	3.94	85.99	nr	**285.12**
Sum of two sides 2850mm	153.72	203.68	3.94	85.99	nr	**289.67**
Sum of two sides 2900mm	157.14	208.21	3.94	85.99	nr	**294.20**
Sum of two sides 2950mm	160.57	212.76	3.94	85.99	nr	**298.74**
Sum of two sides 3000mm	164.00	217.30	4.83	105.41	nr	**322.71**
Sum of two sides 3050mm	166.82	221.04	4.83	105.41	nr	**326.45**
Sum of two sides 3100mm	169.64	224.77	4.83	105.41	nr	**330.19**
Sum of two sides 3150mm	172.45	228.50	4.83	105.41	nr	**333.91**
Sum of two sides 3200mm	175.27	232.23	4.83	105.41	nr	**337.65**
Sum of two sides 3250mm	178.00	235.85	4.83	105.41	nr	**341.26**
Sum of two sides 3300mm	180.73	239.47	4.83	105.41	nr	**344.88**
Sum of two sides 3350mm	183.46	243.08	4.83	105.41	nr	**348.50**
Sum of two sides 3400mm	186.18	246.69	4.83	105.41	nr	**352.10**
Sum of two sides 3450mm	188.91	250.31	4.83	105.41	nr	**355.72**
Sum of two sides 3500mm	191.64	253.92	4.83	105.41	nr	**359.34**
Sum of two sides 3550mm	194.36	257.53	4.83	105.41	nr	**362.94**
Sum of two sides 3600mm	197.09	261.14	4.83	105.41	nr	**366.56**
Sum of two sides 3650mm	199.82	264.76	4.83	105.41	nr	**370.17**
Sum of two sides 3700mm	202.54	268.37	4.83	105.41	nr	**373.78**
Sum of two sides 3750mm	205.27	271.98	4.83	105.41	nr	**377.40**
Sum of two sides 3800mm	208.00	275.60	4.83	105.41	nr	**381.01**
Sum of two sides 3850mm	210.72	279.20	4.83	105.41	nr	**384.62**
Sum of two sides 3900mm	213.45	282.82	4.83	105.41	nr	**388.23**
Sum of two sides 3950mm	216.18	286.44	4.83	105.41	nr	**391.85**
Sum of two sides 4000mm	218.91	290.06	4.83	105.41	nr	**395.47**

376   *Material Costs/Measured Work Prices - Mechanical Installations*

U:VENTILATION/AIR CONDITIONING SYSTEMS

Item	Net Price £	Material £	Labour hours	Labour £	Unit	Total rate £
U10 : DUCTWORK : RECTANGULAR - CLASS C						
Y30 - AIR DUCTLINES						
Galvanised sheet metal DW144 class C rectangular section ductwork; including all necessary stiffeners, joints, couplers in the running length and duct supports (Continued)						
Grille neck						
Sum of two sides 2500mm	107.41	142.32	3.06	66.78	nr	**209.10**
Sum of two sides 2550mm	110.41	146.29	3.06	66.78	nr	**213.08**
Sum of two sides 2600mm	113.41	150.27	3.08	67.22	nr	**217.49**
Sum of two sides 2650mm	116.41	154.24	3.08	67.22	nr	**221.46**
Sum of two sides 2700mm	119.41	158.22	3.08	67.22	nr	**225.44**
Sum of two sides 2750mm	122.40	162.18	3.08	67.22	nr	**229.40**
Sum of two sides 2800mm	125.40	166.16	4.13	90.14	nr	**256.29**
Sum of two sides 2850mm	128.40	170.13	4.13	90.14	nr	**260.27**
Sum of two sides 2900mm	131.40	174.10	4.13	90.14	nr	**264.24**
Sum of two sides 2950mm	134.40	178.08	4.13	90.14	nr	**268.22**
Sum of two sides 3000mm	137.40	182.06	5.02	109.56	nr	**291.61**
Sum of two sides 3050mm	139.72	185.13	5.02	109.56	nr	**294.69**
Sum of two sides 3100mm	142.03	188.19	5.02	109.56	nr	**297.75**
Sum of two sides 3150mm	144.35	191.26	5.02	109.56	nr	**300.82**
Sum of two sides 3200mm	146.66	194.32	5.02	109.56	nr	**303.88**
Sum of two sides 3250mm	148.87	197.25	5.02	109.56	nr	**306.81**
Sum of two sides 3300mm	151.08	200.18	5.02	109.56	nr	**309.74**
Sum of two sides 3350mm	153.29	203.11	5.02	109.56	nr	**312.67**
Sum of two sides 3400mm	155.50	206.04	5.02	109.56	nr	**315.60**
Sum of two sides 3450mm	157.72	208.98	5.02	109.56	nr	**318.54**
Sum of two sides 3500mm	159.93	211.91	5.02	109.56	nr	**321.47**
Sum of two sides 3550mm	162.14	214.84	5.02	109.56	nr	**324.39**
Sum of two sides 3600mm	164.35	217.76	5.02	109.56	nr	**327.32**
Sum of two sides 3650mm	166.56	220.69	5.02	109.56	nr	**330.25**
Sum of two sides 3700mm	168.77	223.62	5.02	109.56	nr	**333.18**
Sum of two sides 3750mm	170.99	226.56	5.02	109.56	nr	**336.12**
Sum of two sides 3800mm	173.20	229.49	5.02	109.56	nr	**339.05**
Sum of two sides 3850mm	175.41	232.42	5.02	109.56	nr	**341.98**
Sum of two sides 3900mm	177.62	235.35	5.02	109.56	nr	**344.91**
Sum of two sides 3950mm	179.83	238.27	5.02	109.56	nr	**347.83**
Sum of two sides 4000mm	182.05	241.22	5.02	109.56	nr	**350.78**

U:VENTILATION/AIR CONDITIONING SYSTEMS

Item	Net Price £	Material £	Labour hours	Labour £	Unit	Total rate £
U10 : DUCTWORK : RECTANGULAR - CLASS C						
Y30 - ANCILLARIES						
Access doors, hollow steel construction; 25mm mineral wool insulation; removeable; fixed with cams; including sub-frame and integral sealing gaskets						
Rectangular duct						
235 x 90mm	26.73	35.42	1.25	27.28	nr	**62.70**
235 x 140mm	28.79	38.15	1.35	29.49	nr	**67.64**
335 x 235mm	31.45	41.67	1.50	32.77	nr	**74.44**
535 x 235mm	37.75	50.02	1.50	32.77	nr	**82.79**
Access doors, hollow steel construction; 25mm mineral wool insulation;hinged; locked with cams; including sub-frame and integral sealing gaskets						
Rectangular duct						
235 x 90mm	26.73	35.42	1.25	27.28	nr	**62.70**
235 x 140mm	28.79	38.15	1.35	29.49	nr	**67.64**
335 x 235mm	31.45	41.67	1.50	32.77	nr	**74.44**
535 x 235mm	37.75	50.02	1.50	32.77	nr	**82.79**

Material Costs/Measured Work Prices - Mechanical Installations

U:VENTILATION/AIR CONDITIONING SYSTEMS

Item	Net Price £	Material £	Labour hours	Labour £	Unit	Total rate £
U10 : DUCTWORK : VOLUME/FIRE DAMPERS						
Y30 - ANCILLAIRES						
Volume control damper; galvanised steel casing; aluminium aerofoil blades; manually operated						
Rectangular						
Sum of two sides 200mm	13.02	17.25	1.60	34.92	nr	52.17
Sum of two sides 250mm	13.45	17.82	1.60	34.92	nr	52.74
Sum of two sides 300mm	13.90	18.42	1.60	34.92	nr	53.34
Sum of two sides 350mm	15.67	20.76	1.60	34.92	nr	55.68
Sum of two sides 400mm	16.26	21.54	1.60	34.92	nr	56.46
Sum of two sides 450mm	16.63	22.03	1.60	34.92	nr	56.95
Sum of two sides 500mm	18.09	23.97	1.60	34.92	nr	58.89
Sum of two sides 550mm	18.38	24.35	1.60	34.92	nr	59.27
Sum of two sides 600mm	19.20	25.44	1.70	37.12	nr	62.56
Sum of two sides 650mm	20.59	27.28	2.00	43.65	nr	70.93
Sum of two sides 700mm	20.88	27.67	2.10	45.85	nr	73.52
Sum of two sides 750mm	21.25	28.16	2.10	45.85	nr	74.01
Sum of two sides 800mm	23.31	30.89	2.15	46.92	nr	77.81
Sum of two sides 850mm	23.68	31.38	2.20	48.01	nr	79.39
Sum of two sides 900mm	23.97	31.76	2.30	50.20	nr	81.96
Sum of two sides 950mm	25.97	34.41	2.30	50.20	nr	84.61
Sum of two sides 1000mm	26.55	35.18	2.40	52.38	nr	87.56
Sum of two sides 1050mm	27.56	36.52	2.50	54.56	nr	91.08
Sum of two sides 1100mm	28.60	37.90	2.60	56.74	nr	94.64
Sum of two sides 1150mm	29.09	38.54	2.70	58.93	nr	97.47
Sum of two sides 1200mm	29.57	39.18	2.80	61.11	nr	100.29
Sum of two sides 1250mm	30.09	39.87	2.95	64.38	nr	104.25
Sum of two sides 1300mm	30.60	40.55	3.10	67.66	nr	108.20
Sum of two sides 1350mm	32.95	43.66	3.17	69.29	nr	112.95
Sum of two sides 1400mm	35.30	46.77	3.25	70.93	nr	117.70
Sum of two sides 1450mm	35.86	47.51	3.33	72.57	nr	120.08
Sum of two sides 1500mm	36.41	48.24	3.40	74.20	nr	122.45
Sum of two sides 1550mm	52.49	69.55	3.42	74.75	nr	144.30
Sum of two sides 1600mm	68.57	90.86	3.45	75.29	nr	166.15
Sum of two sides 1650mm	69.24	91.74	3.52	76.93	nr	168.67
Sum of two sides 1700mm	69.90	92.62	3.60	78.57	nr	171.19
Sum of two sides 1750mm	74.03	98.09	3.75	81.84	nr	179.93
Sum of two sides 1800mm	78.15	103.55	3.90	85.12	nr	188.66
Sum of two sides 1850mm	78.63	104.18	4.05	88.39	nr	192.57
Sum of two sides 1900mm	79.10	104.81	4.20	91.70	nr	196.51
Sum of two sides 1950mm	80.09	106.12	4.26	93.08	nr	199.20
Sum of two sides 2000mm	81.07	107.42	4.33	94.50	nr	201.92
Sum of two sides 2050mm	81.95	108.58	4.38	95.59	nr	204.18
Sum of two sides 2100mm	82.83	109.75	4.43	96.68	nr	206.43
Sum of two sides 2150mm	87.53	115.98	4.49	97.99	nr	213.97
Sum of two sides 2200mm	92.23	122.20	4.55	99.30	nr	221.51
Sum of two sides 2250mm	93.35	123.69	4.63	100.94	nr	224.63
Sum of two sides 2300mm	94.46	125.16	4.70	102.58	nr	227.74
Sum of two sides 2350mm	95.46	126.48	4.79	104.54	nr	231.02
Sum of two sides 2400mm	96.46	127.81	4.88	106.50	nr	234.31

Material Costs/Measured Work Prices - Mechanical Installations

U:VENTILATION/AIR CONDITIONING SYSTEMS

Item	Net Price £	Material £	Labour hours	Labour £	Unit	Total rate £
Sum of two sides 2450mm	96.69	128.11	4.94	107.81	nr	235.93
Sum of two sides 2500mm	96.91	128.41	5.00	109.12	nr	237.53
Sum of two sides 2550mm	98.44	130.43	5.05	110.21	nr	240.65
Sum of two sides 2600mm	99.97	132.46	5.10	111.35	nr	243.81
Sum of two sides 2650mm	106.13	140.62	5.20	113.49	nr	254.11
Sum of two sides 2700mm	112.28	148.77	5.30	115.67	nr	264.44
Sum of two sides 2750mm	112.77	149.42	5.40	117.85	nr	267.27
Sum of two sides 2800mm	113.25	150.06	5.50	120.04	nr	270.09
Sum of two sides 2850mm	132.84	176.01	5.55	121.13	nr	297.14
Sum of two sides 2900mm	152.43	201.97	5.60	122.22	nr	324.19
Sum of two sides 2950mm	154.12	204.21	5.67	123.85	nr	328.06
Sum of two sides 3000mm	155.81	206.45	5.75	125.49	nr	331.94
Sum of two sides 3050mm	157.15	208.22	5.88	128.22	nr	336.44
Sum of two sides 3100mm	158.48	209.99	6.00	130.95	nr	340.93
Sum of two sides 3150mm	158.48	209.99	6.08	132.58	nr	342.57
Sum of two sides 3200mm	158.48	209.99	6.15	134.22	nr	344.21
Sum of two sides 3250mm	160.24	212.32	6.28	136.95	nr	349.27
Sum of two sides 3300mm	161.99	214.64	6.40	139.68	nr	354.31
Sum of two sides 3350mm	163.79	217.02	6.53	142.41	nr	359.43
Sum of two sides 3400mm	165.59	219.41	6.65	145.13	nr	364.54
Sum of two sides 3450mm	167.43	221.84	6.72	146.77	nr	368.62
Sum of two sides 3500mm	169.27	224.28	6.80	148.41	nr	372.69
Sum of two sides 3550mm	171.15	226.77	6.88	150.04	nr	376.82
Sum of two sides 3600mm	173.03	229.26	6.95	151.68	nr	380.95
Sum of two sides 3650mm	174.95	231.81	7.08	154.41	nr	386.22
Sum of two sides 3700mm	176.87	234.35	7.20	157.14	nr	391.49
Sum of two sides 3750mm	178.84	236.96	7.28	158.77	nr	395.74
Sum of two sides 3800mm	180.80	239.56	7.35	160.41	nr	399.97
Sum of two sides 3850mm	182.81	242.22	7.42	162.05	nr	404.27
Sum of two sides 3900mm	184.81	244.87	7.50	163.68	nr	408.56
Sum of two sides 3950mm	186.86	247.59	7.58	165.32	nr	412.91
Sum of two sides 4000mm	188.91	250.31	7.65	166.96	nr	417.26
Circular						
100mm dia.	16.32	21.62	0.80	17.46	nr	39.08
150mm dia.	18.98	25.15	0.90	19.64	nr	44.79
200mm dia.	20.52	27.19	1.05	22.92	nr	50.11
250mm dia.	21.33	28.26	1.20	26.20	nr	54.46
300mm dia.	26.76	35.46	1.35	29.49	nr	64.95
355mm dia.	31.26	41.42	1.65	36.02	nr	77.43
400mm dia.	31.26	41.42	1.90	41.49	nr	82.91
450mm dia.	34.19	45.30	2.10	45.85	nr	91.15
500mm dia.	37.65	49.89	2.95	64.38	nr	114.27
630mm dia.	46.99	62.26	4.55	99.30	nr	161.56
710mm dia.	54.63	72.38	5.20	113.49	nr	185.87
800mm dia.	58.39	77.37	5.80	126.58	nr	203.95
900mm dia.	67.57	89.53	6.40	139.68	nr	229.21
1000mm dia.	76.77	101.72	7.00	152.77	nr	254.49
Flat oval						
345 x 102mm	27.57	36.53	1.20	26.20	nr	62.73
427 x 102mm	29.19	38.68	1.35	29.49	nr	68.17
508 x 102mm	29.63	39.26	1.60	34.92	nr	74.18
559 x 152mm	30.23	40.05	1.90	41.49	nr	81.55

Material Costs/Measured Work Prices - Mechanical Installations

U:VENTILATION/AIR CONDITIONING SYSTEMS

Item	Net Price £	Material £	Labour hours	Labour £	Unit	Total rate £
U10 : DUCTWORK : VOLUME/FIRE DAMPERS						
Y30 - ANCILLAIRES						
Volume control damper; galvanised steel casing; aluminium aerofoil blades; manually operated (Continued)						
Flat oval						
531 x 203mm	30.23	40.05	1.90	41.49	nr	81.55
851 x 203mm	25.02	33.15	4.55	99.30	nr	132.45
582 x 254mm	33.61	44.53	2.10	45.85	nr	90.38
1303 x 254mm	88.51	117.28	6.40	139.68	nr	256.95
632 x 305mm	36.70	48.63	2.95	64.38	nr	113.01
1275 x 305mm	95.42	126.43	6.40	139.68	nr	266.11
765 x 356mm	41.77	55.35	4.55	99.30	nr	154.65
1247 x 356mm	103.61	137.28	6.40	139.68	nr	276.96
1727 x 356mm	112.03	148.44	8.20	178.89	nr	327.33
737 x 406mm	42.87	56.80	4.55	99.30	nr	156.10
818 x 406mm	45.01	59.64	5.20	113.49	nr	173.13
978 x 406mm	47.50	62.94	5.50	120.04	nr	182.97
1379 x 406mm	109.59	145.21	7.00	152.77	nr	297.98
1699 x 406mm	117.03	155.06	8.20	178.96	nr	334.03
709 x 457mm	45.89	60.80	4.50	98.31	nr	159.11
1671 x 457mm	125.45	166.22	8.10	176.78	nr	343.00
678 x 508mm	49.19	65.18	4.55	99.30	nr	164.48
Fire damper; galvanised steel casing; stainless steel folding shutter; fusible link and 24V d.c. electro-magnetic shutter release mechanisms; spring operated; BS 476 4-hour fire rating						
Rectangular						
Sum of two sides 200mm	185.23	245.43	1.60	34.92	nr	280.35
Sum of two sides 250mm	185.23	245.43	1.60	34.92	nr	280.35
Sum of two sides 300mm	185.23	245.43	1.60	34.92	nr	280.35
Sum of two sides 350mm	185.23	245.43	1.60	34.92	nr	280.35
Sum of two sides 400mm	185.23	245.43	1.60	34.92	nr	280.35
Sum of two sides 450mm	185.23	245.43	1.60	34.92	nr	280.35
Sum of two sides 500mm	186.66	247.32	1.60	34.92	nr	282.24
Sum of two sides 550mm	188.21	249.38	1.60	34.92	nr	284.30
Sum of two sides 600mm	189.87	251.58	1.70	37.12	nr	288.69
Sum of two sides 650mm	197.01	261.04	2.00	43.65	nr	304.69
Sum of two sides 700mm	199.09	263.79	2.10	45.85	nr	309.64
Sum of two sides 750mm	201.00	266.32	2.10	45.85	nr	312.17
Sum of two sides 800mm	203.65	269.84	2.15	46.93	nr	316.77
Sum of two sides 850mm	205.80	272.69	2.20	48.07	nr	320.76
Sum of two sides 900mm	207.39	274.79	2.30	50.29	nr	325.08
Sum of two sides 950mm	210.16	278.46	2.30	50.29	nr	328.75
Sum of two sides 1000mm	211.72	280.53	2.40	52.46	nr	332.99
Sum of two sides 1050mm	213.88	283.39	2.60	56.84	nr	340.23
Sum of two sides 1100mm	216.03	286.24	2.60	56.84	nr	343.08
Sum of two sides 1150mm	217.43	288.09	2.80	61.13	nr	349.23

Material Costs/Measured Work Prices - Mechanical Installations

U:VENTILATION/AIR CONDITIONING SYSTEMS

Item	Net Price £	Material £	Labour hours	Labour £	Unit	Total rate £
Sum of two sides 1200mm	218.83	289.95	2.80	61.13	nr	351.08
Sum of two sides 1250mm	220.47	292.12	3.10	67.66	nr	359.78
Sum of two sides 1300mm	222.11	294.30	3.10	67.66	nr	361.95
Sum of two sides 1350mm	224.54	297.52	3.25	70.93	nr	368.45
Sum of two sides 1400mm	226.97	300.74	3.25	70.93	nr	371.67
Sum of two sides 1450mm	228.39	302.62	3.40	74.23	nr	376.85
Sum of two sides 1500mm	229.81	304.50	3.40	74.23	nr	378.73
Sum of two sides 1550mm	232.18	307.64	3.45	75.29	nr	382.93
Sum of two sides 1600mm	234.55	310.78	3.45	75.29	nr	386.07
Sum of two sides 1650mm	236.83	313.80	3.60	78.57	nr	392.37
Sum of two sides 1700mm	239.11	316.82	3.60	78.57	nr	395.39
Sum of two sides 1750mm	241.40	319.86	3.90	85.12	nr	404.97
Sum of two sides 1800mm	243.68	322.88	3.90	85.12	nr	407.99
Sum of two sides 1850mm	247.28	327.65	4.20	91.70	nr	419.35
Sum of two sides 1900mm	340.44	451.08	4.20	91.70	nr	542.78
Sum of two sides 1950mm	341.72	452.78	4.33	94.50	nr	547.28
Sum of two sides 2000mm	342.99	454.46	4.33	94.50	nr	548.96
Sum of two sides 2050mm	452.16	599.11	4.43	96.68	nr	695.79
Sum of two sides 2100mm	453.50	600.89	4.43	96.68	nr	697.57
Sum of two sides 2150mm	457.51	606.20	4.55	99.30	nr	705.50
Sum of two sides 2200mm	461.52	611.51	4.55	99.30	nr	710.82
Sum of two sides 2250mm	463.16	613.69	4.70	102.58	nr	716.26
Sum of two sides 2300mm	464.80	615.86	4.70	102.58	nr	718.44
Sum of two sides 2350mm	466.52	618.14	4.88	106.50	nr	724.64
Sum of two sides 2400mm	468.23	620.40	4.88	106.50	nr	726.91
Sum of two sides 2450mm	469.80	622.48	5.00	109.12	nr	731.61
Sum of two sides 2500mm	471.36	624.55	5.00	109.12	nr	733.67
Sum of two sides 2550mm	473.23	627.03	5.10	111.35	nr	738.38
Sum of two sides 2600mm	475.10	629.51	5.10	111.35	nr	740.86
Sum of two sides 2650mm	479.66	635.55	5.30	115.67	nr	751.22
Sum of two sides 2700mm	484.22	641.59	5.30	115.67	nr	757.26
Sum of two sides 2750mm	485.94	643.87	5.50	120.04	nr	763.91
Sum of two sides 2800mm	487.65	646.14	5.50	120.04	nr	766.17
Sum of two sides 2850mm	490.14	649.44	5.60	122.22	nr	771.65
Sum of two sides 2900mm	492.63	652.73	5.60	122.22	nr	774.95
Sum of two sides 2950mm	495.07	655.97	5.75	125.49	nr	781.46
Sum of two sides 3000mm	497.51	659.20	5.75	125.49	nr	784.69
Sum of two sides 3050mm	499.78	662.21	6.00	130.95	nr	793.16
Sum of two sides 3100mm	502.05	665.22	6.00	130.95	nr	796.16
Sum of two sides 3150mm	504.51	668.48	6.15	134.22	nr	802.70
Sum of two sides 3200mm	477.52	632.71	6.15	134.22	nr	766.94
Sum of two sides 3250mm	477.52	632.71	6.35	138.59	nr	771.30
Sum of two sides 3300mm	448.07	593.69	6.35	138.59	nr	732.28
Sum of two sides 3350mm	449.35	595.39	6.60	144.04	nr	739.43
Sum of two sides 3400mm	450.62	597.07	6.60	144.04	nr	741.11
Sum of two sides 3450mm	461.52	611.51	6.75	147.32	nr	758.83
Sum of two sides 3500mm	464.80	615.86	6.75	147.32	nr	763.18
Sum of two sides 3550mm	681.78	903.36	6.89	150.37	nr	1053.73
Sum of two sides 3600mm	683.49	905.62	6.89	150.37	nr	1056.00
Sum of two sides 3650mm	685.06	907.70	7.16	156.26	nr	1063.97
Sum of two sides 3700mm	686.62	909.77	7.16	156.26	nr	1066.04
Sum of two sides 3750mm	688.49	912.25	7.43	162.16	nr	1074.41
Sum of two sides 3800mm	690.36	914.73	7.43	162.16	nr	1076.88
Sum of two sides 3850mm	694.92	920.77	7.56	164.99	nr	1085.76

Material Costs/Measured Work Prices - Mechanical Installations

U:VENTILATION/AIR CONDITIONING SYSTEMS

Item	Net Price £	Material £	Labour hours	Labour £	Unit	Total rate £
U10 : DUCTWORK : VOLUME/FIRE DAMPERS						
Y30 - ANCILLAIRES						
Fire damper; galvanised steel casing; stainless steel folding shutter; fusible link and 24V d.c. electro-magnetic shutter release mechanisms; spring operated; BS 476 4-hour fire rating (Continued)						
Rectangular						
Sum of two sides 3900mm	699.48	926.81	7.56	164.99	nr	**1091.80**
Sum of two sides 3950mm	701.20	929.09	7.77	169.58	nr	**1098.67**
Sum of two sides 4000mm	702.91	931.36	7.77	169.58	nr	**1100.93**
Circular						
100mm dia.	187.21	248.05	0.80	17.46	nr	**265.51**
150mm dia.	188.93	250.33	0.90	19.64	nr	**269.97**
200mm dia.	191.23	253.38	1.05	22.92	nr	**276.30**
250mm dia.	194.72	258.00	1.20	26.20	nr	**284.20**
300mm dia.	200.02	265.03	1.35	29.49	nr	**294.52**
355mm dia.	212.01	280.91	1.65	36.02	nr	**316.93**
400mm dia.	212.01	280.91	1.90	41.49	nr	**322.40**
450mm dia.	217.73	288.49	2.10	45.85	nr	**334.34**
500mm dia.	224.25	297.13	2.95	64.38	nr	**361.51**
630mm dia.	241.60	320.12	4.55	99.30	nr	**419.42**
710mm dia.	253.33	335.66	5.20	113.49	nr	**449.15**
800mm dia.	259.85	344.30	5.80	126.58	nr	**470.88**
900mm dia.	273.54	362.44	6.40	139.68	nr	**502.12**
1000mm dia.	288.49	382.25	7.00	152.77	nr	**535.02**
Flat oval						
345 x 102mm	199.09	263.79	1.20	26.20	nr	**289.99**
427 x 102mm	203.46	269.58	1.35	29.49	nr	**299.08**
508 x 102mm	205.46	272.23	1.60	34.92	nr	**307.15**
559 x 152mm	210.66	279.12	1.90	41.49	nr	**320.62**
531 x 203mm	213.24	282.54	1.90	41.49	nr	**324.03**
851 x 203mm	228.74	303.08	4.55	99.30	nr	**402.38**
582 x 254mm	231.74	307.06	2.10	45.85	nr	**352.90**
1303 x 254mm	360.65	477.86	6.40	139.68	nr	**617.54**
632 x 305mm	238.46	315.96	2.95	64.38	nr	**380.34**
1275 x 305mm	369.24	489.24	6.40	139.68	nr	**628.92**
765 x 356mm	250.33	331.69	4.55	99.30	nr	**430.99**
1247 x 356mm	329.93	437.16	6.40	139.68	nr	**576.83**
1727 x 356mm	403.62	534.80	8.20	178.89	nr	**713.69**
737 x 406mm	251.69	333.49	4.55	99.30	nr	**432.79**
818 x 406mm	257.13	340.70	5.20	113.49	nr	**454.19**
978 x 406mm	264.76	350.81	5.50	120.04	nr	**470.84**
1379 x 406mm	390.94	518.00	7.00	152.77	nr	**670.77**
1699 x 406mm	406.76	538.96	8.20	178.89	nr	**717.85**
709 x 457mm	253.33	335.66	4.50	98.31	nr	**433.97**
1671 x 457mm	414.62	549.37	8.10	176.78	nr	**726.15**
678 x 508mm	257.70	341.45	4.55	99.30	nr	**440.75**

Material Costs/Measured Work Prices - Mechanical Installations

383

U:VENTILATION/AIR CONDITIONING SYSTEMS

Item	Net Price £	Material £	Labour hours	Labour £	Unit	Total rate £
U10 : GRILLES/DIFFUSERS/LOUVRES						
Y46 - GRILLES/ DIFFUSERS/LOUVRES						
Supply grilles; single deflection; extruded aluminium alloy frame and adjustable horizontal vanes; silver grey polyester powder coated; screw fixed						
Rectangular; for duct, ceiling and sidewall applications						
150 x 100mm	6.12	8.11	0.60	13.09	nr	**21.20**
150 x 150mm	7.23	9.58	0.60	13.09	nr	**22.67**
200 x 150mm	7.79	10.32	0.65	14.19	nr	**24.50**
200 x 200mm	9.46	12.53	0.72	15.71	nr	**28.25**
300 x 100mm	7.79	10.32	0.72	15.71	nr	**26.03**
300 x 150mm	9.46	12.53	0.80	17.46	nr	**29.99**
300 x 200mm	10.01	13.26	0.88	19.21	nr	**32.47**
300 x 300mm	12.79	16.95	1.04	22.70	nr	**39.64**
400 x 100mm	9.46	12.53	0.88	19.21	nr	**31.74**
400 x 150mm	10.57	14.00	0.94	20.52	nr	**34.52**
400 x 200mm	11.68	15.48	1.04	22.70	nr	**38.18**
400 x 300mm	15.02	19.90	1.12	24.44	nr	**44.35**
600 x 200mm	16.69	22.11	1.26	27.50	nr	**49.61**
600 x 300mm	21.13	28.00	1.40	30.55	nr	**58.55**
600 x 400mm	25.59	33.90	1.61	35.14	nr	**69.04**
600 x 500mm	30.59	40.53	1.76	38.41	nr	**78.94**
600 x 600mm	34.49	45.70	2.17	47.36	nr	**93.06**
800 x 300mm	25.03	33.16	1.76	38.41	nr	**71.57**
800 x 400mm	31.15	41.27	2.17	47.36	nr	**88.63**
800 x 600mm	43.38	57.48	3.00	65.47	nr	**122.95**
1000 x 300mm	30.59	40.53	2.60	56.74	nr	**97.28**
1000 x 400mm	37.82	50.11	3.00	65.47	nr	**115.58**
1000 x 600mm	52.84	70.01	3.80	82.93	nr	**152.95**
1200 x 600mm	61.18	81.07	4.61	100.61	nr	**181.68**
Rectangular; for duct, ceiling and sidewall applications; including opposed blade damper volume regulator						
150 x 100mm	12.24	16.22	0.72	15.72	nr	**31.94**
150 x 150mm	12.24	16.22	0.72	15.72	nr	**31.94**
200 x 150mm	13.35	17.68	0.83	18.13	nr	**35.81**
200 x 200mm	15.57	20.63	0.90	19.64	nr	**40.27**
300 x 100mm	16.69	22.11	0.90	19.64	nr	**41.75**
300 x 150mm	16.69	22.11	0.98	21.39	nr	**43.50**
300 x 200mm	17.80	23.59	1.06	23.14	nr	**46.73**
300 x 300mm	21.69	28.74	1.20	26.20	nr	**54.94**
400 x 100mm	18.35	24.32	1.06	23.14	nr	**47.46**
400 x 150mm	18.35	24.32	1.13	24.66	nr	**48.97**
400 x 200mm	21.13	28.00	1.20	26.20	nr	**54.20**
400 x 300mm	26.69	35.37	1.34	29.26	nr	**64.63**
600 x 200mm	28.37	37.59	1.50	32.77	nr	**70.36**
600 x 300mm	35.04	46.42	1.66	36.25	nr	**82.68**
600 x 400mm	43.38	57.48	1.80	39.32	nr	**96.80**

384   *Material Costs/Measured Work Prices - Mechanical Installations*

U:VENTILATION/AIR CONDITIONING SYSTEMS

Item	Net Price £	Material £	Labour hours	Labour £	Unit	Total rate £
U10 : GRILLES/DIFFUSERS/LOUVRES						
Y46 - GRILLES/ DIFFUSERS/LOUVRES						
Supply grilles; single deflection; extruded aluminium alloy frame and adjustable horizontal vanes; silver grey polyester powder coated; screw fixed (Continued)						
Rectangular; for duct, ceiling and sidewall applications; including opposed blade damper volume regulator						
600 x 500mm	59.52	78.86	2.00	43.65	nr	**122.51**
600 x 600mm	66.74	88.44	2.60	56.84	nr	**145.27**
800 x 300mm	44.50	58.96	2.00	43.65	nr	**102.61**
800 x 400mm	55.06	72.96	2.60	56.84	nr	**129.79**
800 x 600mm	86.77	114.97	3.61	78.79	nr	**193.76**
1000 x 300mm	53.40	70.75	3.00	65.54	nr	**136.29**
1000 x 400mm	65.63	86.96	3.61	78.79	nr	**165.75**
1000 x 600mm	105.12	139.28	4.61	100.57	nr	**239.86**
1200 x 600mm	122.36	162.13	5.62	122.61	nr	**284.74**
Supply grilles; double deflection; extruded aluminium alloy frame and adjustable horizontal and vertical vanes; silver grey polyester powder coated; screw fixed						
Rectangular; for duct, ceiling and sidewall applications						
150 x 100mm	8.90	11.79	0.88	19.21	nr	**31.00**
150 x 150mm	8.90	11.79	0.88	19.21	nr	**31.00**
200 x 150mm	10.01	13.26	1.08	23.57	nr	**36.83**
200 x 200mm	11.68	15.48	1.25	27.28	nr	**42.76**
300 x 100mm	11.68	15.48	1.25	27.28	nr	**42.76**
300 x 150mm	11.68	15.48	1.50	32.74	nr	**48.22**
300 x 200mm	12.79	16.95	1.75	38.19	nr	**55.14**
300 x 300mm	16.69	22.11	2.15	46.92	nr	**69.03**
400 x 100mm	14.46	19.16	1.75	38.19	nr	**57.36**
400 x 150mm	14.46	19.16	1.95	42.56	nr	**61.72**
400 x 200mm	15.57	20.63	2.15	46.92	nr	**67.55**
400 x 300mm	20.02	26.53	2.55	55.65	nr	**82.19**
600 x 200mm	22.25	29.48	3.01	65.69	nr	**95.17**
600 x 300mm	27.81	36.85	3.36	73.33	nr	**110.18**
600 x 400mm	33.93	44.96	3.80	82.93	nr	**127.89**
600 x 500mm	40.60	53.79	4.20	91.66	nr	**145.46**
600 x 600mm	47.84	63.38	4.51	98.43	nr	**161.81**
800 x 300mm	38.93	51.59	4.20	91.66	nr	**143.25**
800 x 400mm	48.39	64.12	4.51	98.43	nr	**162.55**
800 x 600mm	69.53	92.12	5.10	111.31	nr	**203.43**
1000 x 300mm	51.17	67.81	4.80	104.76	nr	**172.56**
1000 x 400mm	63.96	84.75	5.10	111.31	nr	**196.06**
1000 x 600mm	91.22	120.86	5.72	124.84	nr	**245.70**
1200 x 600mm	107.90	142.97	6.33	138.15	nr	**281.12**

Material Costs/Measured Work Prices - Mechanical Installations   385

U:VENTILATION/AIR CONDITIONING SYSTEMS

Item	Net Price £	Material £	Labour hours	Labour £	Unit	Total rate £
Rectangular; for duct, ceiling and sidewall applications; including opposed blade damper volume regulator						
150 x 100mm	13.35	17.68	1.00	21.82	nr	**39.51**
150 x 150mm	13.35	17.68	1.00	21.82	nr	**39.51**
200 x 150mm	15.02	19.90	1.26	27.50	nr	**47.40**
200 x 200mm	17.80	23.59	1.43	31.21	nr	**54.80**
300 x 100mm	18.35	24.32	1.43	31.21	nr	**55.52**
300 x 150mm	18.35	24.32	1.68	36.67	nr	**60.98**
300 x 200mm	21.13	28.00	1.93	42.12	nr	**70.12**
300 x 300mm	23.91	31.68	2.31	50.41	nr	**82.10**
400 x 100mm	22.25	29.48	1.93	42.12	nr	**71.60**
400 x 150mm	22.25	29.48	2.14	46.70	nr	**76.18**
400 x 200mm	25.03	33.16	2.31	50.41	nr	**83.58**
400 x 300mm	31.15	41.27	2.77	60.45	nr	**101.73**
600 x 200mm	33.93	44.96	3.25	70.93	nr	**115.89**
600 x 300mm	42.27	56.01	3.62	79.01	nr	**135.02**
600 x 400mm	51.72	68.53	3.99	87.08	nr	**155.61**
600 x 500mm	70.64	93.60	4.44	96.90	nr	**190.50**
600 x 600mm	81.21	107.60	4.94	107.81	nr	**215.41**
800 x 300mm	58.40	77.38	4.44	96.90	nr	**174.28**
800 x 400mm	71.75	95.07	4.94	107.81	nr	**202.88**
800 x 600mm	113.46	150.34	5.71	124.62	nr	**274.96**
1000 x 300mm	73.42	97.28	5.20	113.49	nr	**210.77**
1000 x 400mm	92.33	122.34	5.71	124.62	nr	**246.96**
1000 x 600mm	142.39	188.67	6.53	142.51	nr	**331.18**
1200 x 600mm	167.98	222.57	7.34	160.19	nr	**382.76**
Exhaust grilles; aluminium						
0 degree fixed blade core						
150 x 150mm	7.24	9.59	0.60	13.14	nr	**22.73**
200 x 200mm	9.79	12.97	0.72	15.72	nr	**28.70**
250 x 250mm	10.79	14.30	0.80	17.46	nr	**31.76**
300 x 300mm	12.79	16.95	1.00	21.82	nr	**38.77**
350 x 350mm	17.05	22.59	1.20	26.19	nr	**48.78**
0 degree fixed blade core; including opposed blade damper volume regulator						
150 x 150mm	11.63	15.41	0.62	13.54	nr	**28.94**
200 x 200mm	15.02	19.90	0.72	15.72	nr	**35.63**
250 x 250mm	17.52	23.22	0.80	17.46	nr	**40.68**
300 x 300mm	20.92	27.71	1.00	21.82	nr	**49.54**
350 x 350mm	27.50	36.44	1.20	26.19	nr	**62.63**
45 degree fixed blade core						
150 x 150mm	7.79	10.32	0.62	13.54	nr	**23.85**
200 x 200mm	9.46	12.53	0.72	15.72	nr	**28.26**
250 x 250mm	10.57	14.00	0.80	17.46	nr	**31.46**
300 x 300mm	15.02	19.90	1.00	21.82	nr	**41.73**
350 x 350mm	17.24	22.85	1.20	26.19	nr	**49.04**

Material Costs/Measured Work Prices - Mechanical Installations

U:VENTILATION/AIR CONDITIONING SYSTEMS

Item	Net Price £	Material £	Labour hours	Labour £	Unit	Total rate £
U10 : GRILLES/DIFFUSERS/LOUVRES						
Y46 - GRILLES/ DIFFUSERS/LOUVRES						
Exhaust grilles; aluminium (Continued)						
45 degree fixed blade core; including opposed blade damper volume regulator						
150 x 150mm	15.48	20.51	0.62	13.54	nr	34.05
200 x 200mm	18.81	24.92	0.72	15.72	nr	40.65
250 x 250mm	21.02	27.84	0.80	17.46	nr	45.30
300 x 300mm	29.87	39.58	1.00	21.82	nr	61.40
350 x 350mm	34.30	45.45	1.20	26.19	nr	71.64
Eggcrate core						
150 x 150mm	6.12	8.11	0.62	13.54	nr	21.65
200 x 200mm	9.46	12.53	1.00	21.82	nr	34.36
250 x 250mm	11.12	14.74	0.80	17.46	nr	32.20
300 x 300mm	13.35	17.68	1.00	21.82	nr	39.51
350 x 350mm	16.13	21.37	1.20	26.19	nr	47.56
Eggcrate core; including opposed blade damper volume regulator						
150 x 150mm	10.01	13.26	0.62	13.54	nr	26.80
200 x 200mm	13.35	17.68	0.72	15.72	nr	33.41
250 x 250mm	16.69	22.11	0.80	17.46	nr	39.57
300 x 300mm	19.47	25.79	1.00	21.82	nr	47.62
350 x 350mm	23.91	31.68	1.20	26.19	nr	57.87
Mesh/perforated plate core						
150 x 150mm	6.12	8.11	0.62	13.53	nr	21.64
200 x 200mm	9.46	12.53	0.72	15.71	nr	28.25
250 x 250mm	11.12	14.74	0.80	17.46	nr	32.20
300 x 300mm	13.35	17.68	1.00	21.82	nr	39.51
350 x 350mm	16.13	21.37	1.20	26.19	nr	47.56
Mesh/perforated plate core; including opposed blade damper volume regulator						
150 x 150mm	12.17	16.12	0.62	13.53	nr	29.65
200 x 200mm	18.81	24.92	0.72	15.71	nr	40.64
250 x 250mm	22.13	29.32	0.80	17.46	nr	46.78
300 x 300mm	26.55	35.18	0.80	17.46	nr	52.64
350 x 350mm	32.09	42.51	1.20	26.19	nr	68.70
Plastic air diffusion system						
Eggcrate grilles						
150 x 150mm	5.82	7.71	0.62	10.92	nr	18.63
200 x 200mm	7.69	10.19	0.72	12.68	nr	22.87
250 x 250mm	9.03	11.96	0.80	14.08	nr	26.04
300 x 300mm	11.71	15.51	1.00	17.60	nr	33.11
310 x 315mm	8.49	11.26	1.05	18.48	nr	29.74

Material Costs/Measured Work Prices - Mechanical Installations

U:VENTILATION/AIR CONDITIONING SYSTEMS

Item	Net Price £	Material £	Labour hours	Labour £	Unit	Total rate £
Single deflection grilles						
150 x 150mm	5.36	7.10	0.62	10.92	nr	**18.02**
200 x 200mm	7.16	9.48	0.72	12.68	nr	**22.16**
250 x 250mm	7.69	10.19	0.80	14.08	nr	**24.27**
300 x 300mm	9.93	13.16	1.00	17.60	nr	**30.76**
315 x 315mm	7.69	10.19	1.05	18.48	nr	**28.67**
Double deflection grilles						
150 x 150mm	6.35	8.41	0.62	10.92	nr	**19.33**
200 x 200mm	8.49	11.26	0.72	12.68	nr	**23.94**
250 x 250mm	10.82	14.33	0.80	14.08	nr	**28.41**
300 x 300mm	14.94	19.80	1.00	17.60	nr	**37.40**
315 x 315mm	10.82	14.33	1.05	18.48	nr	**32.81**
Door transfer grilles						
150 x 150mm	10.82	14.33	0.62	10.92	nr	**25.25**
200 x 200mm	14.31	18.96	0.72	12.68	nr	**31.64**
250 x 250mm	15.29	20.26	0.80	14.08	nr	**34.34**
300 x 300mm	19.77	26.20	1.00	17.60	nr	**43.80**
315 x 315mm	15.29	20.26	1.05	18.48	nr	**38.74**
Opposed blade dampers						
150 x 150mm	4.12	5.46	0.62	10.92	nr	**16.38**
200 x 200mm	5.36	7.10	0.72	12.68	nr	**19.78**
250 x 250mm	6.35	8.41	0.80	14.08	nr	**22.49**
300 x 300mm	7.69	10.19	1.00	17.60	nr	**27.79**
315 x 315mm	6.35	8.41	1.05	18.48	nr	**26.89**
Neck reducers						
150 x 150mm	4.93	6.53	0.62	10.92	nr	**17.45**
200 x 200mm	5.82	7.71	0.72	12.68	nr	**20.39**
250 x 250mm	6.80	9.00	0.80	14.08	nr	**23.09**
300 x 300mm	7.65	10.13	1.00	17.60	nr	**27.74**
350 x 350mm	6.82	9.03	1.05	18.48	nr	**27.52**
Ceiling mounted diffusers; circular aluminium multi-core diffuser						
Circular; for ceiling mounting						
152mm dia. neck	15.66	20.75	0.80	17.46	nr	**38.21**
203mm dia. neck	18.90	25.04	1.10	24.01	nr	**49.05**
305mm dia. neck	29.70	39.35	1.40	30.57	nr	**69.92**
381mm dia. neck	34.02	45.08	1.50	32.77	nr	**77.85**
457mm dia. neck	96.66	128.07	2.00	43.65	nr	**171.72**
Circular; for ceiling mounting; including louvre damper volume control						
152mm dia. neck	35.44	46.96	1.00	21.82	nr	**68.79**
203mm dia. neck	37.59	49.81	1.20	26.20	nr	**76.01**
305mm dia. neck	59.10	78.31	1.60	34.92	nr	**113.23**
381mm dia. neck	67.66	89.65	1.90	41.49	nr	**131.14**
457mm dia. neck	192.26	254.74	2.40	52.46	nr	**307.20**

Material Costs/Measured Work Prices - Mechanical Installations

U:VENTILATION/AIR CONDITIONING SYSTEMS

Item	Net Price £	Material £	Labour hours	Labour £	Unit	Total rate £
U10 : GRILLES/DIFFUSERS/LOUVRES						
Y46 - GRILLES/ DIFFUSERS/LOUVRES						
Ceiling mounted diffusers; rectangular aluminium multi cone diffuser; four way flow						
Rectangular; for ceiling mounting						
150 x 150 mm neck	17.82	23.61	1.80	39.32	nr	**62.94**
300 x 300 mm neck	26.46	35.06	2.80	61.13	nr	**96.19**
450 x 450 mm neck	37.80	50.09	3.40	74.23	nr	**124.32**
600 x 600 mm neck	56.16	74.41	4.00	87.30	nr	**161.71**
Rectangular; for ceiling mounting; including opposed blade damper volume regulator						
150 x 150 mm neck	25.38	33.63	1.80	39.32	nr	**72.95**
300 x 300 mm neck	36.72	48.65	2.80	61.13	nr	**109.79**
450 x 450 mm neck	52.92	70.12	3.51	76.58	nr	**146.70**
600 x 600 mm neck	77.76	103.03	5.62	122.61	nr	**225.64**
Slot diffusers; continuous aluminium slot diffuser with flanged frame						
Diffuser						
1 slot	21.96	29.10	3.76	82.05	m	**111.14**
2 slot	35.28	46.75	3.76	82.03	m	**128.77**
3 slot	47.52	62.96	3.76	82.03	m	**144.99**
4 slot	60.48	80.14	4.50	98.31	m	**178.44**
6 slot	87.12	115.43	4.50	98.31	m	**213.74**
Diffuser; including equalizing deflector						
1 slot	21.96	29.10	5.26	114.87	m	**143.96**
2 slot	35.28	46.75	5.26	114.87	m	**161.61**
3 slot	47.52	62.96	5.26	114.87	m	**177.83**
4 slot	60.48	80.14	6.33	138.13	m	**218.27**
6 slot	87.12	115.43	6.33	138.13	m	**253.56**
Extra over for Ends						
1 slot	6.48	8.59	1.00	21.82	nr	**30.41**
2 slot	7.56	10.02	1.00	21.82	nr	**31.84**
3 slot	8.64	11.45	1.00	21.82	nr	**33.27**
4 slot	9.72	12.88	1.30	28.37	nr	**41.25**
6 slot	11.88	15.74	1.40	30.55	nr	**46.30**
Plenum boxes; 1.0m long; circular spigot; including cord operated flap damper						
1 slot	31.86	42.21	2.75	60.12	nr	**102.34**
2 slot	32.40	42.93	2.75	60.12	nr	**103.05**
3 slot	33.48	44.36	2.75	60.12	nr	**104.48**
4 slot	34.02	45.08	3.51	76.58	nr	**121.65**
6 slot	35.64	47.22	3.51	76.58	nr	**123.80**

Material Costs/Measured Work Prices - Mechanical Installations   389

U:VENTILATION/AIR CONDITIONING SYSTEMS

Item	Net Price £	Material £	Labour hours	Labour £	Unit	Total rate £
Plenum boxes; 2.0m long; circular spigot; including cord operated flap damper						
1 slot	31.86	42.21	3.26	71.09	nr	**113.30**
2 slot	49.68	65.83	3.26	71.09	nr	**136.92**
3 slot	50.76	67.26	3.26	71.09	nr	**138.35**
4 slot	50.76	67.26	3.76	82.05	nr	**149.30**
6 slot	53.46	70.83	3.76	82.05	nr	**152.88**
Perforated diffusers; rectangular face aluminium perforated diffuser; quick release face plate; for integration with rectangular ceiling tiles						
Circular spigot; rectangular diffuser						
150mm dia. spigot; 300 x 300 diffuser	49.14	65.11	1.00	21.82	nr	**86.94**
300mm dia. spigot; 600 x 600 diffuser	90.72	120.20	1.40	30.57	nr	**150.77**
Circular spigot; rectangilar diffuser; including louvre damper volume regulator						
150mm dia. spigot; 300 x 300 diffuser	62.10	82.28	1.00	21.82	nr	**104.11**
300mm dia. spigot; 600 x 600 diffuser	105.30	139.52	1.60	34.92	nr	**174.44**
Rectangular spigot; rectangular diffuser						
150 x 150mm dia. spigot; 300 x 300mm diffuser	52.87	70.05	1.00	17.60	nr	**87.65**
300 x 150mm dia. spigot; 600 x 300mm diffuser	61.14	81.01	1.20	21.12	nr	**102.13**
300 x 300mm dia. spigot; 600 x 600mm diffuser	90.18	119.49	1.40	24.64	nr	**144.13**
600 x 300mm dia. spigot; 1200 x 600mm diffuser	160.17	212.23	1.60	28.16	nr	**240.39**
Rectangular spigot; rectangular diffuser; including opposed blade damper volume regulator						
150 x 150mm dia. spigot; 300 x 300mm diffuser	68.50	90.77	1.20	21.12	nr	**111.89**
300 x 150mm dia. spigot; 600 x 300mm diffuser	101.00	133.82	1.40	24.65	nr	**158.48**
300 x 300mm dia. spigot; 600 x 600mm diffuser	109.55	145.15	1.60	28.16	nr	**173.31**
600 x 300mm dia. spigot; 1200 x 600mm diffuser	185.00	245.13	1.80	31.68	nr	**276.81**
Plastic air diffusion system						
Cellular diffusers						
300 x 300mm	19.06	25.25	2.80	49.30	nr	**74.55**
600 x 600mm	41.33	54.76	4.00	70.40	nr	**125.16**
Multicone diffusers						
300 x 300mm	19.06	25.25	2.80	49.30	nr	**74.55**
450 x 450mm	28.54	37.81	3.40	59.87	nr	**97.68**
500 x 500mm	28.54	37.81	3.80	66.92	nr	**104.73**
600 x 600mm	41.33	54.76	4.00	70.40	nr	**125.16**
625 x 625mm	41.33	54.76	4.26	74.90	nr	**129.66**
Opposed blade dampers						
300 x 300mm	5.36	7.10	1.20	21.13	nr	**28.23**
450 x 450mm	8.49	11.26	1.50	26.43	nr	**37.68**
600 x 600mm	16.99	22.51	2.60	45.84	nr	**68.35**

Material Costs/Measured Work Prices - Mechanical Installations

U:VENTILATION/AIR CONDITIONING SYSTEMS

Item	Net Price £	Material £	Labour hours	Labour £	Unit	Total rate £
U10 : GRILLES/DIFFUSERS/LOUVRES						
Y46 - GRILLES/ DIFFUSERS/LOUVRES						
Plastic air diffusion system (Continued)						
Plenum boxes						
300mm	8.49	11.26	2.80	49.30	nr	60.56
450mm	12.69	16.82	3.40	59.87	nr	76.68
600mm	16.99	22.51	4.00	70.40	nr	92.91
Plenum spigot reducer						
600mm	6.35	8.41	1.00	17.60	nr	26.01
Blanking kits for cellular diffusers						
300mm	4.21	5.57	0.88	15.49	nr	21.07
600mm	6.35	8.41	1.10	19.36	nr	27.78
Blanking kits for multicone diffusers						
300mm	4.21	5.57	0.88	15.49	nr	21.07
450mm	5.36	7.10	0.90	15.84	nr	22.94
600mm	6.35	8.41	1.10	19.36	nr	27.78
Acoustic louvres; opening mounted; 300mm deep steel louvres with blades packed with acoustic infill; 12mm galvanised mesh birdscreen; screw fixing in opening						
Louvre units; self finished galvanised steel						
900 high x 600 wide	172.42	228.46	3.00	52.80	nr	281.26
900 high x 900 wide	258.63	342.68	3.00	52.80	nr	395.49
900 high x 1200 wide	344.84	456.91	3.34	58.79	nr	515.70
900 high x 1500 wide	431.06	571.15	3.34	58.79	nr	629.94
900 high x 1800 wide	517.27	685.38	3.34	58.79	nr	744.17
900 high x 2100 wide	603.48	799.61	3.34	58.79	nr	858.40
900 high x 2400 wide	689.69	913.84	3.68	64.77	nr	978.61
900 high x 2700 wide	775.90	1028.07	3.68	64.77	nr	1092.84
900 high x 3000 wide	862.11	1142.30	3.68	64.77	nr	1207.07
1200 high x 600 wide	229.90	304.62	3.00	52.80	nr	357.42
1200 high x 900 wide	344.84	456.91	3.34	58.79	nr	515.70
1200 high x 1200 wide	459.79	609.22	3.34	58.79	nr	668.01
1200 high x 1500 wide	574.74	761.53	3.34	58.79	nr	820.32
1200 high x 1800 wide	689.69	913.84	3.68	64.77	nr	978.61
1200 high x 2100 wide	804.64	1066.15	3.68	64.77	nr	1130.92
1200 high x 2400 wide	919.58	1218.44	3.68	64.77	nr	1283.21
1500 high x 600 wide	287.37	380.77	3.00	52.80	nr	433.57
1500 high x 900 wide	431.06	571.15	3.34	58.79	nr	629.94
1500 high x 1200 wide	574.74	761.53	3.34	58.79	nr	820.32
1500 high x 1500 wide	718.42	951.91	3.68	64.77	nr	1016.68
1500 high x 1800 wide	862.11	1142.30	3.68	64.77	nr	1207.07
1500 high x 2100 wide	1005.79	1332.67	4.00	70.40	nr	1403.07
1800 high x 600 wide	344.84	456.91	3.34	58.79	nr	515.70
1800 high x 900 wide	517.27	685.38	3.34	58.79	nr	744.17
1800 high x 1200 wide	689.69	913.84	3.68	64.77	nr	978.61
1800 high x 1500 wide	862.11	1142.30	3.68	64.77	nr	1207.07

Material Costs/Measured Work Prices - Mechanical Installations

U:VENTILATION/AIR CONDITIONING SYSTEMS

Item	Net Price £	Material £	Labour hours	Labour £	Unit	Total rate £
Louvre units; polyester powder coated steel						
900 high x 600 wide	223.04	295.53	3.00	52.80	nr	348.33
900 high x 900 wide	342.06	453.23	3.00	52.80	nr	506.03
900 high x 1200 wide	446.07	591.04	3.34	58.79	nr	649.83
900 high x 1500 wide	557.59	738.81	3.34	58.79	nr	797.59
900 high x 1800 wide	669.11	886.57	3.34	58.79	nr	945.36
900 high x 2100 wide	780.63	1034.33	3.34	58.79	nr	1093.12
900 high x 2400 wide	892.14	1182.09	3.68	64.77	nr	1246.86
900 high x 2700 wide	1003.66	1329.85	3.68	64.77	nr	1394.62
900 high x 3000 wide	1115.18	1477.61	3.68	64.77	nr	1542.38
1200 high x 600 wide	297.38	394.03	3.00	52.80	nr	446.83
1200 high x 900 wide	446.07	591.04	3.34	58.79	nr	649.83
1200 high x 1200 wide	594.76	788.06	3.34	58.79	nr	846.84
1200 high x 1500 wide	743.45	985.07	3.34	58.79	nr	1043.86
1200 high x 1800 wide	892.14	1182.09	3.68	64.77	nr	1246.86
1200 high x 2100 wide	1040.84	1379.11	3.68	64.77	nr	1443.88
1200 high x 2400 wide	1189.53	1576.13	3.68	64.77	nr	1640.90
1500 high x 600 wide	371.73	492.54	3.00	52.80	nr	545.34
1500 high x 900 wide	557.59	738.81	3.34	58.79	nr	797.59
1500 high x 1200 wide	743.45	985.07	3.34	58.79	nr	1043.86
1500 high x 1500 wide	929.32	1231.35	3.68	64.77	nr	1296.12
1500 high x 1800 wide	1115.18	1477.61	3.68	64.77	nr	1542.38
1500 high x 2100 wide	1301.04	1723.88	4.00	70.40	nr	1794.28
1800 high x 600 wide	446.07	591.04	3.34	58.79	nr	649.83
1800 high x 900 wide	669.11	886.57	3.34	58.79	nr	945.36
1800 high x 1200 wide	892.14	1182.09	3.68	64.77	nr	1246.86
1800 high x 1500 wide	1115.18	1477.61	3.68	64.77	nr	1542.38
Weather louvres; opening mounted; 300mm deep galvanised steel louvres; screw fixing in position						
Louvre units; including 12mm galvanised mesh birdscreen						
900 x 600mm	172.26	215.33	2.25	39.64	nr	254.97
900 x 900mm	233.07	291.34	2.25	39.64	nr	330.98
900 x 1200mm	288.83	361.04	2.50	44.00	nr	405.05
900 x 1500mm	341.33	426.67	2.50	44.00	nr	470.67
900 x 1800mm	391.22	489.02	2.50	44.00	nr	533.02
900 x 2100mm	538.96	673.69	2.50	44.00	nr	717.70
900 x 2400mm	577.58	721.98	2.76	48.62	nr	770.60
900 x 2700mm	645.87	807.33	2.76	48.62	nr	855.95
900 x 3000mm	682.65	853.31	2.76	48.62	nr	901.93
1200 x 600mm	209.72	262.15	2.25	39.64	nr	301.80
1200 x 900mm	283.97	354.96	2.50	44.00	nr	398.96
1200 x 1200mm	352.16	440.21	2.50	44.00	nr	484.21
1200 x 1500mm	416.42	520.52	2.50	44.00	nr	564.52
1200 x 1800mm	477.56	596.94	2.76	48.62	nr	645.56
1200 x 2100mm	656.79	820.99	2.76	48.62	nr	869.61
1200 x 2400mm	704.33	880.41	2.76	48.62	nr	929.03
1500 x 600mm	244.49	305.61	2.25	39.64	nr	345.25
1500 x 900mm	331.17	413.96	2.50	44.00	nr	457.96
1500 x 1200mm	410.95	513.69	2.50	44.00	nr	557.69
1500 x 1500mm	486.13	607.66	2.76	48.62	nr	656.28

U:VENTILATION/AIR CONDITIONING SYSTEMS

Item	Net Price £	Material £	Labour hours	Labour £	Unit	Total rate £
U10 : GRILLES/DIFFUSERS/LOUVRES						
Y46 - GRILLES/ DIFFUSERS/LOUVRES						
Weather louvres; opening mounted; 300mm deep galvanised steel louvres; screw fixing in position (Continued)						
1500 x 1800mm	557.94	697.42	2.76	48.62	nr	**746.04**
1500 x 2100mm	765.97	957.47	3.00	52.86	nr	**1010.32**
1800 x 600mm	277.17	346.46	2.50	44.00	nr	**390.46**
1800 x 900mm	375.67	469.59	2.50	44.00	nr	**513.59**
1800 x 1200mm	466.47	583.09	2.76	48.62	nr	**631.71**
1800 x 1500mm	552.15	690.18	3.00	52.86	nr	**743.04**

Material Costs/Measured Work Prices - Mechanical Installations   393

U:VENTILATION/AIR CONDITIONING SYSTEMS

Item	Net Price £	Material £	Labour hours	Labour £	Unit	Total rate £
U10 : PLANT/EQUIPMENT						
Y41 - FANS						
Axial flow fan; including ancillaries, anti vibration mountings, mounting feet, matching flanges, flex connectors and clips; 415V, 3 phase, 50Hz motor; includes fixing in position; electrical work elsewhere						
Aerofoil blade fan unit; short duct case						
315mm dia.; 0.47 m3/s duty; 147 Pa	382.13	477.66	4.50	79.20	nr	**556.87**
500mm dia.; 1.89 m3/s duty; 500 Pa	631.39	789.24	5.00	88.00	nr	**877.24**
560mm dia.; 2.36 m3/s duty; 147 Pa	528.39	660.49	5.50	96.80	nr	**757.29**
710mm dia.; 5.67 m3/s duty; 245 Pa	863.14	1078.92	6.00	105.60	nr	**1184.53**
Aerofoil blade fan unit; long duct case						
315mm dia.; 0.47 m3/s duty; 147 Pa	426.42	533.02	4.50	79.20	nr	**612.23**
500mm dia.; 1.89 m3/s duty; 500 Pa	715.85	894.81	5.00	88.00	nr	**982.82**
560mm dia.; 2.36 m3/s duty; 147 Pa	593.28	741.60	5.50	96.80	nr	**838.40**
710mm dia.; 5.67 m3/s duty; 245 Pa	1001.16	1251.45	6.00	105.60	nr	**1357.05**
Aerofoil blade fan unit; two stage; long duct case						
315mm; 0.47m3/s @ 500 Pa	747.78	934.73	4.50	79.20	nr	**1013.93**
355mm; 0.83m3/s @ 147 Pa	695.25	869.06	4.75	83.60	nr	**952.67**
710mm; 3.77m3/s @ 431 Pa	1549.12	1936.40	6.00	105.60	nr	**2042.00**
710mm; 6.61m3/s @ 500 Pa	1860.18	2325.22	6.00	105.60	nr	**2430.83**
Roof mounted extract fan; including ancillaries, fibreglass cowling, fitted shutters and bird guard; 415V, 3 phase, 50Hz motor; includes fixing in position; electrical work elsewhere						
Flat roof installation, fixed to curb						
315mm; 900rpm	320.33	400.41	4.50	79.20	nr	**479.62**
315mm; 1380rpm	327.54	409.43	4.50	79.20	nr	**488.63**
400mm; 900rpm	379.04	473.80	5.50	96.80	nr	**570.60**
400mm; 1360rpm	398.61	498.26	5.50	96.80	nr	**595.07**
800mm; 530rpm	800.31	1000.39	7.00	123.21	nr	**1123.59**
800mm; 700rpm	800.31	1000.39	7.00	123.21	nr	**1123.59**
800mm; 920rpm	848.72	1060.90	7.00	123.21	nr	**1184.11**
1000mm; 470rpm	1027.94	1284.92	8.00	140.81	nr	**1425.73**
1000mm; 570rpm	1027.94	1284.92	8.00	140.81	nr	**1425.73**
1000mm; 710rpm	1281.32	1601.65	8.00	140.81	nr	**1742.46**
Pitched roof installation; including purlin mounting box						
315mm; 900rpm	371.25	464.06	4.50	79.20	nr	**543.27**
315mm; 1380rpm	378.46	473.07	4.50	79.20	nr	**552.28**
400mm; 900rpm	440.57	550.71	5.50	96.80	nr	**647.52**
400mm; 1360rpm	460.14	575.17	5.50	96.80	nr	**671.98**
800mm; 530rpm	922.88	1153.60	7.00	123.21	nr	**1276.81**

Material Costs/Measured Work Prices - Mechanical Installations

U:VENTILATION/AIR CONDITIONING SYSTEMS

Item	Net Price £	Material £	Labour hours	Labour £	Unit	Total rate £
U10 : PLANT/EQUIPMENT						
Y41 - FANS						
Roof mounted extract fan; including ancillaries, fibreglass cowling, fitted shutters and bird guard; 415V, 3 phase, 50Hz motor; includes fixing in position; electrical work elsewhere (Continued)						
Pitched roof installation; including purlin mounting box						
800mm; 700rpm	922.88	1153.60	7.00	123.21	nr	**1276.81**
800mm; 920rpm	971.29	1214.11	7.00	123.21	nr	**1337.32**
1000mm; 470rpm	1209.22	1511.52	8.00	140.81	nr	**1652.33**
1000mm; 570rpm	1209.22	1511.52	8.00	140.81	nr	**1652.33**
1000mm; 710rpm	1462.60	1828.25	8.00	140.81	nr	**1969.06**
Centrifugal fan; single speed for internal domestic kitchens/ utility rooms; fitted with standard overload protection; complete with housing; includes placing in position; electrical work elsewhere						
Window mounted						
245m3/hr	61.67	77.09	0.50	8.80	nr	**85.89**
500m3/hr	98.40	123.00	0.50	8.80	nr	**131.80**
Wall mounted						
245m3/hr	74.47	93.09	0.83	14.67	nr	**107.75**
500m3/hr	111.20	139.00	0.83	14.67	nr	**153.67**
Centrifugal fan; various speeds, simultaneous ventilation from seperate areas fitted with standard overload protection; complete with housing; includes placing in position; ducting and electrical work elsewhere						
Fan unit						
147-300m3/hr	161.39	201.74	1.00	17.60	nr	**219.34**
175-411m3/hr	176.21	220.26	1.00	17.60	nr	**237.86**
Toilet extract units; centrifugal fan; various speeds for internal domestic bathrooms/ W.Cs, with built in filter; complete with housing; includes placing in position; electrical work elsewhere						
Fan unit; fixed to wall; including shutter						
Single speed 85m3/hr	57.29	71.61	0.75	13.20	nr	**84.81**
Two speed 60-85m3/hr	67.90	84.88	0.83	14.67	nr	**99.54**
Humidity controlled; autospeed; fixed to wall; including shutter						
30-60-85m3/hr	115.96	144.95	1.00	17.60	nr	**162.55**

Material Costs/Measured Work Prices - Mechanical Installations   395

U:VENTILATION/AIR CONDITIONING SYSTEMS

Item	Net Price £	Material £	Labour hours	Labour £	Unit	Total rate £
U10 : PLANT/EQUIPMENT						
Y42 - AIR FILTRATION						
High efficiency filters; 99.997% EU13; tested to BS 3928						
Standard; 1700m3/ hr air volume; continuous rating up to 70 C; sealed wood case, aluminium spacers, neoprene gaskets; water repellant filter media; including placing in position						
609 x 609 x 298mm	147.81	184.76	2.00	35.20	nr	**219.96**
High capacity; 3400m3/hr air volume; continuous rating up to 70 C; anti-corrosion coated mild steel frame, polyurethane sealant and neoprene gaskets; water repellant filter media; including fixing in position						
609 x 609 x 298mm	264.75	330.94	2.00	35.20	nr	**366.14**
Chemical resistant; 1700m3/hr air volume; continuous rating up to 66 C; anti-corrosion coated mild steel frame, polyurethane sealant and silicone rubber gaskets; water repellant filter media; including fixing in position						
609 x 609 x 298mm	138.79	173.49	2.00	35.20	nr	**208.69**
Bag Filters; 40/60% EU5; tested to BS 6540						
Bag filter; continuous rating up to 60 C; rigid filter assembly; sealed into one piece coated mild steel header with sealed pocket separators; including placing in position						
6 pocket, 592 x 592 x 25mm frame; pockets 350mm long; 1350m3/hr	31.71	39.64	2.00	35.20	nr	**74.84**
6 pocket, 592 x 592 x 25mm frame; pockets 500mm long; 1900m3/hr	34.86	43.58	2.50	44.00	nr	**87.58**
8 pocket, 592 x 592 x 25mm frame; pockets 900mm long; 3400m3/hr	48.88	61.10	3.00	52.80	nr	**113.90**
Grease filters, washable; minimum 65%						
Double sided extract unit; lightweight stainless steel construction; demountable composite filter media of woven metal mat and expanded metal mesh supports; for mounting on hood and extract systems; including placing in position						
500 x 686 x 565mm, 4080m3/hr	566.81	708.51	2.00	35.20	nr	**743.71**
1000 x 686 x 565mm, 8160m3/hr;	1146.39	1432.99	3.00	52.80	nr	**1485.79**
1500 x 686 x 565mm, 12240m3/hr;	1704.44	2130.55	3.50	61.60	nr	**2192.15**

Material Costs/Measured Work Prices - Mechanical Installations

U:VENTILATION/AIR CONDITIONING SYSTEMS

Item	Net Price £	Material £	Labour hours	Labour £	Unit	Total rate £
U10 : PLANT/EQUIPMENT						
Y42 - AIR FILTRATION						
Panel filters; 82% EU3/G3; tested to BS EN779						
Modular filter panels; continuous rating up to 100 C; graduated density media; rigid cardboard frame; including placing in position						
594 x 594 x 50mm; 2380m3/hr	6.76	8.45	2.50	44.00	nr	**52.45**
594 x 296 x 50mm; 1190m3/hr	5.28	6.60	2.50	44.00	nr	**50.60**
Panel filters; 90% EU4; tested to BS 6540						
Modular filter panels; continuous rating up to 100 C; pleated media with wire support; rigid cardboard frame; including placing in position						
594 x 594 x 50mm; 2380m3/hr	7.86	9.82	3.00	52.80	nr	**62.63**
594 x 296 x 50mm; 1190m3/hr	5.96	7.45	3.00	52.80	nr	**60.25**
Carbon filters; standard duty discarb filters; steel frame with bonded carbon panels; for fixing to ductwork; including placing in position						
12 panels						
597 x 597 x 298mm, 1460m3/hr	346.02	432.52	0.33	5.81	nr	**438.33**
597 x 597 x 451mm, 2200m3/hr	435.37	544.21	0.33	5.81	nr	**550.02**
597 x 597 x 597mm, 2930m3/hr	493.52	616.90	0.33	5.81	nr	**622.71**
8 panels						
451 x 451 x 298mm, 740m3/hr	234.62	293.27	0.29	5.10	nr	**298.38**
451 x 451 x 451mm, 1105m3/hr	284.60	355.75	0.29	5.10	nr	**360.85**
451 x 451 x 597mm, 1460m3/hr	322.50	403.13	0.29	5.10	nr	**408.23**
6 panels						
298 x 298 x 298mm, 365m3/hr	172.27	215.34	0.25	4.40	nr	**219.74**
298 x 298 x 451mm, 550m3/hr	186.69	233.36	0.25	4.40	nr	**237.76**
298 x 298 x 597mm, 780m3/hr	211.92	264.90	0.25	4.40	nr	**269.30**

Material Costs/Measured Work Prices - Mechanical Installations

U:VENTILATION/AIR CONDITIONING SYSTEMS

Item	Net Price £	Material £	Labour hours	Labour £	Unit	Total rate £
U70 - AIR CURTAINS						
The selection of an air curtain requires consideration of the particular conditions involved; such as, climatic conditions, wind influence, construction and position; consultation with a specialist manufacturer is therefore, advisable.						
Commercial grade air curtains; recessed or exposed units with rigid sheet steel casing; aluminium grilles; high quality motor/centrifugal fan assembly						
Ambient temperature 240V single phase supply; mounting height 2.20m						
1000 x 555 x 312mm	1909.62	2387.03	12.05	212.06	nr	**2599.08**
1500 x 555 x 312mm	2448.31	3060.39	12.05	212.06	nr	**3272.45**
2000 x 555 x 312mm	2962.28	3702.85	12.05	212.06	nr	**3914.91**
Ambient temperature 240V single phase supply; mounting height 2.50m						
1000 x 555 x 312mm	2211.41	2764.26	16.13	283.88	nr	**3048.15**
1500 x 555 x 312mm	2822.20	3527.75	16.13	283.88	nr	**3811.63**
2000 x 555 x 312mm	3513.33	4391.66	16.13	283.88	nr	**4675.55**
Ambient temperature 240V single phase supply; mounting height 3.00m						
1000 x 686 x 392mm	2887.09	3608.86	17.24	303.46	nr	**3912.32**
1500 x 686 x 392mm	3902.67	4878.34	17.24	303.46	nr	**5181.80**
2000 x 686 x 392mm	4822.46	6028.07	17.24	303.46	nr	**6331.54**
Water heated 240V single phase supply; mounting height 2.20m						
1000 x 555 x 312mm; 3.80 - 15.60kW output	2010.56	2513.20	12.05	212.06	nr	**2725.26**
1500 x 555 x 312mm; 5.80 - 24.00kW output	2577.06	3221.32	12.05	212.06	nr	**3433.38**
2000 x 555 x 312mm; 7.80 - 31.00kW output	3117.81	3897.26	12.05	212.06	nr	**4109.32**
Water heated 240V single phase supply; mounting height 2.50m						
1000 x 555 x 312mm; 5.80 - 20.70kW output	2327.80	2909.75	16.13	283.88	nr	**3193.63**
1500 x 555 x 312mm; 7.10 - 28.70kW output	2970.52	3713.15	16.13	283.88	nr	**3997.03**
2000 x 555 x 312mm; 11.50 - 40.90kW output	3697.70	4622.13	16.13	283.88	nr	**4906.01**
Water heated 240V single phase supply; mounting height 3.00m						
1000 x 686 x 392mm; 9.90 - 34.20kW output	3038.50	3798.13	17.24	303.46	nr	**4101.59**
1500 x 686 x 392mm; 15.10 - 52.10kW output	4107.64	5134.55	17.24	303.46	nr	**5438.01**
2000 x 686 x 392mm; 20.20 - 67.20kW output	5075.84	6344.80	17.24	303.46	nr	**6648.26**
Electrically heated 415V three phase supply; mounting height 2.20m						
1000 x 555 x 312mm; 4.00 - 8.00kW output	2424.62	3030.78	12.05	212.06	nr	**3242.83**
1500 x 555 x 312mm; 5.40 - 10.70kW output	3076.61	3845.76	12.05	212.06	nr	**4057.82**
2000 x 555 x 312mm; 8.00 - 16.10kW output	3663.71	4579.64	12.05	193.09	nr	**4772.73**

Material Costs/Measured Work Prices - Mechanical Installations

U:VENTILATION/AIR CONDITIONING SYSTEMS

Item	Net Price £	Material £	Labour hours	Labour £	Unit	Total rate £
U70 - AIR CURTAINS						
Commercial grade air curtains; recessed or exposed units with rigid sheet steel casing; aluminium grilles; high quality motor/centrifugal fan assembly (Continued)						
Electrically heated 415V three phase supply; mounting height 2.50m						
1000 x 555 x 312mm; 5.40 - 10.70kW output	2907.69	3634.61	16.13	258.50	nr	**3893.11**
1500 x 555 x 312mm; 7.10 - 14.30kW output	3584.40	4480.50	16.13	283.88	nr	**4764.38**
2000 x 555 x 312mm; 10.70 - 21.40kW output	4515.52	5644.40	16.13	283.88	nr	**5928.28**
Industrial grade air curtains; recessed or exposed units with rigid sheet steel casing; aluminium grilles; high quality motor/centrifugal fan assembly						
Ambient temperature 415V three phase supply; including wiring between multiple units; horizontally or vertically mounted; opening maximum 6.00m						
1106 x 516 x 689mm; 1.2A supply	2323.00	2903.75	17.24	303.46	nr	**3207.21**
1661 x 516 x 689mm; 1.8A supply	3343.00	4178.75	17.24	303.46	nr	**4482.21**
Water heated 415V three phase supply; including wiring between multiple units; horizontally or vertically mounted; opening maximum 6.00m						
1106 x 516 x 689mm; 1.2A supply; 34.80kW output	2499.00	3123.75	17.24	303.46	nr	**3427.21**
1661 x 516 x 689mm; 1.8A supply; 50.70kW output	3596.00	4495.00	17.24	303.46	nr	**4798.46**
Water heated 415V three phase supply; including wiring between multiple units; vertically mounted in single bank for openings maximum 6.00m wide or opposing twin banks for openings maximum 10.00m wide						
1106 x 689mm; 1.2A supply; 34.80kW output	2499.00	3123.75	17.24	303.46	nr	**3427.21**
1661 x 689mm; 1.8A supply; 50.70kW output	3596.00	4495.00	17.24	303.46	nr	**4798.46**
Remote mounted electronic controller unit; 415V three phase supply; excluding wiring to units						
five speed; 7A	786.00	982.50	15.00	264.01	-	**1246.51**

Material Costs/Measured Work Prices - Mechanical Installations

U:VENTILATION/AIR CONDITIONING SYSTEMS

Item	Net Price £	Material £	Labour hours	Labour £	Unit	Total rate £
U10 : SILENCERS/ACOUSTIC TREATMENT						
Y45 - SILENCERS/ACOUSTIC TREATMENT						
Attenuators; DW144 galvanised construction c/w splitters; self securing; fitted to ductwork						
To suit rectangular ducts; unit length 600mm						
100 x 100mm	20.60	27.30	0.05	1.09	nr	**28.39**
100 x 150mm	20.60	27.30	0.05	1.09	nr	**28.39**
100 x 200mm	20.60	27.30	0.07	1.53	nr	**28.82**
100 x 400mm	36.05	47.77	0.07	1.53	nr	**49.29**
150 x 150mm	20.60	27.30	0.07	1.53	nr	**28.82**
150 x 200mm	20.60	27.30	0.07	1.53	nr	**28.82**
150 x 300mm	30.90	40.94	0.07	1.53	nr	**42.47**
200 x 200mm	30.90	40.94	0.07	1.53	nr	**42.47**
200 x 300mm	36.05	47.77	0.07	1.53	nr	**49.29**
200 x 400mm	36.05	47.77	0.10	2.18	nr	**49.95**
300 x 150mm	30.90	40.94	0.10	2.18	nr	**43.13**
300 x 200mm	36.05	47.77	0.10	2.18	nr	**49.95**
300 x 300mm	41.20	54.59	0.10	2.18	nr	**56.77**
To suit rectangular ducts; unit length 1200mm						
100 x 100mm	30.90	40.94	0.08	1.75	nr	**42.69**
100 x 150mm	30.90	40.94	0.08	1.75	nr	**42.69**
100 x 200mm	30.90	40.94	0.11	2.40	nr	**43.34**
100 x 400mm	46.35	61.41	0.11	2.40	nr	**63.81**
150 x 150mm	30.90	40.94	0.11	2.40	nr	**43.34**
150 x 200mm	30.90	40.94	0.11	2.40	nr	**43.34**
150 x 300mm	36.05	47.77	0.11	2.40	nr	**50.17**
200 x 200mm	36.05	47.77	0.11	2.40	nr	**50.17**
200 x 300mm	46.35	61.41	0.11	2.40	nr	**63.81**
200 x 400mm	41.20	54.59	0.15	3.27	nr	**57.86**
300 x 150mm	51.50	68.24	0.11	2.40	nr	**70.64**
300 x 200mm	51.50	68.24	0.11	2.40	nr	**70.64**
300 x 300mm	51.50	68.24	0.15	3.27	nr	**71.51**
To suit circular ducts; unit length 600mm						
100mm dia.	51.50	68.24	0.04	0.87	nr	**69.11**
125mm dia.	51.50	68.24	0.04	0.87	nr	**69.11**
160mm dia.	51.50	68.24	0.04	0.87	nr	**69.11**
200mm dia.	72.10	95.53	0.04	0.87	nr	**96.41**
224mm dia.	72.10	95.53	0.04	0.87	nr	**96.41**
250mm dia.	72.10	95.53	0.05	1.09	nr	**96.62**
315mm dia.	82.40	109.18	0.05	1.09	nr	**110.27**
355mm dia.	82.40	109.18	0.05	1.09	nr	**110.27**
400mm dia.	92.70	122.83	0.07	1.53	nr	**124.36**
450mm dia.	92.70	122.83	0.07	1.53	nr	**124.36**
To suit circular ducts; unit length 1200mm						
100mm dia.	92.70	122.83	0.06	1.31	nr	**124.14**
125mm dia.	92.70	122.83	0.06	1.31	nr	**124.14**

Material Costs/Measured Work Prices - Mechanical Installations

U:VENTILATION/AIR CONDITIONING SYSTEMS

Item	Net Price £	Material £	Labour hours	Labour £	Unit	Total rate £
U10 : SILENCERS/ACOUSTIC TREATMENT						
Y45 - SILENCERS/ACOUSTIC TREATMENT						
Attenuators; DW144 galvanised construction c/w splitters; self securing; fitted to ductwork (Continued)						
To suit circular ducts; unit length 1200mm						
160mm dia.	92.70	122.83	0.06	1.31	nr	**124.14**
200mm dia.	123.60	163.77	0.06	1.31	nr	**165.08**
224mm dia.	123.60	163.77	0.06	1.31	nr	**165.08**
250mm dia.	123.60	163.77	0.06	1.31	nr	**165.08**
315mm dia.	154.50	204.71	0.08	1.75	nr	**206.46**
355mm dia.	154.50	204.71	0.08	1.75	nr	**206.46**
400mm dia.	169.95	225.18	0.11	2.40	nr	**227.58**
450mm dia.	169.95	225.18	0.11	2.40	nr	**227.58**

Material Costs/Measured Work Prices - Mechanical Installations   401

U:VENTILATION/AIR CONDITIONING SYSTEMS

Item	Net Price £	Material £	Labour hours	Labour £	Unit	Total rate £
U10 : THERMAL INSULATION						
Y50 - THERMAL INSULATION						
Concealed Ductwork						
Flexible wrap; 20kg-45kg Bright Class O aluminium foil faced; Bright Class O foil taped joints; 62mm metal pins and washers; aluminium bands						
40mm thick insulation	5.21	6.75	0.40	6.68	m2	**13.42**
Semi-rigid slab; 45kg Bright Class O aluminium foil faced rockwall; Bright Class O foil taped joints; 62mm metal pins and washers; aluminium bands						
40mm thick insulation	6.81	8.82	0.65	10.85	m2	**19.67**
Plantroom Ductwork						
Semi-rigid slab; 45kg Bright Class O aluminium foil faced rockwall; Bright Class O foil taped joints; 62mm metal pins and washers; 22 swg plain/embossed aluminium cladding; pop rivited						
50mm thick insulation	17.19	22.26	1.50	25.04	m2	**47.30**
External Ductwork						
Semi-rigid slab; 45kg Bright Class O aluminium foil faced rockwall; Bright Class O foil taped joints; 62mm metal pins and washers; 0.8mm polyisobutylene sheeting; welded joints						
50mm thick insulation	12.53	16.23	1.25	20.87	m2	**37.09**

Material Costs/Measured Work Prices - Mechanical Installations

U:VENTILATION/AIR CONDITIONING SYSTEMS

Item	Net Price £	Material £	Labour hours	Labour £	Unit	Total rate £
U14 : DUCTWORK : SMOKE EXTRACT						
Y30 - DUCTLINES						
Galvanised sheet metal rectangular section ductwork to BS476 Pt 24 (ISO 6944:1985); 2/4 hours fire resistance at 1100C external; including all necessary stiffeners, joints and supports in the running length						
Ductwork up to 600mm longest side						
Sum of two sides 200mm	34.53	45.75	2.91	63.51	m	**109.26**
Sum of two sides 250mm	36.01	47.71	2.90	63.29	m	**111.00**
Sum of two sides 300mm	37.49	49.67	2.99	65.26	m	**114.93**
Sum of two sides 350mm	38.96	51.62	3.08	67.22	m	**118.84**
Sum of two sides 400mm	40.44	53.58	3.17	69.18	m	**122.77**
Sum of two sides 450mm	41.92	55.54	3.27	71.37	m	**126.91**
Sum of two sides 500mm	43.40	57.51	3.37	73.55	m	**131.05**
Sum of two sides 550mm	44.88	59.47	3.46	75.51	m	**134.98**
Sum of two sides 600mm	46.35	61.41	3.54	77.26	m	**138.67**
Sum of two sides 650mm	47.83	63.37	3.63	79.22	m	**142.60**
Sum of two sides 700mm	49.31	65.34	3.72	81.19	m	**146.52**
Sum of two sides 750mm	50.79	67.30	3.81	83.15	m	**150.45**
Sum of two sides 800mm	52.27	69.26	3.90	85.12	m	**154.37**
Sum of two sides 850mm	53.74	71.21	4.78	104.32	m	**175.53**
Sum of two sides 900mm	55.22	73.17	5.04	110.00	m	**183.16**
Sum of two sides 950mm	56.70	75.13	5.31	115.89	m	**191.02**
Sum of two sides 1000mm	58.18	77.09	5.58	121.78	m	**198.87**
Sum of two sides 1050mm	59.66	79.05	5.58	121.78	m	**200.83**
Sum of two sides 1100mm	61.13	81.00	5.84	127.46	m	**208.45**
Sum of two sides 1150mm	62.61	82.96	5.84	127.46	m	**210.41**
Sum of two sides 1200mm	64.09	84.92	6.11	133.35	m	**218.27**
Extra over fittings; Ductwork up to 600mm longest side						
End Cap						
Sum of two sides 200mm	9.80	12.98	0.81	17.68	nr	**30.66**
Sum of two sides 250mm	10.10	13.38	0.83	18.11	nr	**31.50**
Sum of two sides 300mm	10.40	13.78	0.84	18.33	nr	**32.11**
Sum of two sides 350mm	10.40	13.78	0.86	18.77	nr	**32.55**
Sum of two sides 400mm	11.00	14.57	0.87	18.99	nr	**33.56**
Sum of two sides 450mm	11.30	14.97	0.88	19.21	nr	**34.18**
Sum of two sides 500mm	11.59	15.36	0.90	19.64	nr	**35.00**
Sum of two sides 550mm	11.89	15.75	0.91	19.86	nr	**35.61**
Sum of two sides 600mm	12.19	16.15	0.93	20.30	nr	**36.45**
Sum of two sides 650mm	12.49	16.55	0.94	20.52	nr	**37.06**
Sum of two sides 700mm	12.79	16.95	0.96	20.95	nr	**37.90**
Sum of two sides 750mm	13.09	17.34	0.97	21.17	nr	**38.51**
Sum of two sides 800mm	13.39	17.74	0.98	21.39	nr	**39.13**
Sum of two sides 850mm	13.69	18.14	1.14	24.88	nr	**43.02**
Sum of two sides 900mm	13.99	18.54	1.17	25.53	nr	**44.07**
Sum of two sides 950mm	14.29	18.93	1.19	25.97	nr	**44.91**

Material Costs/Measured Work Prices - Mechanical Installations 403

U:VENTILATION/AIR CONDITIONING SYSTEMS

Item	Net Price £	Material £	Labour hours	Labour £	Unit	Total rate £
Sum of two sides 1000mm	14.58	19.32	1.22	26.63	nr	**45.94**
Sum of two sides 1050mm	14.88	19.72	1.22	26.63	nr	**46.34**
Sum of two sides 1100mm	15.18	20.11	1.25	27.28	nr	**47.39**
Sum of two sides 1150mm	15.48	20.51	1.25	27.28	nr	**47.79**
Sum of two sides 1200mm	15.78	20.91	1.28	27.94	nr	**48.84**
Reducer						
Sum of two sides 200mm	38.89	51.53	2.23	48.67	nr	**100.20**
Sum of two sides 250mm	39.78	52.71	2.30	50.20	nr	**102.91**
Sum of two sides 300mm	40.68	53.90	2.37	51.72	nr	**105.63**
Sum of two sides 350mm	41.57	55.08	2.44	53.25	nr	**108.33**
Sum of two sides 400mm	42.46	56.26	2.51	54.78	nr	**111.04**
Sum of two sides 450mm	43.36	57.45	2.58	56.31	nr	**113.76**
Sum of two sides 500mm	44.25	58.63	2.65	57.84	nr	**116.47**
Sum of two sides 550mm	45.14	59.81	2.73	59.58	nr	**119.39**
Sum of two sides 600mm	46.03	60.99	2.79	60.89	nr	**121.88**
Sum of two sides 650mm	46.93	62.18	2.80	61.11	nr	**123.29**
Sum of two sides 700mm	47.82	63.36	2.87	62.64	nr	**126.00**
Sum of two sides 750mm	48.71	64.54	2.94	64.16	nr	**128.71**
Sum of two sides 800mm	49.61	65.73	3.01	65.69	nr	**131.43**
Sum of two sides 850mm	50.50	66.91	3.01	65.69	nr	**132.60**
Sum of two sides 900mm	51.39	68.09	3.06	66.78	nr	**134.88**
Sum of two sides 950mm	52.29	69.28	3.12	68.09	nr	**137.38**
Sum of two sides 1000mm	53.18	70.46	3.17	69.18	nr	**139.65**
Sum of two sides 1050mm	54.07	71.64	3.17	69.18	nr	**140.83**
Sum of two sides 1100mm	54.96	72.82	3.22	70.28	nr	**143.10**
Sum of two sides 1150mm	55.86	74.01	3.22	70.28	nr	**144.29**
Sum of two sides 1200mm	56.75	75.19	3.28	71.58	nr	**146.78**
Offset						
Sum of two sides 200mm	93.86	124.36	2.95	64.38	nr	**188.75**
Sum of two sides 250mm	95.03	125.91	3.03	66.13	nr	**192.04**
Sum of two sides 300mm	96.19	127.45	3.11	67.87	nr	**195.33**
Sum of two sides 350mm	97.36	129.00	3.18	69.40	nr	**198.40**
Sum of two sides 400mm	98.52	130.54	3.26	71.15	nr	**201.69**
Sum of two sides 450mm	99.69	132.09	3.34	72.89	nr	**204.98**
Sum of two sides 500mm	100.85	133.63	3.42	74.64	nr	**208.27**
Sum of two sides 550mm	102.02	135.18	3.50	76.39	nr	**211.56**
Sum of two sides 600mm	103.18	136.71	3.57	77.91	nr	**214.63**
Sum of two sides 650mm	104.35	138.26	3.65	79.66	nr	**217.92**
Sum of two sides 700mm	105.51	139.80	3.73	81.41	nr	**221.21**
Sum of two sides 750mm	106.68	141.35	3.81	83.15	nr	**224.50**
Sum of two sides 800mm	107.84	142.89	3.23	70.49	nr	**213.38**
Sum of two sides 850mm	109.01	144.44	3.34	72.89	nr	**217.33**
Sum of two sides 900mm	110.17	145.98	3.45	75.29	nr	**221.27**
Sum of two sides 950mm	111.34	147.53	3.56	77.70	nr	**225.22**
Sum of two sides 1000mm	112.50	149.06	3.67	80.10	nr	**229.16**
Sum of two sides 1050mm	113.67	150.61	3.67	80.10	nr	**230.71**
Sum of two sides 1100mm	114.83	152.15	3.78	82.50	nr	**234.65**
Sum of two sides 1150mm	116.00	153.70	3.78	82.50	nr	**236.20**
Sum of two sides 1200mm	117.16	155.24	3.89	84.90	nr	**240.13**

Material Costs/Measured Work Prices - Mechanical Installations

U:VENTILATION/AIR CONDITIONING SYSTEMS

Item	Net Price £	Material £	Labour hours	Labour £	Unit	Total rate £
U14 : DUCTWORK : SMOKE EXTRACT						
Y30 - DUCTLINES						
Galvanised sheet metal rectangular section ductwork to BS476 Pt 24 (ISO 6944:1985); 2/4 hours fire resistance at 1100C external; including all necessary stiffeners, joints and supports in the running length (Continued)						
90 degree radius bend						
Sum of two sides 200mm	38.06	50.43	2.06	44.96	nr	**95.39**
Sum of two sides 250mm	39.50	52.34	2.13	46.49	nr	**98.82**
Sum of two sides 300mm	40.94	54.25	2.21	48.23	nr	**102.48**
Sum of two sides 350mm	42.38	56.15	2.28	49.76	nr	**105.91**
Sum of two sides 400mm	43.82	58.06	2.36	51.51	nr	**109.57**
Sum of two sides 450mm	45.27	59.98	2.44	53.25	nr	**113.23**
Sum of two sides 500mm	46.71	61.89	2.51	54.78	nr	**116.67**
Sum of two sides 550mm	48.15	63.80	2.59	56.53	nr	**120.32**
Sum of two sides 600mm	49.59	65.71	2.66	58.05	nr	**123.76**
Sum of two sides 650mm	51.03	67.61	2.74	59.80	nr	**127.41**
Sum of two sides 700mm	52.47	69.52	2.81	61.33	nr	**130.85**
Sum of two sides 750mm	53.91	71.43	2.89	63.07	nr	**134.50**
Sum of two sides 800mm	55.35	73.34	2.97	64.82	nr	**138.16**
Sum of two sides 850mm	56.79	75.25	2.97	64.82	nr	**140.07**
Sum of two sides 900mm	58.23	77.15	2.99	65.26	nr	**142.41**
Sum of two sides 950mm	59.68	79.08	3.02	65.91	nr	**144.99**
Sum of two sides 1000mm	61.12	80.98	3.04	66.35	nr	**147.33**
Sum of two sides 1050mm	62.56	82.89	3.04	66.35	nr	**149.24**
Sum of two sides 1100mm	64.00	84.80	3.07	67.00	nr	**151.80**
Sum of two sides 1150mm	65.44	86.71	3.07	67.00	nr	**153.71**
Sum of two sides 1200mm	66.88	88.62	3.09	67.44	nr	**156.05**
45 degree radius bend						
Sum of two sides 200mm	46.93	62.18	1.52	33.17	nr	**95.36**
Sum of two sides 250mm	47.51	62.95	1.56	34.05	nr	**97.00**
Sum of two sides 300mm	48.10	63.73	1.59	34.70	nr	**98.43**
Sum of two sides 350mm	48.68	64.50	1.63	35.57	nr	**100.08**
Sum of two sides 400mm	49.26	65.27	1.66	36.23	nr	**101.50**
Sum of two sides 450mm	49.84	66.04	1.69	36.88	nr	**102.92**
Sum of two sides 500mm	50.43	66.82	1.73	37.76	nr	**104.58**
Sum of two sides 550mm	51.01	67.59	1.76	38.41	nr	**106.00**
Sum of two sides 600mm	51.59	68.36	1.80	39.28	nr	**107.64**
Sum of two sides 650mm	52.17	69.13	1.83	39.94	nr	**109.06**
Sum of two sides 700mm	52.76	69.91	1.87	40.81	nr	**110.72**
Sum of two sides 750mm	53.34	70.68	1.90	41.47	nr	**112.14**
Sum of two sides 800mm	53.92	71.44	1.93	42.12	nr	**113.57**
Sum of two sides 850mm	54.50	72.21	1.96	42.78	nr	**114.99**
Sum of two sides 900mm	55.09	72.99	1.99	43.43	nr	**116.43**
Sum of two sides 950mm	55.67	73.76	2.02	44.09	nr	**117.85**
Sum of two sides 1000mm	56.25	74.53	2.05	44.74	nr	**119.27**
Sum of two sides 1050mm	56.83	75.30	2.08	45.40	nr	**120.70**
Sum of two sides 1100mm	57.42	76.08	2.11	46.05	nr	**122.13**
Sum of two sides 1150mm	58.00	76.85	2.14	46.70	nr	**123.55**
Sum of two sides 1200mm	58.58	77.62	2.17	47.36	nr	**124.98**

Material Costs/Measured Work Prices - Mechanical Installations

U:VENTILATION/AIR CONDITIONING SYSTEMS

Item	Net Price £	Material £	Labour hours	Labour £	Unit	Total rate £
90 degree mitre bend						
Sum of two sides 200mm	39.45	52.27	2.47	53.91	nr	**106.18**
Sum of two sides 250mm	42.59	56.43	2.56	55.87	nr	**112.30**
Sum of two sides 300mm	45.73	60.59	2.65	57.84	nr	**118.43**
Sum of two sides 350mm	48.86	64.74	2.75	60.02	nr	**124.76**
Sum of two sides 400mm	52.00	68.90	2.84	61.98	nr	**130.88**
Sum of two sides 450mm	55.14	73.06	2.93	63.95	nr	**137.01**
Sum of two sides 500mm	58.28	77.22	3.02	65.91	nr	**143.13**
Sum of two sides 550mm	61.42	81.38	3.11	67.87	nr	**149.26**
Sum of two sides 600mm	64.55	85.53	3.20	69.84	nr	**155.37**
Sum of two sides 650mm	67.69	89.69	3.29	71.80	nr	**161.49**
Sum of two sides 700mm	70.83	93.85	3.38	73.77	nr	**167.62**
Sum of two sides 750mm	73.97	98.01	3.47	75.73	nr	**173.74**
Sum of two sides 800mm	77.11	102.17	3.56	77.70	nr	**179.87**
Sum of two sides 850mm	80.24	106.32	3.56	77.70	nr	**184.01**
Sum of two sides 900mm	83.38	110.48	3.59	78.35	nr	**188.83**
Sum of two sides 950mm	86.52	114.64	3.62	79.01	nr	**193.64**
Sum of two sides 1000mm	89.66	118.80	3.66	79.88	nr	**198.68**
Sum of two sides 1050mm	92.80	122.96	3.66	79.88	nr	**202.84**
Sum of two sides 1100mm	95.93	127.11	3.69	80.53	nr	**207.64**
Sum of two sides 1150mm	99.07	131.27	3.69	80.53	nr	**211.80**
Sum of two sides 1200mm	102.21	135.43	3.72	81.19	nr	**216.62**
Branch						
Sum of two sides 200mm	14.65	19.41	0.98	21.39	nr	**40.80**
Sum of two sides 250mm	14.69	19.46	1.02	22.26	nr	**41.73**
Sum of two sides 300mm	14.73	19.52	1.06	23.13	nr	**42.65**
Sum of two sides 350mm	14.77	19.57	1.10	24.01	nr	**43.58**
Sum of two sides 400mm	14.81	19.62	1.14	24.88	nr	**44.50**
Sum of two sides 450mm	14.85	19.68	1.18	25.75	nr	**45.43**
Sum of two sides 500mm	14.89	19.73	1.22	26.63	nr	**46.36**
Sum of two sides 550mm	14.93	19.78	1.26	27.50	nr	**47.28**
Sum of two sides 600mm	14.97	19.84	1.30	28.37	nr	**48.21**
Sum of two sides 650mm	15.01	19.89	1.34	29.25	nr	**49.13**
Sum of two sides 700mm	15.05	19.94	1.38	30.12	nr	**50.06**
Sum of two sides 750mm	15.08	19.98	1.42	30.99	nr	**50.97**
Sum of two sides 800mm	15.12	20.03	1.46	31.86	nr	**51.90**
Sum of two sides 850mm	15.16	20.09	1.32	28.81	nr	**48.90**
Sum of two sides 900mm	15.24	20.19	1.37	29.90	nr	**50.09**
Sum of two sides 950mm	15.24	20.19	1.41	30.77	nr	**50.97**
Sum of two sides 1000mm	15.28	20.25	1.46	31.86	nr	**52.11**
Sum of two sides 1050mm	15.32	20.30	1.46	31.86	nr	**52.16**
Sum of two sides 1100mm	15.36	20.35	1.50	32.74	nr	**53.09**
Sum of two sides 1150mm	15.40	20.41	1.50	32.74	nr	**53.14**
Sum of two sides 1200mm	15.44	20.46	1.55	33.83	nr	**54.29**
Ductwork 601 to 800mm longest side						
Sum of two sides 900mm	55.21	73.15	5.04	110.00	m	**183.15**
Sum of two sides 950mm	56.69	75.11	5.31	115.89	m	**191.00**
Sum of two sides 1000mm	58.17	77.08	5.58	121.78	m	**198.86**
Sum of two sides 1050mm	59.65	79.04	5.58	121.78	m	**200.82**
Sum of two sides 1100mm	61.13	81.00	5.84	127.46	m	**208.45**
Sum of two sides 1150mm	62.61	82.96	5.84	127.46	m	**210.41**
Sum of two sides 1200mm	64.09	84.92	6.11	133.35	m	**218.27**

Material Costs/Measured Work Prices - Mechanical Installations

U:VENTILATION/AIR CONDITIONING SYSTEMS

Item	Net Price £	Material £	Labour hours	Labour £	Unit	Total rate £
U14 : DUCTWORK : SMOKE EXTRACT						
Y30 - DUCTLINES						
Galvanised sheet metal rectangular section ductwork to BS476 Pt 24 (ISO 6944:1985); 2/4 hours fire resistance at 1100C external; including all necessary stiffeners, joints and supports in the running length (Continued)						
Ductwork 601 to 800mm longest side						
Sum of two sides 1250mm	65.58	86.89	6.11	133.35	m	**220.24**
Sum of two sides 1300mm	67.06	88.85	6.38	139.24	m	**228.10**
Sum of two sides 1350mm	68.54	90.82	6.38	139.24	m	**230.06**
Sum of two sides 1400mm	70.02	92.78	6.65	145.13	m	**237.91**
Sum of two sides 1450mm	71.50	94.74	6.65	145.13	m	**239.87**
Sum of two sides 1500mm	72.98	96.70	6.91	150.81	m	**247.51**
Sum of two sides 1550mm	74.46	98.66	6.91	150.81	m	**249.47**
Sum of two sides 1600mm	75.94	100.62	7.18	156.70	m	**257.32**
Extra over fittings; Ductwork 601 to 800mm longest side						
End Cap						
Sum of two sides 900mm	13.00	17.23	1.17	25.53	nr	**42.76**
Sum of two sides 950mm	13.46	17.83	1.19	25.97	nr	**43.81**
Sum of two sides 1000mm	13.93	18.46	1.22	26.63	nr	**45.08**
Sum of two sides 1050mm	14.39	19.07	1.22	26.63	nr	**45.69**
Sum of two sides 1100mm	14.86	19.69	1.25	27.28	nr	**46.97**
Sum of two sides 1150mm	15.32	20.30	1.25	27.28	nr	**47.58**
Sum of two sides 1200mm	15.79	20.92	1.28	27.94	nr	**48.86**
Sum of two sides 1250mm	16.25	21.53	1.28	27.94	nr	**49.47**
Sum of two sides 1300mm	16.71	22.14	1.31	28.59	nr	**50.73**
Sum of two sides 1350mm	17.18	22.76	1.31	28.59	nr	**51.35**
Sum of two sides 1400mm	17.64	23.37	1.34	29.25	nr	**52.62**
Sum of two sides 1450mm	18.11	24.00	1.34	29.25	nr	**53.24**
Sum of two sides 1500mm	18.57	24.61	1.36	29.68	nr	**54.29**
Sum of two sides 1550mm	19.04	25.23	1.36	29.68	nr	**54.91**
Sum of two sides 1600mm	19.50	25.84	1.39	30.34	nr	**56.17**
Reducer						
Sum of two sides 900mm	48.64	64.45	3.06	66.78	nr	**131.23**
Sum of two sides 950mm	51.35	68.04	3.12	68.09	nr	**136.13**
Sum of two sides 1000mm	51.35	68.04	3.17	69.18	nr	**137.22**
Sum of two sides 1050mm	52.70	69.83	3.17	69.18	nr	**139.01**
Sum of two sides 1100mm	54.05	71.62	3.22	70.28	nr	**141.89**
Sum of two sides 1150mm	55.41	73.42	3.22	70.28	nr	**143.69**
Sum of two sides 1200mm	56.76	75.21	3.28	71.58	nr	**146.79**
Sum of two sides 1250mm	58.12	77.01	3.28	71.58	nr	**148.59**
Sum of two sides 1300mm	59.47	78.80	3.33	72.68	nr	**151.47**
Sum of two sides 1350mm	60.82	80.59	3.33	72.68	nr	**153.26**
Sum of two sides 1400mm	62.18	82.39	3.38	73.77	nr	**156.16**
Sum of two sides 1450mm	63.53	84.18	3.38	73.77	nr	**157.94**

Material Costs/Measured Work Prices - Mechanical Installations 407

U:VENTILATION/AIR CONDITIONING SYSTEMS

Item	Net Price £	Material £	Labour hours	Labour £	Unit	Total rate £
Sum of two sides 1500mm	64.88	85.97	3.44	75.08	nr	**161.04**
Sum of two sides 1550mm	66.24	87.77	3.44	75.08	nr	**162.84**
Sum of two sides 1600mm	67.59	89.56	3.49	76.17	nr	**165.72**
Offset						
Sum of two sides 900mm	103.45	137.07	3.45	75.29	nr	**212.37**
Sum of two sides 950mm	105.91	140.33	3.56	77.70	nr	**218.03**
Sum of two sides 1000mm	108.38	143.60	3.67	80.10	nr	**223.70**
Sum of two sides 1050mm	110.84	146.86	3.67	80.10	nr	**226.96**
Sum of two sides 1100mm	113.30	150.12	3.78	82.50	nr	**232.62**
Sum of two sides 1150mm	115.76	153.38	3.78	82.50	nr	**235.88**
Sum of two sides 1200mm	118.23	156.65	3.89	84.90	nr	**241.55**
Sum of two sides 1250mm	120.69	159.91	3.89	84.90	nr	**244.81**
Sum of two sides 1300mm	123.15	163.17	4.00	87.30	nr	**250.47**
Sum of two sides 1350mm	125.62	166.45	4.00	87.30	nr	**253.74**
Sum of two sides 1400mm	128.08	169.71	4.11	89.70	nr	**259.41**
Sum of two sides 1450mm	130.54	172.97	4.11	89.70	nr	**262.66**
Sum of two sides 1500mm	133.00	176.22	4.23	92.32	nr	**268.54**
Sum of two sides 1550mm	135.47	179.50	4.23	92.32	nr	**271.82**
Sum of two sides 1600mm	137.93	182.76	4.34	94.72	nr	**277.48**
90 degree radius bend						
Sum of two sides 900mm	51.72	68.53	2.99	65.26	nr	**133.78**
Sum of two sides 950mm	54.26	71.89	3.02	65.91	nr	**137.80**
Sum of two sides 1000mm	56.80	75.26	3.04	66.35	nr	**141.61**
Sum of two sides 1050mm	59.34	78.63	3.04	66.35	nr	**144.97**
Sum of two sides 1100mm	61.88	81.99	3.07	67.00	nr	**148.99**
Sum of two sides 1150mm	64.42	85.36	3.07	67.00	nr	**152.36**
Sum of two sides 1200mm	66.96	88.72	3.09	67.44	nr	**156.16**
Sum of two sides 1250mm	69.50	92.09	3.09	67.44	nr	**159.53**
Sum of two sides 1300mm	72.03	95.44	3.12	68.09	nr	**163.53**
Sum of two sides 1350mm	74.57	98.81	3.12	68.09	nr	**166.90**
Sum of two sides 1400mm	77.11	102.17	3.15	68.75	nr	**170.92**
Sum of two sides 1450mm	79.65	105.54	3.15	68.75	nr	**174.28**
Sum of two sides 1500mm	82.19	108.90	3.17	69.18	nr	**178.09**
Sum of two sides 1550mm	84.73	112.27	3.17	69.18	nr	**181.45**
Sum of two sides 1600mm	87.27	115.63	3.20	69.84	nr	**185.47**
45 degree bend						
Sum of two sides 900mm	51.36	68.05	1.74	37.97	nr	**106.03**
Sum of two sides 950mm	52.56	69.64	1.79	39.07	nr	**108.71**
Sum of two sides 1000mm	53.77	71.25	1.85	40.38	nr	**111.62**
Sum of two sides 1050mm	54.97	72.84	1.85	40.38	nr	**113.21**
Sum of two sides 1100mm	56.18	74.44	1.91	41.69	nr	**116.12**
Sum of two sides 1150mm	57.38	76.03	1.91	41.69	nr	**117.71**
Sum of two sides 1200mm	58.59	77.63	1.96	42.78	nr	**120.41**
Sum of two sides 1250mm	59.79	79.22	1.96	42.78	nr	**122.00**
Sum of two sides 1300mm	60.99	80.81	2.02	44.09	nr	**124.90**
Sum of two sides 1350mm	62.20	82.42	2.02	44.09	nr	**126.50**
Sum of two sides 1400mm	63.40	84.00	2.07	45.18	nr	**129.18**
Sum of two sides 1450mm	64.61	85.61	2.07	45.18	nr	**130.79**
Sum of two sides 1500mm	65.81	87.20	2.13	46.49	nr	**133.68**
Sum of two sides 1550mm	67.02	88.80	2.13	46.49	nr	**135.29**
Sum of two sides 1600mm	68.22	90.39	2.18	47.58	nr	**137.97**

Material Costs/Measured Work Prices - Mechanical Installations

U:VENTILATION/AIR CONDITIONING SYSTEMS

Item	Net Price £	Material £	Labour hours	Labour £	Unit	Total rate £
U14 : DUCTWORK : SMOKE EXTRACT						
Y30 - DUCTLINES						
Galvanised sheet metal rectangular section ductwork to BS476 Pt 24 (ISO 6944:1985); 2/4 hours fire resistance at 1100C external; including all necessary stiffeners, joints and supports in the running length (Continued)						
90 degree mitre bend						
Sum of two sides 900mm	77.44	102.61	3.59	78.35	nr	**180.96**
Sum of two sides 950mm	81.57	108.08	3.62	79.01	nr	**187.09**
Sum of two sides 1000mm	85.70	113.55	3.66	79.88	nr	**193.43**
Sum of two sides 1050mm	89.82	119.01	3.66	79.88	nr	**198.89**
Sum of two sides 1100mm	93.95	124.48	3.69	80.53	nr	**205.02**
Sum of two sides 1150mm	98.08	129.96	3.69	80.53	nr	**210.49**
Sum of two sides 1200mm	102.21	135.43	3.72	81.19	nr	**216.62**
Sum of two sides 1250mm	106.34	140.90	3.72	81.19	nr	**222.09**
Sum of two sides 1300mm	110.46	146.36	3.75	81.84	nr	**228.20**
Sum of two sides 1350mm	114.59	151.83	3.75	81.84	nr	**233.67**
Sum of two sides 1400mm	118.72	157.30	3.78	82.50	nr	**239.80**
Sum of two sides 1450mm	122.85	162.78	3.78	82.50	nr	**245.27**
Sum of two sides 1500mm	126.97	168.24	3.81	83.15	nr	**251.39**
Sum of two sides 1550mm	131.10	173.71	3.81	83.15	nr	**256.86**
Sum of two sides 1600mm	135.23	179.18	3.84	83.81	nr	**262.99**
Branch						
Sum of two sides 900mm	14.37	19.04	1.37	29.90	nr	**48.94**
Sum of two sides 950mm	14.55	19.28	1.41	30.77	nr	**50.05**
Sum of two sides 1000mm	14.73	19.52	1.46	31.86	nr	**51.38**
Sum of two sides 1050mm	14.91	19.76	1.46	31.86	nr	**51.62**
Sum of two sides 1100mm	15.09	19.99	1.50	32.74	nr	**52.73**
Sum of two sides 1150mm	15.27	20.23	1.50	32.74	nr	**52.97**
Sum of two sides 1200mm	15.45	20.47	1.55	33.83	nr	**54.30**
Sum of two sides 1250mm	15.63	20.71	1.55	33.83	nr	**54.54**
Sum of two sides 1300mm	15.81	20.95	1.59	34.70	nr	**55.65**
Sum of two sides 1350mm	15.99	21.19	1.59	34.70	nr	**55.89**
Sum of two sides 1400mm	16.17	21.43	1.63	35.57	nr	**57.00**
Sum of two sides 1450mm	16.35	21.66	1.63	35.57	nr	**57.24**
Sum of two sides 1500mm	15.53	20.58	1.68	36.67	nr	**57.24**
Sum of two sides 1550mm	16.72	22.15	1.68	36.67	nr	**58.82**
Sum of two sides 1600mm	16.89	22.38	1.72	37.54	nr	**59.92**
Ductwork 801 to 1000mm longest side						
Sum of two sides 1100mm	61.13	81.00	5.84	127.46	m	**208.45**
Sum of two sides 1150mm	62.61	82.96	5.84	127.46	m	**210.41**
Sum of two sides 1200mm	64.09	84.92	6.11	133.35	m	**218.27**
Sum of two sides 1250mm	65.57	86.88	6.11	133.35	m	**220.23**
Sum of two sides 1300mm	67.05	88.84	6.38	139.24	m	**228.08**
Sum of two sides 1350mm	68.53	90.80	6.38	139.24	m	**230.04**
Sum of two sides 1400mm	70.01	92.76	6.65	145.13	m	**237.90**
Sum of two sides 1450mm	71.49	94.72	6.65	145.13	m	**239.86**

Material Costs/Measured Work Prices - Mechanical Installations

U:VENTILATION/AIR CONDITIONING SYSTEMS

Item	Net Price £	Material £	Labour hours	Labour £	Unit	Total rate £
Sum of two sides 1500mm	72.97	96.69	6.91	150.81	m	**247.49**
Sum of two sides 1550mm	74.45	98.65	6.91	150.81	m	**249.45**
Sum of two sides 1600mm	75.93	100.61	7.18	156.70	m	**257.31**
Sum of two sides 1650mm	77.41	102.57	7.18	156.70	m	**259.27**
Sum of two sides 1700mm	78.89	104.53	7.45	162.59	m	**267.12**
Sum of two sides 1750mm	80.37	106.49	7.45	162.59	m	**269.08**
Sum of two sides 1800mm	81.85	108.45	7.71	168.27	m	**276.72**
Sum of two sides 1850mm	83.33	110.41	7.71	168.27	m	**278.68**
Sum of two sides 1900mm	84.81	112.37	7.98	174.16	m	**286.53**
Sum of two sides 1950mm	86.29	114.33	7.98	174.16	m	**288.49**
Sum of two sides 2000mm	87.77	116.30	8.25	180.05	m	**296.35**
Extra over fittings; Ductwork 801 to 1000mm longest side						
End Cap						
Sum of two sides 1100mm	14.86	19.69	1.25	27.28	nr	**46.97**
Sum of two sides 1150mm	15.32	20.30	1.25	27.28	nr	**47.58**
Sum of two sides 1200mm	15.79	20.92	1.28	27.94	nr	**48.86**
Sum of two sides 1250mm	16.25	21.53	1.28	27.94	nr	**49.47**
Sum of two sides 1300mm	16.71	22.14	1.31	28.59	nr	**50.73**
Sum of two sides 1350mm	17.18	22.76	1.31	28.59	nr	**51.35**
Sum of two sides 1400mm	17.64	23.37	1.34	29.25	nr	**52.62**
Sum of two sides 1450mm	18.10	23.98	1.34	29.25	nr	**53.23**
Sum of two sides 1500mm	18.57	24.61	1.36	29.68	nr	**54.29**
Sum of two sides 1550mm	19.03	25.21	1.36	29.68	nr	**54.90**
Sum of two sides 1600mm	19.49	25.82	1.39	30.34	nr	**56.16**
Sum of two sides 1650mm	19.96	26.45	1.39	30.34	nr	**56.78**
Sum of two sides 1700mm	20.42	27.06	1.42	30.99	nr	**58.05**
Sum of two sides 1750mm	20.88	27.67	1.42	30.99	nr	**58.66**
Sum of two sides 1800mm	21.35	28.29	1.45	31.65	nr	**59.93**
Sum of two sides 1850mm	21.81	28.90	1.45	31.65	nr	**60.54**
Sum of two sides 1900mm	22.27	29.51	1.48	32.30	nr	**61.81**
Sum of two sides 1950mm	22.74	30.13	1.48	32.30	nr	**62.43**
Sum of two sides 2000mm	23.20	30.74	1.51	32.96	nr	**63.70**
Reducer						
Sum of two sides 1100mm	54.05	71.62	3.22	70.28	nr	**141.89**
Sum of two sides 1150mm	55.41	73.42	3.22	70.28	nr	**143.69**
Sum of two sides 1200mm	56.77	75.22	3.28	71.58	nr	**146.81**
Sum of two sides 1250mm	58.13	77.02	3.28	71.58	nr	**148.61**
Sum of two sides 1300mm	59.49	78.82	3.33	72.68	nr	**151.50**
Sum of two sides 1350mm	60.85	80.63	3.33	72.68	nr	**153.30**
Sum of two sides 1400mm	62.21	82.43	3.38	73.77	nr	**156.20**
Sum of two sides 1450mm	63.57	84.23	3.38	73.77	nr	**158.00**
Sum of two sides 1500mm	64.93	86.03	3.44	75.08	nr	**161.11**
Sum of two sides 1550mm	66.30	87.85	3.44	75.08	nr	**162.92**
Sum of two sides 1600mm	67.66	89.65	3.49	76.17	nr	**165.82**
Sum of two sides 1650mm	69.02	91.45	3.49	76.17	nr	**167.62**
Sum of two sides 1700mm	70.38	93.25	3.54	77.26	nr	**170.51**
Sum of two sides 1750mm	71.74	95.06	3.54	77.26	nr	**172.31**
Sum of two sides 1800mm	73.10	96.86	3.60	78.57	nr	**175.43**
Sum of two sides 1850mm	74.46	98.66	3.60	78.57	nr	**177.23**
Sum of two sides 1900mm	75.82	100.46	3.65	79.66	nr	**180.12**
Sum of two sides 1950mm	77.18	102.26	3.65	79.66	nr	**181.92**
Sum of two sides 2000mm	79.90	105.87	3.70	80.75	nr	**186.62**

410   *Material Costs/Measured Work Prices - Mechanical Installations*

U:VENTILATION/AIR CONDITIONING SYSTEMS

Item	Net Price £	Material £	Labour hours	Labour £	Unit	Total rate £
U14 : DUCTWORK : SMOKE EXTRACT						
Y30 - DUCTLINES						
Extra over fittings; Ductwork 801 to 1000mm longest side (Continued)						
Offset						
Sum of two sides 1100mm	112.41	148.94	3.78	82.50	nr	**231.44**
Sum of two sides 1150mm	114.79	152.10	3.78	82.50	nr	**234.59**
Sum of two sides 1200mm	117.16	155.24	3.89	84.90	nr	**240.13**
Sum of two sides 1250mm	119.54	158.39	3.89	84.90	nr	**243.29**
Sum of two sides 1300mm	121.91	161.53	4.00	87.30	nr	**248.83**
Sum of two sides 1350mm	124.29	164.68	4.00	87.30	nr	**251.98**
Sum of two sides 1400mm	126.66	167.82	4.11	89.70	nr	**257.52**
Sum of two sides 1450mm	129.04	170.98	4.11	89.70	nr	**260.68**
Sum of two sides 1500mm	131.41	174.12	4.23	92.32	nr	**266.44**
Sum of two sides 1550mm	133.79	177.27	4.23	92.32	nr	**269.59**
Sum of two sides 1600mm	136.17	180.43	4.34	94.72	nr	**275.14**
Sum of two sides 1650mm	138.54	183.57	4.34	94.72	nr	**278.28**
Sum of two sides 1700mm	140.92	186.72	4.45	97.12	nr	**283.84**
Sum of two sides 1750mm	143.92	190.69	4.45	97.12	nr	**287.81**
Sum of two sides 1800mm	145.67	193.01	4.56	99.52	nr	**292.53**
Sum of two sides 1850mm	148.04	196.15	4.56	99.52	nr	**295.67**
Sum of two sides 1900mm	150.42	199.31	4.67	101.92	nr	**301.23**
Sum of two sides 1950mm	152.79	202.45	4.67	101.92	nr	**304.37**
Sum of two sides 2000mm	157.55	208.75	5.53	120.69	nr	**329.44**
90 degree radius bend						
Sum of two sides 1100mm	61.78	81.86	3.07	67.00	nr	**148.86**
Sum of two sides 1150mm	64.32	85.22	3.07	67.00	nr	**152.23**
Sum of two sides 1200mm	66.87	88.60	3.09	67.44	nr	**156.04**
Sum of two sides 1250mm	69.41	91.97	3.09	67.44	nr	**159.41**
Sum of two sides 1300mm	71.96	95.35	3.12	68.09	nr	**163.44**
Sum of two sides 1350mm	74.50	98.71	3.12	68.09	nr	**166.81**
Sum of two sides 1400mm	77.05	102.09	3.15	68.75	nr	**170.84**
Sum of two sides 1450mm	79.59	105.46	3.15	68.75	nr	**174.20**
Sum of two sides 1500mm	82.14	108.84	3.17	69.18	nr	**178.02**
Sum of two sides 1550mm	84.68	112.20	3.17	69.18	nr	**181.38**
Sum of two sides 1600mm	87.22	115.57	3.20	69.84	nr	**185.41**
Sum of two sides 1650mm	89.77	118.95	3.20	69.84	nr	**188.78**
Sum of two sides 1700mm	92.31	122.31	3.22	70.28	nr	**192.59**
Sum of two sides 1750mm	94.86	125.69	3.22	70.28	nr	**195.96**
Sum of two sides 1800mm	97.40	129.06	3.25	70.93	nr	**199.99**
Sum of two sides 1850mm	99.95	132.43	3.25	70.93	nr	**203.36**
Sum of two sides 1900mm	102.49	135.80	3.28	71.58	nr	**207.38**
Sum of two sides 1950mm	105.04	139.18	3.28	71.58	nr	**210.76**
Sum of two sides 2000mm	107.58	142.54	3.30	72.02	nr	**214.56**
45 degree bend						
Sum of two sides 1100mm	56.55	74.93	1.91	41.69	nr	**116.61**
Sum of two sides 1150mm	57.73	76.49	1.91	41.69	nr	**118.18**
Sum of two sides 1200mm	58.92	78.07	1.96	42.78	nr	**120.85**
Sum of two sides 1250mm	58.92	78.07	1.96	42.78	nr	**120.85**

Material Costs/Measured Work Prices - Mechanical Installations

U:VENTILATION/AIR CONDITIONING SYSTEMS

Item	Net Price £	Material £	Labour hours	Labour £	Unit	Total rate £
Sum of two sides 1300mm	61.28	81.20	2.02	44.09	nr	**125.28**
Sum of two sides 1350mm	62.46	82.76	2.02	44.09	nr	**126.85**
Sum of two sides 1400mm	63.65	84.34	2.07	45.18	nr	**129.51**
Sum of two sides 1450mm	64.83	85.90	2.07	45.18	nr	**131.08**
Sum of two sides 1500mm	66.01	87.46	2.13	46.49	nr	**133.95**
Sum of two sides 1550mm	67.20	89.04	2.13	46.49	nr	**135.53**
Sum of two sides 1600mm	68.38	90.60	2.18	47.58	nr	**138.18**
Sum of two sides 1650mm	69.56	92.17	2.18	47.58	nr	**139.74**
Sum of two sides 1700mm	70.74	93.73	2.24	48.89	nr	**142.62**
Sum of two sides 1750mm	71.93	95.31	2.24	48.89	nr	**144.19**
Sum of two sides 1800mm	73.11	96.87	2.30	50.20	nr	**147.07**
Sum of two sides 1850mm	74.29	98.43	2.30	50.20	nr	**148.63**
Sum of two sides 1900mm	75.47	100.00	2.35	51.29	nr	**151.29**
Sum of two sides 1950mm	76.66	101.57	2.35	51.29	nr	**152.86**
Sum of two sides 2000mm	77.84	103.14	2.76	60.24	nr	**163.37**
90 degree mitre bend						
Sum of two sides 1100mm	93.95	124.48	3.69	80.53	nr	**205.02**
Sum of two sides 1150mm	98.08	129.96	3.69	80.53	nr	**210.49**
Sum of two sides 1200mm	102.21	135.43	3.72	81.19	nr	**216.62**
Sum of two sides 1250mm	106.34	140.90	3.72	81.19	nr	**222.09**
Sum of two sides 1300mm	110.46	146.36	3.75	81.84	nr	**228.20**
Sum of two sides 1350mm	114.59	151.83	3.75	81.84	nr	**233.67**
Sum of two sides 1400mm	118.72	157.30	3.78	82.50	nr	**239.80**
Sum of two sides 1450mm	122.85	162.78	3.78	82.50	nr	**245.27**
Sum of two sides 1500mm	126.98	168.25	3.81	83.15	nr	**251.40**
Sum of two sides 1550mm	131.11	173.72	3.81	83.15	nr	**256.87**
Sum of two sides 1600mm	135.23	179.18	3.84	83.81	nr	**262.99**
Sum of two sides 1650mm	139.36	184.65	3.84	83.81	nr	**268.46**
Sum of two sides 1700mm	143.49	190.12	3.87	84.46	nr	**274.59**
Sum of two sides 1750mm	147.62	195.60	3.87	84.46	nr	**280.06**
Sum of two sides 1800mm	151.75	201.07	3.90	85.12	nr	**286.18**
Sum of two sides 1850mm	155.88	206.54	3.90	85.12	nr	**291.66**
Sum of two sides 1900mm	160.00	212.00	3.93	85.77	nr	**297.77**
Sum of two sides 1950mm	164.13	217.47	3.93	85.77	nr	**303.24**
Sum of two sides 2000mm	168.26	222.94	3.96	86.43	nr	**309.37**
Branch						
Sum of two sides 1100mm	15.07	19.97	1.50	32.74	nr	**52.70**
Sum of two sides 1150mm	15.25	20.21	1.50	32.74	nr	**52.94**
Sum of two sides 1200mm	15.43	20.44	1.55	33.83	nr	**54.27**
Sum of two sides 1250mm	15.62	20.70	1.55	33.83	nr	**54.52**
Sum of two sides 1300mm	15.80	20.93	1.59	34.70	nr	**55.64**
Sum of two sides 1350mm	15.98	21.17	1.59	34.70	nr	**55.87**
Sum of two sides 1400mm	16.16	21.41	1.63	35.57	nr	**56.99**
Sum of two sides 1450mm	16.35	21.66	1.63	35.57	nr	**57.24**
Sum of two sides 1500mm	16.53	21.90	1.68	36.67	nr	**58.57**
Sum of two sides 1550mm	16.71	22.14	1.68	36.67	nr	**58.81**
Sum of two sides 1600mm	16.89	22.38	1.72	37.54	nr	**59.92**
Sum of two sides 1650mm	17.07	22.62	1.72	37.54	nr	**60.16**
Sum of two sides 1700mm	17.26	22.87	1.77	38.63	nr	**61.50**
Sum of two sides 1750mm	17.44	23.11	1.77	38.63	nr	**61.74**
Sum of two sides 1800mm	17.62	23.35	1.81	39.50	nr	**62.85**
Sum of two sides 1850mm	17.80	23.59	1.81	39.50	nr	**63.09**

412 **Material Costs/Measured Work Prices - Mechanical Installations**

U:VENTILATION/AIR CONDITIONING SYSTEMS

Item	Net Price £	Material £	Labour hours	Labour £	Unit	Total rate £
U14 : DUCTWORK : SMOKE EXTRACT						
Y30 - DUCTLINES						
Extra over fittings; Ductwork 801 to 1000mm longest side (Continued)						
Branch						
Sum of two sides 1900mm	17.99	23.84	1.86	40.59	nr	**64.43**
Sum of two sides 1950mm	18.17	24.08	1.86	40.59	nr	**64.67**
Sum of two sides 2000mm	18.53	24.55	1.90	41.47	nr	**66.02**
Ductwork 1001 to 1250mm longest side						
Sum of two sides 1300mm	75.99	100.69	6.38	139.24	m	**239.93**
Sum of two sides 1350mm	77.63	102.86	6.38	139.24	m	**242.10**
Sum of two sides 1400mm	79.27	105.03	6.65	145.13	m	**250.17**
Sum of two sides 1450mm	80.91	107.21	6.65	145.13	m	**252.34**
Sum of two sides 1500mm	82.55	109.38	6.91	150.81	m	**260.19**
Sum of two sides 1550mm	84.19	111.55	6.91	150.81	m	**262.36**
Sum of two sides 1600mm	85.01	112.64	6.91	150.81	m	**263.45**
Sum of two sides 1650mm	85.82	113.71	6.91	150.81	m	**264.52**
Sum of two sides 1700mm	87.46	115.88	7.45	162.59	m	**278.48**
Sum of two sides 1750mm	89.09	118.04	7.45	162.59	m	**280.64**
Sum of two sides 1800mm	90.73	120.22	7.71	168.27	m	**288.49**
Sum of two sides 1850mm	92.37	122.39	7.71	168.27	m	**290.66**
Sum of two sides 1900mm	94.00	124.55	7.98	174.16	m	**298.71**
Sum of two sides 1950mm	95.94	127.12	7.98	174.16	m	**301.28**
Sum of two sides 2000mm	97.28	128.90	8.25	180.05	m	**308.95**
Sum of two sides 2050mm	98.91	131.06	8.25	180.05	m	**311.11**
Sum of two sides 2100mm	100.55	133.23	9.62	209.95	m	**343.18**
Sum of two sides 2150mm	102.18	135.39	9.62	209.95	m	**345.34**
Sum of two sides 2200mm	103.00	136.47	9.62	209.95	m	**346.43**
Sum of two sides 2250mm	103.82	137.56	9.62	209.95	m	**347.51**
Sum of two sides 2300mm	105.46	139.73	10.07	219.77	m	**359.51**
Sum of two sides 2350mm	107.09	141.89	10.07	219.77	m	**361.67**
Sum of two sides 2400mm	108.73	144.07	10.47	228.50	m	**372.57**
Sum of two sides 2450mm	110.36	146.23	10.47	228.50	m	**374.73**
Sum of two sides 2500mm	112.00	148.40	10.87	237.23	m	**385.63**
Extra over fittings; Ductwork 1001 to 1250mm longest side						
End Cap						
Sum of two sides 1300mm	16.72	22.15	1.31	28.59	nr	**50.74**
Sum of two sides 1350mm	17.18	22.76	1.31	28.59	nr	**51.35**
Sum of two sides 1400mm	17.64	23.37	1.34	29.25	nr	**52.62**
Sum of two sides 1450mm	18.10	23.98	1.34	29.25	nr	**53.23**
Sum of two sides 1500mm	18.56	24.59	1.36	29.68	nr	**54.27**
Sum of two sides 1550mm	19.02	25.20	1.36	29.68	nr	**54.88**
Sum of two sides 1600mm	19.49	25.82	1.39	30.34	nr	**56.16**
Sum of two sides 1650mm	19.95	26.43	1.39	30.34	nr	**56.77**
Sum of two sides 1700mm	20.42	27.06	1.42	30.99	nr	**58.05**
Sum of two sides 1750mm	20.88	27.67	1.42	30.99	nr	**58.66**
Sum of two sides 1800mm	21.34	28.28	1.45	31.65	nr	**59.92**

Material Costs/Measured Work Prices - Mechanical Installations

U:VENTILATION/AIR CONDITIONING SYSTEMS

Item	Net Price £	Material £	Labour hours	Labour £	Unit	Total rate £
Sum of two sides 1850mm	21.81	28.90	1.45	31.65	nr	**60.54**
Sum of two sides 1900mm	22.27	29.51	1.48	32.30	nr	**61.81**
Sum of two sides 1950mm	22.74	30.13	1.48	32.30	nr	**62.43**
Sum of two sides 2000mm	23.20	30.74	1.51	32.96	nr	**63.70**
Sum of two sides 2050mm	23.66	31.35	1.51	32.96	nr	**64.30**
Sum of two sides 2100mm	24.13	31.97	2.66	58.05	nr	**90.03**
Sum of two sides 2150mm	24.59	32.58	2.66	58.05	nr	**90.64**
Sum of two sides 2200mm	25.06	33.20	2.80	61.11	nr	**94.31**
Sum of two sides 2250mm	25.52	33.81	2.80	61.11	nr	**94.92**
Sum of two sides 2300mm	25.98	34.42	2.95	64.38	nr	**98.81**
Sum of two sides 2350mm	26.45	35.05	2.95	64.38	nr	**99.43**
Sum of two sides 2400mm	26.91	35.66	3.10	67.66	nr	**103.31**
Sum of two sides 2450mm	27.38	36.28	3.10	67.66	nr	**103.93**
Sum of two sides 2500mm	27.84	36.89	3.24	70.71	nr	**107.60**
Reducer						
Sum of two sides 1300mm	59.47	78.80	3.33	72.68	nr	**151.47**
Sum of two sides 1350mm	60.81	80.57	3.33	72.68	nr	**153.25**
Sum of two sides 1400mm	62.15	82.35	3.38	73.77	nr	**156.12**
Sum of two sides 1450mm	63.49	84.12	3.38	73.77	nr	**157.89**
Sum of two sides 1500mm	64.83	85.90	3.44	75.08	nr	**160.98**
Sum of two sides 1550mm	66.17	87.68	3.44	75.08	nr	**162.75**
Sum of two sides 1600mm	67.52	89.46	3.49	76.17	nr	**165.63**
Sum of two sides 1650mm	68.86	91.24	3.49	76.17	nr	**167.41**
Sum of two sides 1700mm	70.21	93.03	3.54	77.26	nr	**170.29**
Sum of two sides 1750mm	71.55	94.80	3.54	77.26	nr	**172.06**
Sum of two sides 1800mm	72.90	96.59	3.60	78.57	nr	**175.16**
Sum of two sides 1850mm	74.24	98.37	3.60	78.57	nr	**176.94**
Sum of two sides 1900mm	75.59	100.16	3.65	79.66	nr	**179.82**
Sum of two sides 1950mm	76.93	101.93	3.65	79.66	nr	**181.59**
Sum of two sides 2000mm	78.28	103.72	3.70	80.75	nr	**184.47**
Sum of two sides 2050mm	79.62	105.50	3.70	80.75	nr	**186.25**
Sum of two sides 2100mm	80.96	107.27	3.75	81.84	nr	**189.11**
Sum of two sides 2150mm	82.31	109.06	3.75	81.84	nr	**190.90**
Sum of two sides 2200mm	83.65	110.84	3.80	82.93	nr	**193.77**
Sum of two sides 2250mm	85.00	112.63	3.80	82.93	nr	**195.56**
Sum of two sides 2300mm	86.34	114.40	3.85	84.02	nr	**198.43**
Sum of two sides 2350mm	87.69	116.19	3.85	84.02	nr	**200.21**
Sum of two sides 2400mm	89.03	117.96	3.90	85.12	nr	**203.08**
Sum of two sides 2450mm	90.38	119.75	3.90	85.12	nr	**204.87**
Sum of two sides 2500mm	91.72	121.53	3.95	86.21	nr	**207.74**
Offset						
Sum of two sides 1300mm	121.51	161.00	4.00	87.30	nr	**248.30**
Sum of two sides 1350mm	123.92	164.19	4.00	87.30	nr	**251.49**
Sum of two sides 1400mm	126.33	167.39	4.11	89.70	nr	**257.09**
Sum of two sides 1450mm	128.74	170.58	4.11	89.70	nr	**260.28**
Sum of two sides 1500mm	131.15	173.77	4.23	92.32	nr	**266.09**
Sum of two sides 1550mm	133.56	176.97	4.23	92.32	nr	**269.29**
Sum of two sides 1600mm	135.97	180.16	4.34	94.72	nr	**274.88**
Sum of two sides 1650mm	138.37	183.34	4.34	94.72	nr	**278.06**
Sum of two sides 1700mm	140.78	186.53	4.45	97.12	nr	**283.65**
Sum of two sides 1750mm	143.19	189.73	4.45	97.12	nr	**286.85**
Sum of two sides 1800mm	145.60	192.92	4.56	99.52	nr	**292.44**

Material Costs/Measured Work Prices - Mechanical Installations

U:VENTILATION/AIR CONDITIONING SYSTEMS

Item	Net Price £	Material £	Labour hours	Labour £	Unit	Total rate £
U14 : DUCTWORK : SMOKE EXTRACT						
Y30 - DUCTLINES						
Extra over fittings; Ductwork 1001 to 1250mm longest side (Continued)						
Offset						
Sum of two sides 1850mm	148.01	196.11	4.56	99.52	nr	**295.63**
Sum of two sides 1900mm	150.41	199.29	4.67	101.92	nr	**301.21**
Sum of two sides 1950mm	152.82	202.49	4.67	101.92	nr	**304.41**
Sum of two sides 2000mm	155.23	205.68	5.53	120.69	nr	**326.37**
Sum of two sides 2050mm	157.64	208.87	5.53	120.69	nr	**329.56**
Sum of two sides 2100mm	160.05	212.07	5.76	125.71	nr	**337.78**
Sum of two sides 2150mm	162.45	215.25	5.76	125.71	nr	**340.96**
Sum of two sides 2200mm	164.86	218.44	5.99	130.73	nr	**349.17**
Sum of two sides 2250mm	167.27	221.63	5.99	130.73	nr	**352.36**
Sum of two sides 2300mm	169.68	224.83	6.22	135.75	nr	**360.57**
Sum of two sides 2350mm	172.09	228.02	6.22	135.75	nr	**363.77**
Sum of two sides 2400mm	174.49	231.20	6.45	140.77	nr	**371.97**
Sum of two sides 2450mm	176.90	234.39	6.45	140.77	nr	**375.16**
Sum of two sides 2500mm	179.31	237.59	6.68	145.79	nr	**383.37**
90 degree radius bend						
Sum of two sides 1300mm	71.87	95.23	3.12	68.09	nr	**163.32**
Sum of two sides 1350mm	74.42	98.61	3.12	68.09	nr	**166.70**
Sum of two sides 1400mm	79.97	105.96	3.15	68.75	nr	**174.71**
Sum of two sides 1450mm	79.52	105.36	3.15	68.75	nr	**174.11**
Sum of two sides 1500mm	82.07	108.74	3.17	69.18	nr	**177.93**
Sum of two sides 1550mm	84.62	112.12	3.17	69.18	nr	**181.31**
Sum of two sides 1600mm	87.17	115.50	3.20	69.84	nr	**185.34**
Sum of two sides 1650mm	89.72	118.88	3.20	69.84	nr	**188.72**
Sum of two sides 1700mm	92.28	122.27	3.22	70.28	nr	**192.55**
Sum of two sides 1750mm	94.83	125.65	3.22	70.28	nr	**195.92**
Sum of two sides 1800mm	97.38	129.03	3.25	70.93	nr	**199.96**
Sum of two sides 1850mm	99.93	132.41	3.25	70.93	nr	**203.34**
Sum of two sides 1900mm	102.48	135.79	3.28	71.58	nr	**207.37**
Sum of two sides 1950mm	105.03	139.16	3.28	71.58	nr	**210.75**
Sum of two sides 2000mm	107.59	142.56	3.30	72.02	nr	**214.58**
Sum of two sides 2050mm	110.14	145.94	3.30	72.02	nr	**217.96**
Sum of two sides 2100mm	112.69	149.31	3.32	72.46	nr	**221.77**
Sum of two sides 2150mm	115.24	152.69	3.32	72.46	nr	**225.15**
Sum of two sides 2200mm	117.79	156.07	3.34	72.89	nr	**228.97**
Sum of two sides 2250mm	120.34	159.45	3.34	72.89	nr	**232.34**
Sum of two sides 2300mm	122.89	162.83	3.36	73.33	nr	**236.16**
Sum of two sides 2350mm	125.45	166.22	3.36	73.33	nr	**239.55**
Sum of two sides 2400mm	128.00	169.60	3.38	73.77	nr	**243.37**
Sum of two sides 2450mm	130.55	172.98	3.38	73.77	nr	**246.75**
Sum of two sides 2500mm	133.10	176.36	3.40	74.20	nr	**250.56**
45 degree bend						
Sum of two sides 1300mm	58.21	77.13	2.02	44.09	nr	**121.21**
Sum of two sides 1350mm	59.32	78.60	2.02	44.09	nr	**122.68**
Sum of two sides 1400mm	60.83	80.60	2.07	45.18	nr	**125.78**

Material Costs/Measured Work Prices - Mechanical Installations

U:VENTILATION/AIR CONDITIONING SYSTEMS

Item	Net Price £	Material £	Labour hours	Labour £	Unit	Total rate £
Sum of two sides 1450mm	62.14	82.34	2.07	45.18	nr	**127.51**
Sum of two sides 1500mm	63.45	84.07	2.13	46.49	nr	**130.56**
Sum of two sides 1550mm	64.76	85.81	2.13	46.49	nr	**132.29**
Sum of two sides 1600mm	66.07	87.54	2.18	47.58	nr	**135.12**
Sum of two sides 1650mm	67.38	89.28	2.18	47.58	nr	**136.86**
Sum of two sides 1700mm	68.69	91.01	2.24	48.89	nr	**139.90**
Sum of two sides 1750mm	70.00	92.75	2.24	48.89	nr	**141.64**
Sum of two sides 1800mm	71.31	94.49	2.30	50.20	nr	**144.68**
Sum of two sides 1850mm	72.62	96.22	2.30	50.20	nr	**146.42**
Sum of two sides 1900mm	73.93	97.96	2.35	51.29	nr	**149.25**
Sum of two sides 1950mm	75.24	99.69	2.35	51.29	nr	**150.98**
Sum of two sides 2000mm	76.56	101.44	2.76	60.24	nr	**161.68**
Sum of two sides 2050mm	77.87	103.18	2.76	60.24	nr	**163.41**
Sum of two sides 2100mm	79.18	104.91	2.89	63.07	nr	**167.99**
Sum of two sides 2150mm	80.49	106.65	2.89	63.07	nr	**169.72**
Sum of two sides 2200mm	81.80	108.39	3.01	65.69	nr	**174.08**
Sum of two sides 2250mm	83.11	110.12	3.01	65.69	nr	**175.81**
Sum of two sides 2300mm	84.42	111.86	3.13	68.31	nr	**180.17**
Sum of two sides 2350mm	85.73	113.59	3.13	68.31	nr	**181.90**
Sum of two sides 2400mm	87.04	115.33	3.26	71.15	nr	**186.48**
Sum of two sides 2450mm	88.35	117.06	3.26	71.15	nr	**188.21**
Sum of two sides 2500mm	89.66	118.80	3.37	73.55	nr	**192.35**
90 degree mitre bend						
Sum of two sides 1300mm	110.21	146.03	3.75	81.84	nr	**227.87**
Sum of two sides 1350mm	114.38	151.55	3.75	81.84	nr	**233.40**
Sum of two sides 1400mm	118.55	157.08	3.78	82.50	nr	**239.58**
Sum of two sides 1450mm	122.72	162.60	3.78	82.50	nr	**245.10**
Sum of two sides 1500mm	126.89	168.13	3.81	83.15	nr	**251.28**
Sum of two sides 1550mm	131.06	173.65	3.81	83.15	nr	**256.81**
Sum of two sides 1600mm	135.22	179.17	3.84	83.81	nr	**262.97**
Sum of two sides 1650mm	139.39	184.69	3.84	83.81	nr	**268.50**
Sum of two sides 1700mm	143.55	190.20	3.87	84.46	nr	**274.67**
Sum of two sides 1750mm	147.72	195.73	3.87	84.46	nr	**280.19**
Sum of two sides 1800mm	151.88	201.24	3.90	85.12	nr	**286.36**
Sum of two sides 1850mm	156.05	206.77	3.90	85.12	nr	**291.88**
Sum of two sides 1900mm	160.21	212.28	3.93	85.77	nr	**298.05**
Sum of two sides 1950mm	164.38	217.80	3.93	85.77	nr	**303.57**
Sum of two sides 2000mm	168.55	223.33	3.96	86.43	nr	**309.75**
Sum of two sides 2050mm	172.71	228.84	3.96	86.43	nr	**315.27**
Sum of two sides 2100mm	176.88	234.37	3.99	87.08	nr	**321.45**
Sum of two sides 2150mm	181.04	239.88	3.99	87.08	nr	**326.96**
Sum of two sides 2200mm	185.21	245.40	4.02	87.73	nr	**333.14**
Sum of two sides 2250mm	189.37	250.92	4.02	87.73	nr	**338.65**
Sum of two sides 2300mm	193.54	256.44	4.05	88.39	nr	**344.83**
Sum of two sides 2350mm	197.70	261.95	4.05	88.39	nr	**350.34**
Sum of two sides 2400mm	201.87	267.48	4.08	89.04	nr	**356.52**
Sum of two sides 2450mm	206.03	272.99	4.08	89.04	nr	**362.03**
Sum of two sides 2500mm	210.20	278.51	4.11	89.70	nr	**368.21**
Branch						
Sum of two sides 1300mm	15.83	20.97	1.59	34.70	nr	**55.68**
Sum of two sides 1350mm	16.01	21.21	1.59	34.70	nr	**55.91**
Sum of two sides 1400mm	16.19	21.45	1.63	35.57	nr	**57.03**

Material Costs/Measured Work Prices - Mechanical Installations

U:VENTILATION/AIR CONDITIONING SYSTEMS

Item	Net Price £	Material £	Labour hours	Labour £	Unit	Total rate £
U14 : DUCTWORK : SMOKE EXTRACT						
Y30 - DUCTLINES						
Extra over fittings; Ductwork 1001 to 1250mm longest side (Continued)						
Branch						
Sum of two sides 1450mm	16.37	21.69	1.63	35.57	nr	**57.26**
Sum of two sides 1500mm	16.55	21.93	1.68	36.67	nr	**58.59**
Sum of two sides 1550mm	16.73	22.17	1.68	36.67	nr	**58.83**
Sum of two sides 1600mm	16.91	22.41	1.72	37.54	nr	**59.94**
Sum of two sides 1650mm	17.09	22.64	1.72	37.54	nr	**60.18**
Sum of two sides 1700mm	17.28	22.90	1.77	38.63	nr	**61.53**
Sum of two sides 1750mm	17.46	23.13	1.77	38.63	nr	**61.76**
Sum of two sides 1800mm	17.64	23.37	1.81	39.50	nr	**62.88**
Sum of two sides 1850mm	17.82	23.61	1.81	39.50	nr	**63.11**
Sum of two sides 1900mm	18.00	23.85	1.86	40.59	nr	**64.44**
Sum of two sides 1950mm	18.18	24.09	1.86	40.59	nr	**64.68**
Sum of two sides 2000mm	18.37	24.34	1.90	41.47	nr	**65.81**
Sum of two sides 2050mm	18.55	24.58	1.90	41.47	nr	**66.05**
Sum of two sides 2100mm	18.73	24.82	2.58	56.31	nr	**81.12**
Sum of two sides 2150mm	18.91	25.06	2.58	56.31	nr	**81.36**
Sum of two sides 2200mm	19.09	25.29	2.61	56.96	nr	**82.26**
Sum of two sides 2250mm	19.27	25.53	2.61	56.96	nr	**82.50**
Sum of two sides 2300mm	19.45	25.77	2.64	57.62	nr	**83.39**
Sum of two sides 2350mm	19.64	26.02	2.64	57.62	nr	**83.64**
Sum of two sides 2400mm	19.82	26.26	2.88	62.85	nr	**89.12**
Sum of two sides 2450mm	20.00	26.50	2.88	62.85	nr	**89.35**
Sum of two sides 2500mm	20.18	26.74	2.91	63.51	nr	**90.25**
Ductwork 1251 to 2000mm longest side						
Sum of two sides 1700mm	83.34	110.43	7.45	162.59	m	**273.02**
Sum of two sides 1750mm	86.01	113.96	7.45	162.59	m	**276.56**
Sum of two sides 1800mm	88.68	117.50	7.71	168.27	m	**285.77**
Sum of two sides 1850mm	91.35	121.04	7.71	168.27	m	**289.31**
Sum of two sides 1900mm	94.02	124.58	7.98	174.16	m	**298.74**
Sum of two sides 1950mm	96.69	128.11	7.98	174.16	m	**302.27**
Sum of two sides 2000mm	99.36	131.65	8.25	180.05	m	**311.70**
Sum of two sides 2050mm	102.03	135.19	8.25	180.05	m	**315.24**
Sum of two sides 2100mm	104.70	138.73	9.62	209.95	m	**348.68**
Sum of two sides 2150mm	107.37	142.27	9.62	209.95	m	**352.22**
Sum of two sides 2200mm	110.04	145.80	9.66	210.83	m	**356.63**
Sum of two sides 2250mm	112.71	149.34	9.66	210.83	m	**360.17**
Sum of two sides 2300mm	115.37	152.87	10.07	219.77	m	**372.64**
Sum of two sides 2350mm	118.04	156.40	10.07	219.77	m	**376.18**
Sum of two sides 2400mm	120.71	159.94	10.47	228.50	m	**388.44**
Sum of two sides 2450mm	123.38	163.48	10.47	228.50	m	**391.98**
Sum of two sides 2500mm	126.05	167.02	10.87	237.23	m	**404.25**
Sum of two sides 2550mm	128.72	170.55	10.87	237.23	m	**407.79**
Sum of two sides 2600mm	131.39	174.09	11.27	245.96	m	**420.05**
Sum of two sides 2650mm	134.06	177.63	11.27	245.96	m	**423.59**
Sum of two sides 2700mm	136.73	181.17	11.67	254.69	m	**435.86**
Sum of two sides 2750mm	139.40	184.71	11.67	254.69	m	**439.40**

Material Costs/Measured Work Prices - Mechanical Installations

U:VENTILATION/AIR CONDITIONING SYSTEMS

Item	Net Price £	Material £	Labour hours	Labour £	Unit	Total rate £
Sum of two sides 2800mm	142.07	188.24	12.08	263.64	m	451.88
Sum of two sides 2850mm	144.74	191.78	12.08	263.64	m	455.42
Sum of two sides 2900mm	147.41	195.32	12.48	272.37	m	467.69
Sum of two sides 2950mm	150.08	198.86	12.48	272.37	m	471.23
Sum of two sides 3000mm	152.75	202.39	12.88	281.10	m	483.49
Sum of two sides 3050mm	155.42	205.93	12.88	281.10	m	487.03
Sum of two sides 3100mm	158.09	209.47	13.26	289.39	m	498.86
Sum of two sides 3150mm	160.76	213.01	13.26	289.39	m	502.40
Sum of two sides 3200mm	163.43	216.54	13.69	298.78	m	515.32
Sum of two sides 3250mm	166.10	220.08	13.69	298.78	m	518.86
Sum of two sides 3300mm	168.77	223.62	14.09	307.51	m	531.13
Sum of two sides 3350mm	171.44	227.16	14.09	307.51	m	534.67
Sum of two sides 3400mm	174.11	230.70	14.49	316.24	m	546.93
Sum of two sides 3450mm	176.77	234.22	14.49	316.24	m	550.46
Sum of two sides 3500mm	179.44	237.76	14.89	324.97	m	562.73
Sum of two sides 3550mm	182.11	241.30	14.89	324.97	m	566.26
Sum of two sides 3600mm	184.78	244.83	15.29	333.70	m	578.53
Sum of two sides 3650mm	187.45	248.37	15.29	333.70	m	582.07
Sum of two sides 3700mm	190.12	251.91	15.69	342.43	m	594.34
Sum of two sides 3750mm	192.79	255.45	15.69	342.43	m	597.87
Sum of two sides 3800mm	195.46	258.98	16.09	351.16	m	610.14
Sum of two sides 3850mm	198.13	262.52	16.09	351.16	m	613.68
Sum of two sides 3900mm	200.80	266.06	16.49	359.89	m	625.95
Sum of two sides 3950mm	203.47	269.60	16.49	359.89	m	629.49
Sum of two sides 4000mm	206.14	273.14	16.89	368.62	m	641.75
Extra over fittings; Ductwork 1251 to 2000mm longest side						
End Cap						
Sum of two sides 1700mm	30.34	40.20	1.42	30.99	nr	71.19
Sum of two sides 1750mm	31.32	41.50	1.42	30.99	nr	72.49
Sum of two sides 1800mm	32.29	42.78	1.45	31.65	nr	74.43
Sum of two sides 1850mm	33.27	44.08	1.45	31.65	nr	75.73
Sum of two sides 1900mm	34.24	45.37	1.48	32.30	nr	77.67
Sum of two sides 1950mm	35.22	46.67	1.48	32.30	nr	78.97
Sum of two sides 2000mm	36.19	47.95	1.51	32.96	nr	80.91
Sum of two sides 2050mm	37.17	49.25	1.51	32.96	nr	82.21
Sum of two sides 2100mm	38.14	50.54	2.66	58.05	nr	108.59
Sum of two sides 2150mm	39.12	51.83	2.66	58.05	nr	109.89
Sum of two sides 2200mm	40.09	53.12	2.80	61.11	nr	114.23
Sum of two sides 2250mm	41.07	54.42	2.80	61.11	nr	115.53
Sum of two sides 2300mm	42.04	55.70	2.95	64.38	nr	120.09
Sum of two sides 2350mm	43.02	57.00	2.95	64.38	nr	121.38
Sum of two sides 2400mm	43.99	58.29	3.10	67.66	nr	125.94
Sum of two sides 2450mm	44.97	59.59	3.10	67.66	nr	127.24
Sum of two sides 2500mm	45.94	60.87	3.24	70.71	nr	131.58
Sum of two sides 2550mm	46.92	62.17	3.24	70.71	nr	132.88
Sum of two sides 2600mm	47.89	63.45	3.39	73.99	nr	137.44
Sum of two sides 2650mm	48.87	64.75	3.39	73.99	nr	138.74
Sum of two sides 2700mm	49.84	66.04	3.54	77.26	nr	143.30
Sum of two sides 2750mm	50.82	67.34	3.54	77.26	nr	144.60
Sum of two sides 2800mm	51.79	68.62	3.68	80.31	nr	148.94
Sum of two sides 2850mm	52.77	69.92	3.68	80.31	nr	150.23

Material Costs/Measured Work Prices - Mechanical Installations

U:VENTILATION/AIR CONDITIONING SYSTEMS

Item	Net Price £	Material £	Labour hours	Labour £	Unit	Total rate £
U14 : DUCTWORK : SMOKE EXTRACT						
Y30 - DUCTLINES						
Extra over fittings; Ductwork 1251 to 2000mm longest side (Continued)						
End Cap						
Sum of two sides 2900mm	53.75	71.22	3.83	83.59	nr	**154.81**
Sum of two sides 2950mm	54.72	72.50	3.83	83.59	nr	**156.09**
Sum of two sides 3000mm	55.70	73.80	3.98	86.86	nr	**160.66**
Sum of two sides 3050mm	56.67	75.09	3.98	86.86	nr	**161.95**
Sum of two sides 3100mm	57.65	76.39	4.12	89.92	nr	**166.30**
Sum of two sides 3150mm	58.62	77.67	4.12	89.92	nr	**167.59**
Sum of two sides 3200mm	59.60	78.97	4.27	93.19	nr	**172.16**
Sum of two sides 3250mm	60.57	80.26	4.27	93.19	nr	**173.45**
Sum of two sides 3300mm	61.55	81.55	4.42	96.46	nr	**178.02**
Sum of two sides 3350mm	62.52	82.84	4.42	96.46	nr	**179.30**
Sum of two sides 3400mm	63.50	84.14	4.57	99.74	nr	**183.88**
Sum of two sides 3450mm	64.47	85.42	4.57	99.74	nr	**185.16**
Sum of two sides 3500mm	65.11	86.27	4.72	103.01	nr	**189.28**
Sum of two sides 3550mm	65.45	86.72	4.72	103.01	nr	**189.73**
Sum of two sides 3600mm	66.42	88.01	4.87	106.29	nr	**194.29**
Sum of two sides 3650mm	67.40	89.31	4.87	106.29	nr	**195.59**
Sum of two sides 3700mm	68.37	90.59	5.02	109.56	nr	**200.15**
Sum of two sides 3750mm	69.35	91.89	5.02	109.56	nr	**201.45**
Sum of two sides 3800mm	72.27	95.76	5.17	112.83	nr	**208.59**
Sum of two sides 3850mm	72.27	95.76	5.17	112.83	nr	**208.59**
Sum of two sides 3900mm	73.25	97.06	5.32	116.11	nr	**213.16**
Sum of two sides 3950mm	74.22	98.34	5.32	116.11	nr	**214.45**
Sum of two sides 4000mm	75.20	99.64	5.47	119.38	nr	**219.02**
Reducer						
Sum of two sides 1700mm	65.51	86.80	3.54	77.26	nr	**164.06**
Sum of two sides 1750mm	69.50	92.09	3.54	77.26	nr	**169.35**
Sum of two sides 1800mm	73.49	97.37	3.60	78.57	nr	**175.94**
Sum of two sides 1850mm	77.47	102.65	3.60	78.57	nr	**181.22**
Sum of two sides 1900mm	81.46	107.93	3.65	79.66	nr	**187.59**
Sum of two sides 1950mm	85.45	113.22	3.65	79.66	nr	**192.88**
Sum of two sides 2000mm	89.44	118.51	3.70	80.75	nr	**199.26**
Sum of two sides 2050mm	93.43	123.79	3.70	80.75	nr	**204.55**
Sum of two sides 2100mm	97.42	129.08	2.92	63.73	nr	**192.81**
Sum of two sides 2150mm	101.40	134.35	2.92	63.73	nr	**198.08**
Sum of two sides 2200mm	105.39	139.64	3.10	67.66	nr	**207.30**
Sum of two sides 2250mm	109.38	144.93	3.10	67.66	nr	**212.58**
Sum of two sides 2300mm	113.37	150.22	3.29	71.80	nr	**222.02**
Sum of two sides 2350mm	117.36	155.50	3.29	71.80	nr	**227.30**
Sum of two sides 2400mm	121.35	160.79	3.48	75.95	nr	**236.74**
Sum of two sides 2450mm	125.33	166.06	3.48	75.95	nr	**242.01**
Sum of two sides 2500mm	129.32	171.35	3.66	79.88	nr	**251.23**
Sum of two sides 2550mm	133.31	176.64	3.66	79.88	nr	**256.51**
Sum of two sides 2600mm	137.30	181.92	3.85	84.02	nr	**265.95**
Sum of two sides 2650mm	141.29	187.21	3.85	84.02	nr	**271.23**
Sum of two sides 2700mm	145.28	192.50	4.03	87.95	nr	**280.45**

Material Costs/Measured Work Prices - Mechanical Installations

U:VENTILATION/AIR CONDITIONING SYSTEMS

Item	Net Price £	Material £	Labour hours	Labour £	Unit	Total rate £
Sum of two sides 2750mm	149.26	197.77	4.03	87.95	nr	**285.72**
Sum of two sides 2800mm	153.25	203.06	4.22	92.10	nr	**295.16**
Sum of two sides 2850mm	157.24	208.34	4.22	92.10	nr	**300.44**
Sum of two sides 2900mm	161.23	213.63	4.40	96.03	nr	**309.66**
Sum of two sides 2950mm	165.22	218.92	4.40	96.03	nr	**314.94**
Sum of two sides 3000mm	173.19	229.48	4.59	100.17	nr	**329.65**
Sum of two sides 3050mm	169.20	224.19	4.59	100.17	nr	**324.36**
Sum of two sides 3100mm	173.19	229.48	4.78	104.32	nr	**333.80**
Sum of two sides 3150mm	177.18	234.76	4.78	104.32	nr	**339.09**
Sum of two sides 3200mm	181.17	240.05	4.96	108.25	nr	**348.30**
Sum of two sides 3250mm	189.15	250.62	4.96	108.25	nr	**358.87**
Sum of two sides 3300mm	193.13	255.90	5.15	112.40	nr	**368.29**
Sum of two sides 3350mm	197.12	261.18	5.15	112.40	nr	**373.58**
Sum of two sides 3400mm	201.11	266.47	5.34	116.54	nr	**383.01**
Sum of two sides 3450mm	205.10	271.76	5.34	116.54	nr	**388.30**
Sum of two sides 3500mm	209.09	277.04	5.53	120.69	nr	**397.73**
Sum of two sides 3550mm	213.08	282.33	5.53	120.69	nr	**403.02**
Sum of two sides 3600mm	217.06	287.60	5.72	124.84	nr	**412.44**
Sum of two sides 3650mm	221.05	292.89	5.72	124.84	nr	**417.73**
Sum of two sides 3700mm	225.04	298.18	5.91	128.98	nr	**427.16**
Sum of two sides 3750mm	229.03	303.46	5.91	128.98	nr	**432.45**
Sum of two sides 3800mm	233.02	308.75	6.10	133.13	nr	**441.88**
Sum of two sides 3850mm	237.01	314.04	6.10	133.13	nr	**447.17**
Sum of two sides 3900mm	240.99	319.31	6.29	137.28	nr	**456.59**
Sum of two sides 3950mm	244.98	324.60	6.29	137.28	nr	**461.88**
Sum of two sides 4000mm	248.97	329.89	6.48	141.42	nr	**471.31**
Offset						
Sum of two sides 1700mm	65.51	86.80	4.45	97.12	nr	**183.92**
Sum of two sides 1750mm	68.21	90.38	4.45	97.12	nr	**187.50**
Sum of two sides 1800mm	70.91	93.96	4.56	99.52	nr	**193.48**
Sum of two sides 1850mm	73.61	97.53	4.56	99.52	nr	**197.05**
Sum of two sides 1900mm	76.31	101.11	4.67	101.92	nr	**203.03**
Sum of two sides 1950mm	79.00	104.67	4.67	101.92	nr	**206.60**
Sum of two sides 2000mm	81.70	108.25	5.53	120.69	nr	**228.94**
Sum of two sides 2050mm	84.40	111.83	5.53	120.69	nr	**232.52**
Sum of two sides 2100mm	87.10	115.41	5.76	125.71	nr	**241.12**
Sum of two sides 2150mm	89.80	118.98	5.76	125.71	nr	**244.69**
Sum of two sides 2200mm	92.50	122.56	5.99	130.73	nr	**253.29**
Sum of two sides 2250mm	95.20	126.14	5.99	130.73	nr	**256.87**
Sum of two sides 2300mm	97.90	129.72	6.22	135.75	nr	**265.47**
Sum of two sides 2350mm	100.60	133.29	6.22	135.75	nr	**269.04**
Sum of two sides 2400mm	103.29	136.86	6.45	140.77	nr	**277.63**
Sum of two sides 2450mm	105.99	140.44	6.45	140.77	nr	**281.21**
Sum of two sides 2500mm	108.69	144.01	6.68	145.79	nr	**289.80**
Sum of two sides 2550mm	111.39	147.59	6.68	145.79	nr	**293.38**
Sum of two sides 2600mm	114.09	151.17	6.91	150.81	nr	**301.98**
Sum of two sides 2650mm	116.79	154.75	6.91	150.81	nr	**305.55**
Sum of two sides 2700mm	119.49	158.32	7.14	155.83	nr	**314.15**
Sum of two sides 2750mm	122.19	161.90	7.14	155.83	nr	**317.73**
Sum of two sides 2800mm	124.89	165.48	7.37	160.85	nr	**326.33**
Sum of two sides 2850mm	127.59	169.06	7.37	160.85	nr	**329.90**
Sum of two sides 2900mm	130.28	172.62	7.60	165.87	nr	**338.49**
Sum of two sides 2950mm	132.98	176.20	7.60	165.87	nr	**342.07**

420 **Material Costs/Measured Work Prices - Mechanical Installations**

U:VENTILATION/AIR CONDITIONING SYSTEMS

Item	Net Price £	Material £	Labour hours	Labour £	Unit	Total rate £
U14 : DUCTWORK : SMOKE EXTRACT						
Y30 - DUCTLINES						
Extra over fittings; Ductwork 1251 to 2000mm longest side (Continued)						
Offset						
Sum of two sides 3000mm	135.68	179.78	7.83	170.89	nr	**350.66**
Sum of two sides 3050mm	138.38	183.35	7.83	170.89	nr	**354.24**
Sum of two sides 3100mm	141.08	186.93	8.06	175.91	nr	**362.84**
Sum of two sides 3150mm	143.78	190.51	8.06	175.91	nr	**366.41**
Sum of two sides 3200mm	146.48	194.09	8.29	180.93	nr	**375.01**
Sum of two sides 3250mm	149.18	197.66	8.29	180.93	nr	**378.59**
Sum of two sides 3300mm	151.88	201.24	8.52	185.95	nr	**387.19**
Sum of two sides 3350mm	154.57	204.81	8.52	185.95	nr	**390.75**
Sum of two sides 3400mm	157.27	208.38	8.75	190.97	nr	**399.35**
Sum of two sides 3450mm	159.97	211.96	8.75	190.97	nr	**402.93**
Sum of two sides 3500mm	162.67	215.54	8.98	195.98	nr	**411.52**
Sum of two sides 3550mm	165.37	219.12	8.98	195.98	nr	**415.10**
Sum of two sides 3600mm	168.07	222.69	9.21	201.00	nr	**423.70**
Sum of two sides 3650mm	170.77	226.27	9.21	201.00	nr	**427.27**
Sum of two sides 3700mm	173.47	229.85	9.44	206.02	nr	**435.87**
Sum of two sides 3750mm	176.17	233.43	9.44	206.02	nr	**439.45**
Sum of two sides 3800mm	178.86	236.99	9.67	211.04	nr	**448.03**
Sum of two sides 3850mm	181.56	240.57	9.67	211.04	nr	**451.61**
Sum of two sides 3900mm	184.26	244.14	9.90	216.06	nr	**460.21**
Sum of two sides 3950mm	186.96	247.72	9.90	216.06	nr	**463.79**
Sum of two sides 4000mm	189.66	251.30	10.13	221.08	nr	**472.38**
90 degree radius bend						
Sum of two sides 1700mm	91.17	120.80	3.22	70.28	nr	**191.08**
Sum of two sides 1750mm	95.29	126.26	3.22	70.28	nr	**196.53**
Sum of two sides 1800mm	99.41	131.72	3.25	70.93	nr	**202.65**
Sum of two sides 1850mm	103.53	137.18	3.25	70.93	nr	**208.11**
Sum of two sides 1900mm	107.65	142.64	3.28	71.58	nr	**214.22**
Sum of two sides 1950mm	111.77	148.10	3.28	71.58	nr	**219.68**
Sum of two sides 2000mm	115.89	153.55	3.30	72.02	nr	**225.58**
Sum of two sides 2050mm	120.01	159.01	3.30	72.02	nr	**231.03**
Sum of two sides 2100mm	124.13	164.47	3.32	72.46	nr	**236.93**
Sum of two sides 2150mm	128.25	169.93	3.32	72.46	nr	**242.39**
Sum of two sides 2200mm	132.37	175.39	3.34	72.89	nr	**248.28**
Sum of two sides 2250mm	136.49	180.85	3.34	72.89	nr	**253.74**
Sum of two sides 2300mm	140.61	186.31	3.36	73.33	nr	**259.64**
Sum of two sides 2350mm	144.73	191.77	3.36	73.33	nr	**265.10**
Sum of two sides 2400mm	148.85	197.23	3.38	73.77	nr	**270.99**
Sum of two sides 2450mm	152.97	202.69	3.38	73.77	nr	**276.45**
Sum of two sides 2500mm	157.09	208.14	3.40	74.20	nr	**282.35**
Sum of two sides 2550mm	161.21	213.60	3.40	74.20	nr	**287.81**
Sum of two sides 2600mm	165.33	219.06	3.42	74.64	nr	**293.70**
Sum of two sides 2650mm	169.45	224.52	3.42	74.64	nr	**299.16**
Sum of two sides 2700mm	173.57	229.98	3.44	75.08	nr	**305.06**
Sum of two sides 2750mm	177.69	235.44	3.44	75.08	nr	**310.52**
Sum of two sides 2800mm	181.81	240.90	3.46	75.51	nr	**316.41**

Material Costs/Measured Work Prices - Mechanical Installations

U:VENTILATION/AIR CONDITIONING SYSTEMS

Item	Net Price £	Material £	Labour hours	Labour £	Unit	Total rate £
Sum of two sides 2850mm	185.93	246.36	3.46	75.51	nr	**321.87**
Sum of two sides 2900mm	190.04	251.80	3.48	75.95	nr	**327.75**
Sum of two sides 2950mm	194.16	257.26	3.48	75.95	nr	**333.21**
Sum of two sides 3000mm	198.28	262.72	3.50	76.39	nr	**339.11**
Sum of two sides 3050mm	202.40	268.18	3.50	76.39	nr	**344.57**
Sum of two sides 3100mm	206.52	273.64	3.52	76.82	nr	**350.46**
Sum of two sides 3150mm	210.64	279.10	3.52	76.82	nr	**355.92**
Sum of two sides 3200mm	214.76	284.56	3.54	77.26	nr	**361.82**
Sum of two sides 3250mm	218.88	290.02	3.54	77.26	nr	**367.28**
Sum of two sides 3300mm	223.00	295.48	3.56	77.70	nr	**373.17**
Sum of two sides 3350mm	227.12	300.93	3.56	77.70	nr	**378.63**
Sum of two sides 3400mm	231.34	306.53	3.58	78.13	nr	**384.66**
Sum of two sides 3450mm	235.36	311.85	3.58	78.13	nr	**389.98**
Sum of two sides 3500mm	239.48	317.31	3.60	78.57	nr	**395.88**
Sum of two sides 3550mm	243.60	322.77	3.60	78.57	nr	**401.34**
Sum of two sides 3600mm	247.72	328.23	3.62	79.01	nr	**407.23**
Sum of two sides 3650mm	251.84	333.69	3.62	79.01	nr	**412.69**
Sum of two sides 3700mm	255.96	339.15	3.64	79.44	nr	**418.59**
Sum of two sides 3750mm	260.08	344.61	3.64	79.44	nr	**424.05**
Sum of two sides 3800mm	264.20	350.06	3.66	79.88	nr	**429.94**
Sum of two sides 3850mm	268.32	355.52	3.66	79.88	nr	**435.40**
Sum of two sides 3900mm	272.44	360.98	3.68	80.31	nr	**441.30**
Sum of two sides 3950mm	276.56	366.44	3.68	80.31	nr	**446.76**
Sum of two sides 4000mm	280.68	371.90	3.70	80.75	nr	**452.65**
45 degree bend						
Sum of two sides 1700mm	67.59	89.56	2.24	48.89	nr	**138.44**
Sum of two sides 1750mm	71.32	94.50	2.24	48.89	nr	**143.39**
Sum of two sides 1800mm	75.06	99.45	2.30	50.20	nr	**149.65**
Sum of two sides 1850mm	78.79	104.40	2.30	50.20	nr	**154.59**
Sum of two sides 1900mm	82.52	109.34	2.35	51.29	nr	**160.63**
Sum of two sides 1950mm	86.26	114.29	2.35	51.29	nr	**165.58**
Sum of two sides 2000mm	89.99	119.24	2.76	60.24	nr	**179.47**
Sum of two sides 2050mm	93.72	124.18	2.76	60.24	nr	**184.41**
Sum of two sides 2100mm	124.13	164.47	2.89	63.07	nr	**227.55**
Sum of two sides 2150mm	101.19	134.08	2.89	63.07	nr	**197.15**
Sum of two sides 2200mm	104.92	139.02	3.01	65.69	nr	**204.71**
Sum of two sides 2250mm	108.65	143.96	3.01	65.69	nr	**209.65**
Sum of two sides 2300mm	112.39	148.92	3.13	68.31	nr	**217.23**
Sum of two sides 2350mm	116.12	153.86	3.13	68.31	nr	**222.17**
Sum of two sides 2400mm	119.85	158.80	3.26	71.15	nr	**229.95**
Sum of two sides 2450mm	123.59	163.76	3.26	71.15	nr	**234.91**
Sum of two sides 2500mm	127.32	168.70	3.37	73.55	nr	**242.25**
Sum of two sides 2550mm	131.05	173.64	3.37	73.55	nr	**247.19**
Sum of two sides 2600mm	134.78	178.58	3.49	76.17	nr	**254.75**
Sum of two sides 2650mm	138.52	183.54	3.49	76.17	nr	**259.71**
Sum of two sides 2700mm	142.25	188.48	3.62	79.01	nr	**267.49**
Sum of two sides 2750mm	145.98	193.42	3.62	79.01	nr	**272.43**
Sum of two sides 2800mm	149.72	198.38	3.74	81.62	nr	**280.00**
Sum of two sides 2850mm	153.45	203.32	3.74	81.62	nr	**284.95**
Sum of two sides 2900mm	157.18	208.26	3.86	84.24	nr	**292.51**
Sum of two sides 2950mm	160.92	213.22	3.86	84.24	nr	**297.46**
Sum of two sides 3000mm	164.65	218.16	3.98	86.86	nr	**305.02**
Sum of two sides 3050mm	168.38	223.10	3.98	86.86	nr	**309.97**

Material Costs/Measured Work Prices - Mechanical Installations

U:VENTILATION/AIR CONDITIONING SYSTEMS

Item	Net Price £	Material £	Labour hours	Labour £	Unit	Total rate £
U14 : DUCTWORK : SMOKE EXTRACT						
Y30 - DUCTLINES						
Extra over fittings; Ductwork 1251 to 2000mm longest side (Continued)						
45 degree bend						
Sum of two sides 3100mm	172.12	228.06	4.10	89.48	nr	**317.54**
Sum of two sides 3150mm	175.85	233.00	4.10	89.48	nr	**322.48**
Sum of two sides 3200mm	179.58	237.94	4.22	92.10	nr	**330.04**
Sum of two sides 3250mm	183.31	242.89	4.22	92.10	nr	**334.99**
Sum of two sides 3300mm	187.05	247.84	4.34	94.72	nr	**342.56**
Sum of two sides 3350mm	190.78	252.78	4.34	94.72	nr	**347.50**
Sum of two sides 3400mm	194.51	257.73	4.46	97.34	nr	**355.06**
Sum of two sides 3450mm	198.25	262.68	4.46	97.34	nr	**360.02**
Sum of two sides 3500mm	201.98	267.62	4.58	99.96	nr	**367.58**
Sum of two sides 3550mm	205.71	272.57	4.58	99.96	nr	**372.52**
Sum of two sides 3600mm	209.45	277.52	4.70	102.58	nr	**380.10**
Sum of two sides 3650mm	213.18	282.46	4.70	102.58	nr	**385.04**
Sum of two sides 3700mm	216.91	287.41	4.82	105.19	nr	**392.60**
Sum of two sides 3750mm	220.64	292.35	4.82	105.19	nr	**397.54**
Sum of two sides 3800mm	224.38	297.30	4.94	107.81	nr	**405.12**
Sum of two sides 3850mm	228.11	302.25	4.94	107.81	nr	**410.06**
Sum of two sides 3900mm	231.84	307.19	5.06	110.43	nr	**417.62**
Sum of two sides 3950mm	235.58	312.14	5.06	110.43	nr	**422.58**
Sum of two sides 4000mm	239.31	317.09	5.18	113.05	nr	**430.14**
90 degree mitre bend						
Sum of two sides 1700mm	136.55	180.93	3.87	84.46	nr	**265.39**
Sum of two sides 1750mm	148.35	196.56	3.87	84.46	nr	**281.02**
Sum of two sides 1800mm	160.15	212.20	3.90	85.12	nr	**297.31**
Sum of two sides 1850mm	171.95	227.83	3.90	85.12	nr	**312.95**
Sum of two sides 1900mm	183.75	243.47	3.93	85.77	nr	**329.24**
Sum of two sides 1950mm	195.55	259.10	3.93	85.77	nr	**344.87**
Sum of two sides 2000mm	207.34	274.73	3.96	86.43	nr	**361.15**
Sum of two sides 2050mm	219.14	290.36	3.96	86.43	nr	**376.79**
Sum of two sides 2100mm	230.94	306.00	3.82	83.37	nr	**389.37**
Sum of two sides 2150mm	242.74	321.63	3.82	83.37	nr	**405.00**
Sum of two sides 2200mm	254.54	337.27	3.83	83.59	nr	**420.85**
Sum of two sides 2250mm	266.34	352.90	3.83	83.59	nr	**436.49**
Sum of two sides 2300mm	278.14	368.54	3.83	83.59	nr	**452.12**
Sum of two sides 2350mm	289.94	384.17	3.83	83.59	nr	**467.76**
Sum of two sides 2400mm	301.74	399.81	3.84	83.81	nr	**483.61**
Sum of two sides 2450mm	313.54	415.44	3.84	83.81	nr	**499.25**
Sum of two sides 2500mm	325.34	431.08	3.85	84.02	nr	**515.10**
Sum of two sides 2550mm	337.14	446.71	3.85	84.02	nr	**530.74**
Sum of two sides 2600mm	348.93	462.33	3.85	84.02	nr	**546.36**
Sum of two sides 2650mm	360.73	477.97	3.85	84.02	nr	**561.99**
Sum of two sides 2700mm	372.53	493.60	3.86	84.24	nr	**577.85**
Sum of two sides 2750mm	384.33	509.24	3.86	84.24	nr	**593.48**
Sum of two sides 2800mm	396.13	524.87	3.86	84.24	nr	**609.12**
Sum of two sides 2850mm	407.93	540.51	3.86	84.24	nr	**624.75**
Sum of two sides 2900mm	419.73	556.14	3.87	84.46	nr	**640.60**

Material Costs/Measured Work Prices - Mechanical Installations

U:VENTILATION/AIR CONDITIONING SYSTEMS

Item	Net Price £	Material £	Labour hours	Labour £	Unit	Total rate £
Sum of two sides 2950mm	431.53	571.78	3.87	84.46	nr	656.24
Sum of two sides 3000mm	443.33	587.41	3.87	84.46	nr	671.87
Sum of two sides 3050mm	455.13	603.05	3.87	84.46	nr	687.51
Sum of two sides 3100mm	466.93	618.68	3.88	84.68	nr	703.36
Sum of two sides 3150mm	478.72	634.30	3.88	84.68	nr	718.98
Sum of two sides 3200mm	490.52	649.94	3.88	84.68	nr	734.62
Sum of two sides 3250mm	502.32	665.57	3.88	84.68	nr	750.25
Sum of two sides 3300mm	514.12	681.21	3.89	84.90	nr	766.11
Sum of two sides 3350mm	525.92	696.84	3.89	84.90	nr	781.74
Sum of two sides 3400mm	537.72	712.48	3.90	85.12	nr	797.59
Sum of two sides 3450mm	549.52	728.11	3.90	85.12	nr	813.23
Sum of two sides 3500mm	561.32	743.75	3.91	85.33	nr	829.08
Sum of two sides 3550mm	573.12	759.38	3.91	85.33	nr	844.72
Sum of two sides 3600mm	584.92	775.02	3.92	85.55	nr	860.57
Sum of two sides 3650mm	596.72	790.65	3.92	85.55	nr	876.21
Sum of two sides 3700mm	608.52	806.29	3.93	85.77	nr	892.06
Sum of two sides 3750mm	620.31	821.91	3.93	85.77	nr	907.68
Sum of two sides 3800mm	632.11	837.55	3.94	85.99	nr	923.53
Sum of two sides 3850mm	643.91	853.18	3.94	85.99	nr	939.17
Sum of two sides 3900mm	655.71	868.82	3.95	86.21	nr	955.02
Sum of two sides 3950mm	667.51	884.45	3.95	86.21	nr	970.66
Sum of two sides 4000mm	679.31	900.09	3.96	86.43	nr	986.51
Branch						
Sum of two sides 1700mm	15.48	20.51	1.77	38.63	nr	59.14
Sum of two sides 1750mm	17.46	23.13	1.77	38.63	nr	61.76
Sum of two sides 1800mm	16.92	22.42	1.81	39.50	nr	61.92
Sum of two sides 1850mm	17.64	23.37	1.81	39.50	nr	62.88
Sum of two sides 1900mm	18.36	24.33	1.86	40.59	nr	64.92
Sum of two sides 1950mm	19.08	25.28	1.86	40.59	nr	65.87
Sum of two sides 2000mm	19.80	26.23	1.90	41.47	nr	67.70
Sum of two sides 2050mm	20.52	27.19	1.90	41.47	nr	68.66
Sum of two sides 2100mm	21.24	28.14	2.58	56.31	nr	84.45
Sum of two sides 2150mm	21.96	29.10	2.58	56.31	nr	85.40
Sum of two sides 2200mm	22.68	30.05	2.61	56.96	nr	87.01
Sum of two sides 2250mm	23.40	31.00	2.61	56.96	nr	87.97
Sum of two sides 2300mm	24.12	31.96	2.65	57.84	nr	89.79
Sum of two sides 2350mm	24.84	32.91	2.65	57.84	nr	90.75
Sum of two sides 2400mm	25.56	33.87	2.68	58.49	nr	92.36
Sum of two sides 2450mm	26.28	34.82	2.68	58.49	nr	93.31
Sum of two sides 2500mm	27.00	35.77	2.71	59.14	nr	94.92
Sum of two sides 2550mm	27.72	36.73	2.71	59.14	nr	95.87
Sum of two sides 2600mm	28.44	37.68	2.75	60.02	nr	97.70
Sum of two sides 2650mm	29.16	38.64	2.75	60.02	nr	98.65
Sum of two sides 2700mm	29.88	39.59	2.78	60.67	nr	100.26
Sum of two sides 2750mm	30.60	40.55	2.78	60.67	nr	101.22
Sum of two sides 2800mm	31.32	41.50	2.81	61.33	nr	102.83
Sum of two sides 2850mm	32.04	42.45	2.81	61.33	nr	103.78
Sum of two sides 2900mm	32.76	43.41	2.84	61.98	nr	105.39
Sum of two sides 2950mm	33.48	44.36	2.84	61.98	nr	106.34
Sum of two sides 3000mm	34.20	45.31	2.87	62.64	nr	107.95
Sum of two sides 3050mm	34.92	46.27	2.87	62.64	nr	108.91
Sum of two sides 3100mm	35.64	47.22	2.90	63.29	nr	110.51
Sum of two sides 3150mm	36.36	48.18	2.90	63.29	nr	111.47

Material Costs/Measured Work Prices - Mechanical Installations

U:VENTILATION/AIR CONDITIONING SYSTEMS

Item	Net Price £	Material £	Labour hours	Labour £	Unit	Total rate £
U14 : DUCTWORK : SMOKE EXTRACT						
Y30 - DUCTLINES						
Extra over fittings; Ductwork 1251 to 2000mm longest side (Continued)						
Branch						
Sum of two sides 3200mm	37.08	49.13	2.93	63.95	nr	**113.08**
Sum of two sides 3250mm	37.80	50.09	2.93	63.95	nr	**114.03**
Sum of two sides 3300mm	38.52	51.04	2.93	63.95	nr	**114.99**
Sum of two sides 3350mm	39.24	51.99	2.97	64.82	nr	**116.81**
Sum of two sides 3400mm	39.96	52.95	3.00	65.47	nr	**118.42**
Sum of two sides 3450mm	40.68	53.90	3.00	65.47	nr	**119.37**
Sum of two sides 3500mm	41.40	54.85	3.03	66.13	nr	**120.98**
Sum of two sides 3550mm	42.12	55.81	3.03	66.13	nr	**121.94**
Sum of two sides 3600mm	42.84	56.76	3.06	66.78	nr	**123.55**
Sum of two sides 3650mm	43.56	57.72	3.06	66.78	nr	**124.50**
Sum of two sides 3700mm	44.28	58.67	3.09	67.44	nr	**126.11**
Sum of two sides 3750mm	45.00	59.63	3.09	67.44	nr	**127.06**
Sum of two sides 3800mm	45.72	60.58	3.12	68.09	nr	**128.67**
Sum of two sides 3850mm	46.44	61.53	3.12	68.09	nr	**129.63**
Sum of two sides 3900mm	47.16	62.49	3.15	68.75	nr	**131.23**
Sum of two sides 3950mm	47.88	63.44	3.15	68.75	nr	**132.19**
Sum of two sides 4000mm	48.60	64.39	3.18	69.40	nr	**133.80**

Electrical Installations

MATERIAL COSTS/MEASURED WORK PRICES

DIRECTIONS

The following explanations are given for each of the column headings and letter codes.

Unit	Prices for each unit are given as singular (1 metre, 1 nr).
Net price	Industry tender prices, plus nominal allowance for fixings, waste and applicable trade discounts.
Material cost	Net price plus percentage allowance for overheads and profit and preliminaries.
Labour constant	Gang norm (in man-hours) for each operation.
Labour cost	Labour constant multiplied by the appropriate all-in man-hour cost.(See also relevant Rates of Wages Section)
Measured work price	Material cost plus Labour cost.

MATERIAL COSTS
The Material Costs given are based at Second Quarter 2001 but exclude any charges in respect of VAT.

MEASURED WORK PRICES
These prices are intended to apply to new work in the London area. The prices are for reasonable quantities of work and the user should make suitable adjustments if the quantities are especially small or especially large. Adjustments may also be required for locality (e.g. outside London) and for the market conditions (e.g. volume of work on hand or on offer) at the time of use.

426   *Rates of Wages and Working Rules - Electrical Installations*

DIRECTIONS

LABOUR RATE
The labour rate has been based on average tender rates for Third Quarter 2001 plus allowances for all other emoluments and expenses. To this rate has been added 14% to cover site and head office overheads and preliminary items together with a further 3% for profit, resulting in an inclusive rate of £ 18.53 per man hour. The rate per man hour has been calculated on a working year of 2025 hours; a detailed build-up of the rate is given at the end of these Directions.

In calculating the 'Measured Work Prices' the following assumptions have been made:
  (a)   That the work is carried out as a sub-contract under the Standard Form of Building Contract.
  (b)   That, unless otherwise stated, the work is being carried out in open areas at a height which would not require more than simple scaffolding.
  (c)   That the building in which the work is being carried out is no more than six storey's high.

Where these assumptions are not valid, as for example where work is carried out in ducts and similar confined spaces or in multi-storey structures when additional time is needed to get to and from upper floors, then an appropriate adjustment must be made to the prices. Such adjustment will normally be to the labour element only.

Material Costs/Measured Work Prices - Electrical Installations

DIRECTIONS

LABOUR RATE - ELECTRICAL

The annual cost of notional eleven man gang.

		TECHNICIAN	APPROVED ELECTRICIANS	ELECTRICIANS	LABOURERS	SUB-TOTALS
		1 NR	4 NR	4 NR	2 NR	
Hourly Rate from 8 January 2001		11.20	10.02	9.28	7.57	
Working hours per annum per man		1710	1710	1710	1710	
x Hourly rate x nr of men = £ per annum		19,152	68,537	63,475	25,889	
Overtime Rate		16.8	15.03	13.92	11.36	
Overtime hours per annum per man		315	315	315	315	
x Hourly rate x nr of men = £ per annum		5292	18938	17539	7154	
Total		24444	87475	81014	33043	225,976.05
Incentive schemes (insert percentage)	5.00%	1,222	4,374	4,051	1,652	11,298.00
Daily Travel Time Allowance (15-20 miles each way)		3.62	3.62	3.62	3.62	
Days per annum per man		225	225	225	225	
x nr of men = £ per annum		814.50	3,258.00	3,258.00	1,629.00	8,959.50
Daily Travel Allowance (15-20 miles each way)		2.50	2.50	2.50	2.50	
Days per annum per man		225	225	225	225	
x nr of men = £ per annum		562.50	2,250.00	2,250.00	1,125.00	6,187.50
JIB Pension Scheme		1,041.04	3,729.89	3,457.42	1,414.10	9,642.44
JIB combined benefits scheme (nr of weeks per man)		52	52	52	52	
Benefit Credit		36.98	33.31	31.03	25.71	
x nr of men = £ per annum		1,922.96	6,928.48	6,454.24	2,673.84	17,979.52
Holiday Top-up Funding including overtime		33.26	30.32	28.46	24.21	
x nr of men = £ per annum		1,729.52	6,306.56	5,919.68	2,517.84	16,473.60
National Insurance Contributions:						
Weekly gross pay (subject to NI) each		26480.70	95106.33	88323.12	36324.20	
% of NI Contributions		11.9	11.9	11.9	11.9	
£ Contributions/annum		2584.11	9049.28	8242.07	3188.39	23,063.85

	SUB-TOTAL		319,581.26
	TRAINING (INCLUDING ANY TRADE REGISTRATIONS) - SAY	1.00%	3,195.81
	SEVERANCE PAY AND SUNDRY COSTS - SAY	1.50%	4,841.66
	EMPLOYER'S LIABILITY AND THIRD PARTY INSURANCE - SAY	2.00%	6,552.37
	ANNUAL COST OF NOTIONAL GANG		334,171.11
MEN ACTUALLY WORKING = 10.5	THEREFORE ANNUAL COST PER PRODUCTIVE MAN		31,825.82
AVERAGE NR OF HOURS WORKED PER MAN 2025	THEREFORE ALL IN MAN HOUR		15.72
	PRELIMINARY ITEMS SAY	8.00%	1.26
	SITE AND HEAD OFFICE OVERHEADS SAY	6.00%	1.02
	PROFIT SAY	3.00%	0.54
	THEREFORE INCLUSIVE MAN HOUR		18.53

Notes:
(1)   The following assumptions have been made in the above calculations:-
        (a)   Hourly rates are based on London rate and job reporting own transport.
        (b)   The working week of 37.5 hours is made up of 7.5 hours Monday to Friday.
        (c)   Five days in the year are lost through sickness or similar reason.
        (d)   A working year of 2025 hours.
(2)   The incentive scheme addition of 5% is intended to reflect bonus schemes typically in use.
(3)   National insurance contributions are those effective from 6 April 2001.
(4)   Weekly Holiday Credit/Welfare Stamp values are those effective from 25 September 2000.

Material Costs/Measured Work Prices – Electrical Installations

V:ELECTRICAL SUPPLY/POWER/LIGHTING

Item	Net Price £	Material £	Labour hours	Labour £	Unit	Total rate £
V11 : HV SUPPLY - SWITCHGEAR						
Y70 - HV SWITCHGEAR						
H.V. Circuit Breakers; installed on prepared foundations including all supports, fixings and inter panel connections (Excluding main and multi core cabling and heat shrink cable termination kits.)						
Three phase 11kV, 630 Amp, SF6 insulated vacuum circuit breaker panels; hand charged spring closing operation; prospective fault level up to 20 kA; feeders include ammeter with phase selector switch, 3 pole IDMT, overcurrent and earth fault relays with necessary current relays with necessary current transformers; incomers include 3 phase voltage transformer, voltmeter and phase selector switch; vacuum insulation						
Single panel with outgoing cable box	5000.00	5991.50	31.70	587.52	nr	**6579.02**
Three panel with one incomer and two feeders; with cable boxes	16000.00	19172.80	67.83	1257.15	nr	**20429.95**
Five panel with two incoming, two feeders and a bus section; with cable boxes	28000.00	33552.40	99.17	1838.01	nr	**35390.41**
Three phase 11kV, 630 Amp, SF6 insulated ring main unit with vacuum circuit breaker, 200 Amp tee off, 3 pole IDMT, overcurrent and earth fault relays with necessary current relays with necessary current transformers; Cable boxes to two ring circuits and tee off.						
3 way Ring Main Unit	5650.00	6770.40	67.83	1257.15	nr	**8027.55**
Extra for HV Switchgear circuit breakers						
Remote actuator to ring switch portion of ring main unit; per switch	750.00	898.73	-	-	nr	**898.73**
Remote tripping of circuit breaker	250.00	299.57	-	-	nr	**299.57**
Neon indicators; per switch	350.00	419.40	-	-	nr	**419.40**
Castell interlocks	350.00	419.40	-	-	nr	**419.40**
Pressure gauge with alarm contacts	150.00	179.75	-	-	nr	**179.75**

Material Costs/Measured Work Prices – Electrical Installations

V:ELECTRICAL SUPPLY/POWER/LIGHTING

Item	Net Price £	Material £	Labour hours	Labour £	Unit	Total rate £
V11 : HV SUPPLY - SWITCHGEAR						
Y70 - HV SWITCHGEAR						
Step Down Transformers; mounted on skids; provision for inserting axles and wheels; all necessary connections to equipment. (Provision and fixing of plates, discs, etc., for identification is included).						
500 kVA; Three Phase 11 kV/433 Volt 50 Hz and LV cable boxes, with glands and silica gel breathers.						
Naturally cooled, oil filled type ODGM; offload tap changer excluding conservator, dial thermometer or Bucholzrelay.	4848.00	5809.36	38.91	721.15	nr	**6530.51**
Naturally cooled, midel fluid filled, type ODGM in ventilated steel tank.	7777.00	9319.18	38.91	721.15	nr	**10040.33**
500 kVA; Three Phase 11 kV/433 Volt 50 Hz and LV cable boxes, with glands						
Dry type; class C insulated air cooled; type AN in ventilated steel tank; temperature indicator.	13635.00	16338.82	38.91	721.15	nr	**17059.97**
Dry type; cast resin filled, type AN in ventilated steel tank; temperature indicator.	14140.00	16943.96	38.91	721.15	nr	**17665.12**
Extra For						
Conservator	252.50	302.57	8.00	148.27	nr	**450.84**
Dial thermometer	212.10	254.16	2.00	37.07	nr	**291.23**
Bucholz relay	252.50	302.57	3.00	55.66	nr	**358.23**
800 kVA; Three Phase 11 kV/433 Volt 50 Hz and LV cable boxes, with glands and silica gel breathers.						
Naturally cooled, oil filled type ODGM; offload tap changer excluding conservator, dial thermometer or Bucholzrelay	5454.00	6535.53	44.15	818.27	nr	**7353.80**
Naturally cooled, midel fluid filled, type ODGM in ventilated steel tank.	8585.00	10287.41	44.15	818.27	nr	**11105.68**
800 kVA; Three Phase 11 kV/433 Volt 50 Hz and LV cable boxes, with glands						
Dry type; class C insulated air cooled; type AN in ventilated steel tank; temperature indicator.	15655.00	18759.39	44.15	818.27	nr	**19577.66**
Dry type; cast resin filled, type AN in ventilated steel tank; temperature indicator.	16160.00	19364.53	44.15	818.27	nr	**20182.80**

Material Costs/Measured Work Prices – Electrical Installations

V:ELECTRICAL SUPPLY/POWER/LIGHTING

Item	Net Price £	Material £	Labour hours	Labour £	Unit	Total rate £
Extra For:						
Conservator	252.50	302.57	10.00	185.34	nr	**487.91**
Dial thermometer	212.10	254.16	2.00	37.07	nr	**291.23**
Bucholz relay	252.50	302.57	3.00	55.66	nr	**358.23**
1000 kVA Three Phase 11KV/433 volt 50 Hz; HV and LV cable boxes, with glands and silica gel breathers.						
Naturally cooled, oil filled type ODGM; offload tap changer excluding conservator, dial thermometer or Bucholzrelay	5605.50	6717.07	79.75	1478.08	nr	**8195.15**
Naturally cooled, midel fluid filled, type ODGM in ventilated steel tank.	8686.00	10408.43	79.75	1478.08	nr	**11886.51**
1000 kVA Three Phase 11KV/433 volt 50 Hz; HV and LV cable boxes, with glands						
Dry type; class C insulated air cooled; type AN in ventilated steel tank; temperature indicator.	17675.00	21179.95	79.75	1478.08	nr	**22658.03**
Dry type; cast resin filled, type AN in ventilated steel tank; temperature indicator.	17675.00	21179.95	79.75	1478.08	nr	**22658.03**
Extra For:						
Conservator	252.50	302.57	10.00	185.34	nr	**487.91**
Dial thermometer	212.10	254.16	2.00	37.07	nr	**291.23**
Bucholz relay	252.50	302.57	3.00	55.66	nr	**358.23**
1500 kVA; Three Phase 11KV/433 Volt 50 Hz; HV and LV cable boxes, with glands and silica gel breathers.						
Naturally cooled, oil filled type ODGM; offload tap changer excluding conservator, dial thermometer or Bucholzrelay	8029.50	9621.75	99.44	1843.01	nr	**11464.76**
Naturally cooled, midel fluid filled, type ODGM in ventilated steel tank.	11615.00	13918.25	99.44	1843.01	nr	**15761.27**
1500 kVA; Three Phase 11KV/433 Volt 50 Hz; HV and LV cable boxes, with glands						
Dry type; class C insulated air cooled; type AN in ventilated steel tank; temperature indicator.	24240.00	29046.79	99.44	1843.01	nr	**30889.80**
Dry type; cast resin filled; type AN in ventilated steel tank; temperature indicator.	24240.00	29046.79	99.44	1843.01	nr	**30889.80**
Extra For:						
Conservator	303.00	363.08	12.05	223.30	nr	**586.38**
Dial thermometer	212.10	254.16	2.00	37.07	nr	**291.23**
Bucholz relay	252.50	302.57	3.00	55.66	nr	**358.23**

Material Costs/Measured Work Prices – Electrical Installations

V:ELECTRICAL SUPPLY/POWER/LIGHTING

Item	Net Price £	Material £	Labour hours	Labour £	Unit	Total rate £
V11 : HV SUPPLY - SWITCHGEAR						
Y70 - HV SWITCHGEAR						
Step Down Transformers; mounted on skids; provision for inserting axles and wheels; all necessary connections to equipment. (Provision and fixing of plates, discs, etc., for identification is included) (Continued)						
2000 kVA; Three Phase 11KV/433 volt 50Hz; HV and LV cable boxes, with glands and silica gel breathers.						
Naturally cooled, oil filled type ODGM; offload tap changer excluding conservator, dial thermometer or Bucholzrelay	10908.00	13071.06	122.50	2270.40	nr	**15341.46**
Naturally cooled, midel fluid filled, type ODGM in ventilated steel tank.	14847.00	17791.16	122.50	2270.40	nr	**20061.56**
2000 kVA; Three Phase 11KV/433 volt 50Hz; HV and LV cable boxes, with glands						
Dry type; class C insulated air cooled; type AN in ventilated steel tank; temperature indicator	29290.00	35098.21	122.50	2270.40	nr	**37368.61**
Dry type; cast resin filled, type AN in ventilated steel tank; temperature indicator	28280.00	33887.92	122.50	2270.40	nr	**36158.33**
Extra For:						
Conservator	333.30	399.39	12.05	223.30	nr	**622.69**
Dial thermometer	212.10	254.16	2.00	37.07	nr	**291.23**
Bucholz relay	252.50	302.57	3.00	55.66	nr	**358.23**

Material Costs/Measured Work Prices – Electrical Installations

V:ELECTRICAL SUPPLY/POWER/LIGHTING

Item	Net Price £	Material £	Labour hours	Labour £	Unit	Total rate £
V11 : HV SUPPLY - CABLES						
Y61 - HV CABLES						
Cable; XLPE insulated, LSOH sheathed copper stranded conductors to BS 6724; laid in trench/duct including marker tape (cable tiles measured elsewhere)						
1900/3300 Volt grade; single core (Aluminium wire armour)						
50 mm2	3.87	4.64	0.16	2.97	m	**7.60**
70 mm2	4.47	5.36	0.16	2.97	m	**8.32**
95 mm2	5.09	6.10	0.18	3.34	m	**9.44**
120 mm2	5.74	6.88	0.18	3.34	m	**10.21**
150 mm2	6.26	7.50	0.20	3.71	m	**11.21**
185 mm2	7.54	9.04	0.20	3.71	m	**12.74**
240 mm2	8.03	9.62	0.22	4.08	m	**13.70**
300 mm2	10.27	12.31	0.22	4.08	m	**16.38**
400 mm2	12.66	15.17	0.26	4.82	m	**19.99**
1900/3300 Volt grade; Three core (Galvanised steel wire armour)						
25 mm2	3.49	4.18	0.16	2.97	m	**7.15**
35 mm2	4.27	5.12	0.16	2.97	m	**8.08**
50 mm2	5.81	6.96	0.18	3.34	m	**10.30**
70 mm2	7.34	8.80	0.18	3.34	m	**12.13**
95 mm2	9.20	11.02	0.21	3.89	m	**14.92**
120 mm2	11.25	13.48	0.21	3.89	m	**17.37**
150 mm2	13.25	15.88	0.23	4.26	m	**20.14**
185 mm2	15.96	19.12	0.23	4.26	m	**23.39**
240 mm2	20.23	24.24	0.24	4.45	m	**28.69**
300 mm2	24.46	29.31	0.24	4.45	m	**33.76**
400 mm2	30.62	36.69	0.28	5.19	m	**41.88**
Cable; XLPE insulated, LSOH sheathed copper stranded conductors to BS 6724; clipped direct to backgrounds including cleat						
1900/3300 Volt grade; single core (Aluminium wire armour)						
50 mm2	4.24	5.08	0.33	6.12	m	**11.20**
70 mm2	4.90	5.87	0.33	6.12	m	**11.99**
95 mm2	5.86	7.02	0.34	6.30	m	**13.32**
120 mm2	6.60	7.91	0.36	6.67	m	**14.58**
150 mm2	7.19	8.62	0.37	6.86	m	**15.47**
185 mm2	8.67	10.39	0.42	7.78	m	**18.17**
240 mm2	9.24	11.07	0.48	8.90	m	**19.97**
300 mm2	11.81	14.15	0.54	10.01	m	**24.16**
400 mm2	14.56	17.45	0.60	11.12	m	**28.57**

V:ELECTRICAL SUPPLY/POWER/LIGHTING

Item	Net Price £	Material £	Labour hours	Labour £	Unit	Total rate £
V11 : HV SUPPLY - CABLES						
Y61 - HV CABLES						
Cable; XLPE insulated, LSOH sheathed copper stranded conductors to BS 6724; clipped direct to backgrounds including cleat (Continued)						
1900/3300 Volt grade; Three core (Galvanised steel wire armour)						
25 mm2	3.82	4.58	0.35	6.49	m	11.06
35 mm2	4.68	5.61	0.37	6.86	m	12.47
50 mm2	6.36	7.62	0.40	7.41	m	15.03
70 mm2	8.44	10.11	0.41	7.60	m	17.71
95 mm2	10.57	12.67	0.45	8.34	m	21.01
120 mm2	12.94	15.51	0.48	8.90	m	24.40
150 mm2	15.24	18.26	0.51	9.45	m	27.71
185 mm2	18.35	21.99	0.53	9.82	m	31.81
240 mm2	23.27	27.88	0.60	11.12	m	39.00
300 mm2	28.14	33.72	0.66	12.23	m	45.95
400 mm2	35.21	42.19	0.75	13.90	m	56.09
Cable Termination; including drilling and cutting gland plate						
1900/3300 Volt grade; single core (Aluminium wire armour)						
25 mm2	8.84	10.59	1.25	23.17	nr	33.76
35 mm2	8.84	10.59	1.35	25.02	nr	32.74
50 mm2	9.52	11.41	1.57	29.10	nr	40.51
70 mm2	9.52	11.41	2.11	39.11	nr	50.51
95 mm2	9.52	11.41	2.37	43.93	nr	55.33
120 mm2	9.52	11.41	2.65	49.11	nr	60.52
150 mm2	12.02	14.40	3.00	55.60	nr	70.01
185 mm2	12.02	14.40	3.45	63.94	nr	78.35
240 mm2	16.33	19.57	3.75	69.50	nr	89.07
300 mm2	16.33	19.57	4.36	80.81	nr	100.38
400 mm2	23.21	27.81	5.10	94.52	nr	122.34
1900/3300 Volt grade; Three core (Galvanised steel wire armour)						
25 mm2	5.56	6.66	2.50	46.33	nr	53.00
35 mm2	10.31	12.35	2.68	49.67	nr	62.03
50 mm2	10.31	12.35	3.14	58.20	nr	70.55
70 mm2	10.31	12.35	4.22	78.21	nr	90.57
95 mm2	15.60	18.69	4.75	88.04	nr	106.73
120 mm2	15.60	18.69	5.30	98.23	nr	116.92
150 mm2	15.60	18.69	6.00	111.20	nr	129.90
185 mm2	15.60	18.69	6.90	127.88	nr	146.58
240 mm2	27.65	33.13	7.43	137.71	nr	170.84
300 mm2	27.65	33.13	8.75	162.17	nr	195.30
400 mm2	38.23	45.81	10.20	189.05	nr	234.86

Material Costs/Measured Work Prices – Electrical Installations

V:ELECTRICAL SUPPLY/POWER/LIGHTING

Item	Net Price £	Material £	Labour hours	Labour £	Unit	Total rate £
Cable; XLPE insulated, LSOH sheathed copper stranded conductors to BS 6724; laid in trench/duct including marker tape (cable tiles measured elsewhere)						
6350/11000 Volt grade; single core (Aluminium wire armour)						
95 mm2	6.15	7.37	0.20	3.71	m	**11.08**
120 mm2	6.56	7.86	0.20	3.71	m	**11.57**
150 mm2	7.21	8.64	0.22	4.08	m	**12.72**
185 mm2	8.23	9.86	0.22	4.08	m	**13.94**
240 mm2	9.97	11.95	0.24	4.45	m	**16.40**
300 mm2	11.03	13.22	0.26	4.82	m	**18.04**
400 mm2	16.93	20.29	0.28	5.19	m	**25.48**
6350/11000 Volt grade; Three core (Galvanised steel wire armour)						
95 mm2	13.60	16.30	0.23	4.26	m	**20.56**
120 mm2	16.21	19.42	0.23	4.26	m	**23.69**
150 mm2	18.07	21.65	0.25	4.63	m	**26.29**
185 mm2	20.32	24.35	0.25	4.63	m	**28.98**
240 mm2	24.83	29.75	0.27	5.00	m	**34.76**
300 mm2	29.78	35.69	0.29	5.37	m	**41.06**
400 mm2	35.63	42.70	0.31	5.75	m	**48.44**
Cable; XLPE insulated, LSOH sheathed copper stranded conductors to BS 6724; clipped direct to backgrounds including cleat						
6350/11000 Volt grade; single core (Aluminium wire armour)						
95 mm2	7.07	8.47	0.36	6.67	m	**15.14**
120 mm2	7.54	9.04	0.38	7.04	m	**16.08**
150 mm2	8.29	9.93	0.39	7.23	m	**17.16**
185 mm2	9.46	11.34	0.44	8.15	m	**19.49**
240 mm2	11.47	13.74	0.50	9.27	m	**23.01**
300 mm2	12.68	15.19	0.56	10.38	m	**25.57**
400 mm2	19.47	23.33	0.62	11.49	m	**34.82**
6350/11000 Volt grade; Three core (Galvanised steel wire armour)						
95 mm2	15.64	18.74	0.47	8.71	m	**27.45**
120 mm2	18.64	22.34	0.50	9.27	m	**31.60**
150 mm2	20.78	24.90	0.53	9.82	m	**34.72**
185 mm2	23.37	28.00	0.55	10.19	m	**38.20**
240 mm2	23.37	28.00	0.60	11.12	m	**39.12**
300 mm2	34.25	41.04	0.68	12.60	m	**53.64**
400 mm2	40.97	49.09	0.77	14.27	m	**63.37**

436

Material Costs/Measured Work Prices – Electrical Installations

V:ELECTRICAL SUPPLY/POWER/LIGHTING

Item	Net Price £	Material £	Labour hours	Labour £	Unit	Total rate £
V11 : HV SUPPLY - CABLES						
Y61 - HV CABLES						
Cable; XLPE insulated, LSOH sheathed copper stranded conductors to BS 6724; laid in trench/duct including marker tape (cable tiles measured elsewhere) (continued)						
Cable Termination; including drilling and cutting gland plate						
6350/11000 Volt grade; single core (Aluminium wire armour)						
95 mm2	292.34	350.31	2.37	43.93	nr	**394.24**
120 mm2	295.28	353.83	2.65	49.11	nr	**402.95**
150 mm2	296.53	355.33	3.00	55.60	nr	**410.93**
185 mm2	301.12	360.83	3.45	63.94	nr	**424.77**
240 mm2	304.45	364.82	3.75	69.50	nr	**434.32**
300 mm2	317.89	380.93	4.36	80.81	nr	**461.74**
400 mm2	319.11	382.39	5.10	94.52	nr	**476.91**
6350/11000 Volt grade; Three core (Galvanised steel wire armour)						
95 mm2	230.56	276.28	4.75	88.04	nr	**364.32**
120 mm2	233.50	279.80	5.30	98.23	nr	**378.03**
150 mm2	234.75	281.30	6.00	111.20	nr	**392.50**
185 mm2	236.72	283.66	6.90	127.88	nr	**411.55**
240 mm2	240.05	287.65	7.43	137.71	nr	**425.36**
300 mm2	251.55	301.43	8.75	162.17	nr	**463.60**
400 mm2	252.77	302.89	10.20	189.05	nr	**491.94**
Cable; XLPE insulated, LSOH sheathed copper stranded conductors to BS 6724; laid in trench/duct including marker tape (cable tiles measured elsewhere)						
19000/33000 Volt grade; single core (Aluminium wire armour)						
95 mm2	8.90	10.66	0.20	3.71	m	**14.37**
120 mm2	9.70	11.62	0.25	4.63	m	**16.26**
150 mm2	10.79	12.93	0.28	5.19	m	**18.12**
185 mm2	11.94	14.31	0.33	6.12	m	**20.42**
240 mm2	13.82	16.56	0.40	7.41	m	**23.97**
300 mm2	15.51	18.59	0.47	8.71	m	**27.30**
400 mm2	18.00	21.57	0.54	10.01	m	**31.58**
19000/33000 Volt grade; Three core (Galvanised steel wire armour)						
95 mm2	27.65	33.13	0.35	6.49	m	**39.62**
120 mm2	30.26	36.26	0.42	7.78	m	**44.04**
150 mm2	32.97	39.51	0.54	10.01	m	**49.52**
185 mm2	37.54	44.98	0.67	12.42	m	**57.40**
300 mm2	47.71	57.17	0.98	18.16	m	**75.33**
400 mm2	55.24	66.19	1.25	23.17	m	**89.36**

Material Costs/Measured Work Prices – Electrical Installations 437

V:ELECTRICAL SUPPLY/POWER/LIGHTING

Item	Net Price £	Material £	Labour hours	Labour £	Unit	Total rate £
Cable; XLPE insulated, LSOH sheathed copper stranded conductors to BS 6724; clipped direct to backgrounds including cleat						
19000/33000 Volt grade; single core (Aluminium wire armour)						
95 mm2	10.24	12.27	0.20	3.71	m	**15.98**
120 mm2	11.15	13.36	0.25	4.63	m	**17.99**
150 mm2	12.41	14.87	0.28	5.19	m	**20.06**
185 mm2	13.73	16.45	0.33	6.12	m	**22.57**
240 mm2	15.89	19.04	0.40	7.41	m	**26.45**
300 mm2	17.84	21.38	0.47	8.71	m	**30.09**
400 mm2	20.70	24.80	0.54	10.01	m	**34.81**
19000/33000 Volt grade; Three core (Galvanised steel wire armour)						
95 mm2	31.80	38.11	0.35	6.49	m	**44.59**
120 mm2	34.80	41.70	0.42	7.78	m	**49.48**
150 mm2	37.92	45.44	0.54	10.01	m	**55.45**
185 mm2	43.17	51.73	0.67	12.42	m	**64.15**
300 mm2	47.71	57.17	0.98	18.16	m	**75.33**
400 mm2	55.24	66.19	1.25	23.17	m	**89.36**
Cable Termination; including drilling and cutting gland plate						
19000/33000 Volt grade; single core (Aluminium wire armour)						
95 mm2	376.20	450.80	2.37	43.93	nr	**494.73**
120 mm2	379.14	454.32	2.65	49.11	nr	**503.44**
150 mm2	380.39	455.82	3.00	55.60	nr	**511.42**
185 mm2	381.33	456.95	3.45	63.94	nr	**520.89**
240 mm2	384.66	460.94	3.75	69.50	nr	**530.44**
300 mm2	385.38	461.80	4.36	80.81	nr	**542.61**
400 mm2	386.60	463.26	5.10	94.52	nr	**557.79**
Extra over single core 19000/33000 terminations						
Right angled boot	605.52	725.59	6.00	111.20	nr	**836.80**
Pressure testing per cable	500.00	599.15	-	-	nr	**599.15**
19000/33000 Volt grade; Three core (Galvanised steel wire armour)						
95 mm2	365.01	437.39	4.75	88.04	nr	**525.43**
120 mm2	448.65	537.62	5.30	98.23	nr	**635.85**
150 mm2	452.40	542.11	6.00	111.20	nr	**653.31**
185 mm2	455.22	545.49	6.90	127.88	nr	**673.37**
300 mm2	604.08	723.87	8.75	162.17	nr	**886.04**
400 mm2	607.74	728.25	10.20	189.05	nr	**917.30**

438   *Material Costs/Measured Work Prices – Electrical Installations*

V:ELECTRICAL SUPPLY/POWER/LIGHTING

Item	Net Price £	Material £	Labour hours	Labour £	Unit	Total rate £
V11 : HV SUPPLY - CABLES						
Y61 - HV CABLES						
Cable; XLPE insulated, LSOH sheathed copper stranded conductors to BS 6724; clipped direct to backgrounds including cleat (Continued)						
Extra over Three core 19000/33000 terminations						
Right angled boot	605.52	725.59	10.00	185.34	nr	910.93
Pressure testing per cable	500.00	599.15	-	-	nr	599.15
Cable; Paper Insulated; lead sheathed; copper stranded conductors;to BS 6480; laid in trench/duct with marker tape (Cable tiles measured elsewhere).						
6350/11000 Volt grade; Single core (Unarmoured)						
95 mm2	4.13	4.95	0.22	4.08	m	9.03
120 mm2	4.75	5.69	0.22	4.08	m	9.77
150 mm2	5.33	6.39	0.24	4.45	m	10.84
185 mm2	6.38	7.65	0.26	4.82	m	12.46
240 mm2	7.82	9.37	0.34	6.30	m	15.67
300 mm2	9.44	11.31	0.40	7.41	m	18.73
400 mm2	11.59	13.89	-	-	m	13.89
6350/11000 Volt Grade, Three core (Armoured)						
95 mm2	14.28	17.11	0.55	10.19	m	27.31
120 mm2	16.29	19.52	0.63	11.68	m	31.20
150 mm2	19.63	23.52	0.66	12.23	m	35.76
185 mm2	21.28	25.50	0.72	13.34	m	38.84
240 mm2	25.22	30.22	0.82	15.20	m	45.42
300 mm2	29.38	35.21	0.94	17.42	m	52.63
Cable; Paper Insulated; lead sheathed; copper stranded conductors to BS 6480; clipped direct to backgrounds including cleats.						
6350/11000 Volt Grade, Three core (Armoured)						
95 mm2	16.42	19.68	0.55	10.19	m	29.87
120 mm2	18.73	22.44	0.63	11.68	m	34.12
150 mm2	22.57	27.05	0.66	12.23	m	39.28
185 mm2	24.47	29.32	0.72	13.34	m	42.67
240 mm2	29.00	34.75	0.82	15.20	m	49.95
300 mm2	33.79	40.49	0.94	17.42	m	57.91

Material Costs/Measured Work Prices – Electrical Installations

V:ELECTRICAL SUPPLY/POWER/LIGHTING

Item	Net Price £	Material £	Labour hours	Labour £	Unit	Total rate £
Cable Termination; including drilling and cutting gland plate						
6350/11000 Volt grade; Three core (Armoured)						
95 mm2	244.21	292.64	4.75	88.04	nr	**380.67**
120 mm2	248.12	297.32	5.30	98.23	nr	**395.55**
185 mm2	250.31	299.95	6.90	127.88	nr	**427.83**
240 mm2	257.75	308.86	7.43	137.71	nr	**446.57**
300 mm2	273.43	327.65	8.75	162.17	nr	**489.82**
Cable Tiles; single width; laid in trench above cables on prepared sand bed. (Costs of excavation and filling sand excluded.)						
Reinforced concrete covers; concave/ convex ends						
914 x 152 x 63/38 mm	5.40	6.47	0.11	2.04	m	**8.51**
914 x 229 x 63/38 mm	6.88	8.24	0.11	2.04	m	**10.28**
914 x 305 x 63/38 mm	8.82	10.57	0.11	2.04	m	**12.61**

Material Costs/Measured Work Prices – Electrical Installations

V:ELECTRICAL SUPPLY/POWER/LIGHTING

Item	Net Price £	Material £	Labour hours	Labour £	Unit	Total rate £
V12 : LV SUPPLY - STANDBY GENERATORS						
STANDBY GENERATORS						
Standby Diesel Generating Sets; Supply and installation fixing to base; all supports and fixings; all necessary connections to equipment; including fixing of plates, discs etc. for identification						
Three phase, 440 Volt, four wire 50 Hz packaged standby diesel generating set, complete with radio and television suppressors, daily service fuel tank and associated piping, 4 metres of exhaust pipe and primary exhaust silencer, control panel, mains failure relay, starting battery with charger, all internal wiring, interconnections, earthing and labels rated for standby duty; delivery, installation and commissioning is included						
50 kVA	14009.00	16786.98	20.00	370.68	nr	17157.66
100 kVA	14509.00	17386.13	25.00	463.35	nr	17849.48
150 kVA	16261.00	19485.56	27.00	500.42	nr	19985.97
300 kVA	25688.00	30781.93	30.00	556.02	nr	31337.95
500 kVA	36430.00	43654.07	35.00	648.69	nr	44302.76
750 kVA	63164.00	75689.42	37.00	685.75	nr	76375.18
1000 kVA	84497.00	101252.76	40.00	741.36	nr	101994.11
1500kVA	99740.00	119518.44	45.00	834.03	nr	120352.47
2000kVA	154536.00	185180.49	50.00	926.70	nr	186107.18
2500kVA	212196.00	254274.47	55.00	1019.36	nr	255293.83
Extra for:						
Residential Silencer; including connection to exhaust pipe						
50 kVA	435.00	521.26	-	-	nr	521.26
100 kVA	717.00	859.18	-	-	nr	859.18
150 kVA	823.00	986.20	-	-	nr	986.20
300 kVA	1237.00	1482.30	-	-	nr	1482.30
500 kVA	2497.00	2992.16	-	-	nr	2992.16
750 kVA	3049.00	3653.62	-	-	nr	3653.62
1000kVA	4107.00	4921.42	-	-	nr	4921.42
1500kVA	6542.00	7839.28	-	-	nr	7839.28
2000kVA	7086.00	8491.15	-	-	nr	8491.15
2500kVA	8102.00	9708.63	-	-	nr	9708.63
Synchronisation Panel; including interconnecting cables; commissioning and testing; fixing to backgrounds						
2 x 50 kVA	4413.00	5288.10	-	-	nr	5288.10
2 x 100 kVA	5376.00	6442.06	-	-	nr	6442.06
2 x 150 kVA	6783.00	8128.07	-	-	nr	8128.07
2 x 300 kVA	10987.00	13165.72	-	-	nr	13165.72
2 x 500 kVA	16598.00	19889.38	-	-	nr	19889.38

Material Costs/Measured Work Prices – Electrical Installations

V:ELECTRICAL SUPPLY/POWER/LIGHTING

Item	Net Price £	Material £	Labour hours	Labour £	Unit	Total rate £
2 x 750kVA	18730.00	22444.16	-	-	nr	**22444.16**
2 x 1000kVA	20186.00	24188.88	-	-	nr	**24188.88**
2 x 1500kVA	23897.00	28635.78	-	-	nr	**28635.78**
2 x 2000kVA	25892.00	31026.38	-	-	nr	**31026.38**
2 x 2500kVA	30186.00	36171.88	-	-	nr	**36171.88**
Prefabricated acoustic housing						
50kVa	2142.00	2566.76	-	-	nr	**2566.76**
100kVa	2235.00	2678.20	-	-	nr	**2678.20**
150kVa	3629.00	4348.63	-	-	nr	**4348.63**
300kVa	6675.00	7998.65	-	-	nr	**7998.65**
500kVA	12788.00	15323.86	-	-	nr	**15323.86**
750kVA	20153.00	24149.34	-	-	nr	**24149.34**
1000kVA	25496.00	30551.86	-	-	nr	**30551.86**
1500kVA	50574.00	60602.82	-	-	nr	**60602.82**
2000kVA	60789.00	72843.46	-	-	nr	**72843.46**
2500kVA	84546.00	101311.47	-	-	nr	**101311.47**

Material Costs/Measured Work Prices – Electrical Installations

V:ELECTRICAL SUPPLY/POWER/LIGHTING

Item	Net Price £	Material £	Labour hours	Labour £	Unit	Total rate £
V12 : LV SUPPLY - SWITCHGEAR						
Y71 - LV SWITCHGEAR						
L.V. Circuit Breakers; Non-automatic air circuit breakers for purpose made switchboard including overload protection - adjustable instantaneous and inverse overload and short circuit						
Up to 65kA at 415/500V with instantaneous trip;						
1 second; Fixed Manual; Triple Pole						
630 Amp (35kA fault rating)	1683.18	2016.95	2.10	38.92	nr	**2055.88**
1000 Amp	1945.52	2331.32	2.20	40.77	nr	**2372.09**
1250 Amp	1965.07	2354.74	2.29	42.44	nr	**2397.19**
1600 Amp	2181.41	2613.98	2.29	42.44	nr	**2656.43**
2000 Amp	2775.77	3326.21	3.15	58.38	nr	**3384.59**
2500 Amp	3422.49	4101.17	3.15	58.38	nr	**4159.55**
1 Second; Fixed Manual; Four Pole						
630 Amp (35 kA fault protection)	1929.19	2311.75	2.96	54.86	nr	**2366.61**
1000 Amp	2220.69	2661.05	2.96	54.86	nr	**2715.91**
1250 Amp	2241.02	2685.41	3.15	58.38	nr	**2743.80**
1600 Amp	2505.09	3001.85	3.15	58.38	nr	**3060.23**
2000 Amp	3209.53	3845.98	4.01	74.32	nr	**3920.30**
2500 Amp	4102.85	4916.45	4.20	77.84	nr	**4994.29**
1 Second; Drawout Manual; Triple Pole						
630 Amp (35kA fault protection)	2244.93	2690.10	3.15	58.38	nr	**2748.48**
1000 Amp	2442.73	2927.12	3.15	58.38	nr	**2985.51**
1250 Amp	2486.66	2979.76	3.25	60.24	nr	**3040.00**
1600 Amp	2838.46	3401.33	3.53	65.42	nr	**3466.75**
2000 Amp	3541.59	4243.89	4.20	77.84	nr	**4321.73**
2500 Amp	4548.69	5450.70	4.49	83.22	nr	**5533.91**
1 Second; Drawout Manual; Four Pole						
630 Amp (35kA fault protection)	2721.00	3260.57	4.20	77.84	nr	**3338.42**
1000 Amp	3067.70	3676.02	4.20	77.84	nr	**3753.87**
1250 Amp	3118.92	3737.40	4.39	81.36	nr	**3818.77**
1600 Amp	3590.42	4302.40	4.39	81.36	nr	**4383.76**
2000 Amp	4467.12	5352.95	5.35	99.16	nr	**5452.11**
2500 Amp	5677.14	6802.92	5.54	102.68	nr	**6905.59**
Extra For						
Closing Coil	57.11	68.43	2.00	37.07	nr	**105.50**
Shunt Trip Coil	51.86	62.14	2.00	37.07	nr	**99.21**
Motorised Charging Unit	445.41	533.73	2.00	37.07	nr	**570.80**
Under voltage release device - instant	223.86	268.25	2.00	37.07	nr	**305.32**
Under voltage release device - time delay	223.86	268.25	2.00	37.07	nr	**305.32**
Carriage Position Switch	65.28	78.22	3.00	55.66	nr	**133.88**

Material Costs/Measured Work Prices – Electrical Installations

V:ELECTRICAL SUPPLY/POWER/LIGHTING

Item	Net Price £	Material £	Labour hours	Labour £	Unit	Total rate £
Operations Counter	58.57	70.18	3.00	55.66	nr	**125.84**
Breaker Insertion Interlock	23.45	28.10	3.00	55.66	nr	**83.76**
IP54 Door Panel	107.24	128.51	0.50	9.27	nr	**137.77**
Earthing Device						
3 Pole up to 65kA	1440.49	1726.14	2.00	37.07	nr	**1763.21**
4 Pole up to 65kA	1873.15	2244.60	2.00	37.07	nr	**2281.66**
Protection Units						
Unrestricted Earth Fault Protection	205.28	245.99	2.00	37.07	nr	**283.05**
Load Monitoring - Pre-trip alarm	56.56	67.78	2.00	37.07	nr	**104.84**
Load Monitoring - Load shedding warning	56.56	67.78	2.00	37.07	nr	**104.84**
Communication - Remote signalling	252.66	302.76	2.00	37.07	nr	**339.83**
Communication - Data reception / transmission	262.30	314.31	2.00	37.07	nr	**351.38**
Communication - Electronic Interlocks / interface	321.18	384.87	2.00	37.07	nr	**421.94**
Moulded Case Circuit Breakers; in sheet steel enclosure with thermal/magnetic trips; including supports connections jointing to equipment; BS 4572/IEC947; (provision and fixing of plates, discs etc, for identification is included.)						
25 -200 Amp at 25kV Icu range in nominal ratings indicated; T.P.&S.N.						
25-63 Amp	85.01	101.87	2.28	42.26	nr	**144.12**
80 - 100 Amp	86.38	103.51	2.48	45.96	nr	**149.47**
125 Amp	119.02	142.62	2.63	48.74	nr	**191.37**
160 Amp	245.59	294.29	2.88	53.38	nr	**347.67**
200 Amp	245.60	294.30	3.41	63.20	nr	**357.50**
250 Amp	310.34	371.88	3.48	64.50	nr	**436.38**
250-800 Amp at 32kV Icu range in nominal ratings indicated; T.P.&S.N.						
250 Amp	275.87	330.57	3.60	66.72	nr	**397.30**
315 Amp	371.82	445.55	3.94	73.02	nr	**518.58**
400 Amp	371.82	445.55	4.36	80.81	nr	**526.36**
250-800 Amp at 50kV Icu range in nominal ratings indicated; T.P.&S.N.						
400 Amp	487.54	584.22	4.48	83.03	nr	**667.25**
630 Amp	487.54	584.22	5.14	95.26	nr	**679.48**
800 Amp	577.62	692.16	5.68	105.27	nr	**797.43**

Material Costs/Measured Work Prices – Electrical Installations

V:ELECTRICAL SUPPLY/POWER/LIGHTING

Item	Net Price £	Material £	Labour hours	Labour £	Unit	Total rate £
V12 : LV SUPPLY - SWITCHGEAR						
Y71 - LV SWITCHGEAR						
AUTOMATIC TRANSFER SWITCHES						
Automatic transfer switches; steel enclosure; solenoid operating; programmable controller, keypad and LCD display; fixed to backgrounds; including commissioning and testing						
Panel Mounting type 3 pole						
100 amp	1807.05	2165.39	2.60	48.19	nr	**2213.58**
260amp	2538.90	3042.36	3.30	61.16	nr	**3103.53**
400amp	2688.00	3221.03	4.30	79.70	nr	**3300.73**
600amp	4562.25	5466.94	5.30	98.23	nr	**5565.17**
800amp	4856.25	5819.24	5.50	101.94	nr	**5921.18**
1000amp	9772.35	11710.21	5.83	108.05	nr	**11818.26**
1600amp	10849.65	13001.14	6.20	114.91	nr	**13116.05**
2000amp	11043.90	13233.91	6.90	127.88	nr	**13361.79**
3000amp	15777.30	18905.94	7.90	146.42	nr	**19052.36**
4000amp	18198.60	21807.38	7.90	146.42	nr	**21953.80**
Panel Mounting type 4 pole						
100 amp	2012.85	2412.00	2.93	54.30	nr	**2466.30**
260amp	2712.15	3249.97	3.63	67.28	nr	**3317.25**
400amp	2871.75	3441.22	4.63	85.81	nr	**3527.03**
600amp	5170.20	6195.45	5.83	108.05	nr	**6303.50**
800amp	5504.10	6595.56	6.53	121.03	nr	**6716.59**
1000amp	11410.35	13673.02	6.53	121.03	nr	**13794.05**
1600amp	12169.50	14582.71	7.20	133.44	nr	**14716.16**
2000amp	12382.65	14838.13	8.20	151.98	nr	**14990.11**
3000amp	17498.25	20968.15	9.20	170.51	nr	**21138.66**
4000amp	23831.85	28557.71	9.20	170.51	nr	**28728.22**
Enclosed type 3 pole						
100 amp	2077.95	2490.01	2.60	48.19	nr	**2538.20**
260amp	2801.40	3356.92	3.30	61.16	nr	**3418.08**
400amp	2966.25	3554.46	4.30	79.70	nr	**3634.15**
600amp	5191.20	6220.61	4.84	89.70	nr	**6310.32**
800amp	5807.55	6959.19	5.12	94.89	nr	**7054.08**
1000amp	11520.60	13805.14	5.50	101.94	nr	**13907.07**
1600amp	12518.10	15000.44	6.20	114.91	nr	**15115.35**
2000amp	12973.80	15546.50	6.90	127.88	nr	**15674.39**
3000amp	17946.60	21505.41	7.70	142.71	nr	**21648.12**
4000amp	21249.90	25463.76	7.70	142.71	nr	**25606.47**
Enclosed type 4 pole; overlapping neutral						
100 amp	2158.80	2586.89	2.93	54.30	nr	**2641.19**
260amp	2973.60	3563.26	3.63	67.28	nr	**3630.54**
400amp	3147.90	3772.13	4.63	85.81	nr	**3857.94**
600amp	5800.20	6950.38	4.97	92.11	nr	**7042.49**
800amp	6487.95	7774.51	5.47	101.38	nr	**7875.89**
1000amp	13015.80	15596.83	5.83	108.05	nr	**15704.89**
1600amp	14084.70	16877.70	6.53	121.03	nr	**16998.72**
2000amp	14579.25	17470.32	7.20	133.44	nr	**17603.76**

Material Costs/Measured Work Prices – Electrical Installations 445

V:ELECTRICAL SUPPLY/POWER/LIGHTING

Item	Net Price £	Material £	Labour hours	Labour £	Unit	Total rate £
3000amp	19885.95	23829.33	8.08	149.75	nr	**23979.09**
4000amp	26306.70	31523.32	8.08	149.75	nr	**31673.07**
Y71 - DISTRIBUTION BOARDS						
MCB distribution boards; IP3X external Protection enclosure; removable earth and neutral bars and DIN rail; 125/250amp incomers; including fixing to backgrounds						
SP & N						
6 way	62.26	74.60	2.00	37.07	nr	**111.67**
8 way	66.56	79.76	2.50	46.33	nr	**126.09**
12 way	74.66	89.46	3.00	55.60	nr	**145.06**
16 way	84.12	100.80	4.00	74.14	nr	**174.94**
24 way	132.42	158.68	5.00	92.67	nr	**251.35**
TP & N						
4 way	256.98	307.94	3.00	55.60	nr	**363.55**
6 way	263.34	315.57	3.50	64.87	nr	**380.43**
8 way	270.64	324.31	4.00	74.14	nr	**398.44**
12 way	288.26	345.42	4.00	74.14	nr	**419.55**
16 way	340.58	408.11	5.00	92.67	nr	**500.78**
24 way	479.76	574.90	6.40	118.62	nr	**693.51**
Miniature Circuit Breakers for Distribution Boards; BS EN 60 898; DIN rail mounting; including connecting to circuit						
SP&N; including connecting of wiring						
6 Amp	6.66	7.98	0.10	1.85	nr	**9.83**
10 - 40 Amp	5.74	6.88	0.10	1.85	nr	**8.74**
50 - 63 Amp	6.66	7.98	0.14	2.59	nr	**10.57**
TP&N; including connecting of wiring						
6 Amp	26.65	31.93	0.30	5.56	nr	**37.49**
10 - 40 Amp	24.95	29.90	0.45	8.34	nr	**38.24**
50 - 63 Amp	28.25	33.85	0.45	8.34	nr	**42.19**
Residual circuit breakers for Distribution Boards; DIN rail mounting; including connecting to circuit						
SP&N						
10mA						
6 Amp	42.51	50.94	0.21	3.89	nr	**54.83**
10 - 32 Amp	42.51	50.94	0.26	4.82	nr	**55.76**
45 Amp	43.09	51.63	0.26	4.82	nr	**56.45**
30mA						
6 Amp	40.88	48.99	0.21	3.89	nr	**52.88**
10 - 40 Amp	40.88	48.99	0.21	3.89	nr	**52.88**
50 -63 Amp	46.16	55.31	0.26	4.82	nr	**60.13**

Material Costs/Measured Work Prices – Electrical Installations

V:ELECTRICAL SUPPLY/POWER/LIGHTING

Item	Net Price £	Material £	Labour hours	Labour £	Unit	Total rate £
V12 : LV SUPPLY - SWITCHGEAR						
Y71 - DISTRIBUTION BOARDS						
Residual circuit breakers for Distribution Boards; DIN rail mounting; including connecting to circuit (Continued)						
100mA						
6 Amp	40.88	48.99	0.21	3.89	nr	**52.88**
10 - 40 Amp	40.88	48.99	0.21	3.89	nr	**52.88**
50 -63 Amp	50.70	60.76	0.26	4.82	nr	**65.58**
TP&N						
10mA						
6 Amp	42.51	50.94	0.21	3.89	nr	**54.83**
10 - 40 Amp	42.51	50.94	0.21	3.89	nr	**54.83**
50 -63 Amp	43.09	51.63	0.26	4.82	nr	**56.45**
30mA						
6 Amp	40.88	48.99	0.21	3.89	nr	**52.88**
10 - 40 Amp	40.88	48.99	0.21	3.89	nr	**52.88**
50 -63 Amp	46.16	55.31	0.26	4.82	nr	**60.13**
100mA						
6 Amp	40.88	48.99	0.21	3.89	nr	**52.88**
10 - 40 Amp	40.88	48.99	0.21	3.89	nr	**52.88**
50 -63 Amp	50.70	60.76	0.26	4.82	nr	**65.58**
HRC fused distribution boards; IP4X external protection enclosure; including earth and neutral bars; fixing to backgrounds						
SP&N						
20 Amp incomer						
4 way	81.55	97.72	1.00	18.53	nr	**116.26**
6 way	98.46	117.98	1.20	22.24	nr	**140.22**
8 way	115.42	138.31	1.40	25.95	nr	**164.26**
12 way	149.36	178.98	1.80	33.36	nr	**212.34**
32 Amp incomer						
4 way	98.19	117.66	1.00	18.53	nr	**136.20**
6 way	129.13	154.73	1.20	22.24	nr	**176.97**
8 way	151.92	182.05	1.40	25.95	nr	**207.99**
12 way	197.38	236.52	1.80	33.36	nr	**269.88**
TP&N						
20 Amp incomer						
4 way	146.18	175.16	1.50	27.80	nr	**202.96**
6 way	184.89	221.55	2.10	38.92	nr	**260.47**
8 way	574.14	687.99	2.70	50.04	nr	**738.03**
12 way	306.68	367.49	3.90	72.28	nr	**439.78**

Material Costs/Measured Work Prices – Electrical Installations

V:ELECTRICAL SUPPLY/POWER/LIGHTING

Item	Net Price £	Material £	Labour hours	Labour £	Unit	Total rate £
32 Amp incomer						
4 way	174.88	209.56	1.50	27.80	nr	**237.36**
6 way	234.78	281.34	2.10	38.92	nr	**320.26**
8 way	286.81	343.68	2.70	50.04	nr	**393.72**
12 way	397.79	476.67	3.90	72.28	nr	**548.96**
63 Amp incomer						
4 way	371.74	445.45	2.17	40.22	nr	**485.67**
6 way	476.74	571.28	2.83	52.45	nr	**623.73**
8 way	574.14	687.99	2.57	47.63	nr	**735.62**
100 Amp incomer						
4 way	587.90	704.49	2.40	44.48	nr	**748.97**
6 way	768.44	920.82	2.73	50.60	nr	**971.42**
8 way	939.61	1125.93	3.87	71.73	nr	**1197.66**
200 Amp incomer						
4 way	1456.18	1744.95	5.36	99.34	nr	**1844.29**
6 way	1924.65	2306.31	6.17	114.35	nr	**2420.66**
HRC fuse; including connection of wiring						
SP&N						
2-30 Amp	6.23	7.47	0.10	1.85	nr	**9.32**
35 - 63 Amp	6.23	7.47	0.12	2.22	nr	**9.69**
80 Amp	6.23	7.47	0.15	2.78	nr	**10.25**
100 Amp	6.23	7.47	0.15	2.78	nr	**10.25**
125 Amp	10.67	12.79	0.15	2.78	nr	**15.57**
160 Amp	11.64	13.95	0.15	2.78	nr	**16.73**
200 Amp	12.17	14.58	0.15	2.78	nr	**17.36**
TP&N						
2-30 Amp	18.70	22.40	0.15	2.78	nr	**25.18**
35 - 63 Amp	18.70	22.40	0.20	3.71	nr	**26.11**
80 Amp	18.70	22.40	0.20	3.71	nr	**26.11**
100 Amp	18.70	22.40	0.20	3.71	nr	**26.11**
125 Amp	32.02	38.36	0.20	3.71	nr	**42.07**
160 Amp	34.92	41.84	0.25	4.63	nr	**46.48**
200 Amp	36.50	43.74	0.25	4.63	nr	**48.38**
Consumer Units; fixed to backgrounds; including supports, fixings, connections and jointing to equipment. (Provision and fixing of plates, discs etc, for identification is included.)						
Switched and insulated; moulded plastic case, 63 Amp 230 Volt SP&N; earth and neutral bars; 30mA RCCB protection; fitted MCB's						
2 way	63.02	75.52	1.67	30.95	nr	**106.47**
4 way	74.23	88.95	1.59	29.47	nr	**118.42**
6 way	85.48	102.43	2.50	46.33	nr	**148.77**
8 way	97.47	116.80	3.00	55.60	nr	**172.40**
12 way	121.13	145.15	4.00	74.14	nr	**219.28**
16 way	160.78	192.66	5.50	101.94	nr	**294.59**

448
Material Costs/Measured Work Prices – Electrical Installations

V:ELECTRICAL SUPPLY/POWER/LIGHTING

Item	Net Price £	Material £	Labour hours	Labour £	Unit	Total rate £
V12 : LV SUPPLY - SWITCHGEAR						
Y71 - DISTRIBUTION BOARDS						
Consumer Units; fixed to backgrounds; including supports, fixings, connections and jointing to equipment. (Provision and fixing of plates, discs etc, for identification is included) (Continued)						
Switched and insulated; moulded plastic case, 100 Amp 230 Volt SP&N; earth and neutral bars; 30mA RCCB protection; fitted MCB's						
2 way	67.18	80.50	1.67	30.95	nr	**111.45**
4 way	78.30	93.83	1.59	29.47	nr	**123.30**
6 way	89.55	107.31	2.50	46.33	nr	**153.64**
8 way	101.54	121.68	3.00	55.60	nr	**177.28**
12 way	125.20	150.03	4.00	74.14	nr	**224.16**
16 way	164.85	197.54	5.50	101.94	nr	**299.47**
Residual current device; double pole; 230 volt/30mA tripping current						
16 Amp	36.58	43.83	0.22	4.08	nr	**47.91**
30 Amp	39.82	47.72	0.22	4.08	nr	**51.80**
40 Amp	41.44	49.66	0.22	4.08	nr	**53.74**
63 Amp	46.16	55.31	0.22	4.08	nr	**59.39**
80 Amp	50.11	60.05	0.22	4.08	nr	**64.13**
100 Amp	57.08	68.40	0.25	4.63	nr	**73.03**
Residual current device; double pole; 230 volt/100mA tripping current						
63 Amp	51.18	61.33	0.22	4.08	nr	**65.41**
80 Amp	51.18	61.33	0.22	4.08	nr	**65.41**
100 Amp	60.78	72.84	0.25	4.63	nr	**77.47**
Residual current device; four pole; 230-400volt/30mA tripping current						
25 Amp	51.22	61.37	0.35	6.49	nr	**67.86**
40 Amp	61.62	73.84	0.35	6.49	nr	**80.33**
63 Amp	81.34	97.46	0.35	6.49	nr	**103.95**
Heavy duty fuse switches; with HRC fuses BS 5419; short circuit rating 65kVA, 500 volt; including retractable operating switches						
SP&N						
63 Amp	109.08	130.71	1.30	24.09	nr	**154.80**
100 Amp	185.32	222.07	1.95	36.14	nr	**258.21**
TP&N						
63 Amp	149.97	179.71	1.83	33.92	nr	**213.63**
100 Amp	219.37	262.87	2.48	45.96	nr	**308.84**
200 Amp	345.37	413.86	3.13	58.01	nr	**471.87**
300 Amp	596.17	714.39	4.45	82.48	nr	**796.87**
400 Amp	653.59	783.20	4.45	82.48	nr	**865.67**

Material Costs/Measured Work Prices – Electrical Installations   449

V:ELECTRICAL SUPPLY/POWER/LIGHTING

Item	Net Price £	Material £	Labour hours	Labour £	Unit	Total rate £
600 Amp	1271.95	1524.18	5.72	106.01	nr	**1630.19**
800 Amp	1531.04	1834.65	7.88	146.05	nr	**1980.69**
Isolating switches; in sheet steel case; including plates/discs for indentification; fixed to backgrounds						
Double Pole						
20 Amp	39.74	47.62	1.02	18.90	nr	**66.52**
32 Amp	47.95	57.46	1.02	18.90	nr	**76.36**
63 Amp	69.82	83.67	1.21	22.43	nr	**106.09**
100 Amp	107.58	128.91	1.86	34.47	nr	**163.39**
TP&N						
20 Amp	43.47	52.09	1.29	23.91	nr	**76.00**
32 Amp	55.11	66.04	1.83	33.92	nr	**99.96**
63 Amp	85.76	102.77	2.48	45.96	nr	**148.73**
100 Amp	131.78	157.91	2.48	45.96	nr	**203.88**
125 Amp	197.38	243.35	2.48	45.96	nr	**289.51**
Busbar Chambers; fixed to background including all supports, fixings, connections and jointing to equipment. (Provision and fixing of plates, discs etc, for identification is included.)						
Sheet steel case enclosing 4 pole 550 Volt copper bars, detachable metal end plates;						
600mm long						
200 Amp	189.58	227.17	2.62	48.56	nr	**275.73**
300 Amp	243.60	291.91	3.03	56.16	nr	**348.06**
500 Amp	475.91	570.28	4.48	83.03	nr	**653.31**
Sheet steel case enclosing 4 pole 550 Volt copper bars, detachable metal end plates;						
900mm long						
200 Amp	273.05	327.20	3.04	56.34	nr	**383.54**
300 Amp	321.80	385.61	3.59	66.54	nr	**452.15**
500 Amp	475.99	570.38	4.42	81.92	nr	**652.30**
Sheet steel case enclosing 4 pole 550 Volt copper bars, detachable metal end plates;						
1350mm long						
200 Amp	372.96	446.92	3.38	62.64	nr	**509.56**
300 Amp	439.02	526.08	3.94	73.02	nr	**599.10**
500 Amp	702.67	842.01	4.82	89.33	nr	**931.34**

450 *Material Costs/Measured Work Prices – Electrical Installations*

V:ELECTRICAL SUPPLY/POWER/LIGHTING

Item	Net Price £	Material £	Labour hours	Labour £	Unit	Total rate £
V12 : LV SUPPLY - SWITCHGEAR						
Y71 - DISTRIBUTION BOARDS						
Contactor Relays; pressed steel enclosure; fixed to backgrounds including supports, fixings, connections/jointing to equipment. (Provision and fixing of plates, discs, etc, for identification is included.)						
Relays						
6 Amp, 415/240 Volt, 4 pole N/O	31.94	38.27	0.52	9.64	nr	**47.91**
6 Amp, 415/240 Volt, 8 pole N/O	39.05	46.79	0.85	15.75	nr	**62.55**
Push Button Stations; Heavy gauge pressed steel enclosure; polycarbonate cover; IP65; fixed to backgrounds including supports, fixings,connections/joining to equipment. (plates, discs, etc, for identification is included)						
Standard Units						
One button (start or stop)	31.20	37.39	0.39	7.23	nr	**44.62**
Two button (start or stop)	35.47	42.50	0.47	8.71	nr	**51.21**
Three button (forward-reverse-stop)	51.85	62.13	0.57	10.56	nr	**72.70**
Weatherproof Junction Boxes; enclosures with rail mounted terminal blocks; side hung door to receive padlock; fixed to backgrounds, including all supports and fixings. (Suitable for cable up to 2.5mm2; including glandplates and gaskets.)						
Sheet steel with zinc spray finish enclosure						
Overall Size 229 x 152; suitable to receive 3 nr 20(A) glands per gland plate	46.45	55.66	1.43	26.50	nr	**82.16**
Overall Size 306 x 306; suitable to receive 14 nr 20(A) glands per gland plate	62.31	74.67	2.17	40.22	nr	**114.88**
Overall Size 458 x 382; suitable to receive 18 nr 20(A) glands per gland plate	90.72	108.71	3.51	65.05	nr	**173.76**
Overall Size 762 x 508; suitable to receive 26 nr 20(A) glands per gland plate	95.95	114.98	4.85	89.89	nr	**204.87**
Overall Size 914 x 610; suitable to receive 45 nr 20(A) glands per gland plate	106.05	127.08	7.01	129.92	nr	**257.00**

Material Costs/Measured Work Prices – Electrical Installations   451

V:ELECTRICAL SUPPLY/POWER/LIGHTING

Item	Net Price £	Material £	Labour hours	Labour £	Unit	Total rate £
Weatherproof Junction Boxes; enclosures with rail mounted terminal blocks; screw fixed lid; fixed to backgrounds, including all supports and fixings. (Suitable for cable up to 2.5mm2; including glandplates and gaskets.)						
In Fex glassfibre reinforced polycarbonate enclosure						
Overall Size 190 x 190 x 130	65.97	79.05	1.43	26.50	nr	**105.56**
Overall Size 190 x 190 x 180	96.58	115.73	1.53	28.36	nr	**144.09**
Overall Size 280 x 190 x 130	109.08	130.71	2.17	40.22	nr	**170.93**
Overall Size 280 x 190 x 180	122.21	146.44	2.37	43.93	nr	**190.37**
Overall Size 380 x 190 x 130	136.35	163.39	3.33	61.72	nr	**225.11**
Overall Size 380 x 190 x 180	146.45	175.49	3.30	61.16	nr	**236.65**
Overall Size 380 x 280 x 130	156.55	187.59	4.66	86.37	nr	**273.96**
Overall Size 380 x 280 x 180	168.67	202.12	5.36	99.34	nr	**301.46**
Overall Size 560 x 280 x 130	203.01	243.27	7.01	129.92	nr	**373.19**
Overall Size 560 x 380 x 180	209.07	250.53	7.67	142.16	nr	**392.68**

Material Costs/Measured Work Prices – Electrical Installations

V:ELECTRICAL SUPPLY/POWER/LIGHTING

Item	Net Price £	Material £	Labour hours	Labour £	Unit	Total rate £
V20 : LV DISTRIBUTION - CABLE SUPPORTS						
Y60/63 - CABLE SUPPORTS						
LADDER RACK						
Light duty Galvanised Steel Ladder Rack; fixed to backgrounds; including supports, fixings and brackets; earth continuity straps.						
Straight lengths						
150 mm wide ladder	10.44	12.51	0.69	12.79	m	**25.30**
225 mm wide ladder	10.70	12.82	0.75	13.90	m	**26.72**
300 mm wide ladder	11.18	13.39	0.88	16.31	m	**29.70**
450 mm wide ladder	12.63	15.13	1.26	23.35	m	**38.48**
600 mm wide ladder	13.48	16.15	1.51	27.99	m	**44.13**
Extra over; (Cutting and jointing racking to fittings is included.)						
Insider riser bend						
150 mm wide ladder	40.43	48.41	0.33	6.12	nr	**54.52**
225 mm wide ladder	13.70	16.41	0.40	7.41	nr	**23.82**
300 mm wide ladder	42.02	50.31	0.56	10.38	nr	**60.69**
450 mm wide ladder	42.78	51.22	0.85	15.75	nr	**66.97**
600 mm wide ladder	44.57	53.36	0.99	18.35	nr	**71.71**
Outside riser bend						
225 mm wide ladder	41.12	49.23	0.35	6.49	nr	**55.72**
300 mm wide ladder	42.02	50.31	0.43	7.97	nr	**58.28**
450 mm wide ladder	42.78	51.22	0.73	13.53	nr	**64.75**
600 mm wide ladder	44.57	53.36	0.86	15.94	nr	**69.30**
Equal tee						
225 mm wide ladder	49.96	59.81	0.48	8.90	nr	**68.70**
300 mm wide ladder	49.47	59.23	0.62	11.49	nr	**70.72**
450 mm wide ladder	49.13	58.82	1.09	20.20	nr	**79.02**
600 mm wide ladder	59.27	70.96	1.12	20.76	nr	**91.72**
Unequal tee						
225 mm wide ladder	49.96	59.81	0.48	8.90	nr	**68.70**
300 mm wide ladder	49.47	59.23	0.57	10.56	nr	**69.79**
450 mm wide ladder	49.13	58.82	1.17	21.68	nr	**80.50**
600 mm wide ladder	59.27	70.96	1.17	21.68	nr	**92.64**
4 way cross overs						
225 mm wide ladder	79.00	94.58	0.56	10.38	nr	**104.96**
300 mm wide ladder	83.90	100.45	0.72	13.34	nr	**113.79**
450 mm wide ladder	91.36	109.37	1.13	20.94	nr	**130.31**
600 mm wide ladder	109.64	131.26	1.29	23.91	nr	**155.17**

Material Costs/Measured Work Prices – Electrical Installations   453

V:ELECTRICAL SUPPLY/POWER/LIGHTING

Item	Net Price £	Material £	Labour hours	Labour £	Unit	Total rate £
Heavy Duty Galvanised Steel Ladder Rack; fixed to backgrounds; including supports, fixings and brackets; earth continuity straps.						
Straight lengths						
150 mm wide ladder	12.07	14.47	0.68	12.60	m	**27.07**
300 mm wide ladder	13.18	15.79	0.79	14.64	m	**30.43**
450 mm wide ladder	14.26	17.08	1.07	19.83	m	**36.91**
600 mm wide ladder	15.15	18.16	1.24	22.98	m	**41.14**
750 mm wide ladder	16.75	20.07	1.49	27.62	m	**47.68**
900 mm wide ladder	17.89	21.44	1.67	30.95	m	**52.39**
Extra over; (Cutting and jointing racking to fittings is included)						
Flat bend						
150 mm wide ladder	28.84	34.56	0.34	6.30	nr	**40.86**
300 mm wide ladder	30.50	36.55	0.39	7.23	nr	**43.77**
450 mm wide ladder	34.64	41.51	0.43	7.97	nr	**49.48**
600 mm wide ladder	39.54	47.38	0.61	11.31	nr	**58.68**
750 mm wide ladder	45.47	54.49	0.82	15.20	nr	**69.69**
900 mm wide ladder	51.13	61.27	0.97	17.98	nr	**79.25**
Insider riser bend						
150 mm wide ladder	43.68	52.29	0.27	5.00	nr	**57.29**
300 mm wide ladder	45.33	54.27	0.45	8.34	nr	**62.61**
450 mm wide ladder	47.82	57.25	0.65	12.05	nr	**69.29**
600 mm wide ladder	65.83	78.81	0.81	15.01	nr	**93.82**
750 mm wide ladder	56.03	67.08	0.92	17.05	nr	**84.13**
900 mm wide ladder	60.10	71.95	1.06	19.65	nr	**91.60**
Outside riser bend						
150 mm wide ladder	43.68	52.29	0.27	5.00	nr	**57.29**
300 mm wide ladder	45.33	54.27	0.33	6.12	nr	**60.39**
450 mm wide ladder	47.82	57.25	0.61	11.31	nr	**68.55**
600 mm wide ladder	65.83	78.81	0.76	14.09	nr	**92.89**
750 mm wide ladder	56.03	67.08	0.94	17.42	nr	**84.50**
900 mm wide ladder	60.10	71.95	1.05	19.46	nr	**91.41**
Equal tee						
150mm wide ladder	51.06	61.13	0.37	6.86	nr	**67.99**
300 mm wide ladder	56.03	67.08	0.57	10.56	nr	**77.64**
450 mm wide ladder	60.17	72.03	0.83	15.38	nr	**87.42**
600 mm wide ladder	66.72	79.88	0.92	17.05	nr	**96.93**
750 mm wide ladder	88.25	105.65	1.13	20.94	nr	**126.60**
900 mm wide ladder	91.56	109.62	1.20	22.24	nr	**131.86**
Unequal tee						
300 mm wide ladder	56.03	67.08	0.57	10.56	nr	**77.64**
450 mm wide ladder	60.17	72.03	1.17	21.68	nr	**93.72**
600 mm wide ladder	66.72	79.88	1.17	21.68	nr	**101.57**
750 mm wide ladder	88.25	105.65	1.25	23.17	nr	**128.82**
900 mm wide ladder	91.56	109.62	1.33	24.65	nr	**134.27**

454   *Material Costs/Measured Work Prices – Electrical Installations*

V:ELECTRICAL SUPPLY/POWER/LIGHTING

Item	Net Price £	Material £	Labour hours	Labour £	Unit	Total rate £
V20 : LV DISTRIBUTION - CABLE SUPPORTS						
Y60/63 - CABLE SUPPORTS						
Heavy Duty Galvanised Steel Ladder Rack; fixed to backgrounds; including supports, fixings and brackets; earth continuity straps. (Continued)						
4 way cross overs						
150 mm wide ladder	79.00	94.58	0.50	9.27	nr	**103.85**
300 mm wide ladder	82.32	98.55	0.67	12.42	nr	**110.97**
450 mm wide ladder	87.35	104.58	0.92	17.05	nr	**121.63**
600 mm wide ladder	93.15	111.52	1.07	19.83	nr	**131.35**
750 mm wide ladder	119.44	142.99	1.25	23.17	nr	**166.16**
900 mm wide ladder	124.48	149.02	1.36	25.21	nr	**174.23**
Extra Heavy Duty Galvanised Steel Ladder Rack; fixed to backgrounds; including supports, fixings and brackets; earth continuity straps.						
Straight lengths						
150 mm wide ladder	14.99	17.97	0.63	11.68	m	**29.64**
300 mm wide ladder	18.72	22.43	0.70	12.97	m	**35.41**
450 mm wide ladder	20.03	24.00	0.83	15.38	m	**39.39**
600 mm wide ladder	21.71	26.02	0.89	16.50	m	**42.52**
750 mm wide ladder	24.45	29.29	1.22	22.61	m	**51.91**
900 mm wide ladder	25.85	30.98	1.44	26.69	m	**57.67**
Extra over; (Cutting and jointing racking to fittings is included.)						
Flat bend						
150 mm wide ladder	35.47	42.50	0.36	6.67	nr	**49.17**
300 mm wide ladder	37.88	45.39	0.39	7.23	nr	**52.62**
450 mm wide ladder	42.92	51.43	0.43	7.97	nr	**59.40**
600 mm wide ladder	55.55	66.56	0.61	11.31	nr	**77.87**
750 mm wide ladder	61.48	73.67	0.82	15.20	nr	**88.87**
900 mm wide ladder	68.45	82.02	0.97	17.98	nr	**100.00**
Insider riser bend						
150 mm wide ladder	64.10	76.74	0.36	6.67	nr	**83.41**
300 mm wide ladder	64.86	77.65	0.39	7.23	nr	**84.88**
450 mm wide ladder	68.31	81.78	0.43	7.97	nr	**89.75**
600 mm wide ladder	59.41	71.12	0.61	11.31	nr	**82.43**
750 mm wide ladder	61.48	73.60	0.82	15.20	nr	**88.80**
900 mm wide ladder	68.45	81.95	0.97	17.98	nr	**99.92**
Outside riser bend						
150 mm wide ladder	64.10	76.74	0.36	6.67	nr	**83.41**
300 mm wide ladder	64.86	77.65	0.39	7.23	nr	**84.88**
450 mm wide ladder	68.31	81.78	0.41	7.60	nr	**89.38**
600 mm wide ladder	59.41	71.12	0.57	10.56	nr	**81.69**
750 mm wide ladder	61.48	73.60	0.82	15.20	nr	**88.80**
900 mm wide ladder	68.45	81.95	0.93	17.24	nr	**99.18**

Material Costs/Measured Work Prices – Electrical Installations

V:ELECTRICAL SUPPLY/POWER/LIGHTING

Item	Net Price £	Material £	Labour hours	Labour £	Unit	Total rate £
Equal tee						
150 mm wide ladder	71.48	85.58	0.37	6.86	nr	**92.44**
300 mm wide ladder	67.48	80.79	0.57	10.56	nr	**91.35**
450 mm wide ladder	73.35	87.81	0.83	15.38	nr	**103.19**
600 mm wide ladder	80.94	96.90	0.92	17.05	nr	**113.95**
750 mm wide ladder	101.15	121.10	1.13	20.94	nr	**142.04**
900 mm wide ladder	109.64	131.26	1.20	22.24	nr	**153.50**
Unequal tee						
150 mm wide ladder	68.86	82.44	0.37	6.86	nr	**89.30**
300 mm wide ladder	68.86	82.44	0.57	10.56	nr	**93.01**
450 mm wide ladder	73.35	87.81	1.17	21.68	nr	**109.50**
600 mm wide ladder	80.94	96.90	1.17	21.68	nr	**118.58**
750 mm wide ladder	101.15	121.10	1.25	23.17	nr	**144.27**
900 mm wide ladder	109.64	131.26	1.33	24.65	nr	**155.91**
4 way cross overs						
150 mm wide ladder	93.84	112.35	0.50	9.27	nr	**121.61**
300 mm wide ladder	98.81	118.29	0.67	12.42	nr	**130.71**
450 mm wide ladder	123.65	148.03	0.92	17.05	nr	**165.08**
600 mm wide ladder	131.24	157.12	1.07	19.83	nr	**176.95**
750 mm wide ladder	138.48	165.79	1.25	23.17	nr	**188.96**
900 mm wide ladder	145.04	173.64	1.36	25.21	nr	**198.85**
CABLE TRAY						
Galvanised Steel Cable Tray to BS 729; including standard coupling joints, fixings and earth continuity straps. (Supports and hangers are excluded.)						
Light duty tray						
50 mm wide	2.42	2.89	0.19	3.52	m	**6.42**
75 mm wide	3.03	3.63	0.23	4.26	m	**7.89**
100 mm wide	3.76	4.51	0.31	5.75	m	**10.25**
150 mm wide	4.95	5.93	0.33	6.12	m	**12.04**
225 mm wide	9.01	10.80	0.39	7.23	m	**18.03**
300 mm wide	12.75	15.28	0.49	9.08	m	**24.36**
450 mm wide	19.29	23.11	0.60	11.12	m	**34.23**
600 mm wide	32.36	38.77	0.79	14.64	m	**53.41**
750 mm wide	40.02	47.96	1.04	19.28	m	**67.23**
900 mm wide	47.38	56.78	1.26	23.35	m	**80.13**
Extra over; (Cutting and jointing tray to fittings is included.)						
Straight reducer						
75 mm wide	15.11	18.11	0.22	4.08	nr	**22.18**
100 mm wide	16.51	19.78	0.25	4.63	nr	**24.42**
150 mm wide	22.21	26.61	0.27	5.00	nr	**31.62**
225 mm wide	30.11	36.08	0.34	6.30	nr	**42.38**
300 mm wide	38.71	46.39	0.39	7.23	nr	**53.61**
450 mm wide	63.01	75.50	0.49	9.08	nr	**84.59**
600 mm wide	80.61	96.59	0.54	10.01	nr	**106.60**
750 mm wide	103.51	124.04	0.61	11.31	nr	**135.34**
900 mm wide	118.31	141.77	0.69	12.79	nr	**154.56**

Material Costs/Measured Work Prices – Electrical Installations

V:ELECTRICAL SUPPLY/POWER/LIGHTING

Item	Net Price £	Material £	Labour hours	Labour £	Unit	Total rate £
V20 : LV DISTRIBUTION - CABLE SUPPORTS						
Y60/63 - CABLE SUPPORTS						
Galvanised Steel Cable Tray to BS 729; including standard coupling joints, fixings and earth continuity straps. (Supports and hangers are excluded) (Continued)						
Flat bend; 90 degree						
50mm wide	8.31	9.96	0.19	3.52	nr	**13.48**
75 mm wide	8.61	10.32	0.24	4.45	nr	**14.77**
100 mm wide	9.11	10.92	0.28	5.19	nr	**16.11**
150 mm wide	10.01	11.99	0.30	5.56	nr	**17.56**
225 mm wide	14.91	17.87	0.36	6.67	nr	**24.54**
300 mm wide	22.91	27.45	0.44	8.15	nr	**35.61**
450 mm wide	40.31	48.30	0.57	10.56	nr	**58.87**
600 mm wide	53.41	64.00	0.69	12.79	nr	**76.79**
750 mm wide	90.51	108.46	0.81	15.01	nr	**123.47**
900 mm wide	120.11	143.93	0.94	17.42	nr	**161.35**
Adjustable riser						
50 mm wide	13.61	16.31	0.26	4.82	nr	**21.13**
75 mm wide	15.11	18.11	0.29	5.37	nr	**23.48**
100 mm wide	17.41	20.86	0.32	5.93	nr	**26.79**
150 mm wide	22.81	27.33	0.36	6.67	nr	**34.01**
225 mm wide	29.71	35.60	0.44	8.15	nr	**43.76**
300 mm wide	33.11	39.68	0.52	9.64	nr	**49.31**
450 mm wide	59.81	71.67	0.66	12.23	nr	**83.90**
600 mm wide	76.01	91.08	0.79	14.64	nr	**105.72**
750 mm wide	99.91	119.72	1.03	19.09	nr	**138.81**
900 mm wide	120.11	143.93	1.10	20.39	nr	**164.32**
Inside riser; 90 degree						
50 mm wide	13.61	16.31	0.28	5.19	nr	**21.50**
75 mm wide	17.41	20.86	0.31	5.75	nr	**26.61**
100 mm wide	17.41	20.86	0.33	6.12	nr	**26.98**
150 mm wide	22.91	27.45	0.37	6.86	nr	**34.31**
225 mm wide	29.71	35.60	0.44	8.15	nr	**43.76**
300 mm wide	33.11	39.68	0.53	9.82	nr	**49.50**
450 mm wide	59.81	71.67	0.67	12.42	nr	**84.09**
600 mm wide	76.01	91.08	0.79	14.64	nr	**105.72**
750 mm wide	99.91	119.72	0.95	17.61	nr	**137.33**
900 mm wide	120.11	143.93	1.11	20.57	nr	**164.50**
Outside riser; 90 degree						
50 mm wide	13.61	16.31	0.28	5.19	nr	**21.50**
75 mm wide	15.11	18.11	0.31	5.75	nr	**23.85**
100 mm wide	17.41	20.86	0.33	6.12	nr	**26.98**
150 mm wide	22.91	27.45	0.37	6.86	nr	**34.31**
225 mm wide	29.71	35.60	0.44	8.15	nr	**43.76**
300 mm wide	33.11	39.68	0.53	9.82	nr	**49.50**
450 mm wide	59.81	71.67	0.67	12.42	nr	**84.09**
600 mm wide	76.01	91.08	0.79	14.64	nr	**105.72**
750 mm wide	99.91	119.72	0.95	17.61	nr	**137.33**
900 mm wide	120.11	143.93	1.11	20.57	nr	**164.50**

Material Costs/Measured Work Prices – Electrical Installations

V:ELECTRICAL SUPPLY/POWER/LIGHTING

Item	Net Price £	Material £	Labour hours	Labour £	Unit	Total rate £
Equal Tee						
50 mm wide	13.12	15.72	0.30	5.56	nr	**21.28**
75 mm wide	13.62	16.32	0.31	5.75	nr	**22.07**
100 mm wide	14.42	17.28	0.35	6.49	nr	**23.77**
150 mm wide	17.82	21.35	0.36	6.67	nr	**28.03**
225 mm wide	23.62	28.30	0.74	13.72	nr	**42.02**
300 mm wide	35.02	41.96	0.54	10.01	nr	**51.97**
450 mm wide	61.42	73.60	0.71	13.16	nr	**86.76**
600 mm wide	66.82	80.07	0.92	17.05	nr	**97.12**
750 mm wide	124.72	149.45	1.19	22.06	nr	**171.51**
900 mm wide	187.82	225.06	1.44	26.69	nr	**251.75**
Unequal Tee						
75 mm wide	13.62	16.32	0.38	7.04	nr	**23.36**
100 mm wide	14.42	17.28	0.39	7.23	nr	**24.51**
150 mm wide	17.82	21.35	0.43	7.97	nr	**29.32**
225 mm wide	24.42	29.26	0.50	9.27	nr	**38.53**
300 mm wide	35.02	41.96	0.63	11.68	nr	**53.64**
450 mm wide	61.42	73.60	0.80	14.83	nr	**88.43**
600 mm wide	86.82	104.04	1.02	18.90	nr	**122.94**
750 mm wide	124.72	149.45	1.12	20.76	nr	**170.21**
900 mm wide	187.82	225.06	1.35	25.02	nr	**250.09**
4 way crossovers						
50 mm wide	20.13	24.12	0.38	7.04	nr	**31.16**
75 mm wide	20.43	24.48	0.40	7.41	nr	**31.89**
100 mm wide	21.63	25.92	0.40	7.41	nr	**33.33**
150 mm wide	26.13	31.31	0.44	8.15	nr	**39.47**
225 mm wide	36.03	43.17	0.53	9.82	nr	**53.00**
300 mm wide	51.13	61.27	0.64	11.86	nr	**73.13**
450 mm wide	90.03	107.88	0.84	15.57	nr	**123.45**
600 mm wide	118.83	142.39	1.03	19.09	nr	**161.48**
750 mm wide	163.33	195.72	1.13	20.94	nr	**216.66**
900 mm wide	250.83	300.57	1.36	25.21	nr	**325.78**
Medium duty tray with return flange						
75 mm wide	4.97	5.96	0.33	6.12	m	**12.08**
100 mm wide	5.47	6.55	0.35	6.49	m	**13.04**
150 mm wide	6.73	8.07	0.39	7.23	m	**15.30**
225 mm wide	8.57	10.27	0.45	8.34	m	**18.61**
300 mm wide	11.99	14.37	0.57	10.56	m	**24.93**
450 mm wide	18.69	22.40	0.69	12.79	m	**35.19**
600 mm wide	24.10	28.88	0.91	16.87	m	**45.74**
Extra over; (Cutting and jointing tray to fittings is included.)						
Straight reducer						
100 mm wide	25.34	30.36	0.25	4.63	nr	**35.00**
150 mm wide	27.64	33.12	0.27	5.00	nr	**38.13**
225 mm wide	32.44	38.87	0.34	6.30	nr	**45.17**
300 mm wide	41.54	49.78	0.39	7.23	nr	**57.01**
450 mm wide	55.94	67.03	0.49	9.08	nr	**76.11**
600 mm wide	67.84	81.29	0.54	10.01	nr	**91.30**

Material Costs/Measured Work Prices – Electrical Installations

V:ELECTRICAL SUPPLY/POWER/LIGHTING

Item	Net Price £	Material £	Labour hours	Labour £	Unit	Total rate £
V20 : LV DISTRIBUTION - CABLE SUPPORTS						
Y60/63 - CABLE SUPPORTS						
Galvanised Steel Cable Tray to BS 729; including standard coupling joints, fixings and earth continuity straps. (Supports and hangers are excluded) (Continued)						
Flat bend; 90 degree						
75 mm wide	40.74	48.82	0.24	4.45	nr	**53.27**
100 mm wide	43.44	52.05	0.28	5.19	nr	**57.24**
150 mm wide	47.04	56.37	0.30	5.56	nr	**61.93**
225 mm wide	54.84	65.71	0.36	6.67	nr	**72.39**
300 mm wide	67.24	80.57	0.44	8.15	nr	**88.73**
450 mm wide	92.34	110.65	0.57	10.56	nr	**121.22**
600 mm wide	121.14	145.16	0.69	12.79	nr	**157.95**
Adjustable bend						
75 mm wide	42.74	51.22	0.29	5.37	nr	**56.59**
100 mm wide	43.44	52.05	0.32	5.93	nr	**57.98**
150 mm wide	47.04	56.37	0.36	6.67	nr	**63.04**
225 mm wide	54.84	65.71	0.44	8.15	nr	**73.87**
300 mm wide	57.24	68.59	0.52	9.64	nr	**78.23**
Adjustable riser						
75 mm wide	25.34	30.36	0.29	5.37	nr	**35.74**
100 mm wide	28.64	34.32	0.32	5.93	nr	**40.25**
150 mm wide	31.14	37.32	0.36	6.67	nr	**43.99**
225 mm wide	36.04	43.19	0.44	8.15	nr	**51.34**
300 mm wide	41.14	49.30	0.52	9.64	nr	**58.94**
450 mm wide	63.74	76.38	0.66	12.23	nr	**88.61**
600 mm wide	79.94	95.79	0.79	14.64	nr	**110.43**
Inside riser; 90 degree						
75 mm wide	20.94	25.09	0.31	5.75	nr	**30.84**
100 mm wide	23.24	27.85	0.33	6.12	nr	**33.96**
150 mm wide	27.94	33.48	0.37	6.86	nr	**40.34**
225 mm wide	36.04	43.19	0.44	8.15	nr	**51.34**
300 mm wide	40.14	48.10	0.53	9.82	nr	**57.92**
450 mm wide	63.74	76.38	0.67	12.42	nr	**88.80**
600 mm wide	79.94	95.79	0.79	14.64	nr	**110.43**
Outside riser; 90 degree						
75 mm wide	20.94	25.09	0.31	5.75	nr	**30.84**
100 mm wide	23.24	27.85	0.33	6.12	nr	**33.96**
150 mm wide	27.94	33.48	0.37	6.86	nr	**40.34**
225 mm wide	36.04	43.19	0.44	8.15	nr	**51.34**
300 mm wide	40.14	48.10	0.53	9.82	nr	**57.92**
450 mm wide	63.74	76.38	0.67	12.42	nr	**88.80**
600 mm wide	79.94	95.79	0.79	14.64	nr	**110.43**

Material Costs/Measured Work Prices – Electrical Installations

V:ELECTRICAL SUPPLY/POWER/LIGHTING

Item	Net Price £	Material £	Labour hours	Labour £	Unit	Total rate £
Equal Tee						
75 mm wide	59.58	71.39	0.31	5.75	nr	**77.14**
100 mm wide	62.38	74.75	0.35	6.49	nr	**81.24**
150 mm wide	66.08	79.18	0.36	6.67	nr	**85.86**
225 mm wide	72.18	86.49	0.74	13.72	nr	**100.21**
300 mm wide	90.88	108.90	0.54	10.01	nr	**118.91**
450 mm wide	120.38	144.25	0.71	13.16	nr	**157.41**
600 mm wide	176.28	211.24	0.92	17.05	nr	**228.29**
Unequal Tee						
100 mm wide	62.38	74.75	0.39	7.23	nr	**81.98**
150 mm wide	62.38	74.75	0.43	7.97	nr	**82.72**
225 mm wide	72.18	86.49	0.50	9.27	nr	**95.76**
300 mm wide	90.88	108.90	0.63	11.68	nr	**120.58**
450 mm wide	120.38	144.25	0.80	14.83	nr	**159.08**
600 mm wide	176.28	211.24	1.02	18.90	nr	**230.14**
4 way crossovers						
75 mm wide	82.62	99.00	0.40	7.41	nr	**106.42**
100 mm wide	91.22	109.31	0.40	7.41	nr	**116.72**
150 mm wide	95.32	114.22	0.44	8.15	nr	**122.38**
225 mm wide	116.32	139.39	0.53	9.82	nr	**149.21**
300 mm wide	133.42	159.88	0.64	11.86	nr	**171.74**
450 mm wide	200.13	239.82	0.84	15.57	nr	**255.38**
600 mm wide	266.82	319.73	1.03	19.09	nr	**338.82**
Heavy duty tray with return flange						
75 mm	9.53	11.42	0.34	6.30	m	**17.72**
100 mm	9.68	11.60	0.36	6.67	m	**18.27**
150 mm	10.77	12.91	0.40	7.41	m	**20.32**
225 mm	11.20	13.42	0.46	8.53	m	**21.94**
300 mm	12.95	15.52	0.58	10.75	m	**26.27**
450 mm	25.78	30.89	0.70	12.97	m	**43.86**
600 mm	35.59	42.65	0.92	17.05	m	**59.70**
750 mm	44.72	53.59	1.01	18.72	m	**72.30**
900 mm	46.99	56.31	1.14	21.13	m	**77.44**
Extra over; (Cutting and jointing tray to fittings is included.)						
Straight reducer						
100 mm wide	38.60	46.25	0.25	4.63	nr	**50.89**
150 mm wide	42.90	51.41	0.27	5.00	nr	**56.41**
225 mm wide	48.80	58.48	0.34	6.30	nr	**64.78**
300 mm wide	58.10	69.62	0.39	7.23	nr	**76.85**
450 mm wide	86.20	103.29	0.49	9.08	nr	**112.38**
600 mm wide	96.40	115.52	0.54	10.01	nr	**125.52**
750 mm wide	131.30	157.34	0.60	11.12	nr	**168.46**
900 mm wide	148.90	178.43	0.66	12.23	nr	**190.66**

460   *Material Costs/Measured Work Prices – Electrical Installations*

V:ELECTRICAL SUPPLY/POWER/LIGHTING

Item	Net Price £	Material £	Labour hours	Labour £	Unit	Total rate £
V20 : LV DISTRIBUTION - CABLE SUPPORTS						
Y60/63 - CABLE SUPPORTS						
Galvanised Steel Cable Tray to BS 729; including standard coupling joints, fixings and earth continuity straps. (Supports and hangers are excluded) (Continued)						
Flat bend; 90 degree						
75 mm wide	56.10	67.22	0.24	4.45	nr	**71.67**
100 mm wide	61.90	74.17	0.28	5.19	nr	**79.36**
150 mm wide	68.10	81.60	0.30	5.56	nr	**87.16**
225 mm wide	75.50	90.47	0.36	6.67	nr	**97.14**
300 mm wide	79.70	95.50	0.44	8.15	nr	**103.66**
450 mm wide	132.80	159.13	0.57	10.56	nr	**169.70**
600 mm wide	183.60	220.01	0.69	12.79	nr	**232.80**
750 mm wide	236.30	283.16	0.83	15.38	nr	**298.54**
900 mm wide	245.20	293.82	1.01	18.72	nr	**312.54**
Adjustable bend						
75 mm wide	56.10	67.22	0.29	5.37	nr	**72.60**
100 mm wide	61.90	74.17	0.32	5.93	nr	**80.11**
150 mm wide	68.10	81.60	0.36	6.67	nr	**88.28**
225 mm wide	75.50	90.47	0.44	8.15	nr	**98.63**
300 mm wide	79.70	95.50	0.52	9.64	nr	**105.14**
Adjustable riser						
75 mm wide	45.20	54.16	0.29	5.37	nr	**59.54**
100 mm wide	47.00	56.32	0.32	5.93	nr	**62.25**
150 mm wide	51.40	61.59	0.36	6.67	nr	**68.26**
225 mm wide	55.80	66.87	0.44	8.15	nr	**75.02**
300 mm wide	57.50	68.90	0.52	9.64	nr	**78.54**
450 mm wide	87.30	104.61	0.66	12.23	nr	**116.84**
600 mm wide	109.10	130.73	0.79	14.64	nr	**145.38**
750 mm wide	131.90	158.06	1.03	19.09	nr	**177.15**
900 mm wide	148.00	177.35	1.10	20.39	nr	**197.74**
Inside riser; 90 degree						
75 mm wide	42.00	50.33	0.31	5.75	nr	**56.07**
100 mm wide	43.00	51.53	0.33	6.12	nr	**57.64**
150 mm wide	47.10	56.44	0.37	6.86	nr	**63.30**
225 mm wide	50.00	59.91	0.44	8.15	nr	**68.07**
300 mm wide	51.70	61.95	0.53	9.82	nr	**71.78**
450 mm wide	87.30	104.61	0.67	12.42	nr	**117.03**
600 mm wide	109.10	130.73	0.79	14.64	nr	**145.38**
750 mm wide	131.80	157.94	0.95	17.61	nr	**175.54**
900 mm wide	148.00	177.35	1.11	20.57	nr	**197.92**
Outside riser; 90 degree						
75 mm wide	42.00	50.33	0.31	5.75	nr	**56.07**
100 mm wide	43.00	51.53	0.33	6.12	nr	**57.64**
150 mm wide	47.10	56.44	0.37	6.86	nr	**63.30**
225 mm wide	50.00	59.91	0.44	8.15	nr	**68.07**
300 mm wide	51.70	61.95	0.53	9.82	nr	**71.78**
450 mm wide	87.30	104.61	0.67	12.42	nr	**117.03**

Material Costs/Measured Work Prices – Electrical Installations

V:ELECTRICAL SUPPLY/POWER/LIGHTING

Item	Net Price £	Material £	Labour hours	Labour £	Unit	Total rate £
600 mm wide	109.10	130.73	0.79	14.64	nr	**145.38**
750 mm wide	131.80	157.94	0.95	17.61	nr	**175.54**
900 mm wide	148.00	177.35	1.11	20.57	nr	**197.92**
Equal Tee						
75 mm wide	77.00	92.27	0.31	5.75	nr	**98.01**
100 mm wide	83.30	99.82	0.35	6.49	nr	**106.31**
150 mm wide	91.10	109.17	0.36	6.67	nr	**115.84**
225 mm wide	102.20	122.47	0.74	13.72	nr	**136.18**
300 mm wide	110.20	132.05	0.54	10.01	nr	**142.06**
450 mm wide	163.80	196.28	0.71	13.16	nr	**209.44**
600 mm wide	227.00	272.01	0.92	17.05	nr	**289.07**
750 mm wide	299.20	358.53	1.19	22.06	nr	**380.59**
900 mm wide	342.40	410.30	1.45	26.87	nr	**437.17**
Unequal Tee						
75 mm wide	77.00	92.27	0.38	7.04	nr	**99.31**
100 mm wide	83.30	99.82	0.39	7.23	nr	**107.05**
150 mm wide	91.10	109.17	0.43	7.97	nr	**117.13**
225 mm wide	102.20	122.47	0.50	9.27	nr	**131.73**
300 mm wide	110.20	132.05	0.63	11.68	nr	**143.73**
450 mm wide	163.80	196.28	0.80	14.83	nr	**211.11**
600 mm wide	227.00	272.01	1.02	18.90	nr	**290.92**
750 mm wide	299.20	358.53	1.12	20.76	nr	**379.29**
900 mm wide	342.40	410.30	1.35	25.02	nr	**435.32**
4 way crossovers						
75 mm wide	113.70	136.25	0.40	7.41	nr	**143.66**
100 mm wide	122.90	147.27	0.40	7.41	nr	**154.68**
150 mm wide	136.80	163.93	0.44	8.15	nr	**172.08**
225 mm wide	159.10	190.65	0.53	9.82	nr	**200.47**
300 mm wide	168.70	202.15	0.64	11.86	nr	**214.01**
450 mm wide	244.40	292.86	0.84	15.57	nr	**308.43**
600 mm wide	347.80	416.77	1.03	19.09	nr	**435.86**
750 mm wide	375.00	449.36	1.13	20.94	nr	**470.31**
900 mm wide	397.10	475.84	1.36	25.21	nr	**501.05**
GRP Cable Tray including standard coupling joints and fixings; (Supports and hangers excluded).						
Tray						
100 mm wide	12.44	14.91	0.34	6.30	m	**21.21**
200 mm wide	15.74	18.86	0.39	7.23	m	**26.09**
400 mm wide	25.32	30.34	0.53	9.82	m	**40.16**
Cover						
100 mm wide	6.88	8.24	0.10	1.85	m	**10.10**
200 mm wide	9.13	10.94	0.11	2.04	m	**12.98**
400 mm wide	15.58	18.67	0.14	2.59	m	**21.26**

Material Costs/Measured Work Prices – Electrical Installations

V:ELECTRICAL SUPPLY/POWER/LIGHTING

Item	Net Price £	Material £	Labour hours	Labour £	Unit	Total rate £
V20 : LV DISTRIBUTION - CABLE SUPPORTS						
Y60/63 - CABLE SUPPORTS						
GRP Cable Tray including standard coupling joints and fixings; (Supports and hangers excluded) (Continued)						
Extra for; (Cutting and jointing to fittings included).						
Reducer						
200 mm wide	23.48	28.14	0.23	4.26	nr	**32.40**
400 mm wide	34.92	41.84	0.30	5.56	nr	**47.40**
Reducer Cover						
200 mm wide	20.33	24.36	0.25	4.63	nr	**28.99**
400 mm wide	29.72	35.61	0.28	5.19	nr	**40.80**
Bend						
100 mm wide	19.71	23.62	0.34	6.30	nr	**29.92**
200 mm wide	35.51	42.55	0.40	7.41	nr	**49.97**
400 mm wide	33.03	39.58	0.32	5.93	nr	**45.51**
Bend Cover						
100 mm wide	13.50	16.18	0.10	1.85	nr	**18.03**
200 mm wide	18.04	21.62	0.10	1.85	nr	**23.47**
400 mm wide	26.72	32.02	0.13	2.41	nr	**34.43**
Tee						
100 mm wide	23.32	27.94	0.37	6.86	nr	**34.80**
200 mm wide	26.87	32.20	0.43	7.97	nr	**40.17**
400 mm wide	36.05	43.20	0.56	10.38	nr	**53.58**
Tee Cover						
100 mm wide	17.42	20.87	0.27	5.00	nr	**25.88**
200 mm wide	20.97	25.13	0.31	5.75	nr	**30.87**
400 mm wide	29.58	35.45	0.37	6.86	nr	**42.30**
BASKET TRAY						
Mild Steel Cable Basket; Zinc Plated Including Standard Coupling Joints, Fixings and Earth Continuity Straps (supports and hangers are excluded)						
Basket						
100 mm wide	1.78	2.13	0.22	4.08	m	**6.21**
150 mm wide	1.95	2.34	0.25	4.63	m	**6.97**
200 mm wide	2.19	2.62	0.28	5.19	m	**7.81**
300 mm wide	2.55	3.06	0.34	6.30	m	**9.36**
450 mm wide	3.59	4.30	0.44	8.15	m	**12.46**
600 mm wide	5.90	7.07	0.70	12.97	m	**20.04**

Material Costs/Measured Work Prices – Electrical Installations 463

V:ELECTRICAL SUPPLY/POWER/LIGHTING

Item	Net Price £	Material £	Labour hours	Labour £	Unit	Total rate £
Extra for; (cutting and jointing to fittings is included)						
Reducer						
150 mm wide	7.25	8.69	0.25	4.63	nr	**13.32**
200 mm wide	7.48	8.96	0.28	5.19	nr	**14.15**
300 mm wide	8.67	10.39	0.38	7.04	nr	**17.43**
450 mm wide	10.82	12.97	0.48	8.90	nr	**21.86**
600 mm wide	13.94	16.70	0.48	8.90	nr	**25.60**
Bend						
100 mm wide	7.13	8.54	0.23	4.26	nr	**12.81**
150 mm wide	7.76	9.30	0.26	4.82	nr	**14.12**
200 mm wide	8.41	10.08	0.30	5.56	nr	**15.64**
300 mm wide	8.94	10.71	0.35	6.49	nr	**17.20**
450 mm wide	11.58	13.88	0.50	9.27	nr	**23.14**
600 mm wide	16.01	19.18	0.58	10.75	nr	**29.93**
Tee						
100 mm wide	9.36	11.22	0.28	5.19	nr	**16.41**
150 mm wide	9.62	11.53	0.30	5.56	nr	**17.09**
200 mm wide	10.40	12.46	0.33	6.12	nr	**18.58**
300 mm wide	11.82	14.16	0.39	7.23	nr	**21.39**
450 mm wide	13.34	15.99	0.56	10.38	nr	**26.36**
600 mm wide	19.50	23.37	0.65	12.05	nr	**35.41**
Cross Over						
100 mm wide	14.36	17.21	0.40	7.41	nr	**24.62**
150 mm wide	14.55	17.44	0.42	7.78	nr	**25.22**
200 mm wide	16.22	19.44	0.46	8.53	nr	**27.96**
300 mm wide	17.86	21.40	0.51	9.45	nr	**30.85**
450 mm wide	20.38	24.42	0.74	13.72	nr	**38.14**
600 mm wide	27.06	32.43	0.82	15.20	nr	**47.62**
Mild Steel Cable Basket; Epoxy coated Including Standard Coupling Joints, Fixings and Earth Continuity Straps (supports and hangers are excluded)						
Basket						
100 mm wide	1.80	2.16	0.22	4.08	m	**6.23**
150 mm wide	1.97	2.36	0.25	4.63	m	**6.99**
200 mm wide	2.21	2.65	0.28	5.19	m	**7.84**
300 mm wide	2.58	3.09	0.34	6.30	m	**9.39**
450 mm wide	3.63	4.35	0.44	8.15	m	**12.50**
600 mm wide	5.96	7.14	0.70	12.97	m	**20.12**
Extra for; (cutting and jointing to fittings is included)						
Reducer						
150 mm wide	7.32	8.77	0.28	5.19	nr	**13.96**
200 mm wide	7.55	9.05	0.28	5.19	nr	**14.24**
300 mm wide	8.76	10.50	0.38	7.04	nr	**17.54**
450 mm wide	10.93	13.10	0.48	8.90	nr	**21.99**
600 mm wide	14.08	16.87	0.48	8.90	nr	**25.77**

Material Costs/Measured Work Prices – Electrical Installations

V:ELECTRICAL SUPPLY/POWER/LIGHTING

Item	Net Price £	Material £	Labour hours	Labour £	Unit	Total rate £
V20 : LV DISTRIBUTION - CABLE SUPPORTS						
Y60/63 - CABLE SUPPORTS						
Mild Steel Cable Basket; Epoxy coated Including Standard Coupling Joints, Fixings and Earth Continuity Straps (supports and hangers are excluded)						
Bend						
100 mm wide	7.20	8.63	0.23	4.26	nr	**12.89**
150 mm wide	7.84	9.39	0.26	4.82	nr	**14.21**
200 mm wide	8.49	10.17	0.30	5.56	nr	**15.73**
300 mm wide	9.03	10.82	0.35	6.49	nr	**17.31**
450 mm wide	11.70	14.02	0.50	9.27	nr	**23.29**
600 mm wide	16.17	19.38	0.58	10.75	nr	**30.13**
Tee						
100 mm wide	9.45	11.32	0.28	5.19	nr	**16.51**
150 mm wide	9.72	11.65	0.30	5.56	nr	**17.21**
200 mm wide	10.50	12.58	0.33	6.12	nr	**18.70**
300 mm wide	11.94	14.31	0.39	7.23	nr	**21.54**
450 mm wide	13.47	16.14	0.56	10.38	nr	**26.52**
600 mm wide	19.70	23.61	0.65	12.05	nr	**35.65**
Cross Over						
100 mm wide	14.50	17.38	0.40	7.41	nr	**24.79**
150 mm wide	14.70	17.61	0.42	7.78	nr	**25.40**
200 mm wide	16.38	19.63	0.46	8.53	nr	**28.15**
300 mm wide	18.04	21.62	0.51	9.45	nr	**31.07**
450 mm wide	20.58	24.66	0.74	13.72	nr	**38.38**
600 mm wide	27.33	32.75	0.82	15.20	nr	**47.95**
CONDUIT						
Heavy Gauged, Screwed Welded Steel; surface fixed on saddles to backgrounds, with standard pattern boxes and fittings including all fixings and supports. (Forming holes, conduit entry, draw wires etc. and components for earth continuity are included.)						
Black Enameled						
20 mm dia.	1.15	1.38	0.49	9.08	m	**10.46**
25 mm dia.	1.65	1.98	0.56	10.38	m	**12.36**
32 mm dia.	2.98	3.57	0.64	11.86	m	**15.43**
38 mm dia.	5.87	7.03	0.73	13.53	m	**20.56**
50 mm dia.	12.33	14.78	1.04	19.28	m	**34.05**
Galvanised						
20 mm dia.	1.51	1.81	0.49	9.08	m	**10.89**
25 mm dia.	2.22	2.66	0.56	10.38	m	**13.04**
32 mm dia.	3.65	4.37	0.64	11.86	m	**16.24**
38 mm dia.	7.42	8.89	0.73	13.53	m	**22.42**
50 mm dia.	14.29	17.12	1.04	19.28	m	**36.40**

Material Costs/Measured Work Prices – Electrical Installations

V:ELECTRICAL SUPPLY/POWER/LIGHTING

Item	Net Price £	Material £	Labour hours	Labour £	Unit	Total rate £
Heavy Duty; Galvanised Steel Core; IP67 Standards; grey or black PVC covering; surface fixed to backgrounds with standard connectors and components for earth continuity.						
Temperature Range -10°C to +70°C						
16mm	3.27	3.59	0.40	7.41	m	**11.01**
20mm	3.75	4.13	0.40	7.41	m	**11.54**
25mm	4.99	5.49	0.40	7.41	m	**12.90**
32mm	7.44	8.18	0.40	7.41	m	**15.60**
40mm	9.22	10.14	0.40	7.41	m	**17.56**
Temperature Range -25°C to +105°C						
10mm	3.17	3.49	0.40	7.41	m	**10.90**
12mm	3.17	3.49	0.40	7.41	m	**10.90**
16mm	3.89	4.28	0.40	7.41	m	**11.69**
20mm	4.57	5.02	0.40	7.41	m	**12.44**
25mm	1.30	1.43	0.40	7.41	m	**8.84**
32mm	8.98	9.88	0.40	7.41	m	**17.30**
40mm	10.95	12.05	0.40	7.41	m	**19.46**
50mm	15.78	17.36	0.40	7.41	m	**24.77**
63mm	18.85	20.74	0.40	7.41	m	**28.15**
Heavy Duty; Galvanised Steel Core; IP67 Standards; Copper packed; grey or black PVC covering; surface fixed to backgrounds; jointed with standard connectors and components for earth continuity						
Temperature Range -10°C to +60°C						
16mm	5.55	6.11	0.45	8.34	m	**14.45**
20mm	6.74	7.41	0.45	8.34	m	**15.75**
25mm	8.97	9.86	0.45	8.34	m	**18.21**
32mm	12.47	13.72	0.45	8.34	m	**22.06**
40mm	15.42	16.96	0.45	8.34	m	**25.30**
50mm	23.32	25.65	0.45	8.34	m	**33.99**
63mm	29.42	32.37	0.45	8.34	m	**40.71**
Heavy Duty; Galvanised Steel Core Fittings; to IP67 standards; grey or black; including surface fixing.						
Male Connector						
10mm	2.30	2.76	0.08	1.54	nr	**4.30**
12mm	2.30	2.76	0.08	1.54	nr	**4.30**
16mm	2.75	3.30	0.08	1.54	nr	**4.84**
20mm	2.89	3.46	0.08	1.54	nr	**5.00**
25mm	4.13	4.95	0.08	1.54	nr	**6.49**
32mm	6.72	8.06	0.08	1.54	nr	**9.60**
40mm	11.15	13.36	0.08	1.54	nr	**14.90**
50mm	17.64	21.14	0.08	1.54	nr	**22.68**
63mm	32.50	38.95	0.08	1.54	nr	**40.49**

Material Costs/Measured Work Prices – Electrical Installations

V:ELECTRICAL SUPPLY/POWER/LIGHTING

Item	Net Price £	Material £	Labour hours	Labour £	Unit	Total rate £
V20 : LV DISTRIBUTION - CABLE SUPPORTS						
Y60/63 - CABLE SUPPORTS						
Heavy Duty; Galvanised Steel Core Fittings; to IP67 standards; grey or black; including surface fixing (Continued)						
Female Connector						
16mm	2.39	2.86	0.10	1.85	nr	4.72
20mm	2.63	3.15	0.10	1.85	nr	5.01
25mm	3.77	4.51	0.10	1.85	nr	6.37
32mm	14.81	17.75	0.10	1.85	nr	19.60
40mm	9.80	11.74	0.10	1.85	nr	13.59
50mm	14.59	17.48	0.10	1.85	nr	19.33
63mm	28.07	33.64	0.10	1.85	nr	35.49
Elbows						
16mm	4.91	5.88	0.11	2.04	nr	7.92
20mm	4.99	5.98	0.12	2.22	nr	8.20
25mm	7.49	8.98	0.12	2.22	nr	11.20
32mm	13.29	15.93	0.14	2.59	nr	18.53
40mm	23.58	28.25	0.14	2.59	nr	30.85
50mm	27.54	33.00	0.19	3.52	nr	36.52
63mm	41.06	49.21	0.21	3.89	nr	53.10
Medium Duty; Nylon to IP66 Standards; Grey or Black; Surface Fixed to backgrounds with standard components for earth continuity						
Temperature Range -40°C to +100°C						
10mm	2.57	3.38	0.40	7.41	m	10.80
12mm	2.84	3.74	0.40	7.41	m	11.15
16mm	3.24	4.27	0.40	7.41	m	11.68
20mm	3.38	4.45	0.40	7.41	m	11.87
25mm	3.93	5.17	0.40	7.41	m	12.59
32mm	6.61	8.71	0.40	7.41	m	16.12
40mm	8.93	11.77	0.45	8.34	m	20.11
50mm	11.97	15.78	0.45	8.34	m	24.12
Temperature Range -40°C to +120°C						
12mm	1.77	2.34	0.40	7.41	m	9.75
16mm	2.05	2.70	0.40	7.41	m	10.12
20mm	2.39	3.15	0.40	7.41	m	10.56
25mm	2.98	3.93	0.40	7.41	m	11.35
32mm	4.26	5.62	0.40	7.41	m	13.04
Medium Duty; Nylon Fittings to IP66 Standards; Grey or black including surface fixing						
Male Connector						
10mm	1.11	1.33	0.08	1.54	nr	2.86
12mm	1.10	1.31	0.08	1.54	nr	2.85
16mm	1.14	1.36	0.08	1.54	nr	2.90

Material Costs/Measured Work Prices – Electrical Installations

V:ELECTRICAL SUPPLY/POWER/LIGHTING

Item	Net Price £	Material £	Labour hours	Labour £	Unit	Total rate £
20mm	1.42	1.70	0.08	1.54	nr	**3.24**
25mm	2.12	2.54	0.08	1.54	nr	**4.08**
32mm	2.97	3.56	0.08	1.54	nr	**5.10**
40mm	6.20	7.43	0.08	1.54	nr	**8.97**
50mm	9.60	11.50	0.08	1.54	nr	**13.04**
Female Connector						
12mm	1.93	2.32	0.09	1.67	nr	**3.98**
16mm	3.66	4.39	0.09	1.67	nr	**6.06**
20mm	4.71	5.64	0.09	1.67	nr	**7.31**
25mm	8.43	10.10	0.09	1.67	nr	**11.77**
Elbows						
10mm	1.80	2.16	0.12	2.22	nr	**4.38**
12mm	1.84	2.21	0.12	2.22	nr	**4.43**
16mm	1.85	2.22	0.12	2.22	nr	**4.45**
20mm	2.15	2.57	0.12	2.22	nr	**4.80**
25mm	3.68	4.40	0.12	2.22	nr	**6.63**
32mm	5.00	5.99	0.12	2.22	nr	**8.22**
40mm	9.36	11.22	0.12	2.22	nr	**13.44**
50mm	16.78	20.11	0.12	2.22	nr	**22.33**
High Impact Unscrewed PVC; surface fixed on saddles to backgrounds; with standard pattern boxes and fittings; including all fixings and supports.						
Light Gauge						
16 mm dia.	0.75	0.90	0.27	5.00	m	**5.90**
20 mm dia.	0.94	1.13	0.28	5.19	m	**6.32**
25 mm dia.	1.60	1.92	0.33	6.12	m	**8.03**
32 mm dia.	2.12	2.54	0.38	7.04	m	**9.58**
38 mm dia.	2.70	3.24	0.44	8.15	m	**11.39**
50 mm dia.	4.43	5.31	0.48	8.90	m	**14.20**
Heavy Gauge						
16 mm dia.	1.16	1.39	0.27	5.00	m	**6.39**
20 mm dia.	1.38	1.65	0.28	5.19	m	**6.84**
25 mm dia.	1.87	2.24	0.33	6.12	m	**8.36**
32 mm dia.	3.02	3.62	0.38	7.04	m	**10.66**
38 mm dia.	3.91	4.69	0.44	8.15	m	**12.84**
50 mm dia.	6.53	7.82	0.48	8.90	m	**16.72**
Flexible Conduits; including adaptors and locknuts. (For connections to equipment.)						
Metallic, PVC covered conduit; not exceeding 1m long; including zinc plated mild steel adaptors, lock nuts and earth conductor						
16 mm dia.	4.01	4.81	0.46	8.53	nr	**13.33**
20 mm dia.	4.32	5.18	0.42	7.78	nr	**12.96**
25 mm dia.	6.44	7.72	0.43	7.97	nr	**15.69**
32 mm dia.	10.29	12.33	0.51	9.45	nr	**21.78**
38 mm dia.	12.99	15.57	0.56	10.38	nr	**25.94**
50 mm dia.	31.57	37.83	0.82	15.20	nr	**53.03**

Material Costs/Measured Work Prices – Electrical Installations

V:ELECTRICAL SUPPLY/POWER/LIGHTING

Item	Net Price £	Material £	Labour hours	Labour £	Unit	Total rate £
V20 : LV DISTRIBUTION - CABLE SUPPORTS						
Y60/63 - CABLE SUPPORTS						
Flexible Conduits; including adaptors and locknuts. (For connections to equipment) (Continued)						
PVC Conduit; not exceeding 1m long; including nylon adaptors, lock nuts						
16 mm dia.	3.11	3.73	0.46	8.53	nr	**12.25**
20 mm dia.	3.20	3.83	0.48	8.90	nr	**12.73**
25 mm dia.	4.74	5.68	0.50	9.27	nr	**14.95**
32 mm dia.	7.07	8.47	0.58	10.75	nr	**19.22**
PVC Adaptable Boxes; fixed to backgrounds; including all supports and fixings. (Cutting and connecting conduit to boxes is included.)						
Square Pattern						
75 x 75 x 53 mm	1.68	2.01	0.69	12.79	nr	**14.80**
100 x 100 x 75 mm	2.84	3.40	0.71	13.16	nr	**16.56**
150 x 150 x 75 mm	3.63	4.35	0.80	14.83	nr	**19.18**
Terminal Strips; (To be fixed in metal or polythene adaptable boxes.)						
20 Amp High Density Polythene						
2 way	0.68	0.81	0.23	4.26	nr	**5.08**
3 way	0.74	0.89	0.23	4.26	nr	**5.15**
4 way	0.80	0.96	0.23	4.26	nr	**5.22**
5 way	0.86	1.03	0.23	4.26	nr	**5.29**
6 way	0.92	1.10	0.25	4.63	nr	**5.74**
7 way	0.98	1.17	0.25	4.63	nr	**5.81**
8 way	1.07	1.28	0.29	5.37	nr	**6.66**
9 way	1.15	1.38	0.30	5.56	nr	**6.94**
10 way	1.31	1.57	0.34	6.30	nr	**7.87**
11 way	1.36	1.63	0.34	6.30	nr	**7.93**
12 way	1.49	1.79	0.34	6.30	nr	**8.09**
13 way	1.65	1.98	0.37	6.86	nr	**8.83**
14 way	1.85	2.22	0.37	6.86	nr	**9.07**
15 way	1.90	2.28	0.39	7.23	nr	**9.51**
16 way	1.95	2.34	0.45	8.34	nr	**10.68**
18 way	2.10	2.52	0.45	8.34	nr	**10.86**
TRUNKING **Galvanised Steel Trunking; fixed to backgrounds; jointed with standard connectors (including plates for air gap between trunking and background; earth continuity straps included).**						
Single Compartment						
50 x 50 mm	2.52	3.02	0.39	7.23	m	**10.25**
75 x 50 mm	3.74	4.48	0.44	8.15	m	**12.64**
75 x 75 mm	4.70	5.63	0.47	8.71	m	**14.34**

Material Costs/Measured Work Prices – Electrical Installations

V:ELECTRICAL SUPPLY/POWER/LIGHTING

Item	Net Price £	Material £	Labour hours	Labour £	Unit	Total rate £
100 x 50 mm	4.58	5.49	0.50	9.27	m	**14.76**
100 x 75 mm	5.40	6.47	0.57	10.56	m	**17.04**
100 x 100 mm	5.39	6.46	0.62	11.49	m	**17.95**
150 x 50 mm	6.11	7.32	0.78	14.46	m	**21.78**
150 x 100 mm	7.50	8.99	0.78	14.46	m	**23.44**
150 x 150 mm	9.22	11.05	0.86	15.94	m	**26.99**
225 x 75 mm	8.82	10.57	0.88	16.31	m	**26.88**
225 x 150 mm	12.65	15.16	0.84	15.57	m	**30.73**
225 x 225 mm	15.72	18.84	0.99	18.35	m	**37.19**
300 x 75 mm	12.41	14.87	0.96	17.79	m	**32.66**
300 x 100 mm	12.96	15.53	0.99	18.35	m	**33.88**
300 x 150 mm	15.41	18.47	0.99	18.35	m	**36.81**
300 x 225 mm	16.54	19.82	1.09	20.20	m	**40.02**
300 x 300 mm	19.50	23.37	1.16	21.50	m	**44.87**
Double Compartment						
50 x 50 mm	3.90	4.67	0.41	7.60	m	**12.27**
75 x 50 mm	5.00	5.99	0.47	8.71	m	**14.70**
75 x 75 mm	5.38	6.45	0.50	9.27	m	**15.71**
100 x 50 mm	5.50	6.59	0.54	10.01	m	**16.60**
100 x 75 mm	6.26	7.50	0.62	11.49	m	**18.99**
100 x 100 mm	6.59	7.90	0.66	12.23	m	**20.13**
150 x 50 mm	6.97	8.35	0.70	12.97	m	**21.33**
150 x 100 mm	8.70	10.43	0.83	15.38	m	**25.81**
150 x 150 mm	10.98	13.16	0.92	17.05	m	**30.21**
Triple Compartment						
75 x 50 mm	5.93	7.11	0.54	10.01	m	**17.11**
75 x 75 mm	6.43	7.71	0.58	10.75	m	**18.45**
100 x 50 mm	6.35	7.61	0.61	11.31	m	**18.91**
100 x 75 mm	7.56	9.06	0.70	12.97	m	**22.03**
100 x 100 mm	7.94	9.51	0.74	13.72	m	**23.23**
150 x 50 mm	7.79	9.33	0.79	14.64	m	**23.98**
150 x 100 mm	10.02	12.01	0.78	14.46	m	**26.46**
150 x 150 mm	12.94	15.51	1.01	18.72	m	**34.23**
Four Compartment						
100 x 50 mm	5.50	6.59	0.64	11.86	m	**18.45**
100 x 75 mm	6.40	7.67	0.72	13.34	m	**21.01**
100 x 100 mm	6.99	8.38	0.77	14.27	m	**22.65**
150 x 50 mm	6.92	8.29	0.82	15.20	m	**23.49**
150 x 100 mm	8.59	10.29	0.95	17.61	m	**27.90**
150 x 150 mm	11.42	13.68	1.04	19.28	m	**32.96**
Galvanised Steel Trunking Fittings; (Cutting and jointing trunking to fittings is included.)						
Additional Connector or Stop End						
50 x 50 mm	0.43	0.52	0.19	3.52	nr	**4.04**
75 x 50 mm	0.46	0.55	0.20	3.71	nr	**4.26**
75 x 75 mm	0.49	0.59	0.21	3.89	nr	**4.48**
100 x 50 mm	0.52	0.62	0.31	5.75	nr	**6.37**
100 x 75 mm	0.67	0.80	0.27	5.00	nr	**5.81**
100 x 100 mm	0.91	1.09	0.27	5.00	nr	**6.09**
150 x 50 mm	0.97	1.16	0.28	5.19	nr	**6.35**

Material Costs/Measured Work Prices – Electrical Installations

V:ELECTRICAL SUPPLY/POWER/LIGHTING

Item	Net Price £	Material £	Labour hours	Labour £	Unit	Total rate £
V20 : LV DISTRIBUTION - CABLE SUPPORTS						
Y60/63 - CABLE SUPPORTS						
Galvanised Steel Trunking Fittings;						
(Cutting and jointing trunking to fittings						
is included) (Continued)						
Additional Connector or Stop End						
150 x 100 mm	1.07	1.28	0.30	5.56	nr	6.84
150 x 150 mm	1.19	1.43	0.32	5.93	nr	7.36
225 x 75 mm	1.50	1.80	0.35	6.49	nr	8.28
225 x 150 mm	1.87	2.24	0.37	6.86	nr	9.10
225 x 225 mm	2.43	2.91	0.38	7.04	nr	9.95
300 x 75 mm	1.89	2.26	0.42	7.78	nr	10.05
300 x 100 mm	2.08	2.49	0.42	7.78	nr	10.28
300 x 150 mm	2.32	2.78	0.43	7.97	nr	10.75
300 x 225 mm	2.58	3.09	0.45	8.34	nr	11.43
300 x 300 mm	2.81	3.37	0.48	8.90	nr	12.26
Flanged Connector or Stop End						
50 x 50 mm	0.49	0.59	0.19	3.52	nr	4.11
75 x 50 mm	0.49	0.59	0.20	3.71	nr	4.29
75 x 75 mm	0.54	0.65	0.21	3.89	nr	4.54
100 x 50 mm	0.51	0.61	0.26	4.82	nr	5.43
100 x 75 mm	0.61	0.73	0.27	5.00	nr	5.74
100 x 100 mm	0.89	1.07	0.27	5.00	nr	6.07
150 x 50 mm	0.72	0.86	0.28	5.19	nr	6.05
150 x 100 mm	1.01	1.21	0.30	5.56	nr	6.77
150 x 150 mm	1.26	1.51	0.32	5.93	nr	7.44
225 x 75 mm	1.23	1.47	0.35	6.49	nr	7.96
225 x 150 mm	1.72	2.06	0.37	6.86	nr	8.92
225 x 225 mm	2.08	2.49	0.38	7.04	nr	9.54
300 x 75 mm	1.46	1.75	0.42	7.78	nr	9.53
300 x 100 mm	1.75	2.10	0.42	7.78	nr	9.88
300 x 150 mm	2.05	2.46	0.43	7.97	nr	10.43
300 x 225 mm	2.30	2.76	0.45	8.34	nr	11.10
300 x 300 mm	2.64	3.16	0.48	8.90	nr	12.06
Bends 90 Degree; Single Compartment						
50 x 50 mm	2.68	3.21	0.42	7.78	nr	11.00
75 x 50 mm	3.18	3.81	0.45	8.34	nr	12.15
75 x 75 mm	3.30	3.95	0.48	8.90	nr	12.85
100 x 50 mm	3.52	4.22	0.53	9.82	nr	14.04
100 x 75 mm	3.54	4.24	0.56	10.38	nr	14.62
100 x 100 mm	3.70	4.43	0.58	10.75	nr	15.18
150 x 50 mm	4.21	5.04	0.64	11.86	nr	16.91
150 x 100 mm	5.45	6.53	0.91	16.87	nr	23.40
150 x 150 mm	5.93	7.11	0.89	16.50	nr	23.60
225 x 75 mm	8.13	9.74	0.76	14.09	nr	23.83
225 x 150 mm	10.02	12.01	0.82	15.20	nr	27.20
225 x 225 mm	11.40	13.66	0.83	15.38	nr	29.04
300 x 75 mm	9.82	11.77	0.85	15.75	nr	27.52
300 x 100 mm	10.07	12.07	0.90	16.68	nr	28.75
300 x 150 mm	11.43	13.70	0.96	17.79	nr	31.49
300 x 225 mm	13.16	15.77	0.98	18.16	nr	33.93

Material Costs/Measured Work Prices – Electrical Installations

V:ELECTRICAL SUPPLY/POWER/LIGHTING

Item	Net Price £	Material £	Labour hours	Labour £	Unit	Total rate £
300 x 300 mm	15.15	18.15	1.06	19.65	nr	**37.80**
Bends 90 Degree; Double Compartment						
50 x 50 mm	4.06	4.87	0.42	7.78	nr	**12.65**
75 x 50 mm	4.59	5.50	0.45	8.34	nr	**13.84**
75 x 75 mm	4.83	5.79	0.49	9.08	nr	**14.87**
100 x 50 mm	4.39	5.26	0.53	9.82	nr	**15.08**
100 x 75 mm	5.12	6.14	0.56	10.38	nr	**16.51**
100 x 100 mm	5.98	7.17	0.58	10.75	nr	**17.92**
150 x 50 mm	5.74	6.88	0.65	12.05	nr	**18.93**
150 x 100 mm	7.20	8.63	0.69	12.79	nr	**21.42**
150 x 150 mm	8.01	9.60	0.73	13.53	nr	**23.13**
Bends 90 Degree; Triple Compartment						
75 x 50 mm	5.89	7.06	0.47	8.71	nr	**15.77**
75 x 75 mm	6.34	7.60	0.51	9.45	nr	**17.05**
100 x 50 mm	6.27	7.51	0.56	10.38	nr	**17.89**
100 x 75 mm	6.64	7.96	0.59	10.94	nr	**18.89**
100 x 100 mm	7.05	8.45	0.61	11.31	nr	**19.75**
150 x 50 mm	5.85	7.01	0.68	12.60	nr	**19.61**
150 x 100 mm	8.89	10.65	0.73	13.53	nr	**24.18**
150 x 150 mm	10.35	12.40	0.77	14.27	nr	**26.67**
Bends 90 Degree; Four Compartments						
100 x 50 mm	6.25	7.49	0.56	10.38	nr	**17.87**
100 x 75 mm	6.86	8.22	0.59	10.94	nr	**19.16**
100 x 100 mm	7.65	9.17	0.61	11.31	nr	**20.47**
150 x 50 mm	6.04	7.24	0.69	12.79	nr	**20.03**
150 x 100 mm	9.42	11.29	0.73	13.53	nr	**24.82**
150 x 150 mm	11.32	13.56	0.77	14.27	nr	**27.84**
Tees; Single Compartment						
50 x 50 mm	3.20	3.83	0.56	10.38	nr	**14.21**
75 x 50 mm	3.81	4.57	0.57	10.56	nr	**15.13**
75 x 75 mm	3.76	4.51	0.60	11.12	nr	**15.63**
100 x 50 mm	3.91	4.69	0.65	12.05	nr	**16.73**
100 x 75 mm	4.40	5.27	0.71	13.16	nr	**18.43**
100 x 100 mm	4.49	5.38	0.72	13.34	nr	**18.72**
150 x 50 mm	5.17	6.20	0.82	15.20	nr	**21.39**
150 x 100 mm	6.77	8.11	0.84	15.57	nr	**23.68**
150 x 150 mm	7.23	8.66	0.91	16.87	nr	**25.53**
225 x 75 mm	10.55	12.64	0.94	17.42	nr	**30.06**
225 x 150 mm	13.90	16.66	1.01	18.72	nr	**35.38**
225 x 225 mm	14.75	17.67	1.02	18.90	nr	**36.58**
300 x 75 mm	13.44	16.11	1.07	19.83	nr	**35.94**
300 x 100 mm	14.11	16.91	1.07	19.83	nr	**36.74**
300 x 150 mm	16.19	19.40	1.14	21.13	nr	**40.53**
300 x 225 mm	18.57	22.25	1.19	22.06	nr	**44.31**
300 x 300mm	21.47	25.73	1.26	23.35	nr	**49.08**
Tees; Double Compartment						
50 x 50 mm	6.50	7.79	0.56	10.38	nr	**18.17**
75 x 50 mm	7.12	8.53	0.57	10.56	nr	**19.10**
75 x 75 mm	7.63	9.14	0.60	11.12	nr	**20.26**
100 x 50 mm	6.79	8.14	0.65	12.05	nr	**20.18**
100 x 75 mm	7.90	9.47	0.71	13.16	nr	**22.63**

Material Costs/Measured Work Prices – Electrical Installations

V:ELECTRICAL SUPPLY/POWER/LIGHTING

Item	Net Price £	Material £	Labour hours	Labour £	Unit	Total rate £
V20 : LV DISTRIBUTION - CABLE SUPPORTS						
Y60/63 - CABLE SUPPORTS						
Galvanised Steel Trunking Fittings;						
(Cutting and jointing trunking to fittings						
is included) (Continued)						
Tees; Double Compartment						
100 x 100 mm	8.76	10.50	0.72	13.34	nr	**23.84**
150 x 50 mm	9.19	11.01	0.82	15.20	nr	**26.21**
150 x 100 mm	11.38	13.64	0.85	15.75	nr	**29.39**
150 x 150 mm	12.20	14.62	0.91	16.87	nr	**31.49**
Tees; Triple Compartment						
75 x 50 mm	9.47	11.35	0.60	11.12	nr	**22.47**
75 x 75 mm	10.28	12.32	0.63	11.68	nr	**23.99**
100 x 50 mm	9.99	11.97	0.68	12.60	nr	**24.57**
100 x 75 mm	11.08	13.28	0.74	13.72	nr	**26.99**
100 x 100 mm	11.86	14.21	0.75	13.90	nr	**28.11**
150 x 50 mm	9.09	10.89	0.87	16.12	nr	**27.02**
150 x 100 mm	14.50	17.38	0.89	16.50	nr	**33.87**
150 x 150 mm	16.04	19.22	0.96	17.79	nr	**37.01**
Tees; Four Compartment						
100 x 50 mm	13.38	16.03	0.66	12.23	nr	**28.27**
100 x 75 mm	14.93	17.89	0.72	13.34	nr	**31.23**
100 x 100 mm	16.53	19.81	0.72	13.34	nr	**33.15**
150 x 50 mm	10.41	12.47	0.83	15.38	nr	**27.86**
150 x 100 mm	19.64	23.53	0.85	15.75	nr	**39.29**
150 x 150 mm	21.59	25.87	0.92	17.05	nr	**42.92**
Crossovers; Single Compartment						
50 x 50 mm	4.03	4.83	0.65	12.05	nr	**16.88**
75 x 50 mm	5.12	6.14	0.66	12.23	nr	**18.37**
75 x 75 mm	5.92	7.09	0.69	12.79	nr	**19.88**
100 x 50 mm	6.65	7.97	0.74	13.72	nr	**21.68**
100 x 75 mm	6.85	8.21	0.80	14.83	nr	**23.04**
100 x 100 mm	6.92	8.29	0.81	15.01	nr	**23.30**
150 x 50 mm	7.42	8.89	0.91	16.87	nr	**25.76**
150 x 100 mm	8.95	10.72	0.94	17.42	nr	**28.15**
150 x 150 mm	9.54	11.43	0.99	18.35	nr	**29.78**
225 x 75 mm	13.72	16.44	1.01	18.72	nr	**35.16**
225 x 150 mm	16.86	20.20	1.08	20.02	nr	**40.22**
225 x 225 mm	19.05	22.83	1.09	20.20	nr	**43.03**
300 x 75 mm	16.63	19.93	1.14	21.13	nr	**41.06**
300 x 100 mm	17.10	20.49	1.16	21.50	nr	**41.99**
300 x 150 mm	19.27	23.09	1.19	22.06	nr	**45.15**
300 x 225 mm	21.30	25.52	1.21	22.43	nr	**47.95**
300 x 300mm	26.69	31.98	1.29	23.91	nr	**55.89**
Crossovers; Double Compartment						
50 x 50 mm	8.24	9.87	0.66	12.23	nr	**22.11**
75 x 50 mm	9.33	11.18	0.66	12.23	nr	**23.41**
75 x 75 mm	10.58	12.68	0.70	12.97	nr	**25.65**
100 x 50 mm	10.64	12.75	0.74	13.72	nr	**26.46**

Material Costs/Measured Work Prices – Electrical Installations

V:ELECTRICAL SUPPLY/POWER/LIGHTING

Item	Net Price £	Material £	Labour hours	Labour £	Unit	Total rate £
100 x 75 mm	11.48	13.76	0.80	14.83	nr	**28.58**
100 x 100 mm	11.82	14.16	0.81	15.01	nr	**29.18**
150 x 50 mm	12.46	14.93	0.86	15.94	nr	**30.87**
150 x 100 mm	13.97	16.74	0.94	17.42	nr	**34.16**
150 x 150 mm	14.77	17.70	1.00	18.53	nr	**36.23**
Crossovers; Triple Compartment						
75 x 50 mm	12.34	14.79	0.70	12.97	nr	**27.76**
75 x 75 mm	13.79	16.52	0.73	13.53	nr	**30.05**
100 x 50 mm	13.78	16.51	0.79	14.64	nr	**31.15**
100 x 75 mm	14.93	17.89	0.85	15.75	nr	**33.64**
100 x 100 mm	15.55	18.63	0.85	15.75	nr	**34.39**
150 x 50 mm	13.46	16.13	0.97	17.98	nr	**34.11**
150 x 100 mm	18.59	22.28	0.99	18.35	nr	**40.63**
150 x 150 mm	19.43	23.28	1.06	19.65	nr	**42.93**
Crossovers; Four Compartments						
100 x 50 mm	18.45	22.11	0.79	14.64	nr	**36.75**
100 x 75 mm	19.79	23.71	0.85	15.75	nr	**39.47**
100 x 100 mm	20.87	25.01	0.86	15.94	nr	**40.95**
150 x 50 mm	16.94	20.30	0.97	17.98	nr	**38.28**
150 x 100 mm	24.85	29.78	1.00	18.53	nr	**48.31**
150 x 150 mm	26.48	31.73	1.06	19.65	nr	**51.38**
Galvanised Steel Flush Floor Trunking; fixed to backgrounds; supports and fixings; standard coupling joints; (including plates for air gap between trunking and background; earth continuity straps included.)						
Triple Compartment						
350 x 60mm	18.72	22.43	1.32	24.46	m	**46.90**
Four Compartment						
350 x 60mm	21.47	25.73	1.32	24.46	m	**50.19**
Galvanised Steel Flush Floor Trunking; Fittings (Cutting and jointing trunking to fittings is included.)						
Stop End; Triple Compartment						
350 x 60mm	2.28	2.73	0.53	9.82	nr	**12.56**
Stop End; Four Compartment						
350 x 60mm	2.50	3.00	0.53	9.82	nr	**12.82**
Rising Bend; Standard; Triple Compartment						
350 x 60mm	13.60	16.30	1.30	24.09	nr	**40.39**
Rising Bend; Standard; Four Compartment						
350 x 60mm	15.52	18.60	1.30	24.09	nr	**42.69**
Rising Bend; Skirting; Triple Compartment						
350 x 60mm	28.26	33.86	1.33	24.65	nr	**58.51**

474   *Material Costs/Measured Work Prices – Electrical Installations*

V:ELECTRICAL SUPPLY/POWER/LIGHTING

Item	Net Price £	Material £	Labour hours	Labour £	Unit	Total rate £
V20 : LV DISTRIBUTION - CABLE SUPPORTS						
Y60/63 - CABLE SUPPORTS						
Galvanised Steel Flush Floor Trunking; Fittings (Cutting and jointing trunking to fittings is included) (Continued)						
Rising Bend; Skirting; Four Compartment						
350 x 60mm	36.72	44.00	1.33	24.65	nr	**68.65**
Junction Box; Triple Compartment						
350 x 60mm	21.42	25.67	1.16	21.50	nr	**47.17**
Junction Box; Four Compartment						
350 x 60mm	25.92	31.06	1.16	21.50	nr	**52.56**
Body Coupler (pair)						
3 and 4 Compartment	13.04	15.63	0.16	2.97	nr	**18.59**
Service Outlet Module comprising flat lid with flanged carpet trim; twin 13 A outlet and drilled plate for mounting 2 telephone outlets; One blank plate; Triple compartment						
3 Compartment	28.71	34.40	0.47	8.71	nr	**43.11**
Service Outlet Module comprising flat lid with flanged carpet trim; twin 13 A outlet and drilled plate for mounting 2 telephone outlets; Two blank plates; Four compartments						
4 Compartment	34.60	41.46	0.47	8.71	nr	**50.17**
Single Compartment PVC Trunking; grey finish; clip on lid; fixed to backgrounds; including supports and fixings (standard coupling joints)						
Single Compartment						
50 x 50mm	7.95	9.53	0.27	5.00	m	**14.53**
75 x 50mm	9.11	10.92	0.28	5.19	m	**16.11**
75 x 75mm	11.05	13.24	0.29	5.37	m	**18.62**
100 x 50mm	12.90	15.46	0.34	6.30	m	**21.76**
100 x 75mm	13.89	16.64	0.37	6.86	m	**23.50**
100 x 100mm	15.92	19.08	0.37	6.86	m	**25.93**
150 x 75mm	17.19	20.60	0.44	8.15	m	**28.75**
150 x 100mm	20.61	24.70	0.44	8.15	m	**32.85**
150 x 50mm	26.91	32.25	0.48	8.90	m	**41.14**

Material Costs/Measured Work Prices – Electrical Installations

V:ELECTRICAL SUPPLY/POWER/LIGHTING

Item	Net Price £	Material £	Labour hours	Labour £	Unit	Total rate £
Single Compartment PVC Trunking Fittings; (cutting and jointing trunking to fittings is included)						
Crossover						
50 x 50mm	15.25	18.27	0.29	5.37	nr	**23.65**
75 x 50mm	16.79	20.12	0.30	5.56	nr	**25.68**
75 x 75mm	18.23	21.84	0.31	5.75	nr	**27.59**
100 x 50mm	28.42	34.06	0.35	6.49	nr	**40.54**
100 x 75mm	28.82	34.53	0.36	6.67	nr	**41.21**
100 x 100mm	30.00	35.95	0.40	7.41	nr	**43.36**
150 x 75mm	42.47	50.89	0.45	8.34	nr	**59.23**
150 x 100mm	57.32	68.69	0.46	8.53	nr	**77.21**
150 x 150mm	59.34	71.11	0.47	8.71	nr	**79.82**
Stop End						
50 x 50mm	0.72	0.86	0.12	2.22	nr	**3.09**
75 x 50mm	1.03	1.23	0.12	2.22	nr	**3.46**
75 x 75mm	1.37	1.64	0.13	2.41	nr	**4.05**
100 x 50mm	1.76	2.11	0.16	2.97	nr	**5.07**
100 x 75mm	2.68	3.21	0.16	2.97	nr	**6.18**
100 x 100mm	2.81	3.37	0.18	3.34	nr	**6.70**
150 x 75mm	5.64	6.76	0.20	3.71	nr	**10.47**
150 x 100mm	8.14	9.75	0.21	3.89	nr	**13.65**
150 x 150mm	10.80	12.94	0.22	4.08	nr	**17.02**
Flanged Coupling						
50 x 50mm	3.18	3.81	0.32	5.93	nr	**9.74**
75 x 50mm	3.72	4.46	0.33	6.12	nr	**10.57**
75 x 75mm	4.76	5.70	0.34	6.30	nr	**12.01**
100 x 50mm	5.88	7.05	0.44	8.15	nr	**15.20**
100 x 75mm	7.67	9.19	0.45	8.34	nr	**17.53**
100 x 100mm	6.91	8.28	0.46	8.53	nr	**16.81**
150 x 75mm	9.77	11.71	0.57	10.56	nr	**22.27**
150 x 100mm	11.72	14.04	0.57	10.56	nr	**24.61**
150 x 150mm	14.19	17.00	0.59	10.94	nr	**27.94**
Internal Coupling						
50 x 50mm	1.28	1.53	0.07	1.30	nr	**2.83**
75 x 50mm	1.59	1.91	0.07	1.30	nr	**3.20**
75 x 75mm	1.53	1.83	0.07	1.30	nr	**3.13**
100 x 50mm	2.08	2.49	0.08	1.48	nr	**3.98**
100 x 75mm	2.60	3.12	0.08	1.48	nr	**4.60**
100 x 100mm	2.75	3.30	0.08	1.48	nr	**4.78**
External Coupling						
50 x 50mm	1.91	2.29	0.09	1.67	nr	**3.96**
75 x 50mm	2.63	3.15	0.09	1.67	nr	**4.82**
75 x 75mm	2.94	3.52	0.09	1.67	nr	**5.19**
100 x 50mm	5.21	6.24	0.10	1.85	nr	**8.10**
100 x 75mm	5.29	6.34	0.10	1.85	nr	**8.19**
100 x 100mm	6.48	7.76	0.10	1.85	nr	**9.62**
150 x 75mm	7.41	8.88	0.11	2.04	nr	**10.92**
150 x 100mm	8.86	10.62	0.11	2.04	nr	**12.66**
150 x 150mm	10.90	13.06	0.11	2.04	nr	**15.10**

Material Costs/Measured Work Prices – Electrical Installations

V:ELECTRICAL SUPPLY/POWER/LIGHTING

Item	Net Price £	Material £	Labour hours	Labour £	Unit	Total rate £
V20 : LV DISTRIBUTION - CABLE SUPPORTS						
Y60/63 - CABLE SUPPORTS						
Single Compartment PVC Trunking Fittings; (cutting and jointing trunking to fittings is included) Continued)						
Angle, Flat Cover						
50 x 50mm	4.08	4.89	0.18	3.34	nr	8.23
75 x 50mm	5.47	6.55	0.19	3.52	nr	10.08
75 x 75mm	6.53	7.82	0.20	3.71	nr	11.53
100 x 50mm	11.10	13.30	0.23	4.26	nr	17.56
100 x 75mm	14.86	17.81	0.26	4.82	nr	22.63
100 x 100mm	15.74	18.86	0.26	4.82	nr	23.68
150 x 75mm	27.77	33.28	0.30	5.56	nr	38.84
150 x 100mm	31.66	37.94	0.33	6.12	nr	44.05
150 x 150mm	41.91	50.22	0.34	6.30	nr	56.52
Angle, Internal or External Cover						
50 x 50mm	4.47	5.36	0.18	3.34	nr	8.69
75 x 50mm	8.12	9.73	0.19	3.52	nr	13.25
75 x 75mm	10.65	12.76	0.20	3.71	nr	16.47
100 x 50mm	11.80	14.14	0.23	4.26	nr	18.40
100 x 75mm	19.31	23.14	0.26	4.82	nr	27.96
100 x 100mm	20.39	24.43	0.26	4.82	nr	29.25
150 x 75mm	27.02	32.38	0.30	5.56	nr	37.94
150 x 100mm	30.58	36.64	0.33	6.12	nr	42.76
150 x 150mm	41.49	49.72	0.34	6.30	nr	56.02
Tee, Flat Cover						
50 x 50mm	7.85	9.41	0.24	4.45	nr	13.85
75 x 50mm	9.24	11.07	0.25	4.63	nr	15.71
75 x 75mm	10.34	12.39	0.26	4.82	nr	17.21
100 x 50mm	14.99	17.96	0.32	5.93	nr	23.89
100 x 75mm	19.23	23.04	0.33	6.12	nr	29.16
100 x 100mm	20.44	24.49	0.34	6.30	nr	30.79
150 x 75mm	29.86	35.78	0.41	7.60	nr	43.38
150 x 100mm	37.42	44.84	0.42	7.78	nr	52.62
150 x 150mm	47.12	56.46	0.44	8.15	nr	64.62
Tee, Internal or External Cover						
50 x 50mm	11.59	13.89	0.24	4.45	nr	18.34
75 x 50mm	12.77	15.30	0.25	4.63	nr	19.94
75 x 75mm	14.74	17.66	0.26	4.82	nr	22.48
100 x 50mm	18.07	21.65	0.32	5.93	nr	27.58
100 x 75mm	23.79	28.51	0.33	6.12	nr	34.62
100 x 100mm	23.79	28.51	0.34	6.30	nr	34.81
150 x 75mm	39.15	46.91	0.41	7.60	nr	54.51
150 x 100mm	49.54	59.36	0.42	7.78	nr	67.15
150 x 150mm	49.54	59.36	0.44	8.15	nr	67.52
Division Strip (1.8m long)						
50mm	0.72	0.86	0.07	1.30	nr	2.16
75mm	0.94	1.13	0.07	1.30	nr	2.42
100mm	1.21	1.45	0.08	1.48	nr	2.93

Material Costs/Measured Work Prices – Electrical Installations

V:ELECTRICAL SUPPLY/POWER/LIGHTING

Item	Net Price £	Material £	Labour hours	Labour £	Unit	Total rate £
PVC Miniature Trunking; white finish; fixed to backgrounds; including supports and fixing; standard coupling joints						
Single Compartment						
16 x 16mm	0.91	1.09	0.20	3.71	m	**4.80**
25 x 16mm	1.15	1.38	0.21	3.89	m	**5.27**
38 x 16mm	1.39	1.67	0.24	4.45	m	**6.11**
38 x 25mm	1.58	1.89	0.25	4.63	m	**6.53**
Compartmented						
38 x 16mm	1.37	1.64	0.24	4.45	m	**6.09**
38 x 25mm	1.64	1.97	0.25	4.63	m	**6.60**
PVC Miniature Trunking Fittings; single compartment; white finish; (cutting and jointing trunking to fittings is included)						
Coupling						
16 x 16mm	0.29	0.35	0.10	1.85	nr	**2.20**
25 x 16mm	0.29	0.35	0.09	1.67	nr	**2.02**
38 x 16mm	0.32	0.38	0.12	2.22	nr	**2.61**
38 x 25mm	0.65	0.78	0.14	2.59	nr	**3.37**
Stop End						
16 x 16mm	0.29	0.35	0.12	2.22	nr	**2.57**
25 x 16mm	0.29	0.35	0.12	2.22	nr	**2.57**
38 x 16mm	0.30	0.36	0.15	2.78	nr	**3.14**
38 x 25mm	0.36	0.43	0.17	3.15	nr	**3.58**
Bend; Flat, Internal or External						
16 x 16mm	0.29	0.35	0.18	3.34	nr	**3.68**
25 x 16mm	0.29	0.35	0.18	3.34	nr	**3.68**
38 x 16mm	0.32	0.38	0.21	3.89	nr	**4.28**
38 x 25mm	0.73	0.87	0.23	4.26	nr	**5.14**
Tee						
16 x 16mm	0.51	0.61	0.23	4.26	nr	**4.87**
25 x 16mm	0.51	0.61	0.19	3.52	nr	**4.13**
38 x 16mm	0.51	0.61	0.26	4.82	nr	**5.43**
38 x 25mm	0.74	0.89	0.29	5.37	nr	**6.26**
PVC Bench Trunking; White or Grey Finish; fixed to backgrounds; including supports and fixings; standard coupling joints						
Trunking						
90 x 90mm	26.12	31.30	0.33	6.12	m	**37.42**
PVC Bench Trunking Fittings; White or Grey Finish. (Cutting and jointing trunking to fittings is included.)						
Stop End						
90 x 90mm	9.22	11.05	0.09	1.67	nr	**12.72**

Material Costs/Measured Work Prices – Electrical Installations

V:ELECTRICAL SUPPLY/POWER/LIGHTING

Item	Net Price £	Material £	Labour hours	Labour £	Unit	Total rate £
V20 : LV DISTRIBUTION - CABLE SUPPORTS						
Y60/63 - CABLE SUPPORTS						
PVC Bench Trunking Fittings; White or Grey Finish. (Cutting and jointing trunking to fittings is included) (Continued)						
Coupling						
90 x 90mm	5.48	6.57	0.09	1.67	nr	**8.23**
Internal or External Bend						
90 x 90mm	30.84	36.96	0.28	5.19	nr	**42.15**
Socket Plate						
90 x 90mm - 1 gang	8.28	9.92	0.10	1.85	nr	**11.78**
90 x 90mm - 2 gang	10.38	12.44	0.10	1.85	nr	**14.29**
PVC Underfloor Trunking; Single Compartment; fitted in floor screed; standard coupling joints						
Trunking						
60 x 25mm	1.84	2.20	0.22	4.08	m	**6.28**
90 x 35mm	2.85	3.42	0.27	5.00	m	**8.42**
PVC Underfloor Trunking Fittings; Single Compartment; fitted in floor screed; (Cutting and jointing trunking to fittings is included.)						
Jointing Sleeve						
60 x 25mm	0.29	0.35	0.08	1.48	nr	**1.83**
90 x 35mm	0.41	0.49	0.10	1.85	nr	**2.34**
Duct Connector						
90 x 35mm	4.51	5.40	0.17	3.15	nr	**8.56**
Socket Reducer						
90 x 35mm	0.63	0.75	0.12	2.22	nr	**2.98**
Vertical Access Box; 2 compartment						
Shallow	26.94	32.28	0.37	6.86	nr	**39.14**
Duct Bend; Vertical						
60 x 25mm	7.07	8.47	0.27	5.00	nr	**13.48**
90 x 35mm	7.99	9.57	0.35	6.49	nr	**16.06**
Duct Bend; Horizontal						
60 x 25mm	7.41	8.88	0.30	5.56	nr	**14.44**
90 x 35mm	8.13	9.74	0.37	6.86	nr	**16.60**

Material Costs/Measured Work Prices – Electrical Installations

V:ELECTRICAL SUPPLY/POWER/LIGHTING

Item	Net Price £	Material £	Labour hours	Labour £	Unit	Total rate £
Zinc Coated Steel Underfloor Ducting; fixed to backgrounds; standard coupling joints; earth continuity straps; (Including supports and fixing, packing shims where required)						
Double Compartment						
150 x 25mm	8.55	10.25	0.57	10.56	m	**20.81**
Triple Compartment						
225 x 25mm	11.92	14.28	0.93	17.24	m	**31.52**
Zinc Coated Steel Underfloor Ducting Fittings; (cutting and jointing to fittings is included.)						
Stop End; Double Compartment						
150 x 25mm	1.14	1.37	0.31	5.75	nr	**7.11**
Stop End; Triple Compartment						
225 x 25mm	1.28	1.53	0.37	6.86	nr	**8.39**
Rising Bend; Double Compartment; Standard Trunking						
150 x 25mm	11.43	13.70	0.71	13.16	nr	**26.86**
Rising Bend; Triple Compartment; Standard Trunking						
225 x 25mm	11.67	13.98	0.85	15.75	nr	**29.74**
Rising Bend; Double Compartment; To Skirting						
150 x 25	10.34	12.39	0.90	16.68	nr	**29.07**
Rising Bend; Triple Compartment; To Skirting						
225 x 25	13.61	16.31	0.95	17.61	nr	**33.92**
Horizontal Bend; Double Compartment						
150 x 25mm	12.01	14.39	0.64	11.86	nr	**26.25**
Horizontal Bend; Triple Compartment						
225 x 25mm	14.28	17.11	0.77	14.27	nr	**31.38**
Junction or Service Outlet Boxes; Terminal; Double Compartment						
150mm	24.79	29.71	0.91	16.87	nr	**46.57**
Junction or Service Outlet Boxes; Terminal; Triple Compartment						
225mm	31.26	37.46	1.11	20.57	nr	**58.03**
Junction or Service Outlet Boxes; Through or Angle; Double Compartment						
150mm	19.10	22.89	0.97	17.98	nr	**40.87**

Material Costs/Measured Work Prices – Electrical Installations

V:ELECTRICAL SUPPLY/POWER/LIGHTING

Item	Net Price £	Material £	Labour hours	Labour £	Unit	Total rate £
V20 : LV DISTRIBUTION - CABLE SUPPORTS						
Y60/63 - CABLE SUPPORTS						
Zinc Coated Steel Underfloor Ducting Fittings; (cutting and jointing to fittings is included) (Continued)						
Junction or Service Outlet Boxes; Through or Angle; Triple Compartment						
225mm	24.58	29.45	1.17	21.68	nr	**51.14**
Junction or Service Outlet Boxes; Tee; Double Compartment						
150mm	24.79	29.71	1.02	18.90	nr	**48.61**
Junction or Service Outlet Boxes; Tee; Triple Compartment						
225mm	26.10	31.28	1.22	22.61	nr	**53.89**
Junction or Service Outlet Boxes; Cross;Double Compartment						
up to 150mm	21.66	25.96	1.03	19.09	nr	**45.05**
Junction or Service Outlet Boxes; Cross;Triple Compartment						
225mm	26.49	31.74	1.23	22.80	nr	**54.54**
Plates for Junction / Inspection Boxes; Double and Triple Compartment						
Blank Plate	6.06	7.26	0.92	17.05	nr	**24.31**
Conduit Entry Plate	7.58	9.08	0.86	15.94	nr	**25.02**
Trunking Entry Plate	7.58	9.08	0.86	15.94	nr	**25.02**
Service outlet box comprising flat lid with flanged carpet trim; twin 13 A outlet and drilled plate for mounting 2 telephone outlets and terminal blocks; terminal outlet box; double compartment						
150 x 25mm trunking	37.15	44.52	1.68	31.14	nr	**75.65**
Service outlet box comprising flat lid with flanged carpet trim; twin 13 A outlet and drilled plate for mounting 2 telephone outlets and terminal blocks; terminal outlet box; triple compartment						
225 x 25mm trunking	53.17	63.71	1.93	35.77	nr	**99.48**
PVC Skirting/Dado Modular Trunking; White. (Cutting and jointing trunking to fittings and backplates for fixing to walls is included)						
Main carrier/backplate						
50 x 170mm	13.61	16.31	2.02	37.44	m	**53.75**
Extension carrier/backplate						
50 x 42mm	8.34	9.99	0.58	10.75	m	**20.74**

Material Costs/Measured Work Prices – Electrical Installations

V:ELECTRICAL SUPPLY/POWER/LIGHTING

Item	Net Price £	Material £	Labour hours	Labour £	Unit	Total rate £
Carrier/backplate						
Including cover seal	5.80	6.95	0.53	9.82	m	**16.77**
Chamfered covers for fixing to backplates						
50 x 42mm	3.12	3.74	0.33	6.12	m	**9.85**
Square covers for fixing to backplates						
50 x 42 mm	3.12	3.74	0.33	6.12	m	**9.85**
Plain covers for fixing to backplates						
85 mm	1.80	2.16	0.34	6.30	m	**8.46**
Retainers-clip to backplates to hold cables						
For chamfered covers	0.53	0.64	0.07	1.30	m	**1.93**
For square-recessed covers	0.53	0.64	0.07	1.30	m	**1.93**
For plain covers	0.53	0.64	0.07	1.30	m	**1.93**
Prepackaged corner assemblies						
Internal ; for 170 x 50 Assy	5.04	6.04	0.51	9.45	m	**15.49**
Internal; for 215 x 50 Assy	6.32	7.57	0.53	9.82	m	**17.40**
Internal; for 254 x 50 Assy	9.03	10.82	0.53	9.82	m	**20.64**
External; for 170 x 50 Assy	5.04	6.04	0.56	10.38	m	**16.42**
External ; for 215 x 50 Assy	6.32	7.57	0.58	10.75	m	**18.32**
External ; for 254 x 50 Assy	9.03	10.82	0.58	10.75	m	**21.57**
Clip on end caps						
170 x 50 Assy	2.49	2.98	0.11	2.04	m	**5.02**
215 x 50 Assy	2.77	3.32	0.11	2.04	m	**5.36**
254 x 50 Assy	2.87	3.44	0.11	2.04	m	**5.48**
Outlet box						
1 Gang; in horizontal trunking; clip in	2.34	2.80	0.34	6.30	m	**9.11**
2 Gang; in horizontal trunking; clip in	2.87	3.44	0.34	6.30	m	**9.74**
1 Gang; in vertical trunking; clip in	11.46	13.73	0.34	6.30	m	**20.03**
Sheet Steel Adaptable Boxes; with plain or knockout sides; fixed to backgrounds; including supports and fixings. (Cutting and connecting conduit to boxes is included.)						
Square Pattern - Black						
75 x 75 x 37 mm	1.87	2.24	0.69	12.79	nr	**15.03**
75 x 75 x 50 mm	1.93	2.31	0.69	12.79	nr	**15.10**
75 x 75 x 75 mm	2.32	2.78	0.69	12.79	nr	**15.57**
100 x 100 x 50 mm	2.13	2.55	0.71	13.16	nr	**15.71**
150 x 150 x 50 mm	3.12	3.74	0.79	14.64	nr	**18.38**
150 x 150 x 75 mm	3.55	4.25	0.80	14.83	nr	**19.08**
150 x 150 x 100 mm	4.73	5.67	0.80	14.83	nr	**20.50**
225 x 225 x 50 mm	5.81	6.96	0.93	17.24	nr	**24.20**
225 x 225 x 100 mm	6.96	8.34	0.94	17.42	nr	**25.76**
300 x 300 x 100 mm	11.81	14.15	0.99	18.35	nr	**32.50**

Material Costs/Measured Work Prices – Electrical Installations

V:ELECTRICAL SUPPLY/POWER/LIGHTING

Item	Net Price £	Material £	Labour hours	Labour £	Unit	Total rate £
V20 : LV DISTRIBUTION - CABLE SUPPORTS						
Y60/63 - CABLE SUPPORTS						
Sheet Steel Adaptable Boxes; with plain or knockout sides; fixed to backgrounds; including supports and fixings. (Cutting and connecting conduit to boxes is included) (Continued)						
Square Pattern - Galvanised						
75 x 75 x 37 mm	2.88	3.45	0.69	12.79	nr	**16.24**
75 x 75 x 50 mm	3.00	3.59	0.69	12.79	nr	**16.38**
75 x 75 x 75 mm	3.62	4.34	0.70	12.97	nr	**17.31**
100 x 100 x 50 mm	3.30	3.95	0.71	13.16	nr	**17.11**
150 x 150 x 50 mm	4.19	5.02	0.84	15.57	nr	**20.59**
150 x 150 x 75 mm	4.88	5.85	0.80	14.83	nr	**20.67**
150 x 150 x 100 mm	6.53	7.82	0.80	14.83	nr	**22.65**
225 x 225 x 50 mm	7.64	9.15	0.93	17.24	nr	**26.39**
225 x 225 x 100 mm	9.51	11.40	0.94	17.42	nr	**28.82**
300 x 300 x 100 mm	16.91	20.26	0.99	18.35	nr	**38.61**
Rectangular Pattern - Black						
100 x 75 x 50 mm	2.20	2.64	0.69	12.79	nr	**15.42**
150 x 75 x 50 mm	2.38	2.85	0.70	12.97	nr	**15.83**
150 x 75 x 75 mm	2.69	3.22	0.71	13.16	nr	**16.38**
150 x 100 x 75 mm	3.28	3.93	0.71	13.16	nr	**17.09**
225 x 75 x 50 mm	3.11	3.73	0.78	14.46	nr	**18.18**
225 x 150 x 75 mm	5.13	6.15	0.81	15.01	nr	**21.16**
225 x 150 x 100 mm	5.69	6.82	0.81	15.01	nr	**21.83**
300 x 150 x 50 mm	6.13	7.35	0.93	17.24	nr	**24.58**
300 x 150 x 75 mm	6.31	7.56	0.94	17.42	nr	**24.98**
300 x 150 x 100 mm	7.56	9.06	0.96	17.79	nr	**26.85**
Rectangular Pattern - Galvanised						
100 x 75 x 50 mm	2.97	3.56	0.69	12.79	nr	**16.35**
150 x 75 x 50 mm	2.84	3.40	0.70	12.97	nr	**16.38**
150 x 75 x 75 mm	3.49	4.18	0.71	13.16	nr	**17.34**
150 x 100 x 75 mm	3.48	4.17	0.71	13.16	nr	**17.33**
225 x 75 x 50 mm	3.22	3.86	0.89	16.50	nr	**20.35**
225 x 150 x 75 mm	5.97	7.15	0.81	15.01	nr	**22.17**
225 x 150 x 100 mm	6.59	7.90	0.81	15.01	nr	**22.91**
300 x 150 x 50 mm	6.51	7.80	0.93	17.24	nr	**25.04**
300 x 150 x 75 mm	7.33	8.78	0.94	17.42	nr	**26.21**
300 x 150 x 100 mm	8.73	10.46	0.96	17.79	nr	**28.25**

Material Costs/Measured Work Prices – Electrical Installations

V:ELECTRICAL SUPPLY/POWER/LIGHTING

Item	Net Price £	Material £	Labour hours	Labour £	Unit	Total rate £
V20 : LV DISTRIBUTION - CABLES/BUSBAR						
Y61 - LV CABLES AND WIRING						
ARMOURED CABLE						
Cable; XLPE insulated; PVC sheathed; copper stranded conductors to BS 5467; laid in trench/duct including marker tape. (Cable tiles measured elsewhere.)						
600/1000 Volt Grade; Single Core (Aluminium wire armour)						
25 mm2	2.12	2.54	0.15	2.78	m	**5.32**
35 mm2	2.23	2.67	0.15	2.78	m	**5.45**
50 mm2	2.31	2.77	0.17	3.15	m	**5.92**
70 mm2	2.49	2.98	0.18	3.34	m	**6.32**
95 mm2	2.85	3.42	0.20	3.71	m	**7.12**
120 mm2	3.39	4.06	0.22	4.08	m	**8.14**
150 mm2	4.15	4.97	0.24	4.45	m	**9.42**
185 mm2	4.60	5.51	0.26	4.82	m	**10.33**
240 mm2	5.66	6.78	0.30	5.56	m	**12.34**
300 mm2	6.60	7.91	0.31	5.75	m	**13.65**
400 mm2	8.43	10.10	0.38	7.04	m	**17.14**
600/1000 Volt Grade; Two Core (Galvanised steel wire armour)						
1.5 mm2	0.37	0.44	0.06	1.11	m	**1.56**
2.5 mm2	0.42	0.50	0.06	1.11	m	**1.62**
4 mm2	0.60	0.72	0.08	1.48	m	**2.20**
6 mm2	0.75	0.90	0.08	1.48	m	**2.38**
10 mm2	1.12	1.34	0.10	1.85	m	**3.20**
16 mm2	1.39	1.67	0.10	1.85	m	**3.52**
25 mm2	1.86	2.23	0.15	2.78	m	**5.01**
35 mm2	2.26	2.71	0.15	2.78	m	**5.49**
50 mm2	2.81	3.37	0.17	3.15	m	**6.52**
70 mm2	3.49	4.18	0.18	3.34	m	**7.52**
95 mm2	5.24	6.28	0.20	3.71	m	**9.99**
120 mm2	5.78	6.93	0.22	4.08	m	**11.00**
150 mm2	7.47	8.95	0.24	4.45	m	**13.40**
185 mm2	9.47	11.35	0.26	4.82	m	**16.17**
240 mm2	11.93	14.30	0.30	5.56	m	**19.86**
300 mm2	15.70	18.81	0.31	5.75	m	**24.56**
400 mm2	22.07	26.45	0.35	6.49	m	**32.93**
600/1000 Volt Grade; Three Core (Galvanised steel wire armour)						
1.5 mm2	0.41	0.49	0.07	1.30	m	**1.79**
2.5 mm2	0.50	0.60	0.07	1.30	m	**1.90**
4 mm2	0.69	0.83	0.09	1.67	m	**2.49**
6 mm2	0.88	1.05	0.10	1.85	m	**2.91**
10 mm2	1.57	1.88	0.11	2.04	m	**3.92**
16 mm2	1.95	2.34	0.11	2.04	m	**4.38**
25 mm2	2.29	2.74	0.16	2.97	m	**5.71**
35 mm2	2.92	3.50	0.16	2.97	m	**6.46**
50 mm2	3.71	4.45	0.19	3.52	m	**7.97**

Material Costs/Measured Work Prices – Electrical Installations

V:ELECTRICAL SUPPLY/POWER/LIGHTING

Item	Net Price £	Material £	Labour hours	Labour £	Unit	Total rate £
V20 : LV DISTRIBUTION - CABLES/BUSBAR						
Y61 - LV CABLES AND WIRING						
Cable; XLPE insulated; PVC sheathed; copper stranded conductors to BS 5467; laid in trench/duct including marker tape. (Cable tiles measured elsewhere) (Continued)						
600/1000 Volt Grade; Three Core (Galvanised steel wire armour						
70 mm2	4.99	5.98	0.21	3.89	m	**9.87**
95 mm2	6.87	8.23	0.23	4.26	m	**12.50**
120 mm2	8.80	10.54	0.24	4.45	m	**14.99**
150 mm2	10.93	13.10	0.27	5.00	m	**18.10**
185 mm2	13.05	15.64	0.30	5.56	m	**21.20**
240 mm2	16.89	20.24	0.33	6.12	m	**26.36**
300 mm2	21.45	25.70	0.35	6.49	m	**32.19**
400 mm2	26.52	31.78	0.41	7.60	m	**39.38**
600/1000 Volt Grade; Four Core (Galvanised steel wire armour)						
1.5 mm2	0.45	0.54	0.08	1.48	m	**2.02**
2.5 mm2	0.58	0.69	0.09	1.67	m	**2.36**
4 mm2	0.83	0.99	0.10	1.85	m	**2.85**
6 mm2	1.13	1.35	0.10	1.85	m	**3.21**
10 mm2	1.84	2.20	0.12	2.22	m	**4.43**
16 mm2	2.16	2.59	0.12	2.22	m	**4.81**
25 mm2	2.63	3.15	0.18	3.34	m	**6.49**
35 mm2	3.38	4.05	0.19	3.52	m	**7.57**
50 mm2	4.45	5.33	0.21	3.89	m	**9.22**
70 mm2	6.17	7.39	0.23	4.26	m	**11.66**
95 mm2	8.28	9.92	0.26	4.82	m	**14.74**
120 mm2	10.55	12.64	0.28	5.19	m	**17.83**
150 mm2	13.08	15.67	0.32	5.93	m	**21.60**
185 mm2	15.77	18.90	0.35	6.49	m	**25.38**
240 mm2	20.75	24.86	0.36	6.67	m	**31.54**
300 mm2	25.83	30.95	0.40	7.41	m	**38.37**
400 mm2	32.88	39.40	0.45	8.34	m	**47.74**
600/1000 Volt Grade; Seven Core (Galvanised steel wire armour)						
1.5 mm2	0.82	0.98	0.10	1.85	m	**2.84**
2.5 mm2	1.10	1.32	0.10	1.85	m	**3.17**
4 mm2	1.63	1.95	0.11	2.04	m	**3.99**
600/1000 Volt Grade; Twelve Core (Galvanised steel wire armour)						
1.5 mm2	1.32	1.58	0.11	2.04	m	**3.62**
2.5 mm2	1.77	2.12	0.11	2.04	m	**4.16**
600/1000 Volt Grade; Nineteen Core (Galvanised steel wire armour)						
1.5 mm2	1.99	2.38	0.13	2.41	m	**4.79**
2.5 mm2	2.77	3.32	0.14	2.59	m	**5.91**

Material Costs/Measured Work Prices – Electrical Installations

485

V:ELECTRICAL SUPPLY/POWER/LIGHTING

Item	Net Price £	Material £	Labour hours	Labour £	Unit	Total rate £
600/1000 Volt Grade; Twenty seven Core (Galvanised steel wire armour)						
1.5 mm2	2.93	3.51	0.14	2.59	m	**6.11**
2.5 mm2	3.76	4.51	0.16	2.97	m	**7.47**
600/1000 Volt Grade; Thirty seven Core (Galvanised steel wire armour)						
1.5 mm2	3.76	4.51	0.15	2.78	m	**7.29**
2.5 mm2	4.98	5.97	0.17	3.15	m	**9.12**
Cable; XLPE Insulated; PVC sheathed copper stranded conductors to BS 5467; clipped direct to backgrounds including cleat.						
600/1000 Volt Grade; Single Core (Aluminium wire armour)						
25 mm2	2.42	2.90	0.35	6.49	m	**9.39**
35 mm2	2.54	3.04	0.36	6.67	m	**9.72**
50 mm2	2.64	3.16	0.37	6.86	m	**10.02**
70 mm2	2.84	3.40	0.39	7.23	m	**10.63**
95 mm2	3.25	3.89	0.42	7.78	m	**11.68**
120 mm2	3.88	4.65	0.47	8.71	m	**13.36**
150 mm2	4.74	5.68	0.51	9.45	m	**15.13**
185 mm2	5.26	6.30	0.59	10.94	m	**17.24**
240 mm2	6.47	7.75	0.68	12.60	m	**20.36**
300 mm2	7.55	9.05	0.74	13.72	m	**22.76**
400 mm2	9.64	11.55	0.88	16.31	m	**27.86**
600/1000 Volt Grade; Two Core (Galvanised steel wire armour)						
1.5 mm2	0.42	0.50	0.20	3.71	m	**4.21**
2.5 mm2	0.48	0.58	0.20	3.71	m	**4.28**
4.0 mm2	0.68	0.81	0.21	3.89	m	**4.71**
6.0 mm2	0.85	1.02	0.22	4.08	m	**5.10**
10.0 mm2	1.28	1.53	0.24	4.45	m	**5.98**
16.0 mm2	1.58	1.89	0.25	4.63	m	**6.53**
25 mm2	2.12	2.54	0.35	6.49	m	**9.03**
35 mm2	2.58	3.09	0.36	6.67	m	**9.76**
50 mm2	3.22	3.86	0.37	6.86	m	**10.72**
70 mm2	3.98	4.77	0.39	7.23	m	**12.00**
95 mm2	5.99	7.18	0.42	7.78	m	**14.96**
120 mm2	6.60	7.91	0.47	8.71	m	**16.62**
150 mm2	8.53	10.22	0.51	9.45	m	**19.67**
185 mm2	11.04	13.23	0.59	10.94	m	**24.16**
240 mm2	13.63	16.33	0.68	12.60	m	**28.94**
300 mm2	17.94	21.50	0.74	13.72	m	**35.21**
400 mm2	25.22	30.22	0.88	16.31	m	**46.53**
600/1000 Volt Grade; Three Core (Galvanised Steel wire armour)						
1.5 mm2	0.47	0.56	0.20	3.71	m	**4.27**
2.5 mm2	0.58	0.69	0.21	3.89	m	**4.59**
4.0 mm2	0.79	0.95	0.22	4.08	m	**5.02**
6.0 mm2	1.01	1.21	0.22	4.08	m	**5.29**
10.0 mm2	1.80	2.16	0.25	4.63	m	**6.79**
16.0 mm2	2.23	2.67	0.26	4.82	m	**7.49**

Material Costs/Measured Work Prices – Electrical Installations

V:ELECTRICAL SUPPLY/POWER/LIGHTING

Item	Net Price £	Material £	Labour hours	Labour £	Unit	Total rate £
V20 : LV DISTRIBUTION - CABLES/BUSBAR						
Y61 - LV CABLES AND WIRING						
Cable; XLPE Insulated; PVC sheathed copper stranded conductors to BS 5467; clipped direct to backgrounds including cleat. (Continued)						
600/1000 Volt Grade; Three Core (Galvanised Steel wire armour)						
25 mm2	2.62	3.14	0.37	6.86	m	**10.00**
35 mm2	2.62	3.14	0.39	7.23	m	**10.37**
50 mm2	4.24	5.08	0.40	7.41	m	**12.49**
70 mm2	5.70	6.83	0.42	7.78	m	**14.61**
95 mm2	7.85	9.41	0.45	8.34	m	**17.75**
120 mm2	7.85	9.41	0.52	9.64	m	**19.04**
150 mm2	12.49	14.97	0.55	10.19	m	**25.16**
185 mm2	14.92	17.88	0.63	11.68	m	**29.55**
240 mm2	19.31	23.14	0.71	13.16	m	**36.30**
300 mm2	24.52	29.38	0.78	14.46	m	**43.84**
400 mm2	30.31	36.32	0.87	16.12	m	**52.45**
600/1000 Volt Grade; Four Core (Galvanised steel wire armour)						
1.5 mm2	0.52	0.62	0.21	3.89	m	**4.52**
2.5 mm2	0.66	0.79	0.22	4.08	m	**4.87**
4.0 mm2	0.95	1.14	0.22	4.08	m	**5.22**
6.0 mm2	1.30	1.56	0.23	4.26	m	**5.82**
10.0 mm2	2.10	2.52	0.26	4.82	m	**7.34**
16.0 mm2	2.47	2.96	0.26	4.82	m	**7.78**
25 mm2	3.00	3.59	0.39	7.23	m	**10.82**
35 mm2	3.86	4.63	0.40	7.41	m	**12.04**
50 mm2	5.09	6.10	0.41	7.60	m	**13.70**
70 mm2	7.06	8.46	0.45	8.34	m	**16.80**
95 mm2	9.47	11.35	0.50	9.27	m	**20.61**
120 mm2	12.06	14.45	0.54	10.01	m	**24.46**
150 mm2	14.95	17.91	0.60	11.12	m	**29.03**
185 mm2	18.02	21.59	0.67	12.42	m	**34.01**
240 mm2	23.71	28.41	0.75	13.90	m	**42.31**
300 mm2	29.52	35.37	0.83	15.38	m	**50.76**
400 mm2	37.57	45.02	0.91	16.87	m	**61.89**
600/1000 Volt Grade; Seven Core (Galvanised steel wire armour)						
1.5 mm2	0.94	1.13	0.20	3.71	m	**4.83**
2.5 mm2	1.26	1.51	0.20	3.71	m	**5.22**
4.0 mm2	1.85	2.22	0.23	4.26	m	**6.48**
600/1000 Volt Grade; Twelve Core (Galvanised steel wire armour)						
1.5 mm2	2.28	2.73	0.23	4.26	m	**6.99**
2.5 mm2	2.03	2.43	0.24	4.45	m	**6.88**

Material Costs/Measured Work Prices – Electrical Installations

V:ELECTRICAL SUPPLY/POWER/LIGHTING

Item	Net Price £	Material £	Labour hours	Labour £	Unit	Total rate £
600/1000 Volt Grade; Nineteen Core						
(Galvanised steel wire armour)						
1.5 mm2	2.28	2.73	0.26	4.82	m	**7.55**
2.5 mm2	3.17	3.80	0.28	5.19	m	**8.99**
600/1000 Volt Grade; Twenty seven Core						
(Galvanised steel wire armour)						
1.5 mm2	3.35	4.01	0.29	5.37	m	**9.39**
2.5 mm2	4.30	5.15	0.30	5.56	m	**10.71**
600/1000 Volt Grade; Thirty seven Core						
(Galvanised steel wire armour)						
1.5 mm2	4.30	5.15	0.32	5.93	m	**11.08**
2.5 mm2	5.69	6.82	0.33	6.12	m	**12.93**
Cable Termination; brass weatherproof gland						
with inner and outer seal, shroud, brass						
locknut and earth ring (including drilling and						
cutting mild steel gland plate).						
600/1000 Volt grade; Single core (Aluminium wire						
armour)						
25 mm2	6.55	7.85	1.70	31.51	nr	**39.36**
35 mm2	6.55	7.85	1.79	33.18	nr	**41.02**
50 mm2	6.55	7.85	2.06	38.18	nr	**46.03**
70 mm2	7.05	8.45	2.12	39.29	nr	**47.74**
95 mm2	7.05	8.45	2.39	44.30	nr	**52.74**
120 mm2	7.05	8.45	2.47	45.78	nr	**54.23**
150 mm2	8.84	10.59	2.73	50.60	nr	**61.19**
185 mm2	8.84	10.59	3.05	56.53	nr	**67.12**
240 mm2	11.98	14.36	3.45	63.94	nr	**78.30**
300 mm2	16.29	19.52	3.84	71.17	nr	**90.69**
400 mm2	16.29	19.52	4.21	78.03	nr	**97.55**
600/1000 Volt grade; Two core (Galvanised						
steel wire armour)						
1.5 mm2	0.80	0.96	0.58	10.75	nr	**11.71**
2.5 mm2	0.80	0.96	0.58	10.75	nr	**11.71**
4 mm2	0.80	0.96	0.58	10.75	nr	**11.71**
6 mm2	1.01	1.21	0.67	12.42	nr	**13.63**
10 mm2	1.01	1.21	1.00	18.53	nr	**19.74**
16 mm2	1.52	1.82	1.11	20.57	nr	**22.39**
25 mm2	1.52	1.82	1.70	31.51	nr	**33.33**
35 mm2	2.07	2.48	1.79	33.18	nr	**35.66**
50 mm2	2.07	2.48	2.06	38.18	nr	**40.66**
70 mm2	2.07	2.48	2.12	39.29	nr	**41.77**
95 mm2	2.07	2.48	2.39	44.30	nr	**46.78**
120 mm2	4.32	5.18	2.47	45.78	nr	**50.96**
150 mm2	4.32	5.18	2.73	50.60	nr	**55.77**
185 mm2	9.59	11.49	3.05	56.53	nr	**68.02**
240 mm2	9.67	11.59	3.45	63.94	nr	**75.53**
300 mm2	9.67	11.59	3.84	71.17	nr	**82.76**
400 mm2	16.08	19.27	4.21	78.03	nr	**97.30**

488   *Material Costs/Measured Work Prices – Electrical Installations*

V:ELECTRICAL SUPPLY/POWER/LIGHTING

Item	Net Price £	Material £	Labour hours	Labour £	Unit	Total rate £
V20 : LV DISTRIBUTION - CABLES/BUSBAR						
Y61 - LV CABLES AND WIRING						
Cable Termination; brass weatherproof gland with inner and outer seal, shroud, brass locknut and earth ring (including drilling and cutting mild steel gland plate) (Continued)						
600/1000 Volt grade; Three core (Galvanised steel wire armour)						
1.5 mm2	0.80	0.96	0.62	11.49	nr	**12.45**
2.5 mm2	0.80	0.96	0.62	11.49	nr	**12.45**
4 mm2	0.80	0.96	0.62	11.49	nr	**12.45**
6 mm2	1.01	1.21	0.71	13.16	nr	**14.37**
10 mm2	1.01	1.21	1.06	19.65	nr	**20.86**
16 mm2	1.52	1.82	1.19	22.06	nr	**23.88**
25 mm2	1.52	1.82	1.81	33.55	nr	**35.37**
35 mm2	2.07	2.48	1.99	36.88	nr	**39.36**
50 mm2	2.07	2.48	2.23	41.33	nr	**43.81**
70 mm2	4.32	5.18	2.40	44.48	nr	**49.66**
95 mm2	4.32	5.18	2.63	48.74	nr	**53.92**
120 mm2	9.59	11.49	2.83	52.45	nr	**63.94**
150 mm2	9.59	11.49	3.22	59.68	nr	**71.17**
185 mm2	9.67	11.59	3.44	63.76	nr	**75.34**
240 mm2	16.08	19.27	3.83	70.98	nr	**90.25**
300 mm2	16.08	19.27	4.28	79.33	nr	**98.59**
400 mm2	24.83	29.75	5.00	92.67	nr	**122.42**
600/1000 Volt Grade; Four Core (Galvanised steel wire armour)						
1.5 mm2	0.80	0.96	0.67	12.42	nr	**13.38**
2.5 mm2	0.80	0.96	0.67	12.42	nr	**13.38**
4 mm2	1.01	1.21	0.71	13.16	nr	**14.37**
6 mm2	1.01	1.21	0.76	14.09	nr	**15.30**
10 mm2	1.52	1.82	1.14	21.13	nr	**22.95**
16 mm2	1.52	1.82	1.29	23.91	nr	**25.73**
25 mm2	2.07	2.48	1.99	36.88	nr	**39.36**
35 mm2	2.07	2.48	2.16	40.03	nr	**42.51**
50 mm2	4.32	5.18	2.49	46.15	nr	**51.33**
70 mm2	4.32	5.18	2.65	49.11	nr	**54.29**
95 mm2	9.59	11.49	2.98	55.23	nr	**66.72**
120 mm2	9.67	11.59	3.15	58.38	nr	**69.97**
150 mm2	9.67	11.59	3.50	64.87	nr	**76.46**
185 mm2	16.08	19.27	3.72	68.95	nr	**88.21**
240 mm2	24.83	29.75	4.33	80.25	nr	**110.01**
300 mm2	24.83	29.75	4.86	90.07	nr	**119.83**
400 mm2	24.83	29.75	5.46	101.20	nr	**130.95**
600/1000 Volt grade; Seven Core (Galvanised steel wire armour)						
1.5 mm2	0.80	0.96	0.81	15.01	nr	**15.97**
2.5 mm2	0.80	0.96	0.85	15.75	nr	**16.71**
4 mm2	0.80	0.96	0.93	17.24	nr	**18.20**

Material Costs/Measured Work Prices – Electrical Installations

V:ELECTRICAL SUPPLY/POWER/LIGHTING

Item	Net Price £	Material £	Labour hours	Labour £	Unit	Total rate £
600/1000 Volt grade; Twelve Core (Galvanised steel wire armour)						
1.5 mm2	1.01	1.21	1.14	21.13	nr	**22.34**
2.5 mm2	1.52	1.82	1.13	20.94	nr	**22.76**
600/1000 Volt grade; Nineteen Core (Galvanised steel wire armour)						
1.5 mm2	1.52	1.82	1.54	28.54	nr	**30.36**
2.5 mm2	1.52	1.82	1.54	28.54	nr	**30.36**
600/1000 Volt grade; Twenty Seven Core (Galvanised steel wire armour)						
1.5 mm2	1.52	1.82	1.94	35.96	nr	**37.78**
2.5 mm2	2.07	2.48	2.31	42.81	nr	**45.29**
600/1000 Volt grade; Thirty Seven Core (Galvanised steel wire armour)						
1.5 mm2	2.07	2.48	2.53	46.89	nr	**49.37**
2.5 mm2	2.07	2.48	2.87	53.19	nr	**55.67**
Cable; XLPE insulated; LSOH sheathed; copper stranded conductors to BS 6724; laid in trench/duct including marker tape. (Cable tiles measured elsewhere)						
600/1000 Volt Grade; Single Core (Aluminium wire armour)						
50 mm2	2.93	3.51	0.17	3.15	m	**6.66**
70 mm2	3.06	3.67	0.18	3.34	m	**7.00**
95 mm2	3.76	4.51	0.20	3.71	m	**8.21**
120 mm2	4.00	4.79	0.22	4.08	m	**8.87**
150 mm2	4.60	5.51	0.24	4.45	m	**9.96**
185 mm2	5.18	6.21	0.26	4.82	m	**11.03**
240 mm2	6.50	7.79	0.30	5.56	m	**13.35**
300 mm2	7.62	9.13	0.31	5.75	m	**14.88**
400 mm2	9.73	11.66	0.35	6.49	m	**18.15**
600/1000 Volt Grade; Two Core (Galvanised steel wire armour)						
1.5 mm2	0.60	0.72	0.06	1.11	m	**1.83**
2.5 mm2	0.68	0.81	0.06	1.11	m	**1.93**
4 mm2	0.87	1.04	0.08	1.48	m	**2.53**
6 mm2	1.00	1.20	0.08	1.48	m	**2.68**
10 mm2	1.35	1.62	0.10	1.85	m	**3.47**
16 mm2	1.87	2.24	0.10	1.85	m	**4.09**
25 mm2	2.12	2.54	0.15	2.78	m	**5.32**
35 mm2	2.55	3.06	0.15	2.78	m	**5.84**
50 mm2	3.29	3.94	0.17	3.15	m	**7.09**
70 mm2	4.37	5.24	0.18	3.34	m	**8.57**
95 mm2	5.92	7.09	0.20	3.71	m	**10.80**
120 mm2	7.43	8.90	0.22	4.08	m	**12.98**
150 mm2	8.89	10.65	0.24	4.45	m	**15.10**
185 mm2	10.89	13.05	0.26	4.82	m	**17.87**
240 mm2	14.06	16.85	0.30	5.56	m	**22.41**
300 mm2	17.62	21.11	0.31	5.75	m	**26.86**
400 mm2	23.29	27.91	0.35	6.49	m	**34.40**

490   *Material Costs/Measured Work Prices – Electrical Installations*

V:ELECTRICAL SUPPLY/POWER/LIGHTING

Item	Net Price £	Material £	Labour hours	Labour £	Unit	Total rate £
V20 : LV DISTRIBUTION - CABLES/BUSBAR						
Y61 - LV CABLES AND WIRING						
Cable; XLPE insulated; LSOH sheathed; copper stranded conductors to BS 6724; laid in trench/duct including marker tape. (Cable tiles measured elsewhere) (Continued)						
600/1000 Volt Grade; Three Core (Galvanised steel wire armour)						
1.5 mm2	0.65	0.78	0.07	1.30	m	**2.08**
2.5 mm2	0.72	0.86	0.07	1.30	m	**2.16**
4 mm2	0.98	1.17	0.09	1.67	m	**2.84**
6 mm2	1.19	1.43	0.10	1.85	m	**3.28**
10 mm2	1.64	1.97	0.11	2.04	m	**4.00**
16 mm2	2.18	2.61	0.11	2.04	m	**4.65**
25 mm2	2.61	3.13	0.16	2.97	m	**6.09**
35 mm2	3.25	3.89	0.16	2.97	m	**6.86**
50 mm2	4.51	5.40	0.19	3.52	m	**8.93**
70 mm2	5.99	7.18	0.21	3.89	m	**11.07**
95 mm2	7.98	9.56	0.23	4.26	m	**13.83**
120 mm2	9.69	11.61	0.24	4.45	m	**16.06**
150 mm2	11.99	14.37	0.27	5.00	m	**19.37**
185 mm2	14.36	17.21	0.30	5.56	m	**22.77**
240 mm2	18.70	22.41	0.33	6.12	m	**28.52**
300 mm2	23.03	27.60	0.35	6.49	m	**34.08**
400 mm2	28.93	34.67	0.41	7.60	m	**42.27**
600/1000 Volt Grade; Four Core (Galvanised steel wire armour)						
1.5 mm2	0.72	0.86	0.08	1.48	m	**2.35**
2.5 mm2	0.87	1.04	0.09	1.67	m	**2.71**
4 mm2	1.14	1.37	0.10	1.85	m	**3.22**
6 mm2	1.46	1.75	0.10	1.85	m	**3.60**
10 mm2	1.90	2.28	0.12	2.22	m	**4.50**
16 mm2	2.67	3.20	0.12	2.22	m	**5.42**
25 mm2	3.13	3.75	0.18	3.34	m	**7.09**
35 mm2	3.93	4.71	0.19	3.52	m	**8.23**
50 mm2	4.97	5.96	0.21	3.89	m	**9.85**
70 mm2	6.96	8.34	0.23	4.26	m	**12.60**
95 mm2	9.19	11.01	0.26	4.82	m	**15.83**
120 mm2	11.94	14.31	0.28	5.19	m	**19.50**
150 mm2	14.03	16.81	0.32	5.93	m	**22.74**
185 mm2	17.41	20.86	0.35	6.49	m	**27.35**
240 mm2	22.42	26.87	0.36	6.67	m	**33.54**
300 mm2	28.47	34.12	0.40	7.41	m	**41.53**
400 mm2	36.44	43.67	0.45	8.34	m	**52.01**
600/1000 Volt Grade; Seven Core (Galvanised steel wire armour)						
1.5 mm2	1.26	1.51	0.10	1.85	m	**3.36**
2.5 mm2	1.60	1.92	0.10	1.85	m	**3.77**
4 mm2	2.67	3.20	0.11	2.04	m	**5.24**

Material Costs/Measured Work Prices – Electrical Installations

V:ELECTRICAL SUPPLY/POWER/LIGHTING

Item	Net Price £	Material £	Labour hours	Labour £	Unit	Total rate £
600/1000 Volt Grade; Twelve Core (Galvanised steel wire armour)						
1.5 mm2	1.86	2.23	0.11	2.04	m	**4.27**
2.5 mm2	2.31	2.77	0.11	2.04	m	**4.81**
600/1000 Volt Grade; Nineteen Core (Galvanised steel wire armour)						
1.5 mm2	2.71	3.25	0.13	2.41	m	**5.66**
2.5 mm2	3.61	4.33	0.14	2.59	m	**6.92**
600/1000 Volt Grade; Twenty seven Core (Galvanised steel wire armour)						
1.5 mm2	3.80	4.55	0.14	2.59	m	**7.15**
2.5 mm2	4.87	5.84	0.16	2.97	m	**8.80**
600/1000 Volt Grade; Thirty seven Core (Galvanised steel wire armour)						
1.5 mm2	5.28	6.33	0.15	2.78	m	**9.11**
2.5 mm2	6.68	8.00	0.17	3.15	m	**11.16**
Cable; XLPE Insulated; LSOH sheathed copper stranded conductors to BS 6724; clipped direct to backgrounds including cleat.						
600/1000 Volt Grade; Single Core (Aluminium wire armour)						
50 mm2	3.35	4.01	0.37	6.86	m	**10.87**
70 mm2	3.49	4.18	0.39	7.23	m	**11.41**
95 mm2	4.30	5.15	0.42	7.78	m	**12.94**
120 mm2	4.57	5.48	0.47	8.71	m	**14.19**
150 mm2	5.26	6.30	0.51	9.45	m	**15.76**
185 mm2	5.92	7.09	0.59	10.94	m	**18.03**
240 mm2	7.43	8.90	0.68	12.60	m	**21.51**
300 mm2	8.70	10.43	0.74	13.72	m	**24.14**
400 mm2	11.12	13.33	0.81	15.01	m	**28.34**
600/1000 Volt Grade; Two Core (Galvanised steel wire armour)						
1.5 mm2	0.68	0.81	0.20	3.71	m	**4.52**
2.5 mm2	0.78	0.93	0.20	3.71	m	**4.64**
4.0 mm2	1.00	1.20	0.21	3.89	m	**5.09**
6.0 mm2	1.14	1.37	0.22	4.08	m	**5.44**
10.0 mm2	1.55	1.86	0.24	4.45	m	**6.31**
16.0 mm2	2.14	2.56	0.25	4.63	m	**7.20**
25 mm2	2.42	2.90	0.35	6.49	m	**9.39**
35 mm2	2.92	3.50	0.36	6.67	m	**10.17**
50 mm2	3.76	4.51	0.37	6.86	m	**11.36**
70 mm2	4.99	5.98	0.39	7.23	m	**13.21**
95 mm2	6.77	8.11	0.42	7.78	m	**15.90**
120 mm2	8.50	10.19	0.47	8.71	m	**18.90**
150 mm2	10.16	12.17	0.51	9.45	m	**21.63**
185 mm2	12.44	14.91	0.59	10.94	m	**25.84**
240 mm2	16.07	19.26	0.68	12.60	m	**31.86**
300 mm2	20.14	24.13	0.74	13.72	m	**37.85**
400 mm2	26.62	31.90	0.81	15.01	m	**46.91**

492   *Material Costs/Measured Work Prices – Electrical Installations*

V:ELECTRICAL SUPPLY/POWER/LIGHTING

Item	Net Price £	Material £	Labour hours	Labour £	Unit	Total rate £
V20 : LV DISTRIBUTION - CABLES/BUSBAR						
Y61 - LV CABLES AND WIRING						
Cable; XLPE Insulated; LSOH sheathed copper stranded conductors to BS 6724; clipped direct to backgrounds including cleat. (Continued)						
600/1000 Volt Grade; Three Core (Galvanised Steel wire armour)						
1.5 mm2	0.74	0.89	0.20	3.71	m	**4.59**
2.5 mm2	0.83	0.99	0.21	3.89	m	**4.89**
4.0 mm2	1.12	1.34	0.22	4.08	m	**5.42**
6.0 mm2	1.36	1.63	0.22	4.08	m	**5.71**
10.0 mm2	1.87	2.24	0.25	4.63	m	**6.87**
16.0 mm2	2.50	3.00	0.26	4.82	m	**7.81**
25 mm2	2.99	3.58	0.37	6.86	m	**10.44**
35 mm2	3.72	4.46	0.39	7.23	m	**11.69**
50 mm2	5.16	6.18	0.40	7.41	m	**13.60**
70 mm2	6.84	8.20	0.42	7.78	m	**15.98**
95 mm2	9.12	10.93	0.45	8.34	m	**19.27**
120 mm2	11.08	13.28	0.52	9.64	m	**22.91**
150 mm2	13.70	16.42	0.55	10.19	m	**26.61**
185 mm2	16.42	19.68	0.63	11.68	m	**31.35**
240 mm2	21.37	25.61	0.71	13.16	m	**38.77**
300 mm2	26.32	31.54	0.78	14.46	m	**46.00**
400 mm2	33.06	39.62	0.87	16.12	m	**55.74**
600/1000 Volt Grade; Four Core (Galvanised steel wire armour)						
1.5 mm2	0.83	0.99	0.21	3.89	m	**4.89**
2.5 mm2	1.00	1.20	0.22	4.08	m	**5.28**
4.0 mm2	1.31	1.57	0.22	4.08	m	**5.65**
6.0 mm2	1.67	2.00	0.23	4.26	m	**6.26**
10.0 mm2	2.17	2.60	0.26	4.82	m	**7.42**
16.0 mm2	3.05	3.65	0.26	4.82	m	**8.47**
25 mm2	3.58	4.29	0.39	7.23	m	**11.52**
35 mm2	4.49	5.38	0.40	7.41	m	**12.79**
50 mm2	5.68	6.81	0.41	7.60	m	**14.41**
70 mm2	7.96	9.54	0.45	8.34	m	**17.88**
95 mm2	10.50	12.58	0.50	9.27	m	**21.85**
120 mm2	13.64	16.34	0.54	10.01	m	**26.35**
150 mm2	16.03	19.21	0.60	11.12	m	**30.33**
185 mm2	19.90	23.85	0.67	12.42	m	**36.26**
240 mm2	25.62	30.70	0.75	13.90	m	**44.60**
300 mm2	32.53	38.98	0.83	15.38	m	**54.36**
400 mm2	41.64	49.90	0.91	16.87	m	**66.76**
600/1000 Volt Grade; Seven Core (Galvanised steel wire armour)						
1.5 mm2	1.44	1.73	0.20	3.71	m	**5.43**
2.5 mm2	1.82	2.18	0.20	3.71	m	**5.89**
4.0 mm2	3.05	3.65	0.23	4.26	m	**7.92**

Material Costs/Measured Work Prices – Electrical Installations

V:ELECTRICAL SUPPLY/POWER/LIGHTING

Item	Net Price £	Material £	Labour hours	Labour £	Unit	Total rate £
600/1000 Volt Grade; Twelve Core (Galvanised steel wire armour)						
1.5 mm2	2.12	2.54	0.23	4.26	m	**6.80**
2.5 mm2	2.64	3.16	0.24	4.45	m	**7.61**
600/1000 Volt Grade; Nineteen Core (Galvanised steel wire armour)						
1.5 mm2	3.10	3.71	0.26	4.82	m	**8.53**
2.5 mm2	4.13	4.95	0.28	5.19	m	**10.14**
600/1000 Volt Grade; Twenty seven Core (Galvanised steel wire armour)						
1.5 mm2	4.34	5.20	0.29	5.37	m	**10.58**
2.5 mm2	5.57	6.67	0.30	5.56	m	**12.23**
600/1000 Volt Grade; Thirty seven Core (Galvanised steel wire armour)						
1.5 mm2	6.04	7.24	0.32	5.93	m	**13.17**
2.5 mm2	7.63	9.14	0.33	6.12	m	**15.26**
Cable Termination; brass weatherproof gland with inner and outer seal, shroud, brass locknut and earth ring (including drilling and cutting mild steel gland plate)						
600/1000 Volt grade; Single core (Aluminium wire armour)						
50 mm2	6.55	7.85	2.06	38.18	nr	**46.03**
70 mm2	7.05	8.45	2.12	39.29	nr	**47.74**
95 mm2	7.05	8.45	2.39	44.30	nr	**52.74**
120 mm2	7.05	8.45	2.47	45.78	nr	**54.23**
150 mm2	8.84	10.59	2.73	50.60	nr	**61.19**
185 mm2	8.84	10.59	3.05	56.53	nr	**67.12**
240 mm2	11.98	14.36	3.45	63.94	nr	**78.30**
300 mm2	16.29	19.52	3.84	71.17	nr	**90.69**
400 mm2	16.29	19.52	4.21	78.03	nr	**97.55**
600/1000 Volt grade; Two core (Galvanised steel wire armour)						
1.5 mm2	0.80	0.96	0.58	10.75	nr	**11.71**
2.5 mm2	0.80	0.96	0.58	10.75	nr	**11.71**
4 mm2	0.80	0.96	0.58	10.75	nr	**11.71**
6 mm2	1.01	1.21	0.67	12.42	nr	**13.63**
10 mm2	1.01	1.21	1.00	18.53	nr	**19.74**
16 mm2	1.52	1.82	1.11	20.57	nr	**22.39**
25 mm2	1.52	1.82	1.70	31.51	nr	**33.33**
35 mm2	2.07	2.48	1.79	33.18	nr	**35.66**
50 mm2	2.07	2.48	2.06	38.18	nr	**40.66**
70 mm2	2.07	2.48	2.12	39.29	nr	**41.77**
95 mm2	2.07	2.48	2.39	44.30	nr	**46.78**
120 mm2	4.32	5.18	2.47	45.78	nr	**50.96**
150 mm2	4.32	5.18	2.73	50.60	nr	**55.77**
185 mm2	9.59	11.49	3.05	56.53	nr	**68.02**
240 mm2	9.67	11.59	3.45	63.94	nr	**75.53**
300 mm2	9.67	11.59	3.84	71.17	nr	**82.76**
400 mm2	16.08	19.27	4.21	78.03	nr	**97.30**

Material Costs/Measured Work Prices – Electrical Installations

V:ELECTRICAL SUPPLY/POWER/LIGHTING

Item	Net Price £	Material £	Labour hours	Labour £	Unit	Total rate £
V20 : LV DISTRIBUTION - CABLES/BUSBAR						
Y61 - LV CABLES AND WIRING						
Cable Termination; brass weatherproof gland with inner and outer seal, shroud, brass locknut and earth ring (including drilling and cutting mild steel gland plate) (Continued)						
600/1000 Volt grade; Three core (Galvanised steel wire armour)						
1.5 mm2	0.80	0.96	0.62	11.49	nr	12.45
2.5 mm2	0.80	0.96	0.62	11.49	nr	12.45
4 mm2	0.80	0.96	0.62	11.49	nr	12.45
6 mm2	1.01	1.21	0.71	13.16	nr	14.37
10 mm2	1.01	1.21	1.06	19.65	nr	20.86
16 mm2	1.52	1.82	1.19	22.06	nr	23.88
25 mm2	1.52	1.82	1.81	33.55	nr	35.37
35 mm2	2.07	2.48	1.99	36.88	nr	39.36
50 mm2	2.07	2.48	2.23	41.33	nr	43.81
70 mm2	4.32	5.18	2.40	44.48	nr	49.66
95 mm2	4.32	5.18	2.63	48.74	nr	53.92
120 mm2	9.59	11.49	2.83	52.45	nr	63.94
150 mm2	9.59	11.49	3.22	59.68	nr	71.17
185 mm2	9.67	11.59	3.44	63.76	nr	75.34
240 mm2	16.08	19.27	3.83	70.98	nr	90.25
300 mm2	16.08	19.27	4.28	79.33	nr	98.59
400 mm2	24.83	29.75	5.00	92.67	nr	122.42
600/1000 Volt Grade; Four Core (Galvanised steel wire armour)						
1.5 mm2	0.80	0.96	0.67	12.42	nr	13.38
2.5 mm2	0.80	0.96	0.67	12.42	nr	13.38
4 mm2	1.01	1.21	0.71	13.16	nr	14.37
6 mm2	1.01	1.21	0.76	14.09	nr	15.30
10 mm2	1.52	1.82	1.14	21.13	nr	22.95
16 mm2	1.52	1.82	1.29	23.91	nr	25.73
25 mm2	2.07	2.48	1.99	36.88	nr	39.36
35 mm2	2.07	2.48	2.16	40.03	nr	42.51
50 mm2	4.32	5.18	2.49	46.15	nr	51.33
70 mm2	4.32	5.18	2.65	49.11	nr	54.29
95 mm2	9.59	11.49	2.98	55.23	nr	66.72
120 mm2	9.67	11.59	3.15	58.38	nr	69.97
150 mm2	9.67	11.59	3.50	64.87	nr	76.46
185 mm2	16.08	19.27	3.72	68.95	nr	88.21
240 mm2	24.83	29.75	4.33	80.25	nr	110.01
300 mm2	24.83	29.75	4.86	90.07	nr	119.83
400 mm2	24.83	29.75	5.46	101.20	nr	130.95
600/1000 Volt grade; Seven Core (Galvanised steel wire armour)						
1.5 mm2	0.80	0.96	0.81	15.01	nr	15.97
2.5 mm2	0.80	0.96	0.85	15.75	nr	16.71
4 mm2	0.80	0.96	0.93	17.24	nr	18.20

Material Costs/Measured Work Prices – Electrical Installations

V:ELECTRICAL SUPPLY/POWER/LIGHTING

Item	Net Price £	Material £	Labour hours	Labour £	Unit	Total rate £
600/1000 Volt grade; Twelve Core (Galvanised steel wire armour)						
1.5 mm2	1.01	1.21	1.14	21.13	nr	**22.34**
2.5 mm2	1.52	1.82	1.13	20.94	nr	**22.76**
600/1000 Volt grade; Nineteen Core (Galvanised steel wire armour)						
1.5 mm2	1.52	1.82	1.54	28.54	nr	**30.36**
2.5 mm2	1.52	1.82	1.54	28.54	nr	**30.36**
600/1000 Volt grade; Twenty Seven Core (Galvanised steel wire armour)						
1.5 mm2	1.52	1.82	1.94	35.96	nr	**37.78**
2.5 mm2	2.07	2.48	2.31	42.81	nr	**45.29**
600/1000 Volt grade; Thirty Seven Core (Galvanised steel wire armour)						
1.5 mm2	2.07	2.48	2.53	46.89	nr	**49.37**
2.5 mm2	2.07	2.48	2.87	53.19	nr	**55.67**
UN-ARMOURED CABLE						
Cable: XLPE Insulated; PVC sheathed 90c copper to BS 6181e; for internal wiring; clipped to backgrounds; (Supports and fixings included.)						
300/500 Volt Grade; Single Core						
6.0 mm2	0.23	0.28	0.09	1.67	m	**1.94**
10 mm2	0.33	0.40	0.10	1.85	m	**2.25**
16 mm2	0.45	0.54	0.12	2.22	m	**2.76**
Cable; LSF insulated 6491B; non-sheathed copper to; laid/drawn in trunking/conduit						
450/700 Volt Grade; single core						
1.5 mm2	0.06	0.07	0.03	0.56	m	**0.63**
2.5 mm2	0.09	0.11	0.03	0.56	m	**0.66**
4.0 mm2	0.14	0.17	0.03	0.56	m	**0.72**
6.0 mm2	0.20	0.24	0.04	0.74	m	**0.98**
10.0 mm2	0.35	0.42	0.04	0.74	m	**1.16**
16.0 mm2	0.49	0.59	0.05	0.93	m	**1.51**
25.0 mm2	0.80	0.96	0.06	1.11	m	**2.07**
35.0 mm2	1.04	1.25	0.06	1.11	m	**2.36**
50.0 mm2	1.37	1.64	0.07	1.30	m	**2.94**
70.0 mm2	1.85	2.22	0.08	1.48	m	**3.70**
95.0 mm2	2.32	2.78	0.08	1.48	m	**4.26**
120.0 mm2	3.23	3.87	0.10	1.85	m	**5.72**
150.0 mm2	4.20	5.03	0.13	2.41	m	**7.44**

Material Costs/Measured Work Prices – Electrical Installations

V:ELECTRICAL SUPPLY/POWER/LIGHTING

Item	Net Price £	Material £	Labour hours	Labour £	Unit	Total rate £
V20 : LV DISTRIBUTION - CABLES/BUSBAR						
Y61 - LV CABLES AND WIRING						
EARTH CABLE						
Cable; LSF insulated 6491B; non-sheathed copper; laid/drawn in trunking/conduit						
450/700 Volt Grade; single core						
1.5 mm2	0.06	0.07	0.03	0.56	m	**0.63**
2.5 mm2	0.09	0.11	0.03	0.56	m	**0.66**
4.0 mm2	0.14	0.17	0.03	0.56	m	**0.72**
6.0 mm2	0.20	0.24	0.04	0.74	m	**0.98**
10.0 mm2	0.35	0.42	0.04	0.74	m	**1.16**
16.0 mm2	0.49	0.59	0.05	0.93	m	**1.51**
25.0 mm2	0.80	0.96	0.06	1.11	m	**2.07**
35.0 mm2	1.04	1.25	0.06	1.11	m	**2.36**
50.0 mm2	1.37	1.64	0.07	1.30	m	**2.94**
70.0 mm2	1.86	2.23	0.08	1.48	m	**3.71**
95.0 mm2	2.32	2.78	0.08	1.48	m	**4.26**
120.0 mm2	3.23	3.87	0.10	1.85	m	**5.72**
150.0 mm2	4.20	5.03	0.13	2.41	m	**7.44**
185 mm2	5.04	6.04	0.16	2.97	m	**9.00**
240 mm2	5.82	6.97	0.20	3.71	m	**10.68**
FLEXIBLE CABLE						
Flexible cord; PVC insulated; PVC sheathed; copper stranded to BS 218*Y (laid loose)						
300/500 Volt grade; Two Core						
0.50 mm2	0.08	0.08	0.07	1.30	m	**1.38**
0.75 mm2	0.13	0.13	0.07	1.30	m	**1.43**
300/500 Volt grade; Three Core						
0.50 mm2	0.11	0.11	0.07	1.30	m	**1.41**
0.75 mm2	0.14	0.14	0.07	1.30	m	**1.44**
1.0 mm2	0.17	0.20	0.07	1.30	m	**1.50**
1.5 mm2	0.25	0.30	0.07	1.30	m	**1.60**
2.5 mm2	0.35	0.42	0.08	1.48	m	**1.90**
Flexible cord; PVC insulated; PVC sheathed; copper stranded to BS 318*Y (laid loose)						
300/500 Volt grade; Two Core						
0.75 mm2	0.09	0.11	0.07	1.30	m	**1.41**
1.0 mm2	0.11	0.13	0.07	1.30	m	**1.43**
1.5 mm2	0.16	0.19	0.07	1.30	m	**1.49**
2.5 mm2	0.33	0.40	0.07	1.30	m	**1.69**
300/500 Volt grade; Three Core						
0.75 mm2	0.09	0.11	0.07	1.30	m	**1.41**
1.0 mm2	0.11	0.13	0.07	1.30	m	**1.43**
1.5 mm2	0.16	0.19	0.07	1.30	m	**1.49**
2.5 mm2	0.28	0.34	0.08	1.48	m	**1.82**

Material Costs/Measured Work Prices – Electrical Installations

V:ELECTRICAL SUPPLY/POWER/LIGHTING

Item	Net Price £	Material £	Labour hours	Labour £	Unit	Total rate £
300/500 Volt grade; Four Core						
0.75 mm2	0.20	0.24	0.08	1.48	m	**1.72**
1.0 mm2	0.24	0.29	0.08	1.48	m	**1.77**
1.5 mm2	0.34	0.41	0.08	1.48	m	**1.89**
2.5 mm2	0.52	0.62	0.09	1.67	m	**2.29**
Flexible cord; PVC insulated; PVC sheathed for use in high temperature zones; copper stranded to BS 309*Y (laid loose)						
300/500 Volt grade; Two Core						
0.50 mm2	0.14	0.17	0.07	1.30	m	**1.47**
0.75 mm2	0.17	0.20	0.07	1.30	m	**1.50**
1.0 mm2	0.25	0.30	0.07	1.30	m	**1.60**
1.5 mm2	0.36	0.43	0.07	1.30	m	**1.73**
2.5 mm2	0.56	0.67	0.07	1.30	m	**1.97**
300/500 Volt grade; Three Core						
0.50 mm2	0.19	0.23	0.07	1.30	m	**1.53**
0.75 mm2	0.21	0.25	0.07	1.30	m	**1.55**
1.0 mm2	0.28	0.34	0.07	1.30	m	**1.63**
1.5 mm2	0.39	0.47	0.07	1.30	m	**1.76**
2.5 mm2	0.58	0.69	0.07	1.30	m	**1.99**
Flexible cord; Rubber insulated; Rubber sheathed; copper stranded to BS 318 (laid loose)						
300/500 Volt grade; Two Core						
0.50 mm2	0.17	0.20	0.07	1.30	m	**1.50**
0.75 mm2	0.12	0.14	0.07	1.30	m	**1.44**
1.0 mm2	0.15	0.18	0.07	1.30	m	**1.48**
1.5 mm2	0.21	0.25	0.07	1.30	m	**1.55**
2.5 mm2	0.30	0.36	0.07	1.30	m	**1.66**
300/500 Volt grade; Three Core						
0.50 mm2	0.21	0.25	0.07	1.30	m	**1.55**
0.75 mm2	0.25	0.30	0.07	1.30	m	**1.60**
1.0 mm2	0.28	0.34	0.07	1.30	m	**1.63**
1.5 mm2	0.32	0.38	0.07	1.30	m	**1.68**
2.5 mm2	0.37	0.44	0.07	1.30	m	**1.74**
300/500 Volt grade; Four Core						
0.50 mm2	0.71	0.85	0.08	1.48	m	**2.33**
0.75 mm2	0.22	0.26	0.08	1.48	m	**1.75**
1.0 mm2	0.27	0.32	0.08	1.48	m	**1.81**
1.5 mm2	0.33	0.40	0.08	1.48	m	**1.88**
2.5 mm2	0.50	0.60	0.08	1.48	m	**2.08**

Material Costs/Measured Work Prices – Electrical Installations

V:ELECTRICAL SUPPLY/POWER/LIGHTING

Item	Net Price £	Material £	Labour hours	Labour £	Unit	Total rate £
V20 : LV DISTRIBUTION - CABLES/BUSBAR						
Y61 - LV CABLES AND WIRING						
Flexible cord; Rubber insulated; Rubber sheathed; for 90C operation; copper stranded to BS 318*TQ (laid loose)						
450/750 Volt grade; Two Core						
0.50 mm2	0.26	0.31	0.07	1.30	m	**1.61**
0.75 mm2	0.27	0.32	0.07	1.30	m	**1.62**
1.0 mm2	0.31	0.37	0.07	1.30	m	**1.67**
1.5 mm2	0.34	0.41	0.07	1.30	m	**1.70**
2.5 mm2	0.53	0.64	0.07	1.30	m	**1.93**
450/750 Volt grade; Three Core						
0.50 mm2	0.23	0.28	0.07	1.30	m	**1.57**
0.75 mm2	0.25	0.30	0.07	1.30	m	**1.60**
1.0 mm2	0.30	0.36	0.07	1.30	m	**1.66**
1.5 mm2	0.37	0.44	0.07	1.30	m	**1.74**
2.5 mm2	0.53	0.64	0.07	1.30	m	**1.93**
450/750 Volt grade; Four Core						
0.75 mm2	0.45	0.54	0.08	1.48	m	**2.02**
1.0 mm2	0.52	0.62	0.08	1.48	m	**2.11**
1.5 mm2	0.66	0.79	0.08	1.48	m	**2.27**
2.5 mm2	0.85	1.02	0.08	1.48	m	**2.50**
Heavy Flexible cable; Rubber insulated; Rubber sheathed; copper stranded to BS638*P (laid loose)						
450/750 Volt grade; Two Core						
1.0 mm2	0.25	0.30	0.08	1.48	m	**1.78**
1.5 mm2	0.29	0.35	0.08	1.48	m	**1.83**
2.5 mm2	0.42	0.50	0.08	1.48	m	**1.99**
450/750 Volt grade; Three Core						
1.0 mm2	0.30	0.36	0.08	1.48	m	**1.84**
1.5 mm2	0.35	0.42	0.08	1.48	m	**1.90**
2.5 mm2	0.49	0.59	0.08	1.48	m	**2.07**
450/750 Volt grade; Four Core						
1.0 mm2	0.37	0.44	0.08	1.48	m	**1.93**
1.5 mm2	0.44	0.53	0.08	1.48	m	**2.01**
2.5 mm2	0.60	0.72	0.08	1.48	m	**2.20**

Material Costs/Measured Work Prices – Electrical Installations

499

V:ELECTRICAL SUPPLY/POWER/LIGHTING

Item	Net Price £	Material £	Labour hours	Labour £	Unit	Total rate £
FIRE RATED CABLE						
Cable, Mineral Insulated; copper sheathed with copper conductors; fixed with clips to backgrounds. BASEC approval to BS 6207 Part 1 1995; complies with BS 6387 Category CWZ						
Light duty 500 Volt grade; Bare						
2L 1.0	0.72	0.86	0.23	4.26	m	**5.13**
2L 1.5	0.85	1.02	0.23	4.26	m	**5.28**
2L 2.5	1.10	1.32	0.25	4.63	m	**5.95**
2L 4.0	1.66	1.99	0.25	4.63	m	**6.62**
3L 1.0	0.88	1.05	0.24	4.45	m	**5.50**
3L 1.5	1.11	1.33	0.25	4.63	m	**5.96**
3L 2.5	1.74	2.08	0.25	4.63	m	**6.72**
4L 1.0	1.06	1.27	0.25	4.63	m	**5.90**
4L 1.5	1.31	1.57	0.25	4.63	m	**6.20**
4L 2.5	2.12	2.54	0.26	4.82	m	**7.36**
7L 1.5	1.86	2.23	0.28	5.19	m	**7.42**
7L 2.5	2.44	2.92	0.27	5.00	m	**7.93**
Light duty 500 Volt grade; LSF sheathed						
2L 1.0	0.84	1.01	0.23	4.26	m	**5.27**
2L 1.5	0.92	1.10	0.23	4.26	m	**5.37**
2L 2.5	1.17	1.40	0.25	4.63	m	**6.04**
2L 4.0	1.76	2.11	0.25	4.63	m	**6.74**
3L 1.0	1.01	1.21	0.24	4.45	m	**5.66**
3L 1.5	1.23	1.47	0.25	4.63	m	**6.11**
3L 2.5	1.84	2.20	0.25	4.63	m	**6.84**
4L 1.0	1.19	1.43	0.25	4.63	m	**6.06**
4L 1.5	1.45	1.74	0.25	4.63	m	**6.37**
4L 2.5	2.20	2.64	0.26	4.82	m	**7.46**
7L 1.5	2.12	2.54	0.28	5.19	m	**7.73**
7L 2.5	2.72	3.26	0.27	5.00	m	**8.26**
Cable, Mineral Insulated; copper sheathed with copper conductors; fixed with clips to backgrounds; BASEC approval to BS 6207 Part 1 1995; complies with BS 6387 Category CWZ						
Heavy duty 750 Volt grade; Bare						
1H 10	1.75	2.10	0.25	4.63	m	**6.73**
1H 16	2.40	2.88	0.26	4.82	m	**7.69**
1H 25	3.32	3.98	0.27	5.00	m	**8.98**
1H 35	4.46	5.34	0.32	5.93	m	**11.28**
1H 50	5.66	6.78	0.35	6.49	m	**13.27**
1H 70	7.36	8.82	0.38	7.04	m	**15.86**
1H 95	9.67	11.59	0.41	7.60	m	**19.19**
1H 120	11.82	14.16	0.46	8.53	m	**22.69**
1H 150	14.64	17.54	0.50	9.27	m	**26.81**
1H 185	17.82	21.35	0.56	10.38	m	**31.73**
1H 240	23.14	27.73	0.69	12.79	m	**40.52**
2H 1.5	1.49	1.79	0.25	4.63	m	**6.42**
2H 2.5	1.84	2.20	0.26	4.82	m	**7.02**
2H 4	2.30	2.76	0.26	4.82	m	**7.57**

Material Costs/Measured Work Prices – Electrical Installations

V:ELECTRICAL SUPPLY/POWER/LIGHTING

Item	Net Price £	Material £	Labour hours	Labour £	Unit	Total rate £
V20 : LV DISTRIBUTION - CABLES/BUSBAR						
Y61 - LV CABLES AND WIRING						
Cable, Mineral Insulated; copper sheathed with copper conductors; fixed with clips to backgrounds; BASEC approval to BS 6207 Part 1 1995; complies with BS 6387 Category CWZ (Continued)						
Heavy duty 750 Volt grade; Bare						
2H 6	3.07	3.68	0.29	5.37	m	**9.05**
2H 10	3.98	4.77	0.34	6.30	m	**11.07**
2H 16	5.72	6.85	0.40	7.41	m	**14.27**
2H 25	8.03	9.62	0.44	8.15	m	**17.78**
3H 1.5	1.66	1.99	0.25	4.63	m	**6.62**
3H 2.5	2.08	2.49	0.25	4.63	m	**7.13**
3H 4	2.63	3.15	0.27	5.00	m	**8.16**
3H 6	3.40	4.07	0.30	5.56	m	**9.63**
3H 10	4.91	5.88	0.35	6.49	m	**12.37**
3H 16	6.89	8.26	0.41	7.60	m	**15.86**
3H 25	10.56	12.65	0.47	8.71	m	**21.36**
4H 1.5	2.05	2.46	0.24	4.45	m	**6.90**
4H 2.5	2.58	3.09	0.26	4.82	m	**7.91**
4H 4	3.22	3.86	0.29	5.37	m	**9.23**
4H 6	4.30	5.15	0.31	5.75	m	**10.90**
4H 10	6.10	7.31	0.37	6.86	m	**14.17**
4H 16	8.89	10.65	0.44	8.15	m	**18.81**
4H 25	12.90	15.46	0.52	9.64	m	**25.10**
7H 1.5	2.84	3.40	0.30	5.56	m	**8.96**
7H 2.5	3.85	4.61	0.32	5.93	m	**10.54**
12H 2.5	6.71	8.04	0.39	7.23	m	**15.27**
19H 1.5	10.36	12.41	0.42	7.78	m	**20.20**
Heavy duty 750 Volt grade; LSF Sheathed						
1H 10	1.91	2.29	0.25	4.63	m	**6.92**
1H 16	2.60	3.12	0.26	4.82	m	**7.93**
1H 25	3.60	4.31	0.27	5.00	m	**9.32**
1H 35	4.72	5.66	0.32	5.93	m	**11.59**
1H 50	5.95	7.13	0.35	6.49	m	**13.62**
1H 70	7.74	9.27	0.38	7.04	m	**16.32**
1H 95	10.15	12.16	0.41	7.60	m	**19.76**
1H 120	12.38	14.84	0.46	8.53	m	**23.36**
1H 150	15.25	18.27	0.50	9.27	m	**27.54**
1H 185	18.72	22.43	0.56	10.38	m	**32.81**
1H 240	24.13	28.91	0.68	12.60	m	**41.52**
2H 1.5	1.68	2.01	0.25	4.63	m	**6.65**
2H 2.5	1.99	2.38	0.26	4.82	m	**7.20**
2H 4	2.52	3.02	0.26	4.82	m	**7.84**
2H 6	3.32	3.98	0.29	5.37	m	**9.35**
2H 10	4.31	5.16	0.34	6.30	m	**11.47**
2H 16	6.06	7.26	0.40	7.41	m	**14.68**
2H 25	8.57	10.27	0.44	8.15	m	**18.42**
3H 1.5	1.86	2.23	0.25	4.63	m	**6.86**
3H 2.5	2.28	2.73	0.25	4.63	m	**7.37**
3H 4	2.90	3.48	0.27	5.00	m	**8.48**

Material Costs/Measured Work Prices – Electrical Installations

V:ELECTRICAL SUPPLY/POWER/LIGHTING

Item	Net Price £	Material £	Labour hours	Labour £	Unit	Total rate £
3H 6	3.65	4.37	0.30	5.56	m	**9.93**
3H 10	5.23	6.27	0.35	6.49	m	**12.75**
3H 16	7.39	8.86	0.41	7.60	m	**16.45**
3H 25	11.14	13.35	0.47	8.71	m	**22.06**
4H 1.5	2.24	2.68	0.24	4.45	m	**7.13**
4H 2.5	2.81	3.37	0.26	4.82	m	**8.19**
4H 4	3.48	4.17	0.29	5.37	m	**9.54**
4H 6	4.57	5.48	0.31	5.75	m	**11.22**
4H 10	6.46	7.74	0.37	6.86	m	**14.60**
4H 16	9.41	11.28	0.44	8.15	m	**19.43**
4H 25	13.72	16.44	0.52	9.64	m	**26.08**
7H 1.5	3.10	3.71	0.30	5.56	m	**9.27**
7H 2.5	4.15	4.97	0.32	5.93	m	**10.90**
12H 2.5	7.18	8.60	0.39	7.23	m	**15.83**
19H 1.5	10.91	13.07	0.42	7.78	m	**20.86**
Cable terminations for M.I. Cable; Polymeric one piece moulding; containing grey sealing compound; testing; phase marking and connection						
Light Duty 500 Volt grade; Brass gland; polymeric one moulding containing grey sealing compound; coloured conductor sleeving; Earth tag; plastic gland shroud						
2L 1.5	2.09	2.50	0.27	5.00	m	**7.51**
2L 2.5	2.10	2.52	0.27	5.00	m	**7.52**
3L 1.5	2.41	2.89	0.27	5.00	m	**7.89**
4L 1.5	2.44	2.92	0.27	5.00	m	**7.93**
Cable Terminations; for MI copper sheathed cable. Certified for installation in potentially explosive atmospheres; testing; phase marking and connection; BS 6207 Part 2 1995						
Light duty 500 Volt grade; brass gland; brass pot with earth tail; pot closure; sealing compound; conductor sleving; plastic gland shroud; identification markers						
2L 1.0	2.30	2.76	0.39	7.23	nr	**9.98**
2L 1.5	2.30	2.76	0.41	7.60	nr	**10.36**
2L 2.5	2.42	2.90	0.41	7.60	nr	**10.50**
2L 4.0	2.42	2.90	0.46	8.53	nr	**11.43**
3L 1.0	2.41	2.89	0.43	7.97	nr	**10.86**
3L 1.5	2.41	2.89	0.43	7.97	nr	**10.86**
3L 2.5	2.41	2.89	0.44	8.15	nr	**11.04**
4L 1.0	2.44	2.92	0.47	8.71	nr	**11.63**
4L 1.5	2.44	2.92	0.47	8.71	nr	**11.63**
4L 2.5	2.44	2.92	0.50	9.27	nr	**12.19**
7L 1.0	3.68	4.41	0.69	12.79	nr	**17.20**
7L 1.5	3.68	4.41	0.70	12.97	nr	**17.38**
7L 2.5	3.68	4.41	0.74	13.72	nr	**18.12**

Material Costs/Measured Work Prices – Electrical Installations

V:ELECTRICAL SUPPLY/POWER/LIGHTING

Item	Net Price £	Material £	Labour hours	Labour £	Unit	Total rate £
V20 : LV DISTRIBUTION - CABLES/BUSBAR						
Y61 - LV CABLES AND WIRING						
Cable Terminations; for MI copper sheathed cable. Certified for installation in potentially explosive atmospheres; testing; phase marking and connection; BS 6207 Part 2 1995 (Continued)						
Heavy duty 750 Volt grade; brass gland; brass pot with earth tail; pot closure; sealing compound; conductor sleeving; plastic gland shroud; identification markers						
1H 10	2.86	3.43	0.37	6.86	nr	**10.28**
1H 16	2.86	3.43	0.39	7.23	nr	**10.66**
1H 25	4.66	5.58	0.56	10.38	nr	**15.96**
1H 35	4.66	5.58	0.57	10.56	nr	**16.15**
1H 50	9.06	10.86	0.60	11.12	nr	**21.98**
1H 70	3.80	4.55	0.67	12.42	nr	**16.97**
1H 95	3.80	4.55	0.75	13.90	nr	**18.45**
1H 120	7.72	9.25	0.94	17.42	nr	**26.67**
1H 150	7.72	9.25	0.99	18.35	nr	**27.60**
1H 185	7.72	9.25	1.26	23.35	nr	**32.60**
1H 240	15.71	18.83	1.37	25.39	nr	**44.22**
2H 1.5	2.27	2.72	0.42	7.78	nr	**10.50**
2H 2.5	2.27	2.72	0.42	7.78	nr	**10.50**
2H 4	2.69	3.22	0.47	8.71	nr	**11.93**
2H 6	3.32	3.98	0.54	10.01	nr	**13.99**
2H 10	7.16	8.58	0.58	10.75	nr	**19.33**
2H 16	10.62	12.73	0.69	12.79	nr	**25.51**
2H 25	11.62	13.92	0.77	14.27	nr	**28.20**
3H 1.5	2.27	2.72	0.44	8.15	nr	**10.88**
3H 2.5	2.69	3.22	0.44	8.15	nr	**11.38**
3H 4	3.32	3.98	0.57	10.56	nr	**14.54**
3H 6	3.80	4.55	0.61	11.31	nr	**15.86**
3H 10	7.16	8.58	0.65	12.05	nr	**20.63**
3H 16	10.62	12.73	0.78	14.46	nr	**27.18**
3H 25	14.43	17.29	0.85	15.75	nr	**33.05**
4H 1.5	2.27	2.72	0.52	9.64	nr	**12.36**
4H 2.5	3.32	3.98	0.53	9.82	nr	**13.80**
4H 4	3.80	4.55	0.60	11.12	nr	**15.67**
4H 6	6.37	7.63	0.65	12.05	nr	**19.68**
4H 10	6.37	7.63	0.69	12.79	nr	**20.42**
4H 16	11.62	13.92	0.88	16.31	nr	**30.23**
4H 25	14.43	17.29	0.93	17.24	nr	**34.53**
7H 1.5	5.36	6.42	0.71	13.16	nr	**19.58**
7H 2.5	4.58	5.49	0.74	13.72	nr	**19.20**
12H 1.5	7.37	8.83	0.85	15.75	nr	**24.59**
12H 2.5	8.16	9.78	1.00	18.53	nr	**28.31**
19H 2.5	14.43	17.29	1.11	20.57	nr	**37.86**

Material Costs/Measured Work Prices – Electrical Installations

V:ELECTRICAL SUPPLY/POWER/LIGHTING

Item	Net Price £	Material £	Labour hours	Labour £	Unit	Total rate £
Cable: FP100; LOSH insulated; non sheathed fire resistant to LPCB Approved to BS 6387 Catergory CWZ; in conduit including terminations						
450/750 volt grade; single core						
1.0 mm	0.85	1.02	0.13	2.41	m	**3.43**
1.5 mm	0.90	1.08	0.13	2.41	m	**3.49**
2.5 mm	1.08	1.29	0.13	2.41	m	**3.70**
4.0 mm	1.30	1.56	0.13	2.41	m	**3.97**
6.0 mm	1.38	1.65	0.13	2.41	m	**4.06**
10 mm	2.10	2.52	0.16	2.97	m	**5.48**
16 mm	2.33	2.79	0.16	2.97	m	**5.76**
Cable: FP200; Insudite insulated; LSOH sheathed screened fire resistant BASEC Approved to BS 7629; fixed with clips to backgrounds						
300/500 volt grade; two core						
1.5 mm	0.79	0.95	0.13	2.41	m	**3.36**
2.5 mm	1.03	1.23	0.13	2.41	m	**3.64**
4.0 mm	1.61	1.93	0.13	2.41	m	**4.34**
300/500 volt grade; three core						
1.5 mm	1.06	1.27	0.13	2.41	m	**3.68**
2.5 mm	1.31	1.57	0.13	2.41	m	**3.98**
4.0 mm	2.03	2.43	0.13	2.41	m	**4.84**
300/500 volt grade; four core						
1.5 mm	1.28	1.53	0.13	2.41	m	**3.94**
2.5 mm	1.79	2.15	0.13	2.41	m	**4.55**
4.0 mm	2.81	3.37	0.13	2.41	m	**5.78**
Terminations; including sealing, connection to equipment, phase marking						
Two Core						
1.5 mm	0.24	0.29	0.35	6.49	nr	**6.77**
2.5 mm	0.24	0.29	0.35	6.49	nr	**6.77**
4.0 mm	0.24	0.29	0.35	6.49	nr	**6.77**
Three Core						
1.5 mm	0.24	0.29	0.35	6.49	nr	**6.77**
2.5 mm	0.24	0.29	0.35	6.49	nr	**6.77**
4.0 mm	0.24	0.29	0.35	6.49	nr	**6.77**
Four Core						
1.5 mm	0.24	0.29	0.35	6.49	nr	**6.77**
2.5 mm	0.24	0.29	0.35	6.49	nr	**6.77**
4.0 mm	0.55	0.66	0.35	6.49	nr	**7.15**

Material Costs/Measured Work Prices – Electrical Installations

V:ELECTRICAL SUPPLY/POWER/LIGHTING

Item	Net Price £	Material £	Labour hours	Labour £	Unit	Total rate £
V20 : LV DISTRIBUTION - CABLES/BUSBAR						
Y61 - LV CABLES AND WIRING						
Cable: FP400; Polymeric insulated; LSOH sheathed fire resistant; armoured; with copper stranded copper conductors; BASEC Approved to BS 7846; fixed with clips to backgrounds						
600/1000 volt grade; two core						
1.5 mm	1.56	1.87	0.16	2.97	m	**4.83**
2.5 mm	1.74	2.08	0.16	2.97	m	**5.05**
4.0 mm	1.99	2.38	0.16	2.97	m	**5.35**
6.0 mm	2.29	2.74	0.16	2.97	m	**5.71**
10 mm	2.69	3.22	0.20	3.71	m	**6.93**
16 mm	3.67	4.40	0.20	3.71	m	**8.10**
25 mm	4.46	5.34	0.20	3.71	m	**9.05**
600/1000 volt grade; three core						
1.5 mm	1.73	2.07	0.16	2.97	m	**5.04**
2.5 mm	2.09	2.50	0.16	2.97	m	**5.47**
4.0 mm	2.40	2.88	0.16	2.97	m	**5.84**
6.0 mm	2.76	3.31	0.16	2.97	m	**6.27**
10 mm	3.47	4.16	0.20	3.71	m	**7.86**
16 mm	5.21	6.24	0.20	3.71	m	**9.95**
25 mm	6.24	7.48	0.20	3.71	m	**11.18**
600/1000 volt grade; four core						
1.5 mm	2.03	2.43	0.16	2.97	m	**5.40**
2.5 mm	2.11	2.53	0.16	2.97	m	**5.49**
4.0 mm	2.39	2.86	0.16	2.97	m	**5.83**
6.0 mm	2.98	3.57	0.16	2.97	m	**6.54**
10 mm	4.27	5.12	0.20	3.71	m	**8.82**
16 mm	5.70	6.83	0.20	3.71	m	**10.54**
25 mm	6.96	8.34	0.20	3.71	m	**12.05**
Terminations; including sealing, connection to equipment, phase marking						
Two Core						
1.5 mm	1.61	1.93	0.35	6.49	nr	**8.42**
2.5 mm	1.61	1.93	0.35	6.49	nr	**8.42**
4.0 mm	1.61	1.93	0.35	6.49	nr	**8.42**
6.0 mm	2.02	2.42	0.35	6.49	nr	**8.91**
10 mm	2.02	2.42	0.35	6.49	nr	**8.91**
16 mm	2.85	3.42	0.35	6.49	nr	**9.90**
25 mm	2.85	3.42	0.35	6.49	nr	**9.90**
Three Core						
1.5 mm	1.61	1.93	0.35	6.49	nr	**8.42**
2.5 mm	1.61	1.93	0.35	6.49	nr	**8.42**
4.0 mm	1.61	1.93	0.35	6.49	nr	**8.42**
6.0 mm	2.02	2.42	0.35	6.49	nr	**8.91**
10 mm	2.02	2.42	0.35	6.49	nr	**8.91**
16 mm	2.85	3.42	0.35	6.49	nr	**9.90**
25 mm	2.85	3.42	0.35	6.49	nr	**9.90**

Material Costs/Measured Work Prices – Electrical Installations

V:ELECTRICAL SUPPLY/POWER/LIGHTING

Item	Net Price £	Material £	Labour hours	Labour £	Unit	Total rate £
Four Core						
1.5 mm	1.61	1.93	0.35	6.49	nr	**8.42**
2.5 mm	1.61	1.93	0.35	6.49	nr	**8.42**
4.0 mm	2.02	2.42	0.35	6.49	nr	**8.91**
6.0 mm	2.02	2.42	0.35	6.49	nr	**8.91**
10 mm	2.85	3.42	0.35	6.49	nr	**9.90**
16 mm	2.85	3.42	0.35	6.49	nr	**9.90**
25 mm	4.70	5.63	0.35	6.49	nr	**12.12**
Cable; Firetuff fire resistant to BS 6387; including terminations to equipment; fixed with clips to backgrounds						
Two Core						
1.5 mm	0.17	0.20	0.16	2.97	m	**3.17**
2.5 mm	0.22	0.26	0.16	2.97	m	**3.23**
4.0 mm	0.36	0.43	0.16	2.97	m	**3.40**
Three Core						
1.5 mm	0.17	0.20	0.16	2.97	m	**3.17**
2.5 mm	0.22	0.26	0.16	2.97	m	**3.23**
4.0 mm	0.36	0.43	0.16	2.97	m	**3.40**
Four Core						
1.5 mm	0.22	0.26	0.16	2.97	m	**3.23**
2.5 mm	0.22	0.26	0.16	2.97	m	**3.23**
4.0 mm	0.43	0.52	0.16	2.97	m	**3.48**
Cable; Calflam fire resistant to BS EN50265-2-1; including terminations to equipment; fixed with clips to backgrounds						
Two Core						
1.5 mm	0.17	0.20	0.20	3.71	m	**3.91**
2.5 mm	0.22	0.26	0.20	3.71	m	**3.97**
4.0 mm	0.36	0.43	0.20	3.71	m	**4.14**
Three Core						
1.5 mm	0.17	0.20	0.20	3.71	m	**3.91**
2.5 mm	0.22	0.26	0.20	3.71	m	**3.97**
4.0 mm	0.36	0.43	0.20	3.71	m	**4.14**
Four Core						
1.5 mm	0.22	0.26	0.20	3.71	m	**3.97**
2.5 mm	0.22	0.26	0.20	3.71	m	**3.97**
4.0 mm	0.43	0.52	0.20	3.71	m	**4.22**

Material Costs/Measured Work Prices – Electrical Installations

V:ELECTRICAL SUPPLY/POWER/LIGHTING

Item	Net Price £	Material £	Labour hours	Labour £	Unit	Total rate £
V20 : LV DISTRIBUTION - CABLES/BUSBAR						
Y62 - BUSBAR TRUNKING						
MAINS BUSBAR						
Low impedance busbar trunking; fixed to backgrounds including supports, fixings and connections/jointing to equipment; including fixing of plates, discs etc, for identification						
Straight copper busbar						
1000 Amp TP&N	216.00	258.83	3.41	63.20	m	**322.03**
1350 Amp TP&N	265.50	318.15	3.58	66.35	m	**384.50**
2000 Amp TP&N	333.00	399.03	5.00	92.67	m	**491.70**
2500 Amp TP&N	503.10	602.86	5.90	109.35	m	**712.21**
Extra for fittings Mains Bus Bar						
IP54 Protection						
1000 Amp TP&N	18.00	21.57	2.16	40.03	m	**61.60**
1350 Amp TP&N	19.35	23.19	2.61	48.37	m	**71.56**
2000 Amp TP&N	26.84	32.16	3.51	65.05	m	**97.21**
2500 Amp TP&N	31.87	38.19	3.96	73.39	m	**111.58**
End Cover						
1000 Amp TP&N	20.70	24.80	0.56	10.38	nr	**35.18**
1350 Amp TP&N	21.60	25.88	0.56	10.38	nr	**36.26**
2000 Amp TP&N	34.65	41.52	0.66	12.23	nr	**53.75**
2500 Amp TP&N	35.10	42.06	0.66	12.23	nr	**54.29**
Edge Elbow						
1000 Amp TP&N	258.30	309.52	2.01	37.25	nr	**346.77**
1350 Amp TP&N	290.70	348.35	2.01	37.25	nr	**385.60**
2000 Amp TP&N	450.90	540.31	2.40	44.48	nr	**584.79**
2500 Amp TP&N	592.20	709.63	2.40	44.48	nr	**754.11**
Flat Elbow						
1000 Amp TP&N	224.10	268.54	2.01	37.25	nr	**305.79**
1350 Amp TP&N	243.00	291.19	2.01	37.25	nr	**328.44**
2000 Amp TP&N	344.70	413.05	2.40	44.48	nr	**457.54**
2500 Amp TP&N	429.30	514.43	2.40	44.48	nr	**558.91**
Offset						
1000 Amp TP&N	450.00	539.24	3.00	55.60	nr	**594.84**
1350 Amp TP&N	553.50	663.26	3.00	55.60	nr	**718.86**
2000 Amp TP&N	872.10	1045.04	3.50	64.87	nr	**1109.91**
2500 Amp TP&N	992.70	1189.55	3.50	64.87	nr	**1254.42**
Edge Z Unit						
1000 Amp TP&N	674.10	807.77	3.00	55.60	nr	**863.38**
1350 Amp TP&N	863.10	1034.25	3.00	55.60	nr	**1089.85**
2000 Amp TP&N	1293.30	1549.76	3.50	64.87	nr	**1614.63**
2500 Amp TP&N	1476.90	1769.77	3.50	64.87	nr	**1834.64**

Material Costs/Measured Work Prices – Electrical Installations

V:ELECTRICAL SUPPLY/POWER/LIGHTING

Item	Net Price £	Material £	Labour hours	Labour £	Unit	Total rate £
Flat Z Unit						
1000 Amp TP&N	578.70	693.46	3.00	55.60	nr	**749.06**
1350 Amp TP&N	729.00	873.56	3.00	55.60	nr	**929.16**
2000 Amp TP&N	1085.40	1300.63	3.50	64.87	nr	**1365.50**
2500 Amp TP&N	1305.00	1563.78	3.50	64.87	nr	**1628.65**
Edge Tee						
1000 Amp TP&N	674.10	807.77	2.20	40.77	nr	**848.55**
1350 Amp TP&N	863.10	1034.25	2.20	40.77	nr	**1075.03**
2000 Amp TP&N	1293.30	1549.76	2.60	48.19	nr	**1597.95**
2500 Amp TP&N	1477.80	1770.85	2.60	48.19	nr	**1819.04**
Tap Off; TP&N Integral Contactor/Breaker						
18 Amp	130.53	156.41	0.82	15.20	nr	**171.61**
Tap Off; TP&N Fusable with on-load switch; excludes fuses						
32 Amp	369.49	442.76	0.82	15.20	nr	**457.95**
63 Amp	376.89	451.63	0.88	16.31	nr	**467.94**
100 Amp	461.17	552.62	1.18	21.87	nr	**574.49**
160 Amp	524.54	628.55	1.41	26.13	nr	**654.69**
250 Amp	677.15	811.43	1.76	32.62	nr	**844.05**
315 Amp	799.40	957.92	2.06	38.18	nr	**996.10**
Tap Off; TP&N MCCB						
63 Amp	466.89	559.48	0.88	16.31	nr	**575.79**
125 Amp	559.58	670.55	1.18	21.87	nr	**692.42**
160 Amp	611.00	732.16	1.41	26.13	nr	**758.30**
250 Amp	785.82	941.64	1.76	32.62	nr	**974.26**
400 Amp	1001.74	1200.38	2.06	38.18	nr	**1238.56**
RISING MAINS BUSBAR						
Rising mains busbar; insulated supports, earth continuity bar; including couplers; fixed to backgrounds						
Straight Aluminium bar						
200 Amp TP&N	90.90	108.93	2.13	39.48	m	**148.40**
315 Amp TP&N	101.70	121.87	2.15	39.85	m	**161.72**
400 Amp TP&N	117.90	141.28	2.15	39.85	m	**181.13**
630 Amp TP&N	145.80	174.71	2.47	45.78	m	**220.49**
800 Amp TP&N	218.70	262.07	2.88	53.38	m	**315.45**
Extra for fittings Rising Busbar						
End feed unit						
200 Amp TP&N	189.00	226.48	2.57	47.63	nr	**274.11**
315 Amp TP&N	189.00	226.48	2.76	51.15	nr	**277.63**
400 Amp TP&N	211.50	253.44	2.76	51.15	nr	**304.59**
630 Amp TP&N	211.50	253.44	3.64	67.46	nr	**320.90**
800 Amp TP&N	238.50	285.79	4.54	84.14	nr	**369.94**

Material Costs/Measured Work Prices – Electrical Installations

V:ELECTRICAL SUPPLY/POWER/LIGHTING

Item	Net Price £	Material £	Labour hours	Labour £	Unit	Total rate £
V20 : LV DISTRIBUTION - CABLES/BUSBAR						
Y62 - BUSBAR TRUNKING						
Extra for fittings Rising Busbar (Continued)						
Top feeder unit						
200 Amp TP&N	189.00	226.48	2.57	47.63	nr	**274.11**
315 Amp TP&N	189.00	226.48	2.76	51.15	nr	**277.63**
400 Amp TP&N	211.50	253.44	2.76	51.15	nr	**304.59**
630 Amp TP&N	211.50	253.44	3.64	67.46	nr	**320.90**
800 Amp TP&N	238.50	285.79	4.54	84.14	nr	**369.94**
End Cap						
200 Amp TP&N	16.20	19.41	0.18	3.34	nr	**22.75**
315 Amp TP&N	16.20	19.41	0.27	5.00	nr	**24.42**
400 Amp TP&N	18.00	21.57	0.27	5.00	nr	**26.57**
630 Amp TP&N	18.00	21.57	0.41	7.60	nr	**29.17**
800 Amp TP&N	52.20	62.55	0.41	7.60	nr	**70.15**
Edge Elbow						
200 Amp TP&N	22.50	26.96	0.55	10.19	nr	**37.16**
315 Amp TP&N	22.50	26.96	0.94	17.42	nr	**44.38**
400 Amp TP&N	169.20	202.75	0.94	17.42	nr	**220.17**
630 Amp TP&N	169.20	202.75	1.45	26.87	nr	**229.63**
800 Amp TP&N	160.20	191.97	1.45	26.87	nr	**218.84**
Flat Elbow						
200 Amp TP&N	72.90	87.36	0.55	10.19	nr	**97.55**
315 Amp TP&N	72.90	87.36	0.94	17.42	nr	**104.78**
400 Amp TP&N	99.00	118.63	0.94	17.42	nr	**136.05**
630 Amp TP&N	99.00	118.63	1.45	26.87	nr	**145.51**
800 Amp TP&N	136.80	163.93	1.45	26.87	nr	**190.80**
Edge Tee						
200 Amp TP&N	102.60	122.95	0.61	11.31	nr	**134.25**
315 Amp TP&N	102.60	122.95	1.02	18.90	nr	**141.85**
400 Amp TP&N	144.00	172.56	1.02	18.90	nr	**191.46**
630 Amp TP&N	144.00	172.56	1.57	29.10	nr	**201.65**
800 Amp TP&N	205.20	245.89	1.57	29.10	nr	**274.99**
Flat Tee						
200 Amp TP&N	131.40	157.46	0.61	11.31	nr	**168.76**
315 Amp TP&N	102.60	122.95	1.02	18.90	nr	**141.85**
400 Amp TP&N	207.00	248.05	1.02	18.90	nr	**266.95**
630 Amp TP&N	207.00	248.05	1.57	29.10	nr	**277.15**
800 Amp TP&N	288.90	346.19	1.57	29.10	nr	**375.29**
Tap Off Units						
TP&N Fusable with on-load switch; excludes fuses						
32 Amp	103.50	124.02	0.82	15.20	nr	**139.22**
63 Amp	136.80	163.93	0.88	16.31	nr	**180.24**
100 Amp	183.60	220.01	1.18	21.87	nr	**241.88**
250 Amp	275.40	330.01	1.41	26.13	nr	**356.14**

Material Costs/Measured Work Prices – Electrical Installations

V:ELECTRICAL SUPPLY/POWER/LIGHTING

Item	Net Price £	Material £	Labour hours	Labour £	Unit	Total rate £
400 Amp	401.40	481.00	2.06	38.18	nr	**519.18**
TP&N MCCB						
32 Amp	105.30	126.18	0.82	15.20	nr	**141.38**
63 Amp	145.80	174.71	0.88	16.31	nr	**191.02**
100 Amp	-	-	1.18	21.87	nr	**21.87**
250 Amp	-	474.53	1.41	26.13	nr	**500.66**
400 Amp	694.80	832.58	2.06	38.18	nr	**870.76**
LIGHTING BUSBAR						
Pre-Wired Busbar, Plug-In Trunking for Lighting; Galvanised Sheet Steel Housing (PE); Tin-Plated Copper Conducters with Tap-Off units at 1m intervals.						
Straight Lengths - 25 Amp						
2 Pole & PE	9.76	11.70	0.16	2.97	m	**14.66**
4 Pole & PE	10.77	12.91	0.16	2.97	m	**15.88**
Straight Lengths - 40 Amp						
2 Pole & PE	10.77	12.91	0.16	2.97	m	**15.88**
4 Pole & PE	13.13	15.73	0.16	2.97	m	**18.70**
Components for Pre-Wired Busbars, Plug-In Trunking for Lighting.						
Plug-In Tap-Off Units						
10 Amp with phase selection, 2P & PE; 2m of cable	7.70	9.23	0.10	1.85	nr	**11.08**
10 Amp 4 Pole & PE; 3m of cable	10.81	12.95	0.10	1.85	nr	**14.81**
16 Amp 4 Pole & PE; 3m of cable	10.43	12.50	0.10	1.85	nr	**14.35**
16 Amp with phase selection, 2P & PE; no cable	9.06	10.86	0.10	1.85	nr	**12.71**
Trunking Components						
End Feed Unit & Cover; 4P & PE	18.91	22.66	0.23	4.26	nr	**26.92**
Centre Feed Unit	104.47	125.19	0.29	5.37	nr	**130.56**
Right hand, Intermediate Terminal Box Feed unit	19.74	23.65	0.23	4.26	nr	**27.92**
End Cover (for R/hand feed)	3.58	4.29	0.06	1.11	nr	**5.40**
Flexible Elbow Unit	49.50	59.32	0.12	2.22	nr	**61.54**
Fixing Bracket - Universal	1.16	1.39	0.10	1.85	nr	**3.24**
Suspension Bracket - Flat	0.82	0.98	0.10	1.85	nr	**2.84**
UNDERFLOOR BUSBAR						
Pre-Wired Busbar, Plug-In Trunking for Underfloor Power Distribution; Galvanised Sheet Steel Housing (PE); Copper Conductors with Tap-Off units at 300mm intervals.						
Straight Lengths - 63 Amp						
2 Pole & PE	11.45	13.72	0.28	5.19	m	**18.91**
3 Pole & PE; Clean Earth System	14.48	17.35	0.28	5.19	m	**22.54**

510   *Material Costs/Measured Work Prices – Electrical Installations*

V:ELECTRICAL SUPPLY/POWER/LIGHTING

Item	Net Price £	Material £	Labour hours	Labour £	Unit	Total rate £
V20 : LV DISTRIBUTION - CABLES/BUSBAR						
Y62 - BUSBAR TRUNKING						
Components for Pre-Wired Busbars, Plug-In Trunking for Underfloor Power Distribution.						
Plug-In Tap-Off Units						
32 Amp 2P & PE; 3m metal flexible pre-wired conduit	19.71	23.62	0.25	4.63	nr	**28.25**
32 Amp 3P & PE; clean earth; 3m metal flexible pre-wired conduit	23.72	28.42	0.28	5.19	nr	**33.61**
Trunking Components						
End Feed Unit & Cover; 2P & PE	21.23	25.44	0.35	6.49	nr	**31.93**
End Feed Unit & Cover; 3P & PE; clean earth	23.25	27.86	0.38	7.04	nr	**34.90**
End Cover; 2P & PE	6.61	7.92	0.11	2.04	nr	**9.96**
End Cover; 3P & PE	7.11	8.52	0.11	2.04	nr	**10.56**
Flexible interlink/corner; 2P&PE; 1m long	36.58	43.83	0.34	6.30	nr	**50.14**
Flexible interlink/corner; 3P&PE; 1m long	41.31	49.50	0.35	6.49	nr	**55.99**
Flexible interlink/corner; 2P&PE; 2m long	44.64	53.49	0.37	6.86	nr	**60.35**
Flexible interlink/corner; 3P&PE; 2m long	48.98	58.69	0.37	6.86	nr	**65.55**

Material Costs/Measured Work Prices – Electrical Installations

V:ELECTRICAL SUPPLY/POWER/LIGHTING

Item	Net Price £	Material £	Labour hours	Labour £	Unit	Total rate £
V20 : LV DISTRIBUTION - MODULAR WIRING						
MODULAR WIRING						
Modular wiring systems; including commissioning						
Master Distribution box; steel; fixed to backgrounds; 6 Port						
4.0mm 18 core armoured home run cable	85.34	102.26	0.90	16.68	nr	**118.94**
4.0mm 24 core armoured cable home run cable	85.34	102.26	0.95	17.61	nr	**119.87**
4.0mm 18 core armoured home run cable & data cable	90.66	108.64	0.95	17.61	nr	**126.25**
6.0mm 18 core armoured home run cable	85.34	102.26	1.00	18.53	nr	**120.79**
6.0mm 24 core armoured home run cable	85.34	102.26	1.10	20.39	nr	**122.65**
6.0mm 18 core armoured home run cable & date cable	90.66	108.64	1.10	20.39	nr	**129.03**
Master Distribution box; steel; fixed to backgrounds; 9 Port						
4.0mm 27 core armoured home run cable	106.66	127.82	1.30	24.09	nr	**151.91**
4.0mm 27 core armoured home run cable & data cable	106.66	127.82	1.45	26.87	nr	**154.69**
6.0mm 27 core armoured home run cable	112.00	134.21	1.45	26.87	nr	**161.08**
6.0mm 27 core armoured home run cable & data cable	112.00	134.21	1.55	28.73	nr	**162.94**
Metal Clad cable; BSEN 60439 Part 2 1993; BASEC approved						
4.0mm 18 core	7.46	8.93	0.30	5.56	m	**14.49**
4.0mm 24 core	11.46	13.74	0.32	5.93	m	**19.67**
4.0mm 27 core	11.46	13.74	0.32	5.93	m	**19.67**
6.0mm 18 core	9.94	11.91	0.32	5.93	m	**17.84**
6.0mm 27 core	15.43	18.49	0.35	6.49	m	**24.98**
Metal Clad data cable						
Single twisted pair	1.45	1.74	0.18	3.34	m	**5.07**
Twin twisted pair	2.34	2.80	0.18	3.34	m	**6.14**
Distribution cables; armoured; BSEN 60439 Part 2 1993; BASEC approved						
3 wire; 6.1 metre long	22.55	27.02	0.92	17.05	nr	**44.07**
4 wire; 6.1 metre long	27.09	32.46	0.96	17.79	nr	**50.25**
Extender cables; armoured; BSEN 60439 Part 2 1993; BASEC approved 3 Wire						
0.9 metre long	9.86	11.82	0.13	2.41	nr	**14.22**
1.5 metre long	11.56	13.85	0.23	4.26	nr	**18.12**
2.1 metre long	13.26	15.89	0.31	5.75	nr	**21.63**
2.7 metre long	14.95	17.91	0.40	7.41	nr	**25.33**

Material Costs/Measured Work Prices – Electrical Installations

V:ELECTRICAL SUPPLY/POWER/LIGHTING

Item	Net Price £	Material £	Labour hours	Labour £	Unit	Total rate £
V20 : LV DISTRIBUTION - MODULAR WIRING						
MODULAR WIRING						
Modular wiring systems; including Commissioning (Continued)						
Extender cables; armoured; BSEN 60439 Part 2 1993; BASEC approved						
3 Wire						
3.4 metre long	16.65	19.95	0.51	9.45	nr	**29.40**
4.6 metre long	20.05	24.03	0.69	12.79	nr	**36.81**
6.1 metre long	24.29	29.11	0.92	17.05	nr	**46.16**
7.6 metre long	28.54	34.20	1.14	21.13	nr	**55.33**
9.1 metre long	32.79	39.29	1.37	25.39	nr	**64.68**
10.7 metre long	44.70	53.56	1.61	29.84	nr	**83.40**
4 Wire						
0.9 metre long	10.73	12.86	0.14	2.59	nr	**15.45**
1.5 metre long	12.85	15.40	0.24	4.45	nr	**19.85**
2.1 metre long	14.98	17.95	0.32	5.93	nr	**23.88**
2.7 metre long	17.10	20.49	0.43	7.97	nr	**28.46**
3.4 metre long	19.22	23.03	0.51	9.45	nr	**32.48**
4.6 metre long	23.47	28.12	0.67	12.42	nr	**40.54**
6.1 metre long	28.78	34.49	0.92	17.05	nr	**51.54**
7.6 metre long	34.08	40.84	1.22	22.61	nr	**63.45**
9.1 metre long	39.39	47.20	1.46	27.06	nr	**74.26**
10.7 metre long	44.70	53.56	1.71	31.69	nr	**85.26**
3 Wire; including twisted pair						
0.9 metre long	11.74	14.07	0.13	2.41	nr	**16.48**
1.5 metre long	14.43	17.29	0.23	4.26	nr	**21.55**
2.1 metre long	17.12	20.51	0.31	5.75	nr	**26.26**
2.7 metre long	19.81	23.74	0.40	7.41	nr	**31.15**
3.4 metre long	22.50	26.96	0.51	9.45	nr	**36.41**
4.6 metre long	27.88	33.41	0.69	12.79	nr	**46.20**
6.1 metre long	34.61	41.47	0.92	17.05	nr	**58.52**
7.6 metre long	41.33	49.53	1.14	21.13	nr	**70.65**
9.1 metre long	48.06	57.59	1.37	25.39	nr	**82.98**
10.7 metre long	54.79	65.65	1.61	29.84	nr	**95.49**
Extender whip ended cables; armoured; BSEN 60439 Part 2 1993; BASEC approved						
3 wire; 3.0 metre long	13.69	16.40	0.30	5.56	nr	**21.96**
4 wire; 3.0 metre long	16.01	19.18	0.30	5.56	nr	**24.74**
T Connectors						
3 wire						
Snap fix	8.24	9.87	0.10	1.85	nr	**11.73**
0.3 metre flexible cable	9.34	11.19	0.10	1.85	nr	**13.05**
0.3 metre armoured cable	9.33	11.18	0.15	2.78	nr	**13.96**
0.3 metre armoured cable with twisted pair	10.40	12.46	0.15	2.78	nr	**15.24**

Material Costs/Measured Work Prices – Electrical Installations

V:ELECTRICAL SUPPLY/POWER/LIGHTING

Item	Net Price £	Material £	Labour hours	Labour £	Unit	Total rate £
4 Wire						
Snap fix	8.60	10.31	0.10	1.85	nr	**12.16**
0.3 metre flexible cable	9.67	11.59	0.10	1.85	nr	**13.44**
0.3 metre armoured cable	9.67	11.59	0.18	3.34	nr	**14.92**
Splitters						
5 wire	13.19	15.81	0.20	3.71	nr	**19.51**
5 wire converter	10.22	12.25	0.20	3.71	nr	**15.95**
Switch Modules						
3 wire; 6.1 metre long armoured cable	27.51	32.97	0.75	13.90	nr	**46.87**
4 wire; 6.1 metre long armoured cable	31.48	37.72	0.80	14.83	nr	**52.55**
Distribution cables; unarmoured; IEC 998 DIN/VDE 0628						
3 wire; 6.1 metre long	9.62	11.53	0.70	12.97	nr	**24.50**
4 wire; 6.1 metre long	11.51	13.79	0.75	13.90	nr	**27.69**
Extender cables; unarmoured; IEC 998 DIN/VDE 0628						
3 Wire						
0.9 metre long	6.68	8.00	0.07	1.30	nr	**9.30**
1.5 metre long	7.22	8.65	0.12	2.22	nr	**10.88**
2.1 metre long	7.76	9.30	0.17	3.15	nr	**12.45**
2.7 metre long	9.68	11.60	0.22	4.08	nr	**15.68**
3.4 metre long	8.84	10.59	0.27	5.00	nr	**15.60**
4.6 metre long	10.01	11.99	0.37	6.86	nr	**18.85**
6.1 metre long	11.36	13.61	0.49	9.08	nr	**22.69**
7.6 metre long	12.71	15.23	0.61	11.31	nr	**26.54**
9.1 metre long	14.06	16.85	0.73	13.53	nr	**30.38**
10.7 metre long	15.50	18.57	0.86	15.94	nr	**34.51**
4 Wire						
0.9 metre long	7.64	9.15	0.08	1.48	nr	**10.64**
1.5 metre long	8.32	9.97	0.14	2.59	nr	**12.56**
2.1 metre long	9.00	10.78	0.19	3.52	nr	**14.31**
2.7 metre long	9.68	11.60	0.24	4.45	nr	**16.05**
3.4 metre long	10.36	12.41	0.31	5.75	nr	**18.16**
4.6 metre long	11.83	14.18	0.41	7.60	nr	**21.77**
6.1 metre long	13.53	16.21	0.55	10.19	nr	**26.41**
7.6 metre long	15.23	18.25	0.68	12.60	nr	**30.85**
9.1 metre long	19.93	23.88	0.82	15.20	nr	**39.08**
10.7 metre long	18.75	22.47	0.96	17.79	nr	**40.26**
5 Wire						
0.9 metre long	9.01	10.80	0.09	1.67	nr	**12.46**
1.5 metre long	10.11	12.11	0.15	2.78	nr	**14.89**
2.1 metre long	11.20	13.42	0.21	3.89	nr	**17.31**
2.7 metre long	12.29	14.73	0.27	5.00	nr	**19.73**
3.4 metre long	13.39	16.05	0.34	6.30	nr	**22.35**
4.6 metre long	15.75	18.87	0.46	8.53	nr	**27.40**
6.1 metre long	18.49	22.16	0.61	11.31	nr	**33.46**
7.6 metre long	21.22	25.43	0.76	14.09	nr	**39.51**
9.1 metre long	23.95	28.70	0.91	16.87	nr	**45.57**
10.7 metre long	26.87	32.20	1.07	19.83	nr	**52.03**

514   *Material Costs/Measured Work Prices – Electrical Installations*

V:ELECTRICAL SUPPLY/POWER/LIGHTING

Item	Net Price £	Material £	Labour hours	Labour £	Unit	Total rate £
V20 : LV DISTRIBUTION - MODULAR WIRING						
MODULAR WIRING						
Modular wiring systems; including Commissioning (Continued)						
Extender whip ended cables; armoured; IEC 998 DIN/VDE 0628						
3 wire; 2.5mm; 3.0 metre long	6.81	8.16	0.30	5.56	nr	**13.72**
4 wire; 2.5mm; 3.0 metre long	8.00	9.59	0.30	5.56	nr	**15.15**
T Connectors						
3 wire						
5 pin; direct fix	6.89	8.26	0.10	1.85	nr	**10.11**
5 pin; 1.5mm flexible cable; 0.3 metre long	6.81	8.16	0.15	2.78	nr	**10.94**
4 Wire						
5 pin; direct fix	8.01	9.60	0.20	3.71	nr	**13.31**
5 pin; 1.5mm flexible cable; 0.3 metre long	7.78	9.32	0.20	3.71	nr	**13.03**
5 Wire						
5 pin; direct fix	9.13	10.94	0.20	3.71	nr	**14.65**
Splitters						
3 way; 5 pin	5.26	6.30	0.25	4.63	nr	**10.94**
Switch Modules						
3 wire	14.98	17.95	0.20	3.71	nr	**21.66**
4 wire	15.62	18.72	0.22	4.08	nr	**22.79**

Material Costs/Measured Work Prices – Electrical Installations 515

V:ELECTRICAL SUPPLY/POWER/LIGHTING

Item	Net Price £	Material £	Labour hours	Labour £	Unit	Total rate £
V21 : GENERAL LIGHTING						
Y73 - LUMINAIRES						
LUMINAIRES						
Fluorescent Luminaires; surface fixed to backgrounds.						
Batten type; surface mounted						
600 mm Single - 18 W	6.00	7.18	0.58	10.75	nr	**17.93**
600 mm Twin - 18 W	10.62	12.72	0.59	10.94	nr	**23.66**
1200 mm Single - 36 W	7.93	9.50	0.76	14.09	nr	**23.59**
1200 mm Twin - 36 W	15.19	18.21	0.77	14.27	nr	**32.48**
1500 mm Single - 58 W	8.97	10.75	0.84	15.57	nr	**26.31**
1500 mm Twin - 58 W	18.05	21.63	0.85	15.75	nr	**37.38**
1800 mm Single - 70 W	10.83	12.97	1.05	19.46	nr	**32.43**
1800 mm Twin - 70 W	19.79	23.72	1.06	19.65	nr	**43.36**
2400 mm Single - 100 W	14.82	17.75	1.25	23.17	nr	**40.92**
2400 mm Twin - 100 W	25.92	31.07	1.27	23.54	nr	**54.60**
Surface mounted, opal diffuser						
600 mm Twin - 18 W	14.93	17.89	0.62	11.49	nr	**29.38**
1200 mm Single - 36 W	12.41	14.87	0.79	14.64	nr	**29.51**
1200 mm Twin - 36 W	24.38	29.22	0.80	14.83	nr	**44.04**
1500 mm Single - 58 W	14.33	17.17	0.88	16.31	nr	**33.48**
1500 mm Twin - 58 W	29.48	35.33	0.90	16.68	nr	**52.01**
1800 mm Single - 70 W	16.63	19.93	1.09	20.20	nr	**40.13**
1800 mm Twin - 70 W	22.56	27.04	1.10	20.39	nr	**47.43**
2400 mm Single - 100 W	23.82	28.55	1.30	24.09	nr	**52.64**
2400 mm Twin - 100 W	44.35	53.15	1.31	24.28	nr	**77.43**
Surface mounted; high frequency control gear; low brightness cross blade reflector						
CIBSE LG3 VDT Cat 2						
600mm 4 x 18 watt	106.69	127.85	0.87	16.12	nr	**143.97**
Extra for						
Emergency pack	67.08	80.39	0.25	4.63	nr	**85.02**
1200mm 1 x 36 watt	55.01	65.92	1.09	20.20	nr	**86.12**
1200mm 2 x 36 watt	59.52	71.33	1.09	20.20	nr	**91.53**
Extra for						
Emergency pack	63.00	75.49	0.25	4.63	nr	**80.13**
1500mm 1 x 58 watt	62.41	74.79	0.90	16.68	nr	**91.47**
1500mm 2 x 58 watt	68.13	81.65	0.90	16.68	nr	**98.33**
Extra for						
Emergency pack	63.00	75.49	0.25	4.63	nr	**80.13**
1800mm 1 x 70 watt	66.75	79.98	0.90	16.68	nr	**96.67**
1800mm 2 x 70 watt	78.83	94.47	0.90	16.68	nr	**111.15**
Extra for						
Emergency pack	65.10	78.01	0.25	4.63	nr	**82.64**

516 **Material Costs/Measured Work Prices – Electrical Installations**

V:ELECTRICAL SUPPLY/POWER/LIGHTING

Item	Net Price £	Material £	Labour hours	Labour £	Unit	Total rate £
V21 : GENERAL LIGHTING						
Y73 - LUMINAIRES						
Modular recessed; high frequency control gear; including louvre; fitted to exposed T grid ceiling						
CIBSE LG3 VDT Cat 2						
600 x 600 mm 3 x 18 watt	57.99	69.49	0.84	15.57	nr	**85.06**
600 x 600 mm 4 x 18 watt	52.27	62.63	0.87	16.12	nr	**78.76**
Extra for						
Emergency pack	42.62	51.07	0.25	4.63	nr	**55.70**
600 x 600 mm 2 x 36 watt	67.59	80.99	0.82	15.20	nr	**96.19**
600 x 600 mm 3 x 36 watt	103.69	124.25	0.84	15.57	nr	**139.82**
600 x 600 mm 4 x 36 watt	114.82	137.59	0.87	16.12	nr	**153.71**
Extra for						
Emergency pack	42.62	51.07	0.25	4.63	nr	**55.70**
300 x 1200 mm 2 x 36 watt	52.08	62.41	0.87	16.12	nr	**78.53**
Extra for						
Emergency pack	35.61	42.67	0.25	4.63	nr	**47.30**
600 x 1200 mm 3 x 36 watt	68.31	81.86	0.89	16.50	nr	**98.35**
600 x 1200 mm 4 x 36 watt	75.23	90.15	0.91	16.87	nr	**107.02**
Extra for						
Emergency pack	33.73	40.41	0.25	4.63	nr	**45.05**
CIBSE LG3 VDT Cat 3						
600 x 600 mm 3 x 18 watt	37.52	44.96	0.84	15.57	nr	**60.52**
600 x 600 mm 4 x 18 watt	50.77	60.83	0.87	16.12	nr	**76.96**
Extra for						
Emergency pack	42.62	51.07	0.25	4.63	nr	**55.70**
Modular recessed; low energy; high power factor control gear; fitted into exposed T grid ceiling						
300 x 300 mm 2D 4 pin 38 watt; opal diffuser	52.28	62.65	0.75	13.90	nr	**76.55**
300 x 300 mm 2D 4 pin 38 watt; nibe cell louvre	78.32	93.85	0.75	13.90	nr	**107.75**
Extra for						
Emergency pack	68.25	81.78	0.25	4.63	nr	**86.42**
Downlighter recessed; low voltage; mirror reflector with white/chrome bezel; dimmable transformer; for dichroic lamps						
85mm dia x 20/50 watt	56.95	68.25	0.66	12.23	nr	**80.48**
118mm dia x 50 watt	61.82	74.08	0.66	12.23	nr	**86.32**
165mm dia x 100 watt	100.45	120.37	0.66	12.23	nr	**132.61**

Material Costs/Measured Work Prices – Electrical Installations   517

V:ELECTRICAL SUPPLY/POWER/LIGHTING

Item	Net Price £	Material £	Labour hours	Labour £	Unit	Total rate £
High/Low Bay luminaires						
Compact discharge; aluminium reflector						
150 watt	55.32	66.30	1.50	27.80	nr	**94.10**
250 watt	55.58	66.60	1.50	27.80	nr	**94.40**
400 watt	70.83	84.88	1.50	27.80	nr	**112.68**
Sealed discharge; aluminium reflector						
150 watt	171.12	205.05	1.50	27.80	nr	**232.85**
250 watt	180.42	216.20	1.50	27.80	nr	**244.00**
400 watt	205.11	245.78	1.50	27.80	nr	**273.58**
Corrosion resistant GRP body; gasket sealed; acrylic diffuser						
600 mm Single - 18 W	24.97	29.92	0.49	9.08	nr	**39.00**
600 mm Twin - 18 W	34.25	41.04	0.49	9.08	nr	**50.12**
1200 mm Single - 36 W	28.11	33.68	0.64	11.86	nr	**45.55**
1200 mm Twin - 36 W	39.59	47.44	0.64	11.86	nr	**59.30**
1500 mm Single - 58 W	31.33	37.54	0.72	13.34	nr	**50.89**
1500 mm Twin - 58 W	45.26	54.24	0.72	13.34	nr	**67.58**
1800 mm Single - 70 W	46.36	55.55	0.94	17.42	nr	**72.98**
1800 mm Twin - 70 W	62.87	75.34	0.94	17.42	nr	**92.76**
Flameproof to IIA/IIB,I.P. 64; Aluminium Body; BS 229 and 899						
600 mm Single - 18 W	256.37	307.21	1.04	19.28	nr	**326.48**
600 mm Twin - 18 W	319.38	382.71	1.04	19.28	nr	**401.99**
1200 mm Single - 36 W	280.82	336.51	1.31	24.28	nr	**360.79**
1200 mm Twin - 36 W	347.19	416.04	1.18	21.87	nr	**437.91**
1500 mm Single - 58 W	300.64	360.26	1.64	30.40	nr	**390.65**
1500 mm Twin - 58 W	363.03	435.02	1.64	30.40	nr	**465.41**
1800 mm Single - 70 W	329.57	394.92	1.97	36.51	nr	**431.44**
1800 mm Twin - 70 W	380.22	455.62	1.97	36.51	nr	**492.13**
External Lighting						
Ground mounted 50 watt	164.55	197.17	2.25	41.70	nr	**238.88**
Ceiling mounted 50 watt	164.96	197.67	2.25	41.70	nr	**239.37**
Bulkhead; aluminium body and polycarbonate bowl; vandel resistant; IP65						
60 watt	26.60	31.87	0.75	13.90	nr	**45.77**
Extra for						
Emergency version	13.90	16.66	0.25	4.63	nr	**21.29**
2D 2 pin 16 watt	22.24	26.65	0.66	12.23	nr	**38.88**
2D 2 pin 28 watt	22.90	27.44	0.66	12.23	nr	**39.67**
Extra for						
Emergency version	70.83	84.88	0.25	4.63	nr	**89.51**
Photocell	15.76	18.89	0.75	13.90	nr	**32.79**

Material Costs/Measured Work Prices – Electrical Installations

V:ELECTRICAL SUPPLY/POWER/LIGHTING

Item	Net Price £	Material £	Labour hours	Labour £	Unit	Total rate £
V21 : GENERAL LIGHTING						
Y73 - LUMINAIRES						
External Lighting (Continued)						
1500 mm high circular bollard; polycarbonate visor; vandel resistant; IP54						
50 watt	203.81	244.22	1.75	32.43	nr	**276.65**
70 watt	203.81	244.22	1.75	32.43	nr	**276.65**
80 watt	195.74	234.56	1.75	32.43	nr	**266.99**
Floodlight; enclosed high performance Discharge light; integral control gear; reflector; toughened glass; IP65						
70 watt	71.12	85.22	1.25	23.17	nr	**108.39**
100 watt	109.87	131.66	1.25	23.17	nr	**154.83**
150 watt	112.64	134.98	1.25	23.17	nr	**158.15**
250 watt	212.61	254.78	1.25	23.17	nr	**277.94**
400 watt	233.66	279.99	1.25	23.17	nr	**303.16**
Extra for						
Photocell	16.30	19.53	0.75	13.90	nr	**33.43**
Lighting Track						
Single circuit; extruded aluminium white finish; low voltage with copper conductors; including couplers and supports; fixed to backgrounds						
Straight track	13.58	16.27	0.50	9.27	m	**25.54**
Live end	4.60	5.51	0.33	6.12	nr	**11.63**
Dead end	1.51	1.81	0.25	4.63	nr	**6.45**
Elbow	9.17	10.98	0.33	6.12	nr	**17.10**
Tee	13.79	16.52	0.33	6.12	nr	**22.64**
Cross	18.35	21.99	0.50	9.27	nr	**31.26**
Flexible couplers	22.26	26.67	0.33	6.12	nr	**32.79**
Three circuit; extruded aluminium white finish; low voltage with copper conductors; including couplers and supports; fixed to backgrounds						
Straight track	21.97	26.32	0.75	13.90	m	**40.22**
Live end	9.95	11.93	0.50	9.27	nr	**21.19**
Dead end	1.09	1.31	0.45	8.34	nr	**9.65**
Elbow	13.71	16.43	0.40	7.41	nr	**23.85**
Tee	17.49	20.96	0.55	10.19	nr	**31.16**
Cross	20.71	24.81	0.88	16.31	nr	**41.12**
Flexible couplers	25.54	30.60	0.45	8.34	nr	**38.94**

Material Costs/Measured Work Prices – Electrical Installations

V:ELECTRICAL SUPPLY/POWER/LIGHTING

Item	Net Price £	Material £	Labour hours	Labour £	Unit	Total rate £
Y74 - ACCESSORIES						
SWITCHES						
6 Amp metal clad surface mounted switch,gridswitch; one way						
1 Gang	4.01	4.81	0.43	7.97	nr	**12.77**
2 Gang	5.40	6.47	0.55	10.19	nr	**16.66**
3 Gang	8.63	10.34	0.77	14.27	nr	**24.61**
4 Gang	10.02	12.01	0.88	16.31	nr	**28.32**
6 Gang	17.10	20.49	1.10	20.39	nr	**40.88**
8 Gang	20.79	24.91	1.28	23.72	nr	**48.64**
12 Gang	29.62	35.49	1.67	30.95	nr	**66.45**
Extra for						
6 Amp - Two way switch	1.21	1.45	0.03	0.56	nr	**2.01**
20 Amp - Two way switch	1.77	2.12	0.04	0.74	nr	**2.86**
20 Amp - Intermediate	4.08	4.89	0.08	1.48	nr	**6.37**
20 Amp - One way SP switch	1.98	2.37	0.08	1.48	nr	**3.86**
Steel blank plate; 1 Gang	0.77	0.92	0.07	1.30	nr	**2.22**
Steel blank plate; 2 Gang	1.26	1.51	0.08	1.48	nr	**2.99**
6 Amp modular type switch; galvanised steel box, bronze or satin chrome coverplate; metalclad switches; flush mounting; one way						
1 Gang	8.98	10.76	0.43	7.97	nr	**18.73**
2 Gang	12.97	15.54	0.55	10.19	nr	**25.74**
3 Gang	20.27	24.29	0.77	14.27	nr	**38.56**
4 Gang	24.26	29.07	0.88	16.31	nr	**45.38**
6 Gang	39.63	47.49	1.18	21.87	nr	**69.36**
8 Gang	49.71	59.57	1.63	30.21	nr	**89.78**
9 Gang	57.78	69.24	1.83	33.92	nr	**103.15**
12 Gang	74.14	88.84	2.29	42.44	nr	**131.28**
6 Amp modular type switch; galvanised steel box; bronze or satin chrome cover plate; flush mounting; two way						
1 Gang	11.21	13.43	0.43	7.97	nr	**21.40**
2 Gang	14.55	17.44	0.55	10.19	nr	**27.63**
3 Gang	24.04	28.81	0.77	14.27	nr	**43.08**
4 Gang	27.37	32.80	0.88	16.31	nr	**49.11**
6 Gang	47.03	56.36	1.18	21.87	nr	**78.23**
8 Gang	58.15	69.68	1.63	30.21	nr	**99.89**
9 Gang	68.65	82.26	1.83	33.92	nr	**116.18**
12 Gang	87.53	104.89	2.22	41.15	nr	**146.03**
Plate switches; 10 Amp flush mounted, white plastic fronted; 16mm metal box; fitted brass earth terminal						
1 Gang 1 Way, Single Pole	2.44	2.92	0.28	5.19	nr	**8.11**
1 Gang 2 Way, Single Pole	3.14	3.76	0.33	6.12	nr	**9.88**
2 Gang 2 Way, Single Pole	5.04	6.04	0.44	8.15	nr	**14.19**
3 Gang 2 Way, Single Pole	7.07	8.47	0.56	10.38	nr	**18.85**
1 Gang Intermediate	5.67	6.79	0.43	7.97	nr	**14.76**
1 Gang 1 Way, Double Pole	4.59	5.50	0.33	6.12	nr	**11.62**
1 Gang Single Pole with bell symbol	3.33	3.99	0.23	4.26	nr	**8.25**

520   *Material Costs/Measured Work Prices – Electrical Installations*

V:ELECTRICAL SUPPLY/POWER/LIGHTING

Item	Net Price £	Material £	Labour hours	Labour £	Unit	Total rate £
V21 : GENERAL LIGHTING						
Y74 – ACCESSORIES (Continued)						
Plate switches; 10 Amp flush mounted, white plastic fronted; 16mm metal box; fitted brass earth terminal						
1 Gang Single Pole marked "PRESS"	4.61	5.52	0.23	4.26	nr	**9.79**
Time delay switch, suppressed	26.38	31.61	0.49	9.08	nr	**40.69**
Plate switches; 6 Amp flush mounted white plastic fronted; 25mm metal box; fitted brass earth terminal						
4 Gang 2 Way, Single Pole	9.72	11.65	0.42	7.78	nr	**19.43**
6 Gang 2 Way, Single Way	17.17	20.57	0.47	8.71	nr	**29.29**
Architrave plate switches; 6 Amp flush mounted, white plastic fronted; 27mm metal box; brass earth terminal						
1 Gang 2 Way, Single Pole	2.42	2.90	0.30	5.56	nr	**8.46**
Ceiling switches, white moulded plastic, pull cord; standard unit						
6 Amp, 1 Way, Single Pole	3.05	3.65	0.32	5.93	nr	**9.59**
6 Amp, 2 Way, Single Pole	3.49	4.18	0.34	6.30	nr	**10.48**
16 Amp, 1 Way, Double Pole	6.24	7.48	0.37	6.86	nr	**14.33**
45 Amp, 1 Way, Double Pole with neon indicator	8.50	10.19	0.47	8.71	nr	**18.90**
10 Amp splash proof moulded switch with plain, threaded or PVC entry						
1 Gang,2 Way Single Pole	9.94	11.91	0.34	6.30	nr	**18.21**
2 Gang, 1 Way Single Pole	11.41	13.67	0.36	6.67	nr	**20.34**
2 Gang, 2 Way Single Pole	14.47	17.34	0.40	7.41	nr	**24.75**
6 Amp Watertight switch; metalclad; BS 3676; ingress protected to IP65 surface mounted						
1 Gang, 2 Way; terminal entry	8.88	10.64	0.41	7.60	nr	**18.24**
1 Gang, 2 Way; through entry	23.29	27.91	0.42	7.78	nr	**35.69**
2 Gang, 2 Way; terminal entry	28.84	34.56	0.54	10.01	nr	**44.57**
2 Gang, 2 Way; through entry	28.84	34.56	0.53	9.82	nr	**44.38**
2 Way replacement switch	9.09	10.89	0.10	1.85	nr	**12.75**
15 Amp Watertight switch; metalclad; BS 3676; ingress protected to IP65; surface mounted						
1 Gang 2 Way, terminal entry	12.10	14.50	0.42	7.78	nr	**22.28**
1 Gang 2 Way, through entry	12.10	14.50	0.43	7.97	nr	**22.47**
2 Gang 2 Way, terminal entry	31.15	37.33	0.55	10.19	nr	**47.52**
2 Gang 2 Way, through entry	31.15	37.33	0.54	10.01	nr	**47.34**
Intermediate interior only	9.09	10.89	0.11	2.04	nr	**12.93**
2 way interior only	9.09	10.89	0.11	2.04	nr	**12.93**
Double pole interior only	9.09	10.89	0.11	2.04	nr	**12.93**

Material Costs/Measured Work Prices – Electrical Installations

V:ELECTRICAL SUPPLY/POWER/LIGHTING

Item	Net Price £	Material £	Labour hours	Labour £	Unit	Total rate £
Electrical Accessories; fixed to backgrounds (Including fixings.)						
Dimmer switches; Rotary action; for individual lights; moulded plastic case; metal backbox; flush mounted						
1 Gang, 1 Way; 250 Watt	11.03	13.22	0.28	5.19	nr	**18.41**
1 Gang, 1 Way; 400 Watt	15.65	18.75	0.28	5.19	nr	**23.94**
Dimmer switches; Push on/off action; for individual lights; moulded plastic case; metal backbox; flush mounted						
1 Gang, 2 Way; 250 Watt	15.60	18.69	0.34	6.30	nr	**25.00**
3 Gang, 2 Way; 250 Watt	76.65	91.85	0.48	8.90	nr	**100.75**
4 Gang, 2 Way; 250 Watt	100.74	120.72	0.57	10.56	nr	**131.28**
Dimmer switches; Rotary action; metal cald; metal backbox; BS 5518 and BS 800; flush mounted						
1 Gang, 1 Way; 400 Watt	24.82	29.74	0.33	6.12	nr	**35.86**
CEILING ROSES						
Ceiling Rose: white moulded plastic; flush fixed to conduit box						
Plug in type; ceiling socket with 2 terminals , loop-in and ceiling plug with 3 terminals and cover	4.07	4.88	0.34	6.30	nr	**11.18**
BC Lampholder; white moulded plastic; Heat resistent PVC insulated and sheathed cable; flush fixed						
2 Core; 0.75mm2	2.32	2.78	0.33	6.12	nr	**8.90**
Batten Holder: white moulded plastic; 3 terminals; BS 5042; fixed to conduit						
Straight pattern; 2 terminals with loop-in and Earth	2.12	2.54	0.29	5.37	nr	**7.92**
Angled pattern ; looped in terminal	2.03	2.43	0.29	5.37	nr	**7.81**

Material Costs/Measured Work Prices – Electrical Installations

V:ELECTRICAL SUPPLY/POWER/LIGHTING

Item	Net Price £	Material £	Labour hours	Labour £	Unit	Total rate £
V21 : GENERAL LIGHTING						
LIGHTING CONTROL						
Lighting control system; including software, commissioning and testing. Typical component parts indicated. System requirements dependant on final lighting design						
Y61 - CABLES						
Cable; Twin twisted bus; LSF sheathed; aluminium conductors	1.00	1.20	0.08	1.48	m	**2.68**
Cable; ELV 4 core 7/0.2; LSF sheathed; alumimium screened; copper conductor	1.30	1.56	0.15	2.78	m	**4.34**
EQUIPMENT						
Central supervisor controller	5000.00	5991.50	12.00	222.41	nr	**6213.91**
Area control unit	1052.00	1260.61	4.00	74.14	nr	**1334.75**
Lighting control module; plug in; 5m flying leads and plug						
Base and lid assembly	195.50	234.27	2.05	37.99	nr	**272.26**
Lighting control module; hard wired						
Base and lid assembly	172.50	206.71	1.85	34.29	nr	**240.99**
Phase control dimmer unit						
1 kVA	230.00	275.61	1.00	18.53	nr	**294.14**
Presence detectors; flush mounted	46.00	55.12	0.60	11.12	nr	**66.24**
Scene switch plate; anodised aluminium finish						
4 way	109.25	130.91	1.20	22.24	nr	**153.16**

Material Costs/Measured Work Prices – Electrical Installations

523

V:ELECTRICAL SUPPLY/POWER/LIGHTING

Item	Net Price £	Material £	Labour hours	Labour £	Unit	Total rate £
V22 : SMALL POWER						
Y74 - ACCESSORIES						
OUTLETS						
Socket outlet: unswitched; 13 Amp metal clad; BS 1363; galvanised steel box and coverplate with white plastic inserts; fixed surface mounted						
1 Gang	3.86	4.63	0.41	7.60	nr	**12.22**
2 Gang	7.77	9.31	0.41	7.60	nr	**16.91**
Socket outlet: switched; 13 Amp metal clad; BS 1363; galvanised steel box and coverplate with white plastic inserts; fixed surface mounted						
1 Gang	4.98	5.97	0.43	7.97	nr	**13.94**
2 Gang	9.77	11.71	0.45	8.34	nr	**20.05**
Socket outlet: switched with neon indicator; 13 Amp metal clad; BS 1363; galvanised steel box and coverplate withwhite plastic inserts; fixed surface mounted						
1 Gang	7.77	9.31	0.43	7.97	nr	**17.28**
2 Gang	14.11	16.91	0.45	8.34	nr	**25.25**
Socket outlet: unswitched; 13 Amp; BS 1363; white moulded plastic box and coverplate; fixed surface mounted						
1 Gang	2.71	3.25	0.41	7.60	nr	**10.85**
2 Gang	5.05	6.05	0.41	7.60	nr	**13.65**
Socket outlet; switched; 13 Amp; BS 1363; white moulded plastic box and coverplate; fixed surface mounted						
1 Gang	2.93	3.51	0.43	7.97	nr	**11.48**
2 Gang	5.38	6.45	0.45	8.34	nr	**14.79**
Socket outlet: switched with neon indicator; 13 Amp; BS 1363; white moulded plastic box and coverplate; fixed surface mounted						
1 Gang	6.20	7.43	0.43	7.97	nr	**15.40**
2 Gang	10.79	12.93	0.45	8.34	nr	**21.27**
Socket outlet: switched; 13 Amp; BS 1363; galvanised steel box, white moulded coverplate; flush fitted						
1 Gang	2.93	3.51	0.43	7.97	nr	**11.48**
2 Gang	5.11	6.12	0.45	8.34	nr	**14.46**
Socket outlet: switched with neon indicator; 13 Amp; BS 1363; galvanised steel box, white moulded coverplate; flush fixed						
1 Gang	6.20	7.43	0.43	7.97	nr	**15.40**
2 Gang	10.51	12.59	0.45	8.34	nr	**20.93**

Material Costs/Measured Work Prices – Electrical Installations

V:ELECTRICAL SUPPLY/POWER/LIGHTING

Item	Net Price £	Material £	Labour hours	Labour £	Unit	Total rate £
V22 : SMALL POWER						
Y74 – ACCESSORIES (Continued)						
Socket outlet: switched; 13 Amp; BS 1363; galvanised steel box, satin chrome coverplate; BS 4662; flush fixed						
1 Gang	6.55	7.85	0.43	7.97	nr	**15.82**
2 Gang	11.31	13.55	0.45	8.34	nr	**21.89**
Socket outlet: switched with neon indicator; 13 Amp; BS 1363; steel backbox, satin chrome coverplate; BS 4662; flush fixed						
1 Gang	9.33	11.18	0.43	7.97	nr	**19.15**
2 Gang	18.02	21.59	0.45	8.34	nr	**29.93**
Weatherproof socket outlet: 13 Amp; switched; single gang; RCD protected; water and dust protected to I.P.66; surface mounted						
40A 30mA tripping current protecting 1 socket	159.42	191.03	0.52	9.64	nr	**200.67**
40A 30mA tripping current protecting 2 sockets	205.94	246.78	0.64	11.86	nr	**258.64**
Plug for weatherproof socket outlet: protected to I.P.66						
13Amp plug	14.76	17.69	0.21	3.89	nr	**21.58**
Floor service outlet box; comprising flat lid with flanged carpet trim; twin 13A switched socket outlets; punched plate for mounting 2 telephone outlets; one blank plate; triple compartment						
3 compartment	30.82	36.93	0.88	16.31	nr	**53.24**
Floor service outlet box; comprising flat lid with flanged carpet trim; twin 13A switched socket outlets; punched plate for mounting 2 telephone outlets; two blank plates; four compartment						
4 compartment	36.21	43.39	0.88	16.31	nr	**59.70**
Floor service outlet box; comprising flat lid with flanged carpet trim; single 13A unswitched socket outlet; single compartment; circular						
1 compartment	22.42	26.87	0.79	14.64	nr	**41.51**
Floor service grommet, comprising flat lid with flanged carpet trim; circular						
Floor Grommet	12.40	14.86	0.49	9.08	nr	**23.94**
POWER POSTS/POLES/PILLARS						
Power Post						
Power post; aluminium painted body; PVC-U cover; 5 nr outlets	87.05	104.21	4.00	74.14	nr	**178.35**
Power Pole						
Power pole; 3.6 metres high; aluminium painted body; PVC-U cover; 6 nr outlets	245.82	294.29	4.00	74.14	nr	**368.43**

Material Costs/Measured Work Prices – Electrical Installations   525

V:ELECTRICAL SUPPLY/POWER/LIGHTING

Item	Net Price £	Material £	Labour hours	Labour £	Unit	Total rate £
Extra for						
Power pole extension bar; 900mm long	22.73	27.22	1.50	27.80	nr	**55.02**
Vertical multi compartment pillar; PVC-U; BS 4678 Part4 EN60529; excludes accessories						
Single						
630mm long	49.50	59.26	2.00	37.07	nr	**96.33**
3000mm long	117.00	140.07	2.00	37.07	nr	**177.14**
Double						
630mm long	81.00	96.97	3.00	55.60	nr	**152.57**
3000mm long	198.90	238.12	3.00	55.60	nr	**293.72**
CONNECTION UNITS						
Connection units: Moulded pattern; BS 5733; moulded plastic box; white coverplate; knockout for flex outlet; surface mounted - standard fused						
DP Switched	4.85	5.81	0.49	9.08	nr	**14.89**
Unswitched	4.92	5.90	0.49	9.08	nr	**14.98**
DP Switched with neon indicator	6.70	8.03	0.49	9.08	nr	**17.11**
Connection units: moulded pattern; BS 5733; galvanised steel box; white coverplate; knockout for flex outlet; surface mounted						
DP Switched	4.85	5.81	0.49	9.08	nr	**14.89**
DP Unswitched	6.06	7.26	0.49	9.08	nr	**16.34**
DP Switched with neon indicator	7.21	8.64	0.49	9.08	nr	**17.72**
Connection units: galvanised pressed steel pattern; galvanised steel box; satin chrome or satin brass finish; white moulded plastic inserts; flush mounted - standard fused						
DP Switched	8.76	10.50	0.49	9.08	nr	**19.58**
Unswitched	9.22	11.05	0.49	9.08	nr	**20.13**
DP Switched with neon indicator	11.53	13.82	0.49	9.08	nr	**22.90**
Connection units: galvanised steel box; satin chrome or satin brass finish; white moulded plastic inserts; flex outlet; flush mounted - standard fused						
Switched	12.40	14.86	0.49	9.08	nr	**23.94**
Unswitched	13.36	16.01	0.49	9.08	nr	**25.09**
Switched with neon indicator	13.99	16.76	0.49	9.08	nr	**25.85**
SHAVER SOCKET						
Shaver Unit: self setting overload device; 200/250 voltage supply; white moulded plastic faceplate; unswitched						
Surface type with moulded plastic box	12.18	14.60	0.55	10.19	nr	**24.79**
Flush type with galvanised steel box	11.53	13.82	0.57	10.56	nr	**24.38**

Material Costs/Measured Work Prices – Electrical Installations

V:ELECTRICAL SUPPLY/POWER/LIGHTING

Item	Net Price £	Material £	Labour hours	Labour £	Unit	Total rate £
V22 : SMALL POWER						
Y74 – ACCESSORIES (Continued)						
Shaver Unit: dual voltage supply unit; white moulded plastic faceplate; unswitched						
Surface type with moulded plastic box	18.79	22.52	0.62	11.49	nr	**34.01**
Flush type with galvanised steel box	17.85	21.39	0.64	11.86	nr	**33.25**
COOKER CONTROL UNIT						
Cooker Control Unit: BS 4177; 45 amp D.P. main switch; 13 Amp switched socket outlet; metal coverplate; plastic inserts; neon indicators						
Surface mounted with mounting box	18.60	22.29	0.61	11.31	nr	**33.59**
Flush mounted with galvanised steel box	16.84	20.18	0.61	11.31	nr	**31.49**
Cooker Control Unit: BS 4177; 45 Amp D.P. main switch; 13 Amp switched socket outlet; moulded plastic box and coverplate; surface mounted						
Standard	11.73	14.06	0.61	11.31	nr	**25.36**
With neon indicators	15.41	18.47	0.61	11.31	nr	**29.77**
CONTROL COMPONANTS						
Connector unit : moulded white plastic cover and block; galvanised steel back box; to immersion heaters						
3Kw up to 915mm long; fitted to thermostat	16.14	17.76	0.75	13.90	nr	**31.66**
Water heater switch : 20 Amp; switched with neon indicator						
DP Switched with neon indicator	9.28	10.34	0.45	8.34	nr	**18.68**
SWITCH DISCONNECTORS						
Switch disconnectors; moulded plastic enclosure; fixed to backgrounds						
3 pole; IP54; Grey						
16 Amp	15.41	18.45	0.80	14.83	nr	**33.27**
25 Amp	18.12	21.69	0.80	14.83	nr	**36.52**
40 Amp	28.69	34.35	0.80	14.83	nr	**49.17**
63 Amp	43.14	51.64	1.00	18.53	nr	**70.18**
80 Amp	73.84	88.40	1.25	23.17	nr	**111.57**
6 pole; IP54; Grey						
25 Amp	23.34	27.95	1.00	18.53	nr	**46.48**
63 Amp	58.44	69.96	1.25	23.17	nr	**93.13**
80 Amp	75.74	90.67	1.80	33.36	nr	**124.03**

Material Costs/Measured Work Prices – Electrical Installations 527

V:ELECTRICAL SUPPLY/POWER/LIGHTING

Item	Net Price £	Material £	Labour hours	Labour £	Unit	Total rate £
3 pole; IP54; Yellow						
16 Amp	16.95	20.29	0.80	14.83	nr	**35.12**
25 Amp	19.94	23.87	0.80	14.83	nr	**38.69**
40 Amp	31.56	37.78	0.80	14.83	nr	**52.61**
63 Amp	47.45	56.80	1.00	18.53	nr	**75.34**
6 pole; IP54; Yellow						
25 Amp	25.68	30.74	1.00	18.53	nr	**49.28**
INDUSTRIAL SOCKETS/PLUGS						
Plugs; Splashproof; 100-130 volts, 50-60 Hz; IP44 (Yellow)						
2 pole and earth						
16 Amp	2.10	2.52	0.55	10.19	nr	**12.71**
32 Amp	7.03	8.42	0.60	11.12	nr	**19.54**
3 pole and earth						
16 Amp	7.80	9.34	0.65	12.05	nr	**21.39**
32 Amp	10.49	12.56	0.72	13.34	nr	**25.90**
3 pole; neutral and earth						
16 Amp	8.33	9.97	0.72	13.34	nr	**23.31**
32 Amp	12.62	15.10	0.78	14.46	nr	**29.56**
Connectors; Splashproof; 100-130 volts, 50-60 Hz; IP 44 (Yellow)						
2 pole and earth						
16 Amp	3.19	3.82	0.42	7.78	nr	**11.61**
32 Amp	9.30	11.13	0.50	9.27	nr	**20.40**
3 pole and earth						
16 Amp	9.33	11.17	0.48	8.90	nr	**20.06**
32 Amp	13.54	16.21	0.58	10.75	nr	**26.96**
3 pole; neutral and earth						
16 Amp	12.16	14.56	0.52	9.64	nr	**24.20**
32 Amp	15.41	18.45	0.73	13.53	nr	**31.98**
Angles sockets; surface mounted; Splashproof; 100-130 volts, 50-60 Hz; IP44 (Yellow)						
2 pole and earth						
16 Amp	4.58	5.49	0.55	10.19	nr	**15.68**
32 Amp	10.66	12.77	0.60	11.12	nr	**23.89**
3 pole and earth						
16 Amp	10.17	12.17	0.65	12.05	nr	**24.22**
32 Amp	19.09	22.85	0.72	13.34	nr	**36.20**
3 pole; neutral and earth						
16 Amp	14.10	16.89	0.72	13.34	nr	**30.23**
32 Amp	18.45	22.09	0.78	14.46	nr	**36.54**

Material Costs/Measured Work Prices – Electrical Installations

V:ELECTRICAL SUPPLY/POWER/LIGHTING

Item	Net Price £	Material £	Labour hours	Labour £	Unit	Total rate £
V22 : SMALL POWER						
Y74 – ACCESSORIES (Continued)						
Plugs; Watertight; 100-130 volts, 50-60 Hz; IP67 (Yellow)						
2 pole and earth						
16 Amp	8.09	9.68	0.55	10.19	nr	**19.88**
32 Amp	15.18	18.17	0.60	11.12	nr	**29.29**
63 Amp	33.59	40.22	0.75	13.90	nr	**54.12**
Connectors; Watertight; 100-130 volts, 50-60 Hz; IP 67 (Yellow)						
2 pole and earth						
16 Amp	12.18	14.58	0.42	7.78	nr	**22.36**
32 Amp	20.89	25.01	0.50	9.27	nr	**34.27**
63 Amp	35.70	42.74	0.67	12.42	nr	**55.16**
Angles sockets; surface mounted; Watertight; 100-130 volts, 50-60 Hz; IP 67 (Yellow)						
2 pole and earth						
16 Amp	21.84	26.15	0.55	10.19	nr	**36.34**
32 Amp	32.16	38.50	0.60	11.12	nr	**49.62**
Plugs; Splashproof; 200-250 volts, 50-60 Hz; IP 44 (Blue)						
2 pole and earth						
16 Amp	2.10	2.52	0.55	10.19	nr	**12.71**
32 Amp	5.41	6.47	0.60	11.12	nr	**17.59**
63 Amp	26.06	31.20	0.75	13.90	nr	**45.10**
3 pole and earth						
16 Amp	6.56	7.85	0.65	12.05	nr	**19.90**
32 Amp	12.43	14.88	0.72	13.34	nr	**28.23**
63 Amp	26.17	31.33	0.83	15.38	nr	**46.71**
3 pole; neutral and earth						
16 Amp	7.21	8.63	0.72	13.34	nr	**21.97**
32 Amp	10.88	13.03	0.78	14.46	nr	**27.48**
Connectors; Splashproof; 200-250 volts, 50-60 Hz; IP 44 (Blue)						
2 pole and earth						
16 Amp	3.48	4.17	0.42	7.78	nr	**11.95**
32 Amp	8.72	10.44	0.50	9.27	nr	**19.71**
63 Amp	31.65	37.89	0.67	12.42	nr	**50.31**
3 pole and earth						
16 Amp	9.33	11.17	0.48	8.90	nr	**20.06**
32 Amp	12.43	14.88	0.58	10.75	nr	**25.63**
63 Amp	25.98	31.11	0.75	13.90	nr	**45.01**
3 pole; neutral and earth						
16 Amp	10.89	13.04	0.52	9.64	nr	**22.67**
32 Amp	15.41	18.45	0.73	13.53	nr	**31.98**

Material Costs/Measured Work Prices – Electrical Installations

V:ELECTRICAL SUPPLY/POWER/LIGHTING

Item	Net Price £	Material £	Labour hours	Labour £	Unit	Total rate £
Angle sockets; surface mounted; Splashproof; 200-250 volts, 50-60 Hz; IP 44 (Blue)						
2 pole and earth						
16 Amp	4.58	5.49	0.55	10.19	nr	**15.68**
32 Amp	7.27	8.71	0.60	11.12	nr	**19.83**
63 Amp	27.86	33.35	0.75	13.90	nr	**47.25**
3 pole and earth						
16 Amp	8.92	10.68	0.65	12.05	nr	**22.73**
32 Amp	16.59	19.86	0.72	13.34	nr	**33.21**
63 Amp	40.36	48.32	0.83	15.38	nr	**63.70**
3 pole; neutral and earth						
16 Amp	10.50	12.57	0.72	13.34	nr	**25.91**
32 Amp	16.48	19.73	0.78	14.46	nr	**34.19**
Plugs; Watertight; 200-250 volts, 50-60 Hz; IP67 (Blue)						
2 pole and earth						
16 Amp	8.56	10.25	0.41	7.60	nr	**17.85**
32 Amp	14.90	17.84	0.50	9.27	nr	**27.11**
63 Amp	38.94	46.62	0.66	12.23	nr	**58.86**
125 Amp	99.81	119.49	0.86	15.94	nr	**135.43**
Connectors; Watertight; 200-250 volts, 50-60 Hz; IP 67 (Blue)						
2 pole and earth						
16 Amp	12.02	14.39	0.42	7.78	nr	**22.17**
32 Amp	19.23	23.02	0.50	9.27	nr	**32.29**
63 Amp	40.92	48.99	0.67	12.42	nr	**61.41**
125 Amp	122.50	146.65	0.87	16.12	nr	**162.78**
Angles sockets; surface mounted; Watertight; 200-250 volts, 50-60 Hz; IP67 (Blue)						
2 pole and earth						
16 Amp	13.77	16.48	0.55	10.19	nr	**26.68**
32 Amp	26.88	32.18	0.60	11.12	nr	**43.30**
125 Amp	201.91	241.73	1.00	18.53	nr	**260.26**
Plugs; Splashproof; 380-415 volts, 50-60 Hz; IP 44 (Red)						
2 pole and earth						
16 Amp	7.55	9.04	0.41	7.60	nr	**16.64**
3 pole and earth						
16 Amp	4.89	5.85	0.48	8.90	nr	**14.75**
32 Amp	7.00	8.38	0.58	10.75	nr	**19.13**
63 Amp	22.66	27.12	0.75	13.90	nr	**41.02**
3 pole; neutral and earth						
16 Amp	4.78	5.72	0.52	9.64	nr	**15.36**
32 Amp	6.21	7.43	0.73	13.53	nr	**20.96**
63 Amp	22.30	26.70	0.87	16.12	nr	**42.83**
Connectors; Splashproof; 380-415 volts, 50-60 Hz; IP 44 (Red)						
2 pole and earth						
16 Amp	8.77	10.50	0.42	7.78	nr	**18.28**

Material Costs/Measured Work Prices – Electrical Installations

V:ELECTRICAL SUPPLY/POWER/LIGHTING

Item	Net Price £	Material £	Labour hours	Labour £	Unit	Total rate £
V22 : SMALL POWER						
Y74 – ACCESSORIES (Continued)						
3 pole and earth						
16 Amp	6.74	8.06	0.48	8.90	nr	**16.96**
32 Amp	9.94	11.90	0.58	10.75	nr	**22.65**
63 Amp	28.31	33.90	0.75	13.90	nr	**47.80**
3 pole; neutral and earth						
16 Amp	8.58	10.28	0.52	9.64	nr	**19.91**
32 Amp	9.77	11.69	0.73	13.53	nr	**25.22**
63 Amp	28.82	34.51	0.87	16.12	nr	**50.63**
Angles sockets; surface mounted; Splashproof; 380-415 volts, 50-60 Hz; IP 44 (Red)						
2 pole and earth						
16 Amp	10.12	12.12	0.55	10.19	nr	**22.31**
3 pole and earth						
16 Amp	6.71	8.04	0.65	12.05	nr	**20.08**
32 Amp	9.70	11.61	0.72	13.34	nr	**24.95**
63 Amp	31.79	38.06	0.83	15.38	nr	**53.44**
3 pole; neutral and earth						
16 Amp	6.76	8.09	0.72	13.34	nr	**21.44**
32 Amp	8.72	10.44	0.78	14.46	nr	**24.90**
63 Amp	33.58	40.20	0.90	16.68	nr	**56.88**
Plugs; Watertight; 380-415 volts, 50-60 Hz; IP67 (Red)						
3 pole and earth						
16 Amp	13.82	16.55	0.48	8.90	nr	**25.45**
32 Amp	18.86	22.58	0.58	10.75	nr	**33.33**
63 Amp	27.86	33.35	0.75	13.90	nr	**47.25**
125 Amp	113.09	135.39	1.08	20.02	nr	**155.41**
3 pole; neutral and earth						
16 Amp	13.97	16.72	0.52	9.64	nr	**26.36**
32 Amp	24.36	29.16	0.73	13.53	nr	**42.69**
63 Amp	30.83	36.91	0.87	16.12	nr	**53.04**
125 Amp	90.94	108.87	1.17	21.68	nr	**130.55**
Connectors; Watertight; 380-415 volts, 50-60 Hz; IP 67 (Red)						
3 pole and earth						
16 Amp	14.94	17.89	0.48	8.90	nr	**26.79**
32 Amp	24.04	28.78	0.58	10.75	nr	**39.53**
63 Amp	35.53	42.53	0.75	13.90	nr	**56.43**
125 Amp	131.71	157.69	1.08	20.02	nr	**177.70**
3 pole; neutral and earth						
16 Amp	18.45	22.09	0.52	9.64	nr	**31.72**
32 Amp	22.78	27.28	0.73	13.53	nr	**40.81**
63 Amp	36.97	44.26	0.87	16.12	nr	**60.38**
125 Amp	133.34	159.64	1.17	21.68	nr	**181.32**

Material Costs/Measured Work Prices – Electrical Installations

V:ELECTRICAL SUPPLY/POWER/LIGHTING

Item	Net Price £	Material £	Labour hours	Labour £	Unit	Total rate £
Angles sockets; surface mounted; Watertight; 380-415 volts, 50-60 Hz; IP 67 (Red)						
3 pole and earth						
16 Amp	32.03	38.35	0.65	12.05	nr	**50.40**
32 Amp	34.32	41.09	0.72	13.34	nr	**54.43**
125 Amp	193.06	231.13	1.17	21.68	nr	**252.81**
3 pole; neutral and earth						
16 Amp	25.20	30.17	0.72	13.34	nr	**43.51**
32 Amp	34.32	41.09	0.78	14.46	nr	**55.54**
125 Amp	193.07	231.15	1.33	24.65	nr	**255.80**

Material Costs/Measured Work Prices – Electrical Installations

V:ELECTRICAL SUPPLY/POWER/LIGHTING

Item	Net Price £	Material £	Labour hours	Labour £	Unit	Total rate £
V32 : UNINTERRUPTED POWER SUPPLY						
Uninterruptable Power Supply; 230 Volt AC input and output; standard 13 Amp socket outlet connection; sheet steel enclosure; self contained battery pack; including testing and commissioning.						
Single Phase; I/P and O/P						
1.2kVA (7 minute supply)	1196.00	1433.17	0.31	5.79	nr	1438.96
1.2kVA (18 minute supply)	1454.00	1742.33	3.21	59.40	nr	1801.73
1.6kVA (7 minute supply)	1458.00	1747.12	3.40	63.04	nr	1810.16
1.6kVA (15 minute supply)	1711.00	2050.29	3.40	63.04	nr	2113.33
2.2kVA (10 minute supply)	1842.00	2207.27	3.61	66.91	nr	2274.18
2.2kVA (15 minute supply)	2350.00	2816.01	3.61	66.91	nr	2882.91
3.0kVA (7 minute supply)	1849.00	2215.66	4.00	74.14	nr	2289.79
3.0kVA (24 minute supply)	2484.00	2976.58	4.00	74.14	nr	3050.71
5.0kVA (5 minute supply)	3255.00	3900.47	4.50	83.49	nr	3983.95
5.0kVA (15 minute supply)	3255.00	3900.47	4.50	83.49	nr	3983.95
7.5kVA (6 minute supply)	3771.00	4518.79	4.90	90.85	nr	4609.64
7.5kVA (18 minute supply)	4406.00	5279.71	4.90	90.85	nr	5370.56
10.0kVA (5 minute supply)	4280.00	5128.72	5.00	92.67	nr	5221.39
10.0kVA (12 minute supply)	4915.00	5889.64	5.00	92.67	nr	5982.31
Uninterruptable Power Supply; including final connections and testing/ commissioning.						
Medium size static; three phase input, single phase output; integral sealed battery.						
7.5 kVA (32 minutes supply)	6829.00	8183.19	30.30	561.63	nr	8744.82
10.0 kVA (30 minutes supply)	7432.00	8905.77	30.30	561.63	nr	9467.40
15.0 kVA (20 minutes supply)	7335.00	8789.53	35.71	661.93	nr	9451.46
20.0 kVA (12 minutes supply)	7699.00	9225.71	40.00	741.36	nr	9967.07
25.0 kVA (6 minutes supply)	8334.00	9986.63	40.00	741.36	nr	10727.99
Medium size static; three phase input, three phase output; integral sealed battery;						
10.0 kVA (30 minutes supply)	8258.00	9895.56	35.71	661.93	nr	10557.49
15.0 kVA (20 minutes supply)	8639.00	10352.11	35.71	661.93	nr	11014.04
20.0 kVA (12 minutes supply)	9122.00	10930.89	40.00	741.36	nr	11672.25
30.0 kVA (6 minutes supply)	10067.00	12063.29	30.30	561.63	nr	12624.92
Large size static; three phase input/output; lead acid batteries on separate open racks.						
40 kVA (16 minutes supply)	13478.00	16150.69	43.48	805.82	nr	16956.51
60 kVA (9 minutes supply)	14783.00	17714.47	45.45	842.45	nr	18556.92
80 kVA (16 minutes supply)	18594.00	22281.19	45.45	842.45	nr	23123.64
100 kVA (12 minutes supply)	19310.00	23139.17	50.00	926.70	nr	24065.87
200 kVA (10 minutes supply)	35287.00	42284.41	55.56	1029.66	nr	43314.07
300 kVA (10 minutes supply)	45572.00	54608.93	62.50	1158.37	nr	55767.30

Material Costs/Measured Work Prices – Electrical Installations

V:ELECTRICAL SUPPLY/POWER/LIGHTING

Item	Net Price £	Material £	Labour hours	Labour £	Unit	Total rate £
Integral diesel rotary; three phase input/output; no break supply including ventilation and accoustic attenuation; oil day tank and interconnecting pipework						
100 kVA	102036.00	122269.74	-	-	nr	**122269.74**
125 kVA	115844.00	138815.87	-	-	nr	**138815.87**
150 kVA	124555.00	149254.26	-	-	nr	**149254.26**
180 kVA	133467.00	159933.51	-	-	nr	**159933.51**
200 kVA	196642.00	235636.11	-	-	nr	**235636.11**
250 kVA	202705.00	242901.40	-	-	nr	**242901.40**
300 kVA	208570.00	249929.43	-	-	nr	**249929.43**
400 kVA	245952.00	294724.28	-	-	nr	**294724.28**
500 kVA	270094.00	323653.64	-	-	nr	**323653.64**
630 kVA	326749.00	391543.33	-	-	nr	**391543.33**
800 kVA	394508.00	472738.94	-	-	nr	**472738.94**
1000 kVA	443334.00	531247.13	-	-	nr	**531247.13**
1125 kVA	499932.00	599068.52	-	-	nr	**599068.52**
1250 kVA	526239.00	630592.19	-	-	nr	**630592.19**
1500 kVA	577570.00	692102.13	-	-	nr	**692102.13**
1750 kVA	655349.00	785304.71	-	-	nr	**785304.71**

Material Costs/Measured Work Prices – Electrical Installations

V:ELECTRICAL SUPPLY/POWER/LIGHTING

Item	Net Price £	Material £	Labour hours	Labour £	Unit	Total rate £
V40 : EMERGENCY LIGHTING						
Y73 - LUMINAIRES						
Self contained; polycarbonate base and diffuser; LED charging light 3 hour duration; as European Signs Directive						
Non Maintained						
8 watt	46.20	55.36	0.80	14.83	nr	**70.19**
8 watt; vandel resistant; weatherproof; IP65	49.98	59.89	0.80	14.83	nr	**74.72**
Maintained						
8 watt	47.97	57.49	0.80	14.83	nr	**72.31**
8 watt; vandel resistant; weatherproof; IP65	49.98	59.89	0.80	14.83	nr	**74.72**
Self contained; luxury illuminated gold finish; LED charging light 3 hour duration; as European Signs Directive						
8 watt non maintained	73.71	88.33	1.00	18.53	nr	**106.86**
8 watt maintained	76.55	91.72	1.00	18.53	nr	**110.26**
Extra for						
Ceiling bracket	17.22	20.63	0.25	4.63	nr	**25.27**
Conversion Equipment						
Module and battery charger for non maintained or maintained; 3 hour duration	26.77	32.08	0.25	4.63	nr	**36.72**
12 volt low voltage lighting; non maintained or maintained						
1 x 20 watt lamp load; 10 hour duration	115.50	138.40	1.00	18.53	nr	**156.94**
2 x 20 watt lamp load; 3 hour duration	81.90	98.14	1.20	22.24	nr	**120.38**
1 x 50 watt lamp load; 3 hour duration	81.90	98.14	1.00	18.53	nr	**116.67**
2 x 50 watt lamp load; 1 hour duration	84.00	100.66	1.20	22.24	nr	**122.90**
EQUIPMENT						
Central battery system; cubilcle with combined charger/battery (10 year life span); solid state constant voltage charge control module; 24 hour recharging with current limiting facility; LED display, audible alarm, volt free changeover contacts; including commissioning and testing						
Sealed Lead acid batteries						
24 volt, 1000 watt, 50 Amp; 1 hour cubilcle	2465.40	2954.29	10.00	185.34	nr	**3139.63**
24 volt, 1000 watt, 45 Amp; 2 hour cubilcle	2775.15	3325.46	15.00	278.01	nr	**3603.47**
24 volt, 900 watt, 40 Amp; 3 hour cubilcle	2699.55	3234.87	20.00	370.68	nr	**3605.55**
50 volt, 1500 watt, 30 Amp; 1 hour cubilcle	2250.15	2696.35	10.00	185.34	nr	**2881.69**
50 volt, 1000 watt, 20 Amp; 2 hour cubilcle	2525.25	3026.01	15.00	278.01	nr	**3304.02**
50 volt, 1000 watt, 20 Amp; 3 hour cubilcle	3137.40	3759.55	20.00	370.68	nr	**4130.22**
110 volt, 1000 watt, 10 Amp; 1 hour cubilcle	2457.00	2944.22	10.00	185.34	nr	**3129.56**
110 volt, 1000 watt, 10 Amp; 2 hour cubilcle	2763.60	3311.62	15.00	278.01	nr	**3589.63**
110 volt, 1000 watt, 10 Amp; 3 hour cubilcle	3279.15	3929.41	20.00	370.68	nr	**4300.08**

Material Costs/Measured Work Prices – Electrical Installations

V:ELECTRICAL SUPPLY/POWER/LIGHTING

Item	Net Price £	Material £	Labour hours	Labour £	Unit	Total rate £
Fluoresant slave lumiaires; fixed to backgrounds						
24 volt						
8 watt; indoor use	35.17	42.15	0.80	14.83	nr	56.98
Exit sign box; indoor use	43.78	52.47	0.80	14.83	nr	67.29
8 watt; weatherproof	39.90	47.81	0.80	14.83	nr	62.64
Conversion module	42.52	50.96	0.25	4.63	nr	55.59
50 volt						
8 watt; indoor use	35.17	42.15	0.80	14.83	nr	56.98
Exit sign box; indoor use	43.78	52.47	0.80	14.83	nr	67.29
8 watt; weatherproof	39.90	47.81	0.80	14.83	nr	62.64
Conversion module	42.52	50.96	0.25	4.63	nr	55.59
110 volt						
8 watt; indoor use	35.17	42.15	0.80	14.83	nr	56.98
Exit sign box; indoor use	43.78	52.47	0.80	14.83	nr	67.29
8 watt; weatherproof	39.90	47.81	0.80	14.83	nr	62.64
Conversion module	42.52	50.96	0.25	4.63	nr	55.59
Static inverter central battery system; cubilcle with combined charger/battery (10 year life span); solid state contant voltage charge control module; 24 hour recharging with current limiting facility; LED display, audible alarm, volt free changeover contacts; including commissioning and testing						
1 hour Sealed Lead acid batteries						
1.25 kVA	2415.00	2893.89	2.00	37.07	nr	2930.96
3.75 kVA	3337.95	3999.87	5.00	92.67	nr	4092.53
5.0 kVA	5466.30	6550.27	6.00	111.20	nr	6661.47
7.5 kVA	6622.35	7935.56	8.00	148.27	nr	8083.83
10 kVA	8507.10	10194.06	10.00	185.34	nr	10379.40
13 kVA	11067.00	13261.59	12.00	222.41	nr	13483.99
15 kVA	11826.15	14171.28	13.00	240.94	nr	14412.22
20 kVA	14708.40	17625.08	17.00	315.08	nr	17940.15
3 hour Sealed Lead acid batteries						
1.25 kVA	2845.50	3409.76	3.00	55.60	nr	3465.36
3.75 kVA	4604.25	5517.27	7.50	139.00	nr	5656.28
5.0 kVA	6724.20	8057.61	9.00	166.81	nr	8224.41
7.5 kVA	9479.40	11359.17	12.00	222.41	nr	11581.57
10 kVA	11482.80	13759.84	15.00	278.01	nr	14037.85
13 kVA	15444.45	18507.08	18.00	333.61	nr	18840.69
15 kVA	16376.85	19624.38	19.50	361.41	nr	19985.79
20 kVA	21490.35	25751.89	25.50	472.61	nr	26224.50

Material Costs/Measured Work Prices – Electrical Installations

V:ELECTRICAL SUPPLY/POWER/LIGHTING

Item	Net Price £	Material £	Labour hours	Labour £	Unit	Total rate £
V40 : EMERGENCY LIGHTING						
Battery chargers; switchgear tripping and closing; double wound transformer and earth screen; including comissioning and testing						
Valve regulated lead acid battery						
30 volt; 19 Ah; 3A	1208.55	1448.21	6.50	120.47	nr	**1568.68**
30 volt; 29 Ah; 3A	1278.90	1532.51	8.50	157.54	nr	**1690.04**
110 volt; 19 Ah; 3A	2205.00	2642.25	6.50	120.47	nr	**2762.72**
110 volt; 29 Ah; 3A	2413.95	2892.64	8.50	157.54	nr	**3050.17**
110 volt; 38 Ah; 3A	2683.80	3216.00	10.00	185.34	nr	**3401.34**

Material Costs/Measured Work Prices – Electrical Installations

W:COMMUNICATIONS/SECURITY/CONTROL

Item	Net Price £	Material £	Labour hours	Labour £	Unit	Total rate £
W10 : TELECOMMUNICATIONS						
Y61 – CABLES						
Multipair Internal Telephone cable; BS 6746; loose laid on tray or drawn in conduit or trunking.(Including cable sleeves)						
0.5 millimetre diameter conductor p.v.c.insulated and sheathed multipair cables; BT specification CW 1308						
3 pair	0.11	0.13	0.05	0.93	m	**1.06**
4 pair	0.13	0.16	0.06	1.11	m	**1.27**
6 pair	0.20	0.24	0.06	1.11	m	**1.35**
10 pair	0.34	0.41	0.07	1.30	m	**1.70**
12 pair	0.43	0.52	0.07	1.30	m	**1.81**
15 pair	0.38	0.46	0.07	1.30	m	**1.75**
20 pair + 1 wire	0.61	0.73	0.09	1.67	m	**2.40**
25 pair	0.70	0.84	0.10	1.85	m	**2.69**
40 pair + earth	0.93	1.11	0.13	2.41	m	**3.52**
60 pair + earth	1.41	1.69	0.15	2.78	m	**4.47**
80 pair + earth	1.78	2.13	0.20	3.71	m	**5.84**
Telephone Undercarpet Cable; Low Profile; laid loose						
0.5 millimetre; PVC Insulated; PVC Sheathed multicore cable; BT CW 1316						
6 Core	0.48	0.58	0.05	0.93	m	**1.50**
Telephone Drop Wire Cable; drawn in conduit or trunking						
0.5 millimetre conductor; PVC insulate twisted pairs; Polyethylene sheathed; BT CW 1378						
Drop Wire 10	0.44	0.53	0.06	1.11	m	**1.64**
Y74 - ACCESSORIES						
Telephone outlet: moulded plastic plate with box; fitted and connected; flush or surface mounted						
Single Master outlet	5.63	6.75	0.35	6.49	nr	**13.23**
Single Secondary outlet	4.16	4.98	0.35	6.49	nr	**11.47**
Telephone outlet: bronze or satin chromeplate; with box; fitted and connected; flush or surface mounted						
Single Master outlet	9.16	10.98	0.35	6.49	nr	**17.46**
Single Secondary outlet	10.01	11.99	0.35	6.49	nr	**18.48**

Material Costs/Measured Work Prices – Electrical Installations

W:COMMUNICATIONS/SECURITY/CONTROL

Item	Net Price £	Material £	Labour hours	Labour £	Unit	Total rate £
W20 : RADIO/TELEVISION						
RADIO						
Y61 - CABLES						
Radio Frequency Cable; BS 2316 ; PVC sheathed; laid loose						
7/0.41mm tinned copper inner conductor; solid polyethylene dielectric insulation; bare copper wire braid; PVC sheath; 75 ohm impedance						
Cable	0.51	0.61	0.05	0.93	m	**1.54**
Twin 1/0.58mm copper covered steel solid core wire conductor; solid polyethylene dielectric insulation; barecopper wire braid; PVC sheath; 75 ohm impedance						
Cable	1.15	1.38	0.05	0.93	m	**2.30**
TELEVISION						
Y61 - CABLES						
Television Aerial Cable; coaxial; PVC sheathed; fixed to backgrounds						
General purpose TV aerial downlead; copper stranded inner conductor; cellular polythene insulation; copper braid outer conductor; 75 ohm impedance						
7/0.25mm	0.53	0.64	0.06	1.11	m	**1.75**
Low loss TV aerial downlead; solid copper inner conductor; cellular polythene insulation; copper braid outer; conductor; 75 ohm impedance						
1/1.12mm	0.59	0.71	0.06	1.11	m	**1.82**
Low loss air spaced; solid copper inner conductor; air spaced polythene insulation; copper braid outer conductor; 75 ohm impedance						
1/1.00mm	0.46	0.55	0.06	1.11	m	**1.66**
Satellite aerial downlead; solid copper inner conductor; air spaced polythene insulation; copper tape and braid outer conductor; 75 ohm impedance						
1/1.00mm	0.46	0.55	1.00	18.53	m	**19.09**
Satellite TV coaxial; solid copper inner conductor; semi air spaced polyethylene dielectric insulation; plain annealed copper foil and copper braid screen in outer conductor; PVC sheath; 75 ohm impedance						
1/1.25mm	0.64	0.77	0.08	1.48	m	**2.25**

Material Costs/Measured Work Prices – Electrical Installations

W:COMMUNICATIONS/SECURITY/CONTROL

Item	Net Price £	Material £	Labour hours	Labour £	Unit	Total rate £
Satellite TV coaxial; solid copper inner conductor; air spaced polyethylene dielectric insulation; plain annealed copper foil and copper braid screen in outer conductor; PVC sheath; 75 ohm impedance						
1/1.67mm	0.85	1.02	0.09	1.67	m	**2.69**
Video Cable; PVC Fame Retardant sheath; laid loose						
7/0.1mm silver coated copper covered annealed steel wire conductor; polyethylene dielectric insulation with tin coated copper wire braid; 75 ohms impedance						
Cable	0.72	0.86	0.05	0.93	m	**1.79**
Y74 - ACCESSORIES						
TV Co-Axial socket outlet: moulded plastic box: flush or surface mounted						
One way Direct Connection	4.30	5.15	0.35	6.49	nr	**11.64**
Two way Direct Connection	6.20	7.43	0.35	6.49	nr	**13.92**
One way Isolated UHF/VHF	6.77	8.11	0.35	6.49	nr	**14.60**
Two way Isolated UHF/VHF	9.30	11.14	0.35	6.49	nr	**17.63**

540　**Material Costs/Measured Work Prices – Electrical Installations**

W:COMMUNICATIONS/SECURITY/CONTROL

Item	Net Price £	Material £	Labour hours	Labour £	Unit	Total rate £
W23 : CLOCKS						
Clock Timing Systems; Master and Slave Units; fixed to background; (Excluding supports and fixings).						
Quartz master clock with solid state digital readout for parallel loop operation; one minute, half minute and one second pulse; maximum of 80 clocks						
Over two loops only	510.05	611.19	4.40	81.55	nr	**692.74**
Power supplies for above, giving 24 hours power reserve						
2 6 Amp hour batteries	252.50	302.57	3.00	55.60	nr	**358.17**
2 15 Amp hour batteries	328.25	393.34	5.00	92.67	nr	**486.01**
Radio receiver to accept BBC Rugby Transmitter MSF signal						
To synchronise time of above Quartz master clock	207.05	248.11	7.04	130.52	nr	**378.63**
Wall clocks for slave (impulse) systems;24V DC, white dial with black numerals fitted with axxispolycarbonate disc; BS 467.7 Class O						
227mm diameter 1 minute impulse	50.50	60.51	0.62	11.49	nr	**72.01**
305mm diameter 1 minute impulse	50.50	60.51	1.37	25.39	nr	**85.91**
227mm diameter 1/2 minute impulse	50.50	60.51	1.37	25.39	nr	**85.91**
305mm diameter 1/2 minute impulse	50.50	60.51	1.37	25.39	nr	**85.91**
227mm diameter 1 second impulse	70.19	84.11	1.37	25.39	nr	**109.50**
305mm diameter 1 second impulse	70.19	84.11	1.37	25.39	nr	**109.50**
Quartz Battery Movement; BS 467.7 Class O; white dial with black numerals and sweep second hand; fitted with axxispolycarbonate disc; stove enamel case						
227mm diameter	23.23	27.84	0.77	14.27	nr	**42.11**
305mm diameter	25.25	30.26	0.77	14.27	nr	**44.53**
Internal wall mounted electric clock; white dial with black numerals; 240v, 50 Hz						
227mm diameter	27.77	33.28	1.00	18.53	nr	**51.81**
305mm diameter	29.80	35.71	1.00	18.53	nr	**54.24**
458mm diameter	131.30	157.34	1.00	18.53	nr	**175.87**

Material Costs/Measured Work Prices – Electrical Installations

W:COMMUNICATIONS/SECURITY/CONTROL

Item	Net Price £	Material £	Labour hours	Labour £	Unit	Total rate £
Elapsed time clock; BS 467.7 Class O; 240V AC, 50 Hz mains supply; 12 hour duration; dial with 0-55 and 1-12 duration; remote control facility; IP66; axxispolycarbonate disc; spun metal movement cover; 6 point fixing bezel						
227mm diameter	555.50	665.66	0.62	11.49	nr	**677.15**
Matching clock; BS 467.7 Class O; 240V AC, 50/60 Hz mains supply; 12 hour duration; dial with 1-12; IP 66; axxispolycarbonate disc; spun metal movement cover; semi flush mount on 6 point fixing bezel						
227mm diameter	196.95	236.01	0.62	11.49	nr	**247.50**
Remote Control Unit; BS 1363; integral Stop/Start and Reset; 2 gang; flush or surface mounted						
Remote Control Unit	60.60	72.62	0.77	14.27	nr	**86.89**
Digital clocks; 240V, 50 hz supply; with/without synchronisation from Masterclock;12/24 hour display;stand alone operation; 50mm digits						
Type hours/minutes/seconds or minutes/seconds/10th seconds	227.25	272.31	0.57	10.56	nr	**282.88**
Type hours/minutes or minutes/seconds	202.00	242.06	0.57	10.56	nr	**252.62**

542

Material Costs/Measured Work Prices – Electrical Installations

W:COMMUNICATIONS/SECURITY/CONTROL

Item	Net Price £	Material £	Labour hours	Labour £	Unit	Total rate £
W30 : DATA TRANSMISSION						
Cabinets						
Floor standing; suitable for 19" patch panels with glass lockable doors, metal rear doors, side panels, vertical cable management, 2 x 4 way PDU's, 4 way fan, earth bonding kit; installed on raised floor						
600 wide x 800 deep - 18U	600.00	711.12	4.00	60.68	nr	771.80
600 wide x 800 deep - 24U	625.00	740.75	4.00	60.68	nr	801.43
600 wide x 800 deep - 33U	670.00	794.08	4.00	60.68	nr	854.77
600 wide x 800 deep - 42U	720.00	853.34	4.00	60.68	nr	914.03
600 wide x 800 deep - 47U	760.00	900.75	4.00	60.68	nr	961.44
800 wide x 800 deep - 42U	815.00	965.94	4.00	60.68	nr	1026.62
800 wide x 800 deep - 47U	850.00	1007.42	4.00	60.68	nr	1068.10
Traffolyte cabinet label	2.00	2.37	0.25	3.79	nr	6.16
Wall mounted; suitable for 19" patch panels with glass lockable doors, side panels, vertical cable management, 2 x 4 way PDU's, 4 way fan, earth bonding kit; fixed to wall						
19 wide x 500 deep - 9U	400.00	474.08	3.00	45.51	nr	519.59
19 wide x 500 deep - 12U	410.00	485.93	3.00	45.51	nr	531.44
19 wide x 500 deep - 15U	425.00	503.71	3.00	45.51	nr	549.22
19 wide x 500 deep - 18U	500.00	592.60	3.00	45.51	nr	638.11
19 wide x 500 deep - 21U	520.00	616.30	3.00	45.51	nr	661.82
Cabinet label	1.90	2.25	0.25	4.63	nr	6.89
Frames						
Floor standing; suitable for 19" patch panels with supports, vertical cable management, earth bonding kit; installed on raised floor						
19 wide x 500 deep - 25U	450.00	533.34	2.50	37.93	nr	571.27
19 wide x 500 deep - 39U	550.00	651.86	2.50	37.93	nr	689.79
19 wide x 500 deep - 42U	600.00	711.12	2.50	37.93	nr	749.05
19 wide x 500 deep - 47U	650.00	770.38	2.50	37.93	nr	808.31
Frame label	1.90	2.25	0.25	3.79	nr	6.04
Patch Panels						
Category 5; 19" wide fully loaded, finished in black including termination and forming of cables						
16 port - RJ45 UTP - Krone / 110	70.00	82.96	3.25	49.31	nr	132.27
24 port - RJ45 UTP - Krone / 110	90.00	106.67	4.75	72.06	nr	178.73
32 port - RJ45 UTP - Krone / 110	130.00	154.08	6.30	95.58	nr	249.65
48 port - RJ45 UTP - Krone / 110	175.00	207.41	9.35	141.85	nr	349.26
Patch panel labeling per port	0.20	0.24	0.17	2.58	nr	2.82

Material Costs/Measured Work Prices – Electrical Installations

W:COMMUNICATIONS/SECURITY/CONTROL

Item	Net Price £	Material £	Labour hours	Labour £	Unit	Total rate £
Category 6; 19" wide fully loaded, finished in black including termination and forming of cables						
16 port - RJ45 UTP - Krone / 110	95.00	112.59	3.40	51.58	nr	**164.17**
24 port - RJ45 UTP - Krone / 110	120.00	142.22	5.00	75.85	nr	**218.08**
32 port - RJ45 UTP - Krone / 110	170.00	201.48	6.60	100.13	nr	**301.61**
48 port - RJ45 UTP - Krone / 110	230.00	272.60	9.80	148.67	nr	**421.27**
Patch panel labeling per port	0.20	0.24	0.17	2.58	nr	**2.82**
Cat 3 / Voice; 19" wide fully loaded finished in black including termination and forming of cables (assuming 2 pairs per port)						
16 port - RJ45 UTP - Krone	75.00	88.89	2.00	30.34	nr	**119.23**
24 port - RJ45 UTP - Krone	90.00	106.67	2.60	39.44	nr	**146.11**
32 port - RJ45 UTP - Krone	130.00	154.08	3.25	49.31	nr	**203.38**
48 port - RJ45 UTP - Krone	140.00	165.93	4.65	70.54	nr	**236.47**
900 pair fully loaded PB type frame including forming and termination of 9 x 100 pair cables	490.00	580.75	25.00	379.27	nr	**960.02**
Installation and termination of Krone strip (10 pair block - 237A) including designation label	5.50	6.52	0.50	7.59	nr	**14.10**
Patch panel labeling per port	0.20	0.24	0.17	2.58	nr	**2.82**
Fibre; 19" wide fully loaded, labeled, aluminum alloy c/w couplers, fibre management and glands. (excludes termination of fibre cores)						
8 way ST; fixed drawer	60.00	71.11	0.50	7.59	nr	**78.70**
16 way ST; fixed drawer	90.00	106.67	0.50	7.59	nr	**114.25**
24 way ST; fixed drawer	125.00	148.15	0.50	7.59	nr	**155.74**
8 way ST; sliding drawer	80.00	94.82	0.50	7.59	nr	**102.40**
16 way ST; sliding drawer	110.00	130.37	0.50	7.59	nr	**137.96**
24 way ST; sliding drawer	145.00	171.85	0.50	9.27	nr	**181.12**
8 way (4 duplex) SC; fixed drawer	75.00	88.89	0.50	7.59	nr	**96.48**
16 way (8 duplex) SC; fixed drawer	110.00	130.37	0.50	7.59	nr	**137.96**
24 way (12 duplex) SC; fixed drawer	135.00	160.00	0.50	7.59	nr	**167.59**
8 way (4 duplex) SC; sliding drawer	95.00	112.59	0.50	7.59	nr	**120.18**
16 way (8 duplex) SC; sliding drawer	130.00	154.08	0.50	7.59	nr	**161.66**
24 way (12 duplex) SC; sliding drawer	155.00	183.71	0.50	7.59	nr	**191.29**
8 way (4 duplex) MTRJ; fixed drawer	85.00	100.74	0.50	7.59	nr	**108.33**
16 way (8 duplex) MTRJ; fixed drawer	130.00	154.08	0.50	7.59	nr	**161.66**
24 way (12 duplex) MTRJ; fixed drawer	155.00	183.71	0.50	7.59	nr	**191.29**
8 way (4 duplex) FC/PC; fixed drawer	95.00	112.59	0.50	7.59	nr	**120.18**
16 way (8 duplex) FC/PC; fixed drawer	145.00	171.85	0.50	7.59	nr	**179.44**
24 way (12 duplex) FC/PC; fixed drawer	175.00	207.41	0.50	7.59	nr	**215.00**
Patch panel label per way	0.20	0.24	0.17	2.58	nr	**2.82**

544 **Material Costs/Measured Work Prices – Electrical Installations**

W:COMMUNICATIONS/SECURITY/CONTROL

Item	Net Price £	Material £	Labour hours	Labour £	Unit	Total rate £
W30 : DATA TRANSMISSION (Continued)						
Patch leads						
Category 5; straight through booted RJ45 UTP - RJ45 UTP						
Patch lead 1m length	2.50	2.96	0.09	1.29	nr	**4.25**
Patch lead 3m length	3.60	4.09	0.09	1.29	nr	**5.38**
Patch lead 5m length	5.30	6.28	0.10	1.52	nr	**7.80**
Patch lead 7m length	7.00	8.30	0.10	1.52	nr	**9.81**
Category 6; straight through booted RJ45 UTP - RJ45 UTP						
Patch lead 1 m length	5.00	5.93	0.09	1.29	nr	**7.22**
Patch lead 3 m length	6.25	7.09	0.09	1.29	nr	**8.38**
Patch lead 5 m length	7.50	8.89	0.10	1.52	nr	**10.41**
Patch lead 7 m length	8.75	10.37	0.10	1.52	nr	**11.89**
Simplex 62.5/125 ST - ST						
Fibre patch lead 1 m length	9.50	11.26	0.08	1.21	nr	**12.47**
Fibre patch lead 3 m length	10.80	12.80	0.08	1.21	nr	**14.01**
Fibre patch lead 5 m length	12.10	14.34	0.10	1.52	nr	**15.86**
Simplex 62.5/125 ST - SC						
Fibre patch lead 1m length	11.90	14.10	0.08	1.21	nr	**15.32**
Fibre patch lead 3m length	13.20	15.64	0.08	1.21	nr	**16.86**
Fibre patch lead 5m length	14.50	17.19	0.10	1.52	nr	**18.70**
Duplex 62.5/125 MTRJ -MTRJ						
Fibre patch lead 1 m length	22.50	26.67	0.08	1.21	nr	**27.88**
Fibre patch lead 3 m length	24.50	29.04	0.08	1.21	nr	**30.25**
Fibre patch lead 5 m length	26.50	31.41	0.10	1.52	nr	**32.92**
Duplex 62.5/125 MTRJ -ST						
Fibre patch lead 1 m length	31.30	37.10	0.08	1.21	nr	**38.31**
Fibre patch lead 3 m length	33.30	39.47	0.08	1.21	nr	**40.68**
Fibre patch lead 5 m length	35.30	41.84	0.10	1.52	nr	**43.35**
Duplex 62.5/125 ST - ST						
Fibre patch lead 1 m length	13.00	15.41	0.08	1.21	nr	**16.62**
Fibre patch lead 3 m length	16.00	18.96	0.08	1.21	nr	**20.18**
Fibre patch lead 5 m length	15.00	17.78	0.10	1.52	nr	**19.30**
Duplex 62.5/125 ST - SC						
Fibre patch lead 1 m length	14.50	17.19	0.08	1.21	nr	**18.40**
Fibre patch lead 3 m length	16.00	18.96	0.08	1.21	nr	**20.18**
Fibre patch lead 5 m length	17.00	20.15	0.10	1.52	nr	**21.67**
Duplex 62.5/125 SC - SC						
Fibre patch lead 1 m length	17.00	20.15	0.08	1.21	nr	**21.36**
Fibre patch lead 3 m length	19.00	22.52	0.08	1.21	nr	**23.73**
Fibre patch lead 5 m length	21.00	24.89	0.10	1.52	nr	**26.41**

Material Costs/Measured Work Prices – Electrical Installations

W:COMMUNICATIONS/SECURITY/CONTROL

Item	Net Price £	Material £	Labour hours	Labour £	Unit	Total rate £
Data Cabling						
Unshielded twisted pair; solid copper conductors; PVC insulation; nominal impedance 100 Ohm; Cat 5e to ISO 11801, EIA / TIA 568-A and EN 50173 standards to the current revisions						
4 pair 24AWG; nominal outside diameter 5.6mm; installed above ceiling	0.18	0.21	0.02	0.23	m	**0.44**
4 pair 24AWG; nominal outside diameter 5.6mm; installed in riser	0.18	0.21	0.02	0.23	m	**0.44**
4 pair 24AWG; nominal outside diameter 5.6mm; installed below floor	0.18	0.21	0.01	0.12	m	**0.33**
4 pair 24AWG; nominal outside diameter 5.6mm; installed in trunking	0.18	0.21	0.02	0.30	m	**0.52**
Category 5 cable test	-	-	0.20	3.03	nr	**3.03**
Unshielded twisted pair; solid copper conductors; LSOH sheathed; nominal impedance 100 Ohm; Cat 5e to ISO 11801, EIA / TIA 568-A and EN 50173 standards to the current revisions						
4 pair 24AWG; nominal outside diameter 5.6mm; installed above ceiling	0.23	0.27	0.02	0.23	m	**0.50**
4 pair 24AWG; nominal outside diameter 5.6mm; installed in riser	0.23	0.27	0.02	0.23	m	**0.50**
4 pair 24AWG; nominal outside diameter 5.6mm; installed below floor	0.23	0.27	0.01	0.12	m	**0.39**
4 pair 24AWG; nominal outside diameter 5.6mm; installed in trunking	0.23	0.27	0.02	0.30	m	**0.58**
Category 5 cable test	-	-	0.20	3.03	nr	**3.03**
Unshielded twisted pair; solid copper conductors; PVC insulation; nominal impedance 100 Ohm; Cat 6 to ISO 11801, EIA / TIA 568-A and EN 50173 standards to the current revisions						
4 pair 24AWG; nominal outside diameter 5.6mm; installed above ceiling	0.22	0.26	0.02	0.23	m	**0.49**
4 pair 24AWG; nominal outside diameter 5.6mm; installed in riser	0.22	0.26	0.02	0.23	m	**0.49**
4 pair 24AWG; nominal outside diameter 5.6mm; installed below floor	0.22	0.26	0.01	0.12	m	**0.38**
4 pair 24AWG; nominal outside diameter 5.6mm; installed in trunking	0.22	0.26	0.02	0.30	m	**0.56**
Category 6 cable test	-	-	0.22	3.34	nr	**3.34**
Unshielded twisted pair; solid copper conductors; LSOH sheathed; nominal impedance 100 Ohm; Cat 6 to ISO 11801, EIA / TIA 568-A and EN 50173 standards to the current revisions						
4 pair 24AWG; nominal outside diameter 5.6mm; installed above ceiling	0.27	0.32	0.02	0.23	m	**0.55**
4 pair 24AWG; nominal outside diameter 5.6mm; installed in riser	0.27	0.32	0.02	0.23	m	**0.55**

Material Costs/Measured Work Prices – Electrical Installations

W:COMMUNICATIONS/SECURITY/CONTROL

Item	Net Price £	Material £	Labour hours	Labour £	Unit	Total rate £
W30 : DATA TRANSMISSION (Continued)						
Unshielded twisted pair; solid copper conductors; LSOH sheathed; nominal impedance 100 Ohm; Cat 6 to ISO 11801, EIA / TIA 568-A and EN 50173 standards to the current revisions						
4 pair 24AWG; nominal outside diameter 5.6mm; installed below floor	0.27	0.32	0.01	0.12	m	**0.44**
4 pair 24AWG; nominal outside diameter 5.6mm; installed in trunking	0.27	0.32	0.02	0.30	m	**0.62**
Category 6 cable test	-	-	0.22	3.34	nr	**3.34**
Fibre optic cable, tight buffered, internal/external application, singlemode, LSOH sheathed						
4 core fibre optic cable	1.65	1.96	0.16	2.43	m	**4.38**
8 core fibre optic cable	2.60	3.08	0.16	2.43	m	**5.51**
12 core fibre optic cable	3.25	3.85	0.16	2.43	m	**6.28**
16 core fibre optic cable	4.00	4.74	0.16	2.43	m	**7.17**
24 core fibre optic cable	4.75	5.63	0.16	2.43	m	**8.06**
Singlemode core test per core	-	-	0.25	3.79	nr	**3.79**
Fibre optic cable, tight buffered, internal/external application, 62.5/125 multimode fibre, LSOH sheathed						
4 core fibre optic cable	1.75	2.07	0.16	2.43	m	**4.50**
8 core fibre optic cable	2.75	3.26	0.16	2.43	m	**5.69**
12 core fibre optic cable	3.50	4.15	0.16	2.43	m	**6.58**
16 core fibre optic cable	4.30	5.10	0.16	2.43	m	**7.52**
24 core fibre optic cable	5.00	5.93	0.16	2.43	m	**8.35**
Multimode core test per core	-	-	0.25	3.79	nr	**3.79**
Fibre optic single and multimode connectors and couplers						
ST singlemode booted connector	-	7.35	0.25	3.79	nr	**11.14**
ST multimode booted connector	-	3.56	0.25	3.79	nr	**7.35**
SC simplex singlemode booted connector	8.80	10.43	0.25	3.79	nr	**14.22**
SC simplex multimode booted connector	3.20	3.79	0.33	5.01	nr	**8.80**
SC duplex multimode booted connector	6.40	7.59	0.33	5.01	nr	**12.59**
ST - SC duplex adaptor	14.00	16.59	-	-	nr	**16.59**
ST inline bulkhead coupler	3.00	3.56	-	-	nr	**3.56**
SC duplex coupler	5.60	6.64	-	-	nr	**6.64**
Voice cabling						
Low speed data; unshielded twisted pair; solid copper conductors; PVC insulation; nominal impedance 100 Ohm; Category 3 to ISO IS 11801/EIA / TIA 568-A and EN50173 standards to current revisions						
Installed in riser						
25 pair 24AWG	1.00	1.19	0.03	0.46	m	**1.64**

Material Costs/Measured Work Prices – Electrical Installations

W:COMMUNICATIONS/SECURITY/CONTROL

Item	Net Price £	Material £	Labour hours	Labour £	Unit	Total rate £
50 pair 24AWG	2.00	2.37	0.06	0.91	m	**3.28**
100 pair 24AWG	3.50	4.15	0.10	1.52	m	**5.67**
Installed below floor						
25 pair 24AWG	1.00	1.19	0.02	0.30	m	**1.49**
50 pair 24AWG	2.00	2.37	0.05	0.76	m	**3.13**
100 pair 24AWG	3.50	4.15	0.08	1.21	m	**5.36**
Cat 3 cable circuit test per pair	-	-	0.08	1.21	nr	**1.21**
Low speed data; unshielded twisted pair; solid copper conductors; LSOH sheath; nominal impedance 100 Ohm; Category 3 to ISO IS 11801/EIA / TIA 568-A and EN50173 standards to current revisions						
Installed in riser						
25 pair 24AWG	1.25	1.48	0.03	0.46	m	**1.94**
50 pair 24AWG	2.50	2.96	0.06	0.91	m	**3.87**
100 pair 24AWG	4.20	4.98	0.10	1.52	m	**6.49**
Installed below floor						
25 pair 24AWG	1.25	1.48	0.02	0.30	m	**1.78**
50 pair 24AWG	2.50	2.96	0.05	0.76	m	**3.72**
100 pair 24AWG	4.20	4.98	0.08	1.21	m	**6.19**
Cat 3 cable circuit test per pair	-	-	0.08	1.21	nr	**1.21**
Accessories						
Category 5 RJ45 data outlet plate and multiway outlet boxes for wall, ceiling and below floor installations including label to ISO 11801 standards						
Wall mounted; fully loaded						
One gang LSOH PVC plate	6.00	7.11	0.25	3.79	nr	**10.90**
Two gang LSOH PVC plate	8.50	10.07	0.33	5.01	nr	**15.08**
Four gang LSOH PVC plate	14.60	17.30	0.66	10.01	nr	**27.32**
One gang satin brass plate	11.00	13.04	0.25	3.79	nr	**16.83**
Two gang satin brass plate	13.50	16.00	0.33	5.01	nr	**21.01**
Ceiling mounted; fully loaded						
One gang metal clad plate	7.00	8.30	0.33	5.01	nr	**13.30**
Two gang metal clad plate	9.50	11.26	0.45	6.83	nr	**18.09**
Below floor; fully loaded						
Four way outlet box, 5 m length 20mm flexible conduit with glands and starin relief bracket	21.50	25.48	0.50	7.59	nr	**33.07**
Six way outlet box, 5 m length 25mm flexible conduit with glands and strain relief bracket	27.80	32.95	0.75	11.38	nr	**44.33**
Eight way outlet box, 5 m length 25mm flexible conduit with glands and strain relief bracket	35.00	41.48	1.00	15.17	nr	**56.65**
Ten way outlet box, 5 m length 32mm flexible conduit with glands and strain relief bracket	60.00	71.11	1.25	18.96	nr	**90.08**
Installation of outlet boxes to desks	-	-	0.40	6.07	nr	**6.07**

Material Costs/Measured Work Prices – Electrical Installations

W:COMMUNICATIONS/SECURITY/CONTROL

Item	Net Price £	Material £	Labour hours	Labour £	Unit	Total rate £
W30 : DATA TRANSMISSION (Continued)						
Category 6 RJ45 data outlet plate and multiway outlet boxes for wall, ceiling and below floor installations including label to ISO 11801 standards						
Wall mounted; fully loaded						
One gang LSOH PVC plate	7.50	8.89	0.25	3.79	nr	**12.68**
Two gang LSOH PVC plate	11.00	13.04	0.33	5.01	nr	**18.04**
Four gang LSOH PVC plate	19.80	23.47	0.66	10.01	nr	**33.48**
One gang satin brass plate	12.50	14.81	0.25	3.79	nr	**18.61**
Two gang satin brass plate	16.00	18.96	0.33	5.01	nr	**23.97**
Ceiling mounted; fully loaded						
One gang metal clad plate	8.50	10.07	0.33	5.01	nr	**15.08**
Two gang metal clad plate	12.00	14.22	0.45	6.83	nr	**21.05**
Below floor; fully loaded						
Four way outlet box, 5 m length 20mm flexible conduit with glands and starin relief bracket	26.50	31.41	0.50	7.59	nr	**38.99**
Six way outlet box, 5 m length 25mm flexible conduit with glands and strain relief bracket	35.30	41.84	0.75	11.38	nr	**53.22**
Eight way outlet box, 5 m length 25mm flexible conduit with glands and strain relief bracket	45.00	53.33	1.00	15.17	nr	**68.50**
Ten way outlet box, 5 m length 32mm flexible conduit with glands and strain relief bracket	66.00	78.22	1.25	18.96	nr	**97.19**
Installation of outlet boxes to desks	-	-	0.40	6.07	nr	**6.07**

Material Costs/Measured Work Prices – Electrical Installations

W:COMMUNICATIONS/SECURITY/CONTROL

Item	Net Price £	Material £	Labour hours	Labour £	Unit	Total rate £
W50 : FIRE DETECTION AND ALARM						
STANDARD FIRE DETECTION CONTROL PANEL						
Zone control panel; 2 x 12 volt batteries/charge up to 48 hours standby; mild steel case; flush or surface mounting						
1 zone	135.00	161.77	3.00	55.60	nr	**217.37**
2 zone	145.00	173.75	3.51	65.03	nr	**238.79**
4 zone	220.00	263.63	4.00	74.14	nr	**337.76**
8 zone	350.00	419.40	5.00	92.67	nr	**512.07**
12 zone	457.00	547.62	6.00	111.20	nr	**658.83**
16 zone	645.00	772.90	6.00	111.20	nr	**884.11**
24 zone	868.00	1040.12	6.00	111.20	nr	**1151.33**
Repeater Panels						
8 zone	237.00	284.00	4.00	74.14	nr	**358.13**
EQUIPMENT						
Manual call point units: plastic covered						
Surface mounted						
Call point	7.61	9.12	0.50	9.27	nr	**18.39**
Call point; Weatherproof	54.78	65.64	0.80	14.83	nr	**80.47**
Flush mounted						
Call point	6.63	7.94	0.56	10.38	nr	**18.33**
Call point; Weatherproof	51.51	61.72	0.86	15.95	nr	**77.67**
Detectors						
Smoke, ionisation type with mounting base	26.60	31.87	0.75	13.90	nr	**45.78**
Smoke, optical type with mounting base	26.60	31.87	0.75	13.90	nr	**45.78**
Fixed temperature heat detector with mounting base (60 degrees)	21.41	25.66	0.75	13.90	nr	**39.56**
Rate of Rise heat detector with mounting base (90 degrees)	20.07	24.05	0.75	13.90	nr	**37.95**
Duct detector including optical smoke detector and base	169.51	203.12	2.00	37.07	nr	**240.19**
Remote smoke detector LED indicator with base	6.85	8.21	0.50	9.27	nr	**17.48**
Sounders						
6" bell, conduit box	13.69	16.40	0.75	13.90	nr	**30.31**
6" bell, conduit box; weatherproof	18.74	22.46	0.75	13.90	nr	**36.36**
Siren; 230V	37.62	45.08	1.25	23.17	nr	**68.25**
Magnetic Door Holder; 230V ; surface fixed	38.13	45.69	1.50	27.83	nr	**73.52**

550   *Material Costs/Measured Work Prices – Electrical Installations*

W:COMMUNICATIONS/SECURITY/CONTROL

Item	Net Price £	Material £	Labour hours	Labour £	Unit	Total rate £
W50 : FIRE DETECTION AND ALARM						
ADDRESSABLE FIRE DETECTION						
CONTROL PANEL						
Analogue addressable panel; BS EN54 Part 2 and 4 1998; incorporating 120 addresses per loop (maximum 1-2km length); sounders wired on loop; sealed lead acid integral battery standby providing 48 hour standby; 24 volt DC; mild steel case; surface fixed						
1 loop; 4 x 12 volt batteries	771.84	771.84	6.00	111.20	nr	**883.04**
Extra for 1 loop panel						
Loop card	154.44	154.44	1.00	18.53	nr	**172.97**
Repeater panel	392.40	392.40	6.00	111.20	nr	**503.60**
Network nodes	823.32	823.32	6.00	111.20	nr	**934.52**
Interface unit; for other systems						
Mains powered	138.60	138.60	1.50	27.80	nr	**166.40**
Loop powered	101.52	101.52	1.00	18.53	nr	**120.05**
Single channel I/O	35.64	35.64	1.00	18.53	nr	**54.17**
Zone module	50.40	50.40	1.50	27.80	nr	**78.20**
4 loop; 4 x 12 volt batteries; 24 hour standby; 30 minute alarm	1234.44	1234.44	8.00	148.27	nr	**1382.71**
Extra for 4 loop panel						
Loop card	154.44	154.44	1.00	18.53	nr	**172.97**
Repeater panel	823.32	823.32	6.00	111.20	nr	**934.52**
Mimic panel	1242.00	1242.00	5.00	92.67	nr	**1334.67**
Network nodes	823.20	823.20	6.00	111.20	nr	**934.40**
Interface unit; for other systems						
Mains powered	138.60	138.60	1.50	27.80	nr	**166.40**
Loop powered	101.52	101.52	1.00	18.53	nr	**120.05**
Single channel I/O	35.64	35.64	1.00	18.53	nr	**54.17**
Zone module	50.40	50.40	1.50	27.80	nr	**78.20**
Line modules	5.15	5.15	1.00	18.53	nr	**23.68**
8 loop; 4 x 12 volt batteries; 24 hour standby; 30 minute alarm	2468.88	2468.88	12.00	222.41	nr	**2691.29**
Extra for 8 loop panel						
Loop card	154.44	154.44	1.00	18.53	nr	**172.97**
Repeater panel	823.32	823.32	6.00	111.20	nr	**934.52**
Mimic panel	1242.00	1242.00	5.00	92.67	nr	**1334.67**
Network nodes	823.20	823.20	6.00	111.20	nr	**934.40**
Interface unit; for other systems						
Mains powered	138.60	138.60	1.50	27.80	nr	**166.40**
Loop powered	101.52	101.52	1.00	18.53	nr	**120.05**
Single channel I/O	35.64	35.64	1.00	18.53	nr	**54.17**
Zone module	50.40	50.40	1.50	27.80	nr	**78.20**
Line modules	5.15	5.15	1.00	18.53	nr	**23.68**

Material Costs/Measured Work Prices – Electrical Installations

W:COMMUNICATIONS/SECURITY/CONTROL

Item	Net Price £	Material £	Labour hours	Labour £	Unit	Total rate £
EQUIPMENT						
Manual Call Point						
Surface mounted						
Call point	31.91	38.24	1.00	18.53	nr	**56.77**
Call point; Weather proof	66.95	80.23	1.25	23.17	nr	**103.39**
Flush mounted						
Call point	31.24	37.43	1.00	18.53	nr	**55.97**
Call point; Weather proof	47.72	57.18	1.25	23.17	nr	**80.35**
Detectors						
Smoke, ionisation type with mounting base	44.96	53.88	0.75	13.90	nr	**67.78**
Smoke, optical type with mounting base	38.59	46.24	0.75	13.90	nr	**60.15**
Fixed temperature heat detector with mounting base (60 degrees)	41.36	49.56	0.75	13.90	nr	**63.47**
Rate of Rise heat detector with mounting base (90 degrees)	41.36	49.56	0.75	13.90	nr	**63.47**
Duct Detector including optical smoke detector and addressable base	180.46	216.25	2.00	37.07	nr	**253.31**
Beam smoke detector with transmitter and receiver unit	376.98	451.74	2.00	37.07	nr	**488.80**
Zone short circuit isolator	31.40	37.63	0.75	13.90	nr	**51.53**
Plant interface unit	23.13	27.72	0.50	9.27	nr	**36.98**
Sounders						
Xenon flasher, 24 volt, conduit box	19.28	23.10	0.50	9.27	nr	**32.37**
Xenon flasher, 24 volt, conduit box; weatherproof	32.78	39.28	0.50	9.27	nr	**48.55**
6" bell, conduit box	13.69	16.40	0.75	13.90	nr	**30.31**
6" bell, conduit box; weatherproof	18.74	22.46	0.75	13.90	nr	**36.36**
Siren; 24V polarised	13.26	15.89	1.00	18.53	nr	**34.42**
Siren; 240V	37.62	45.08	1.25	23.17	nr	**68.25**
Magnetic Door Holder; 240V ; surface fixed	38.13	45.69	1.50	27.83	nr	**73.52**

Material Costs/Measured Work Prices – Electrical Installations

W:COMMUNICATIONS/SECURITY/CONTROL

Item	Net Price £	Material £	Labour hours	Labour £	Unit	Total rate £
W51 : EARTHING AND BONDING						
EARTH BAR						
Earth bar; polymer insulators and base mounting; including connections						
Non disconnect link						
6 way	93.69	112.00	0.81	15.01	nr	**127.01**
8 way	123.21	147.29	0.81	15.01	nr	**162.30**
10 way	126.41	151.11	0.81	15.01	nr	**166.12**
Disconnect link						
6 way	105.11	125.65	1.01	18.72	nr	**144.37**
8 way	124.01	148.24	1.01	18.72	nr	**166.96**
10 way	137.21	164.02	1.01	18.72	nr	**182.74**
Solid earth bar; including connections						
150 x 50 x 6 mm	29.75	35.56	1.01	18.72	nr	**54.28**
Extra for earthing						
Disconnecting link						
300 x 50 x 6mm	31.34	37.46	1.16	21.50	nr	**58.96**
500 x 50 x 6mm	31.34	37.46	1.16	21.50	nr	**58.96**
Crimp lugs; including screws and connections to cable						
25 mm	0.33	0.39	0.31	5.75	nr	**6.14**
35 mm	0.50	0.60	0.31	5.75	nr	**6.34**
50 mm	0.66	0.79	0.32	5.93	nr	**6.72**
70 mm	0.99	1.18	0.32	5.93	nr	**7.11**
95 mm	1.19	1.42	0.46	8.53	nr	**9.95**
120 mm	1.25	1.49	1.25	23.17	nr	**24.66**
Earth clamps; connection to pipework						
15mm to 32mm dia	0.66	0.79	0.15	2.78	nr	**3.57**
32mm to 50mm dia	0.82	0.98	0.18	3.34	nr	**4.32**
50mm to 75mm dia	0.97	1.16	0.20	3.71	nr	**4.87**

Material Costs/Measured Work Prices – Electrical Installations

W:COMMUNICATIONS/SECURITY/CONTROL

Item	Net Price £	Material £	Labour hours	Labour £	Unit	Total rate £
W52 : LIGHTNING PROTECTION						
CONDUCTOR TAPE						
PVC sheathed copper tape						
25 x 3 mm	4.47	5.62	0.30	5.56	m	**11.18**
25 x 6 mm	7.71	9.70	0.30	5.56	m	**15.26**
50 x 6 mm	16.47	20.72	0.30	5.56	m	**26.28**
PVC sheathed copper solid circular conductor						
8mm	2.86	3.43	0.50	9.27	m	**12.69**
Bare copper tape						
20 x 3 mm	2.89	3.46	0.30	5.56	m	**9.02**
25 x 3 mm	3.00	3.78	0.30	5.56	m	**9.34**
25 x 6 mm	6.00	7.55	0.40	7.41	m	**14.96**
50 x 6 mm	11.99	14.36	0.50	9.27	m	**23.63**
Bare copper solid circular conductor						
8mm	2.04	2.44	0.50	9.27	m	**11.71**
Tape fixings; flat; metallic						
PVC sheathed copper						
25 x 3 mm	1.10	1.39	0.33	6.12	nr	**7.50**
25 x 6 mm	1.73	2.18	0.33	6.12	nr	**8.30**
50 x 6 mm	3.77	4.75	0.33	6.12	nr	**10.86**
8mm	3.24	4.07	0.50	9.27	nr	**13.34**
Bare copper						
20 x 3 mm	2.89	3.64	0.30	5.56	nr	**9.20**
25 x 3 mm	3.00	3.78	0.30	5.56	nr	**9.34**
25 x 6 mm	6.00	7.55	0.40	7.41	nr	**14.96**
50 x 6 mm	11.99	14.36	0.50	9.27	nr	**23.63**
8mm	3.24	4.07	0.50	9.27	nr	**13.34**
Tape fixings; flat; non-metallic; PVC sheathed copper						
25 x 3 mm	0.39	0.50	0.30	5.56	nr	**6.06**
Tape fixings; flat; non-metallic; Bare copper						
20 x 3 mm	0.37	0.47	0.30	5.56	nr	**6.03**
25 x 3 mm	0.37	0.47	0.30	5.56	nr	**6.03**
50 x 6 mm	0.94	1.13	0.30	5.56	nr	**6.69**
Puddle flanges; copper						
600 mm long	24.25	29.06	0.93	17.24	nr	**46.30**
AIR RODS						
Pointed air rod fixed to structure; copper						
10 mm diameter						
500 mm long	6.09	7.29	1.00	18.53	nr	**25.83**
1000mm long	9.23	11.06	1.50	27.80	nr	**38.86**

554   **Material Costs/Measured Work Prices – Electrical Installations**

W:COMMUNICATIONS/SECURITY/CONTROL

Item	Net Price £	Material £	Labour hours	Labour £	Unit	Total rate £
W52 : LIGHTNING PROTECTION						
AIR RODS (Continued)						
Extra for 10mm dia air rod						
Air terminal base	8.25	9.88	0.35	6.49	nr	**16.37**
Strike Pad	11.90	14.26	0.35	6.49	nr	**20.75**
16 mm diameter						
500 mm long	8.68	10.40	0.91	16.87	nr	**27.26**
1000mm long	15.86	19.01	1.75	32.43	nr	**51.44**
2000mm long	28.96	34.71	2.50	46.33	nr	**81.04**
Extra for 16mm dia air rod						
Multiple point	16.83	20.17	0.35	6.49	nr	**26.65**
Air terminal base	8.75	10.48	0.35	6.49	nr	**16.97**
Ridge saddle	15.74	18.86	0.35	6.49	nr	**25.35**
Side mounting bracket	15.66	18.76	0.50	9.27	nr	**28.03**
Rod to tape coupling	7.30	8.75	0.50	9.27	nr	**18.01**
Strike Pad	11.90	14.26	0.35	6.49	nr	**20.75**
AIR TERMINALS						
16 mm diameter						
500 mm long	8.68	10.40	0.65	12.05	nr	**22.45**
1000mm long	15.86	19.01	0.78	14.46	nr	**33.46**
2000mm long	28.96	34.71	1.50	27.80	nr	**62.51**
Extra for 16mm dia air terminal						
Multiple point	16.83	20.17	0.35	6.49	nr	**26.65**
Flat saddle	15.74	18.86	0.35	6.49	nr	**25.35**
Side bracket	15.66	18.76	0.50	9.27	nr	**28.03**
Rod to cable coupling	7.30	8.75	0.50	9.27	nr	**18.01**
BONDS AND CLAMPS						
Bond to flat surface; copper						
26 mm	1.72	2.07	0.45	8.34	nr	**10.41**
8 mm diameter	7.01	8.40	0.33	6.12	nr	**14.51**
Pipe bond						
26 mm	4.75	5.69	0.45	8.34	nr	**14.03**
8 mm diameter	14.48	17.35	0.33	6.12	nr	**23.47**
Rod to tape clamp						
26 mm	13.19	15.81	0.45	8.34	nr	**24.15**
Square clamp; copper						
25 x 3 mm	2.99	3.59	0.33	6.12	nr	**9.70**
50 x 6 mm	13.63	16.33	0.50	9.27	nr	**25.60**
8 mm diameter	3.55	4.26	0.33	6.12	nr	**10.37**
Test clamp; copper						
26 x 8 mm; oblong	4.72	5.66	0.50	9.27	nr	**14.93**
26 x 8 mm; plate type	3.04	3.64	0.50	9.27	nr	**12.91**
26 x 8 mm; screw down	9.30	11.14	0.50	9.27	nr	**20.41**

Material Costs/Measured Work Prices – Electrical Installations

W:COMMUNICATIONS/SECURITY/CONTROL

Item	Net Price £	Material £	Labour hours	Labour £	Unit	Total rate £
Cast in earth points						
2 hole	9.30	11.14	0.75	13.90	nr	**25.05**
4 hole	14.78	17.71	1.00	18.53	nr	**36.25**
Extra for csat in earth points						
Cover plate; 25 x 3 mm	10.43	12.49	0.25	4.63	nr	**17.13**
Cover plate; 8 mm	10.43	12.49	0.25	4.63	nr	**17.13**
Rebar clamp; 8 mm	22.13	26.52	0.25	4.63	nr	**31.15**
Static earth receptacle	50.79	60.87	0.50	9.27	nr	**70.13**
Copper braided bonds						
25 x 3 mm						
200 mm hole centres	6.39	7.65	0.33	6.12	nr	**13.77**
400 mm holes centres	9.25	11.09	0.40	7.41	nr	**18.50**
U bolt clamps						
16 mm	3.64	4.36	0.33	6.12	nr	**10.48**
20 mm	4.13	4.95	0.33	6.12	nr	**11.07**
25 mm	5.06	6.07	0.33	6.12	nr	**12.18**
EARTH PITS/MATS						
Earth inspection pit; hand to others for fixing						
Concrete	22.31	26.73	1.00	18.53	nr	**45.27**
Polypropylene	22.02	26.38	1.00	18.53	nr	**44.92**
Extra for earth pit						
5 hole copper earth bar; concrete pit	16.51	19.78	0.35	6.49	nr	**26.27**
5 hole earth bar; polypropylene	14.08	16.87	0.35	6.49	nr	**23.36**
Water proof electrode seal						
Single flange	143.47	171.93	0.93	17.24	nr	**189.16**
Double flange	241.42	289.30	0.93	17.24	nr	**306.53**
Earth electrode mat; laid in ground and connected						
Copper tape lattice						
600 x 600 x 3 mm	39.10	46.85	0.93	17.24	nr	**64.09**
900 x 900 x 3 mm	70.25	84.17	0.93	17.24	nr	**101.41**
Copper tape plate						
600 x 600 x 1.5 mm	37.48	44.91	0.93	17.24	nr	**62.15**
600 x 600 x 3 mm	74.54	89.33	0.93	17.24	nr	**106.56**
900 x 900 x 1.5 mm	83.51	100.07	0.93	17.24	nr	**117.30**
900 x 900 x 3 mm	167.45	200.66	0.93	17.24	nr	**217.90**
EARTH RODS						
Solid cored copper earth electrodes driven into ground and connected						
15 mm diameter						
1200 mm long	10.66	12.78	0.93	17.24	nr	**30.01**

Material Costs/Measured Work Prices – Electrical Installations

W:COMMUNICATIONS/SECURITY/CONTROL

Item	Net Price £	Material £	Labour hours	Labour £	Unit	Total rate £
W52 : LIGHTNING PROTECTION						
EARTH ROD (Continued)						
Extra for 15mm dia earth rod						
Coupling	0.65	0.77	0.06	1.11	nr	1.89
Driving stud	0.80	0.95	0.06	1.11	nr	2.07
Spike	0.75	0.90	0.06	1.11	nr	2.01
Rod Clamp; flat tape	7.27	8.71	0.25	4.63	nr	13.34
Rod Clamp; solid conductor	1.44	1.73	0.25	4.63	nr	6.36
20 mm diameter						
1200 mm long	19.14	22.94	0.98	18.16	nr	41.10
Extra for 20mm dia earth rod						
Coupling	0.65	0.77	0.06	1.11	nr	1.89
Driving stud	1.32	1.58	0.06	1.11	nr	2.69
Spike	1.21	1.45	0.06	1.11	nr	2.57
Rod Clamp; flat tape	9.72	11.64	0.25	4.63	nr	16.28
Rod Clamp; solid conductor	1.62	1.94	0.25	4.63	nr	6.58
Stainless steel earth electrodes driven into ground and connected						
16 mm diameter						
1200 mm long	21.89	26.23	0.93	17.24	nr	43.47
Extra for 16mm dia earth rod						
Coupling	0.86	1.03	0.06	1.11	nr	2.14
Driving head	0.80	0.95	0.06	1.11	nr	2.07
Spike	0.75	0.90	0.06	1.11	nr	2.01
Rod Clamp; flat tape	7.27	8.71	0.25	4.63	nr	13.34
Rod Clamp; solid conductor	1.44	1.73	0.25	4.63	nr	6.36
SURGE PROTECTION						
Single Phase; including connection to equipment						
90 - 150v	184.28	220.82	5.00	92.67	nr	313.49
200 - 280v	184.28	220.82	5.00	92.67	nr	313.49
Three Phase; including connection to equipment						
156 - 260v	368.55	441.63	10.00	185.34	nr	626.97
346 - 484v	368.55	441.63	10.00	185.34	nr	626.97
349 - 484v; remote display	415.80	498.25	10.00	185.34	nr	683.59
346 - 484v; 60kA	708.75	849.30	10.00	185.34	nr	1034.63
346 - 484v; 120kA	1370.25	1641.97	10.00	185.34	nr	1827.31

PART THREE

Rates of Wages and Working Rules

Mechanical Installations, *page 557*
Electrical Installations, *page 595*

Mechanical Installations
RATES OF WAGES AND WORKING RULES

RATES OF WAGES

HEATING, VENTILATING, AIR CONDITIONING, PIPING AND
DOMESTIC ENGINEERING INDUSTRY

Extracts from National Agreement made between:

Heating and Ventilating and	Manufacturing Science Finance
Contractor's Association	Union (MSF)
ESCA House,	Park House,
34 Palace Court,	64-66 Wandsworth Common
Bayswater,	North Side
London W2 4JG	London SW18 2SH
Telephone: (020) 73134900	Telephone: (020) 8871 2100
Internet: www.hvac.org.uk	

WAGE RATES, ALLOWANCES AND OTHER PROVISIONS

Hourly rates of wages
All districts of the United Kingdom

Main Grades	*From* *3 September 2001* *p/hr*
Foreman	10.00
Senior Craftsman (+2nd welding skill)	8.96
Senior Craftsman	8.26
Craftsman (+2nd welding skill)	7.91
Craftsman	7.56
Operative	6.87
Adult Trainee	5.79
Mate (over 18)	5.79
Mate (17-18)	3.72
Mate (up to 17)	2.68
Traditional NJIC-based Craft Apprentices	
Year 3	5.70
Year 4	6.87
Modern Apprentices	
Junior	3.77
Intermediate	5.33
Senior	6.87

Note: Ductwork Erection Operatives are entitled to the same rates and allowances as the parallel Fitter grades shown.

Rates of Wages and Working Rules - Mechanical Installations

RATES OF WAGES

Trainee Rates of Pay

Junior Ductwork Trainees (Probationary)

Age at entry	From 3 September 2001 p/hr
17	3.09
18	3.72
19	4.75
20	5.79

Junior Ductwork Erectors (Year of Training)

	From 3 September 2001		
	1 yr	2 yr	3 yr
Age at entry	p/h	p/hr	p/hr
17	3.72	4.75	5.79
18	4.75	5.79	6.17
19	5.79	5.83	6.45
20	5.79	6.13	6.46

Responsibility Allowance (Craftsmen)	From 3 September 2001 p/hr
Second welding skill or supervisory responsibility (one unit)	0.35
Second welding skill and supervisory responsibility (two units)	0.70

Responsibility Allowance (Senior Craftsmen)	From 3 September 2001 p/hr
Second welding skill	0.35
Supervising responsibility	0.70
Second welding skill and supervisory responsibility	1.05

Daily travelling allowance

C: Craftsmen including Installers
M&A: Mates, Apprentices and Adult Trainees
Direct distance from centre to job in miles

		From 3 September 2001	
		C	M&A
Over	Not exceeding	p/hr	p/hr
10	20	3.85	3.31
20	30	6.90	5.95
30	40	9.08	7.83
40	50	11.33	9.71

558   *Rates of Wages and Working Rules - Mechanical Installations*

RATES OF WAGES

Weekly Holiday Credit and Welfare Contributions

	£ a	£ b	£ c	£ d	£ e	£ f	£ g	£ h
Weekly Holiday Credit	43.61	39.16	36.46	34.15	31.08	25.83	17.40	12.82
Combined Weekly/Welfare Holiday Credit and Contribution	48.16	43.71	41.01	38.70	35.63	30.38	21.95	17.37
Total	91.77	82.87	77.47	72.85	66.71	56.21	39.35	30.19

From
2 October 2000 (subject to review)

From
3 September 2001

Daily abnormal conditions money
per day   2.99 (No change)

Lodging allowance
£ per night   23.70

Weekly Sickness and Accident Benefit - Payable in accordance with the Rates of the WELPLAN Welfare and Holiday Scheme Supplement to the National Agreement.

From 3 September 2001 (No change)

	Weeks 1-28	Weeks 29-52
Category a	179.13	86.90
Category b	148.75	74.41
Category c	135.80	67.90
Category d	122.85	61.39
Category e	98.14	49.07
Category f	71.54	35.77
Category g	26.88	-
Category h	4.90	-
Category i		-

From 2 October 2000 (subject to review)

Death Benefit for Dependants	33,600
Accidental Dismemberment	17,000
Permanent Total Disability Benefit	
- Up to and including age 54	17,000
- Between ages 55-59	11,300
- Between ages 60-64	5,700
Index Benefits	
- Loss of four fingers or thumb	3,400
- Loss of index finger	2,300
- Loss of any other finger	500
- Loss of big toe	1,100
- Loss of any other toe	280

Rates of Wages and Working Rules - Mechanical Installations 559

RATES OF WAGES

Notes

1 From 2 October 2000, the grades of H&V Operatives covered by the range of credit values and entitled to the different rates of Sickness and Accident Benefit, Weekly Holiday Credit and Welfare Contribution are as follows:

a	*b*	*c*	*d*
Foreman	Senior Craftsman	Craftsman	Craftsman
Senior Craftsman	(RAS)	(+1st welding skill)	
(RAS and RAW)	Senior Craftsman		
	(RAW)		
	Senior Craftsman		
	Craftsman (+2 RA)		

e	*f*	*g*	*h*
Installer	Adult trainee	Craft Apprentice	Mate (under age 17)
Craft Apprentice	Mate over 18	- Year 2	
- Year 4	Craft Apprentice	Intermediate	
Senior Modern	- Year 3	Junior Modern Apprentice	
Apprentice	Intermediate	Mate (17-18 inclusive)	
	Modern Apprentice		

5 Payment of death benefit is subject to Inland Revenue requirements which currently provide that it may not exceed four times annual earnings of the deceased, subject to a minimum of □5,000.

6 Payment of sick pay, accidental dismemberment, benefits, death benefit and permanent total disability benefit is discretional and the amounts stated are the maxima.

560   *Rates of Wages and Working Rules - Mechanical Installations*

RATES OF WAGES

Authorised rates of wages agreed by the Joint Industry Board for the Plumbing Mechanical Engineering Services Industry in England and Wales

The Joint Industry Board for Plumbing Mechanical
Engineering Services in England and Wales,
Brook House, Brook Street,
St Neots, Huntingdon, Cambs. PE19 2HW
Telephone: 01480 476925
e-mail: postmaster@jib-pmes.compulink.co.uk

WAGE RATES, ALLOWANCES AND OTHER PROVISIONS

EFFECTIVE FROM 21 AUGUST 2000

Hourly rates of pay - inclusive of tool allowance

	Hourly rate £
Operatives	
Technical plumber and gas service technician	8.40
Advanced plumber and gas service engineer	7.56
Trained plumber and gas service fitter	6.48
Apprentices	
4th year of training with NVQ level 3*	6.28
4th year of training with NVQ level 2*	5.69
4th year of training	5.00
3rd year of training with NVQ level 2*	4.94
3rd year of training	4.06
2nd year of training	3.60
1st year of training	3.15
Adult Trainees	
3rd 6 months of employment	5.66
2nd 6 months of employment	5.42
1st 6 months of employment	5.07

* Where Apprentices have achieved NVQ's, the appropriate rate is payable from the date of attainment except that it shall not be any earlier than the commencement of the promulgated year of training in which it applies.

Wages for August 2001 agreement not promulgated at date of print.

Overtime
Overtime premium rates shall be paid after 39 hours are worked. The working week shall be 37.5 hours, after 8pm overtime shall be paid at double time.

Daily travelling allowance plus return fares
All daily travel allowances are to be paid at the daily rate as follows :

Over	*Not exceeding*	*All* *Operative*	*3rd & 4th* *Year* *Apprentices*	*1st & 2nd* *Year* *Apprentices*
10	20	£4.88	£3.08	£1.99
20	30	£6.54	£4.14	£2.63
30	40	£7.70	£4.95	£3.14
40	50	£8.82	£5.24	£3.32

Rates of Wages and Working Rules - Mechanical Installations

RATES OF WAGES

Notes on Daily Travel Allowances
(i)   Daily travel allowances are payable in addition to fares.
(ii)   The above allowances are paid at a daily rate with the distance calculated for the journey one way.
(iii)   For all distances over 50 miles operatives are to be paid Lodging Allowance.
(iv)   The Daily Travel Allowances as set above are to be paid when public transport is used.
(v)   When the employer provides transport or if alternative means of transport are used, then the employer and operative shall agree an appropriate allowance based on the actual time taken.

Mileage allowance .... £0.30 per mile

Responsibility money (Band 1) .... Up to £0.24 per hour

Responsibility money (Band 2) .... £0.25 to £0.44 per hour

Responsibility money (Band 3) .... £0.45 to £0.64 per hour

Responsibility money (Band 4) .... £0.65 to £0.84 per hour

Lodging allowance (Note 1) .... £18.00 per night

Subsistence Allowance (London Only) (Note 1) .... £4.18 per night

NOTES on Lodging and Subsistence Allowances

1)   When convenient lodgings cannot be secured or where the **Lodging Allowance** is found to be inadequate, an operative shall, with the prior approval of the employer, be reimbursed for the actual expenditure incurred for which a proper receipt shall be produced.
2)   Please note that by way of **concession** from the Inland Revenue the **Lodging Allowance** as shown above is **payable without the deduction of income tax**.
3)   The **Subsistence Allowance** is subject to Schedule E Income Tax through the PAYE System.

Plumbers welding supplement
Possession of Gas or Arc Certificate .... £0.27 per hour
Possession of Gas and Arc Certificate.. £0.46 per hour

Sickness with Pay and Accident Benefits
Effective from 6 April 1998

	Weeks 1-28		Weeks 29-52	
	Daily £	Weekly £	Daily £	Weekly £
Technical Plumber and Gas Service Technician	10.20	71.40	7.15	50.05
Advanced Plumber and Gas Service Engineer	8.70	60.90	5.65	39.55
Trained Plumber and Gas Service Fitter	7.50	52.50	4.40	30.80
Adult Trainee	7.50	52.50	4.40	30.80
Apprentice in last year of training	7.50	52.50	4.40	30.80
Apprentice 2nd to 3rd year of training	6.00	42.00	3.00	21.00
1st year Apprentice	1.00	7.00	N/A	N/A
Ancillary Operative	6.50	45.50	3.60	25.20

Accident Permanent Total Disability and Dismemberment Benefit   £3,500

Rates of Wages and Working Rules - Mechanical Installations

RATES OF WAGES

Death Benefit (effective 24 August 1998)   £

(a)   Operatives

Technical Plumber and Gas Service Technician   29,172
Advanced Plumber and Gas Service Engineer   26,130
Trained Plumber and Gas Service Fitter   23,322

(b)   Apprentices

Apprentice 4th year of Training with NVQ level 3   21,216
Apprentice 4th year of Training with NVQ Level 2   19,227
Apprentice 4th year of Training   16,887
Apprentice 3rd year of Training with NVQ Level 2   16,692
Apprentice 3rd year of Training   13,728
Apprentice 2nd year of Training   12,168
Apprentice 1st year of Training   10,218

(c)   Adult Trainee

Adult Trainee third 6 months of training   20,397
Adult Trainee second 6 months of training   19,539
Adult Trainee first 6 months of training   18,291

Annual Holiday with Pay (HWP)

	Holiday with Pay £
Technical Plumber and Gas Service Technician	22.77
Advance Plumber and Gas Service Engineer	20.30
Trained Plumber and Gas Service Fitter	18.16
Adult Trainee	18.16
Apprentice in last year of training	18.16
Apprentice 2nd and 3rd year	7.76
Apprentice 1st year	5.42
Ancillary Employee	14.86

Additional Holiday Pay (AHP)

PMES Operatives, Adult Trainees, Apprentices and Ancillary Employees who are in current membership of the AEEU at the time a holiday is taken, shall also be entitled to receive payment of AHP, the amount of which is to be based on the number of JIB for PMES stamps that are properly affixed to their card for a particular holiday.

Operatives, Adult Trainees & Ancillary Employee   £1.15
Apprentices   £1.00

Rates of Wages and Working Rules - Mechanical Installations 563

WORKING RULES

Extracts from National Agreement Working Rules (HVCA) incorporating amendments agreed up to May 1999 and published in the National Agreement handbook.

HOURS OF WORK (CLAUSE 3)

(a)　The normal working week shall consist of 38 hours to be worked in five days from Monday to Friday inclusive. The length of each normal working day shall be determined by the Employer but shall not be less than six hours or more than eight hours unless otherwise agreed between the Employer and the Operative concerned.

(b)　The Employer and the Operative concerned may agree to extend the working hours to more than 38 hours per week for particular jobs, provided that overtime shall be paid in accordance with Clause 9. (Attention is also drawn to the provisions for containing overtime).

MEAL AND TEA BREAKS (CLAUSE 4)

(a)　The normal break for lunch shall be one hour except when such a break would make it impossible for the normal working day to be worked, in which case the break may be reduced to not less than half an hour.

(b)　An Operative directed to start work before his normal starting time or to continue work after his normal finishing time shall be entitled to a quarter of an hour meal interval with pay at the appropriate overtime rate for each two hours of working (or part thereof exceeding one hour) in excess of the normal working day, which on Saturdays and Sundays shall mean eight hours. Where an Operative is entitled to a morning and/or evening meal interval under this clause, the meal interval shall replace the morning and/or afternoon tea break referred to in clause 4(c).

(c)　A tea break shall, subject to 4(b), be allowed in the morning and in the afternoon without loss of pay, provided that the Operative co-operates with the Employer in minimising the interruption to production. To this end the duration of the tea break shall be limited to the time necessary to drink tea and the tea shall be drunk at the Operatives workplace wherever possible.

GUARANTEED WEEK (CLAUSE 5)

(a)　Subject to the provisions of this clause an Operative who has been continuously employed by the same Employer for not less than two weeks is guaranteed wages equivalent to his inclusive hourly normal time earnings for 38 hours in any normal working week; provided that during working hours he is capable of, available for, and willing to perform satisfactorily the work associated with his usual occupation, or reasonable alternative work if his usual work is not available.

(b)　In the case of a week in which holidays recognised by agreement, custom or practice occur, the guaranteed week shall be reduced for each day of holiday by the normal working day as determined in Clause 3(a).

(c)　In the event of a dislocation of production as a result of industrial action the guarantee shall be automatically suspended. In the event of such dislocation being caused by Operatives working under other Agreements and the Operatives covered by this Agreement not being parties to the dislocation, the Employers shall, in accordance with Clause 5(d), endeavour to provide other work or if not able to do so will provide for the return of the Operatives to the shop or office from which they were sent. The Operatives will receive instructions as soon as is practicable as to proceedings to other work or return to shop.

(d)　The basis upon which the Employer shall endeavour to provide alternative work as required in Clause 5(c) shall be as follows:

(i)　where possible the Employer shall try to organise work on each job so as to provide a normal day's work for five days, Monday to Friday.

564   *Rates of Wages and Working Rules - Mechanical Installations*

WORKING RULES

GUARANTEED WEEK (CLAUSE 5) (CONTD)

    (ii)   where this is not possible on any particular job, the Employer shall endeavour to arrange to transfer Operatives to other sites to make up working hours to a normal day's work for five days, Monday to Friday.

    (iii)   where an Employer finds it impossible to provide a normal day's work for five days, Monday to Friday, he should rearrange the working hours in agreement with the Operatives concerned so that normal time earnings for 38 hours in the normal working week can be earned but in less than five days

    (iv)   where it is not possible to provide Operatives with a minimum of 38 hours during the week, rather than resort to dismissals a reduced working week may be agreed.

(e)   In the event of dislocation of production as a result of civil commotion, the guarantee shall be automatically suspended at the termination of the pay week after the dislocation first occurs and the Operative may be required by the Employer to register as an unemployed person.

GRADING DEFINITIONS (CLAUSE 6)

(a)   Operatives covered by the Agreement shall be graded in accordance with definitions in Clause 6e.

(b)   The rates of wages for each grade shall be agreed from time to time between the Association and the Union and shall be enumerated in an Appendix to this Agreement.

(c)   When an Operative is re-graded, the rate for the new job shall apply from the date of the re-grading.

(d)   Any dispute between an Employer and an Operative as to the grade which is appropriate for the Operative or in connection with any other matter relating to grading including a refusal or delay on the part of the Employer to consider re-grading shall be dealt with in accordance with Clause 25, Conciliation, of the Agreement.

(e)   The definitions of the grades shall be:

Mate

In order to secure maximum utilisation of manpower and optimum economic production, Mates shall be required to provide a range of support activities for Craftsmen and Senior Craftsmen. However, the work of a Mate shall not be confined to the manual work of fetching and carrying. Mates shall within their capability carry out semi-skilled tasks with one objective of improving productivity and the other objective of permitting those who wish to do so to qualify for consideration for appointment as an Adult Trainee. While not required to demonstrate developed technical skills, Mates shall be able to undertake semi-skilled repetitive tasks, including the use of power tools.

A Mate must also be aware of the basic safety requirements of the job, having had appropriate health and safety training.

All Mates will carry the same grade title, irrespective of age, although the grade will have three age-related pay rates.

Rates of Wages and Working Rules - Mechanical Installations 565

WORKING RULES

Adult Trainee

An Adult Trainee shall be graded the same as a Mate but shall be undergoing recognised training or pursuing accreditation of previous learning and/or experience with a view to achieving National Vocational Qualification/Scottish Vocational Qualification (NVQ/SVQ) Level 2 in H&V Installation.

Installer

An Installer shall be able, under close, but not constant supervision, to carry out the installation of domestic or industrial/commercial pipework and/or ductwork, and associated components and systems. An Installer shall be able to:

(i)   demonstrate a basic knowledge of how the components within a system relate to each other;

(ii)   plan the installation of system components;

(iii)   install and test system components;

(iv)   carry out pre-commissioning testing; and

(v)   de-commission systems

An Installer shall also be able to demonstrate competence in a health and safety, human interaction, quality control and environmental requirements appropriate to their scope of work.

All Installers entering the grade other than by re-grading from the Assistant and Improver grades shall demonstrate that they have satisfactorily completed training or received formal accreditation for the skills and experience they possess, howsoever acquired, in accordance with the requirements of NVQ/SVQ Level 2 H&V Installation (whether Industrial and Commercial, Domestic or Ductwork options) which may be amended from time to time - as approved by the appropriate national accreditation bodies.

On 3 April 2000

All Assistants and Improvers (as defined within the terms of the National Agreement as at 19 October 1997) employed in either grade prior to 5 October 1998 shall automatically be re-graded as an Installer.

The qualifications for the former Assistant and Improver grades set out in the National Agreement as at 19 October 1997 are reproduced for information in the Note for Guidance at 6e(1).

There should be no further recruitment to either of these grades after 5 October 1998.

All Installers are required to register their grade with the UK Register of HVACR Operatives by providing on an approved application form, their National Insurance number, specimen signature, photograph and a number of other details of their background and experience, to be sent to:

566   *Rates of Wages and Working Rules - Mechanical Installations*

WORKING RULES

UK Register of HVACR Operatives,
Old Mansion House
Eamont Bridge
Penrith
Cumbria CA10 2BX
Tel: 01768 864771

Employers will assist where practically possible with the registration of an Installer's grade.

Upon payment of the relevant registration fee, Installers will be sent a credit card-sized Register card certifying their grade and qualifications.

Apprentice/Trainee

An Apprentice/Trainee - as distinct from an Adult Trainee - shall be undertaking an approved course of training as follows:

(i)   in accordance with the Agreement on Modern Apprenticeships at Appendix G: or

(ii)   in accordance with a duly executed Agreement of Service in the form prescribed by the National Joint Industry Council (NJIC). Details of these arrangements can be found in Clause 23; or

(iii)   in the case of Ductwork Installation Trainees, an approved in-company scheme of training (further details are given in Clause 23).

Craftsman

A Craftsman shall be able without supervision to carry out the installation of domestic or industrial/commercial pipework and/or ductwork, and associated components and systems. A Craftsman shall be able to:

(i)   demonstrate a greater depth of technical knowledge and level of responsibility than an Installer: in particular, a Craftsman shall be able to demonstrate detailed knowledge of a system operating principles;

(ii)   set, identify and establish the requirements of the job, whether from drawings or customers other instructions;

(iii)   liase with other trades, suppliers and customers, as appropriate;

(iv)   solve problems within the scope of the work carried out;

(v)   ensure compliance with all relevant standards;

(vi)   specify and monitor programmes for installing and commissioning systems;

(vii)   commission and test systems.

A Craftsman shall also be able to demonstrate competence in the health and safety, human interaction, quality control and environmental requirements appropriate to their scope of work. Where work is undertaken on gas systems, it shall comply with the requirements of relevant regulations and the nationally accredited scheme for the certification of gas-fitting personnel.

Rates of Wages and Working Rules - Mechanical Installations 567

WORKING RULES

A Craftsman shall have one of the following qualifications:

before 24 August 1998

(i)   have successfully completed an apprenticeship approved by the NJIC, and have passed the practical examination of an appropriate City and Guilds of London Institute basic craft course which has been recognised by the NJIC and approved by the Parties. Approved City and Guilds of London Institute courses include:

City and Guilds H&V Fitting 337 (until 1969)

City and Guilds H&V Fitting 618 (until 1975)

City and Guilds H&V Fitting 597 (until 1991)

City and Guilds H&V Fitting 604 (until introduction of NVQ's/SVQ's)

City and Guilds Gas Fitting 598; or

(ii)   have already been employed as a 'tradesman' in the industry, within the terms of the National Agreement on 24 February 1969; or

(iii)   have already been employed as a Ductwork Erector in the industry (within the terms of the National Agreement as at 19 October 1997). The qualifications for the former Ductwork Erector grade set out in the National Agreement as at 19 October 1997 are reproduced for information in the Note for Guidance at 6e(2); or

Before 3 April 2000

A Craftsman wishing to progress to the Senior Craftsman grade must meet all the criteria previously listed, but additionally must hold a current certificate of competence in one of the welding skills acknowledged in the industry or before 3 April 2000

(iv)   have successfully completed an Improvership of one year under Qualification A or 18 months under Qualification B within the terms of the National Agreement as at 19 October 1997. The conditions required to be fulfilled under both Qualifications are in the Note for Guidance at 6e(1) See also the Note for Guidance at 6e(3); or

Before 31 May 2000

(v)   have worked in the industry for four consecutive years; and have successfully completed an apprenticeship approved by the NJIC; or have completed a Modern Apprenticeship; or have completed a training programme as a Ductwork Installation Trainee; or have received formal accreditation for the skills and experience they possess, howsoever acquired; In accordance with the requirements of:

NVQ/SVQ Level 2 and/or 3 - H&V Domestic Installation;

NVQ/SVQ Level 2 and/or 3 - H&V Industrial and Commercial Installations;

568   *Rates of Wages and Working Rules - Mechanical Installations*

WORKING RULES

HVCA Interim Certificate and/or NVQ/SVQ Level 3 - H&V Ductwork Installation; or

On or after 1 June 1999

(vi) have worked in the industry for four consecutive years; and have successfully completed an apprenticeship approved by the NJIC; or have completed a Modern Apprenticeship; or have completed a training programme as a Ductwork Installation Trainee; or have received formal accreditation for the skills and experience they possess, howsoever acquired;

In accordance with the requirements of NVQ/SVQ Level 3 in H&V Installation (whether Industrial and Commercial, Domestic or Ductwork options) which may be amended from time to time - as approved by the appropriate national accreditation bodies.

All Craftsmen are required to register their grade with the UK Register of HVACR Operatives by providing on an approved application form their National Insurance number, specimen signature, photograph and a number of other details of their background and experience, to be sent to:

UK Register of HVACR Operatives
Old Mansion House
Eamont Bridge
Penrith
Cumbria CA10 2BX

Employers will assist where practically possible with the registration of a Craftsman's grade.

Upon payment of the relevant registration fee, Craftsmen will be sent a credit card-sized Register card certifying their grade and qualifications.

Senior Craftsman

A Senior Craftsman shall have at least the same qualifications as a Craftsman, except that a Senior Craftsman shall have gained not less than five years experience of working in the industry after achieving status as a Craftsman. Grading as a Senior Craftsman shall be in accordance with the following:

on 24 August 1998

All Advanced Fitters or Advanced Ductwork Erectors, (as defined within the National Agreement as at 19 October 1997) employed in either grade shall automatically be re-graded as a Senior Craftsman.

The qualifications for the former Advanced Fitter and Ductwork Advanced Erector grades set out in the National Agreement as at 19 October 1997 are reproduced for information in the Note for Guidance at 6e.4.

Rates of Wages and Working Rules - Mechanical Installations 569

WORKING RULES

After 24 August 1998

A Senior Craftsman shall:

(i) have experience beyond that of a Craftsman by virtue of additional proficiency, speed and flexibility, and have other special skills over and above that detailed in the definition of a Craftsman; and

(ii) agree to undertake the day-to-day on-the-job training and instruction of Adult Trainees, Craft Apprentices, Modern Apprentices and other trainees/candidates undergoing, for example, the accreditation of their prior learning aimed at the achievement of industry-recognised Vocational Qualifications; and

(iii) be able to take responsibility for the day-to-day supervision of work squads with an average labour force of three other Craftsmen/Senior Craftsmen.

Re-grading as a Senior Craftsman shall be on the basis of the Craftsman having the capabilities required by the grade, rather than the employers requirements for such level of work to be performed.

A Senior Craftsman shall retain his grade as a Senior Craftsman while in the employment of the employer re-grading him and when moving to any subsequent employer.

All Craftsmen are required to register their grade with the UK Register of HVACR Operatives by providing on an approved application form their National Insurance number, specimen signature, photograph and a number of other details of their background and experience, to be sent to:

> UK Register of HVACR Operatives
> Old Mansion House
> Eamont Bridge
> Penrith
> Cumbria CA10 2BX

Employers will assist where practically possible with the registration of a Senior Craftsman's grade.

Upon payment of the relevant registration fee, Senior Craftsmen will be sent a credit card-sized Register card certifying their grade and qualifications.

Foreman

On 24 August 1998

All Foremen and Foremen (Ductwork) (as defined within the National Agreement as at 19 October 1997) employed in either grade shall be automatically re-graded as a Foreman. From this date, the distinction and pay differential between Foreman and Foreman (Ductwork) will be discontinued: both former grades will be assimilated into a single grade with the same pay rate.

570   *Rates of Wages and Working Rules - Mechanical Installations*

WORKING RULES

After 24 August 1998

Either:

A Craftsman who satisfies the qualifications of a Senior Craftsman may be designated by the Employer as a Foreman, provided he is competent to perform all the duties listed below (or the vast majority of them as appropriate to and in accordance with the requirements of the Employer):

(i)   assign tasks to Senior Craftsmen with supervisory responsibilities and other Operatives under his direct control.

(ii)   redeploy Senior Craftsmen with supervisory responsibilities or other Operatives under his direct control, in order to achieve the optimum productivity including on-site batch production and fabrication.

(iii)   decide methods to be used for individual operations and instruct other Operatives accordingly.

(iv)   ensure variation work does not proceed without authority from the office.

(v)   maintain site contract control procedure.

(vi)   requisition and progress supply of necessary equipment and materials to other Operatives when required.

(vii)   ensure that other Operatives take all reasonable steps to safeguard, maintain and generally take care of Employers tools and materials.

(viii)   maintain day-to-day liaison and programme of work with main contractor and other sub-contractors.

(ix)   inspect and review progress of work of sub-contractors.

(x)   monitor progress to main contractors, in order that agreed programme is met.

(xi)   measure and record progress of work.

(xii)   inspect the work of other Operatives for quality, progress and satisfactory completion.

(xiii)   check weekly progress against programme and identify deviations therefrom.

(xiv)   verify bookings on time and job cards and despatch them promptly to the office.

(xv)   notify office of impending delays likely to affect progress or give rise to a claim.

(xvi)   establish reasons for delays to work and notify office.

(xvii)   provide information for cost variation investigations when necessary.

(xviii) forecast labour requirements.

(xix)   ensure company instructions and standards of discipline, workmanship and safety are maintained on site.

Rates of Wages and Working Rules - Mechanical Installations   571

WORKING RULES

(xx)   ensure that the conditions of the National Agreement and any other conditions of employment are complied with.

(xxi)   supervise training of Apprentices assigned to his control.

(xxii)   take overall charge of all his Employers labour on site and act where necessary as the Employers site agent.

(xxiii)   evolve and/or agree order of work within overall programme and control its progress.

(xxiv)   decide or agree locations of site office, site stores, site workshop and other work stations and adjust same to suit site progress and changing conditions.

(xxv)   ensure compliance of all work, whether executed by own Operatives or sub-contractors with drawings and specifications.

(xxvi)   organise, supervise and record such tests (e.g. hydraulic) and/or inspections as are required during progress of contract.

(xxvii)   requisition or otherwise procure such attendances and facilities as are required of the main contractors and/or of other sub-contractors.

(xxviii)   attend site meetings (if so required by Employer).

(xxix)   ensure that safe methods of work are adopted by other Operatives under his direct control.

(xxx)   ensure clearance of rubbish as specified

(xxxi)   arrange and supervise testing on completion, including compliance with specifications, snagging and operational handing over as directed and final site clearance.

(xxxii)   such other details as are reasonably required by the Employer.

Alternatively:

A parallel route to demonstrating the capabilities required of a Foreman will also exist through the achievement of the NVQ/SVQ Level 3 Building Services Engineering Supervision qualification - which may be amended from time to time as approved by the appropriate national accreditation bodies.

The Employer is not obliged to re-grade every Senior Craftsman as Foremen who may have the capabilities required by the Foreman grade or the NVQ/SVQ Level 3 Supervision. A Senior Craftsman who is graded as a Foreman does not necessarily carry the Foreman grade with him to a new Employer.

BALANCE OF GANGS (CLAUSE 7)

The balance of gangs as between Craftsmen, Assistants, Mates and Craft Apprentices shall be on the basis that:

(i)   Assistants may do semi-skilled tasks including support work for craftsmen without constant direct supervision and shall perform the manual work of fetching and carrying, receiving and checking materials as required by the Employer.

572   *Rates of Wages and Working Rules - Mechanical Installations*

WORKING RULES

(ii)   Support work for skilled men may be done by the skilled men themselves or by Assistants, Mates or Craft Apprentices in order to secure the maximum utilisation of labour and the optimum economic production; thus one Mate can be used to do the support work for two or more Craftsmen or conversely two or more Mates may work with one Craftsman.

(iii)   Mates shall not be confined to the manual work of fetching and carrying. They shall within their capacity, carry out semi-skilled tasks; one object being to improve productivity and the other being to permit those who wish to do so to qualify for consideration for regarding as Assistants.

(iv)   In order to provide Craft Apprentices with appropriate practical experience and to permit them to make the fullest possible contribution to production, they shall be permitted to work with the tools with the maximum of supervision necessary but always on work which is under the control of a recognised Craftsman. In the case of welding, this shall mean that Craft Apprentices shall not weld until they have completed the appropriate City and Guilds welding course. Craft Apprentices shall not be employed solely and continuously on heavy labouring work.

WAGES AND ALLOWANCES (CLAUSE 8)

(a)   Rates of wages, all allowances and holiday credits referred to in this Agreement shall be agreed from time to time between the Association and the Union, and shall be enumerated in an Appendix to this Agreement.

(b)   The rates of wages and other conditions of employment for Operatives engaged on HVAC work on 'Nominated' Engineering Construction Projects shall be as set out in the Agreement between the Association and the Union in Appendix C to this Agreement.

Payment of Wages

(c)   Unless otherwise agreed between the Employer and the Operative, payment of wages shall be by credit transfer into a bank or building society account in the name of the Operative concerned.

(d)   Unless otherwise agreed between the Employer and the Operative, the pay week shall normally end at midnight Friday and wages shall be paid on the following Thursday.

(e)   The Employer at his discretion may pay each Operative to the nearest £1 upwards each week, carrying the credit forward, deducting it from the next wage payment which is again paid to the nearest £1 upwards.

(f)   Where wages due cannot be calculated on time sheets, the Employer shall make assessed payment for the days worked. Any necessary corrections shall appear in wages payable for the following week.

Former Chargehands and Ductwork Erector Chargehands

(g)   With effect from 24 August 1998 the Chargehand and Ductwork Erector Chargehand grades will be discontinued. All Operatives who before this effective date were graded Chargehand or Ductwork Erector Chargehand under the previous conditions of the National Agreement will be entitled to retain their hourly rate as at 24 August 1998 as a personal lead rate on a mark time basis until this is overtaken by the hourly rate of the Senior Craftsman grade.

Rates of Wages and Working Rules - Mechanical Installations 573

WORKING RULES

(h)   The previous National Agreement conditions for the Chargehand and Ductwork Erector Chargehand grades are reproduced for information in the Note for Guidance to Clause 6. It is emphasised that Operatives who were so graded, and who retain the previous Chargehand rate on the basis described above, are not necessarily entitled to carry the grade or the mark-time allowance with them to a new Employer.

Responsibility Allowance

(i)   From 24 August 1998 a general Responsibility Allowance is to be introduced. The rate of the Allowance shall be agreed between the Association and the Union and enumerated in an Appendix to the Agreement. The Allowance can be paid in multiples of one or two to Craftsmen or Senior Craftsmen, at the discretion of the Employer, depending upon the nature and level of the responsibilities involved, taking account of the following considerations:

Conditions for Payment of the Allowance to Craftsmen and Senior Craftsmen with day-to-day supervisory responsibilities

(j)   From 24 August 1998 Senior Craftsmen will take responsibility for the day-to-day supervision of work squads with an average labour force of three other Craftsmen/Senior Craftsmen. Payment for this responsibility is included in the hourly rate for the grade.

(k)   However, where there is a requirement for a Senior Craftsman to take sole responsibility for the day-to-day supervision of larger work squads, a Responsibility Allowance shall be payable.

(l)   Where there is a requirement for a Craftsman to take responsibility for the day-to-day supervision of work squads with an average labour force of three other Craftsmen, a Responsibility Allowance shall be payable.

(m)   Where a Responsibility Allowance is paid in respect of supervisory responsibilities, it may be paid on a permanent, temporary or short term basis for as long as the supervisory requirement continues, provided the Employer informs the Operative concerned of the likely length of the period of payment of the Allowance and gives due notice of its cessation when the supervisory requirement has come to an end.

(n)   A Responsibility Allowance paid on this basis shall be reckonable for overtime insofar as the supervisory responsibilities are carried out during overtime working.

Conditions for Payment of the Allowance to Craftsmen and Senior Craftsmen with welding skills

(o)   A Responsibility Allowance may also be payable to a Craftsman or Senior Craftsman who holds a second current Certificate of Competency in oxy-acetylene or metal arc welding to the standards set out in the Welding of Carbon Steel Pipework - Code of Practice' (informally known as the 'Grey Book'), provided that Craftsmen and Senior Craftsmen shall keep both Welding Certificates current. Payment for the first welding skill is included in the hourly rate for the Craftsman and Senior Craftsman.

General

(p)   Where a Craftsman or Senior Senior Craftsman has responsibilities for supervision and holds a current welding certificate at the same time, the Responsibility Allowance shall be paid in units of one or two, depending on the particular circumstances, as shown in the matrix in the agreed Note for Guidance.

574   *Rates of Wages and Working Rules - Mechanical Installations*

WORKING RULES

Certification of Welding Skills

(q) Craftsmen and Senior Craftsmen are required to ensure their welding skills are updated and properly certificated by registering their welding competency through the UK Register of HVACR Operatives. To do this, Operatives shall provide on an approved application form their National Insurance number, specimen signature, photograph and a number of other details of their background and experience, to be sent to:

UK Register of HVACR Operatives
Old Mansion House
Eamont Bridge
Penrith
Cumbria CA10 2BX
Tel: 01768 864771

Employers will assist where practically possible with registration of an Operative's grade/welding skills.

(r) Upon payment of the relevant registration fee, Craftsmen and Senior Craftsmen will be sent a credit card-sized Register card certifying their grade and qualifications, including certification of their welding competency.

(s) Information about the availability of welding test facilities can be obtained from:

Heating, Ventilating and Domestic Engineers' National Joint Industrial Council
Esca House
34 Palace Court
Bayswater
LONDON W2 4JG

(t) From August 1999, welding skills will be incorporated as optional units into the Level 3 NVQ/SVQ in H&V Installation. Whether they have attained this qualification or not Craftsmen and Senior Craftsmen will receive recognition of their competence in respect of their first welding skill achieved through the NVQ/SVQ by means of a consolidated allowance contained within their hourly rate for the grade in respect of the welding skill contained within their NVQ/SVQ.

(u) Payment in respect of the second current Certificate of Competency will be made as above.

Welding Certificates and the Senior Craftsman Grade

The 1998/99 - 1999/2000 Wage Agreement made a number of changes in the way welding skills are paid for within the terms of the National Agreement. Payment for the first welding skill was incorporated into the basic hourly rate for the Craftsman and Senior Craftsman grades. This change covered all existing Craftsmen and Senior Craftsmen, irrespective of whether they actually undertook welding or not.

The new NVQ/SVQ Level 3 in H & V Installation - which becomes the basic qualification for the Craftsman grade on 1 June 1999 - includes an optional welding unit, which means that in the future all Operatives completing this qualification (including the welding unit), and subsequently becoming a Craftsman, will be able to weld.

WORKING RULES

However, these changes notwithstanding it is recognised there will be a sizeable number of Operatives who have completed a qualification entitling them to Craftsman status, but who gained their qualification at a time before welding was formally incorporated in the qualification. It is quite possible that some of these people have never undertaken welding training, nor held a certificate of welding competence, but - following consolidation of the welding supplements into basic rates - they are paid the same as a Craftsman or Senior Craftsman who actually does undertake welding.

In order to ensure that there is a continuing incentive for non-welding Craftsmen to acquire welding skills, the following has been agreed by the parties.

From 31 December 1999

In order to progress from the Craftsman grade to the Senior Craftsman grade, an Operative must meet all the previously established criteria and must, in addition, hold (and maintain) a current certificate of competence in at least one welding skill. A Craftsman meeting the other criteria for progression, but not holding a welding certificate, will not be allowed to progress to the Senior Craftsman grade, but will remain as a Craftsman until such time as he gains a current welding certificate.

Employers not requiring welding skills

It is recognised by the parties that employers in some sectors of the H & V industry may not actually require their Senior Craftsmen to weld. In these circumstances, a Craftsman can progress to the Senior Craftsman grade without holding a current welding certificate (provided that they meet all the other criteria for progression) as long as their employer expressly states that welding competence is not required. The card issued by the UK Register of HVACR Operatives would state that the individual had attained Senior Craftsman status, without possessing welding skills, in line with the requirements of their current employer.

Senior Craftsman status acquired in these circumstances would not automatically transfer, however, with the Operative into new employment, unless the new employer also expressly stated that welding skills were not required.

Merit Money

(v) Payment of merit money to an Operative may be made at the option of the Employer for mobility, loyalty, long service etc and for special skill over and above that detailed in the definition of the relevant grade at Clause 6.

Abnormal Conditions

(w) Operatives engaged on exceptionally dirty work, or work under abnormal conditions, of such a character as to be equally onerous, shall receive an allowance extra per day or part of a day. The determination of the conditions to which this allowance shall apply shall be agreed between the Employer and the Operative concerned in each case. The allowance shall be agreed from time to time by the Association and the Union and shall be enumerated in an Appendix to this Agreement.

576   *Rates of Wages and Working Rules - Mechanical Installations*

WORKING RULES

Target Incentive Schemes

(x)   Where it has been agreed between the Employer and the majority of his workforce, that target incentive schemes shall be operated in connection with works on which they are or are to be employed, such schemes shall be operated in accordance with general principles established by the Association and the Union for their operation, which are set out in Appendix B.

Notes for guidance for the above Clause are available under JCC letter 67

OVERTIME (CLAUSE 9)

(a)   It is accepted by the Parties that overtime must be contained. To this end, except in cases of urgency or emergency, actual working hours should not exceed:

(i)   an average of 45 per week in the case of travelling men who should only work on Saturdays and/or Sundays in cases of urgency or emergency.

(ii)   an average of 55 per week in the case of lodging men whose work on Saturdays and/or Sundays should be reasonably contained.

(b)   It is also accepted by the Parties that the reference period for calculating compliance with the provisions of the European Directive on Working Time (93/104/EC) concerning certain aspects of working time and UK legislation deriving therefrom shall be a period of 12 months.

(c)   The Parties agree that where temporary redundancies threaten in an area, e.g. during the winter months, overtime should be cut or eliminated in order to spread the available work over as many relevant Operatives as possible; any reduction in overtime in accordance with this principle shall be for negotiation between the Union and the Employer concerned, having regard to the difference between lodging and travelling jobs.

(d)   The difference between the normal hourly rate and the overtime rate shall be known as the 'premium' payment.

Overtime during the Normal Working Week

(e)   For the purposes of calculating overtime, and regardless of the length of the normal working day as determined under Clause 3, time worked in excess of the normal working day on Monday to Friday inclusive shall be paid for as follows:

(i)   first 8 hours worked after normal starting time - normal hourly rate.

(ii)   thereafter until 4 hours after normal finishing time - time-and-a-half.

Rates of Wages and Working Rules - Mechanical Installations

WORKING RULES

(iii) thereafter until normal starting time next morning.

(iv) if time is lost through the fault of the Operative, the time lost shall be added to the normal starting time and the resultant time shall be used for the purpose of calculating overtime payable at time-and-a-half.

(v) an Operative directed to start work before the normal starting time shall be paid the appropriate overtime rates for all hours worked before the normal starting time, but if through the action of the Operative the normal working day is not worked, the normal hourly rate shall be paid for all hours worked.

(vi) the calculation of overtime for any day shall not be affected by any hours of absence arising from:

certified sickness

absence with the concurrence of the Employer

absence for which the Operative can provide to the satisfaction of the Employer that his absence was due to causes beyond his control

(vii) an Operative called back to work at any time between the period commencing two hours after normal finishing time and until two hours before normal starting time, shall be paid such overtime rates as would apply had work been continuous from normal finishing time and shall be paid a minimum of two hours at the appropriate rate.

Time Worked outside the Normal Working Week

(f) Time worked, outside the normal working week, shall be paid for as follows:

(i) Saturday - first 5 hours, time-and-a-half; after the first 5 hours, double time, but if time is lost through the fault of the Operative, the double time rate shall not apply until time lost has been made up.

(ii) Sunday - double time for all hours worked until starting time on Monday morning.

(g) An Operative who is required to work continuously from normal finishing time until after midnight shall normally not continue working beyond normal starting time the following day. If, however, the Employer considers it to be an emergency, work may continue after normal starting time by agreement between the Operative and the Employer.

After ceasing work, the Operative shall take a break of at least 8 hours before restarting work. Where the break falls on a normal working day the Operative shall be entitled to one hour off with pay at the normal hourly rate for each hour worked after midnight until normal starting time.

Notes for guidance for the above Clause are available under JCC letter 75

This clause is to be amended as JCC letter 79

PAYMENT FOR HOLIDAYS WORKED (CLAUSE 10)

(a) This clause applies to all recognised holidays as defined in Clause 18, and in Scotland - three days of the Winter Holiday period as defined in Clause 10(c).

(b) An Operative who works on any of the days in Clause 10(a) shall be paid a minimum of two hours at the appropriate rate. In addition an Operative shall be granted a day's holiday with pay for each holiday day worked as provided in Clause 18(c).

578   *Rates of Wages and Working Rules - Mechanical Installations*

WORKING RULES

(c)   Time worked on such days shall be paid as follows:

In England and Wales
New Year's Day, Good Friday, Easter Monday, May Bank Holiday, Spring Bank Holiday, Late Summer Bank Holiday, Christmas Day, Boxing Day:
                  - Double time for all hours worked.

In Scotland
Three consecutive days of the Winter Holiday period including New Year's Day and the one or two holiday days which immediately follow it (if any), Spring Bank Holiday, Friday before Spring Bank Holiday, May Bank Holiday, Autumn Holiday (one day): Boxing Day.
                  - Double time for all hours worked.
Christmas Day and the one day of recognised holiday to be agreed locally.
The normal working day as determined in Clause 3(a), time and a half,   thereafter double time.
Friday before Autumn Holiday.
The normal working day as determined in Clause 3(a), normal hourly rates; thereafter overtime rates in accordance with Clause 9.

In Northern Ireland
The eight days of recognised holidays agreed by the Northern Ireland Branches of the Association and Union. Double time for all hours worked.

(d)   The general conditions of the Agreement shall apply to men called back to work on these holidays.

NIGHT SHIFTS AND NIGHT WORK (CLAUSE 11)

(a)   For an Operative who works for at least five consecutive nights:
   (i)   The basic rate, called the night shift rate, shall be one and a third times the normal rate.
   (ii)   Overtime rates and conditions shall be as for normal working days (provided in clause 9) but the basic rate shall be the night shift.
(b)   An Operative who works for less than five nights and does not work during the day shall be paid at overtime rates as if the normal day had already been worked.

CONTINUOUS SHIFT WORK (CLAUSE 12)

Where jobs have to be continuously operated the work shall be carried out in two or three shifts of eight hours each according to requirements. The Operatives concerned shall be paid time and a third in cases where a six day shift is worked and time and a half in cases where a seven day shift is worked, overtime and night shift rates being compounded in these rates. Arrangements shall be made to change the shifts worked by each Operative.

PROVISION OF TOOLS (CLAUSE 13)

(a)   The Operative shall provide a rule and spirit level. Other tools shall be provided by the Employer but the Operative shall take all reasonable steps to safeguard, maintain and generally take care of the Employer's tools.

(b)   The Operative shall co-operate in the implementation of reasonable procedures properly designed to prevent loss of or damage to tools.

WORKING RULES

ALLOWANCES TO OPERATIVES WHO TRAVEL DAILY (CLAUSE 15)

(a) Except where his centre is the job, an Operative who is required by the Employer to travel daily up to 50 miles to the job shall be paid fares and travelling time as stated in (i) and (ii) below:

(i) Return daily travelling fares from his centre to the job. Where cheap daily or period fares or other cheap travel arrangements by public transport are available the Employer may pay fares on that basis. Where, however, a change in such travel arrangements results from a change in the working arrangements the Employer must pay the Operative for any additional cost. The Employer at his option may provide suitable conveyance for the Operative to and from the job in which case fares shall not be paid.

(ii) Allowances for travelling time, provided that the normal hours are worked on the job. The allowances for travelling time shall be agreed from time to time by the Association and the Union and shall be enumerated in an Appendix to this Agreement. When a reasonably direct journey is not possible, a claim for special consideration may be made by the Operative and in case of dispute the matter shall be referred to the Chief Officials of the parties, whose decision shall be final.

(b) Except where his centre is the job, payment to the Operative of allowances for travelling time and fares for journeys beyond fifty miles daily from his centre to the job will be for agreement between the Employer and the Operative concerned.

ALLOWANCES TO OPERATIVES WHO LODGE (CLAUSE 16)

(a) Where an Operative is sent to a job to which it is impracticable to travel daily and where the Operative lodges away from his place of residence he shall (except if he is engaged at the job or if his centre is the job) be paid the items in (i) to (v) below where appropriate:

(i) A nightly lodging allowance including the night of the day of return and when on week-end leaves in accordance with Clause 17(a). The nightly lodging allowance shall be agreed from time to time by the Association and the Union and shall be enumerated in an Appendix to this Agreement. The lodging allowance shall not be paid when an Operative is absent from work without the concurrence of the Employer, nor when suitable lodging is arranged by the Employer at no expense to the Operative, nor during the annual holidays defined in Clause 20 including the week of Winter Holiday. The Operative shall provide the Employer with a statement signed by himself to the effect that he is in lodgings for the period of payment of lodging allowance under this clause. Without such evidence, the Employer shall deduct tax on lodging allowance paid.

(ii) Any VAT charged on the cost of lodgings, subject to the provision by the Operative of a valid tax invoice on which the Employer can claim input credit.

(iii) When suitable lodgings are not available within two miles of the job, daily return fares from lodgings to job. The Employer at his option may provide suitable conveyance for the Operative between the lodgings and the job, in which case the fares shall not be paid.

(iv) Time spent in travelling to and from the centre at the commencement and completion of the job at the normal hourly rate but when an excessive number of hours of travelling is necessarily incurred, a claim for special consideration may be made by the Operative to the Employer or by the Employer to the Operative and in case of dispute the matter shall be referred to the Chief Officials of the parties, whose decision shall be final.

(v) Fares between his centre and the job at the commencement and the completion of the job. Return fares shall be used when available.

(vi) Weekend leaves in accordance with Clause 17(a).

580   *Rates of Wages and Working Rules - Mechanical Installations*

WORKING RULES

(b)   An Operative whose employment is terminated in accordance with Clause 2a during the course of a job, shall be entitled to travelling time and a single fare for the journey from the job to his centre. This condition shall not apply to an Operative who is discharged for misconduct or who leaves the job without the concurrence of his Employer.

WEEK-END LEAVES (CLAUSE 17)

(a)   An Operative who is in receipt of lodging allowance in accordance with Clause 16 shall be allowed a week-end leave every two weeks. Such Operative shall be entitled to return to his respective centre for the recognised holidays prescribed in Clause 18 and to facilitate this, the nearest normal week-end leave shall, where necessary, be deferred or brought forward to coincide with the holiday.

(b)   Unless the Employer and the Operative agree otherwise the week-end leave shall be from normal finishing time on Friday to normal starting time on Monday.

(c)   An Operative shall not normally be required to start his return journey before 6.00a.m. on the appropriate day of return to the job but shall, where the return journey makes it impossible to commence work at the normal starting time, agree with his Employer the working arrangements for the day.

(d)   Weekend return fares shall be paid for weekend leaves. If an Operative does not elect to return to his centre a single fare from the job to his centre shall be paid.

(e)   The following travelling time arrangements shall apply to an operative on weekend leave for journeys to and from his centre:

(i)   Where the job is up to 150 miles from his centre, he shall travel in his own time from the job to his centre, but travelling time from his centre to the job shall be paid at the normal hourly rate.

(ii)   Where the job is 150 miles or more from his centre, he shall be paid four hours at the normal hourly rate from the job to his centre, and travelling time from his centre to the job shall be paid at the normal hourly rate.

If an Operative elects to stay at the job travelling time shall not be paid.

(f)   When a reasonably direct journey is not possible or when an excessive number of hours travelling is necessarily incurred on jobs more than 150 miles from an Operative's centre, a claim for special consideration in respect of travelling time may be made by the Operative to the Employer or by the Employer to the Operative and in case of dispute the matter shall be referred to the Chief Officials of the parties, whose decision shall be final.

(g)   An Operative on week-end leaves (including holidays, provided under Clause 18), shall be paid the nightly lodging allowance, provided that the leave is within this Agreement or is agreed with the Employer.

Rates of Wages and Working Rules - Mechanical Installations 581

WORKING RULES

RECOGNISED HOLIDAYS (CLAUSE 18)

(a) The following days have been designated as recognised holidays and shall be paid in accordance with WELPLAN the HVACR Welfare and Holiday Scheme. The pay shall consist of the appropriate holiday credits standing to the credit of the Operative

If any of these days comes within the annual holidays as provided in Clause 20, mutual arrangements shall be made to substitute some other day for the day or days included.

In England and Wales
New Year's Day; Good Friday; Easter Monday; May Bank Holiday; Spring Bank Holiday; Late Summer Bank Holiday; Christmas Day; Boxing Day.

In Scotland
New Years Day; Spring Holiday; May Holiday; Autumn Holiday (two days); Christmas Day; Boxing Day; plus one other day to be agreed locally.

In Northern Ireland
Note: The days when recognised holidays are to be taken in Northern Ireland are subject to discussion between the Northern Ireland Branches of the Association and the Union.

(b) Any Operative who has insufficient credits in his WELPLAN account to pay for the three days of recognised holiday included in the winter holiday period, because he entered the industry after the commencement of the appropriate accounting period, shall be entitled to three days pay at the normal hourly rate for eight hours. The Employer shall be responsible for paying the difference between this sum and the value of any holiday credits that may have been accrued in the appropriate stamping period.

(c) Operatives who work on a recognised holiday as set out in Clause 18(a) shall be paid in accordance with Clause 10(c) and shall be entitled to a day's holiday in lieu, at a mutually agreed time, payment for which shall be the sum of the appropriate holiday credits for that day of recognised holiday.

(d) The general conditions of the Agreement shall apply to Operatives called back to work on these holidays.

582   *Rates of Wages and Working Rules - Mechanical Installations*

WORKING RULES

WELPLAN THE HVACR WELFARE AND HOLIDAY SCHEME (CLAUSE 19)

(a) The Employer shall notify WELPLAN of all Operatives to be included in the Scheme. The rules of the Scheme which are incorporated into and form part of this Agreement are set out in a separate Supplement. Operatives are entitled to a weekly credit subject to the rules of the scheme to be purchased by the Employer by means of a four weekly return to WELPLAN. The credit shall cover:

(i)   A weekly credit in respect of annual and recognised holidays (the value of the credit shall be agreed from time to time between the Association and the Union and shall be enumerated in an Appendix to this Agreement).

(ii)   A weekly premium in respect of welfare benefits (the value of the premium shall be determined from time to time by the Association and shall be enumerated in an Appendix to this Agreement).

(b) Variation or Amendment: Clauses 19, 20, 21 and 22 of this Agreement may be varied or amended by agreement of the Parties but any variation or amendment shall, subject to the rules of the WELPLAN, only become operative at the beginning of a new accounting period. Notice of any proposed variation must be given in writing to each of the other Parties at least six months prior to the commencement of any accounting period.

(c) Termination: Either of the Parties to this Agreement may terminate Clauses 19, 20, 21 or 22 at the end of any accounting period by giving notice in writing to the other Party at least 12 months before the end of the accounting period. In the event of termination of the `Annual and Recognised Holidays Provision' the Parties agree to provide the holiday facilities and holiday payments until such time as the rights acquired by the Operatives in respect of holiday credits under this section have been met.

ANNUAL HOLIDAYS (CLAUSE 20)

During 2000 and thereafter, annual holidays shall consist of :

England and Wales

(i)   four days of Spring Holiday
(ii)   two weeks of Summer Holiday
(iii) seven days of Winter Holiday

Scotland

(i)   four days of Spring Holiday
(ii)   eleven days of Summer Holiday
(iii) six days of Winter Holiday

Northern Ireland

(i)   three days of Spring Holiday
(ii)   two weeks of Summer Holiday
(iii) one day of Autumn Holiday
(iv)   seven days of Winter Holiday

Rates of Wages and Working Rules - Mechanical Installations   583

WORKING RULES

Extracts from National Working Rules (JIB PMES)

WORKING HOURS (CLAUSE 1)

(1.1)   Working Week

The normal working week shall be 37½ hours consisting of five working days, Monday to Friday inclusive, each day to consist of 7½ working hours. The normal starting time shall be 8.00 am but this may be varied nationally by the National JIB or locally by a Regional JIB, according to the circumstances.

(1.2)   Utilisation of Working Hours

The hours set out in Rule 1.1 are working hours which shall be fully utilised and shall not be subject to unauthorised 'breaks'. Time permitted for tea breaks shall not be exceeded. Bad time-keeping and/or unauthorised absence from the place of work during working hours shall be the subject of disciplinary action.

Meetings of Operatives shall not be held during working hours except by arrangement with the Job, Shop or Site Representative and with the Employer or the Employer's Representative.

(1.3)   Meal Intervals

Meal intervals in each such working day shall be as follows:

(1.3.1)   a midday break of not less than half an hour, to be taken at a time fixed by the Employer, and shall be unpaid.

(1.3.2)   two paid tea-breaks, not exceeding ten minutes each in duration, which shall be taken at times determined by the Employer.

GUARANTEED WEEK (CLAUSE 2)

(2.1)   Guarantee

An Operative who has been continually employed by the same Employer for not less than one full week is guaranteed wages equivalent to his inclusive normal graded earnings for 37½ hours in any normal working week, provided that during working hours he is capable of, available for, and willing to perform satisfactorily the work associated with his usual occupation, or reasonable alternative work if his usual work is not available, as determined by the Employer and his Representative.

(2.2)   Holidays

In the case of a week in which recognised holidays occur, the guaranteed week shall be reduced by 7½hours for each day of holiday.

584   *Rates of Wages and Working Rules - Mechanical Installations*

WORKING RULES

(2.3)   Industrial Action

In the event of an interruption of work as a result of industrial action by the Operatives in the employ of an Employer who is a party to the Agreement, the guarantee shall be automatically suspended in respect of Operatives affected on the site, job or shop where the industrial action is taking place.

In the event of such interruption being caused by Operatives working under other Agreements and the Operatives covered by this Agreement not being parties to the interruption, the Employers will try to provide other work or if not able to do so will provide for the return of the Operatives to the shop or office from which they were sent. The Operatives will receive instructions as soon as is practicable as to proceeding to other work or return to the shop. If other work is not available the Employer may suspend Operatives for a temporary period and the provisions of Rule 2.1 shall be suspended until normal working is restored or alternative work becomes available.

(2.4)   Temporary Lay-Off

Where an Operative is not provided with work for a complete payweek, although remaining available for work, he shall be paid in accordance with the Guarantee set out in Rule 2.1. Thereafter, following consultation with a written confirmation of the agreement from the Representatives of the AEEU, the Employer may require the Operative to register as an unemployed person and Rule 2.1 will not apply for the duration of the stoppage of work. Where an Operative, who has been subject to temporary layoff, is restarted, his employment will be deemed to have been continuous and no break in service will apply.

OVERTIME (CLAUSE 3)

(3.1)   Hours

Overtime is strongly discouraged by the Board and systematic overtime should only be introduced to meet specific circumstances and be limited to the period to which the circumstances apply, i.e. breakdown, urgent maintenance and repairs or country jobs.

(3.2)   Payment

(3.2.1)   39 hours shall be worked at normal rates in any one week (Monday to Friday) before any overtime is calculated. Overtime shall be paid at time-and-a-half. Overtime after 8.00 p.m. shall be paid at double time. After midnight a 7☐ hour rest period, starting at the time at which the work ceased and paid at normal rates of pay, shall follow night work. All hours worked between 1.00 p.m. on Saturday until normal starting time on Monday shall be paid at double time. To qualify for the rest period the night work must be within the twenty-four hour period in which the normal shift occurs.

           Exceptions: For the purpose of overtime payment, an Operative shall be deemed to have worked normal hours on days when, although no payment is made by the Employer, the Operative:-

      (a)   has lost time through certified sickness;
      (b)   was absent with the Employer's permission;
           or when the Operative
      (c)   was on statutory holiday;
      (d)   was on a rest period for the day following continuous working all the previous night.

Rates of Wages and Working Rules - Mechanical Installations   585

WORKING RULES

(3.2.2) Any Operative who has not worked five days from Monday to Friday taking into account the exceptions detailed under 3.2.1 is precluded from working the following Saturday or Sunday unless specifically required to do so by his Employer due to a shortage of labour. In such a case overtime rates as in 3.2.1 shall be payable.

(3.2.3) Payday: Wages shall be paid on Thursday of the week following the week in which hours were worked unless a statutory holiday intervenes or if strikes or other circumstances beyond the Employer's control cause delay in payment, when payday may be Friday.

Payment of wages by credit transfer should be encouraged to avoid delays.

(3.3) Call Out

When an Operative is called to return to work after his normal finishing time and before his next normal starting time, he shall be guaranteed the equivalent of 3 hours payment at the Operative's JIB Graded Rate of Pay for any Call Out commencing before 8.00 p.m. and 4 hours payment at the Operative's JIB Graded Rate of Pay for any Call Out commencing after 8.00 p.m. in accordance with the overtime provisions contained in Rule 3.2.1, provided the Operative has completed 39 hours work in the relevant week.

Payment for call out will be for all hours inclusive home to home.

NIGHT SHIFT WORK (CLAUSE 4)

(4.1) Hours

Night shift is where Operatives, other than day shift Operatives, work throughout the night for not less than three consecutive nights.

A full night shift pattern shall consist of 37☐ working hours worked on 4 or 5 nights by mutual agreement with breaks for meals each night to be mutually arranged. The Employer shall agree the working hours (including breaks) on each contract.

(4.2) Payment

Night shift shall be paid at the rate of time-and-a-third for all hours worked up to 39 hours in any one week (Monday to Friday inclusive). Overtime shall be at the rate of double basic time.

586   *Rates of Wages and Working Rules - Mechanical Installations*

WORKING RULES

HOLIDAYS (CLAUSE 5)

(5.1) Annual Holiday and Stamp Credit Card Scheme

Operative Plumbers and Apprentice Plumbers shall be entitled to 21 working days paid annual holiday in each year to be taken in conformity with Rules 5.1 and 5.1.4. Employers of Operative Plumbers and Apprentice Plumbers **must in all** circumstances purchase, from the Board on behalf of their Operatives, weekly credit stamps in respect of annual holidays and covering for the sickness and other benefits referred to in Rule 9. The values of credit stamps for each contribution period (from the beginning of April to the end of March in the succeeding year) shall be determined from time to time by the Board (see Appendix B). The credit stamps shall be affixed weekly to a card supplied by the Board and the stamped cards shall be regarded as the property of the Operative but retained in safe keeping by the Employer during the currency of employment. Upon termination of that employment, the cards, stamped up to the date of termination, shall be handed to the Operative who shall provide acknowledgement of receipt of the stamped cards. To qualify for a holiday stamp in any one week an Operative must work a minimum of four full days. Statutory holidays or authorised absence including Jury Service, count as worked days.

Excepting Apprentices (see below), at the time of the Operative taking his holiday the Employer shall pay to the Operative the holiday credit due, calculable by reference to the "Holiday Credit Calculator" and the number of stamps affixed to the stamped card. In the case of Apprentices, the Employer shall pay to the Apprentice the normal wages due to the Apprentice according to the wage for year of training scale for each day of annual holiday.

The Board will, upon receipt of the stamped cards from the Employer, reimburse the Employer the amount of holiday credit which has been properly paid on the total of stamps affixed thereon, in accordance with the terms of this agreement, provided the Board is satisfied that the claim has been made in accordance with the object and rules of the Scheme.

The object of the Scheme is to secure a rest period with pay and it is not a means by which an Operative can achieve double pay by not taking all or part of his prescribed holiday. According to the obligations of membership of the Board, participation in the Scheme is obligatory upon all Employers. HOLIDAYS MUST BE TAKEN.

(5.1.1) Operatives Covered by the Scheme

The Scheme shall apply to all Operatives whose rates of wages and working conditions are determined by the Board and to Apprentices in accordance with the terms of their Board Apprentice Training Agreement and failure to comply will be a breach of the Agreement.
Any Operative in the Plumbing Mechanical Engineering Services Industry may be admitted to the Scheme at the Director's discretion.

Rates of Wages and Working Rules - Mechanical Installations   587

WORKING RULES

(5.1.2) Annual Holiday Periods

Annual holidays cannot be carried forward from one Holiday Period to the next and no Operatives shall in any circumstances be eligible to receive in any Holiday Period payment of Holiday Credits except those made on his behalf during the relevant Contribution Period. "Contribution Period" means the period of one year from the commencement of the week in which 1st April falls.

"Holiday Period" means the period 1st April to 31st March immediately following the Contribution Period.

(5.1.3) Extent of Annual Holidays

All Operatives covered by this Scheme shall be entitled to Annual Holidays consisting of 21 working days for which payment shall be made as provided in Rule 5.1. Ten days of holiday are to be taken in the period 1st May to 30th September, the balance of the holiday to be taken at a time to be agreed with the Employer.

Where possible, the holidays shall commence at normal finishing time on Friday. For the purpose of this Scheme no Saturday or Sunday shall be considered as a working day.

(5.1.4) Notice

Operatives shall give the Employer at least six weeks notice of the dates on which a holiday is to be taken. Where the dates are mutually agreed a shorter period of notice is permissible.

(5.2) Statutory and Other Holidays

(5.2.1) Holiday Dates and Qualification for Payment

Seven-and-a-half hours pay at the appropriate JIB graded rate of wages shall be paid for 8 days of holiday per annum additional to the annual holidays in accordance with Rule 5.1. In general, the following shall constitute such paid holidays:

New Years Day
Good Friday
Easter Monday
May Day
Spring Bank Holiday
Late Summer Bank Holiday
Christmas Day
Boxing Day

In areas where any of these days are not normally observed as holidays in the Industry, traditional local holidays may be substituted by the appropriate Regional Board.

In order to qualify payment, Operatives must work full-time for the normal day on the working days preceding and following the holiday, except where the Operative is absent with the Employer's permission or has lost time through certified sickness. In no case shall the Operative lose pay in excess of a single day of the holiday.

588   *Rates of Wages and Working Rules - Mechanical Installations*

WORKING RULES

For the purpose of this Rule Operatives shall be deemed to have worked on one or both of the qualifying days if they comply with the following conditions:

(i)   were absent through certified sickness;
(ii)   were on a rest period for the day following continuous working all the previous night;
(iii)   were absent with the Employer's permission;
(iv)   except where Rule 6.4 applies;
(v)   where the preceding working day was a statutory holiday when the Operatives would have to be available for work only on the day following the holiday;
(vi)   were absent on Jury Service.

(5.2.2)   Payment for Working During Statutory and/or Public Holidays

When Operatives are required to work on a paid holiday within the scope of this Agreement, they shall receive wages at the following rates for all hours worked.

Christmas Day: Double time and a day or shift off in lieu for which they shall be paid wages at bare time rates for the hours constituting a normal working day. The alternative day hereunder shall be mutually agreed between the Employer and the Operative concerned.

In respect of all other days of holidays:
Time-and-a-half plus a day or shift off in lieu for which they shall be paid wages at bare time rates for the hours constituting a normal working day. The alternative day hereunder shall be mutually agreed between the Employer and the Operative concerned. In the case of night shift workers required to work on a holiday, the overtime rates mentioned above shall be calculated upon the night shift rate. Time off in lieu of statutory holidays shall be paid at bare time day rates.

(5.2.3)   Coincidence of Annual and Other Holidays

Where a holiday under Rule 5.2 coincides with part of an Operative's annual holiday, the annual holiday shall be paid at the rate laid down in Rule 8.1 and time off in lieu of the day or days of statutory or other holiday shall be granted at a later date mutually agreed between the Employer and the Operative concerned, and shall be paid for at ordinary time rates. Payment in lieu of the holiday shall not be permitted.

WAGES AND ALLOWANCES (CLAUSE 8)

(8.1) Graded Rates of Wages

The national standard rates of wages for each grade of Operative and Apprentices, shall be those made by the Board and currently applicable.

Employers are not permitted to pay, nor Operatives to receive, any rate other than the national standard rate. Additional payments and deductions shall be as permitted by these Working Rules.

The current national standard graded rates of wages shall be separately published. (See Appendix A, Section 1).

Rates of Wages and Working Rules - Mechanical Installations 589

WORKING RULES

(8.2) Responsibility Money

Advanced Plumbers designated by the Employer to be in charge of other Operatives shall be paid responsibility money at an hourly rate for the period of such responsibility. (See Appendix A, Section 1).

(8.3) Incentive Bonus Schemes

Bona fide incentive bonus schemes only may be operated, subject to the approval of the Board. (See Appendix C).

(8.4) Daily Travelling Allowances

(8.4.1) Journeys of less than 50 miles

Except where his Centre is the Job, an Operative who is required by his Employer to travel daily up to 50 miles to the Job, shall be paid fares and travelling time as stated below.

Fares

Return travelling fares (cheapest available) from his Centre to the Job. The Employer at his option may provide suitable conveyance for the Operative to and from the Job in which case fares shall not be paid.

Travelling allowance

Allowances for travelling time, provided that the normal hours are worked on the Job, shall be agreed from time to time by the Board and shall be promulgated and separately published by the Board. When a reasonably direct journey is not possible, a claim for special consideration may be made by the Operative and in case of dispute the matter shall be referred to the Board whose decision shall be final. (See Appendix A, Section 1).

(8.4.2) Employers Own Transport

(a)   It is the Employer's responsibility, if he or the main contractor provides suitable free transport to a site, to get the Operatives to and from the job on time. The Employer's liability for late arrival of the provided transport shall be agreed between the parties.

(b)   Mileage Allowance - Use of Private Vehicles on Company Business

The use of Operatives private vehicles on employer business will be subject to compliance with the following:-

(i)   that prior agreement regarding use of such vehicles be reached between employer and operative

(ii)   it is the responsibility of the operative to provide insurance for the vehicle and to ascertain that the insurance is adequate for such use

(iii)   the rate of payment for use of vehicle for distance travelled will be as promulgated and that this payment be made in lieu of fares

(iv)   use of private vehicle is not a condition of employment

590   *Rates of Wages and Working Rules - Mechanical Installations*

WORKING RULES

(8.4.3) Definition of Centre

The Centre for determining distances under this Rule must be agreed between the Employer and the Operative and must be either:

(a) the Job on which the Operative is for the time being employed, if the Operative is engaged on the understanding that his Centre will be the Job,
or
(b) a convenient Centre near the Operative's place of residence. Unless otherwise agreed such Centre must be the nearest convenient public transport boarding point to the Operative's place of residence.

(8.4.4) Change of Centre

When an Operative whose Centre is the Job, as in (a) above, is an Operative in Regular Employment as defined below, his Centre may, by agreement between Employer and Operative, be transferred to one located in accordance with his place of residence as in (b) above.

An Operative after 28 days continuous employment or who, after his first engagement, is transferred to another Job or who is re-engaged by the same firm or any of its subsidiaries within 28 days of ceasing work for any reason with the said firm, shall be regarded as an Operative in Regular Employment for the purpose of this Rule.

Any change of residence after an Operative has been engaged and a Centre established in accordance with Rule 8.4.3 which substantially varies the journey time/distance shall not alter the Centre without prior agreement.

(8.4.5) Measurement of Travelling Distances

All distances referred to in this Rule are to be calculated in a straight line, point to point. When a reasonably direct journey is not possible, a claim for special consideration may be made by the Operative to the Employer.

(8.4.6) Allowances to Operatives who Lodge

(a)   When an Operative is sent to a Job to which it is impracticable to travel daily and where the Operative lodges away from his place of residence, he shall (except if he is engaged at the Job or if his Centre is the Job) be paid the items (i) to (v) below, where appropriate.

Rates of Wages and Working Rules - Mechanical Installations　　591

WORKING RULES

(i)　　A nightly lodging allowance including the night of the day of return and when on week-end leaves in accordance with Rule 8.4.7.(i). The nightly lodging allowance shall enumerated in a separate appendix to this Agreement. (See Appendix A, Section 1). When convenient lodgings cannot be secured or where the Lodging Allowance is found to be inadequate, an operative shall, with the prior approval of the employer, be reimbursed for the actual expenditure incurred for which a proper receipt shall be produced. The lodging allowance shall not be paid when an Operative is absent from work without the concurrence of the Employer nor when suitable lodging is arranged by the Employer at no expense to the Operative, nor during the annual holidays defined in Rule 5 including the week of Winter Holiday. The Operative shall provide the Employer with a statement signed by himself to the effect that he is in lodgings for the period of payment of lodging allowance under this Rule. Without such evidence the Employer shall deduct tax on lodging allowance paid.

(ii)　　When suitable lodgings are not available within two miles from the Job, daily return fares from lodgings to Job. The Employer at his option may provide suitable conveyance for the Operative between lodgings and the Job, in which case fares shall not be paid.

(iii)　　Travelling time for the time spent in travelling from the Centre at the commencement and completion of the Job at the normal time rates but when an excessive number of hours of travelling is necessarily incurred, a claim for special consideration may be made. In case of dispute, the matter shall be referred to the Board.

(iv)　　Fares between his Centre and the Job at the commencement and the completion of the job. Return fares shall be used when available.

(v)　　Week-end leaves in accordance with Rule 8.4.7.

(b)　　An Operative whose employment is terminated by proper notice on either side during the course of a Job, shall be entitled to travelling time back to his Centre and a single fare for the journey from the Job to his Centre. This condition shall not apply to an Operative who is discharged for misconduct or who leaves the Job without the concurrence of his Employer.

(8.4.7)　　Week-End Leaves

(i)　　An Operative who is in receipt of lodging allowance in accordance with Rule 8.4.6 shall be allowed a week-end leave every two weeks. Such Operative shall be entitled to return to his respective Centre for the recognised holidays prescribed in Rule 5 and to facilitate this, the nearest normal week-end leave shall, where necessary, be deferred or brought forward to coincide with the holiday.

(ii)　　Unless the Employer and the Operative agree otherwise, the week-end leave shall be from normal finishing time on Friday to normal starting time on Monday.

(iii)　　An Operative shall not be required to start his return journey before 6.00 am on the appropriate day of return to the Job.

(iv)　　Week-end return fares shall be paid for week-end leaves. If an Operative does not elect to return to his Centre, a single fare from the Job to his Centre shall be paid.

Rates of Wages and Working Rules - Mechanical Installations

Rates of Wages and Working Rules - Mechanical Installations

WORKING RULES

(v)   An Operative on a week-end leave whose work is up to 150 miles from his Centre shall travel home in his own time, but travelling time from the Centre to the Job shall be paid at the normal time rate. An Operative whose work is 150 miles or more from his Centre shall be paid travelling time for four hours at normal rate from the Job to the Centre; travelling time for the journey back to the Job to be paid at normal time rate. If an Operative elects to stay at the Job, travelling time shall not be paid.

(vi)   When a reasonably direct journey is not possible or when an excessive number of hours travelling is necessarily incurred on jobs more than 150 miles from an Operative's Centre, a claim for special consideration in respect of travelling time may be made. In case of dispute, the matter shall be referred to the Board.

(vii)   An Operative on week-end leave (including statutory holidays provided under Rule 5) shall be paid the nightly lodging allowance provided that the leave is within this agreement or is agreed with the Employer.

(8.5) Tool Allowance

Tool allowance being in respect of the provision, maintenance and upkeep of tools provided by the Operatives, is not deemed to be wage-payment. Where employment starts after Monday, or is terminated in accordance with Rule 6.3 otherwise than on a Friday, the amount of tool allowance paid shall be the appropriate proportion of the weekly allowance for each day worked. Tool allowance (see Appendix A, Section 1) is payable to all Operatives who provide and maintain the following set of tools:

Allen Keys - 1 set	Hand Drill/Brace
Adjustable Spanner - up to 6"	Junior Hacksaw/Saw
Adjustable Spanner - up to 12"	Mole Wrench
Basin Key	Padsaw/Compass Saw
Bending Spring - 15mm	Pipewrench -
Bending Spring - 22mm	Stillsons -
Blow Torch and Nozzle - complete	up to type 14"
(similar to Primus Type B)	Pliers - Insulated -
Bolster Chisel - 2□"	General Type
Bossing Mallet	Pocket Knife -
Bossing Stick	(Stanley type)
Bradawl	Putty Knife*
Chisel - Brick - up to 20"	Rasp
Chisel - Wood	Rule - 3m Tape
Dresser for Lead	Screwdriver - Large
Flooring Chisel	Screwdriver - Small
Footprints - 9"	Shave Hook
Gas or Adjustable Pliers	Snips
Glass Cutter*	Spirit Level - 600mm
Hacking Knife*	Tank Cutter
Hacksaw Frame	Tool Bag
Hammer - Large	Trowel
Hammer - Small	Tube Cutter
Wiping Cloth - (one)	

* Only if glazing is normally done by plumbers in the district.

Rates of Wages and Working Rules - Mechanical Installations 593

WORKING RULES

(8.6) Storage Accommodation for Tools

Where reasonably practical a lock-fast and weather-proof place shall be provided on all jobs where tools can be left at the owner's risk. The Board provides for financial assistance to be given to Operatives in replacing lost or stolen tools, subject to the conditions of the scheme. (See Appendix A, Section 8).

(8.7) Abnormal Conditions

Operatives required to work in exceptionally dirty conditions or under other abnormal conditions as listed in the Schedule attached to these Rules, (see Appendix D) shall be paid a daily supplement, as shown in Appendix A, Section 1. This supplement is a fixed daily amount and does not vary with the hours worked or with different combinations of abnormal conditions.

(8.8) Recognition of Certificates of Competency in Welding

Graded Operatives who hold one or more current JIB Certificates of Competency in Welding or such other qualifications as the Board may require, shall be paid a differential rate as determined by the Board, in addition to the graded rate of wages. (See Appendix E).

For the purposes of this Rule the holder of JIB Certificate of Competency in Oxy-acetylene Welding of Mild Steel Pipework and/or in Bronze Welding of Copper Sheet and Tube shall be classified as a certified "Gas Welder". The holder of JIB Certificate of Competency in Metal Arc Welding of Mild Steel Pipework shall be classified as a certified "Arc Welder".

GRADING DEFINITIONS (CLAUSE 14)

(14.1) Trained Plumber, Mechanical Pipe Fitter and Gas Service Fitter

(i) Must have obtained a National Vocational Qualification (NVQ) - MES/Plumbing Level 2 or City & Guilds of London Institute (CGLI) 603/1 Craft Certificate or such other qualifications as are acceptable to the Board.

(ii) Entry by way of a recognised and registered form of training (usually a 4 year term as a JIB registered apprentice) or other accepted method of entry into the Industry.

(iii) Must be at least 20 years of age.

(iv) Must be able to carry out all such installation and maintenance work to the recognised standard and level of productivity expected from an Operative working under minimum supervision.

(14.2) Advanced Plumber, Mechanical Pipe Fitter and Gas Service Engineer

(i) Must have obtained a National Vocational Qualification (NVQ) - MES/Plumbing Level 3 or City & Guilds of London Institute (CGLI) 603/2 Advanced Craft Certificate or such other qualifications as are acceptable to the Board.

(ii) After achieving **NVQ-MES/Plumbing Level 3** must have **at least one year's experience** working as a Trained Plumber, Mechanical Pipe Fitter or Gas Service Fitter and must be **at least 21 years old**.

594 *Rates of Wages and Working Rules - Mechanical Installations*

WORKING RULES

(iii)   After obtaining the CGLI **603/2 Advanced Craft Certificate** or other such qualifications as are acceptable to the Board must have **at least two years experience** working as a Trained Plumber, Mechanical Pipe Fitter or Gas Service Fitter and must be **at least 22 years old**.

(iv)   Entry by way of a recognised and registered form of training (usually a 4 year term as a JIB registered apprentice) or other accepted method of entry into the industry.

(v)   Must possess particular and productive skills and be able to work without supervision in the most efficient and economical manner and must be able to set out jobs from working drawings and specifications and requisition the necessary installation materials and/or have technical and supervisory knowledge and skill beyond that expected of a Trained Plumber, Mechanical Pipe Fitter or Gas Service Fitter.

(14.3)   Technical Plumber, Mechanical Pipe Fitter and Gas Service Technician

(i)   Entry by way of a recognised and registered form of training or other accepted method of entry into the Industry.

(ii)   Must have obtained such academic qualifications or such other qualifications as are acceptable to the Board.

(iii)   Must have superior technical skill, ability and experience beyond that expected of an Advanced Plumber, Mechanical Pipe Fitter or Gas Service Engineer, and be able to lay out and prepare contract work in accordance with the Building Regulations and Water Bye-laws, take off quantities and measure work, assess labour requirements and control and supervise all manner of plumbing or other relevant installations in the most economic and effective way and achieve a high level of productivity.

AND EITHER

(iv)   Must be at least 27 years of age.

(v)   Must have had at least five years experience as an Advanced Plumber, Mechanical Pipe Fitter or Gas Service Engineer, with a minimum of three years in a supervisory capacity in charge of plumbing or other relevant installations of such a complexity and size as to require wide technical experience and organised ability.

OR

(vi)   May not have reached 27 years of age as in (v) or have the full experience as required in (v) but is otherwise fully qualified in accordance with (i), (ii) and (iii) and his present Employer wishes to have him graded as a Technical Plumber, Mechanical Pipe Fitter or Gas Service Technician in which event he may be granted this grade by the Board.

Electrical Installations
RATES OF WAGES AND WORKING RULES

RATES OF WAGES

ELECTRICAL CONTRACTING INDUSTRY

Extracts from National Working Rules determined by:

The Joint Industry Board for the Electrical Contracting Industry
Kingswood House
47/51 Sidcup Hill, Sidcup, Kent, DA14 6HP
Telephone : 020 8302 0031
Internet : www.jib.org.uk

WAGES (Graded Operatives)

Rates

Up to 7th February 2000 there was one wage rate per grade.
    From 7th February 2000 three different wage rates apply to JIB Graded operatives, depending on whether they report directly to the shop and their method of transport to their place of work. The three categories are:

Shop Reporting
    Payable to an operative who is required to start and finish at the shop.

Job Reporting (Transport Provided)
    Payable to an operative who is required to start and finish at normal starting and finishing times on jobs, travelling in their own time in transport provided by the employer. The operative shall also be entitled to payment for Travel Time, where eligible, as detailed in the appropriate scale.
Job Reporting (Own Transport)
    Payable to an operative who is required to start and finish at normal starting and finishing times on jobs, travelling in their own time by their own means. The operative shall be entitled to payment for Travel Time and Travel Allowance, where eligible, as detailed in the appropriate scale.

    Productivity and Incentive Schemes
    Employers and operatives, with the involvement of the local full time AEEU official, may agree on any job, arrangements for the maximum utilisation of working hours, a bonus payment or payments related to progress of the work and productivity levels or any other related matters in addition to the normal hourly rates of pay.

From and including **7th February 2000** there is no set London Weighting figure.

596   *Rates of Wages and Working Rules - Electrical Installations*

WORKING RULES

The JIB rates of wages are set out below:

From and including **7th February 2000**, the JIB hourly rates of wages shall be set out as below:

(i)   National Standard Rates:

Grade	Shop Reporting	Job Reporting	
		Transport Provided	Own Transport
Technician (or equivalent specialist grade)	£ 8.94	£ 9.27	£ 9.49
Approved Electrician (or equivalent specialist grade)	£ 7.82	£ 8.15	£ 8.37
Electrician (or equivalent specialist grade)	£ 7.13	£ 7.46	£ 7.68
Labourer	£ 5.52	£ 5.85	£ 6.07

(ii)   London Rate:

Grade	Shop Reporting	Job Reporting	
		Transport Provided	Own Transport
Technician (or equivalent specialist grade)	£ 9.46	£ 9.81	£ 10.04
Approved Electrician (or equivalent specialist grade)	£ 8.34	£ 8.69	£ 8.92
Electrician (or equivalent specialist grade)	£ 7.65	£ 8.00	£ 8.22
Labourer	£ 6.04	£ 6.39	£ 6.62

From and including **8th January 2001**, the JIB hourly rates of wages shall be set out below:

(i)   National Standard Rate:

Grade	Shop Reporting	Job Reporting	
		Transport Provided	Own Transport
Technician (or equivalent specialist grade)	£ 9.45	£ 10.10	£ 10.53
Approved Electrician (or equivalent specialist grade)	£ 8.27	£ 8.91	£ 9.35
Electrician (or equivalent specialist grade)	£ 7.54	£ 8.18	£ 8.62
Labourer	£ 5.83	£ 6.49	£ 6.92

Rates of Wages and Working Rules - Electrical Installations

WORKING RULES

(ii)   London Rate:

| | | Job Reporting | |
Grade	Shop Reporting	Transport Provided	Own Transport
Technician (or equivalent specialist grade)	£ 10.07	£ 10.74	£ 11.20
Approved Electrician (or equivalent specialist grade)	£ 8.89	£ 9.56	£ 10.02
Electrician (or equivalent specialist grade)	£ 8.16	£ 8.84	£ 9.28
Labourer	£ 6.45	£ 7.13	£ 7.57

1983 Joint Industry Board Apprentice Training Scheme
Apprentice rates effective from and including October 1999

Junior Apprentice ....…....…… Training allowance of £ 2.35 per hour
Senior Apprentice (stage 1) ……………………………………...……………… £ 3.44 per hour
Senior Apprentice (stage 2) …………………………………….....……………… £ 5.20 per hour

1999 Joint Industry Board Apprentice Training Scheme Apprentice rates effective from and including October 1999.

Stage 1 ……………………………………………………………………….. £ 2.35 per hour
Stage 2 ……………………………………………………………………….... £ 3.44 per hour
Stage 3 ………………………………………………………………………….. £ 5.20 per hour
Stage 4 ……………………………………………………………………….... £ 5.41 per hour

Note:
Apprentices may be affected by the DTI guidance on the National Minimum Wage.

DTI Guidance on the National Minimum Wage
Apprentices are entitled to the National Minimum Wage twelve months after reaching the age of 18 years if the apprenticeship commenced at 16, 17 or 18 years of age or, if commencing between 19 and 25 years of age, twelve months after commencement of the apprenticeship. Apprentices aged 26 and above must be paid the National Minimum Wage from the commencement of their apprenticeship.

The National Minimum Wage rates for apprentices are:
18 - 21 years old …………………………………………………………....…… £ 3.00 per hour
22 - 25 years old …………………………………………………………....……… £ 3.60 per hour
26 years and over ………..………………………………………………....……….. £ 3.00 per hour

Travelling Time and Travel Allowances

Provision of Transport or Mileage Allowance

Where an employer provides transport free of charge, operatives provided with such transport shall not be permitted to travel allowance. An operative who considers that the transport provided by his employer is unsuitable may pursue a complaint through the Procedure for Handling Grievances and Avoiding Disputes. Payment of travel allowance shall be satisfied by the provision by en employer of such as a periodic travel pass (e.g. season ticket).

An employer by agreement with an operative may pay a mileage allowance where the operative agrees to use his own car for business use; the operative shall have appropriate insurance cover.

598   *Rates of Wages and Working Rules - Electrical Installations*

WORKING RULES

On any day when such a mileage allowance is paid in respect of a journey or journeys between an operative's home (or lodgings) and his employer's place of business or a site at which the operative is required by his employer to attend, travel allowance will not also be payable.

From and including 7th February 2000

Operatives who are required to start and finish at the normal starting and finishing time on jobs which are 10 miles and over from the shop - in a straight line - shall receive payment for Travelling Time and, where transport is not provided by the employer, Travel Allowance, as follows:

Distance	Total Daily Travel Allowance	Total Daily Travelling Time
(a) National Standard Rate		
Up to 10 miles	Nil	Nil
Over 10 & up to 15 miles each way	£ 2.13	£ 4.26
Over 15 & up to 20 miles each way	£ 3.82	£ 5.10
Over 20 & up to 25 miles each way	£ 4.61	£ 5.95
Over 25 & up to 35 miles each way	£ 5.56	£ 6.85
Over 35 & up to 55 miles each way	£ 8.25	£ 8.25
Over 55 & up to 75 miles each way	£ 9.70	£ 9.70

For each additional 10 mile band over 75 miles, additional payments of £ 1.25 for Daily Travel Allowance and £ 1.25 for Daily Travel Time will be made.

	Total Daily Travel Allowance	Total Daily Travelling Time
(b) London Rate		
Up to 10 miles	Nil	Nil
Over 10 & up to 15 miles each way	£ 2.13	£ 4.54
Over 15 & up to 20 miles each way	£ 3.82	£ 5.43
Over 20 & up to 25 miles each way	£ 4.61	£ 6.34
Over 25 & up to 35 miles each way	£ 5.56	£ 7.30
Over 35 & up to 55 miles each way	£ 8.25	£ 8.79
Over 55 & up to 75 miles each way	£ 9.70	£ 10.30

For each additional 10 mile band over 75 miles, additional payments of £ 1.25 for Daily Travel Allowance and £ 1.25 for Daily Travel Time will be made.

From and including 8th January 2001

From and including **8th January 2001**, Travelling Time and Travel Allowances shall only be payable to operatives who travel over 15 miles from the shop - in a straight line.

Operatives who are required to start and finish at the normal starting and finishing time on jobs which are 15 miles and over from the shop - in a straight line - shall receive payment for Travelling Time and, where transport is not provided by the employer, Travel Allowance, as follows:

WORKING RULES

	Total Daily Travel Allowance	Total Daily Travelling Time
Distance		
(a) National Standard Rate		
Up to 15 miles	Nil	Nil
Over 15 & up to 20 miles each way	£ 2.50	£ 3.40
Over 20 & up to 25 miles each way	£ 3.30	£ 4.30
Over 25 & up to 35 miles each way	£ 4.35	£ 5.25
Over 35 & up to 55 miles each way	£ 6.95	£ 6.95
Over 55 & up to 75 miles each way	£ 8.50	£ 8.50

For each additional 10 mile band over 75 miles, additional payment of £ 1.50 for Daily Travel Allowance and £ 1.50 for Daily Travel Time will be made.

	Total Daily Travel Allowance	Total Daily Travelling Time
(b) London Rate		
Up to 15 miles	Nil	Nil
Over 15 & up to 20 miles each way	£ 2.50	£ 3.62
Over 20 & up to 25 miles each way	£ 3.30	£ 4.73
Over 25 & up to 35 miles each way	£ 4.35	£ 5.70
Over 35 & up to 55 miles each way	£ 6.95	£ 7.69
Over 55 & up to 75 miles each way	£ 8.50	£ 9.00

For each additional 10 mile band over 75 miles, additional payments of £ 1.50 for Daily Travel Allowance and £ 1.50 for Daily Travel Time will be made.

Travel time and travel allowance from 6 October 1997 - Section 8

Apprentices who are required to start and finish at the normal starting and finishing time on jobs which are up to and including 35 miles from the shop - in a straight line - shall receive travel allowances and payment for travelling time as follows:

	Total Daily Travel Allowance	Total Daily Travelling Time		
		Junior Apprentice	Senior Apprentice (Stage 1)	Senior Apprentice (Stage 2)
(a) National Standard Rate				
Up to 1 mile each way	83p	Nil	Nil	Nil
Up to 2 miles each way	95p	38p	63p	93p
Up to 3 miles each way	109p	63p	105p	155p
Up to 4 miles each way	121p	73p	126p	186p
Up to 5 miles each way	171p	100p	168p	248p
Over 5 miles & up to 10 miles each way	241p	150p	252p	372p
Over 10 miles & up to 15 miles each way	343p	200p	336p	496p
Over 15 miles & up to 20 miles each way	482p	225p	378p	558p
Over 20 miles & up to 15 miles each way	552p	250p	420p	620p
Over 25 miles & up to 35 miles each way	622p	275p	462p	682p
(b) London Area				
Up to 1 mile each way	83p	Nil	Nil	Nil
Up to 2 miles each way	95p	43p	68p	98p
Up to 3 miles each way	109p	71p	114p	164p
Up to 4 miles each way	121p	86p	137p	197p
Up to 5 miles each way	171p	114p	182p	262p
Over 5 miles & up to 10 miles each way	241p	171p	273p	393p
Over 10 miles & up to 15 miles each way	343p	228p	364p	524p
Over 15 miles & up to 20 miles each way	482p	257p	410p	590p
Over 20 miles & up to 15 miles each way	552p	285p	455p	655p
Over 25 miles & up to 35 miles each way	622p	314p	501p	721p

600   *Rates of Wages and Working Rules - Electrical Installations*

WORKING RULES

Please note that whilst Country Allowance has been abolished for Adult Graded Operatives from 7th **February 2000**, *it remains in force for Apprentices undertaking the 1983 Training Scheme (see National Working Rule 12(c)).*

Country Allowance
£ 22.40 from and including 4 January 1999

From 7th February 2000 COUNTRY ALLOWANCE IS ABOLISHED for Adult Graded Operatives.

Please note that Apprentices and Adult Trainees (Under 21) continue to receive Country Allowance when they are required to start and finish at the normal starting and finishing time on jobs over 35 miles from the Shop and travel daily to the job instead of taking lodgings. This remains at £ 22.40 per day.

Lodging Allowances
£ 22.40 from and including 4 January 1999

Lodgings weekend retention fee, maximum reimbursement
£ 22.40 from and including 4 January 1999

Annual Holiday Lodging Allowance Retention
£ 4.60 per night (£ 32.20 per week) from and including 4 January 1999

Responsibility money
From and including 30 March 1998 the minimum payment increased to 10p per hour and the maximum to £1.00 per hour.

From and including 4 January 1992 responsibility payments are enhanced by overtime and shift premiums where appropriate.

Combined JIB Benefits Stamp Value (from week commencing 25 September 2000)

JIB Grade	Weekly JIB combined credit value	Holiday value
Technician	£36.98	£29.42
Approved Electrician	£33.31	£25.75
Electrician	£31.03	£23.47
Senior Graded Electrical Trainee	£28.70	£21.14
Labourer & Adult Trainee	£25.71	£18.15
Adult Trainee(Under 21)	£21.17	£13.61

Rates of Wages and Working Rules - Electrical Installations   601

WORKING RULES

JIB Welfare Benefits

Sick Pay, Death Benefit, Accidental Death Benefit

(i)   The Sickness Benefit is paid at:

£ 65.00 per week for up to 26 weeks in any 52 week period (from and including 4 January 1999).

(ii)   Death Benefit

£17,000.00 for death from any cause.

(iii)   Accidental Death Benefits with effect from 4 January 1992

£12,500.00 for adult operatives (£6,250.00 for apprentices) for death from an accident either at work, or travelling to or from work, making a total of £29,500.00 (£21,250.00 for apprentices) for death due to an accident at work.

602   *Rates of Wages and Working Rules - Electrical Installations*

WORKING RULES

EXTRACTS FROM THE NATIONAL WORKING RULES

(JIB for the Electrical Contracting Industry)

INTRODUCTION

The JIB National working Rules are made under rule 80 of the Rules of the Joint Industry Board (Section 1) as the National Joint Industrial Council for the Electrical Contracting Industry.

The principal objects of the Joint Industry Board are to regulate the relations between employers and employees engaged in the industry and to provide all kinds of benefits for persons concerned with the Industry in such ways as the Joint Industry Board may think fit, for the purpose of stimulating and furthering the improvement and progress of the industry for the mutual advantage of the employers and employees engaged therein, and in particular, for the purpose aforesaid, and in the public interest, to regulate and control employment and productive capacity within the Industry and the levels of skill and proficiency, wages, and welfare benefits of persons concerned in the Industry.

"The Industry" means the Electrical Contracting Industry in all its aspects in England, Wales, Northern Ireland, the Isle of Man and Channel Islands and such other places as may from time to time be determined by the Joint Industry Board, including the design, manufacture, sale, distribution, installation, erection, maintenance, repair and renewal of all kinds of electrical installations, equipment and appliances and ancillary plant activities.

1: GENERAL

These JIB National Working Rules and Industrial Determinations supersede previous Rules and Agreements made between the constituent parties of the National Joint Industrial Council for the Electrical Contracting Industry and shall govern and control the conditions for electrical instrumentation and control engineering, data and communications transmission work, its installation, maintenance and its dismantling and other ancillary activities covered by the Joint Industry Board for the Electrical Contracting Industry ("JIB") and shall come into effect in respect of work performed on and after Monday, 2 March 1970. These Rules apply nationally and in such a manner as may be determined from time to time by the JIB National Board.

2: GRADING

Graded operatives shall comply in all respects with the Grading Definitions set out in Section 4 in carrying out the work of the Industry, erect their own mobile scaffolds and use such power operated and other tools, plant, etc, as may be provided by their Employer and the standard JIB Graded Rates of Wages shall be paid. Grading shall only be valid by the possession of a Grade Card issued by the Joint Industry Board.

Nothing in these rules shall prevent the maximum flexibility in the employment of skilled operatives.

3: WORKING HOURS

(a) Standard Working

The working week shall normally be 37.5 hours worked on five days Monday to Friday inclusive unless another arrangement is agreed between an employer and an operative.

The first 7.5 hours worked on any day shall constitute the normal working hours for that day.

The first 37.5 hours paid in any pay week at an operative's normal hourly rate of pay shall meet an Employer's contractual commitment for the payment of wages for the normal working week in any pay week.

Rates of Wages and Working Rules - Electrical Installations 603

WORKING RULES

Day workers (operatives other than those employed on shift or night work or on any work carried out beyond working hours) shall be paid at the rate of time and a half of the appropriate hourly rate of pay for time worked before 7:30 a.m. or after 6:30 p.m. on any day or at such higher rate as may otherwise apply under these National Working Rules.

Where shifts are required which fall outside these limits, payments and conditions of work shall be determined by the Joint Industry Board.

Meal breaks, including washing time, shall be unpaid of one hour duration or lesser period at the Employers discretion and shall not be exceeded. The Employer shall declare the working days and hours (including breaks) on each job.

(b) Flexible Working

Because of the very wide range of work activities covered by this agreement, in certain situations in the interest of efficiency and productivity a flexible working pattern may be appropriate. Therefore by mutual agreement and following vetting by a full time official of the AEEU the following rule may apply. Any flexible working patterns must not be introduced to circumvent existing overtime provisions.

An employee who agrees to work a flexible working pattern (i.e. to work any five days out of seven) shall:-

(i) be paid his graded rate plus a premium of 15% for working the agreed pattern of work in each week

(ii) receive the appropriate overtime premium calculated on the basic graded pay rate after working 38 hours in any week on the agreed pattern of work

(iii) be paid time and a half at the basic graded rate of pay for the first four hours worked on the first agreed rest day of each week. Thereafter at the rate of double time for the remainder of any rest day or part thereof worked in that week

(iv) not receive overtime premium for any Saturday or Sunday included in the agreed working pattern

(v) forfeit the 15% premium for any week in which there is a failure, without an acceptable reason, to report to duty on any Saturday or Sunday which is included in the agreed pattern of work

Examples of acceptable reasons are when an operative
has lost time through illness certified by a medical practitioner
was on a rest period for the day following continuous working all the previous night
was absent with the employers permission

For the purpose of NWR 9(c) - Call Out - the two agreed rest days in the flexible working sheet shall be deemed to be Saturday and Sunday respectively.

4: UTILISATION OF WORKING HOURS

There shall be full utilisation of working hours which shall not be subject to unauthorised "breaks". Time permitted for tea breaks shall not be exceeded.

Bad timekeeping and/or unauthorised absence from the place of work, during working hours, shall be construed as Industrial Misconduct.

Meeting of operatives shall not be held during working hours except by arrangement with the Job/Shop Representative and with the prior permission of the Employer or the Employers site management or the Employers representative.

604 *Rates of Wages and Working Rules - Electrical Installations*

WORKING RULES

5: TOOLS

The Employer shall provide all power-operated and expendable tools as required; operatives shall act with the greatest possible responsibility in respect of the use, maintenance and safe-keeping of tools and equipment of their Employers.

The operative shall have a kit of hand tools appropriate for carrying out efficiently the work for which he is employed; the kit shall include a lockable tool box.

The Employer shall provide, where practicable, suitable and lockable facilities for storing operatives tool-kits.

6: WAGES (Graded Operatives)

(a) Rates

(i)   National Standard Rate

The National Standard JIB Graded Rates of Wages (hereinafter called "the JIB Rates of Wages") shall be those from time to time determined by the Joint Industry Board pursuant to Rule 80 of the Rules of the Joint Industry Board.

The JIB Rates of Wages appropriate to operatives shall be such rates for their grades as the Joint Industry Board may from time to time determine to be appropriate for their grade in the place where they are working and they shall be paid no more and no less wages.

(ii) Productivity and Incentive Schemes

Employers and operatives, with the involvement of the local full time AEEU official, may agree on any job, arrangements for the maximum utilisation of working hours, a bonus payment or payments related to progress of the work and productivity levels or any other related matters in addition to the normal hourly rates of pay.

(b) London Weighting

(i)   Definition of the London Zone

That area lying within and including the M25 London Orbital Motorway.

(ii) Application

London Weighting as determined from time to time by the Joint Industry Board shall apply to all operatives and (at separate rates) to apprentices working on jobs in the London Zone as defined above in Rule 6 (b) (i) and for avoidance of doubt it is intended that such London Weighting shall be deemed to be part of the standard JIB Rates of Wages.

Such London Weighting shall also apply to any operative or apprentice who has been working from a London based shop in the London Zone for not less than 12 weeks and who is sent by his employer to a job out of the London Zone for a period of not more than 12 weeks or for the duration of one particular contract, whichever is the longer.

London Weighting shall apply to all paid hours including overtime and shift premium payments (National Working Rules 9 and 10), statutory holiday payments (National Working Rule 13) and travelling time payments (National Working Rule 12(b)).

(iii) The amount of London Weighting shall be as determined from time to time by the JIB National Board

Rates of Wages and Working Rules - Electrical Installations 605

WORKING RULES

(c) Responsibility Money

Approved Electricians in charge of work, who undertake the supervision of other operatives, shall be paid "responsibility money" as determined from time to time by the JIB National Board, currently not less than 10p and not more than £1.00 per hour.

The Supervision of apprentices or trainees is a responsibility of all skilled personnel and the supervision of 'other operatives' does not include apprentices/trainees for the purposes of payment of responsibility money.

Responsibility payments shall be enhanced by overtime and shift premiums where appropriate.

7: PAYMENT OF WAGES

Wages shall normally be paid by Credit Transfer. Alternatively, another method of payment may be adopted by mutual arrangement between Employer and Operative.

Wages shall be calculated for weekly periods and paid to the Operative within 5 normal working days of week termination, unless alternative arrangements are agreed.

Each operative shall receive an itemised written pay statement in accordance with the Employment Rights Act 1996.

8: TERMINATION OF EMPLOYMENT

Except for instances of industrial misconduct, an Operative's employment shall be terminated by the statutory period of notice under the Employment Rights Act 1996 (applicable in England, Wales and Scotland) or the Contracts of Employment and Redundancy Payments Act (Northern Ireland) 1965.

The provisions of the appropriate Acts apply to operatives with at least one month's continuous employment. Operatives not covered by these provisions shall be entitled to receive and shall be required to give notice on not less than one working day.

On termination of employment, immediate payment shall be made for all hours worked at the appropriate rate of wages with approximate deductions for Income Tax. The balance of wages due, any outstanding holiday pay received by the employer from the ECIBA and P.A.Y.E. form P.45, parts 2 and 3, shall be forwarded by the Employer to the Operative by registered post within three days. (Where the incidence of a national or local holiday makes this impracticable the balance of wages, etc., shall be dispatched on the first working day after the holiday).

It is recognised that when an operative is instantly dismissed for Industrial Misconduct it may not always be possible to make immediate payment. In these cases, payment will be forwarded by the Employer to the Operative by registered post within three days (where the incidence of a national or local holiday makes this impractical the balance of Wages, etc. shall be dispatched on the first working day after such a holiday).

An operative with six months' continuous service who has been dismissed shall be entitled, on request, to be provided with a written statement by the employer within 14 days, giving the reason(s) for dismissal.

Rates of Wages and Working Rules - Electrical Installations

WORKING RULES

9: OVERTIME

(a) Hours

Overtime is deprecated by the Joint Industry Board; systematic overtime in particular is to be avoided.

A Regional Joint Industry Board may, from time to time, declare permissible hours of overtime in the Region which shall not be exceeded without permission of the Regional Joint Industry Board.

Overtime will not be restricted in the case of Breakdowns or Urgent Maintenance and Repairs.

(b) Payment

(i) The number of hours to be worked at normal rates in any one week (Monday to Friday) before any overtime premium is calculated shall be 38 hours.

Premium time shall be paid at time-and-a-half. All hours worked between 1 p.m. on Saturday and normal starting time on Monday shall be paid at double time. Overtime premium payments shall be calculated on the appropriate standard rate of pay.

Exceptions: For the purpose of premium payment, an Operative shall be deemed to have worked normal hours on days where, although no payment is made by the Employer, the Operative:

(a) has lost time through certified sickness.
(b) was on a rest period for the day following continuous working the previous night.
(c) was absent with the Employers permission.

(ii) Any Operative who has not worked five days (as determined in Rule 3) from Monday to Friday, taking into account the exceptions detailed above, is precluded from working the following Saturday or Sunday.

(c) Call out

Notwithstanding the previous Clause, when an operative is called upon to return to work after his normal finishing time and before his next normal starting time he shall be paid at time-and-a-half for all the hours involved (home-to-home), subject to the guaranteed minimum payment. Between 1 pm on Saturday, and the normal starting time on Monday the appropriate premium rate shall be double time.

The guaranteed minimum payment for a single call out under this clause shall be the equivalent of 4 hours at the Operatives JIB Rates of Wages.

10: SHIFTWORK

Operatives may be required to undertake shiftworking arrangements in order to meet the requirements of the job or client. Operatives may not be so required without reasonable notice.

(a) Permanent Night Shift

(i) Night shift is where operatives (other than as overtime after the end of a day shift) work throughout the night for not less than three consecutive nights.

A full night shift shall consist of 37.5 hours worked on five nights, Monday night to Friday night inclusive, with unpaid breaks for meals each night, to be mutually arranged. The employer shall declare working hours including breaks on each contract.

Rates of Wages and Working Rules - Electrical Installations 607

WORKING RULES

(ii) Payment

Night shifts shall be paid at the rate of time and one-third for all hours worked up to 37.5 in any one week, Monday to Friday.

(b) Double day Shift (Rotating)

(i) The shift week will be from Monday to Friday. Each shift shall be of 7.5 hours worked with an unpaid half hour meal break. The distribution of the hours will be subject to local requirements. Shifts will normally be on an early and late basis.

(ii) Payment

Rotating double-day shift working will be paid at the rate of time plus 20% for normal hours in the early shift and time plus 30% for normal hours worked in the late shift.

(c) Three Shift Working (Rotating)

(i) The shift week will be from Monday to Friday. Each shift shall be of 7.5 hours duration with an unpaid half hour meal break. The distribution of the hours will be subject to local requirements. Shifts will normally be on an early, late and night shift basis.

(ii) Payment

Rotating Three-Shift work will be paid at the rate of time plus 20%, time plus 30% and time plus 33 1/3% for the early, late and night shifts respectively.

(d) Three Shift Working (Seven day continuous)

(i) Occasional

Where continuous shift work is occasionally required to cover both weekdays and weekends, weekend working shall attract the appropriate premiums contained in Rule 9(b) above. Weekday working shall attract the premiums contained in Rule 10(c) above. Generally speaking, "occasional" shiftwork shall be defined as a shiftwork requirement for a period of four weeks or less to meet some short term or emergency exigency.

(ii) Rostered

Where continuous three shift working is required to cover a regular seven day working pattern the following conditions shall be observed:

(i) Unless the requirements for continuous three shift working is specified in the operatives contract of employment or terms of engagement, four weeks prior notice shall be given before the introduction of a rostered three shift working system.

(ii) Prior to the introduction of a rostered three shift working system the employer will discuss and agree with his employees representatives the most suitable pattern of hours to achieve the required cover.

(iii) Subject to the above, rostered three shift working shall not be restricted.

(iv) The normal shift week shall be from Monday to Sunday and will comprise a maximum 37.5 hours in any one week for which the employees shall be paid at time plus thirty per cent.

608   *Rates of Wages and Working Rules - Electrical Installations*

WORKING RULES

> (v)   All hours rostered, or unrostered, in excess of 37.5 hours in any week, Monday to Sunday, shall fall within the terms of Rule 10(e) below.

(e) Overtime on Shifts

The number of hours to be worked at the appropriate shift rates before overtime premium is calculated shall be 38 hrs

Premium payments shall be calculated on the appropriate standard rate of pay and not on the shift rate.

(f) Other Shift Arrangements

Detailed arrangements for any other shift system of those operating on sites covered by the JIB/NJC Treaty Arrangement will be as approved by the JIB.

11: DEFINITION OF 'SHOP'

Employers shall declare the branches of their business as the Shop from which entitlement to travelling time and allowance shall be calculated, subject to the branches fulfilling the following conditions:-

The premises are owned or rented by the Employer. The premises are used for the purpose of general trading or personnel management as a distinct from the management of one contract or one site.

There shall be a full time staff available during normal working hours capable of dealing with, and resolving, enquiries relating to recruitment, payment of wages and other matters affecting employment.

The place of employment is the shop and by custom and practice all shop recruited operatives are transferable from job to job.

Employers are required to notify the Joint Industry Board when they establish a new Branch Office or Shop in any Region.

12: TRAVELLING TIME AND TRAVEL ALLOWANCES, COUNTRY AND LODGING ALLOWANCES

(a) (i) Wages and Allowances

Operatives who are required to book on and off at the Employers Shop shall be entitled to time from booking on until booking off with overtime if the time so booked exceeds the normal working day. They shall also be entitled to be reimbursed, if transport is not provided free of charge, the cost of any actual fare reasonably incurred.

(ii) Provision of Transport or Mileage Allowance

Where an employer provides transport free of charge, operatives provided with such transport shall not be entitled to travel allowance. An operative who considers that the transport provided by his employer is unsuitable may pursue a complaint through the Procedure for Handling Grievances and Avoiding Disputes. Payment of travel allowance shall be satisfied by the provision by an employer of such as a periodic travel pass (e.g. season ticket)

An employer by agreement with an operative may pay a mileage allowance where the operative agrees to use his own car for business use; the operative shall have appropriate insurance cover.

Rates of Wages and Working Rules - Electrical Installations   609

WORKING RULES

On any day when such mileage allowance is paid in respect of a journey or journeys between an operative's home (or lodgings) and his employers place or business or a site at which the operative is required to attend, travel allowance will not also be payable.

(b) Travelling Time and Travel Allowances

(i) Supplementary Travel Payment

Operatives who are required to start and finish at the normal starting and finishing time on jobs which are up to and including 10 miles from the shop - in a straight line - shall receive a Supplementary Travel Payment as determined from time to time by the JIB National Board for all hours worked (i.e. not subject to overtime, shift etc. premium).

(ii) Travelling Time and Travel Allowances

Operatives who are required to start and finish at the normal starting and finishing time on jobs which are over 10 miles and up to and including 55 miles from the shop - in a straight line - shall not receive the Supplementary Travel Payment but will receive both TRAVEL ALLOWANCES AND PAYMENT FOR TRAVELLING TIME as determined from time to time by the JIB National Board.

(c) Country Allowance

Operatives sent from the Employers Shop who are required to start and finish at the normal starting and finishing time on jobs over 55 miles from the Shop and who travel daily to the job instead of taking lodgings will be paid Country Allowance as determined from time to time by the JIB National Board in lieu of travelling time and Travel Allowance. In the case of London jobs this is over and above the JIB Rates of Wages for the London Zone.

Where an employer provides transport free of charge the operative shall receive 50% of the Country Allowance.

(d) Lodging Allowance

(i) Operatives sent from the Employers Shop who are required to start and finish at the normal starting and finishing time on jobs where their Employer requires them to stay away from their normal place of residence and provide proof of lodging will be paid Lodging Allowance as determined from time to time by the JIB National Board (see Appendix C).

The employer may, at his absolute discretion, chose to meet a reasonable bill presented by the operative rather than paying an untaxed lodging allowance.

(ii) Travelling time and travel allowance between the lodgings and the job shall not normally be paid. Where it is proved to the Employers satisfaction that suitable lodging and accommodation is not available near the job, travelling time and travel allowances for any distance of more than 15 miles each way will be paid in accordance with the scale contained in paragraph 12(b) above on the excess distance.

(iii) On being sent to the job the Operative shall receive his actual fare and travelling time at ordinary rates from the Employers Shop and when he returns to the Employers Shop except that when, of his own free will, he leaves the job within one calendar month from the date of his arrival and in cases where he is dismissed by the Employer for proved bad timekeeping, improper work or similar misconduct, no return travelling time or fares shall be paid.

(iv) The payment of Lodging Allowance shall not be made when suitable board and lodging is arranged by the Employer at no cost to the Operative.

WORKING RULES

Where an employer is involved in providing or arranging accommodation, the employer has a responsibility for ensuring beforehand that the accommodation, inclusive of breakfast and evening meal, is adequate and suitable, and shall deal expeditiously with any problem reported by an employee regarding such accommodation.

When circumstances are such that to travel daily to a job is an onerous requirement on an employee, the employer shall not unreasonably refuse to pay lodging allowance or provide accommodation as an alternative to travelling daily.

(v) The payment of Lodging Allowance shall not be made during absence from employment unless a Medical Certificate is produced for the whole of the period claimed. When an operative is sent home by the firm at their cost the payment of Lodging Allowance shall cease.

(vi) No payment for the retention of lodgings during Annual Paid Holiday shall be made by the Employer except in cases where the Operative is required to pay a retention fee during Annual Paid Holiday when reimbursement shall be of the amount actually paid to a maximum, as determined from time to time by the JIB National Board, upon production of proof of payment to the Employers satisfaction.

(vii) Where an Operative is away from his lodgings at a weekend under Rule 12(e) but has to pay a retention fee for his lodgings, reimbursement shall be the amount actually paid, to a maximum as determined from time to time by the JIB National Board upon production of proof of payment to the Employers satisfaction.

(e) Period Return Fares for Operatives who lodge

(i) On jobs up to and including 100 miles from the Employers Shop, return railway fares from the Job to the Employers Shop, without travelling time, shall be paid for every two weeks.

(ii) On jobs over 100 miles and up to and including 250 miles from the Employers Shop, return railway fares from the Job to the Employers Shop, with 4 hours travelling time at ordinary rate time, shall be paid every four weeks.

(iii) On jobs over 250 miles from the Employers Shop, return railway fares from the Job to the Employers Shop, with 7.5 hours travelling time at ordinary rates, shall be paid every four weeks.

(iv) In cases under sub-clauses (ii) and (iii) above, where the Employer, through necessity or expediency, requires his Operatives to work during the specified weekend leave period, he shall arrange that they shall have another period in substitution but this provision shall not apply under sub-clause (i) above.

(v) When an employer arranges to transport operatives between the Job and the Shop then fares shall not be payable under (i) (ii) and (iii) above. When travel between the job and the Shop can only be by air, then a return journey shall be arranged every six weeks, regardless of distance.

N.B. All distances shall be calculated in a straight line (point to point).

When Annual Holidays with pay are taken the period returns may be moved forward or backward from the date upon which they become due, to enable the period returns to coincide with the date of the Annual Paid Holiday.

Special consideration shall be given to Operatives where it is necessary for them to return home on compassionate grounds, e.g. domestic illness.

Rates of Wages and Working Rules - Electrical Installations 611

WORKING RULES

(f) Locally Engaged Labour

Where an Employer does not have a Shop within 25 miles of the Job, he can engage labour domiciled within a 25 miles radius of that Job. Operatives shall receive the JIB Rates of Wages applicable to the Zone of the Job and travelling time and Travel Allowance in accordance with Clause (b), but with the exception of "home" being substituted for "shop" in Clause (b).

Locally engaged labour, domiciled within a 25 miles radius of the Job, can be transferred to other Jobs within that radius without affecting their entitlements under this Rule. Operatives transferred to a Job outside that radius will, by mutual agreement, have their contracts of employment changed to a shop recruited basis.

N.B. (a) Upon engagement, the site upon which the operative is placed is formally his place of work and must be declared (as provided for under the amended JIB statement of particulars of terms of employment) to comply with the requirements of the Employment Rights Act 1996.

(b) When a redundancy situation occurs at 'the place where the employee was so employed' (Employment Rights Act 1996) that place is, for the shop recruited operative, the entire work area covered by the shop of recruitment and, for the site recruited operative, the site of recruitment and selection for redundancy must be made on that basis; within the parameters of the company's agreed selection procedures and the statutory rights of those concerned.

(c) It is advisable that an employer should make the points of law set out under (a) and (b) clear to the operative, particularly the site engaged operative, when he accepts work as a site recruited operative.

In the event of a dispute arising each case will be considered in the light of justice, equity and the merits of the individual case.

13: STATUTORY HOLIDAYS

(a) Qualification

Seven and a half hours pay at the appropriate JIB Rates of Wages shall be paid for a maximum of eight Statutory Holidays per annum. In general, the following shall constitute such paid holidays:-

New Years Day; Good Friday; Easter Monday; May Day; Spring Time Bank Holiday; Late Summer Bank Holiday; Christmas Day; Boxing Day.

In areas where any of these days are not normally observed as holidays in the Electrical Contracting Industry, traditional local holidays may be substituted by mutual agreement and subject to the determination of the appropriate Regional Joint Industry Board.

When Christmas Day and/or Boxing Day or New Years Day falls on a Saturday or Sunday, the following provisions apply:-

Christmas Day
When Christmas Day falls on a Saturday or a Sunday, the Tuesday next following shall be deemed to be a paid holiday.

Boxing Day or New Years Day
When Boxing Day or New Years Day falls on a Saturday or Sunday, the Monday next following shall be deemed to be a paid holiday.

In order to qualify for payment, operatives must work full time for the normal day on the working days preceding and following the holiday.

Rates of Wages and Working Rules - Electrical Installations

WORKING RULES

For the purpose of this Rule, an operative shall be deemed to have worked on one or both of the qualifying days when the Operative
(i) has lost time through certified sickness.
(ii) was on a rest period for the day following continuous working all the previous night.
(iii) was absent with the Employers permission.

(b) Payment for working Statutory Holidays

When operatives are required to work on a Paid Holiday within the scope of this Agreement, they shall receive wages at the following rates for all hours worked:-

CHRISTMAS DAY - Double time and a day or shift off in lieu for which they shall be paid wages at bare time rates for the hours constituting a normal working day. The alternative day hereunder shall be mutually agreed between the Employer and the Operatives concerned.
In respect of all other days:-

Either (a) Time-and-a-half plus a day or shift off in lieu for which they shall be paid wages at bare time rates for the hours constituting a normal working day. The alternative day hereunder shall be mutually agreed between the Employer and the Operatives concerned.

Or (b) at the discretion of the Employer 2.5 times the bare time rate in which event no alternative day is to be given.

In the case of night shift workers required to work on a Statutory Holiday, the premiums mentioned above shall be calculated upon the night shift rate of time-and-a-third. Time off in lieu of Statutory Holidays shall be paid at bare time day rates.

14: ANNUAL HOLIDAYS

Operatives shall be entitled to payment for Annual Holidays as determined from time to time under the JIB Annual Holiday with Pay Scheme, depending upon their continuity of service in the Industry.

The JIB Annual Holiday with Pay Scheme is carried out on behalf of the Joint Industry Board by the Electrical Contracting Industry Benefits Agency (ECIBA) who operate a Benefits Credit collection system. Details of the scheme are shown in the JIB Handbook.

N.B. A person authorised by the Joint Industry Board is entitled to inspect the Benefits Credits records held by any JIB Employer participant at any time. Individual operatives are entitled to inspect their own individual records held by their employer upon request, subject to reasonable notice, and further the operative may consult the ECIBA to establish that the correct Benefits Credits have been purchased in his name.

15: SICKNESS WITH PAY AND GROUP LIFE INSURANCE, ACCIDENTAL DEATH AND PERMANENT AND TOTAL DISABILITY SCHEMES.

With 4 weeks' qualifying service operatives shall be entitled to sickness benefit, excluding the first 2 weeks of absence due to sickness, as may be determined by the JIB National Board from time to time payable up to the 26th week in any 52 week period. Similarly, operatives shall be entitled to Death Benefit, Accidental Death from accident at work or travelling to and from work, Dismemberment and Permanent and Total Disability Schemes

Any Payment made under the JIB Rules to graded operatives, Trainees and Apprentices shall be additional to any Statutory Sick Pay that may be payable in respect of the same day of incapacity for work under the regulations made under the Social Security Contributions and Benefits Act 1992.

Rates of Wages and Working Rules - Electrical Installations 613

WORKING RULES

The obligations of JIB Employer Participants to provide the above benefits for their operatives shall be discharged by the Industry's Insurance arrangements, provided the employer complies with the requirements of the JIB Combined Benefits Scheme, or by other means contributing to the scheme on behalf of each operative in a manner determined by the JIB.

NOTE

The JIB Combined Benefits Scheme provides continuity of entitlement to an operative whilst employed by a JIB Employer Participant and, subject to certain conditions, Death Benefit continues after employment by a JIB Employer Participant provided the operative has registered as unemployed with the JIB and has continued to register as unemployed with the DSS.

Details of the JIB Combined Benefits Insurance Schemes are shown in the JIB Handbook.

N.B. A person authorised by the Joint Industry Board is entitled to inspect the Benefits Credits records held by any JIB Employer Participant at any time. Individual operatives are entitled to inspect their own individual records held by their employer upon request, subject to reasonable notice, and further the operative may consult the ECIBA to establish the correct Benefits Credits have been purchased in his name.

GRADING DEFINITIONS

1.1 Technician

Qualification and Training

Must have been a Registered Apprentice and have had practical training in electrical installation work and must have obtained the City and Guilds of London Institute Electrical Installation Work Part III Course Certificate (or approved equivalent), and either:

(a) Must be at least 27 years of age.
Must have had at least five years' experience as a Approved Electrician with "responsibility money", including a minimum of three years in a supervisory capacity in charge of electrical engineering installations of such a complexity and dimension as to require wide technical experience and organisational ability.

or (b) Must have exceptional technical skill, ability and experience beyond that expected of an Approved Electrician, so that his value to the Employer would be as if he were qualified as a Technician under (a) above and, with the support of his present Employer, may be granted this grade by the Joint Industry Board

Duties

Technicians must have knowledge of the most economical and effective layout of electrical installations together with the ability to achieve a high level of productivity in the work which they control. They must also be able to apply a thorough working knowledge of the National Working Rules for the Electrical Contracting Industry, of the current IEE Regulations for Electrical Installations, of the Electricity at Work Regulations 1989, the Electricity Supply Regulations, Installations (i.e. Regulations 22-29 inclusive and 31), of any Regulations dealing with Consumers Installations which may be issued, relevant British Standards and Codes of Practice, and of the Construction Industry Safety Regulations.

614   *Rates of Wages and Working Rules - Electrical Installations*

WORKING RULES

1.2 Approved Electrician

Qualification and Training

Must satisfy the following three conditions :-

Must have been a Registered Apprentice or undergone some equivalent method of training and have had practical training in electrical installation work.
     and
Must have obtained the NVQ Level 3 - Installing and Commissioning Electrical Systems and Equipment, together with the necessary practical experience
     or
Must have obtained at least the City and Guilds 2360 Electrical Installation Theory Part 2 Course Certificate (or approved equivalent)
     and
have obtained Achievement Measurement 2 or must be able, with the application for Grading and any other relevant supporting evidence (i.e. the City and Guilds Electricians Certificate) which may be required, to satisfy the Grading Committee of his experience and suitability
     and
must have had two years experience working as an Electrician subsequent to the satisfactory completion of training and immediately prior to the application for this grade, or be 22 years of age, whichever is the sooner.

Duties

Approved Electricians must possess particular practical, productive and electrical engineering skills with adequate technical supervisory knowledge so as to be able to work on their own proficiently and carry out electrical installation work without detailed supervision in the most efficient and economical manner; be able to set out jobs from drawings and specifications and requisition the necessary installation materials. They must also have a thorough working knowledge of the National Working Rules for the Electrical Contracting Industry, of the current IEE Regulations for Electrical Installations, of the Electricity Supply Regulations, 1988, issued by the Electricity Commissioners so far as they deal with Consumers' Installations (i.e. Regulations 22-29 inclusive and 31), of any Regulations dealing with Consumers' Installations which may be issued, relevant British Standards and Codes of Practice, and of the Construction Industry Safety Regulations.

1.3. Electrician

Qualification and Training

Must have been a registered Apprentice or undergone some equivalent method of training and have had adequate practical training in electrical installation work
     and
Must have obtained the NVQ Level 3 - Installing and Commissioning Electrical Systems and Equipment, together with the necessary practical experience
     or
Must have completed the City and Guilds 2360 Electrical Installation Work Theory Part 2 Course (or approved equivalent) and have obtained Achievement Measurement 2 or must be able, with the application for grading and any other relevant supporting evidence which may be required, to satisfy the Grading Committee of his experience and suitability.
     and
Must be at least 21 years of age (which requirement may be waived if the applicant has obtained a pass in the City and Guilds 2360 Electrical Installation Theory Part 2 Course or approved equivalent).

Rates of Wages and Working Rules - Electrical Installations

WORKING RULES

Duties

Must be able to carry out electrical installation work efficiently in accordance with the National Working Rules for the Electrical Contracting Industry, the current IEE Regulations for Electrical Installations, and the Construction Industry Safety Regulations.

1.4. Labourer

Labourers may be employed to assist in the installation of cables in accordance with Section 5.1 - Cable Agreement: and to do other unskilled work under supervision provided that they should not be used to re-introduce pair working. Nothing in these rules should be taken to imply that labourers must be employed where there is not sufficient unskilled work to justify their employment, nor to prevent skilled men from doing a complete electrical installation job including the unskilled elements in these circumstances. On any Site at any time there shall be employed in total no more than one Labourer to four skilled JIB Graded Operatives. This particular requirement may be reviewed in the light of the particular circumstances in respect of a particular site upon application, by either Party to the appropriate Regional Joint Industry Board.

SPON'S PRICE BOOKS 2002

with free CD-ROM

Free CD-ROM when you order any Spon's 2002 Price Book.
Use the CD-ROM to:
- produce tender documents
- customise data
- keyword search
- export to other major packages
- perform simple calculations.

Spon's Architects' and Builders' Price Book 2002
Davis Langdon & Everest

"Spon's Price Books have always been a 'Bible' in my work - now they have got even better! The CDs are not only quick but easy to use. The CD ROMs will really help me to get the most from my Spon's in my role as a Freelance Surveyor."
Martin Taylor, Isle of Lewis

New Features for 2002 include:
- A new section on Captial Allowances
- Inclusion of new items within a seperate Measured Works section

September 2001: 1024 pages
Hb & CD-ROM: 0-415-26216-X: £110.00

Spon's Landscapes and External Works Price Book 2002
Davis Langdon & Everest, in association with Landscape Projects

New Features for 2002 include:
- Fees for professional services
- Revised and updated sections on Cost Information and how to use this book
- Revisions and expansions of the Approximate Estimating section, together with direct links into the Measured Works Section

September 2001: 484 pages
Hb & CD-ROM: 0-415-26220-8: £80.00

Spon's Mechanical and Electrical Services Price Book 2002
Mott Green & Wall

"An essential reference for everybody concerned with the calculation of costs of mechanical and electrical works." *Cost Engineer*

New Features for 2002 include:
- New sections on modular wiring, emergency lighting, lighting control, sprinkler pre fabricated pipework, UPVC rainwater and gutters, carbon steel pipework and fittings

September 2001: 584 pages
Hb & CD-ROM: 0-415-26222-4: £110.00

updates available to download from the web
www.pricebooks.co.uk

Spon's Civil Engineering and Highway Works Price Book 2002
Davis Langdon & Everest

New Features for 2002 include:
- A revised and extended section on Land Remediation
- The Rail Track section now includes data on Permanent Way work with fully reviewed pricing
- Fully reviewed pricing for the Geotextiles section

September 2001: 688 pages
Hb & CD-ROM: 0-415-26218-6: £120.00

Return your orders to: Spon Press Customer Service Department, ITPS, Cheriton House, North Way, Andover, Hampshire, SP10 5BE · Tel: +44 (0) 1264 343071 · Fax: + 44 (0) 1264 343005 · Email: book.orders@tandf.co.uk
Postage & Packing: 5% of order value (min. charge £1, max. charge £10) for 3–5 days delivery · Option of next day delivery at an additional £6.50.

PART FOUR

Daywork

Heating and Ventilating Industry, *page 617*
Electrical Industry, *page 620*
Building Industry Plant Hire Costs, *page 624*

When work is carried out in connection with a contract which cannot be valued in any other way, it is usual to assess the value on a cost basis with suitable allowances to cover overheads and profit. The basis of costing is a matter for agreement between the parties concerned but definitions of prime cost for the Heating and Ventilating and Electrical Industries have been prepared and published jointly by the Royal Institution of Chartered Surveyors and the appropriate bodies of the industries concerned for the convenience of those who wish to use them together with a schedule of basic plant charges published by the Royal Institution of Chartered Surveyors. These documents are reproduced by kind permission of the publishers.

Keep your figures up to date, free of charge
Download updates from the web: www.pricebooks.co.uk

This section, and most of the other information in this Price Book, is brought up to date every three months with the Price Book Updates, until the next annual edition. The updates are available free to all Price Book purchasers.

To ensure you receive your free copies, either complete the reply card from the centre of the book and return it to us or register via the website www.pricebooks.co.uk

Daywork

HEATING AND VENTILATING INDUSTRY

DEFINITION OF PRIME COST OF DAYWORK CARRIED OUT UNDER A HEATING, VENTILATING, AIR CONDITIONING, REFRIGERATION, PIPEWORK AND/OR DOMESTIC ENGINEERING CONTRACT (JULY 1980 EDITION)

This Definition of Prime Cost is published by the Royal Institution of Chartered Surveyors and the Heating and Ventilating Contractors Association for convenience, and for use by people who choose to use it. Members of the Heating and Ventilating Contractors Association are not in any way debarred from defining Prime Cost and rendering accounts for work carried out on that basis in any way they choose. Building owners are advised to reach agreement with contractors on the Definition of Prime Cost to be used prior to entering into a contract or sub-contract.

SECTION 1: APPLICATION

1.1 This Definition provides a basis for the valuation of daywork executed under such heating, ventilating, air conditioning, refrigeration, pipework and or domestic engineering contracts as provide for its use.

1.2 It is not applicable in any other circumstances, such as jobbing or other work carried out as a separate or main contract nor in the case of daywork executed after a date of practical completion.

1.3 The terms 'contract' and 'contractor' herein shall be read as 'sub-contract' and 'sub-contractor' as applicable.

SECTION 2: COMPOSITION OF TOTAL CHARGES

2.1 The Prime Cost of daywork comprises the sum of the following costs:
    (a)   Labour as defined in Section 3.
    (b)   Materials and goods as defined in Section 4.
    (c)   Plant as defined in Section 5.

2.2 Incidental costs, overheads and profit as defined in Section 6, as provided in the contract and expressed therein as percentage adjustments, are applicable to each of 2.1 (a)-(c).

SECTION 3: LABOUR

3.1 The standard wage rates, emoluments and expenses referred to below and the standard working hours referred to in 3.2 are those laid down for the time being in the rules or decisions or agreements of the Joint Conciliation Committee of the Heating, Ventilating and Domestic Engineering Industry applicable to the works (or those of such other body as may be appropriate) and to the grade of operative concerned at the time when and the area where the daywork is executed.

3.2 Hourly base rates for labour are computed by dividing the annual prime cost of labour, based upon the standard working hours and as defined in 3.4, by the number of standard working hours per annum. See example.

3.3 The hourly rates computed in accordance with 3.2 shall be applied in respect of the time spent by operatives directly engaged on daywork, including those operating mechanical plant and transport and erecting and dismantling other plant (unless otherwise expressly provided in the contract) and handling and distributing the materials and goods used in the daywork.

3.4 The annual prime cost of labour comprises the following:
    (a)   Standard weekly earnings (i.e. the standard working week as determined at the appropriate rate for the operative concerned).
    (b)   Any supplemental payments.
    (c)   Any guaranteed minimum payments (unless included in Section 6.1(a)-(p)).
    (d)   Merit money.
    (e)   Differentials or extra payments in respect of skill, responsibility, discomfort, inconvenience or risk (excluding those in respect of supervisory responsibility - see 3.5)
    (f)   Payments in respect of public holidays.

618 *Daywork*

HEATING AND VENTILATING INDUSTRY

    (g)   Any amounts which may become payable by the contractor to or in respect of operatives arising from the rules etc. referred to in 3.1 which are not provided for in 3.4 (a)-(f) nor in Section 6.1 (a)-(p).

    (h)   Employers contributions to the WELPLAN, the HVACR Welfare and Holiday Scheme or payments in lieu thereof.

    (i)   Employers National Insurance contributions as applicable to 3.4 (a)-(h).

    (j)   Any contribution, levy or tax imposed by Statute, payable by the contractor in his capacity as an employer.

3.5 Differentials or extra payments in respect of supervisory responsibility are excluded from the annual prime cost (see Section 6). The time of principals, staff, foremen, chargehands and the like when working manually is admissible under this Section at the rates for the appropriate grades.

SECTION 4: MATERIALS AND GOODS

4.1 The prime cost of materials and goods obtained specifically for the daywork is the invoice cost after deducting all trade discounts and any portion of cash discounts in excess of 5%.

4.2 The prime cost of all other materials and goods used in the daywork is based upon the current market prices plus any appropriate handling charges.

4.3 The prime cost referred to in 4.1 and 4.2 includes the cost of delivery to site.

4.4 Any Value Added Tax which is treated, or is capable of being treated, as input tax (as defined by the Finance Act 1972, or any re-enactment or amendment thereof or substitution therefore) by the contractor is excluded.

SECTION 5: PLANT

5.1 Unless otherwise stated in the contract, the prime cost of plant comprises the cost of the following:

    (a)   use or hire of mechanically-operated plant and transport for the time employed on and/or provided or retained for the daywork;

    (b)   use of non-mechanical plant (excluding non-mechanical hand tools) for the time employed on and/or provided or retained for the daywork;

    (c)   transport to and from the site and erection and dismantling where applicable.

5.2 The use of non-mechanical hand tools and of erected scaffolding, staging, trestles or the like is excluded (see Section 6), unless specifically retained for the daywork.

SECTION 6: INCIDENTAL COSTS, OVERHEADS AND PROFIT

6.1 The percentage adjustments provided in the contract which are applicable to each of the totals of Sections 3, 4 and 5 comprise the following:

    (a)   Head office charges.

    (b)   Site staff including site supervision.

    (c)   The additional cost of overtime (other than that referred to in 6.2).

    (d)   Time lost due to inclement weather.

    (e)   The additional cost of bonuses and all other incentive payments in excess of any included in 3.4.

    (f)   Apprentices' study time.

    (g)   Fares and travelling allowances.

    (h)   Country, lodging and periodic allowances.

    (i)   Sick pay or insurances in respect thereof, other than as included in 3.4.

    (j)   Third party and employers' liability insurance.

    (k)   Liability in respect of redundancy payments to employees.

    (l)   Employer's National Insurance contributions not included in 3.4.

    (m) Use and maintenance of non-mechanical hand tools.

    (n)   Use of erected scaffolding, staging, trestles or the like (but see 5.2).

    (o)   Use of tarpaulins, protective clothing, artificial lighting, safety and welfare facilities, storage and the like that may be available on site.

Daywork

619

HEATING AND VENTILATING INDUSTRY

(p)   Any variation to basic rates required by the contractor in cases where the contract provides for the use of a specified schedule of basic plant charges (to the extent that no other provision is made for such variation - see 5.1).

(q)   In the case of a sub-contract which provides that the sub-contractor shall allow a cash discount, such provision as is necessary for the allowance of the prescribed rate of discount.

(r)   All other liabilities and obligations whatsoever not specifically referred to in this Section nor chargeable under any other Section.

(s)   Profit.

6.2   The additional cost of overtime where specifically ordered by the Architect/Supervising Officer shall only be chargeable in the terms of a prior written agreement between the parties.

MECHANICAL INSTALLATIONS

Calculation of Hourly Base Rate of Labour for Typical Main Grades applicable from 3 September 2001.

	FOREMAN	SENIOR CRAFTSMAN (+ 2ND WELDING SKILL)	SENIOR CRAFTSMAN	CRAFTSMAN	INSTALLER	MATE OVER 18
Hourly Rate from 3 September 2001	10.00	8.61	8.26	7.56	6.87	5.79
Annual standard earnings excluding all holidays 45.8 weeks x 38 hours	17,404.00	14,984.84	14,375.70	13,157.42	11,956.55	10,076.92
Employers national insurance contributions from 6 April 2001	1,574.13	1,286.25	1,213.76	1,068.79	925.89	702.21
Weekly holiday credit and welfare contributions (52 weeks) from 1 October 2001	2,757.04	2,583.36	2,583.36	2,407.08	2,319.20	2,230.80
Annual prime cost of labour	21,735.17	18.854.46	18,172.83	16,633.29	15,201.63	13,009.93
£ contributions per annum	12.49	10.83	10.44	9.56	8.73	7.48

Notes:
(1) Annual industry holiday (4.6 weeks x 38 hours) and public holidays (1.6 weeks x 38 hours) are paid through weekly holiday credit and welfare stamp scheme.
(2) Where applicable, Merit money and other variables (e.g. daily abnormal conditions money), which attract Employer's National Insurance contribution, should be included.
(3) Contractors in Northern Ireland should add the appropriate amount of CITB Levy to the annual prime cost of labour prior to calculating the hourly base rate.

620   *Daywork*

ELECTRICAL INDUSTRY

DEFINITION OF PRIME COST OF DAYWORK CARRIED OUT UNDER AN ELECTRICAL CONTRACT (MARCH 1981 EDITION)

This Definition of Prime Cost is published by The Royal Institution of Chartered Surveyors and The Electrical Contractors' Associations for convenience and for use by people who choose to use it. Members of The Electrical Contractors' Association are not in any way debarred from defining Prime Cost and rendering accounts for work carried out on that basis in any way they choose. Building owners are advised to reach agreement with contractors on the Definition of Prime Cost to be used prior to entering into a contract or sub-contract.

SECTION 1: APPLICATION

  1.1 This Definition provides a basis for the valuation of daywork executed under such electrical contracts as provide for its use.

  1.2 It is not applicable in any other circumstances, such as jobbing, or other work carried out as a separate or main contract, nor in the case of daywork executed after the date of practical completion.

  1.3 The terms 'contract' and 'contractor' herein shall be read as 'sub-contract' and 'sub-contractor' as the context may require.

SECTION 2: COMPOSITION OF TOTAL CHARGES

  2.1 The Prime Cost of daywork comprises the sum of the following costs:
      (a) Labour as defined in Section 3.
      (b) Materials and goods as defined in Section 4.
      (c) Plant as defined in Section 5.

  2.2 Incidental costs, overheads and profit as defined in Section 6, as provided in the contract and expressed therein as percentage adjustments, are applicable to each of 2.1 (a)-(c).

SECTION 3: LABOUR

  3.1 The standard wage rates, emoluments and expenses referred to below and the standard working hours referred to in 3.2 are those laid down for the time being in the rules and determinations or decisions of the Joint Industry Board or the Scottish Joint Industry Board for the Electrical Contracting Industry (or those of such other body as may be appropriate) applicable to the works and relating to the grade of operative concerned at the time when and in the area where daywork is executed.

  3.2 Hourly base rates for labour are computed by dividing the annual prime cost of labour, based upon the standard working hours and as defined in 3.4 by the number of standard working hours per annum. See examples.

  3.3 The hourly rates computed in accordance with 3.2 shall be applied in respect of the time spent by operatives directly engaged on daywork, including those operating mechanical plant and transport and erecting and dismantling other plant (unless otherwise expressly provided in the contract) and handling and distributing the materials and goods used in the daywork.

  3.4 The annual prime cost of labour comprises the following:
      (a) Standard weekly earnings (i.e. the standard working week as determined at the appropriate rate for the operative concerned).
      (b) Payments in respect of public holidays.
      (c) Any amounts which may become payable by the Contractor to or in respect of operatives arising from operation of the rules etc. referred to in 3.1 which are not provided for in 3.4(a) and (b) nor in Section 6.
      (d) Employer's National Insurance Contributions as applicable to 3.4 (a)-(c).
      (e) Employer's contributions to the Joint Industry Board Combined Benefits Scheme or Scottish Joint Industry Board Holiday and Welfare Stamp Scheme, and holiday payments made to apprentices in compliance with the Joint Industry Board National Working Rules and Industrial Determinations as an employer.

Daywork

ELECTRICAL INDUSTRY

    (f)   Any contribution, levy or tax imposed by Statute, payable by the Contractor in his capacity as an employer.

3.5   Differentials or extra payments in respect of supervisory responsibility are excluded from the annual prime cost (see Section 6). The time of principals and similar categories, when working manually, is admissible under this Section at the rates for the appropriate grades.

SECTION 4: MATERIALS AND GOODS

4.1   The prime cost of materials and goods obtained specifically for the daywork is the invoice cost after deducting all trade discounts and any portion of cash discounts in excess of 5%.

4.2   The prime cost of all other materials and goods used in the daywork is based upon the current market prices plus any appropriate handling charges.

4.3   The prime cost referred to in 4.1 and 4.2 includes the cost of delivery to site.

4.4   Any Value Added Tax which is treated, or is capable of being treated, as input tax (as defined by the Finance Act 1972, or any re-enactment or amendment thereof or substitution therefore) by the Contractor is excluded.

SECTION 5: PLANT

5.1   Unless otherwise stated in the contract, the prime cost of plant comprises the cost of the following:

    (a)   Use or hire of mechanically-operated plant and transport for the time employed on and/or provided or retained for the daywork;

    (b)   Use of non-mechanical plant (excluding non-mechanical hand tools) for the time employed on and/or provided or retained for the daywork;

    (c)   Transport to and from the site and erection and dismantling where applicable.

5.2   The use of non-mechanical hand tools and of erected scaffolding, staging, trestles or the likes is excluded (see Section 6), unless specifically retained for daywork.

5.3   Note: Where hired or other plant is operated by the Electrical Contractor's operatives, such time is to be included under Section 3 unless otherwise provided in the contract.

SECTION 6: INCIDENTAL COSTS, OVERHEADS AND PROFIT

6.1   The percentage adjustments provided in the contract which are applicable to each of the totals of Sections 3, 4 and 5, compromise the following:

    (a)   Head Office charges.

    (b)   Site staff including site supervision.

    (c)   The additional cost of overtime (other than that referred to in 6.2).

    (d)   Time lost due to inclement weather.

    (e)   The additional cost of bonuses and other incentive payments.

    (f)   Apprentices' study time.

    (g)   Travelling time and fares.

    (h)   Country and lodging allowances.

    (i)   Sick pay or insurance in lieu thereof, in respect of apprentices.

    (j)   Third party and employers' liability insurance.

    (k)   Liability in respect of redundancy payments to employees.

    (l)   Employers' National Insurance Contributions not included in 3.4.

    (m)   Use and maintenance of non-mechanical hand tools.

    (n)   Use of erected scaffolding, staging, trestles or the like (but see 5.2.).

    (o)   Use of tarpaulins, protective clothing, artificial lighting, safety and welfare facilities, storage and the like that may be available on site.

    (p)   Any variation to basic rates required by the Contractor in cases where the contract provides for the use of a specified schedule of basic plant charges (to the extent that no other provision is made for such variation - see 5.1).

    (q)   All other liabilities and obligations whatsoever not specifically referred to in this Section nor chargeable under any other Section.

    (r)   Profit.

Daywork

ELECTRICAL INDUSTRY

(s) In the case of a sub-contract which provides that the sub-contractor shall allow a cash discount, such provision as is necessary for the allowance of the prescribed rate of discount.

6.2 The additional cost of overtime where specifically ordered by the Architect/Supervising Officer shall only be chargeable in the terms of a prior written agreement between the parties.

NEW FROM SPON PRESS

The Architectural Expression of Environmental Control Systems

George Baird, Victoria University of Wellington, New Zealand

*The Architectural Expression of Environmental Control System*s examines the way project teams can approach the design and expression of both active and passive thermal environmental control systems in a more creative way. Using seminal case studies from around the world and interviews with the architects and environmental engineers involved, the book illustrates innovative responses to client, site and user requirements, focusing upon elegant design solutions to a perennial problem.

This book will inspire architects, building scientists and building services engineers to take a more creative approach to the design and expression of environmental control systems - whether active or passive, whether they influence overall building form or design detail.

March 2001: 276x219: 304pp
135 b+w photos, 40 colour, 90 line illustrations
Hb: 0-419-24430-1: £49.95

To Order: Tel: +44 (0) 8700 768853, or +44 (0) 1264 343071 Fax: +44 (0) 1264 343005, or
Post: Spon Press Customer Services, ITPS Andover, Hants, SP10 5BE, UK Email: book.orders@tandf.co.uk.

Postage & Packing: UK: 5% of order value (min. charge £1, max. charge £10) for 3-5 days delivery. Option of next day delivery at an additional £6.50. Europe: 10% of order value (min. charge £2.95, max. charge £20) for delivery surface post. Option of airmail at an additional £6.50. ROW: 15% of order value (min.charge£6.50, max. charge £30) for airmail delivery.

For a complete listing of all our titles visit: www.sponpress.com

Visit www.pricebooks.co.uk

Daywork

ELECTRICAL INDUSTRY

ELECTRICAL INSTALLATIONS

Calculation of Hourly Base Rate of Labour for Typical Main Grades applicable from 8 January 2001

	TECHNICIAN	APPROVED ELECTRICIAN	ELECTRICIAN	LABOURER
Hourly Rate from 8 January 2001 (London Rates)	11.20	10.02	9.28	7.57
Annual standard earnings excluding all holidays 46 weeks x 37.5 hours	19,320.00	17,284.50	16,008.00	13,058.25
Employers national insurance contributions from 6 April 2001	1,802.14	1,559.91	1,408.01	1,056.99
JIB Combined benefits from 25 September 2000	1,922.96	1,732.12	1,613.56	1,336.92
Holiday top up funding	990.08	915.72	867.36	759.20
Annual prime cost of labour	24,035.15	21,492.25	19,896.93	16,211.36
£ contributions per annum	13.93	12.46	11.53	9.40

Notes

(1)   Annual industry holiday (4.4 weeks x 37.5 hours) and public holidays (1.6 weeks x 37.5 hours)

(2)   It should be noted that all labour costs incurred by the Contractor in his capacity as an Employer, other than those contained in the hourly rate above, must be taken into account under Section 6.

(3)   Public Holidays are included in annual earnings.

(4)   Contractors in Northern Ireland should add the appropriate amount of CITB Levy to the annual prime cost of labour prior to calculating the hourly base rate.

624   *Daywork*

BUILDING INDUSTRY PLANT HIRE COSTS

SCHEDULE OF BASIC PLANT CHARGES (MAY 2001)

This Schedule is published by the Royal Institution of Chartered Surveyors and is for use in connection with Dayworks under a Building Contract.

EXPLANATORY NOTES

1   The rates in the Schedule are intended to apply solely to daywork carried out under and incidental to a Building Contract. They are NOT intended to apply
to:
(i)   jobbing or any other work carried out as a main or separate contract; or
(ii)   work carried out after the date of commencement of the Defects Liability Period.

2   The rates apply to plant and machinery already on site, whether hired or owned by the Contractor.

3   The rates, unless otherwise stated, include the cost of fuel and power of every description, lubricating oils, grease, maintenance, sharpening of tools, replacement of spare parts, all consumable stores and for licences and insurances applicable to items of plant.

4   The rates, unless otherwise stated, do not include the costs of drivers and attendants (unless otherwise stated).

5   The rates in the Schedule are base costs and may be subject to an overall adjustment for price movement, overheads and profit, quoted by the Contractor prior to the placing of the Contract.

6   The rates should be applied to the time during which the plant is actually engaged in daywork.

7   Whether or not plant is chargeable on daywork depends on the daywork agreement in use and the inclusion of an item of plant in this schedule does not necessarily indicate that item is chargeable.

8   Rates for plant not included in the Schedule or which is not already on site and is specifically provided or hired for daywork shall be settled at prices which are reasonably related to the rates in the Schedule having regard to any overall adjustment quoted by the Contractor in the Conditions of Contract.

NOTE: All rates in the schedule were calculated during the first quarter of 2001.

Daywork

BUILDING INDUSTRY PLANT HIRE COSTS

MECHANICAL PLANT AND TOOLS

Item of Plant	Size/Rating	Unit	Rate/hr
PUMPS			
Mobile Pumps			
Including pump hoses, valves and strainers etc.			
Diaphragm	50mm dia.	Each	0.87
Diaphragm	76mm dia.	Each	1.29
Submersible	50mm dia.	Each	1.18
Induced flow	50mm dia.	Each	1.54
Induced flow	76mm dia.	Each	2.05
Centrifugal, self priming	50mm dia.	Each	1.96
Centrifugal, self priming	102mm dia.	Each	2.52
Centrifugal, self priming	152mm dia.	Each	3.87
SCAFFOLDING, SHORING, FENCING			
Complete Scaffolding			
Mobile working towers, single width	1.8m x 0.8m base x 7m high	Each	2.00
Mobile working towers, single width	1.8m x 0.8m base x 9m high	Each	2.80
Mobile working towers, double width	1.8m x 1.4m base x 7m high	Each	2.15
Mobile working towers, double width	1.8m x 1.4m base x 15m high	Each	5.10
Chimney scaffold, single unit		Each	1.79
Chimney scaffold, twin unit		Each	2.05
Chimney scaffold, four unit		Each	3.59
Trestles			
Trestle, adjustable	Any height	Pair	0.10
Trestle, painters	1.8m high	Pair	0.21
Trestle, Painters	2.4m high	Pair	0.26
Shoring, Planking and Struting			
'Acrow' adjustable prop	Sizes up to 4.9m (open)	Each	0.10
'Strong boy' support attachment		Each	0.15
Adjustable trench struts	Sizes up to 1.67m (open)	Each	0.10
Trench sheet		Metre	0.01
Backhoe trench box		Each	1.00
Temporary Fencing			
Including block and coupler			
Site fencing steel grid panel	3.5m x 2.0m	Each	0.08
Anti-climb site steel grid fence panel	3.5m x 2.0m	Each	0.08
LIFTING APPLIANCES AND CONVEYORS			
Cranes			
Mobile Cranes			
Rates are inclusive of drivers			
Lorry mounted, telescopic jib			
Two wheel drive	6 tonnes	Each	24.40
Two wheel drive	7 tonnes	Each	25.00
Two wheel drive	8 tonnes	Each	25.62
Two wheel drive	10 tonnes	Each	26.90

626   *Daywork*

BUILDING INDUSTRY PLANT HIRE COSTS

MECHANICAL PLANT AND TOOLS

Item of Plant	Size/Rating	Unit	Rate/hr
Rates are inclusive of drivers			
Lorry mounted, telescopic jib			
Two wheel drive	12 tonnes	Each	28.25
Two wheel drive	15 tonnes	Each	29.66
Two wheel drive	18 tonnes	Each	31.14
Two wheel drive	20 tonnes	Each	32.70
Two wheel drive	25 tonnes	Each	34.33
Four wheel drive	10 tonnes	Each	27.44
Four wheel drive	12 tonnes	Each	28.81
Four wheel drive	15 tonnes	Each	30.25
Four wheel drive	20 tonnes	Each	33.35
Four wheel drive	25 tonnes	Each	35.19
Four wheel drive	30 tonnes	Each	37.12
Four wheel drive	45 tonnes	Each	39.16
Four wheel drive	50 tonnes	Each	41.32

Track-mounted tower crane
Rates inclusive of driver
Note : Capacity equals maximum lift in tonnes times maximum radius at which it can be lifted

	Capacity (metre/tonnes) up to	Height under hook above ground (m) up to17	Unit	Rate/hr
Tower crane	10	17	Each	7.99
Tower crane	15	18	Each	8.59
Tower crane	20	20	Each	9.18
Tower crane	25	22	Each	11.56
Tower crane	30	22	Each	13.78
Tower crane	40	22	Each	18.09
Tower crane	50	22	Each	22.20
Tower crane	60	22	Each	24.32
Tower crane	70	22	Each	23.00
Tower crane	80	22	Each	25.91
Tower crane	110	22	Each	26.45
Tower crane	125	30	Each	29.38
Tower crane	150	30	Each	32.35

Static tower cranes
Rates inclusive of driver
To be charged at 90% of the above rates for track mounted tower cranes

Crane Equipment

Muck tipping skip	Up to 0.25m³	Each	0.56
Muck tipping skip	0.5m³	Each	0.67
Muck tipping skip	0.75m³	Each	0.82
Muck tipping skip	1.0m³	Each	1.03
Muck tipping skip	1.5m³	Each	1.18
Muck tipping skip	2.0m³	Each	1.38

Daywork

BUILDING INDUSTRY PLANT HIRE COSTS

MECHANICAL PLANT AND TOOLS

Item of Plant	Size/Rating		Unit	Rate/hr
Crane Equipment (contd)				
Mortar skips	up to 0.38m³			0.41
Boat skips	1.0m³			1.08
Boat skips	1.5m³			1.33
Boat skips	2.0m³			1.59
Concrete skips, hand levered	0.5m³			1.00
Concrete skips, hand levered	0.75m³			1.10
Concrete skips, hand levered	1.0m³			1.25
Concrete skips, hand levered	1.5m³			1.50
Concrete skips, hand levered	2.0m³			1.65
Concrete skips, geared	0.5m³			1.30
Concrete skips, geared	0.75m³			1.40
Concrete skips, geared	1.0m³			1.55
Concrete skips, geared	1.5m³			1.80
Concrete skips, geared	2.0m³			2.05
Hoists				
Scaffold hoists	200kg			1.92
Rack and pinion (goods only)	500kg			3.31
Rack and pinion (goods only)	1100kg			4.28
Rack and pinion goods and passenger	15 person, 1200kg			5.62
Wheelbarrow chain sling				0.31
Conveyors				
<u>Belt conveyors</u>				
Conveyor	7.5m long x 400mm wide			6.41
Miniveyor, control box and loading hopper	3m unit			3.59
<u>Other Conveying Equipment</u>				
Wheelbarrow				0.21
Hydraulic superlift				2.95
Pavac slab lifter				1.03
Hand pad and hose attachment				0.26
Lifting Trucks				
Fork lift, two wheel drive	Payload	Maximum Lift		
Fork lift, two wheel drive	1100kg	up to 3.0m	Each	4.87
Fork lift, two wheel drive	2540kg	up to 3.7m	Each	5.12
Fork lift, four wheel drive	1524kg	up to 6.0m	Each	6.04
Fork lift, four wheel drive	2600kg	up to 5.4m	Each	7.69
Lifting Platforms				
Hydraulic platform (Cherry picker)	7.5m		Each	4.23
Hydraulic platform (Cherry picker)	13m		Each	9.23
Scissors lift	7.8m		Each	7.56
Telescopic handlers	7m, 2 tonne		Each	7.18
Telescopic handlers	13m, 3 tonne		Each	8.72
Lifting and Jacking Gear				
Pipe winch including gantry	1 tonne		Sets	1.92
Pipe winch including gantry	3 tonnes		Sets	3.21
Chain block	1 tonne		Each	0.45
Chain block	2 tonnes		Each	0.71

628   *Daywork*

BUILDING INDUSTRY PLANT HIRE COSTS

MECHANICAL PLANT AND TOOLS

Item of Plant	Size/Rating	Unit	Rate/hr
Chain block	5 tonnes	Each	1.22
Pull lift (Tirfor winch)	1 tonne	Each	0.64
Pull lift (Tirfor winch)	1.6 tonnes	Each	0.90
Pull lift (Tirfor winch)	3.2 tonnes	Each	1.15
Brother or chain slings, two legs	not exceeding 4.2 tonnes	Set	0.35
Brother or chain slings, two legs	not exceeding 7.5 tonnes	Set	0.45
Brother or chain slings, four legs	not exceeding 3.1 tonnes	Set	0.41
Brother or chain slings, four legs	not exceeding 11.2 tonnes	Set	1.28

CONSTRUCTION VEHICLES
Lorries
Plated lorries
Rates are inclusive of driver

Platform lorries	7.5 tonnes	Each	19.00
Platform lorries	17 tonnes	Each	21.00
Platform lorries	24 tonnes	Each	26.00
Platform lorries with winch and skids	7.5 tonnes	Each	21.40
Platform lorries with crane	17 tonnes	Each	27.50
Platform lorries with crane	24 tonnes	Each	32.10

Tipper Lorries
Rates are inclusive of driver

Tipper lorries	15/17 tonnes	Each	19.50
Tipper lorries	24 tonnes	Each	21.40
Tipper lorries	30 tonnes	Each	27.10

Dumpers
Site use only (excluding tax, insurance and extra
Cost of DERV etc. when operating on highway)

	Makers capacity		
Two wheel drive	0.8 tonnes	Each	1.20
Two wheel drive	1 tonne	Each	1.30
Two wheel drive	1.2 tonnes	Each	1.60
Four wheel drive	2 tonnes	Each	2.50
Four wheel drive	3 tonnes	Each	3.00
Four wheel drive	4 tonnes	Each	3.50
Four wheel drive	5 tonnes	Each	4.00
Four wheel drive	6 tonnes	Each	4.50

Dumper Trucks
Rates are inclusive of drivers

Dumper trucks	10/13 tonnes	Each	20.00
Dumper trucks	18/20 tonnes	Each	20.40
Dumper trucks	22/25 tonnes	Each	26.30
Dumper trucks	35/40 tonnes	Each	36.60

Tractors
<u>Agricultural Type</u>
Wheeled, rubber-clad tyred

Light	48 h.p.	Each	4.65

Daywork

BUILDING INDUSTRY PLANT HIRE COSTS

MECHANICAL PLANT AND TOOLS

Item of Plant	Size/Rating	Unit	Rate/hr
Tractors (contd)			
Heavy	65 h.p.	Each	5.15
<u>Crawler Tractors</u>			
With bull or angle dozer	80/90 h.p.	Each	21.40
With bull or angle dozer	115/130 h.p.	Each	25.10
With bull or angle dozer	130/150 h.p.	Each	26.00
With bull or angle dozer	155/175 h.p.	Each	27.74
With bull or angle dozer	210/230 h.p.	Each	28.00
With bull or angle dozer	300/340 h.p.	Each	31.10
With bull or angle dozer	400/440 h.p.	Each	46.90
With loading shovel	0.8m³	Each	25.00
With loading shovel	1.0m³	Each	28.00
With loading shovel	1.2m³	Each	32.00
With loading shovel	1.4m³	Each	36.00
With loading shovel	1.8m³	Each	45.00
Light Vans			
Ford Escort or the like		Each	4.74
Ford Transit or the like	1.0 tonnes	Each	6.79
Luton Box Van or the like	1.8 tonnes	Each	8.33
Water/Fuel Storage			
Mobile water container	110 litres	Each	0.28
Water bowser	1100 litres	Each	0.55
Water bowser	3000 litres	Each	0.74
Mobile fuel container	110 litres	Each	0.28
Fuel bowser	1100 litres	Each	0.65
Fuel bowser	3000 litres	Each	1.02

EXCAVATORS AND LOADERS

	Size/Rating	Unit	Rate/hr
Excavators			
Wheeled, hydraulic	7/10 tonnes	Each	12.00
Wheeled, hydraulic	11/13 tonnes	Each	12.70
Wheeled, hydraulic	15/16 tonnes	Each	14.80
Wheeled, hydraulic	17/18 tonnes	Each	16.70
Wheeled, hydraulic	20/23 tonnes	Each	16.70
Crawler, hydraulic	12/14 tonnes	Each	12.00
Crawler, hydraulic	15/17.5 tonnes	Each	14.00
Crawler, hydraulic	20/23 tonnes	Each	16.00
Crawler, hydraulic	25/30 tonnes	Each	21.00
Crawler, hydraulic	30/35 tonnes	Each	30.00
Mini excavators	1000/1500kg	Each	4.50
Mini excavators	2150/2400kg	Each	5.50
Mini excavators	2700/3500kg	Each	6.50
Mini excavators	3500/4500kg	Each	8.50
Mini excavators	4500/6000kg	Each	9.50

630   *Daywork*

BUILDING INDUSTRY PLANT HIRE COSTS

MECHANICAL PLANT AND TOOLS

Item of Plant	Size/Rating	Unit	Rate/hr
Loaders			
Wheeled skip loader		Each	4.50
Shovel loaders, four wheel drive	1.6m³	Each	12.00
Shovel loaders, four wheel drive	2.4m³	Each	19.00
Shovel loaders, four wheel drive	3.6m³	Each	22.00
Shovel loaders, four wheel drive	4.4m³	Each	23.00
Shovel loaders, crawlers	0.8m³	Each	11.00
Shovel loaders, crawlers	1.2m³	Each	14.00
Shovel loaders, crawlers	1.6m³	Each	16.00
Shovel loaders, crawlers	2m³	Each	17.00
Skid steer loaders wheeled	300/400kg payload	Each	6.00
Excavator Loaders			
Wheeled tractor type with back-hoe excavator			
Four wheel drive	2.5/3.5 tonnes	Each	7.00
Four wheel drive, 2 wheel steer	7/8 tonnes	Each	9.00
Four wheel drive, 4 wheel steer	7/8 tonnes	Each	10.00
Crawler, hydraulic	12 tonnes	Each	20.00
Crawler, hydraulic	20 tonnes	Each	16.00
Crawler, hydraulic	30 tonnes	Each	35.00
Crawler, hydraulic	40 tonnes	Each	38.00
Attachments			
Breakers for excavators			7.50
Breakers for mini excavators			3.60
Breakers for back-hoe excavator/loaders			6.00
COMPACTION EQUIPMENT			
Rollers			
Vibrating roller	368kg - 420kg	Each	1.68
Single roller	533kg	Each	1.92
Single roller	750kg	Each	2.41
Twin roller	698kg	Each	1.93
Twin roller	851kg	Each	2.41
Twin roller with seat end steering wheel	1067kg	Each	3.03
Twin roller with seat end steering wheel	1397kg	Each	3.17
Pavement rollers	3 - 4 tonnes dead weight	Each	3.18
Pavement rollers	4 - 6 tonnes	Each	4.13
Pavement rollers	6 - 10 tonnes	Each	4.84
Rammers			
Tamper rammer 2 stroke-petrol	225mm - 275mm	Each	1.59
Soil Compactors			
Plate compactor	375mm - 400mm	Each	1.20
Plate compactor rubber pad	375mm - 1400mm	Each	0.33
Plate compactor reversible plate - petrol	400mm	Each	2.20

Daywork

BUILDING INDUSTRY PLANT HIRE COSTS

MECHANICAL PLANT AND TOOLS

Item of Plant	Size/Rating	Unit	Rate/hr
CONCRETE EQUIPMENT			
Concrete/Mortar Mixers			
Open drum without hopper	0.09/0/06m³	Each	0.62
Open drum without hopper	0.12/0.09m³	Each	0.68
Open drum without hopper	0.15/0.10m³	Each	0.72
Open drum with hopper	0.20/0.15m³	Each	0.80
Concrete/Mortar Transport Equipment			
Concrete pump including hose, valve and couplers			
Lorry mounted concrete pump	23m max. distance	Each	36.00
Lorry mounted concrete pump	50m max. distance	Each	46.00
Concrete Equipment			
Vibrator, poker, petrol type	up to 75mm dia.	Each	1.62
Air vibrator (excluding compressor and hose)	up to 75mm dia.	Each	0.79
Extra poker heads	5m	Each	0.77
Vibrating screed unit with beam	3m - 5m	Each	1.77
Vibrating screed unit with adjustable beam	725mm - 900mm	Each	2.18
Power float		Each	1.72
Power grouter		Each	0.92
TESTING EQUIPMENT			
Pipe Testing Equipment			
Pressure testing pump, electric		Sets	1.87
Pipe pressure testing equipment, hydraulic		Sets	2.46
Pressure test pump		Sets	0.64
SITE ACCOMMODATION AND TEMPORARY SERVICES			
Heating Equipment			
Space heaters - propane	80,000Btu/hr	Each	0.77
Space heaters - propane/electric	125,000Btu/hr	Each	1.56
Space heaters - propane/electric	250,000Btu/hr	Each	1.79
Space heaters, propane	125,000Btu/hr	Each	1.33
Space heaters, propane	260,000Btu/hr	Each	1.64
Cabinet heaters		Each	0.41
Cabinet heater catalytic		Each	0.46
Electric halogen heaters		Each	1.28
Ceramic heaters	3kW	Each	0.79
Fan heaters	3kW	Each	0.41
Cooling Fan			1.15
Mobile cooling unit - small			1.38
Mobile cooling unit - large			1.54
Air conditioning unit			2.62
Site Lighting and Equipment			
Tripod floodlight	500W	Each	0.36
Tripod floodlight	1000W	Each	0.34
Towable floodlight	4 x 1000W	Each	2.00
Hand held floodlight	500W	Each	0.22

632   *Daywork*

BUILDING INDUSTRY PLANT HIRE COSTS

MECHANICAL PLANT AND TOOLS

Item of Plant	Size/Rating	Unit	Rate/ hr
Rechargeable light		Each	0.62
Inspection light		Each	0.15
Plasterers light		Each	0.56
Lighting mast		Each	0.92
Festoon light string	33m	Each	0.31
Site Electrical Equipment			
Extension leads	240V/14m	Each	0.20
Extension leafs	110V/14m	Each	0.20
Cable reel	25m 110V/240V	Each	0.28
Cable reel	50m 110V/240V	Each	0.33
4 way junction box	110V	Each	0.17
Power Generating Units			
Generator - petrol	2kVA	Each	1.08
Generator - silenced petrol	2kVA	Each	1.54
Generator - petrol	3VA	Each	1.38
Generator - diesel	5kVA	Each	1.92
Generator - silenced diesel	8kVA	Each	3.59
Generator - silenced diesel	1.5kVA	Each	7.69
Tail adaptor	240V	Each	0.20
Transformers			
Transformer	3kVA	Each	0.36
Transformer	5kVA	Each	0.51
Transformer	7.5kVA	Each	0.82
Transformer	10kVA	Each	0.87
Rubbish Collection and Disposal Equipment			
Rubbish Chutes			
Standard plastic module	1m section	Each	0.18
Steel liner insert		Each	0.26
Steel top hopper		Each	0.20
Plastic side entry hopper		Each	0.20
Plastic side entry hopper liner		Each	0.20
Dust Extraction Plant			
Dust extraction unit, light duty		Each	1.03
Duct extraction unit, heavy duty		Each	1.64

SITE EQUIPMENT
Welding Equipment
Arc-(Electric) Complete with Leads

Welder generator - petrol	200 amp	Each	2.26
Welder generator - diesel	300/350 amp	Each	3.33
Welder generator - diesel	400 amp	Each	4.74
Extra welding lead sets		Each	0.29

Daywork   633

BUILDING INDUSTRY PLANT HIRE COSTS

MECHANICAL PLANT AND TOOLS

Item of Plant	Size/Rating	Unit	Rate/hr
<u>Gas-Oxy Welder</u>			
Welding and cutting set (including oxygen and acetylene, excluding underwater equipment and thermic boring)			
Small		Each	1.41
Large		Each	2.00
Mig welder		Each	1.00
Fume extractor		Each	0.92
Road Works Equipment			
Traffic lights, mains/generator	2-way	Set	4.01
Traffic lights, mains/generator	3-way	Set	7.92
Traffic lights, mains/generator	4-way	Set	9.81
Traffic lights, mains/generator - trailer mounted	2-way	Set	3.98
Flashing lights		Each	0.20
Road safety cone	450mm	10	0.26
Safety cone	750mm	10	0.38
Safety barrier plank	1.25m	Each	0.03
Safety barrier plank	2m	Each	0.04
Road sign		Each	0.26
DPC Equipment			
Damp proofing injection machine		Each	1.49
Cleaning Equipment			
Vacuum cleaner (industrial wet) single motor		Each	0.62
Vacuum cleaner (industrial wet) twin motor		Each	1.23
Vacuum cleaner (industrial wet) triple motor		Each	1.44
Vacuum cleaner (industrial wet) back pack		Each	0.97
Pressure washer, light duty, electric	1450 PSI	Each	0.97
Pressure washer, heavy duty, diesel	2500 PSI	Each	2.69
Cold pressure washer, electric		Each	1.79
Hot pressure washer, petrol		Each	2.92
Cold pressure washer, petrol		Each	2.00
Sandblast attachment to last washer		Each	0.54
Drain cleaning attachment to last washer		Each	0.31
Surface Preparation Equipment			
Rotavators	5 h.p.	Each	1.67
Scabbler, up to three heads		Each	1.15
Scabbler, pole		Each	1.50
Scabbler, multi-headed floor		Each	4.00
Floor preparation machine		Each	2.82
Compressors and Equipment			
<u>Portable Compressors</u>			
Compressors - electric	0.23m³/min	Each	1.59
Compressors - petrol	0.28m³/min	Each	1.74
Compressors - petrol	0.71m³/min	Each	2.00
Compressors - diesel	up to 2.83m³/min	Each	1.24
Compressors - diesel	up to 3.68m³/min	Each	1.49
Compressors - diesel	up to 4.25m³/min	Each	1.60

634
Daywork

BUILDING INDUSTRY PLANT HIRE COSTS

MECHANICAL PLANT AND TOOLS

Item of Plant	Size/Rating	Unit	Rate/hr
Compressors - diesel	up to 4.81m³/min	Each	1.92
Compressors - diesel	up to 7.64m³/min	Each	3.08
Compressors - diesel	up to 11.32m³/min	Each	4.23
Compressors - diesel	up to 18.40m³/min	Each	5.73
<u>Mobile Compressors</u>			
Lorry mounted compressors	2.86-4.24m³/min	Each	12.50
(machine plus lorry only)			
Tractor mounted compressors	2.86-3.40m³/min	Each	13.50
(machine plus rubber tyred tractor)			
<u>Accessories (Pneumatic Tools)</u>			
(with and including up to 15m of air hose)			
Demolition pick		Each	1.03
Breakers (with six steels) light	up to 150kg	Each	0.79
Breakers (with six steels) medium	295kg	Each	1.08
Breakers (with six steels) heavy	386kg	Each	1.44
Rock drill (for use with compressor) hand held		Each	0.90
Additional hoses	15m	Each	0.16
Muffler, tool silencer		Each	0.14
Breakers			
Demolition hammer drill, heavy duty, electric		Each	1.00
Road breaker, electric		Each	1.65
Road breaker, 2 stroke, petrol		Each	2.05
Hydraulic breaker unit, light duty, petrol		Each	2.05
Hydraulic breaker unit, heavy duty, petrol		Each	2.60
Hydraulic breaker unit, heavy duty, diesel		Each	2.95
Quarrying and Tooling Equipment			
Block and stone splitter, hydraulic	600mm x 600mm	Each	1.35
Block and slab splitter, manual		Each	1.10
Steel Reinforcement Equipment			
Bar bending machine - manual	up to 13mm dia. rods	Each	0.90
Bar bending machine - manual	up to 20mm dia. rods	Each	1.28
Bar shearing machine - electric	up to 38mm dia. rods	Each	2.82
Bar shearing machine - electric	up to 40mm dia. rods	Each	3.85
Bar cropper machine - electric	up to 13mm dia. rods	Each	1.54
Bar cropper machine - electric	up to 20mm dia. rods	Each	2.05
Bar cropper machine - electric	up to 40mm dia. rods	Each	2.82
Bar cropper machine - 3 phase	up to 40mm dia. rods	Each	3.85
Dehumidifiers			
110/240v Water	68 litres extraction per 24 hrs	Each	1.28
110/240v Water	90 litres extraction per 24 hrs	Each	1.85

SMALL TOOLS
Saws

Masonry bench saw	350mm - 500mm dia.	Each	2.80
Floor saw	350mm dia., 125mm max. cut	Each	1.90
Floor saw	450mm dia., 150mm max. cut	Each	2.60

Daywork

BUILDING INDUSTRY PLANT HIRE COSTS

MECHANICAL PLANT AND TOOLS

Item of Plant	Size/Rating	Unit	Rate/hr
Saws (contd)			
Floor saw, reversible	Max. cut 300mm	Each	13.00
Chop/cut off saw, electric	350mm dia.	Each	1.33
Circular saw, electric	230mm dia.	Each	0.60
Tyrannosaw		Each	1.20
Reciprocating saw		Each	0.60
Door trimmer		Each	0.90
Chainsaw, petrol	500mm	Each	2.13
Full chainsaw safety kit		Each	0.50
Worktop jig		Each	0.60
Pipework Equipment			
Pipe bender	15mm - 22mm	Each	0.33
Pipe bender, hydraulic	50mm	Each	0.60
Pipe bender, electric	50mm - 150mm dia.	Each	1.35
Pipe cutter, hydraulic		Each	1.84
Tripod pipe vice		Set	0.40
Ratchet threader	12mm - 32mm	Each	0.55
Pipe threading machine, electric	12mm - 75mm	Each	2.40
Pipe threading machine, electric	12mm - 100mm	Each	3.00
Impact wrench, electric		Each	0.54
Impact wrench, two stroke, petrol		Each	4.49
Impact wrench, heavy duty, electric		Each	1.13
Plumber's furnace, calor gas or similar		Each	2.16
Hand-held Drills and Equipment			
Impact or hammer drill	up to 25mm dia.	Each	0.50
Impact or hammer drill	35mm dia.	Each	0.90
Angle head drills		Each	0.70
Stirrer, mixer drills		Each	0.70
Paint, Insulation Application Equipment			
Airless spray unit		Each	4.20
Portaspray unit		Each	1.65
HVLP turbine spray unit		Each	1.65
Compressor and spray gun		Each	2.20
Other Handtools			
Screwing machine	13mm - 50mm dia.	Each	0.77
Screwing machine	25mm - 100mm dia.	Each	1.57
Staple gun		Each	0.33
Air nail gun	110V	Each	3.33
Cartridge hammer		Each	1.00
Tongue and groove nailer complete with mallet		Each	0.93
Chasing machine	152mm	Each	1.72
Chasing machine	76mm - 203mm	Each	5.99
Floor grinder		Each	3.00
Floor plane		Each	3.67
Diamond concrete planer		Each	2.05
Autofeed screwdriver, electric		Each	1.13
Laminate trimmer		Each	0.64

636   *Daywork*

BUILDING INDUSTRY PLANT HIRE COSTS

MECHANICAL PLANT AND TOOLS

Item of Plant	Size/Rating	Unit	Rate/hr
Biscuit jointer		Each	0.87
Random orbital sander		Each	0.72
Floor sander		Each	1.33
Palm, delta, flap or belt sander		Each	0.38
Saw cutter, 2 stroke, petrol	300mm	Each	1.26
Grinder, angle or cutter	up to 225mm	Each	0.60
Grinder, angle or cutter	300mm	Each	1.10
Mortar raking tool attachment		Each	0.15
Floor/polisher scrubber	325mm	Each	1.03
Floor tile stripper		Each	1.74
Wallpaper stripper, electric		Each	0.56
Electric scraper		Each	0.51
Hot air paint stripper		Each	0.38
Electric diamond tile cutter	All sizes	Each	1.38
Hand tile cutter		Each	0.36
Electric needle gun		Each	1.08
Needle chipping gun		Each	0.72
Pedestrian floor sweeper	1.2m wide	Each	0.87

NEW FROM SPON PRESS

Photovoltaics and Architecture

Edited by Randall Thomas, Max Fordham and Partners, UK

With a foreword by Amory Lovins

It has been said that the nineteenth century was the age of coal, the twentieth of oil, and the twenty-first will be the age of solar energy. *Photovoltaics and Architecture* describes vividly how buildings can contribute to this transition.

PVs are changing the form of our communities and buildings and encouraging designers to make the most use of solar energy. The challenges and potential are significant.

Photovoltaics and Architecture is the first book to set out the basic principles of PV design and examines their implications in a largely UK context. It will be of value to designers, clients and students.

March 2001: 297x210: 176pp
105 line illustrations, 63 b+w photos
Pb: 0-415-23182-5: £29.95

To Order: Tel: +44 (0) 8700 768853, or +44 (0) 1264 343071 Fax: +44 (0) 1264 343005, or
Post: Spon Press Customer Services, ITPS Andover, Hants, SP10 5BE, UK Email:
book.orders@tandf.co.uk.

Postage & Packing: UK: 5% of order value (min. charge £1, max. charge £10) for 3-5 days delivery. Option of next day delivery at an additional £6.50. Europe: 10% of order value (min. charge £2.95, max. charge £20) for delivery surface post. Option of airmail at an additional £6.50. ROW: 15% of order value (min.charge£6.50, max. charge £30) for airmail delivery.

NEW FROM SPON PRESS

Understanding the Building Regulations
2nd Edition
Simon Polley, BRCS, UK

This is a new edition of the highly successful introductory guide to current Building Regulations and Approval Documents. Including the major revisions to part B, it is an essential tool for those involved in design and construction and for those who require knowledge of building control.

Thoroughly revised and updated, it will provide all the information necessary to design and build to the building regulations. This is an essential tool for construction professionals requiring a 'pocket book' guide to the regulations.

Reviews of the first edition: *'...covers all the requirements of the Building Regulations as we know them today. It is clear and concise in its explanations...a good book...'* **Clerk of Works Journal**

> **Contents**: Preface. Acknowledgements. Introduction. The Building Regulations 2000. Approved Document to support Regulation 7: Materials and Workmanship. Approved Document A: Structure. Approved Document B: Fire Safety. Approved Document C: Site Preparation and Resistance to Moisture. Approved Document D: Toxic Substances. Approved Document E: Resistance to the Passage of Sound. Approved Document F: Ventilation. Approved Document G: Hygiene. Approved Document H: Drainage and Waste Disposal. Approved Document K: Protection from Falling Collision and Impact. Document L: Conservation of Fuel and Power. Approved Document M: Access and Facilities for Disabled People. Approved Document N: Glazing - Safety in Relation to Impact, Opening and Cleaning. Further Information. Index.

August 2001: 234x156: 208pp :illus.55 line figs. Pb: 0-419-24720-3: £16.99

To Order: Tel: +44 (0) 8700 768853, or +44 (0) 1264 343071 Fax: +44 (0) 1264 343005, or
Post: Spon Press Customer Services, ITPS Andover, Hants, SP10 5BE, UK Email: book.orders@tandf.co.uk.

Postage & Packing: UK: 5% of order value (min. charge £1, max. charge £10) for 3-5 days delivery. Option of next day delivery at an additional £6.50. Europe: 10% of order value (min. charge £2.95, max. charge £20) for delivery surface post. Option of airmail at an additional £6.50. ROW: 15% of order value (min.charge£6.50, max. charge £30) for airmail delivery.

For a complete listing of all our titles visit: www.sponpress.com

Tables and Memoranda

Conversion Tables, *page 638*
Formulae, *page 641*
Fractions, Decimals and Millimetre Equivalents, *page 642*
Imperial Standard Wire Gauge, *page 643*
Water Pressure Due to Height, *page 644*
Table of Weights for Steelwork, *page 645*
Dimension and Weights of Copper and Stainless Steel Pipes, *page 650*
Dimensions of Steel Pipes, *page 651*
Approximate Metres per Tonne of Tubes, *page 652*
Flange Dimension Chart, *page 653*
Minimum Distances Between Supports/Fixings, *page 654*
Litres of Water Storage Required per Person per Building Type, *page 654*
Cold Water Plumbing - Insulation Thickness Required Against Frost, *page 655*
Capacity and Dimensions of Galvanised Mild Steel Cisterns, *page 655*
Capacity of Cold Water Polypropylene Storage Cisterns, *page 655*
Storage Capacity and Recommended Power of Hot Water Storage Boilers, *page 656*
Thickness of Thermal Insulation for Heating Installations, *page 657*
Capacities and Dimensions of : Galvanised Mild Steel Indirect Cylinders, *page 659*
Copper Indirect Cylinders, *page 659*
Recommended Air Conditioning Design Loads, *page 660*

638   *Tables and Memoranda*

CONVERSION TABLES

Example : 1mm = 0.039in - 1in = 25.4mm

1   LINEAR

0.039	in	- 1 -	mm	25.4
0.281	ft	- 1 -	metre	0.305
1.094	yd	- 1 -	metre	0.914

2   WEIGHT

0.020	cwt	- 1 -	kg	50.802
0.984	ton	- 1 -	tonne	1.016
2.205	lb	- 1 -	kg	0.454

3   CAPACITY

1.76	pt	- 1 -	litre	0.568
0.220	gal	- 1 -	litre	4.546

4   AREA

0.002	in^2	- 1 -	mm^2	645.16
10.764	ft^2	- 1 -	m^2	0.093
1.196	yd^2	- 1 -	m^2	0.836
2.471	acre	- 1 -	ha	0.405
0.386	$mile^2$	- 1 -	km^2	2.59

5   VOLUME

0.061	in^3	- 1 -	cm^3	16.387
35.315	ft^3	- 1 -	m^3	0.028
1.308	yd^3	- 1 -	m^3	0.765

6   POWER

1.310	HP	- 1 -	kW	0.746

Tables and Memoranda 639

CONVERSION TABLES

LENGTH Conversion Factors

Centimetre	(cm)	1 in	=	2.54	cm	:	1 cm	=	0.3937	in
Metre	(m)	1 ft	=	0.3048	m	:	1 m	=	3.2808	ft
Kilometre	(km)	1 yd	=	0.9144	m	:	1 m	=	1.0936	yd
Kilometre	(km)	1 mile	=	1.6093	km	:	1 km	=	0.6214	mile

NOTE :	1 cm	=	10	mm	1 ft	=	12	in
	1 m	=	100	cm	1 yd	=	3	ft
	1 km	=	1000	m	1 mile	=	1760	yd

AREA

Square Millimetre	(mm^2)	$1\ in^2$	=	645.2	mm^2	:	$1\ mm^2$	=	0.0016	in^2
Square Centimetre	(cm^2)	$1\ in^2$	=	6.4516	cm^2	:	$1\ cm^2$	=	0.1550	in^2
Square Metre	(m^2)	$1\ ft^2$	=	0.0929	m^2	:	$1\ m^2$	=	10.764	ft^2
Square Metre	(m^2)	$1\ yd^2$	=	0.8361	m^2	:	$1\ m^2$	=	1.1960	yd^2
Square Kilometre	(km^2)	$1\ mile^2$	=	2.590	km^2	:	$1\ km^2$	=	0.3861	$mile^2$

NOTE :	$1\ cm^2$	=	100	mm^2	$1\ ft^2$	=	144	in^2
	$1\ m^2$	=	10000	cm^2	$1\ yd^2$	=	9	ft^2
	$1\ km^2$	=	100	ha	$1\ mile^2$	=	640	acre
					1 acre	=	4840	yd^2

VOLUME

Cubic Centimetre	(cm^3)	$1\ cm^3$	=	0.0610	in^3	:	$1\ in^3$	=	16.387	cm^3
Cubic Decimetre	(dm^3)	$1\ dm^3$	=	0.0353	ft^3	:	$1\ ft^3$	=	28.329	dm^3
Cubic Metre	(m^3)	$1\ m^3$	=	35.315	ft^3	:	$1\ ft^3$	=	0.0283	m^3
Cubic Metre	(m^3)	$1\ m^3$	=	1.3080	yd^3	:	$1\ yd^3$	=	0.7646	m^3
Litre	(L)	1 L	=	1.76	pint	:	1 pint	=	0.5683	L
Litre	(L)	1 L	=	2.113	US pt	:	1 pint	=	0.4733	US L

NOTE :	$1\ dm^3$	=	1000	cm^3	$1\ ft^3$	=	1728	in^3
	$1\ m^3$	=	1000	dm^3	$1\ yd^3$	=	27	ft^3
	1 L	=	1	dm^3	1 pint	=	20	fl oz
	1 HL	=	100	L	1 gal	=	8	pints

MASS

Milligram	(mg)	1 mg	=	0.0154	grain	:	1 grain	=	64.935	mg
Gram	(g)	1 g	=	0.0353	oz	:	1 oz	=	28.35	g
Kilogram	(kg)	1 kg	=	2.2046	lb	:	1 lb	=	0.4536	kg
Tonne	(t)	1 t	=	0.9842	ton	:	1 ton	=	1.016	t

NOTE :	1 g	=	1000	mg	1 oz	=	437.5	grains
	1 kg	=	1000	g	1 lb	=	16	oz
	1 t	=	1000	kg	1 stone	=	14	lb
					1 cwt	=	112	lb
					1 ton	=	20	cwt

640 *Tables and Memoranda*

CONVERSION TABLES

FORCE   Conversion Factors

Newton	(N)	1 lb f	=	4.448	N	:	1 kg f	=	9.807 N
Kilonewton	(kN)	1 lb f	=	0.004448	kN	:	1 ton f	=	9.964 kN
Meganewton	(mN)	100 ton f =		0.9964	mN				

PRESSURE AND STRESS

Kilonewton
per square metre (kN/m^2)
Meganewton
per square metre (mN/m^2)

$1 \text{ lb f/in}^2 = 6.895 \quad kN/m^2$:
$1 \text{ bar} = 100 \quad kN/m^2$:
$1 \text{ ton f/ft}^2 = 107.3 \quad kN/m^2 \qquad = 0.1073 \quad mN/m^2$
$1 \text{ kg f/cm}^2 = 98.07 \quad kN/m^2$:
$1 \text{ lb f/ft}^2 = 0.0479 \quad kN/m^2$:

TEMPERATURE

Degree Celsius ($^{\circ}$C) $^{\circ}$C $= 5 \square 9 \; (^{\circ}F - 32^{\circ}) \qquad ^{\circ}F \qquad = 9 \square 5 \; (^{\circ}C + 32^{\circ})$

Tables and Memoranda 641

FORMULAE

Two dimensional figures

Figure	Area
Triangle	0.5 x base x height, or $N(s(s - a)(s - b)(s - c))$ where s = 0.5 x the sum of the three sides and a, b and c are the lengths of the three sides, or $a^2 = b^2 + c^2 - 2 \times bc \times COS\ A$ where A is the angle opposite side a
Hexagon	$2.6 \times (side)^2$
Octagon	$4.83 \times (side)^2$
Trapezoid	height x 0.5 (base + top)
Circle	$3.142 \times radius^2$ or $0.7854 \times diameter^2$ (circumference = 2 x 3.142 x radius or 3.142 x diameter)
Sector of a circle	0.5 x length of arc x radius
Segment of a circle	area of sector - area of triangle
Ellipse of a circle	3.142 x AB (where A = 0.5 x height and B = 0.5 x length)
Spandrel	$3/14 \times radius^2$

Three dimensional figure

Figure	Volume Surface	Area
Prism x height	Area of base x height	circumference of base
Cube	(length of side) cubed	$6 \times (length\ of\ side)^2$
Cylinder	$3.142 \times radius^2 \times$ height	2 x 3.142 x radius x (height - radius)
Sphere	$4/3 \times 3.142 \times radius^3$	$4 \times 3.142 \times radius^2$
Segment of a sphere	$[(3.142 \times h)/6] \times$ $(3 \times r^2 + h^2)$	$(2 \times 3.142 \times r \times h)$
Pyramid	1/3 x (area of base x height)	0.5 x circumference of base x slant height

FRACTIONS, DECIMALS AND MILLIMETRE EQUIVALENTS

Fractions	Decimals	mm	Fractions	Decimals	mm
1/64	0.015625	0.396875	33/64	0.515625	13.096875
1/32	0.03125	0.79375	17/32	0.53125	13.49375
3/64	0.046875	1.190625	35/64	0.546875	13.890625
1/16	0.0625	1.5875	9/16	0.5625	14.2875
5/64	0.078125	1.984375	37/64	0.578125	14.684375
3/32	0.09375	2.38125	19/32	0.59375	15.08125
7/64	0.109375	2.778125	39/64	0.609375	15.478125
1/8	0.125	3.175	5/8	0.625	15.875
9/64	0.140625	3.571875	41/64	0.640625	16.271875
5/32	0.15625	3.96875	21/32	0.65625	16.66875
11/64	0.171875	4.365625	43/64	0.671875	17.065625
3/16	0.1875	4.7625	11/16	0.6875	17.4625
13/64	0.203125	5.159375	45/64	0.703125	17.859375
7/32	0.21875	5.55625	23/32	0.71875	18.25625
15/64	0.234375	5.953125	47/64	0.734375	18.653125
1/4	0.25	6.35	3/4	0.75	19.05
17/64	0.265625	6.746875	49/64	0.765625	19.446875
9/32	0.28125	7.14375	25/32	0.78125	19.84375
19/64	0.296875	7.540625	51/64	0.796875	20.240625
5/16	0.3125	7.9375	13/16	0.8125	20.6375
21/64	0.328125	8.334375	53/64	0.828125	21.034375
11/32	0.34375	8.73125	27/32	0.84375	21.43125
23/64	0.359375	9.128125	55/64	0.859375	21.828125
3/8	0.375	9.525	7/8	0.875	22.225
25/64	0.390625	9.921875	57/64	0.890625	22.621875
13/32	0.40625	10.31875	29/32	0.90625	23.01875
27/64	0.421875	10.71563	59/64	0.921875	23.415625
7/16	0.4375	11.1125	15/16	0.9375	23.8125
29/64	0.453125	11.50938	61/64	0.953125	24.209375
15/32	0.46875	11.90625	31/32	0.96875	24.60625
31/64	0.484375	12.30313	63/64	0.984375	25.003125
1/2	0.5	12.7	1.0	1	25.4

Tables and Memoranda

IMPERIAL STANDARD WIRE GAUGE (SWG)

SWG No	Diameter in	Diameter mm	SWG No	Diameter in	Diameter mm
7/0	0.5	12.7	23	0.024	0.61
6/0	0.464	11.79	24	0.022	0.559
5/0	0.432	10.97	25	0.02	0.508
4/0	0.4	10.16	26	0.018	0.457
3/0	0.372	9.45	27	0.0164	0.417
2/0	0.348	8.84	28	0.0148	0.376
1/0	0.324	8.23	29	0.0136	0.345
1	0.3	7.62	30	0.0124	0.315
2	0.276	7.01	31	0.0116	0.295
3	0.252	6.4	32	0.0108	0.274
4	0.232	5.89	33	0.01	0.254
5	0.212	5.38	34	0.009	0.234
6	0.192	4.88	35	0.008	0.213
7	0.176	4.47	36	0.008	0.193
8	0.16	4.06	37	0.007	0.173
9	0.144	3.66	38	0.006	0.152
10	0.128	3.25	39	0.005	0.132
11	0.116	2.95	40	0.005	0.122
12	0.104	2.64	41	0.004	0.112
13	0.092	2.34	42	0.004	0.102
14	0.08	2.03	43	0.004	0.091
15	0.072	1.83	44	0.003	0.081
16	0.064	1.63	45	0.003	0.071
17	0.056	1.42	46	0.002	0.061
18	0.048	1.22	47	0.002	0.051
19	0.04	1.016	48	0.002	0.041
20	0.036	0.914	49	0.001	0.031
21	0.032	0.813	50	0.001	0.025
22	0.028	0.711			

Tables and Memoranda

WATER PRESSURE DUE TO HEIGHT

Imperial

Head Feet	Pressure lb/in^2	Head Feet	Pressure lb/in^2
1	0.43	70	30.35
5	2.17	75	32.51
10	4.34	80	34.68
15	6.5	85	36.85
20	8.67	90	39.02
25	10.84	95	41.18
30	13.01	100	43.35
35	15.17	105	45.52
40	17.34	110	47.69
45	19.51	120	52.02
50	21.68	130	56.36
55	23.84	140	60.69
60	26.01	150	65.03
65	28.18		

Metric

Head m	Pressure bar	Head m	Pressure bar
0.5	0.049	18.0	1.766
1.0	0.098	19.0	1.864
1.5	0.147	20.0	1.962
2.0	0.196	21.0	2.06
3.0	0.294	22.0	2.158
4.0	0.392	23.0	2.256
5.0	0.491	24.0	2.354
6.0	0.589	25.0	2.453
7.0	0.687	26.0	2.551
8.0	0.785	27.0	2.649
9.0	0.883	28.0	2.747
10.0	0.981	29.0	2.845
11.0	1.079	30.0	2.943
12.0	1.177	32.5	3.188
13.0	1.275	35.0	3.434
14.0	1.373	37.5	3.679
15.0	1.472	40.0	3.924
16.0	1.57	42.5	4.169
17.0	1.668	45.0	4.415

1 bar	=	14.5038 lbf/in^2
1 metre	=	3.2808 ft or 39.3701 in
1 foot	=	0.3048 metres
1 lbf/in^2	=	0.06895 bar
1 in wg	=	2.5 mbar (249.1 N/m^2)

Tables and Memoranda

TABLE OF WEIGHTS FOR STEELWORK

Mild Steel Bar

Diameter (mm)	Weight (kg/m)	Diameter (mm)	Weight (kg/m)
6	0.22	20	2.47
10	0.62	25	3.85
12	0.89	30	5.55
16	1.58	32	6.31

Mild Steel Flat

Size (mm)	Weight (kg/m)	Size (mm)	Weight (kg/m)
15 x 3	0.36	15 x 5	0.59
20 x 3	0.47	20 x 5	0.79
25 x 3	0.59	25 x 5	0.98
30 x 3	0.71	30 x 5	1.18
40 x 3	0.94	40 x 5	1.57
45 x 3	1.06	45 x 5	1.77
50 x 3	1.18	50 x 5	1.96
20 x 6	0.94	20 x 8	1.26
25 x 6	1.18	25 x 8	1.57
30 x 6	1.41	30 x 8	1.88
40 x 6	1.88	40 x 8	2.51
45 x 6	2.12	45 x 8	2.83
50 x 6	2.36	50 x 8	3.14
55 x 6	2.60	55 x 8	3.45
60 x 6	2.83	60 x 8	3.77
65 x 6	3.06	65 x 8	4.08
70 x 6	3.30	70 x 8	4.40
75 x 6	3.53	75 x 8	4.71
100 x 6	4.71	100 x 8	6.28
20 x 10	1.57	20 x 12	1.88
25 x 10	1.96	25 x 12	2.36
30 x 10	2.36	30 x 12	2.83
40 x 10	3.14	40 x 12	3.77
45 x 10	3.53	45 x 12	4.24
50 x 10	3.93	50 x 12	4.71
55 x 10	4.32	55 x 12	5.12
60 x 10	4.71	60 x 12	5.65
65 x 10	5.10	65 x 12	6.12
70 x 10	5.50	70 x 12	6.59
75 x 10	5.89	75 x 12	7.07
100 x 10	7.85	100 x 12	9.42

646 *Tables and Memoranda*

TABLE OF WEIGHTS FOR STEELWORK

Mild Steel Equal Angle

Size (mm)	Weight (kg/m)	Size (mm)	Weight (kg/m)
13 x 13 x 3	0.56	60 x 60 x 10	8.69
20 x 20 x 3	0.88	70 x 70 x 10	10.30
25 x 25 x 3	1.11	75 x 75 x 10	11.05
30 x 30 x 3	1.36	80 x 80 x 10	11.90
40 x 40 x 3	1.82	90 x 90 x 10	13.40
45 x 45 x 3	2.06	100 x 100 x 10	15.00
50 x 50 x 3	2.30	120 x 120 x 10	18.20
		150 x 156 x 10	23.00
30 x 30 x 6	2.56		
40 x 40 x 6	3.52	75 x 75 x 12	13.07
45 x 45 x 6	4.00	80 x 80 x 12	14.00
50 x 50 x 6	4.47	90 x 90 x 12	15.90
60 x 60 x 6	5.42	100 x 120 x 12	21.90
70 x 70 x 6	6.38	120 x 120 x 12	21.60
75 x 75 x 6	6.82	150 x 150 x 12	27.30
80 x 80 x 6	7.34	200 x 200 x 12	36.74
90 x 90 x 6	8.30		
40 x 40 x 8	4.55		
50 x 50 x 8	5.82		
60 x 60 x 8	7.09		
70 x 70 x 8	8.36		
75 x 75 x 8	8.96		
80 x 80 x 8	9.63		
90 x 90 x 8	10.90		
100 x 100 x 8	12.20		
120 x 120 x 8	14.70		

Mild Steel Unequal Angle

Size (mm)	Weight (kg/m)	Size (mm)	Weight (kg/m)
40 x 25 x 6	2.79	100 x 65 x 10	12.30
50 x 40 x 6	4.24	100 x 75 x 10	13.00
60 x 30 x 6	3.99	125 x 75 x 10	15.00
65 x 50 x 6	5.16	150 x 75 x 10	17.00
75 x 50 x 6	5.65	150 x 90 x 10	18.20
80 x 60 x 6	6.37	200 x 100 x 10	23.00
125 x 75 x 6	9.18		
75 x 50 x 8	7.39	100 x 75 x 12	15.40
80 x 60 x 8	8.34	125 x 75 x 12	17.80
100 x 65 x 8	9.94	150 x 75 x 12	20.20
100 x 75 x 8	10.60	150 x 90 x 12	21.60
125 x 75 x 8	12.20	200 x 100 x 12	27.30
137 x 102 x 8	14.88	200 x 150 x 12	32.00

Tables and Memoranda

TABLE OF WEIGHTS FOR STEELWORK

Rolled Steel Channels

Size (mm)	Weight (kg/m)	Size (mm)	Weight (kg/m)
32 x 27	2.80	178 x 76	20.84
38 x 19	2.49	178 x 79	26.81
51 x 25	4.46	203 x 76	23.82
51 x 38	5.80	203 x 89	29.78
64 x 25	6.70	229 x 76	26.06
76 x 38	7.46	229 x 89	32.76
76 x 51	9.45	254 x 76	28.29
102 x 51	10.42	254 x 89	35.74
127 x 64	14.90	305 x 89	41.67
152 x 76	17.88	305 x 102	46.18
152 x 89	23.84	381 x 102	55.10

Rolled Steel Joists

Size (mm)	Weight (kg/m)	Size (mm)	Weight (kg/m)
76 x 38	6.25	152 x 76	17.86
76 x 76	12.65	152 x 89	17.09
102 x 44	7.44	152 x 127	37.20
102 x 64	9.65	178 x 102	21.54
102 x 102	23.06	203 x 102	25.33
127 x 76	13.36	203 x 152	52.03
127 x 114	26.78	254 x 114	37.20
127 x 114	29.76	254 x 203	81.84
		305 x 203	96.72

Universal Columns

Size (mm)	Weight (kg/m)	Size (mm)	Weight (kg/m)
152 x 152	23.00	254 x 254	89.00
152 x 152	30.00	254 x 254	107.00
152 x 152	37.00	254 x 254	132.00
203 x 203	46.00	254 x 254	167.00
203 x 203	52.00	305 x 305	97.00
203 x 203	60.00	305 x 305	118.00
203 x 203	71.00	305 x 305	137.00
203 x 203	86.00	305 x 305	158.00
254 x 254	73.00	305 x 305	198.00

648 *Tables and Memoranda*

TABLE OF WEIGHTS FOR STEELWORK

Universal Beams

Size (mm)	Weight (kg/m)	Size (mm)	Weight (kg/m)
203 x 133	25.00	305 x 127	48.00
203 x 133	30.00	305 x 165	40.00
254 x 102	22.00	305 x 165	46.00
254 x 102	25.00	305 x 165	54.00
254 x 102	28.00	356 x 127	33.00
254 x 146	31.00	356 x 127	39.00
254 x 146	37.00	356 x 171	45.00
254 x 146	43.00	356 x 171	51.00
305 x 102	25.00	356 x 171	57.00
305 x 102	28.00	356 x 171	67.00
305 x 102	33.00	381 x 152	52.00
305 x 127	37.00	381 x 152	60.00
305 x 127	42.00	381 x 152	67.00

Circular Hollow Sections

Size (mm)	Weight (kg/m)	Size (mm)	Weight (kg/m)
21.3 x 3.2	1.43	76.1 x 3.2	5.75
26.9 x 3.2	1.87	76.1 x 4.0	7.11
33.7 x 2.6	1.99	76.1 x 5.0	8.77
33.7 x 3.2	2.41	88.9 x 3.2	6.76
33.7 x 4.0	2.93	88.9 x 4.0	8.36
42.4 x 2.6	2.55	88.9 x 5.0	10.30
42.4 x 3.2	3.09	114.3 x 3.6	9.83
42.4 x 4.0	3.79	114.3 x 5.0	13.50
48.3 x 3.2	3.56	114.3 x 6.3	16.80
48.3 x 4.0	4.37	139.7 x 5.0	16.60
48.3 x 5.0	5.34	139.7 x 6.3	20.70
60.3 x 3.2	4.51	139.7 x 8.0	26.00
60.3 x 4.0	5.55	139.7 x 10.0	32.00
60.3 x 5.0	6.82	168.3 x 5.0	20.10

Tables and Memoranda

TABLE OF WEIGHTS FOR STEELWORK

Square Hollow Sections

Size (mm)	Weight (kg/m)	Size (mm)	Weight (kg/m)
20 x 20 x 2.0	1.12	90 x 90 x 3.6	9.72
20 x 20 x 2.6	1.39	90 x 90 x 5.0	13.30
30 x 30 x 2.6	2.21	90 x 90 x 6.3	16.40
30 x 30 x 3.2	2.65	100 x 100 x 4.0	12.00
40 x 40 x 2.6	3.03	100 x 100 x 5.0	14.80
40 x 40 x 3.2	3.66	100 x 100 x 6.3	18.40
40 x 40 x 4.0	4.46	100 x 100 x 8.0	22.90
50 x 50 x 3.2	4.66	100 x 100 x 10.0	27.90
50 x 50 x 4.0	5.72	120 x 120 x 5.0	18.00
50 x 50 x 5.0	6.97	120 x 120 x 6.3	22.30
60 x 60 x 3.2	5.67	120 x 120 x 8.0	27.90
60 x 60 x 4.0	6.97	120 x 120 x 10.0	34.20
60 x 60 x 5.0	8.54	150 x 150 x 5.0	22.70
70 x 70 x 3.2	7.46	150 x 150 x 6.3	28.30
70 x 70 x 5.0	10.10	150 x 150 x 8.0	35.40
80 x 80 x 3.6	8.59	150 x 150 x 10.0	43.60
80 x 80 x 5.0	11.70		
80 x 80 x 6.3	14.40		

Rectangular Hollow Sections

Size (mm)	Weight (kg/m)	Size (mm)	Weight (kg/m)
50 x 30 x 2.6	3.03	120 x 80 x 5.0	14.80
50 x 30 x 3.2	3.66	120 x 80 x 6.3	18.40
60 x 40 x 3.2	4.66	120 x 80 x 8.0	22.90
60 x 40 x 4.0	5.72	120 x 80 x 10.0	27.90
80 x 40 x 3.2	5.67	150 x 100 x 5.0	18.70
80 x 40 x 4.0	6.97	150 x 100 x 6.3	23.30
90 x 50 x 3.6	7.46	150 x 100 x 8.0	29.10
90 x 50 x 5.0	10.10	150 x 100 x 10.0	35.70
100 x 50 x 3.2	7.18	160 x 80 x 5.0	18.00
100 x 50 x 4.0	8.86	160 x 80 x 6.3	22.30
100 x 50 x 5.0	10.90	160 x 80 x 8.0	27.90
100 x 60 x 3.6	8.59	160 x 80 x 10.0	34.20
100 x 60 x 5.0	11.70	200 x 100 x 5.0	22.70
100 x 60 x 6.3	14.40	200 x 100 x 6.3	28.30
120 x 60 x 3.6	9.72	200 x 100 x 8.0	35.40
120 x 60 x 5.0	13.30	200 x 100 x 10.0	43.60
120 x 60 x 6.3	16.40		

DIMENSIONS AND WEIGHTS OF COPPER PIPES TO BS 2871 PART 1

Outside Diameter (mm)	Internal Diameter (mm)	Weight per Metre (kg)	Internal Diameter (mm)	Weight per Metre (kg)	Internal Diameter (mm)	Weight per Metre (kg)
	Table X		Table Y		Table Z	
6	4.80	0.0911	4.40	0.1170	5.00	0.0774
8	6.80	0.1246	6.40	0.1617	7.00	0.1054
10	8.80	0.1580	8.40	0.2064	9.00	0.1334
12	10.80	0.1914	10.40	0.2511	11.00	0.1612
15	13.60	0.2796	13.00	0.3923	14.00	0.2031
18	16.40	0.3852	16.00	0.4760	16.80	0.2918
22	20.22	0.5308	19.62	0.6974	20.82	0.3589
28	26.22	0.6814	25.62	0.8985	26.82	0.4594
35	32.63	1.1334	32.03	1.4085	33.63	0.6701
42	39.63	1.3675	39.03	1.6996	40.43	0.9216
54	51.63	1.7691	50.03	2.9052	52.23	1.3343
76.1	73.22	3.1287	72.22	4.1437	73.82	2.5131
108	105.12	4.4666	103.12	7.3745	105.72	3.5834
133	130.38	5.5151	--	--	130.38	5.5151
159	155.38	8.7795	--	--	156.38	6.6056

DIMENSIONS OF STAINLESS STEEL PIPES TO BS 4127.

Outside Diameter (mm)	Maximum Outside Diameter (mm)	Minimum Outside Diameter (mm)	Wall Thickness (mm)	Working Pressure (bar)
6	6.045	5.940	0.6	330
8	8.045	7.940	0.6	260
10	10.045	9.940	0.6	210
12	12.045	11.940	0.6	170
15	15.045	14.940	0.6	140
18	18.045	17.940	0.7	135
22	22.055	21.950	0.7	110
28	28.055	27.950	0.8	121
35	35.070	34.965	1.0	100
42	42.070	41.965	1.1	91
54	54.090	53.940	1.2	77

Tables and Memoranda

DIMENSIONS OF STEEL PIPES TO BS 1387

Nominal Size	Approx. Outside Diamete	Outside Diameter				Thickness		
		Light		Medium and Heavy		Light	Medium	Heavy
		Maximu	Minimu	Maximu	Minimum			
mm	mm	mm	mm	mm	mm	mm	mm	mm
6	10.20	10.10	9.70	10.40	9.80	1.80	2.00	2.65
8	13.50	13.60	13.20	13.90	13.30	1.80	2.35	2.90
10	17.20	17.10	16.70	17.40	16.80	1.80	2.35	2.90
15	21.30	21.40	21.00	21.70	21.10	2.00	2.65	3.25
20	26.90	26.90	26.40	27.20	26.60	2.35	2.65	3.25
25	33.70	33.80	33.20	34.20	33.40	2.65	3.25	4.05
32	42.40	42.50	41.90	42.90	42.10	2.65	3.25	4.05
40	48.30	48.40	47.80	48.80	48.00	2.90	3.25	4.05
50	60.30	60.20	59.60	60.80	59.80	2.90	3.65	4.50
65	76.10	76.00	75.20	76.60	75.40	3.25	3.65	4.50
80	88.90	88.70	87.90	89.50	88.10	3.25	4.05	4.85
100	114.30	113.90	113.00	114.90	113.30	3.65	4.50	5.40
125	139.70	--	--	140.60	138.70	--	4.85	5.40
150	165.1*	--	--	166.10	164.10	--	4.85	5.40

* 165.1mm (6.5in) outside diameter is not generally recommended except where screwing to BS 21 is necessary.

All dimensions are in accordance with ISO R65 except approximate outside diameters which are in accordance with ISO R64.

Light quality is equivalent to ISO R65 Light Series II.

Tables and Memoranda

APPROXIMATE METRES PER TONNE OF TUBES TO BS 1387

Nom. Size mm	BLACK						GALVANISED					
	Plain/screwed ends			Screwed & socketed			Plain/screwed ends			Screwed & socketed		
	Light m	Mediu m	Heavy m	Light m	Mediu m	Heavy m	Light m	Mediu m	Heavy m	Light m	Mediu m	Heavy m
6	2765	2461	2030	2743	2443	2018	2604	2333	1948	2584	2317	1937
8	1936	1538	1300	1920	1527	1292	1826	1467	1254	1811	1458	1247
10	1483	1173	979	1471	1165	974	1400	1120	944	1386	1113	939
15	1050	817	688	1040	811	684	996	785	665	987	779	661
20	712	634	529	704	628	525	679	609	512	673	603	508
25	498	410	336	494	407	334	478	396	327	474	394	325
32	388	319	260	384	316	259	373	308	254	369	305	252
40	307	277	226	303	273	223	296	268	220	292	264	217
50	244	196	162	239	194	160	235	191	158	231	188	157
65	172	153	127	169	151	125	167	149	124	163	146	122
80	147	118	99	143	116	98	142	115	97	139	113	96
100	101	82	69	98	81	68	98	81	68	95	79	67
125	--	62	56	--	60	55	--	60	55	--	59	54
150	--	52	47	--	50	46	--	51	46	--	49	45

The figures for `plain or screwed ends' apply also to tubes to BS 1775 HFW of equivalent size and thickness.

Tables and Memoranda

FLANGE DIMENSION CHART TO BS 4504 & BS 10

Normal Pressure Rating (N.P.6) 6 Bar

Nom. Size	Flang Outsid Diam.	Table 6/2 Forged Weldin Neck	Table 6/3 Plate Slip on	Table 6/4 Forged Bossed Screwe	Table 6/5 Forged Bossed Slip on	Table 6/8 Plate Blank	Raised Face Diam.	T'ness	Pitch Circle Diam.	Nr. Bolt Hol	Size of Bolt
15	80	12	12	12	12	12	40	2	55	4	M10 x 40
20	90	14	14	14	14	14	50	2	65	4	M10 x 45
25	100	14	14	14	14	14	60	2	75	4	M10 x 45
32	120	14	16	14	14	14	70	2	90	4	M12 x 45
40	130	14	16	14	14	14	80	3	100	4	M12 x 45
50	140	14	16	14	14	14	90	3	110	4	M12 x 45
65	160	14	16	14	14	14	110	3	130	4	M12 x 45
80	190	16	18	16	16	16	128	3	150	4	M16 x 55
100	210	16	18	16	16	16	148	3	170	4	M16 x 55
125	240	18	20	18	18	18	178	3	200	8	M16 x 60
150	265	18	20	18	18	18	202	3	225	8	M16 x 60
200	320	20	22	--	20	20	258	3	280	8	M16 x 60
250	375	22	24	--	22	22	312	3	335	12	M16 x 65
300	440	22	24	--	22	22	365	4	395	12	M20 x 70

Normal Pressure Rating (N.P.16) 16 Bar

Nom. Size	Flang Outsid Diam.	Table 6/2 Forged Weldin Neck	Table 6/3 Plate Slip on	Table 6/4 Forged Bossed Screwe	Table 6/5 Forged Bossed Slip on	Table 6/8 Plate Blank	Raised Face Diam.	T'ness	Pitch Circle Diam.	Nr. Bolt Hol	Size of Bolt
15	95	14	14	14	14	14	45	2	55	4	M12 x 45
20	105	16	16	16	16	16	58	2	65	4	M12 x 50
25	115	16	16	16	16	16	68	2	75	4	M12 x 50
32	140	16	16	16	16	16	78	2	90	4	M16 x 55
40	150	16	16	16	16	16	88	3	100	4	M16 x 55
50	165	18	18	18	18	18	102	3	110	4	M16 x 60
65	185	18	18	18	18	18	122	3	130	4	M16 x 60
80	200	20	20	20	20	20	138	3	150	8	M16 x 60
100	220	20	20	20	20	20	158	3	170	8	M16 x 65
125	250	22	22	22	22	22	188	3	200	8	M16 x 70
150	285	22	22	22	22	22	212	3	225	8	M20 x 70
200	340	24	24	--	24	24	268	3	280	12	M20 x 75
250	405	26	26	--	26	26	320	3	335	12	M24 x 90
300	460	28	28	--	28	28	378	4	395	12	M24 x 90

MINIMUM DISTANCES BETWEEN SUPPORTS/FIXINGS

Material	BS Nominal Pipe Size		Pipes - Vertical	Pipes - Horizontal on to low gradients
	inch	mm	support distance in metres	support distance in metres
Copper	0.50	15.00	1.90	1.30
	0.75	22.00	2.50	1.90
	1.00	28.00	2.50	1.90
	1.25	35.00	2.80	2.50
	1.50	42.00	2.80	2.50
	2.00	54.00	3.90	2.50
	2.50	67.00	3.90	2.80
	3.00	76.10	3.90	2.80
	4.00	108.00	3.90	2.80
	5.00	133.00	3.90	2.80
	6.00	159.00	3.90	2.80
muPVC	1.25	32.00	1.20	0.50
	1.50	40.00	1.20	0.50
	2.00	50.00	1.20	0.60
Polypropyle	1.25	32.00	1.20	0.50
	1.50	40.00	1.20	0.50
uPVC	--	82.40	1.20	0.50
	--	110.00	1.80	0.90
	--	160.00	1.80	1.20

LITRES OF WATER STORAGE REQUIRED PER PERSON PER BUILDING TYPE

Type of Building	Storage per person litres
Houses and flats	90
Hostels	90
Hotels	135
Nurses home and medical quarters	115
Offices with canteen	45
Offices without canteen	35
Restaurants, per meal served	7
Boarding schools	90
Day schools	30

Tables and Memoranda

COLD WATER PLUMBING - INSULATION THICKNESS REQUIRED AGAINST FROST

Bore of Tube		Pipework within Buildings - Declared Thermal Conductivity (W/m degrees C)		
inch	mm	Up to 0.040	0.041 to 0.055	0.056 to 0.070
0.50	15	32	50	75
0.75	20	32	50	75
1.00	25	32	50	75
1.25	32	32	50	75
1.50	40	32	50	75
2.00	50	25	32	50
2.50	65	25	32	50
3.00	80	25	32	50
4.00	100	19	25	38

CAPACITY AND DIMENSIONS OF GALVANISED MILD STEEL CISTERNS - BS 417

Capacity (litres)	BS type (SCM)	Dimensions		
		Length (mm)	Width (mm)	Depth (mm)
18	45	457	305	305
36	70	610	305	371
54	90	610	406	371
68	110	610	432	432
86	135	610	457	482
114	180	686	508	508
159	230	736	559	559
191	270	762	584	610
227	320	914	610	584
264	360	914	660	610
327	450/1	1220	610	610
336	450/2	965	686	686
423	570	965	762	787
491	680	1090	864	736
709	910	1070	889	889

CAPACITY OF COLD WATER POLYPROPYLENE STORAGE CISTERNS - BS 4213

Capacity (litres)	BS type (PC)	Maximum Height mm
18	4	310
36	8	380
68	15	430
91	20	510
114	25	530
182	40	610
227	50	660
273	60	660
318	70	660
455	100	760

656   *Tables and Memoranda*

STORAGE CAPACITY AND RECOMMENDED POWER OF HOT WATER STORAGE BOILERS

Type of Building		Storage at 65°C (litres / person)	Boiler power to 65°C (kW / person)
Flats and Dwellings			
(a)	Low Rent Properties	25	0.5
(b)	Medium Rent Properties	30	0.7
(c)	High Rent Properties	45	1.2
Nurses Homes		45	0.9
Hostels		0.7	0.7
Hotels			
(a)	Top Quality - Up Market	1.2	1.2
(b)	Average Quality - Low Market	0.9	0.9
Colleges and Schools			
(a)	Live-in Accommodation	0.7	0.7
(b)	Public Comprehensive	0.1	0.1
Factories		0.1	0.1
Hospitals			
(a)	General	1.5	1.5
(b)	Infectious	1.5	1.5
(c)	Infirmaries	0.6	0.6
(d)	Infirmaries with Laundry	0.9	0.9
(e)	Maternity	2.1	2.1
(f)	Mental	0.7	0.7
Offices		0.1	0.1
Sports Pavilions		0.3	0.3

THICKNESS OF THERMAL INSULATION FOR HEATING INSTALLATIONS

Size of Tube mm	Declared Thermal Conductivity			
	Up to 0.025	0.026 to 0.040	0.041 to 0.055	0.056 to 0.070
LTHW Systems	Minimum thickness of insulation			
15	25	25	38	38
20	25	32	38	38
25	25	38	38	38
32	32	38	38	50
40	32	38	38	50
50	38	38	50	50
65	38	50	50	50
80	38	50	50	50
100	38	50	50	63
125	38	50	50	63
150	50	50	63	63
200	50	50	63	63
250	50	63	63	63
300	50	63	63	63
Flat Surfaces	50	63	63	63

MTHW Systems and Condensate	Minimum thickness of insulation			
15	25	38	38	38
20	32	38	38	50
25	38	38	38	50
32	38	50	50	50
40	38	50	50	50
50	38	50	50	50
65	38	50	50	50
80	50	50	50	63
100	50	63	63	63
125	50	63	63	63
150	50	63	63	63
200	50	63	63	63
250	50	63	63	75
300	63	63	63	75
Flat Surfaces	63	63	63	75

658 *Tables and Memoranda*

THICKNESS OF THERMAL INSULATION FOR HEATING INSTALLATIONS

Size of Tube mm	Declared Thermal Conductivity			
	Up to 0.025	0.026 to 0.040	0.041 to 0.055	0.056 to 0.070
HTHW Systems and Steam	Minimum thickness of insulation			
15	38	50	50	50
20	38	50	50	50
25	38	50	50	50
32	50	50	50	63
40	50	50	50	63
50	50	50	75	75
65	50	63	75	75
80	50	63	75	75
100	63	63	75	100
125	63	63	100	100
150	63	63	100	100
200	63	63	100	100
250	63	75	100	100
300	63	75	100	100
Flat Surfaces	63	75	100	100

Tables and Memoranda

CAPACITIES AND DIMENSIONS OF GALVANISED MILD STEEL INDIRECT CYLINDERS TO BS 1565

Approximate Capacity (litres)	BS Size No.	Internal Diameter (mm)	External height over dome (mm)
109	BSG.1M	457	762
136	BSG.2M	457	914
159	BSG.3M	457	1067
227	BSG.4M	508	1270
273	BSG.5M	508	1473
364	BSG.6M	610	1372
455	BSG.7M	610	1753
123	BSG.8M	457	838

CAPACITIES AND DIMENSIONS OF COPPER INDIRECT CYLINDERS (COIL TYPE) TO BS 1566

Approximate Capacity (litres)	BS Type	External Diameter (mm)	External height over dome (mm)
96	0	300	1600
72	1	350	900
96	2	400	900
114	3	400	1050
84	4	450	675
95	5	450	750
106	6	450	825
117	7	450	900
140	8	450	1050
162	9	450	1200
190	10	500	1200
245	11	500	1500
280	12	600	1200
360	13	600	1500
440	14	600	1800

RECOMMENDED AIR CONDITIONING DESIGN LOADS

Building Type	Design Loading
Computer rooms	500w per m² of floor area
Restaurants	150w per m² of floor area
Banks (main area)	100w per m² of floor area
Supermarkets	25w per m² of floor area
Large Office Buildings (exterior zone)	100w per m² of floor area
Large Office Block (interior zone)	80w per m² of floor area
Small Office Block (interior zone)	80w per m² of floor area

Index

Addressable Call Point	551	flexible	467
Air Conditioning		Connection Units	525
approximate estimating	8	Conversion Tables	639
recommended loads	660	Cost Indices	5
All-in rates	12	Cylinders	
Approximate Metres per tonne of tube	652	copper; direct	149
Arts and Drama Centres	31	copper; indirect	150
Attenuators	399	storage	151
Automatic Air Vents	219		
Automatic Transfer Switches	444	Data Systems	542
		Daywork	618
Batten Lampholders	521	Dimension of Copper and Stainless	
Boilers		Steel Pipes	650
cast iron section	165	Dimension of Steel Pipes	651
domestic water, gas fired	164	Distribution Boards	445
packaged water	167	Domestic Central Heating	13
Bus-Bar, prewired	509	Dry Risers	157
Bus-Bar Systems	506	Ductwork	
		access doors	333
Cable		acoustic louvres	390
Calflam	505	approximate estimating	14
coaxial, television, radio,		dampers	378
frequency video	538	diffusers/grilles	383
Firetuff	505	external louvres	391
flexible	496	rectangular ductwork, class B	286
FP 100	503	rectangular ductwork, class C	334
FP 200	503	circular ductwork, class B	275
FP 400	504	circular ductwork, class C	278
data	545	flat oval ductwork	282
LSF 6491B	495	fire rated ductwork	402
mineral insulated	499		
paper insulated	438	Earthing Systems	552
patch	544	Emergency Lighting	534
Voice	546	Electrical Working Rules	602
XLPE/LOSH/SWA-HV	433	Elemental Rates	7
XLPE/LOSH/SWA-LV	489	Expansion Joints	207
XLPE/PVC	495	External Lighting	20
XLPE/PVC/SWA-LV	483		
Cable Tiles	439	Factories	6
Cable Tray		Fan Coil Systems	8
galvanised steel	445	Fans	
basket	462	axial flow fans	393
GRP	461	roof fans	393
Calorifiers		toilet fans	394
Non storage water	227	kitchen fans	394
Non storage steam	270	multivent fans	394
Chimneys		Filters	395
domestic gas boilers	167	Fire Alarms	549
industrial and commercial		Fire Extinguishers	162
oil and gas appliances	429	Fire Hydrants	162
Circuit Breakers	442	Fire Protection	7
high voltage	445	Flange Dimension Chart	653
low voltage	443	Flats (L.A.)	6
miniature	540	Formulae	641
moulded case		Fractions, Decimals & Millimetric	
Clock Systems	464	Equivalents	642
Conduit	464		
steel, black enamelled	467		
steel, galvanised			
PVC			

INDEX

Galvanised Steel Trunking		carbon steel	199	
single compartment	468	copper	112	
double compartment	469	galvanised steel	102	
triple compartment	469	malleable grooved	192	
four compartment	469	MDPE		
underfloor	473	blue	90	
Generating Sets	440	yellow	153	
Gutters		Pressfit		
half round PVC-U	44	copper	129	
cast iron	52	stainless steel	136	
square PVC-U	45	carbon steel	189	
square high capacity PVC-U	45	PVC-ABS	79	
deep eliptical PVC-U	46	PVC-C	99	
		PVC-U		
Heat Pumps	178	rainwater		
Heaters		- soil, solvent joints	74	
air curtains	397	- soil, ring seal	77	
perimeter heaters	228	waste, solvent joints	64	
radiant panels	230	waste, ring-seal joints	69	
radiant strip heaters	229	water	96	
unit heaters	271	stainless steel	132	
Hose Reels	156	Polypropylene Traps	70	
Houses (L.A.)	6	Pumps		
		accelerator	226	
Imperial Standard Wire Gauge	643	centrifugal	222	
Industrial Sockets	527	glandless	224	
Insulation		sump pump	145	
closed cell	147	pressurisation - cold water	145	
mineral fibre	236	PVC Trunking		
ductwork	401	single compartment	474	
		bench	477	
Ladder Rack	452	miniature	477	
Labour Rates		skirting/dado	480	
ductwork	42	underfloor	478	
electrical	427			
mechanical	40	Regional Variations	5	
Lift and Conveyor Installations	21	Ring Main Units	429	
Lighting		Room Thermostats	234	
Switches	519			
Lightning Protection	553	Shaver Unit	525	
Luminaires	515	Sight Glasses	268	
		Socket Outlets	523	
Mechanical Working Rules	593	Split Units	274	
Minimum Distance between Supports	654	Sprinkler Heads	160	
Modular Wiring	511	Sprinkler Pipework	158	
Museums	6	Sprinkler Valves	160	
		Square Metre Rates	6	
Offices	8	Steam Generators	177	
		Steam Traps		
Pipe Freezing	233	cast iron	267	
Pipe Rollers/Chairs	157	stainless steel	267	
Pipes		Steelwork, weights of	647	
ABS		Support & Fixings, minimum distance	654	
waste	68			
water	94	Tanks		
black steel		fuel storage tanks	155	
screwed	180	moulded glassfibre	146	
welded	192	polypropylene	146	
cast iron		Telephone		
soil	80	cable	537	
rainwater	60	outlet	537	
		Theatres	6	

INDEX

Trace Heating System	272
Transformers	430
Tubes, weights	652
Uninterruptable Power Supplies	532
Universities	6
Valves	
ball valves	211
ball float valves	141
bronze	213
check valves - DZR	144
commissioning valves	215
control valves	218
drain cocks	219
gate valves - DZR	141
globe valves	212
isolating	209
pressure reading	269
radiator valves	220
regulators	217
relief valves	218
safety valves	218
stopcocks	141
strainers	217
Variable Air Volume System	8
Warehouses	6
Water Pressure Due to Height	644
Water Storage Capacities	655
Zinc Underfloor Trunking	
double compartment	479
triple compartment	479

Keep your figures up to date, free of charge
Download updates from the web: www.pricebooks.co.uk

This section, and most of the other information in this Price Book, is brought up to date every three months with the Price Book Updates, until the next annual edition. The updates are available free to all Price Book purchasers.

To ensure you receive your free copies, either complete the reply card from the centre of the book and return it to us or register via the website www.pricebooks.co.uk

CD-Rom Single-User Licence Agreement

We welcome you as a user of this Spon Press CD-ROM and hope that you find it a useful and valuable tool. Please read this document carefully. **This is a legal agreement** between you (hereinafter referred to as the "Licensee") and Taylor and Francis Books Ltd., under the imprint of Spon Press (the "Publisher"), which defines the terms under which you may use the Product. **By breaking the seal and opening the package containing the CD-ROM you agree to these terms and conditions outlined herein. If you do not agree to these terms you must return the Product to your supplier intact, with the seal on the CD case unbroken.**

1. Definition of the Product
The product which is the subject of this Agreement, *Spon's Mechanical and Electrical Services Price Book on CD-ROM* (the "Product") consists of:

1.1	Underlying data comprised in the product (the "Data")
1.2	A compilation of the Data (the "Database")
1.3	Software (the "Software") for accessing and using the Database
1.4	A CD-ROM disk (the "CD-ROM")

2. Commencement and Licence
2.1   This Agreement commences upon the breaking open of the package containing the CD-ROM by the Licensee (the "Commencement Date").

2.2   This is a licence agreement (the "Agreement") for the use of the Product by the Licensee, and not an agreement for sale.

2.3   The Publisher licenses the Licensee on a non-exclusive and non-transferable basis to use the Product on condition that the Licensee complies with this Agreement. The Licensee acknowledges that it is only permitted to use the Product in accordance with this Agreement.

3. Installation and Use
3.1   The Licensee may provide access to the Product for individual study in the following manner: The Licensee may install the Product on a secure local area network on a single site for use by one user. For more than one user or for a wide area network or consortium, use is only permissible with the express permission of the Publisher in writing and requires the payment of the appropriate fee as specified by the Publisher, and signature by the Licensee of a separate multi-user licence agreement.

3.2   The Licensee shall be responsible for installing the Product and for the effectiveness of such installation.

3.3   Text from the Product may be incorporated in a coursepack. Such use is only permissible with the express permission of the Publisher in writing and requires the payment of the appropriate fee as specified by the Publisher and signature of a separate licence agreement.

4. Permitted Activities
4.1   The Licensee shall be entitled:

4.1.1   to use the Product for its own internal purposes;

4.1.2   to download onto electronic, magnetic, optical or similar storage medium reasonable portions of the Database provided that the purpose of the Licensee is to undertake internal research or study and provided that such storage is temporary;

4.1.3   to make a copy of the Database and/or the Software for back-up/archival/disaster recovery purposes.

4.2   The Licensee acknowledges that its rights to use the Product are strictly as set out in this Agreement, and all other uses (whether expressly mentioned in Clause 5 below or not) are prohibited.

5. Prohibited Activities
The following are prohibited without the express permission of the Publisher:

5.1   The commercial exploitation of any part of the Product.

5.2   The rental, loan (free or for money or money's worth) or hire purchase of the product, save with the express consent of the Publisher.

5.3   Any activity which raises the reasonable prospect of impeding the Publisher's ability or opportunities to market the Product.

5.4   Any networking, physical or electronic distribution or dissemination of the product save as expressly permitted by this Agreement.

5.5   Any reverse engineering, decompilation, disassembly or other alteration of the Product save in accordance with applicable national laws.

5.6   The right to create any derivative product or service from the Product save as expressly provided for in this Agreement.

5.7   Any alteration, amendment, modification or deletion from the Product, whether for the purposes of error correction or otherwise.

666 CD-Rom User License Agreement

6. General Responsibilities of the Licensee

6.1 The Licensee will take all reasonable steps to ensure that the Product is used in accordance with the terms and conditions of this Agreement.

6.2 The Licensee acknowledges that damages may not be a sufficient remedy for the Publisher in the event of breach of this Agreement by the Licensee, and that an injunction may be appropriate.

6.3 The Licensee undertakes to keep the Product safe and to use its best endeavours to ensure that the product does not fall into the hands of third parties, whether as a result of theft or otherwise.

6.4 Where information of a confidential nature relating to the product or the business affairs of the Publisher comes into the possession of the Licensee pursuant to this Agreement (or otherwise), the Licensee agrees to use such information solely for the purposes of this Agreement, and under no circumstances to disclose any element of the information to any third party save strictly as permitted under this Agreement. For the avoidance of doubt, the Licensee's obligations under this sub-clause 6.4 shall survive termination of this Agreement.

7. Warrant and Liability

7.1 The Publisher warrants that it has the authority to enter into this Agreement, and that it has secured all rights and permissions necessary to enable the Licensee to use the Product in accordance with this Agreement.

7.2 The Publisher warrants that the CD-ROM as supplied on the Commencement Date shall be free of defects in materials and workmanship, and undertakes to replace any defective CD-ROM within 28 days of notice of such defect being received provided such notice is received within 30 days of such supply. As an alternative to replacement, the Publisher agrees fully to refund the Licensee in such circumstances, if the Licensee so requests, provided that the Licensee returns the Product to the Publisher. The provisions of this sub-clause 7.2 do not apply where the defect results from an accident or from misuse of the product by the Licensee.

7.3 Sub-clause 7.2 sets out the sole and exclusive remedy of the Licensee in relation to defects in the CD-ROM.

7.4 The Publisher and the Licensee acknowledge that the Publisher supplies the Product on an "as is" basis. The Publisher gives no warranties:

7.4.1 that the Product satisfies the individual requirements of the Licensee; or

7.4.2 that the Product is otherwise fit for the Licensee's purpose; or

7.4.3 that the Data are accurate or complete or free of errors or omissions; or

7.4.4 that the Product is compatible with the Licensee's hardware equipment and software operating environment.

7.5 The Publisher hereby disclaims all warranties and conditions, express or implied, which are not stated above.

7.6 Nothing in this Clause 7 limits the Publisher's liability to the Licensee in the event of death or personal injury resulting from the Publisher's negligence.

7.7 The Publisher hereby excludes liability for loss of revenue, reputation, business, profits, or for indirect or consequential losses, irrespective of whether the Publisher was advised by the Licensee of the potential of such losses.

7.8 The Licensee acknowledges the merit of independently verifying Data prior to taking any decisions of material significance (commercial or otherwise) based on such data. It is agreed that the Publisher shall not be liable for any losses which result from the Licensee placing reliance on the Data or on the Database, under any circumstances.

7.9 Subject to sub-clause 7.6 above, the Publisher's liability under this Agreement shall be limited to the purchase price.

8. Intellectual Property Rights

8.1 Nothing in this Agreement affects the ownership of copyright or other intellectual property rights in the Data, the Database or the Software.

8.2 The Licensee agrees to display the Publisher's copyright notice in the manner described in the Product.

8.3 The Licensee hereby agrees to abide by copyright and similar notice requirements required by the Publisher, details of which are as follows:

" © 2002 Spon Press. All rights reserved. All materials in *Spon's Mechanical and Electrical Services Price Book on CD-ROM* are copyright protected. © 2001 Adobe Systems Incorporated. All rights reserved. No such materials may be used, displayed, modified, adapted, distributed, transmitted, transferred, published or otherwise reproduced in any form or by any means now or hereafter developed other than strictly in accordance with the terms of the licence agreement enclosed with the CD-ROM. However, text and images may be printed and copied for research and private study within the preset program limitations. Please note the copyright notice above, and that any text or images printed or copied must credit the source."

8.4 This Product contains material proprietary to and copyrighted by the Publisher and others. Except for the licence granted herein, all rights, title and interest in the Product, in all languages, formats and media throughout the world, including all copyrights therein, are and remain the property of the Publisher or other copyright owners identified in the Product.

9. Non-assignment

This Agreement and the licence contained within it may not be assigned to any other person or entity without the written consent of the Publisher.

10. Termination and Consequences of Termination

10.1 The Publisher shall have the right to terminate this Agreement if:

10.1.1	the Licensee is in material breach of this Agreement and fails to remedy such breach (where capable of remedy) within 14 days of a written notice from the Publisher requiring it to do so; or
10.1.2	the Licensee becomes insolvent, becomes subject to receivership, liquidation or similar external administration; or
10.1.3	the Licensee ceases to operate in business.
10.2	The Licensee shall have the right to terminate this Agreement for any reason upon two months' written notice. The Licensee shall not be entitled to any refund for payments made under this Agreement prior to termination under this sub-clause 10.2.
10.3	Termination by either of the parties is without prejudice to any other rights or remedies under the general law to which they may be entitled, or which survive such termination (including rights of the Publisher under sub-clause 6.4 above).
10.4	Upon termination of this Agreement, or expiry of its terms, the Licensee must:
10.4.1	destroy all back up copies of the product; and
10.4.2	return the Product to the Publisher.

11. General

11.1	*Compliance with export provisions* The Publisher hereby agrees to comply fully with all relevant export laws and regulations of the United Kingdom to ensure that the Product is not exported, directly or indirectly, in violation of English law.
11.2	*Force majeure* The parties accept no responsibility for breaches of this Agreement occurring as a result of circumstances beyond their control.
11.3	*No waiver* Any failure or delay by either party to exercise or enforce any right conferred by this Agreement shall not be deemed to be a waiver of such right.
11.4	*Entire agreement* This Agreement represents the entire agreement between the Publisher and the Licensee concerning the Product. The terms of this Agreement supersede all prior purchase orders, written terms and conditions, written or verbal representations, advertising or statements relating in any way to the Product.
11.5	*Severability* If any provision of this Agreement is found to be invalid or unenforceable by a court of law of competent jurisdiction, such a finding shall not affect the other provisions of this Agreement and all provisions of this Agreement unaffected by such a finding shall remain in full force and effect.
11.6	*Variations* This Agreement may only be varied in writing by means of variation signed in writing by both parties.
11.7	*Notices* All notices to be delivered to: Spon Press, an imprint of Taylor & Francis Books Ltd., 11 New Fetter Lane, London EC4P 4EE, UK.
11.8	*Governing law* This Agreement is governed by English law and the parties hereby agree that any dispute arising under this Agreement shall be subject to the jurisdiction of the English courts.

If you have any queries about the terms of this licence, please contact:

Spon's Price Books
Spon Press
an imprint of Taylor & Francis Books Ltd.
11 New Fetter Lane
London EC4P 4EE
United Kingdom
Tel: +44 (0) 20 7583 9855
Fax: +44 (0) 20 7842 2298
www.sponpress.com

CD-Rom Installation Instructions

System requirements

Minimum

- 66 MhZ processor
- 12 MB of RAM
- 10 MB available hard disk space
- Quad speed CD-ROM drive
- Microsoft Windows 95/98/2000/NT
- VGA or SVGA monitor (256 colours)
- Mouse

Recommended

- 133 MhZ (or better) processor
- 16 MB of RAM
- 10 MB available hard disk space or more
- 12x speed CD-ROM drive
- Microsoft Windows 95/98/2000/NT
- SVGA monitor (256 colours) or better
- Mouse

Microsoft ® is a registered trademark and Windows™ is a trademark of the Microsoft Corporation.

Installation

How to install *Spon's Mechanical and Electrical Services Price Book 2002 CD-ROM*

Windows 95/98/2000/NT

Spon's Mechanical and Electrical Services Price Book 2002 CD-ROM should run automatically when inserted into the CD-ROM drive. If it fails to run, follow the instructions below.

- Click the **Start** button and choose **Run**.
- Click the **Browse** button.
- Select your CD-ROM drive.
- Select the Setup file [setup.exe] then click **Open**.
- Click the OK button.
- Follow the instructions on screen.
- The installation process will create a folder containing an icon for *Spon's Mechanical and Electrical Services Price Book CD-ROM* and also an icon on your desktop.

How to run the *Spon's Mechanical and Electrical Services Price Book CD-ROM*

- Double click the icon (from the folder or desktop) installed by the Setup program.
- Follow the instructions on screen.

© COPYRIGHT ALL RIGHTS RESERVED

All materials in *Spon's Mechanical and Electrical Works Price Book 2002 CD-ROM* are copyright protected. No such materials may be used, displayed, modified, adapted, distributed, transmitted, transferred, published or otherwise reproduced in any form or by any means now or hereafter developed other than strictly in accordance with the terms of the above licence agreement.

The software used in *Spon's Mechanical and Electrical Services Price Book 2002 CD-ROM* is furnished under a licence agreement. The software may be used only in accordance with the terms of the licence agreement.